Water Loss Control Manual

Water Loss Control Manual

JULIAN THORNTON

McGraw-Hill
New York Chicago San Francisco Lisbon London Madrid
Mexico City Milan New Delhi San Juan Seoul
Singapore Sydney Toronto

Library of Congress Cataloging-in-Publication Data

Thornton, Julian.
    Water loss control manual / Julian Thornton.
       p.   cm.
    Includes bibliographical references and index.
    ISBN 0-07-137434-5
    1. Water leakage.   2. Water—Distribution—Management.   3. Loss control.
I. Title.
TD495.T46   2002
628.1'4—dc21   2002070103

Copyright © 2002 by The McGraw-Hill Companies, Inc. All rights reserved. Printed in the United States of America. Except as permitted under the United States Copyright Act of 1976, no part of this publication may be reproduced or distributed in any form or by any means, or stored in a data base or retrieval system, without the prior written permission of the publisher.

2 3 4 5 6 7 8 9 0   KGP/KGP   0 8 7 6 5 4 3

ISBN 0-07-137434-5

The sponsoring editor for this book was Larry S. Hager, the editing supervisor was Stephen M. Smith, and the production supervisor was Sherri Souffrance. It was set in Garamond by Kim Sheran and Joanne Morbit of McGraw-Hill Professional's Hightstown, N.J., composition unit.

Printed and bound by Quebecor/Kingsport.

McGraw-Hill books are available at special quantity discounts to use as premiums and sales promotions, or for use in corporate training programs. For more information, please write to the Director of Special Sales, McGraw-Hill Professional, Two Penn Plaza, New York, NY 10121-2298. Or contact your local bookstore.

This book is printed on acid-free paper.

Information contained in this work has been obtained by The McGraw-Hill Companies, Inc. ("McGraw-Hill") from sources believed to be reliable. However, neither McGraw-Hill nor its authors guarantee the accuracy or completeness of any information published herein and neither McGraw-Hill nor its authors shall be responsible for any errors, omissions, or damages arising out of use of this information. This work is published with the understanding that McGraw-Hill and its authors are supplying information but are not attempting to render engineering or other professional services. If such services are required, the assistance of an appropriate professional should be sought.

Contents

	Preface	vii
	Acknowledgments	ix
Chapter 1	Introduction: Water Supplied and Water Lost	1
Chapter 2	Understanding the Types of Water Loss	13
Chapter 3	Traditional and Progressive Approaches to Water Loss: The Experiences of the United States and England	25
Chapter 4	Evaluating Water Loss: Using Water Audits and Performance Indicators	35
Chapter 5	Data Formatting and Management	59
Chapter 6	Equipment and Techniques: Flow Metering, Pressure Measurement, Control, and Leak Detection	65
Chapter 7	Modeling Water Losses	117
Chapter 8	Completing the Audit and Determining Cost to Benefit	153
Chapter 9	Potential Solutions for Controlling Water Losses	161
Chapter 10	Reducing Real Losses in the Field: Active Leakage Management	171
Chapter 11	Speed and Quality of Leak Repair	241
Chapter 12	Pressure Management	261
Chapter 13	Pipe Maintenance, Rehabilitation, and Replacement	315
Chapter 14	Resolving Apparent Losses	341
Chapter 15	Water Efficiency Programs	381
Chapter 16	Using In-House Staff or a Contractor and Designing a Bid Document	401
Chapter 17	Understanding Basic Hydraulics	417
Chapter 18	Articles and Case Studies	429
Appendix A	Water Accountability	491
Appendix B	Meter Installation and Testing	533
Appendix C	Types of Flowmeters	573
Appendix D	Demand Profiling for Optimal Meter Sizing	597
Appendix E	Pipe Properties	615
	Index	633

Preface

This manual has been written as a comprehensive guide to water auditing and hands-on reduction of losses in a water distribution system. In addition to raising awareness of the extent of the water loss problem and current practices in many systems in North America and around the world, it covers all of the basic tools required to perform an audit both on paper and in the field. This book discusses how to calculate and evaluate losses and cost-to-benefit ratios at variable production costs and at sales cost and how to set up suitable and sustainable field intervention programs.

For each system and water loss problem, different tools will need to be used from the practitioner's tool kit. This manual will aid in selecting the correct tools and methodology for each job.

This manual discusses many new technologies. Technology, however, is changing at such a rapid pace that it is almost impossible to keep up! What may be a new technology for one company, municipality, or utility may be an old technology for another. The emphasis of this manual is to ensure efficient use of the tools we have on hand and to learn to assess the cost and the benefit of improving upon our current practices.

This book is suitable either as an educational tool for the inexperienced operator or as a reference manual for the more experienced operator. The level of detail in the book often jumps from in-depth reasoning to all of the things we either learned or forgot or indeed somehow managed to skip. As a field operator for many years, I have learned that it is often easier to understand a complex problem if we return to basics and try to understand the individual components of the problem.

Various types of measurement units (SI, U.S., Imperial) are mentioned and used throughout this book, as reference material and case studies have been provided from many different countries. In addition, I have adopted recent terminology formulated by the IWA such as *real losses* and *apparent losses*. Real losses have often been termed *physical losses* in the past and apparent losses were often referred to as *nonphysical* losses.

There are other publications which cover some of the topics in this manual. However, this book pulls together ideas, techniques, methodologies, and references from many sources across the world, therefore making it a flexible and comprehensive guide which can be used in a variety of field situations.

The main text is supported throughout with case studies of actual loss management programs. Case studies are an important tool which can be used as a first step when attempting to justify a project or a change in rationale. It always helps to know that somebody else has the same or a similar problem and has dealt with it in an efficient and economical manner. Certain editing changes have been made to some of the case studies and I would urge interested readers to refer to the originals wherever possible.

Throughout the book references are made to types of equipment, techniques, and software, all of which are generally accepted in the industry. The intent of this book is not to promote one particular product, consultant, contractor, or process but to promote awareness of the water loss problems we face and the desire to overcome them and become more efficient.

DISCLAIMER

While every effort has been made not to place emphasis on any particular make or model of equipment, consultant, contractor, software, or process, the author and the publisher accept no responsibility or liability for any omission or claim to loss of revenue caused by the omission of a process type or alternative service provider. The sole intention of this book is to pass on practical field knowledge to end-users with an interest in water loss management.

Julian Thornton

Acknowledgments

SPECIAL THANKS

Special thanks go to George Kunkel, who has done so much more work on this book than any other co-author, as well as bringing about much positive change in our AWWA LDWA committee. Special thanks go also to Allan Lambert, who has supported our industry and my endeavors for the last 10 years, providing many new and innovative tools to assist in water loss management.

I would also like to extend special thanks to Olinda Everret, my coordinator for this project; Larry Hager, my editor; and Colin Murcray and his team from AWWA, who have all given me excellent support throughout this "first book" experience.

CO-AUTHORS

Thanks go to Bill Gauley, who tirelessly prepared Chap. 15 while his new baby was being born, and to Dave Southern, Jack Jackson, Brad Brainard, and Rodney Briar, who all submitted subsections of this book.

PEER REVIEW

Thanks go to Allan Lambert, Ken Brothers, George Kunkel, Mark Shepherd, and Tim Brown, who spent time reviewing the material that went into this manual throughout the time it was being compiled.

CASE STUDIES AND ARTICLES

Thanks go to all of the authors of the original case studies and articles presented in this book. These authors are recognized in their individual sections. I would also like to thank all of the water systems which directly and indirectly allowed their data to be published. (Readers are urged to see the original case studies, as some changes have been made during editing.)

A special note of thanks goes to the AWWA, which allowed the reproduction of many of the cases in this book for which it holds the copyright as well as certain sections of its manuals.

COMMITTEE

Thanks go to all of my colleagues on the AWWA LDWA committee who have given up their time over the years to serve on a volunteer committee dedicated to improving awareness for our industry and meeting the challenges it brings.

COMPANIES WHICH SUBMITTED INFORMATION AND ALLOWED THE USE OF THEIR DATA

BBL (Brazil); Restor (Brazil); Heath Consultants Inc. (United States); Hetek Solutions Inc. (Canada); Pitometer Water Services, a division of Severn Trent Pipeline Services, Inc. (United States); Ductile Iron Pipe Research Association (United States); Invensys Metering Systems (formerly Sensus Technologies) (United States); Invensys Metering Systems (South Africa); BKS Pty Ltd. (South Africa); Watts ACV (United States); Ross Valve Mfg. Co., Inc. (United States); ADS Environmental Services (United States); Veritec Consulting Inc. (Canada); F. S. Brainard & Co. (United States); Effective Fluid Engineering (United Kingdom); and all of the other companies whose information appears directly or indirectly in this manual.

GENERAL

Thanks go to all of my clients, colleagues, and competitors, who have made this industry what it is today.

DEDICATION

I would like to dedicate this book to my beloved daughter and good friend Victoria.

CHAPTER

1

Introduction

Water Supplied and Water Lost

Julian Thornton
George Kunkel

1.1 INTRODUCTION

The world's population exploded during the twentieth century. At the close of 1998 approximately 5.9 billion inhabitants could call the planet Earth home, up from 4 billion in 1975. That such growth could occur is a testament to our unique ability to provide the essentials of clean air, water, food, and health care to our masses. However, during the latter half of the same century, people also recognized that the world's resources could not continue to sustain this rate of growth indefinitely—at least, not if we continue to use the same methods to which we have become accustomed. Our resources are finite. Although efforts are under way to control population growth, similar efforts exist, and are being refined, to ensure that our most vital resources are used wisely to sustain the world's population—which *will* continue to grow even if we succeed in slowing the rate of growth.

The availability of safe water has been a major contributing factor in the growth of the world's population, by serving our drinking water and sanitation needs. The ability to create large water supply systems to abstract or withdraw, treat and transport vital water to whole communities stands as one of history's great engineering marvels. Yet notable caveats exist to this success story. Many developing countries still do not have the water supply infrastructure to provide clean water to individual customers, or to supply it on a continuous basis. In such places, modern water systems are lacking due to the same social, political, and economic complexities that challenge all aspects of development in these lands. While these populations struggle to gain basic levels of service, many highly developed water systems in technologically advanced countries suffer an insidious problem that threatens the long-term sustainability of water resources for the future: water loss. Most of the world's water systems have been highly successful in delivering high-quality water to large populations. However, most of these systems also incur a notable amount of loss in their operations. In years past the seemingly infinite supply of water in expanding "new worlds" allowed water loss to be largely overlooked. With water readily available and relatively inexpensive, losses were ignored by water utilities, or assumed

> **D**id you know that there are almost 6 billion people in the world?

to be naturally inherent in operating a water supply system. With the demands of growing populations, however, realization of the limits on our natural resources and increasing costs from regulations and customer demands have made it increasingly unrealistic to allow water loss to be ignored.

Water loss occurs in two fundamental ways:

1. Water lost from the distribution system through leaks, tank overflows, or improperly open drains or system blow-offs. These have recently been labeled *real losses*.[1]
2. Water that reaches a customer or other end user—including beneficial and unauthorized use—but is not properly measured or tabulated. These are referred to as *apparent losses*.

> **D**id you know that many locations in the United States suffer from water shortages, yet there are no federal regulations governing how much water a supplier can lose?

Throughout the world, water losses are occurring at both the end user's plumbing and the water supplier's distribution piping. In the United States, promoting the wise use of water by customers, including finding and fixing plumbing leaks, has typically fallen under the discipline of *water conservation*. Water loss encountered by water suppliers has been evaluated, rather loosely, under the questionable label of *water accountability*. Unfortunately, there exists no truly reliable measure of the amount of water lost annually in the United States. However, the literature abounds with case studies of water utilities that could not identify where huge portions, sometimes the majority, of their delivered water went.

A new model of water loss management has been developed recently, taking root in England and spreading quickly to a number of other nations. The National Leakage Initiative was an extensive research endeavor carried out by British and Welsh water companies in the early 1990s. Its results formed the basis for the development of a progressive leakage management structure that arguably is now the world's best practice model. The crux of the model is basic applied engineering, stressing a proactive approach toward eliminating and preventing leakage, and contrasting dramatically with the largely reactive models of most water systems worldwide. In less than 10 years, this structure eliminated up to 85 percent of all recoverable leakage in England and Wales.[2] It is highly valuable in that it represents easily transferable technology for nations around the world to employ. It is now evident that the world's water suppliers not only have a need to reduce and manage their losses, but also have the methods and technology to do so effectively.

1.2 THE NEED FOR WATER

> **W**ater is the human body's second most urgent need, after air.

The human body is approximately 50–65 percent water,[3] which must be replenished on a daily basis; a minimum of 8 glasses per day are recommended for each person. Without water we die within three or four days! Water is the body's second most urgent need, after air. Like the human body, many of the fruits and vegetables we eat are also mostly water. Potatoes are approximately 60 percent water, and some fruits are up to 85 percent water.[4] Obviously, water is an extremely important resource, even though people in many developed countries often take its relative abundance and high quality for granted.

The world's surface is approximately 80 percent covered by water, which is basically an indestructible product. Of this water, approximately 97 percent is salt water, 2 percent is frozen in glaciers, and 1 percent is available for use. Through the natural patterns of the world's climate and the hydrologic cycle, the availability of this water varies widely over time and distance. At any point in time some part of the world is enduring severe drought while other parts are experiencing floods. Rarely does this natural cycle coincide with the routine variation in people's use of water.

> Only 1% of the earth's water is readily available for use—we should take better care of it!

1.3 HOW PEOPLE HAVE HISTORICALLY OBTAINED WATER

The amount of water on the earth is fixed and limited. Our predecessors probably drank several times in the past the water we drink today! The water cycle has not really changed much since the beginning of time. The water cycle is essentially evaporation, cloud formation, rainfall, and passage to the sea by rivers and streams. People interfere with the later stages of this cycle and redirect that passage back to the sea through water piping or distribution systems, human bodies, and sewer systems.

Although the water cycle has not changed over time, what we do to the water and how we have to treat it to make it usable have changed considerably. This has been particularly true since the consolidation of populations into major city centers, which has usually led to increased industry, pollution, and demand for services. The more polluted water becomes, the more expensive it is to treat. The farther it is from the source to a population center, the more expensive water becomes to transport. Given continuing worldwide population expansion and relocation, it is inevitable that water is becoming more expensive to provide.

Water distribution systems have been in use for thousands of years. The ancient Egyptians, Greeks, and Romans all captured, treated, and distributed water in ways not dissimilar to those we use today. Although the technology has changed, the basics remain much the same:

- ❑ Source
- ❑ Primary lift stations
- ❑ Storage
- ❑ Pumping or gravity supply
- ❑ Transmission system
- ❑ Distribution system
- ❑ Customer service connection piping, with or without water meters

> There are 55,000 public water systems in the United States, and water losses are suspected to be around 6 billion gallons a day.

1.3.1 Inefficiency and Losses

There are more than 55,000 public water systems in the United States alone, which process nearly 40 billion gallons of water per day.[5] Unfortunately, almost 6 billion gallons per day of this total are approximated to occur as "public uses and losses," with the losses likely much greater than the public uses for most systems. Inaccuracies or inconsistencies in the data reported also contribute to the difference between the total water delivered and total consumed. This amount of water is more than enough to meet the delivery needs of the 10 largest cities in the United States and should be viewed as a considerable concern for the country with the third largest population in the world.

The author's research for this work included referencing the *Reader's Digest* book *How Was It Done?* (1998), which states that around 40 million gallons of water per day were supplied to ancient Rome through a network of 260 mi (420 km) of pipework and channels. The pipelines and channels were made of brick and stone with cement linings, along with some lead pipes. Apparently, service connections were $3/4$ in (20 mm) with simple stopcock arrangements—not so different from what we use today. The first system was apparently installed in 312 B.C. There were approximately 250 reservoir sites, and the system was gravity-fed. A commissioner and a team consisting of engineers, technicians, workers, and clerks administered this system. One of the priority jobs was to locate and repair leaks.

The durability of the workmanship of the ancient aqueducts is evidenced by the fact that one system installed between 98 and 117 A.D. is still in use in Spain. Not many water systems, or infrastructure of any kind, can boast such a history!

Less ancient system management came in more or less the following order:

- 1800s: formulas for unavoidable losses (Kuichling)
- 1800s: pitot rod district measurements
- 1800s: simple wooden sounding rods
- 1900s: simple mechanical geophones
- 1900s: mechanical meter recording devices
- 1940s: electronic geophones and listening devices
- 1970s: computerized leak noise correlators
- 1980s: battery-operated loggers
- 2000: digital equipment and Geographic Information System (GIS)–linked equipment for leak detection
- 2000: International Water Association recommendations for performance indicators for water supply services, including Unavoidable Annual Real Losses and Infrastructure Leakage Index (ILI)

1.4 THE OCCURRENCE AND IMPACT OF LOST WATER

Most water systems in the world experience water losses, and have done so virtually since the systems were new. Water losses cannot be completely avoided, but they can be managed so that they remain within economic limits. People have been attempting to curb water losses since the days of those first Roman water systems, but water distribution systems have often suffered from the "out-of-sight, out-of-mind" syndrome, particularly where water has been inexpensive and plentiful.

The problems associated with water loss are numerous. High real losses indirectly require water suppliers to extract, treat, and transport greater volumes of water than their customer demand requires. The additional energy needed for treatment and transport taxes energy-generating capabilities, which often rely on large quantities of water. Leaks, bursts, and overflows often cause considerable damage and inflate liability for the supplier. Most leakage finds its way into community waste or storm water collection systems and may be treated at the local wastewater treatment plant—two rounds of expensive treatment that provide no beneficial use! Watersheds are taxed unnecessarily by inordinately high withdrawals. Thus, high losses may limit growth in a region due to restrictions on available source water. The full effect of leakage losses has yet to be assessed, but the economics of leakage, discussed elsewhere in this manual, show that its impact is substantial.

Apparent losses do not carry the physical impact that real losses impart. Instead, they exert a significant financial effect on suppliers and customers. (Table 1.1 shows estimated water prices and usage around the world.) These losses represent service rendered without payment received. The economic impact of apparent losses is often relatively much greater than real losses, since the marginal costs of apparent losses occur at the retail rate charged to customers, while the baseline marginal cost of real losses is the production cost. For water suppliers the unit retail cost to customers may be 10–40 times the marginal production costs for treatment and delivery. Apparent losses occur at the "cash register" of the water utility and directly impact the water supplier's revenue stream, yet many systems have such unstructured water accounting and billing practices that they cannot even show that such loss is occurring. It is evident that reducing water loss would not only improve water supply operations but would also result in increased revenue. Sound water loss management, therefore, usually generates a direct and rapid payback to the water utility.

> **M**any water systems around the world cannot account for their lost water. What if banks could not account for the money that was deposited?

TABLE 1.1 Water Prices and Usage around the World

Region	Access to Water within < 200 m	Average Consumption (liters per capita per day)	Average Cost (US$/m³; 1993 data)
Africa	69.08%	53.56	1.30
Arab countries	88.23%	157.86	0.65
Asia	87.46%	160.7	0.54
Industrialized countries	99.60%	262.34	2.24
Latin America	86.87%	182.79	0.91
Transitional countries	99.08%	306.59	0.41
All	84.38%	161.33	1.08

> **M**ore than 1 billion people lack access to safe drinking water, and 3 million die each year from avoidable water-related diseases.

1.5 FORCES DRIVING CHANGE IN THE WAY WATER LOSS IS VIEWED AND MANAGED

Many areas of the world have water shortages and are unable to provide a continuous supply of treated water 24 hours a day. The World Bank reports that over 1 billion people in the world today lack access to safe drinking water, and 3 million people die every year from avoidable water-related diseases.[6] This situation has often been viewed as a problem faced only by developing countries, but even in the United States, 24 percent of water-borne disease outbreaks reported during the 1990s were caused by contaminants that entered the distribution system and not by poorly treated water. The rapidly expanding world population is requiring more treated drinking water. Since 1975 the world's population has expanded by roughly 50 percent, or 1.9 billion people.[7] Much of this additional population has congregated in cities that are already experiencing water stress or in new areas that are far from readily available water sources.

> **I**n the United States alone, 24 percent of water-borne disease outbreaks are caused by contaminants entering the distribution system. Leaks are an ideal place for contaminants to enter. We need to pay more attention to the public health aspects of leak management.

Twenty-eight countries in the world today do not have sufficient water to meet future demand, which means that people will die of thirst or, alternatively, extremely costly solutions to supplying water will need to be addressed. Such costly solutions often applied in arid climates include water desalination and the seeding of clouds with silver iodide to promote precipitation. The latter only works, however, if some cloud formation is present.[8] Both methods are extremely expensive and make the reduction of lost water as a potential new source in these situations much more attractive.

1.6 WHAT IS BEING DONE AROUND THE WORLD TO REDUCE WATER LOSS?

The challenges for us today are the same as they were during the days of the Romans; we just have more advanced methodologies and technologies to apply to the problem. We can look back at past efforts and smile and think that we are so much better, but to be honest we just have better tools. An open mind, unwillingness to accept existing inefficiencies, and a wish to improve are the basic tools a water system operator needs to have today. The rest can be purchased as work progresses. Water audits and loss control programs will only be successful if the operator and the utility are willing to accept what they find and act on it. Therefore it is critical that system operators understand the extent and impact of water loss, and that the control of water loss be considered of paramount importance throughout the entire organization.

> **W**ater system operators are now under pressure from various stakeholder groups to operate systems more efficiently, reduce losses, and improve performance.

Due to a number of dramatic late-twentieth-century changes in the water supply business model worldwide, a new breed of water utility manager has entered the water supply scene—one who strives to increase the performance of the utility, increase profits, and be accountable for the efficient use of one of nature's most precious resources. The need for this new breed of water system operator has been created by pressure from stakeholder groups who will no longer tolerate abuse and inefficient use of natural water resources. These include the environmental community, which has been successful in raising grass-roots consciousness to the level of environmental regulation at the national and international levels. Consumer advocates now carefully monitor the value of service per unit cost paid by the customer, expecting the utility to provide quality service at reasonable cost. Competitive forces have also increased, focusing utilities on improving both technical and business efficiency. The awesome power of the Internet, the media, and other communication forums has helped to accelerate all of these forces, which are mandating that water loss not be tolerated or overlooked as it has been in the past.

The successful structure established in England and Wales was implemented in a relatively short period of time and was driven by a number of the forces mentioned above. British water companies were privatized and reorganized along watershed boundaries in 1989. They also fell under a heavy regulatory burden at that time, which focused on effectiveness and the imparting of company operations and cost to the customer. The ability of water companies to pass costs along to customers is greatly limited by this structure, which ties approval for increased rates or tariffs to company performance. Consequently, innovation was accelerated as companies sought ways to improve performance, cut costs, and increase profits. Environmental concerns and the relatively high density of the population also have elevated support for the wise use of water in the United Kingdom. A notable catalyst was the severe drought that hit the country in the mid-1990s. This event triggered the establishment of new leakage reduction requirements and targets, which the companies were able to implement with the results of the National Leakage Initiative to guide them. While achieving great success in reducing leakage, the United Kingdom water industry still continues to study all aspects of water loss, as well as conservation, reuse, and other water efficiency practices. The relatively sophisticated system that is in place continues to be refined, due largely to the motivation of the government, environmental, and consumer sectors, which have placed a high value on protecting water resources.

The British technology has also had a dramatic effect on other nations as these methods have begun to take hold in perhaps several dozen countries. During the late 1990s, national or regional governments in South Africa, Malaysia, Australia, New Zealand, Brazil, and Canada adopted major new programs that emphasize leakage reduction. Strong programs in Germany and Japan are being refined. Extensive initiatives are under way in Malaysia and Brazil that will extend for 10 years or more, with ongoing investments of over $100 million in each project. The projects include auditing, pressure management, improved leakage monitoring, detection, and repair, and revenue enhancement. A major advantage of this leakage management technology is its transferability. Its techniques can be applied to water systems of varying characteristics and its performance indicators allow comparisons to be drawn with systems around the world. This aspect of the technology is perhaps its most compelling, and is likely a primary reason why it has spread so quickly in its use in the United Kingdom and around the world.

> We need to spend $325 billion on water system upgrades in the United States over the next 20 years.

1.7 PROGRAM NEEDS AND REQUIREMENTS FOR WATER LOSS CONTROL

According to American Water Works Association (AWWA) estimations, approximately $325 billion needs to be spent on upgrading distribution systems in the United States in the next 20 years.[9] Using average demand figures, the annual value of lost water and revenue, and therefore the approximate annual value of the water loss control market in the United States and worldwide, can be approximated. Interestingly, water loss control is estimated at approximately 29 percent of the above AWWA figure, or $94 billion. These estimations can be found in Table 1.2 and are approximations only. However, even if they are in error by 50 percent, the findings represent a huge, virtually untouched potential market for water loss control, which can be approached by water system operators, consultants, contractors, plumbers, and facility managers.

A complete loss control program is often referred to as a *water loss optimization program*. Optimizing basically means doing everything possible to improve the technical and financial performance of the water system, whether it is a public, private, or demand-side system. Optimization usually entails reduction of operating overhead and enhancement of revenue streams. Figure 1.1 shows a typical optimization graph. In this case it can be seen that the profitability in the beginning is low, as the cost of the water loss project is being borne on a performance basis.

Water loss optimization programs are sometimes undertaken on a performance basis. This means that the utility enters a special partnership agreement with a contractor or consultant. The contractor or consultant is paid a portion of the money recovered from the project over a certain

TABLE 1.2 Approximate Value of Water Loss Control Market

U.S. market potential	
U.S. population	250,000,000
Average consumption per person (kgal/year)	36.5
Average loss	16%
Split of real losses	60%
Average cost treated per kgal	$2.50
Average cost sold per kgal	$4.00
Recoverable percentage	75%
Total losses (kgal/year)	1,460,000,000
Total real losses (kgal/year)	876,000,000
Total apparent losses (kgal/year)	584,000,000
Value of recovered product per year	$2,190,000,000
Value of recovered revenue per year	$2,336,000,000
Value of recoverable percentage (product)	$1,642,500,000
Value of recoverable percentage (revenue)	$1,752,000,000
Market size per year	$3,394,500,000
Simple calculation for world market size	
Assumes like numbers, as:	
Other countries use less but with high loss	
U.S. water is cheap compared to others	
Loss value per capita per year in U.S.	$13.58
World population	6,000,000,000
World market size per year	$81,468,000,000

Source: Julian Thornton.

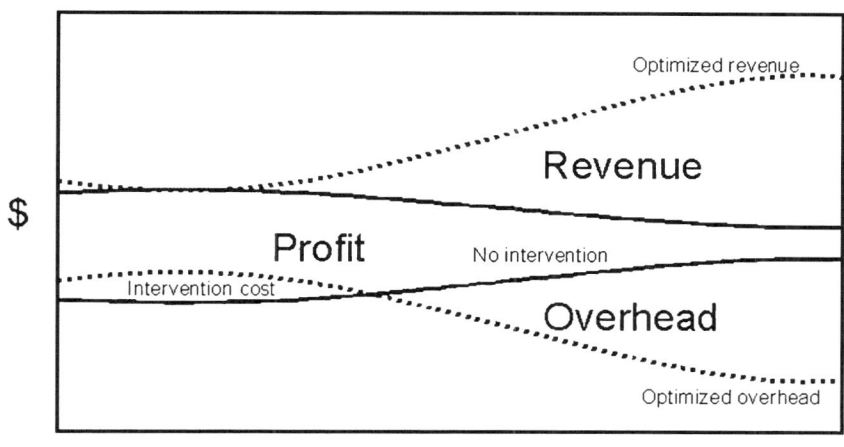

Figure 1.1 Sustainable solutions for improving water system performance. (*Source: Julian Thornton.*)

time frame. This is an excellent way of undertaking a project, especially for utilities that do not have a substantial initial budget to allocate for loss control, but do have an existing operating budget, which includes a fixed cost to operate the system *with* losses. The performance approach allows the utility to continue budgeting for normal allocation, while the actual cost of operation drops and the revenue stream increases as the work continues. At a certain point the contractor can drop out of the equation and the annual operating budget either reduces with increased income, and therefore profitability, or the additional funds can be redirected into other maintenance or training functions as required.

Water loss optimization programs may be implemented in four phases:

1. Water audit and analysis using performance indicators
2. Pilot study to demonstrate initial recommendations of the water audit analysis in the field
3. Global intervention
4. Ongoing maintenance of the loss control mechanism

Budgets may be relatively restricted for phases 1 and 2, until methodologies and techniques have been identified with paybacks in line with the expectations of the utility for their system. Case studies and industry papers are an excellent way of adding credibility to a decision for either phase 1 or phase 2 of a project, with little additional cost.

Operators must learn to be proactive and identify realistic programs and budgets to combat loss. They must learn to identify efficient, inventive methods to reach economic levels of loss, not just apply a minimal budget to loss control and then resolve the rest by way of a "pencil" audit, writing off a major portion of loss as unavoidable. The traditional rule-of-thumb notions of the amount of water loss viewed as "unavoidable" has changed with new methods that calculate system-specific levels of technical unavoidable annual real losses. This level of loss is much smaller than the traditional ways (Kuichling equation) due to the advent of new technologies, which allow us to control losses to much lower levels economically.

Some of the tasks included in a water loss optimization program are the following:

1. Overhead reduction tasks (real losses)
   a. Leakage reduction
   b. Hydraulic controls (pressure management)
   c. Pipe repair and replacement
   d. Customer service pipe replacement
   e. Condition assessment and rehabilitation
   f. Energy management
   g. Resources management
2. Revenue stream enhancement tasks (apparent losses)
   a. Baseline analysis
   b. Meter population management
   c. Meter testing and change-out
   d. Correct meter sizing and change-out
   e. Periodic testing
   f. Automatic meter reading (AMR)
3. Billing structure, analysis, and improvements
   a. Nonpayment actions
      (1) Turn off supply
      (2) Reduce supply to minimum
      (3) Legal action
      (4) Prepayment schemes
      (5) Reduction of fraud and illegal or unregistered connections
      (6) Continuous field inspections and testing
   b. Rate or tariff management
   c. Customer base management
   d. Modeling for efficient installation
   e. Modeling to assure economic efficiency

> In most cases water loss management is extremely cost-effective, with paybacks measured in days, weeks, or months, not years.

Automation is often a common component in an optimization program. Water loss optimization is usually a highly cost-efficient endeavor, since so many water supply systems currently suffer excessive water loss. The greatest challenge for today's progressive water manager is to change dated mindsets that view water as infinite and inexpensive. Once policy and decision makers understand the true value of water, implementing intervention techniques can be a relatively straightforward undertaking.

1.8 HOW THIS MANUAL CAN PROVIDE GUIDANCE

Like other water loss publications in print today, this manual discusses loss control methods and technology in great detail. However, this book also seeks to promote awareness, foster positive attitudes, and pull together not just the ideas of the author, but also those of other specialists in the field. In addition to our ideas and thoughts stemming from many years of hands-on field intervention against water loss and inefficient use, this book also highlights recent case studies and industry-specific papers to reinforce the concepts and methods already being applied successfully in the field.

Case studies are an excellent tool for assisting operators in preparing a master plan that takes an aggressive stance against loss and inefficiency. The fact that somebody else has done it before often makes the job of selling an aggressive program and budget to a board of directors more feasible.

The steps undertaken in a water loss optimization program are discussed and reviewed in detail throughout this manual. The chapters are self-contained and do not need to be read in order, although an operator with no experience in progressive water loss control methods is urged to read the entire book. The manual focuses heavily on the progressive methods pioneered in England and Wales in the 1990s and now used widely on an international basis. It also consistently evaluates the more "traditional" conditions that exist in North America, and other nations, where water loss has not been a top priority. This is done to demonstrate the need to control water loss proactively in even the most developed nations, and that easily transferable technology does now exist to control water loss.

This book includes sections that allow the reader to:

- ❑ Understand the nature and scope of water loss occurring in public water supply systems
- ❑ Assess loss conditions of any system by using a water audit
- ❑ Implement field interventions to control real losses
- ❑ Implement field interventions to control apparent losses
- ❑ Implement demand control
- ❑ Perform cost-to-benefit calculations
- ❑ Identify when and how to use a contractor or consultant

This manual is intended to be a hands-on tool for water system managers who are motivated to understand the nature of water loss and to take meaningful action to reduce it. Its content provides a detailed road map for any water system operator to implement a program that is the appropriate response for the individual water system's needs.

> This book provides many useful case studies, which may be used to justify implementation of a more aggressive loss management program in your utility.

1.9 OUR ROLE AS WATER STEWARD

Upon close evaluation it appears that many of the reasons for water loss from meter error, leakage, or data mishandling are actually based on human failings and lack of maintenance. The problems we face today are not much different from the problems faced 100 and even 2000 years ago by water system operators. The intention of this manual is to explain the reasons why suppliers should reduce water loss and identify how to resolve water loss problems using today's technology in an economically sound manner.

All water utilities, as well as industrial and residential end users, should practice water loss control and water conservation regardless of the size of their system or the nature of their use. The level of water loss management effort that is being exercised by water suppliers worldwide varies considerably. Unfortunately, most of the water industry in the United States and many parts of the world accord water loss only secondary priority, since the true economic and social impact of water loss has not yet been realized by policymakers. Water loss thus continues to suffer from a lack of good auditing practices and a failure to reduce leakage proactively, but instead just waits for the next customer complaint to prompt reactive repair of the next problem leak. However, in a small but quickly growing number of countries throughout the world, comprehensive water efficiency goals have been established. Water conservation, watershed protection, reuse, and the new discipline of leakage management have been implemented as required practice by the highest level of government, and supplier performance is closely monitored and sometimes regulated. This new model of water resources management is the way of the future, as it must be if humanity is to continue to sustain its growth and its environment.

1.10 REFERENCES

1. Lambert, A. O., Myers, S., and Trow. S., *Managing Water Leakage: Economic and Technical Issues,* Financial Times Energy Publications, 1998.
2. Lambert, A. O., International Water Data Comparisons, Ltd., personal communication, October 2000, re interpretation of United Kingdom Office of Water Services (OFWAT) reported leakage results.
3. *The Family Health Medical Encyclopedia,* edited by I. A. G. MacQueen, Book Club Associates/William Collins Sons & Co., 1978.
4. Pearce, E., *Anatomy and Physiology for Nurses,* 16th ed., Faber, London, 1975.
5. U.S. Geological Survey, *Estimated Use of Water in the United States in 1995,* Circular 1200, 1998.
6. World Bank reports.
7. *1999 World Almanac and Book of Facts,* World Almanac Books, Mahwah, N.J., 1998.
8. *The Times Concise Atlas of the World,* 5th ed., Times Books, London, 1986.
9. American Water Works Association, *Stats on Tap,* revised Feb. 15, 2001, www.awwa.org/pressroom/statswp5.htm.

CHAPTER

2

Understanding the Types of Water Loss

Julian Thornton
George Kunkel

2.1 DEFINING WATER SUPPLIER LOSSES

Simply stated, the problems of water and revenue losses are[1]

- *Technical:* Not all water supplied by a water utility reaches the customer.
- *Financial:* Not all of the water that reaches the end user is properly measured or paid for.
- *Terminology:* Standardized definitions of water and revenue losses are lacking.

The International Water Association (IWA) defines two major categories under which all types of supplier water loss occurrences fall:

- *Real losses* are the physical escape of water from the distribution system, and include leakage and overflows prior to the point of end use.
- *Apparent losses* are essentially "paper" losses and consist of customer use which is not recorded due to metering error, incorrect assumptions of unmeasured use, or unauthorized consumption.

While these two definitions are distinguished by a stark physical differentiation, a dramatic economic difference also exists, on a marginal cost basis. Real losses, which are most usually leakage, are typically valued at a marginal production cost of the water. Apparent losses, which occur at the customer destination, penalize the water supplier at the retail cost—a rate usually much higher than the production cost. While the marginal costs are only short-term costs, and other long-term costs must also be assessed, the cost implications of real and apparent losses require that a careful assessment of each be undertaken to design the most appropriate water loss optimization program.

2.1.1 Why Real (Leakage) Losses Occur

A water supplier's real losses are not really losses at all, since these losses (leakage) do not represent water that is destroyed but water that is usually returned to the watershed—albeit by expending considerable costs for transport and treatment without recouping any revenue. Apparent losses include no physical impropriety, as the water has reached the destination of an end user; but this successful supply function lacks

14 CHAPTER 2 UNDERSTANDING THE TYPES OF WATER LOSS

either a full and accurate accounting and revenue capture, or the use of the water was unauthorized.

Leakage is the most common form of real losses for water suppliers, and occurs for a number of reasons, including:

- Poor installation and workmanship
- Poor materials
- Mishandling of materials prior to installation
- Incorrect back-fill
- Pressure transients
- Pressure fluctuations
- Excess pressure
- Corrosion
- Vibration and traffic loading
- Environmental conditions such as cold
- Lack of proper scheduled maintenance

British leakage management terminology distinguishes *reported versus unreported* leaks, or, more literally, *reported bursts* and *unreported leaks*. Dramatic pipe bursts are the most recognizable example of reported leaks, which, due to their damage-causing nature, are usually quickly reported, responded to, and contained. However, unreported leaks, often running at a small rate of flow on underground pipes, frequently escape the attention of the water supplier and the public, but account for larger amounts of lost water since they run undetected for long periods of time. In the United States, the terms *reported* and *unreported* are not employed, so the distinction between a "leak" and a "break" (burst) is rather subjective, and one of a number of examples of inconsistent terminology.

> **S**ervice leaks often cause the largest volumes of loss.

A significant finding of leakage efforts over the past decade has been the large amount of water loss occurring on the customer service piping branching from the water main and supplying water to a single or multiple user premises. For many systems, leaks on these small-diameter pipes represent the greatest number of leaks encountered in water supply operations. Often, supplier policies require customers to own their service lines and to execute repairs or replacement when necessary. Unfortunately, many customers are often unaware of their ownership responsibilities and, when advised to repair known leaks, are neither timely nor effective in getting relatively expensive repairs executed. Consequently, customer service piping leaks can run for considerable periods, even after being reported, and account for substantial water loss. Severe drought in England in the mid-1990s resulted in emergency regulations that required some water suppliers to implement repairs on leaking customer service lines. The resulting savings in lost water were found to be so significant and the repair methods so efficient that national regulations were soon established requiring all water companies to implement policies for company-executed customer service line leak repairs. Two other notable aspects of these regulations are that customers retained their ownership of the lines, and, once high initial backlogs of customer leaks were repaired, the rate of occurrence of new leaks was sufficiently low that the repair policies for the water companies were found to be manageable and cost-effective. This experience demonstrates dramatically the principle that leakage losses depend on two primary variables: rate of flow and time permitted to run. Both parameters must be considered in developing a leakage management strategy. Too often, water suppliers lose track of small-volume leaks, allowing indefinite leak time to occur and losses to mount.

Using data from 1993 to 1996, the American Water Works Association (AWWA) estimated that approximately 75,000 main breaks occur each year in the United States. It is unknown how many leaks occur, but annual leaks likely outnumber main bursts several times over in typical water supply systems, likely resulting in 250,000–500,000 leaks per year. The United States has approximately 880,000 mi of distribution mains, many of which are old, unlined cast iron in need of repair, rehabilitation, or replacement. However, good leakage control practices can help prolong the life of the existing infrastructure by reducing the occurrence of leaks, breaks, and forces leading to water main failures.

> The United States has approximately 880,000 mi of mains.

Another technique employed in recent times by progressive leakage management programs around the world is the science of pressure management. In designing water infrastructure, engineers have frequently specified distribution system pressure levels with the primary objective of providing service above a minimum design pressure. However, local guidelines for providing fire flows, expansion capacity, and "safety factors" have frequently resulted in systems supplying water pressures far above minimum requirements, without consideration of the impact of the "excessive" pressure. By the late 1990s, fundamental relationships between pressure and leakage rates had been established, which showed that certain types of leaks are highly sensitive to changes in pressure. It can now be taken that, while certain minimal levels of pressure should be provided, maximal levels for pressure should also be established and not exceeded. Excessive water pressure not only increases certain types of leakage, but also influences main break rates and the amount of needless energy costs a supplier expends. In progressively managed water systems, water pressure is now controlled within an appropriate range that meets the needs of the customer and the supplier without causing waste or harmful impact to the infrastructure.

> Pressure has a much greater impact on leakage than we originally suspected. System design should take into account maximum pressure limits as well as minimum ones.

Real losses—largely leakage losses—typically account for the greatest volume of water "lost" by suppliers, although not necessarily at the greatest cost. Considerable research work has been conducted in the past decade on the nature and impact of leakage, and highly effective practices and technologies have been implemented around the world. It is in the interest of all water suppliers to evaluate closely the leakage occurring in their systems and to take advantage of these methods which may be considered best practice in controlling leakage losses.

2.1.2 Why Apparent Losses Occur

Apparent losses are significant for two reasons:

- ❑ Relative to real losses, apparent losses typically have a much greater short-run marginal cost effect, since they affect revenue at the retail customer rate.

16 CHAPTER 2   UNDERSTANDING THE TYPES OF WATER LOSS

❑ An inappropriate assessment of apparent loss often results in real losses being overstated in the water audit, potentially misguiding water loss optimization planning by placing inordinate emphasis on leakage while potential revenue recovery goes unattended.

Apparent losses of water can be perceived as occurring in three primary ways:

1. Errors in water flow measurement
2. Errors in water accounting
3. Unauthorized usage

Before discussing the specifics of these losses, it is appropriate to review the typical metering and billing structures used by water suppliers. Most of the world's water supply systems have come into existence in the last 150 years, a period in which the earth's population has more than quadrupled. Water supply systems have multiplied to serve these masses, typically requiring considerable infrastructure including pipelines, pumping facilities, and treatment plants. With the establishment of modern indoor plumbing, customer service pipes have been tapped directly into local water pipes or mains to bring water directly into the homes of the consumer. Figure 2.1 shows a typical direct-feed situation.

Many water suppliers have chosen to incorporate customer water meters at the end-user premises and gather regular meter readings for the purpose of billing per unit volume of actual water used. Customer meters also allow users to monitor their own water usage and provide customers the option to exercise restraint against excessive use and identify waste. Outwardly, this approach seems to follow the norms of typical free-market commodities: charges are based on the volume of product or service delivered. However, the use of customer meters and usage-based billing

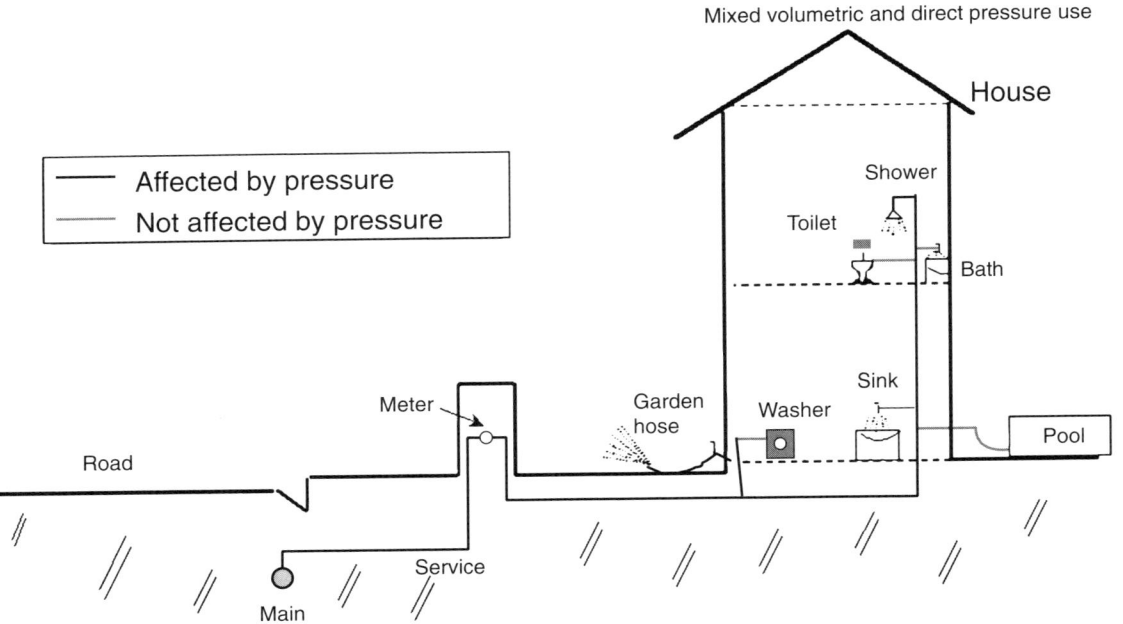

Figure 2.1 Typical direct pressure residential supply situation. (*Source: Julian Thornton.*)

is far from universal in the water industry in the United States or the world at large. For a large portion of public water supply customers, service is provided without any measuring of their actual water usage, and billings are based on flat-rate charges assigned by customer user type. In the United States, perhaps only one-half of all users have water meters, with sentiments regarding metering sharply divided in certain areas of the country. In England and Wales, traditionally only the industrial, commercial, and institutional (ICI) population was metered. Environmentalists and regulators support the establishment of universal residential customer metering, and a slow transition is occurring, with meters being installed in new construction and upon customer request. Approximately 17 percent of all residential properties were metered as of the close of 2000.

2.2 QUANTITY AND VALUE OF WATER

While opinions differ about the use of meters and billing for actual usage, it is difficult to dismiss the argument that humans hold a strong intrinsic value on the relationship between the amount of a commodity and its price. That the use of "volume discounts" that is so prevalent in the business world indicates that the cost–usage relationship is an important business strategy. Discounts and penalties are used as necessary tools to promote increased or decreased sales, respectively. Water rate or tariff structures are designed with this in mind, with decreasing block rate structures tending to promote water sales and increasing block rates promoting conservation. Again, these strategies are effective only if water consumption is measured at the point of usage. Constant or flat rates typically make it more difficult to influence customer usage patterns, although the British experience shows that this is possible in its culture by using a combination of public education, select customer usage profiling, and other techniques. For many water suppliers in the United States that utilize flat billing rates, do not provide customer meters, and lack public education or conservation programs, the tacit message reaching the customer resembles an "all you can eat buffet" advertisement.

While equity issues, disadvantaged population concerns, politics, and other nontechnical issues often influence decisions on the type of rate structure to employ, the purely technical aspects of usage versus cost play a unique and usually overlooked role in the amount of water loss incurred and tolerated in a society. Providing water service without measuring its usage removes the "finite" sense of the resource from the mind of the consumer. If any amount of water is available for the same cost—and the water supplier is not advocating conservation—it is natural for the customer to hold water in lower regard than other vital commodities such as food, health care, transportation, and other utilities. The vast majority of these vital needs are purchased per unit volume of commodity or service, not at flat billing rates for unlimited usage. Using constant or flat rates devalues water in the mind of the consumer, and what we do not value, we waste. This is an unfortunate aspect of human nature.

2.2.1 The Importance of Metering

Metering is an essential component of an effective water loss optimization program. In the words of Lord Kelvin, "If you don't measure it, you can't manage it." Water flows must be accurately measured if they are to be traced in their journey from source to consumer. Flow measurement should occur at multiple points along this journey, to account for valid uses and assess loss that may be occurring between the two metering points. The primary metering points used in water supply systems around the world typically include:

- Production meters
- Meters on district metered areas (DMAs)
- Customer meters

Production or master meters measure the bulk flows that are the input to the distribution system. Having functioning, accurate production meters is the foundation of sound water resource management, yet an alarming number of water utilities have no production meters, or have none that function accurately. Making sure that production metering is established and routinely calibrated and tested is an essential first step for many water suppliers who need to gain a grasp of their water loss status.

Measuring flows into DMAs is a technique that has been in use for over 100 years. However, the application and sophisticated technology used with this technique has been greatly improved. DMAs are established by creating a discrete zone within the water distribution system, by closing valves in "connect-the-dot" fashion, leaving only one or two pipelines to supply water to the zone. Portable or permanent flow meters are used at this input point(s) to measure water supply into the zone on a continuous basis. The data of greatest interest in these measurements is the rate of flow occurring in the early morning hours, when most of the population is asleep and legitimate water usage is at a minimum. Since leaks will continue to waste water on a 24-h basis at a relatively constant rate, leakage is at its maximum percentage of the total flow during these hours. Computer modeling methods now exist to aid the assessment of legitimate night water usage and distinguish it from the derived leakage. This method of night flow leakage measurement is now the standard means of quantifying leakage in most progressive leakage management programs. Absent the use of this technique, many water systems are only able to estimate leakage, often as the remainder of loss after all other (apparent) losses have been quantified.

Customer meters measure flow at the point of use. The definition of "point of use," however, may vary from supplier to supplier. Some systems provide one water meter to each customer service pipe, regardless of whether that service pipe supplies a small dwelling for a family of four or a multi-unit high-rise building serving hundreds of people. Other systems utilize tiers of submetering within large buildings to measure flow to each dwelling unit, the true end-user point. How customer service pipes are configured, metered, and billed is determined most frequently by local regulations and preferences. Regardless of the way they are configured, these functions collectively have many complexities regarding the gather-

> Many utilities still do not have accurate master and supply meters.

ing of metered usage data, execution of equitable billings, and water demand management.

Metered water data is the basis for the establishment of the *water audit* or *water balance*. Similar to a financial account, an accurate balance statement should be drawn up by the water supplier at regular intervals, typically every year. Just as financial balance statements track the flow of money into and out of accounts, the water audit tracks the quantities of flow into and out of the distribution system and identifies all of its uses and losses. In progressive water loss management systems all three levels of metering are utilized to provide a high level of accountability in the auditing process.

In England and Wales, the use of DMAs is required and extensive, and gives a direct measurement of leakage. While most residential customers there do not have permanent meters, representative usage profiles have been obtained for customers by establishing test metering in select customer properties. These profiles provide a good picture of legitimate customer usage as well as assisting in the derivation of nighttime leakage flows. Consequently, the water audits, which are required for all of the 26 large water companies in England and Wales, are reasonably good portrayals of water usage in public water supply systems. Water audits are mandatory in the United States only under certain limited state or regional regulations, and it is likely that most water suppliers in the United States perform no regular water auditing for their systems.

2.2.2 How Errors in Water Flow Measurement Occur

Errors in measurement can occur in several ways. First, water meter readings can be in error due to a variety of mechanical or applications reasons. Due to widely varying water usage patterns among customer populations, a number of different meter sizes, and sometimes types, can be found in any single water utility. Standard displacement or velocity meters provide accurate flow measurement for residential users, while large ICI users may experience dramatic differences in daytime and night time flows, requiring meters that are accurate through a wide range of flow rates, i.e., compound-type meters. Other factors place demands on the water supplier to provide accurate metering. Some of the major reasons why water meters fail to measure water flow accurately include the following:

- ❏ Wear over time
- ❏ Water quality impact
- ❏ Chemical build up
- ❏ Poor finish and workmanship
- ❏ Environmental conditions such as extreme heat or cold
- ❏ Incorrect installation
- ❏ Incorrect sizing
- ❏ Incorrect specification of meter type for the application
- ❏ Tampering
- ❏ Lack of routine testing and maintenance
- ❏ Incorrect repair

Recommended maintenance practices for customer meters include monitoring recorded usage patterns and rotating the meter out of use on a regular basis for testing, calibration, repair, or replacement.

Many systems use estimates for customer usage for accounts where water meters are nonexistent, defective, or unreadable. Estimates, which are used both temporarily or permanently, can be inaccurate if they are not devised in a rational manner or kept up to date with changing customer usage patterns; hence another form of inaccurate water measurement can occur here.

Meter reading is another step in obtaining accurate water usage data. Errors in meter reading are essentially errors in measurement. With the growing use of automatic meter reading (AMR) systems, the opportunity for meter reading error is probably being reduced relative to that occurring in traditional manual meter reading operations. However, all systems seeking to optimize should include at least a brief assessment of the accuracy of meter reading operations in transferring actual measured water usage into the information handling (billing) system.

2.2.3 How Errors in Water Accounting Occur

Errors in the handling of customer accounts can occur in a number of ways, some of which include:

- Customer water usage data is modified during billing adjustments.
- Some customers who use water are inadvertently or intentionally omitted from billing records and go unmonitored.
- Certain users are accorded nonbilled (free or subsidized) status, and usage is not recorded.
- Human errors occur during data analysis and billing.
- Weak policies create loopholes in billing and water accounting.
- Meter reading or billing systems are poorly structured.
- Changes in real estate ownership or other changes in customer account status are tracked poorly or not at all.
- Technical and managerial relationships in assessing, reducing, and preventing water loss are poorly understood or implemented.

In the United States, "water accounting" is not an established practice like financial accounting, which has substantial controls and accountability built into its standardized process. The fact that consistent standards for water accounting do not exist likely results in many water systems understating actual customer usage and failing to capture full billing potential.

> **M**ost errors in water accounting occur as a result of lack of structure and controls in the accounting process.

The Pitfalls of Using the Billing System to Extract Water Accounting Data

The advancement of society into the information age has resulted in the use of computers for almost every application, including public water supply. Likely the most important computer application of any utility—and certainly the most visible one to the customer—is the billing system. Billing systems serve a primary function: to generate a regular bill to the customer for water services. The water bill is the water utility's primary revenue recovery instrument. However, many water suppliers in the United States also use the billing system—without adequate controls—as a water accounting system to track customer water usage. Water bills include charges for water supply as well as administrative or service fees, and sometimes waste or storm water charges. As described previously, the basis for the water charge may be actual customer meter readings, estimates, or flat fees based on user type. In a typical billing system an account is established for each user, although the definition of "user" can vary from system to system. For some, each distinct dwelling or property represents a user. In systems where submetering is used in multi-unit buildings, many user accounts may exist in a single structure. For systems that meter their customers, meter readings should be obtained either manually or automatically on a regular interval, monthly or quarterly being the most common. In this way customers are able to track their usage since the last meter reading cycle. The water utility merely adds billed usage from all accounts to determine the total customer demand for the period.

The collective meter readings for a system over a meter reading cycle may be referred to as *customer metered usage,* which is effectively the input to the billing system. In similar fashion, *customer billed usage* is the output of the billing system. With adequate controls and accountability, customer billed usage should equal customer metered usage. Most water utilities in the United States simply assume that this is so and do not make a distinction, carrying one "usage" or "consumption" parameter in the billing system. However, some billing systems are programmed in a manner that may result in changes to the customer metered usage for some accounts when needed billing adjustments are made.

As an example, envision a customer account that has gone estimated for a period of years due to an inability to obtain actual meter readings. Finally, perhaps a change in ownership of the property results in the water supplier obtaining an actual meter reading. Tracing back to the last known reading, the utility may find that it has either overestimated or underestimated the usage. If the account has been overestimated, then a (possibly large) credit is due to the customer. A credit is often shown as a negative charge on customer bills, and this is standard financial practice. However, some billing systems, in order to calculate the appropriate negative charge, may trigger this charge via the water usage parameter and thus enter a (possibly large) negative water usage amount. It may be argued that, since an overstatement of the customer's usage occurred in small increments for years, including a negative adjustment results in an appropriate correction with no net gain or loss in the water usage measurement. However, including a large negative number in a single billing cycle distorts water usage during that period and for the current year.

A more severe understating of actual customer usage can occur if the billing system triggers strict cost adjustments, such as discounts and waivers, via changes in the actual water usage parameter. In the United States, many water suppliers take the amount of water consumed by the service population as the total customer billed usage figure extracted from the billing system. Water suppliers should take care, however, to ascertain that the customer metered usage that is fed as input into the billing system is not unrealistically modified, as it becomes customer billed usage, or the output of the billing system.

A relatively simple concept can be employed to provide a control to safeguard against inappropriate water usage adjustments. Billing systems can be configured to include separate parameters for customer metered usage and customer billed usage. Any billing cost adjustments triggered by usage data would change only the customer billed usage figure, while the customer metered usage figure would always include the most representative water usage data for the customer.

Other Accounting and Accountability Errors

Other forms of apparent loss arise from an array of accounting errors that occur in practice. All systems likely have some customers who have not been properly documented in the utility's billing system. They use water freely, with never a call or a bill from the local water company. Such situations can occur due to human error, gaps in permitting regulations, or policy shortcomings. Changes in ownership or building renovations which result in new water service pipes and different sized meters can also cause confusion in billing account tracking if sound accountability procedures are not in place.

In many communities throughout the United States, local government requires that the municipal water utility not bill government-owned properties. Unfortunately, since the distinction between water accounting and billing accounting is not understood, many such properties go unmetered and unmonitored as a result. Government use, which often includes water used in water and wastewater treatment processing, public fountains, irrigation for public parks, gardens, and golf courses, and other municipal uses, can be substantial. This practice used by many local governments is also an example of water not being properly valued as a resource and a commodity, but instead as a political instrument.

> **P**olitical influences in municipal water supply settings have likely played a large role in allowing unrecorded water usage.

Devising coordinated metering, billing, and accounting practices for water suppliers is one of the major needs to improve water loss control throughout the world. For water system managers setting out to evaluate the nature and extent of water loss within their operation, careful scrutiny should be given to these activities and their impact on apparent losses.

Unauthorized Water Usage

The last of the three primary occurrences of apparent water loss is unauthorized water usage. While human nature holds a high regard for the quantity–cost relationship, it is also true of human nature that a certain small segment of a population will attempt to obtain service illegally and without making payment. Unauthorized use is likely a more common phenomenon in systems where customer meters are in use and water is billed per unit volume. Where flat rates are charged and usage is not routinely monitored, customers may draw greater quantities of water and lower their own effective unit cost. These customers need to evade inclusion in the billing process altogether in order to obtain water service without paying.

Unauthorized usage can occur in a number of ways. Much unauthorized usage occurs at the point of established end users. Some customers tamper with meters or meter reading equipment in order to lower meter readings. Fortunately, many AMR systems have tamper-detection features that help thwart such activity. Unscrupulous users with large water meters have been known to open valves on unmetered bypass piping, thereby routing their supply around the active water meter. Some users or contractors may consciously or unwittingly connect branch plumbing pipes to customer service lines upstream from the water meter, which also provides supply without passing through the meter.

Urban systems in the northeastern part of the United States have encountered the frequent occurrence of customer restoration of terminated service connections. Closing and locking curb-stop valves on the customer service line is a common means of terminating service used by water utilities in the United States against delinquent customers. Illegal restoration occurs when delinquent customers reactivate their own water service after the water supplier has stopped it because of nonpayment. These situations evidence the need for water suppliers to continue to monitor terminated accounts after they are shut off, to check for resumed, unauthorized usage. The city of Philadelphia provides such monitoring and has achieved success in reducing illegal restorations, lowering the discovery rate from 35 percent of all terminated accounts to roughly 20 percent in less than five years. A city-wide AMR system for residential accounts, completed in 1999, is also assisting this effort in Philadelphia, since meter reading and usage continue to be monitored even if an account has been shut off for nonpayment. In contrast to the U.S. experience, regulations do not allow water companies in England and Wales to terminate water service to customers under any circumstances.

Unauthorized usage has also been known to occur when persons find ways of withdrawing water from a location in the distribution system other than the customer service line. With fire hydrants constructed as aboveground appurtenances in the United States, illegal opening of these devices happens regularly in many cities. In some areas, using fire hydrants to fill street cleaning equipment, landscapers' trucks, and construction vehicles has occurred so casually that upstanding businesses perceive this to be acceptable practice. Water utilities in such places have a public education challenge to instill the value of water as a commodity in the business community. Establishing bulk water dispensaries is now common for water systems that wish to allow—and even promote—water sales outside the normal customer service line connection. Some systems allow water to be used from fire hydrants in an authorized manner with the filing of a permit. With concerns for cross-connection protection and the accountability of water, such a practice is not a preferred one for most water utilities.

All water suppliers should be mindful that the potential for unauthorized usage exists to some degree in their systems. Just as retail establishments must take safeguards against shoplifters, water systems should have appropriate controls to monitor for unauthorized usage and hold it in check.

> Theft of water can be a common occurrence in the United States and is not just a Third World problem.

2.3 THE HYDRA OF WATER LOSS

It is evident that water loss can occur in many ways in today's water utility. Initiating a water loss optimization program may seem akin to Hercules preparing to battle the multiheaded Hydra of legend. Multiple efforts must be carried out to address all the different types of water loss occurring in a water system at the same time. Having a good understanding of these types of loss is a fundamental need that the reader has now met by reviewing this chapter.

Just as Hercules had certain strengths and weapons at his disposal in order to vanquish the Hydra, so too do today's water managers in seeking reductions in lost water. Many methods and tools are now available, the first of which is the water audit or water balance, described in Chap. 4. Compiling a sound water audit allows the manager to identify the type and extent of leakage occurring in the system, and is the first proactive step in identifying the best corrective actions to take.

2.4 REFERENCE

1. International Water Association, *Performance Indicators for Water Supply Services,* Manual of Best Practice, IWA, 2000.

CHAPTER

3

Traditional and Progressive Approaches to Water Loss

The Experiences of the United States and England

Julian Thornton
George Kunkel

3.1 WATER LOSS PERSPECTIVES

Many utilities ignore their water losses due to a perceived lack of resources and the burden of many other priorities of system operation. Others have downplayed their losses through fear of public resentment, especially in cases where the utility is asking the customer to conserve water or pay higher rates or tariffs. In areas where only limited water audit regulations exist, utilities often eliminate their losses on paper using "pencil" audits that are not scrutinized by outside stakeholders. Most of these practices, however, merely reflect the lack of a regional or national agenda for water loss control.

Many developing, and some developed, countries have systems which cannot provide customers a continuous water supply on a 24-h basis, particularly during times of drought. Other systems are faced with rapidly developing communities requiring costly new water sources. Some serve a heavy holiday and tourist trade, resulting in weekend and holiday peaks many times higher than normal operating peak flows. These systems often borrow significant funds and install costly new water sources that are utilized on only a part-time basis. The rest of the time the costly investment sits unused and inefficient. For these and other systems, initiating or accelerating a successful water loss optimization program will defer the cost of loans for capital investments and usually provide a very fast payback.

> **M**any utilities use "pencil" audits as a way of hiding losses. As public awareness of the level of losses grows, operators will be called upon to be more accountable and report real levels of loss.

3.1.1 How Water Loss Is Viewed in the United States

The United States is blessed with bountiful natural resources. Water is a primary resource that has been consistently developed to help the country grow to the level of strength and prosperity that it enjoys today. Unfortunately, the availability of plentiful water during the country's early history may have contributed to a water supply infrastructure and an American psyche that now tolerates significant water loss. Coupled with a general lack of awareness of this fact by the public *and* many water supply professionals, this condition is a cause for concern as many water systems face the challenges of growing populations and limited resources in the near future.

Due to a lack of standard reporting methods, it is difficult to quantify the extent of water loss occurring throughout the United States. Data from the U.S. Geological Survey for 1995 documents a significant difference of 6 billion gallons per day that exists between source water withdrawals or abstractions and water consumed by users in the country. A 1987 American Water Works Association (AWWA) Research Foundation study estimated that the value of water that is lost or not properly accounted for in the United States varies from $158 million to $800 million each year. These numbers were based on different definitions of "unaccounted-for water" and water losses.[1] The literature is also replete with numerous case studies of systems—large and small—that suffered huge water loss before corrective action was taken. While consistent rational methods are not yet employed in the United States to quantify water loss accurately, it is evident that a significant problem exists from the number of documented case studies, the absence of accountability standards, and the casual state of awareness of this condition in the U.S. water supply industry.

The term *water accountability* has been used casually in the United States for the last several decades to label a variety of activities that affect the delivery efficiency of water utilities. Water accountability, however, exists more as art than science, and its methods often generate as much confusion as explanation in interpreting water loss conditions. Symptomatically, this confusion stems from inconsistent terminology, unreliable percentage measures, and a lack of procedures to evaluate and compare water loss performance. On a broader level, however, water accountability is a weak discipline due to the lack of awareness of the extent of water loss occurring in the United States. Lacking a concern for this problem as a major issue, no national agenda exists for the reduction of water lost by suppliers, i.e., supply-side losses.

Conversely, the field of water conservation exists as a better-structured discipline that has achieved certain success in limiting unnecessary water usage, particularly in the arid and semiarid regions of the country, where significant population growth is occurring and water is both limited and expensive. Water conservation focuses largely on water reduction by the end user, by improving usage efficiency and reducing waste. It has achieved recognition at the national level with legislation that sets requirements for household water appliances and other water uses. Unfortunately, supply-side losses occurring due to leakage and poor accounting by water utilities are likely to be many times greater than the end-user savings achieved through conservation.

High water loss occurs in the United States for many reasons that are varied and complex. Some of the root causes are discussed below.

> It seems strange that the public is willing to conserve water when such large levels of loss occur on the supply side. This will most likely change.

Cultural Attitudes

Americans are the world's consumers. This consumption includes water, with relatively high per-capita usage compared to many other industrialized nations.

"Conserving" is sometimes viewed as "doing with less," a notion that runs contrary to the American way of thinking, which is often geared toward building, development, and exploitation of resources.

For many utilities, water is unmetered, thus removing the "finite" sense of the resource from the thinking of both the consumer and the supplier.

As in other parts of the world, water is often undervalued—literally and emotionally—in the United States. Costs to the consumer are often intentionally suppressed for social or political reasons.

Geography and Demographics

Populations are growing in the "Sunbelt" states, where water is often scarce and expensive. The critical role of water in assisting development results in a good appreciation for conservation in these areas, and generally younger infrastructure encounters less loss due to leakage. However, the frequent need to import water over vast distances requires complex planning and negotiations and large infrastructure (reservoirs and pipelines).

Population growth has slowed in the former industrial states where water has been relatively plentiful and inexpensive. Often having still-abundant resources and excess capacity, but a declining customer base and ageing infrastructure, losses are often overlooked in these systems, even as they continue to grow.

Water Utility Organization and Structure

Most of the 55,000 water suppliers in the United States are extremely small utilities existing in rural areas, while a relatively small number of medium- and large-sized systems supply the largest share of consumers in densely populated areas.

The organizational and management structure of water utilities varies widely, with many systems operated by local governments, either as municipalities or authorities; and many large and small privately operated systems exist as well.

System boundaries usually coincide with political boundaries rather than natural (watershed) boundaries.

Frequently, water accountability practitioners are distribution system operators and water conservationists are public affairs or policy professionals. Lacking a national awareness and consensus on the overall water loss problem, these two camps have not interacted widely or integrated their efforts under a single demand management or water efficiency mission.

Establishing standards amid this wide array of conditions is complex, but as the use of water quality mandates under the U.S. Safe Drinking Water Act (1974; amended most recently in 1996) has proven, not insurmountable.

> As federal organizations become aware of the potential for increased efficiency in the public water supply systems, it is likely that regulations will be put into place to enforce performance, as was the case with water quality.

Environmental Perspective

The environmental consciousness has grown steadily in the United States over the past several decades and is now a balancing force in planning and development decisions in this country. The establishment of the Environmental Protection Agency (EPA) confirmed that consideration for the environment must be part of the decision-making process.

However, high losses indirectly require oversized infrastructure, excess energy usage, and unneeded withdrawals or abstractions from source water supplies, all of which have potentially unnecessary—and sometimes damaging—effects on the environment.

It is likely that many new source water abstractions and infrastructure expansions could be avoided if loss reduction was achieved; i.e., water loss reduction might represent one of the largest components of untapped water resources and potential for energy reduction in the United States.

Government Action

In the late twentieth century, significant federal governmental involvement created extensive water quality legislation and rules for clean streams and drinking water. Conversely, federal requirements for auditing water delivery and customer usage have only minimal structure and degree of impact.

Considerable concern has grown for the need to replace aging infrastructure and identify appropriate funding mechanisms. Yet the scope of infrastructure needs is often based on projections that may not include improvements from loss reduction. A more modest estimate of national infrastructure needs might be derived if realistic loss reduction and conservation were consistently included in the analysis.

While a national agenda to combat water loss has yet to develop in the United States, a considerable number of systems have independently addressed lost water via several recognized methods. The AWWA published Manual M36, *Water Audits and Leak Detection,* in 1990,[2] with a revised edition issued in 1999.[2] This document provides excellent instruction to water suppliers on assembling a water audit. (More details can be found in App. A.) Like the financial audits accountants routinely perform to track the use of all monies flowing in and out of an organization, a water audit attempts to define just where all the water a supplier produces goes. Unfortunately, it is believed that most U.S. suppliers do not routinely compile a water audit. Those that do frequently employ their own audit structure or method, without following any established guidelines. Most state and regional water supply oversight agencies require water systems within their jurisdiction to complete reports stating basic water usage and loss data from their operations. The sophistication and usefulness of such reporting is believed to vary widely, however. Dry, heavily populated states such as California employ a consistent reporting structure for suppliers, with results carefully monitored and audited when data are in doubt.[3] Many states, however, use report formats structured only in very general terms—leaving water utilities wide latitude in the type of data and responses that they provide. Similarly, these states do not perform field audits or investigations for water loss unless gross error or malfeasance is suspected.

Many water systems in the United States employ regular efforts to address leakage, but, unfortunately for the vast majority, this consists solely of repairing broken or burst water mains and leaks that have generated customer complaints. This means of addressing leakage in the United States is purely reactive in nature. British leakage management terminology distinguishes *reported* versus *unreported leaks*. Burst water mains are the most recognizable example of reported leaks, which, due to their damage-causing nature, are usually quickly reported, responded to, and contained. However, unreported leaks, which frequently escape the attention of the water supplier and the public, account for larger amounts of lost water, since they run undetected for long periods of time. While most U.S. water suppliers provide reasonable response to reported leaks, those that conduct regular unreported leak searches or surveys (usually at one- to five-year intervals) probably represent a minority of the country's systems. Many systems conduct no surveys to detect unreported leaks. Generally only the larger water systems employ specific "leak detection" personnel and purchase sophisticated leak correlators or other electronic equipment. Smaller systems typically rely on leak detection consultants to provide pinpointing services for hard-to-find leaks and to conduct periodic surveys of their systems to search for unreported leaks.

For over 200 years, plentiful water resources have assisted growth and development in the United States. This same abundance, however, has allowed water loss to be a secondary priority for most water suppliers. Times are changing, however, as expanding and shifting populations stress resources that, due to environmental safeguards, are not exploited as freely as in the past. The demands for stringent water quality, high customer satisfaction, and renewal of the nation's infrastructure indicate that the United States should begin to adopt a more proactive approach toward water loss control. Such approaches have begun to take root around the world, with the most systematic structure for the efficient use of water implemented in England and Wales in the 1990s. Due to a variety of reasons, the fields of leakage and demand management progressed to highly proactive levels in England and Wales, and this technology is being readily grasped in South Africa, Brazil, Canada, and other parts of the world. Since the focus on water loss in the United States has been much less extensive, the U.S. water supply industry can benefit from the experiences of these countries.

> The United States will have to become more proactive in its water accounting and loss management practices.

3.2 A COMPARISON OF WATER LOSS REDUCTION EFFORTS: UNITED STATES VERSUS ENGLAND AND WALES

An interesting contrast can be drawn between the proactive system addressing water loss in England and Wales and current conditions in the United States. A comparison of water loss methods in both countries is given in Table 3.1 for real losses and in Table 3.2 for apparent losses.

A number of factors contributed to the establishment of England's progressive demand and leakage management structure in the 1990s. The reor-

TABLE 3.1 Comparison of Water Loss Methods in the United States and in England and Wales: Real Losses

Method	United States	England and Wales
Metering		
Production	Use advocated; prevalence and upkeep likely vary widely	Prevalent, reliable metering exists
Customer	Roughly half of systems have universal metering	Most residential accounts unmetered, but growing slowly
Automatic meter reading (AMR)	New technology, but being applied rapidly	Limited but likely to grow with population of residential meters
District metered areas (DMAs)	Generally not used to any wide extent	A well-established and required practice
Leakage management		
Monitoring	Reactive: wait for complaints, or only as frequent as leakage surveys	Automated as part of DMAs, polled real time or down loaded at short intervals
Pinpointing	Leak correlators likely used by only a small number of systems and consultants	Leak correlators in use by all companies
Pressure management	Employed rarely	Standard component of leakage strategy and employed whenever applicable
District metered areas (DMAs)	Generally not used to any wide extent	A well-established and required practice
Repair of customer service connections	Typically referred to the customer to arrange repairs	Company-paid or -subsidized repairs for first or subsequent leaks
Economic intervention policy	Not developed or employed	Required use by all companies

ganization, privatization, and regulation of the small number of large water companies in 1989 created an important change in the business model used for water supply. With revenue growth potential limited due to government regulation of customer rates or tariffs, leakage reduction was one of many efficiency improvements targeted by the companies to cut costs and improve their bottom line. The National Leakage Initiative of the early 1990s was a major research project underwritten by the water companies to determine the best methods to employ to reduce leakage. The severe drought of the mid-1990s prompted mandatory targets for leakage reduction from the government regulator Office of Water Services (OFWAT), which most companies have achieved due to their ability to quickly implement the recommendations of their leakage reduction research.

TABLE 3.2 Comparison of Water Loss Methods in the United States and in England and Wales: Apparent Losses

Method	United States	England and Wales
Water auditing Audit guidance document	AWWA Manual M36, *Water Audits and Leak Detection*	Government regulations and IWA's *Performance Indicators for Water Supply Services*
Use of audits	Very limited overall, required by certain states	Required for all companies
Impact of audit information	Believed to be very limited overall, varies by state	Performance targets for leakage and other parameters are set and serve as basis for costs charged to customers
Water accounting Performance indicators	Metered water ratio or similar simplistic percentage ratios are widely quoted, but have proven to be unreliable indicators	Measured leakage required by government regulator; IWA recommends volume and Infrastructure Leakage Index (ILI) as measures of real losses, and financial indicators for non-revenue water cost
Accuracy of water usage data	Questionable due to inconsistencies and lack of controls; usage data may be corrupted during billing adjustments; varies widely	Required auditing and consistent practices among companies ensure accurate water usage data
Water loss standards and regulations	Limited in extent, detail, and, where mandated, level of enforcement; regulations vary widely at the state, regional, and local levels	Extensive and detailed; uniformly enforced by the central government regulator

A glossary of rational and consistent terminology now serves as the foundation of water supplier efficiency in England. *Real losses* represent actual physical losses including leakage and tank overflows, and *apparent losses* represent those that occur on paper, including meter inaccuracies, unauthorized actions, and data error. These definitions are new to the U.S. industry, and only a few water systems apply them. In 2000 the International Water Association (IWA) published a Manual of Best Practice, *Performance Indicators for Water Supply Services*,[4] which explains these and other terms and provides an improved water audit methodology that offers a number of advantages to the AWWA M36 method.

In addition to improved performance measures to assess water loss status accurately, the British water industry has employed a number of innovative techniques in reducing leakage losses. Routine monitoring and quantification of leakage levels inferred from minimum night flows in district metered areas (DMAs) is now a well-established and required

> **S**evere droughts in the United Kingdom prompted mandatory leakage reduction.

practice. The influence of pressure on leakage levels has been mathematically established and incorporated into leakage management strategies, with pressure reduction employed when appropriate to reduce leakage from hard-to-find background leaks and to slow the number of mains ruptures. A major policy initiative also took place when regulations were established requiring water companies to provide free or subsidized repairs of initial and (some) subsequent leaks on customer service connections. In the United Kingdom, the United States, and many other parts of the world, customers usually own their service connections and are responsible for having repairs made when necessary. Research has proven that a substantial portion of leaks occur on customer connections and, since bureaucracy and customer hesitancy often create delays in getting repairs effected, connection leaks are one of the greatest contributors to the overall level of water loss incurred by water suppliers. Results from England and Wales have shown that, once initial backlogs of connections leaks are addressed, the rate of occurrence of new leaks is usually sufficiently slow as to make this approach both effective in reducing lost water and economically viable.[5] Enhanced asset management and other improvements have also helped to reduce water loss in England and Wales.

The British experience in managing water loss in recent years is a dramatic step forward in the realm of water efficiency. The methods, policies, and regulations that exist arguably represent the most systematic supply-side management structure for water loss in the world today. While not every aspect of this structure is applicable to systems in the United States or every part of the world, its many advantages likely render it as the best practice model in the world today, and its primary techniques are readily transferable to other countries.

3.2.1 Possible Futures for the United States

When examined closely, water loss in the United States should be viewed as an important water resources concern due to:

- ❑ Its prevalence and extent, which are believed to be significant
- ❑ Demographic changes that exert increasing pressure on water resources
- ❑ The occurrence of drought and water shortages in wide areas of the country during the 1990s and the year 2000
- ❑ High customer expectations
- ❑ Increasing focus on competition
- ❑ Effects of water loss on energy usage

With still-plentiful water resources and a heavy agenda of water quality and infrastructure improvements to manage, the water loss problem may not garner serious attention from the U.S. water industry in the foreseeable future. One possibility for this country is that the status quo may remain for some years, likely until a series of supply disruptions occur due to drought, infrastructure failure, or other emergency—generating sufficient customer and media outcry to prompt government action.

> **D**rought conditions in parts of the United States may prompt mandatory leakage targets, as leakage is probably one of the largest untapped new sources of water for systems in arid conditions.

It is hoped that a more enlightened future might occur in the United States, with water loss accorded a priority focus in the near future. In this scenario the problem of high loss will be properly defined, recognized, and addressed. Several important steps are needed for this to occur:

- ❏ The nature and impact of high water loss in the United States must be clearly defined.
- ❏ The true value of lost water must be established and trumpeted.
- ❏ The water industry should integrate its separate conservation and distribution system camps under a common, progressive water efficiency mission.
- ❏ The water industry must establish a priority focus on water loss reduction in the government, media, consumer, and environmental arenas.
- ❏ New ideas and values must be embraced, letting go of past perceptions and recognizing both the limits and potential of the country's people and natural resources.

With the ingenuity and determination that Americans have displayed in creating a strong nation, no challenge seems insurmountable. If awareness of this important issue is appropriately raised, losses can be reduced to ensure an abundant water supply in the United States for many generations.

3.3 REFERENCES

1. American Water Works Association Research Foundation, *Water and Revenue Losses: Unaccounted-for Water,* 1987.
2. American Water Works Association, Manual M36, *Water Audits and Leak Detection,* 2d ed., AWWA, 1999.
3. Pike, C., San Juan Water District, personal communication, January 2001, re State of California water loss reporting and monitoring requirements.
4. International Water Association, *Performance Indicators for Water Supply Services,* Manual of Best Practice, IWA, 2000.
5. Lambert, A. O., Myers, S., and Trow, S., *Managing Water Leakage: Economic and Technical Issues,* Financial Times Energy Publications, 1998.

CHAPTER 4

Evaluating Water Loss

Using Water Audits and Performance Indicators

George Kunkel

4.1 INTRODUCTION

Just as businesses routinely prepare statements of debits and credits for their customers and banks provide statements of monies flowing into and out of accounts, the water audit displays how quantities of water flow into and out of the distribution system and to the customer. Yet, as essential and commonplace as financial audits are to the world of commerce, water audits have been surprisingly uncommon in public water supply throughout most of the world. In places where the intrinsic value of water has not been recognized, little motivation has existed to prompt requirements for auditing and sound assessments of water loss performance. As water is becoming a more valued commodity, however, this picture is beginning to change.

Throughout the 1990s, efforts were made to develop a rational, standardized water audit methodology and water loss performance indicators. Part of the motivation for this work was the focus on demand management and the wise use of water in England and Wales, which was driven by competition, drought-related water shortages, and other factors. In the late 1990s the International Water Association (IWA) initiated a large-scale effort to assess water supply operations, which resulted in the publication of *Performance Indicators for Water Supply Services* in 2000.[1] While this initiative included various groups assessing all aspects of water supply operations, the Task Force on Water Loss worked specifically to devise an acceptable water audit format and performance indicators that could be used to make effective comparisons of water loss performance of systems anywhere in the world.

The methods put forth by the IWA Task Force on Water Loss, while new, are offered as the current "best practice" model for water auditing and performance measurement. This is not just because of the multination process used in assembling the results, but primarily because the work was groundbreaking in providing a clear structure for a need that was void of knowledge throughout most of the world. Additionally, the work has been tested thoroughly using data from dozens of countries. Within a very short time a number of countries—including South Africa, Australia,

> Compiling a reliable water audit or water balance is the critical first step in managing losses in public water supplies.

and New Zealand—have adopted these methods as the basis for their national water loss management structures.

What is different about the IWA methods? The best form of explanation is to compare the methods with the practices of the North American water supply industry. In Chap. 3, reasons are given why water suppliers throughout North America do not utilize consistent water auditing practices and measures, and why this should change. Adoption of reliable auditing methods will follow adoption of water loss management as an issue of priority at the national level in the United States and Canada. If the nation decides to make the reduction of water loss a government strategy, then reliable methods must be employed. The following explains why current practices are not sufficient, while the IWA methods offer many advantages.

For most water utilities in the United States there is no requirement to conduct regular water auditing. Therefore, most water utilities do not compile audits. Various state or regional water oversight agencies require suppliers to report various water supply and loss data, but these efforts are typically very limited in detail and the report formats differ from agency to agency. A true water audit is a detailed statement, performed on a regular reporting basis (typically the business or calendar year), that quantifies system input, billed and unbilled customer and supplier usage, apparent losses, and real losses. In North America, the best guide for water auditing is the American Water Works Association (AWWA) Manual M36, *Water Audits and Leak Detection*.[2] (Chapter 2 of M36 is reproduced in App. A of this book; however, it is recommended that readers interested in auditing obtain, and read, the entire manual.) Although it is not known whether any water oversight agencies require use of this particular format as part of their reporting procedures, the following discussion will largely reference the methods of Manual M36 publication and practices of North American water utilities in contrast to the improved structure offered by the IWA format. The IWA structure is further exhibited and explained as the recommended method for water suppliers to assemble water audit and reliable performance measures of their water loss standing.

> Users of the M36 audit procedure can easily transfer data to the IWA audit.

4.2 THE NEED TO IMPROVE WATER LOSS ASSESSMENTS IN NORTH AMERICA

To the outsider, the methods used for assessing, publicizing, and comparing water losses in public water supply systems in North America appear likely to underestimate the true extent of the water loss that is occurring, for several reasons:

❑ There is typically no requirement to publish comparative data on water losses.
❑ There is no standard methodology for defining and calculating water losses.
❑ Though not part of the M36 format, many North American utilities that assemble water audits include various estimates of "unavoidable losses" and "discovered leaks and overflows" as part of "authorized consumption" rather than as real leakage losses.

- ❑ The common practice of expressing water losses as a percentage of system input volume, in a region with one of the highest per-capita consumptions in the world, tends to understate the extent of the problem (10 percent water losses in North America could represent around 25 percent water losses for typical lower per-capita European consumption).
- ❑ There exists no rational definition for the term "unaccounted-for" water.
- ❑ While only a small portion of water suppliers perform any water loss accounting, inconsistency abounds because numerous methodologies are in use in North America.

In contrast to the North American picture, the advantages of the IWA methodology can be summarized as follows.

- ❑ The IWA methods are structured to serve as a standard international "best practice" methodology and terminology for such calculations, based on the conclusions of recent International Water Association Task Forces on Water Losses and Performance Indicators.
- ❑ The IWA methods question the desirability of the common North American practice of counting unavoidable water losses and discovered leaks and overflows as part of "authorized consumption."
- ❑ A system-specific method for calculating "unavoidable real losses" is included.
- ❑ The IWA method counters the deficiencies in the performance indicators most commonly used in North America—percentage of system input volume and losses per mile of mains.
- ❑ The IWA has dropped the term "unaccounted-for water" (UFW) in favor of "nonrevenue water" (NRW), because there is no internationally accepted definition of UFW, and all components of the water audit can be "accounted for" using the IWA methodology.
- ❑ Using the standard international methodology, comparisons of the performance of various North American water utilities with an international data set of 28 systems from 20 countries have been conducted, using both the "traditional" and IWA-recommended performance indicators.

4.3 A ROSETTA STONE FOR WATER LOSS MEASUREMENT?

In 1799, Napoleon's soldiers found an ancient piece of carved black basalt at Rosetta, near the mouth of the Nile River. The Rosetta stone carried a decree of the Egyptian priests of Ptolemy V. Epihanes (205–181 B.C.) written in Egyptian hieroglyphics, Demotic characters, and Greek, permitting a simultaneous translation of these three written texts. As a result, Egyptian hieroglyphics could be read correctly for the first time.

The story of the Rosetta stone could have more to do with water loss accounting in North America than the reader may at first imagine. Remarkably, in North America, there is no single standard terminology, or commonly accepted definitions or methodology for undertaking an annual water audit of the components of a "water balance." The water balance calculation seeks to identify the destinations of all water entering a distribution system, so that the water losses occurring within the distribution system can be assessed. Every state, government organization, professional institution, consultant, or contractor can (and usually does!) define the terminology and undertake the calculations in any way they please. This is perhaps because few states request or require water utilities to report such data on an annual basis. However, water is an important natural resource, and in an increasing number of developed countries similar absences of accountability for demonstrating responsible stewardship of natural resources are being actively addressed.

For example, in England and Wales, since 1992 the privatized water companies have had to produce annual, independently audited calculations of water losses in a standard format, for national publication by their economic regulator. Publication of standardized data raised questions regarding performance and economic levels of water losses, which in turn (spurred by the 1995/1996 drought and political impetus) resulted first in voluntary, and then mandatory, leakage targets. Some five years later, leakage from public water supply systems in England and Wales has been reduced overall by 40 percent,[3] or some 480 million gallons per day (MGD), and U.K. expertise in modern leakage management is now internationally recognized. None of this might have happened if the English and Welsh water utilities had been permitted to choose for themselves:

- ❑ Whether to undertake annual calculations of water losses
- ❑ How the calculations should be carried out and which performance indicators should be used
- ❑ Whether the results should be published

The extent of the problem in North America can be illustrated by the results of a project undertaken in 2000 to compare water losses among utilities in North America.[4] Only seven water utilities that were willing to contribute data were located, despite an outreach effort. Each utility submitted water balance data in a different format, to different degrees of detail, using different definitions and terminology, none of which followed the M36 or any other established format. It was only by reallocating the components of each water balance into a "standard" water balance format, the IWA format, that the volumes of water in the water audits could be defined on a comparable basis.

4.4 THE INTERNATIONAL STANDARD WATER AUDIT

Before attempting the performance comparisons cited in Ref. 4, water balance data were reallocated as standard water balance components based on IWA recommended best practice, as shown in Fig. 4.1. This standard includes terms familiar to North American practice, such as "authorized consumption." However, the IWA has dropped the term "unaccounted-for water" (UFW) in favor of "nonrevenue water" (NRW), because there is no

4.4 THE INTERNATIONAL STANDARD WATER AUDIT

Own Sources	System Input	Water Exported	Authorized Consumption	Billed Authorized Consumption	Revenue Water	Billed Water Exported
						Billed Metered Consumption
		Water Supplied				Billed Unmetered Consumption
Water Imported	(Allow for Known Errors)			Unbilled Authorized Consumption	Nonrevenue Water	Unbilled Metered Consumption
						Unbilled Unmetered Consumption
			Water Losses	Apparent Losses		Unauthorized Consumption
						Customer Metering Inaccuracies
				Real Losses		Leakage on Mains
						Leakage and Overflows at Storages
						Leakage on Service Connections up to Point of Customer Metering

Figure 4.1 An international standard water balance. (*Source: Ref. 4.*)

internationally accepted definition of UFW, and all components of the water balance can be accounted for using the above process.

All data in the water balance should be expressed as a volume per year. Each component of the water balance needs to be specifically defined in a glossary of terms, with supplementary notes where appropriate. For example, the three components of nonrevenue water are:

1. *Unbilled authorized consumption:* The annual volume of unbilled metered and/or unmetered water taken by registered customers, the water supplier, and others who are implicitly or explicitly authorized to do so by the water supplier, for residential, commercial, and industrial purposes.
2. *Apparent losses:* All types of inaccuracies associated with customer metering and billing, plus unauthorized consumption (theft or illegal use).
3. *Real losses:* Physical water losses from the pressurized system, up to the point of measurement of customer use. The annual volume lost through all types of leaks, bursts, and overflows depends on frequencies, flow rates, and average duration of individual leaks, bursts, and overflows.

The calculation procedure is as follows:

- ❑ Obtain system input volume and correct for known errors.
- ❑ Obtain components of revenue water and calculate revenue water, which equals billed authorized consumption.
- ❑ Calculate nonrevenue water (= system input − revenue water).
- ❑ Assess unbilled authorized consumption.
- ❑ Calculate authorized consumption [= (billed + unbilled) authorized consumption].
- ❑ Calculate water losses (= system input − authorized consumption).
- ❑ Assess components of apparent losses and calculate apparent losses.

❏ Calculate real losses (= water losses − apparent losses).
❏ Assess components of real losses from first principles (e.g., burst frequency/flow rate/duration calculations, night flow analysis, modeling) and cross-check with calculated volume of real losses.

Particular points to note: Comparisons are made with the AWWA M36 water audit procedure and the July 1996 recommendations of the AWWA Leak Detection and Water Accountability Committee,[5] and that of the IWA methodology:

❏ *Authorized consumption* is separated into *billed* and *unbilled* components, to allow both financial and operations performance indicators to be calculated.
❏ *Authorized metered consumption* does not include customer meter errors, which are included in apparent losses in the IWA water balance.
❏ In the IWA methodology, "authorized consumption" does not include volumes of losses from known (discovered) leaks, breaks, and storage tank leakage/overflows, or estimates of unavoidable losses—these are all part of the "real losses" in the IWA methodology.

Components of any of the various North American water audits, or from any other countries, can quite easily be reallocated into components of the IWA standard water balance—an example is given in App. 1 of Ref. 4—allowing the IWA standard balance to be considered as the equivalent of an international "Rosetta stone" for standardized calculations of water losses.

4.5 UNAVOIDABLE WATER LOSSES, AND DISCOVERED LEAKS AND OVERFLOWS

Since the last century, it has been common practice in North America to estimate, using various formulas, the "unavoidable" leakage from pressurized pipework systems—those small leaks which are believed to be undetectable, or which are considered uneconomical to repair. The original intention of this practice was presumably to try to define a baseline or lower limit for leakage management, below which it is uneconomical to attempt further leakage control. An outline of the various methods previously used in North America is presented in Ref. 4. The system-specific predictions based on an auditable component-based equation proposed by the IWA Task Force on Water Losses,[6] described later in this chapter, can be regarded as a natural progression of previous North American efforts to predict unavoidable losses.

Because of the simplified nature of some of the formulas previously used in North America, or the very generous allowances given for old pipework (particularly cast-iron pipes), the effect of the "unavoidable leakage" calculation has in practice often resulted in a considerable amount of leakage being written off as beyond control. In fact there are infrastructure and pressure management options that now exist to reduce it.

A similar situation applies regarding "discovered" leaks from pressurized pipework and overflows from service reservoirs. In many North American water balances, the volumes of each one of these events is estimated, from assessed flow rate and duration, and these volumes are then written off as part of authorized consumption so that they do not appear as unaccounted-for water. However, it is of course possible to take management action to reduce the volume of water losses from discovered leaks and overflows, by reducing the runtimes of leaks and installing telemetry or altitude valves to avoid overflows.

The most common practice in countries outside North America is to calculate the annual volume of water losses from the water balance *without* making any deductions for unavoidable leakage or discovered leaks and overflows, and then to calculate the performance indicators. Accordingly, superficial comparisons of North American water losses with water losses from other countries often present a more favorable picture than is actually the case.

The IWA recommended standard methodology for water audit calculations and performance indicators allows unavoidable losses and discovered leaks and overflows to be considered, but only as partial explanations of the total volume of water losses, which should always be stated explicitly before attempting to explain or justify the total volume. The IWA system-specific approach to unavoidable real losses is described below.

The IWA approach is described in detail in the December 1999 issue of the IWA magazine *AQUA*,[6] and can be seen as a natural development of previous North American attempts to take key local factors into account. The component-based approach is based on auditable assumptions for break frequencies/flow rates/durations; background and breaks estimates concepts to calculate the components of unavoidable real losses for a system with well-maintained infrastructure[7]; speedy, good-quality repairs of all detectable leaks and breaks; and efficient active leakage control to locate unreported leaks and breaks. The parameters used in the calculation, taken from Ref. 6 and converted to North American units, are shown in Table 4.1.

Table 4.2 shows these parameters in a more user-friendly format for calculation purposes. The "UARL Total" values, in the units shown in Table 4.2, provide a rational yet flexible basis for predicting UARL values for a wide range of distribution systems. The calculation takes into account length of mains, number of service lines, location of customer meters relative to property line (curb stop), and average operating pressure (leakage rate varies approximately linearly with pressure for most large systems).

An important aspect of Table 4.2 is the value assigned to unavoidable "Background [undetectable real] Losses," shown in col. 2. These figures are based on international data, from analysis of night flows in sectors just after all detectable leaks and breaks have been located and repaired. This component of unavoidable real losses does not appear to have been quantified previously in North American practice, yet it accounts for at least 50 percent of the unavoidable real losses components in Table 4.2. Estimates of background (undetectable) leakage following intensive leak detection surveys in small U.S. systems have been compared with IWA unavoidable background loss predictions based on the second column of Table 4.2. Initial comparisons are encouraging, and more comparisons are being actively sought.

TABLE 4.1 Parameter Values Used for Calculation of Unavoidable Annual Real Losses

Infrastructure Component	Background (Undetectable) Losses	Reported Breaks	Unreported Breaks
Mains	8.5 gal/mi/h	0.20 breaks/mi/year at 50 gal/min for 3 days' duration	0.01 breaks/mi/year at 25 gal/min for 50 days' duration
Service lines, main to curb stop	0.33 gal/service line/h	2.25/1000 service lines/year at 7 gal/min for 8 days' duration	0.75/1000 service lines/year at 7 gal/min for 100 days' duration
Underground pipes, curb stop to meter (for 50 ft avg. length)	0.13 gal/service line/h	1.5/1000 service lines/year at 7 gal/min for 9 days' duration	0.50/1000 service lines/year at 7 gal/min for 101 days' duration

gal = U.S. gallon; all flow rates are at a reference pressure of 70 psi.
Source: Ref. 4.

TABLE 4.2 Components of Unavoidable Annual Real Losses

Infrastructure Component	Background Losses	Reported Bursts	Unreported Bursts	UARL Total	Units
Mains	2.87	1.75	0.77	5.4	gal/mi of mains/day/psi of pressure
Service lines, mains to curb stop	0.112	0.007	0.030	0.15	gal/mi/day/psi of pressure
Underground pipes between curb stop and customer meters	4.78	0.57	2.12	7.5	gal/mi of pipe/day/psi of pressure

Source: Ref. 4.

There are many different ways to present the UARL equation. Figure 4.2 shows UARL in gal/mi/day/psi of pressure (*Y* axis) plotted against density of service lines. The large variation of unavoidable losses per mile of mains for different densities of service lines shows why it is not recommended to use "per mile" for comparisons of real losses. However, Fig. 4.2 can be used to estimate unavoidable annual real losses for any system, as the following example shows.

Example A water supply system has 60,000 service connections and 600 mi of mains (a connection density of 100 service lines per mile of mains), and the average operating pressure is 70 psi. Calculate the unavoidable

4.5 UNAVOIDABLE WATER LOSSES, AND DISCOVERED LEAKS AND OVERFLOWS

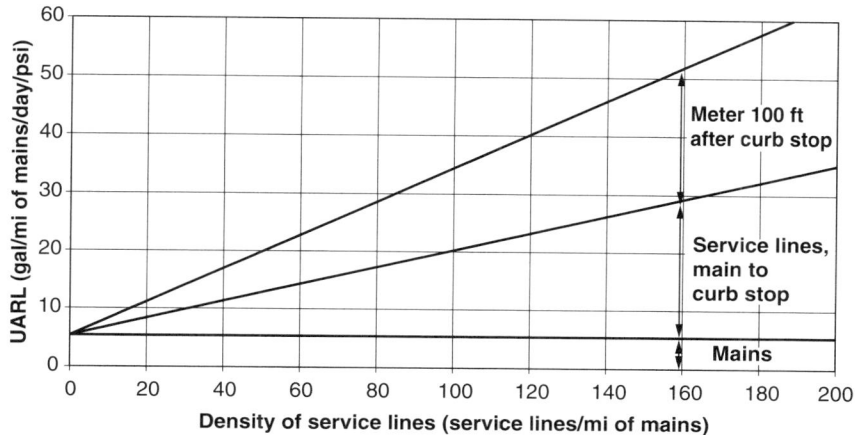

Figure 4.2 Unavoidable annual real losses (gal/mi of mains/day/psi) versus density of service connections. (*Source: Ref. 4.*)

annual real losses from Fig. 4.2 if the average distance of customer meters from the curb stop is (a) 100 ft or (b) 20 ft.

Answer At a connection density of 100 per mile of mains (X axis), from Fig. 4.2 the UARL is: (a) 34 gal/mi/day/psi of pressure $\times$ 70 psi = 2380 gal/mi/day $\times$ 600 mi = 1.43 MGD (for customer meters 100 ft from the curb stop); or (b) 23 gal/mi/day/psi of pressure $\times$ 70 psi = 1610 gal/mi/day $\times$ 600 mi = 0.97 MGD (for customer meters 20 ft from the curb stop).

Comparison of IWA system-specific values of unavoidable annual real losses in gallons per mile of mains per day compare well with the range of 1000–3000 gal/mi/day usually quoted for North American systems. However, the IWA prediction method has the considerable advantage that it allows estimates to be made on a system-specific basis, taking account of density of connections, average operating pressure, and locations of customer meters (relative to the curb stop). The last of these factors is particularly important in a region of diverse climates such as North America, where some customer meters are close to the curb stop and others are in buildings more distant from the curb stop.

The UARL values in Table 4.2 can just as easily be plotted as a graph of gallons per service line per day per psi of pressure versus density of service lines, as shown in Fig. 4.3. In well-run systems worldwide, the greatest annual volume of real losses occurs from long-running, small-to-medium–sized leaks on service connections, except at low densities of service connections. This is why the IWA Task Forces (see Refs. 1 and 6) recommend using "per service connection" instead of "per mile of mains" as the basic performance indicator for real losses, for connection densities exceeding 32 per mile.

Using the previous calculation example, for the system with 60,000 service connections and 600 mi of mains, the UARL derived from Fig. 4.3 would be (a) 0.34 gal/service/day/psi of pressure $\times$ 70 psi = 23.8 gal/service/day $\times$ 60,000 services = 1.43 MGD (for customer meters 100 ft from the curb stop); or (b) 0.23 gal/service/day/psi of pressure $\times$ 70 psi

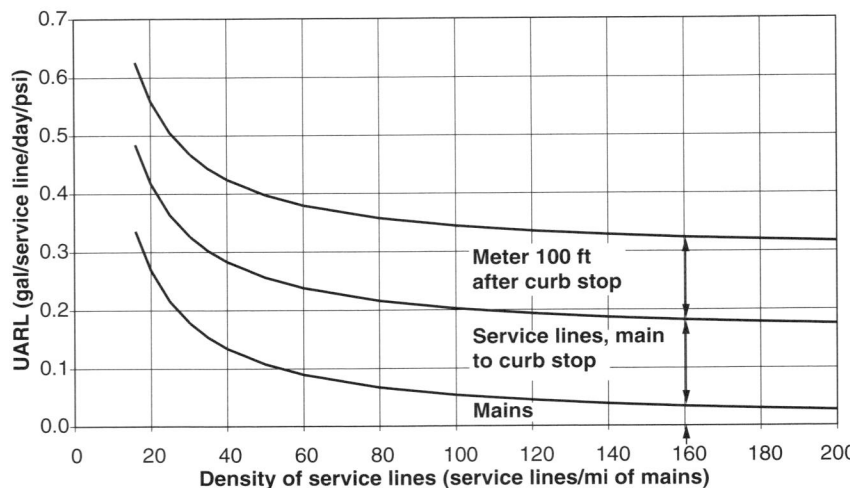

Figure 4.3 Unavoidable annual real losses (gal/service line/day/psi) versus density of service lines. (*Source: Ref. 4.*)

= 16.1 gal/service/day × 60,000 services = 0.97 MGD (for customer meters 20 ft from the curb stop).

The curved lines in Fig. 4.3 are relatively flat for a wide range of connection densities. In calculating unavoidable annual real losses, for example, systems with customer meters 50 ft from the curb stop, and connection densities in the range 80–200 per mile, an acceptable simplification from Fig. 4.3 would be to say that the UARL is 0.25 gal/mi/day/psi of pressure (= ±10%).

4.6 WHICH PERFORMANCE INDICATOR? WHY NOT PERCENT?

Because water utilities are of different sizes, with different characteristics, comparisons of performance in management of water losses need to be made in terms other than volume per year. Traditionally, several different performance indicators are used by North American utilities to compare water losses, percentage of system input volume or metered water ratio, and per mile of mains per day, appear to be the most common. But are these reliable indicators for comparing performance? Why do some countries use "per property per day," or "per service connection per day," or "per kilometer of systems (mains + services length) per day"? The IWA Task Force on Water Losses, with nominated representation from the American Water Works Association, has been considering best practice internationally, and their conclusions (Ref. 6) strongly suggest that there are more reliable and meaningful performance indicators than percentage of system input and per mile of mains.

In emphasizing the importance of the correct choice of measuring units, another example from history is useful. Two thousand years ago, in the first century A.D., Julius Frontinius Sextus, then water commissioner for Rome, was spending the whole of his professional career trying (and failing) to achieve a meaningful balance between the quantities of water

> **E**xpressing losses as a percentage is not the best way to compare loss management performance, as systems with lower demands will never be able to compete with those with larger demands. Instead, the volume of loss per service connection per day should be used.

entering and leaving the aqueducts which served the city. Failure was not due to lack of diligence on his part—he was simply using the wrong measures. The accepted Roman method was to compare areas of flow; because they did not take velocity of flow into account, their calculations could never be reliable for management purposes.

Because per-capita consumption in North America is so high compared to most other countries, the common practice of expressing water losses as a percent of system input volume tends to produce lower figures that would be the case in the other countries. This gives a false impression of true performance when comparisons of performance are made with other countries with lower per-capita consumption.

The same problem occurs when comparisons are made between North American utilities with a high consumption base and North American utilities with a low consumption base. 1996 data showed that 51 water supply systems in California had density of connections varying from 24 per mile to 155 per mile, with an average of 75 per mile. The average metered consumption per connection varied from 136 to 2200 gal/service connection/day, with an average of some 600 gal/service connection/day. Suppose that each of these water utilities was achieving real losses of 60 gal/service line/day, which is around three times the unavoidable annual real losses (21 gal/service line/day) for a system with 75 connections per mile, pressure of 70 psi, and customer meters 50 ft from the curb stop. Table 4.3 and Fig. 4.4 show that the percent real losses for various systems in California would vary from less than 3 percent to almost 30 percent, a 10-fold range, depending on their average consumption per connection, even if all of them had exactly the same actual leakage management performance of 60 gal/service connection/day.

Based on the average consumption of 600 gal/service line/day, a target of 10 percent real losses or less might seem reasonable. However, from the above figures it can be shown that:

❑ For utilities with low consumption per service connection this would be a quite unrealistic target, being almost equal to the unavoidable annual real losses.

❑ For utilities with high consumption this would represent real losses of around 11 times the unavoidable annual real losses.

TABLE 4.3 How Percent Real Losses Vary with Consumption, for Real Losses of 60 gal/service line/day

System Consumption in gal/service line/day	Real Losses in gal/service line/day	System Input in gal/service line/day	Real Losses as Percent of System Input Volume
150	60	210	28.6
300	60	360	16.7
600	60	660	9.1
1200	60	1260	4.8
1800	60	1860	3.2
2400	60	2460	2.4

Source: Allan Lambert.

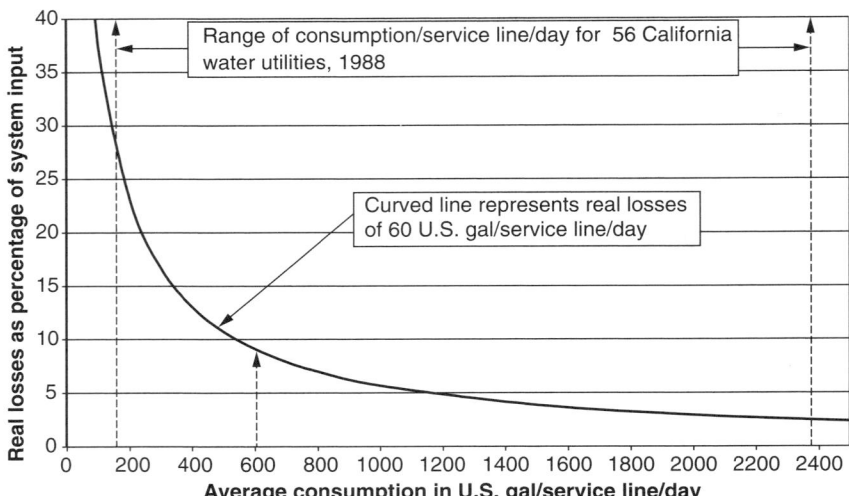

Figure 4.4 How percent real losses vary with consumption, for real losses of 60 gal/service line/day. (*Source: Allan Lambert.*)

If Table 4.3 and Fig. 4.4 were not in themselves sufficient to demonstrate the problem of using percentages for comparisons of performance in managing real losses, there are further serious disadvantages.

❑ When a utility exports water, the percent real losses will be lower if the exported volumes are included in the calculation, and higher if they are excluded.
❑ The problem of expressing water losses in percentage terms is compounded when demand management measures to reduce per-capita consumption (pcc) are applied, as the pcc goes down, and the percent water losses goes up—not a great incentive to demand management in its widest sense, simply because of the choice of an inappropriate performance indicator!

Technical committees worldwide (Germany, U.K., South Africa) have recognized these paradoxes of using percentages, but perhaps most significantly, the England and Wales economic regulator [Office of Water Services (OFWAT)] also recognized it and stopped publishing water losses statistics in percentage terms in 1998. Water system managers who unquestioningly accept percentages as a valid measure of technical performance in management of water losses should consider whether they are falling into the same trap as Julius Frontinius Sextus, 2000 years ago—using a simple (and inappropriate) measure to draw inappropriate conclusions.

4.7 IWA-RECOMMENDED PERFORMANCE INDICATORS FOR NONREVENUE WATER AND REAL LOSSES

During the period 1996–2000, various IWA task forces undertook a detailed study to determine the most appropriate performance indicators (PIs) for different water supply purposes. Table 4.4 shows the PIs for non-

4.7 IWA-RECOMMENDED PERFORMANCE INDICATORS FOR NONREVENUE WATER AND REAL LOSSES

TABLE 4.4 IWA-Recommended Performance Indicators for Nonrevenue Water and Water Losses

Function	Ref.	Level	Performance Indicator	Comments
Financial: nonrevenue water by volume	Fi36	1 (basic)	Volume of nonrevenue water as percent of system input volume	Can be calculated from simple water balance
Financial: nonrevenue water by cost	Fi37	3 (detailed)	Value of nonrevenue water as percent of annual cost of running system	Allows different unit costs for NRW components
Inefficiency of use of water resources	WR1	1 (basic)	Real losses as a percent of system input volume	Unsuitable for assessing efficiency of management of distribution systems
Operational: real losses	Op24	1 (basic)	Gal/service line/day, when system pressurized	Best "traditional" basic performance indicator
Operational: real losses	Op25	3 (detailed)	Infrastructure leakage index	Ratio of current annual real losses to unavoidable annual real losses

Source: Ref. 4.

revenue water and real losses recommended by the IWA (Refs. 1 and 6) converted to North American units. The PIs are categorized by function and by level, defined as follows.

- *Level 1 (basic):* A first layer of indicators that provides a general management overview of the efficiency and effectiveness of the water undertaking.
- *Level 2 (intermediate):* Additional indicators, which provide a better insight than the Level 1 indicators for users who need further depth.
- *Level 3 (detailed):* Indicators that provide the greatest amount of specific detail, but are still relevant at the top management level.

Particular points to note from Table 4.4 are the following.

- Fi36: Percentage of nonrevenue water is the basic *financial PI*.
- Fi37: This detailed financial PI is a development of a 1996 recommendation of the AWWA Leak Detection and Water Accountability Committee (Ref. 5).
- WR1: Real losses as a percentage are unsuitable for assessing efficiency of management of distribution systems for control of real losses (because of the influence of consumption).
- Op24: Gallons/service line/day is the most reliable of the traditional PIs for real losses, for all systems with service line densities of >32/mi.
- To improve on Op24, take account of three key system-specific factors: density of service connections, location of customer meter relative to curb stop, and average operating pressure. *Note:* By expressing Op24 as gallons/service line/day/psi of pressure, the influence of pressure is included.
- Op25, the infrastructure leakage index (ILI), is a measure of how well the system is being managed for the control of real losses, at the current operating pressure.

- ILI is the ratio of current annual real losses to unavoidable annual real losses.
- Unavoidable annual real losses (UARL) are calculated as previously described in this chapter, using the IWA methodology which takes into account average operating pressure, length of mains, number of service lines, and location of customer meters relative to the curb stop.

> The ILI ratio is a great way to demonstrate loss management performance, as each system effectively compares the ratio of its individual best possible performance against how it is actually performing.

The infrastructure leakage index is a new, and potentially very useful, performance indicator. Being a ratio, it has no units, so it facilitates comparisons between countries that use different measurement units (metric, U.S., or imperial). The ILI can perhaps be better envisaged from Fig. 4.5, which shows the four components of leakage management. The large square represents the current annual volume of leakage, which is always tending to increase, as infrastructure systems grow older. This increase, however, can be constrained by an appropriate combination of the four components of a successful leakage management policy. The small square represents the unavoidable annual real losses—the lowest technically achievable value for real losses at the current operating pressure. The ratio of the current annual real losses (the large square) to the unavoidable annual real losses (the small square) is a measure of how well the three infrastructure management functions—repairs, pipe materials management, and active leakage control—are being controlled. We will be seeing more of this diagram in future chapters, where we will be discussing some of the hands-on techniques associated with in-the-field loss reduction programs.

An infrastructure leakage index close to 1.0 demonstrates that all aspects of a successful leakage management policy are being implemented by a water utility. However, typically it will only be economical to

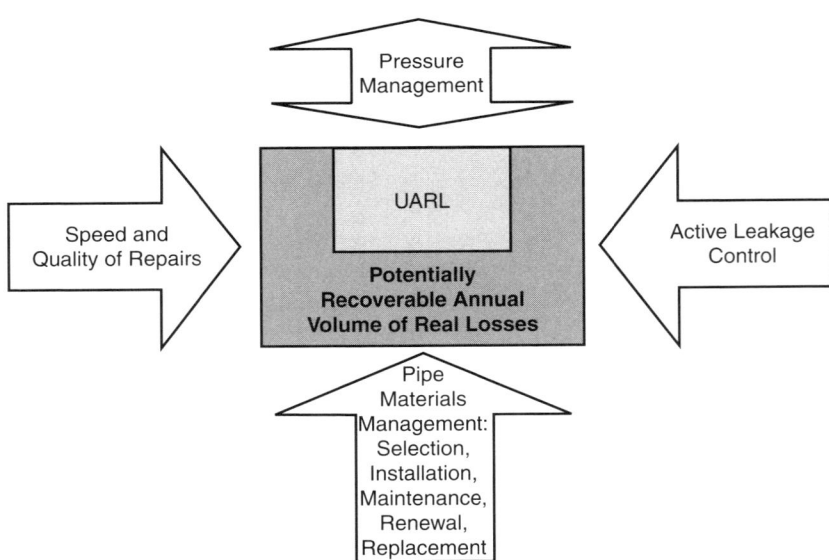

Figure 4.5 The four components of a successful leakage management policy. (*Source: Ref. 4.*)

achieve an ILI close to 1.0 if water is very expensive, scarce, or both. Economic values of ILI depend on the system-specific marginal cost of real losses, and typically lie in the range 1.5–2.5 for most systems.

4.8 SOME PUBLISHED NORTH AMERICAN WATER LOSSES STATISTICS

A 1996 overview of water production and delivery statistics for 469 U.S. water utilities showed metered water ratios averaging 84 percent, implying unaccounted-for water of 16 percent of production, on average. The range of UFW was very large, from over 50 percent down to 1 percent, with a 13 percent modal (most frequently occurring) value. So how does this compare to other countries worldwide?

An international data set of 27 systems from 20 countries, assembled by the IWA Task Force on Water Losses (Ref. 5), showed a range of non-revenue water (NRW) from 2 to 40 percent, with an average of 17.5 percent and a 13 percent modal value.

The IWA Task Force recommendations were mentioned in a presentation to the American Water Works Association Distribution Systems Symposium in Reno, Nevada, in 1999. As a result of the interest this generated, the Leak Detection and Water Accountability Committee of the American Water Works Association encouraged a number of utilities to volunteer water audit and other data, to apply the approach to North American supply systems. Seven of the systems provided sufficient data for the analysis in Table 4.5.

The values of the nonrevenue water percentage and the three new real losses PIs for the seven North American systems in Table 4.5 were entered alongside the values from the 27 systems from 20 countries in the IWA international data set. The North American systems are shown in black in Figs. 4.6 and 4.7. The fact that almost all of the seven North American systems in the data sample have NRW values greater than 16 percent (the average for the large 1996 sample referred to previously) means that these systems cannot be considered as representative for comparison of North American performance with international performance. A wider range of system data is needed for such a comparison.

When expressed as percent NRW (Fig. 4.6), the North American data bunches together in the "worst" 50th percentile of the extended data set; but when the preferred performance measures are used, as shown in Figs. 4.7–4.9, a clearer discrimination of performance is the result. It can be seen that two of the North American systems are consistently just within the "best" 50th percentile of the extended international data set.

In one of the North American systems, the IWA performance measures were calculated for three discrete subsystems known to have very different leakage characteristics. Individual ILI values were calculated for each subsystem and ranged from 1.6 to over 10, reflecting the local situation reliably, but in greater detail than had previously been possible with the cruder, percent NRW approach. The methodology can be used for subsystems down to approximately 5000 service connections.

TABLE 4.5 Summary of Statistics for Seven North American Systems

	A	B	C	D	E	F	G	Average
Traditional Statistics								
Metered water ratio (%)	81.2	84.7	67.6	76.0	78.0	78.6	66.5	76.1
Nonrevenue water (%)	18.8	15.3	32.4	24.0	22.0	21.4	33.5	23.9
Real losses (gal/mi/day)	3,652	3,558	4,975	13,378	13,960	13,105	20,686	10,473
Consumption (gal/service line/day)	337	551	358	466	772	756	444	526
Key System Factors								
Service line density/mi	86	67	40	104	85	80	145	87
Average meter location after property line (ft)	0	23	7	24	25	0	12	13
Average pressure, psi	75	75	85	70	72	65	55	71
Unavoidable Annual Real Losses (IWA Equation)								
gal/service line/day/psi	0.21	0.26	0.29	0.24	0.25	0.22	0.20	0.24
gal/service line/day	16.0	19.7	25.1	16.5	17.9	14.2	11.2	17.2
New PIs for Real Losses								
Basic: gal/service line/day	42	53	124	129	163	165	143	117
Intermediate: gal/service line/day/psi	0.57	0.70	1.46	1.84	2.27	2.53	2.59	1.71
Detailed: Infrastructure leakage index	2.66	2.69	4.96	7.80	9.13	11.63	12.69	7.37

Source: Ref. 4.

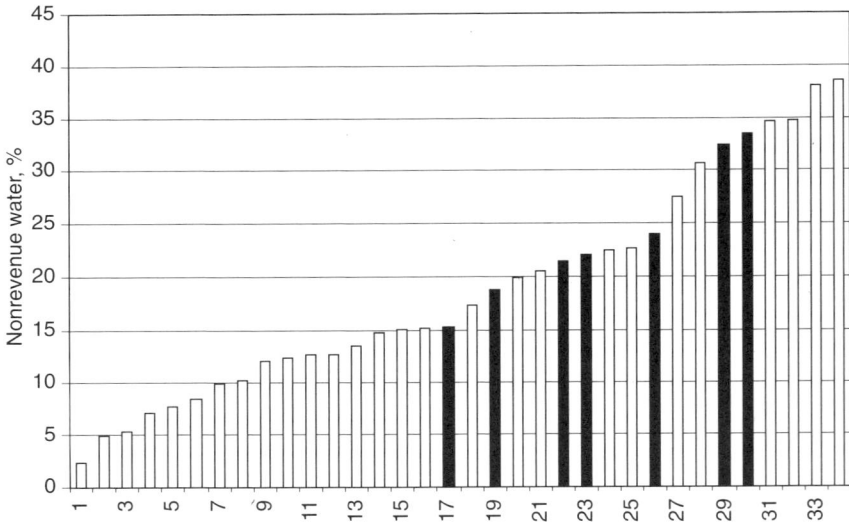

Figure 4.6 Nonrevenue water, percent of system input. (*Source: Ref. 4.*)

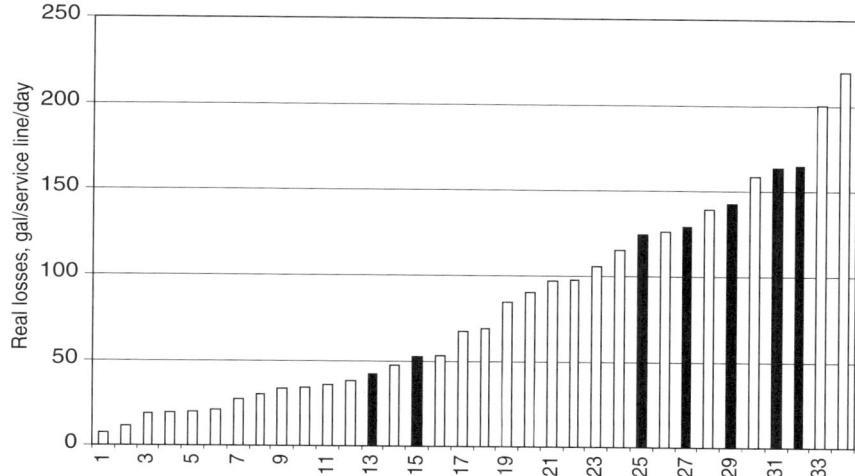

Figure 4.7 Real losses (gal/service line/day). (*Source: Ref. 4.*)

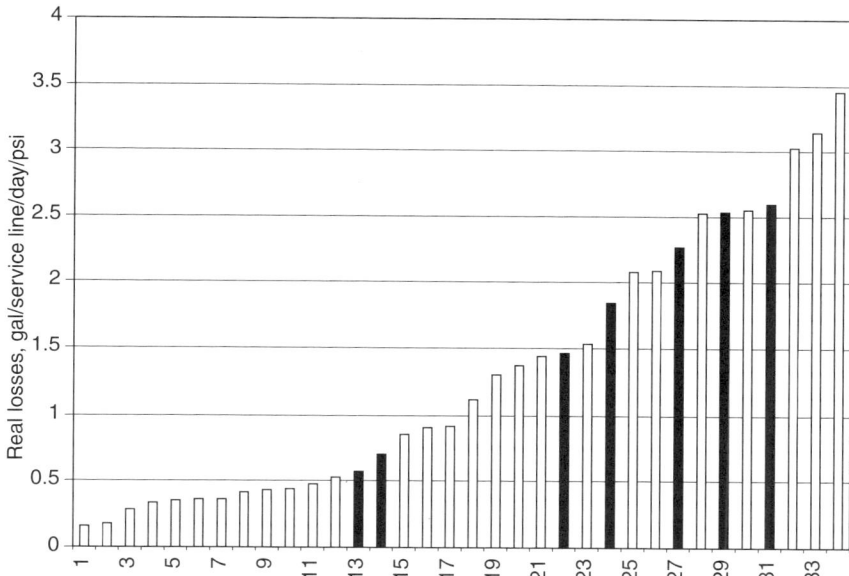

Figure 4.8 Real losses (gal/service line/day/psi). (*Source: Ref. 4.*)

4.9 CONCLUDING COMMENTS

By expressing the outputs of the IWA task forces in units familiar to North American practitioners, it is hoped that this chapter will stimulate further interest in the state of the art for water losses performance indicators. The methodologies may be viewed as a continuation of the work of North American researchers who have been seeking a more rational basis for comparisons, which takes into account key system-specific factors, notably pressure, density of service lines, and location of customer meters relative to the curb stop.

Use of the IWA standard terminology and water balance methodology, together with the new equation for unavoidable annual real losses and

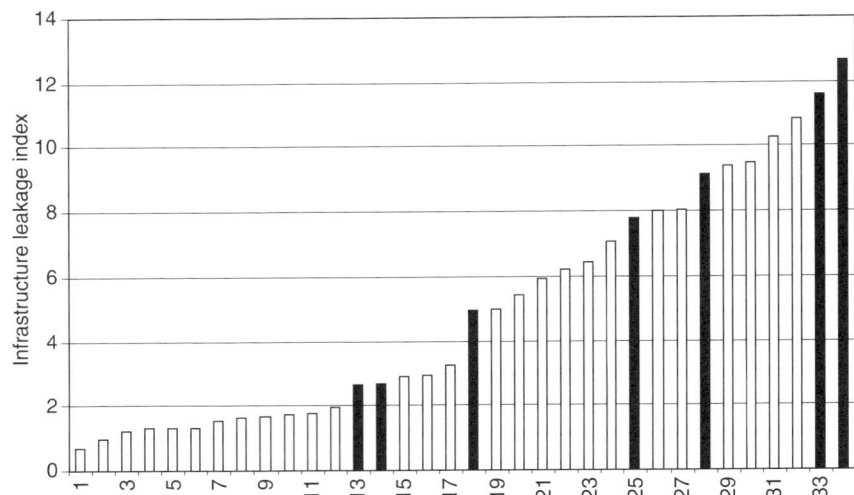

Figure 4.9 Infrastructure leakage index. (*Source: Ref. 4.*)

recommended performance indictors such as the infrastructure leakage index (ILI) could become the basis of a "Rosetta stone" for more rational assessment of water losses volumes and more meaningful comparisons of water losses management.

There is no indication from the applications described here to suggest that the methodology is not applicable to North American systems, although further testing is recommended to provide additional validation. Potential problems likely to be encountered in further applications are the following.

- Changing the mindset of water system managers to recognize water loss as a compelling issue and the need to apply rational, standardized methods
- The wide diversity of formats and terminology used for water balance calculations in North America
- Training in a standard methodology for calculating average pressure

All of these problems have been overcome in developed economies outside North America.

4.9.1 Sample Audit Data

Philadelphia's abbreviated data can be seen in IWA audit format in Fig. 4.10 on pp. 54–55 and Boston's data can be seen in AWWA M36 audit format in Fig. 4.11 on pp. 56–57. A more detailed review of the Philadelphia LMA can be found in Chap. 18, which discusses in detail the compilation of the water balance in the new IWA format.

4.10 REFERENCES

1. Alegre, H., Hirner, W., Baptista, J., and Parena, R., *Performance Indicators for Water Supply Services*, Manual of Best Practice, International Water Association, 2000.
2. American Water Works Association, Manual M36, *Water Audits and Leak Detection*, 2d ed., AWWA, 1999.
3. Office of Water Services, United Kingdom, Leakage and the Efficient Use of Water: 1999–2000 Report.
4. Lambert, A., Huntington, D., and Brown, T. G., "Water Loss Management in North America: Just How Good Is It?" paper presented to Workshop on Progressive Developments in Leakage and Water Loss Management, AWWA Distribution Systems Symposium, New Orleans, La., September 2000 (*Note:* This paper has been reproduced in full in Chap. 18. Much of the data used in Chap. 4 came from this paper. Special thanks go to Allan Lambert, Tim Brown, and Dale Huntington for allowing us to use the paper.)
5. AWWA Leak Detection and Water Accountability Committee, "Committee Report: Water Accountability," *Journal AWWA*, July 1996 (reproduced in App. A of this book).
6. Lambert, A., Brown, T. G., Takizawa, M., and Weimer, D., "A Review of Performance Indicators for Real Losses from Water Supply Systems," *AQUA*, December 1999.
7. Lambert, A. O., Myers, S., and Trow, S. *Managing Water Leakage: Economic and Technical Issues*, Financial Times Energy Publications, 1998.

City of Philadelphia Annual Water Audit in International Water Association Format
Fiscal Year 2000: July 1, 1999 to June 30, 2000
(Water data shown in millions of gallons per day)

Corrected System Input Volume

Water Delivery	280.500
Master Meter Over-Registration	2.805
	277.695

Authorized Water Usage

Billed Metered	185.800
Billed Unmetered	0.000
Unbilled Metered	0.424
Unbilled Unmetered	2.570
	188.794

Water Losses 88.901

Apparent Losses

Customer Meter Under-Registration	2.008
Bypassed Flow to Fire System	0.100
Unauthorized Usage/Illegal Activities	6.150
Data/SCADA System Error	0.000
Customer Meter Malfunction	1.836
Meter Reading/Estimating Error	0.047
Accounts Lacking Proper Billing	2.500
City Properties	4.000
Billing Adjustments/Waivers	2.000
Apparent Losses	**18.641**

Apparent Losses Cost: $13,750,897

Real Losses

Operator Error (Tank Overflows)	0.000
Unavoidable Annual Real Loss	5.724
Recoverable Leakage	
Customer Service Line Leaks	38.633
Main Leaks	24.348
Main Breaks (Bursts)	0.067
Other	1.488
Real Losses	**70.260**

Real Losses Cost: $2,965,228

Fiscal Year 2000 Financial Data

$3,297 Apparent Losses per MG – Small Meter Accounts (5/8" and 3/4")
$2,904 Apparent Losses per MG – Large Meter Accounts (1" and Larger)
$2,828 Apparent Losses per MG for City Property Accounts
$3,120 Apparent Losses – Overall Average Customer Rate
$110.25 Real Losses – Marginal Cost per MG
$130,132 Real Loss Indemnity Costs – Added to Total

$151,181,693 Water Supply Operating Cost – Fiscal Year 1999 Data

Technical and Financial Performance Indicators for Water Supply System Losses

Technical Performance Indicator for Real Losses

Real Losses	70.240 MGD
Unavoidable Annual Real Losses	5.724 MGD
Infrastructure Leakage Index (ILI)	12.30
Recommended Long-Term Targets*	
Long-Term ILI	3.500
Target Long-Term Real Losses	20.000 MGD
Current Excess Real Losses	50.24 MGD
Value of Current Excess	
Real Losses	$2,027,259

* Recommended by Allan Lambert, International Water Data Comparisons, Ltd. An ILI of 3.5 represents the median value of an international data set drawn from over 20 countries. Specific Philadelphia economic influences may dictate a different long-term target ILI.

Financial Performance Indicator for Nonrevenue Water

Nonrevenue water is defined as real and apparent losses and unbilled authorized usage.

The nonrevenue cost ratio is the percentage of the annual cost of nonrevenue water over the annual running costs for the water system.

Nonrevenue water cost = $17,109 + $131,882 + $13,750,897 + $2,965,228 = $16,865,116

Nonrevenue cost ratio = ($16,865,116/$151,181,693) × 100 = 11.15%

Figure 4.10 Annual water audit in International Water Association format. (*Source: City of Philadelphia.*)

56 CHAPTER 4 EVALUATING WATER LOSS

	AUDIT WORKSHEET	(Adapted from AWWA M36)		
For:	Boston Water and Sewer Commission	Study period Jan 1 2000-Dec 31 2000		
			Water Volume	
Line	Item	Subtotal	Total Cumulative	Unit
Task 1	**Measure the Supply**			MG
1	Total Water Supply to Distribution System			
	Uncorrected Total Water Supply to the Distribution System (total of master meters)		30,306.0	MG
2	Adjustments to Total Water Supply			
	a. Source Meter Error (+/-)	0.0		MG
	b. Change in Reservoir and Tank Storage	0.0		MG
	c. Other Contributions or Losses (+/-)	0.0		MG
3	Total Adjustments to Total Water Supply (2a, 2b and 2c)		0.0	MG
4	Adjusted Total Water Supply to the Distribution System (add lines 1 and 3)		30,306.0	MG
Task 2	**Measure Authorized Metered Use**			
5	Uncorrected Total Metered Water Use (sales)	23,601.0		MG
6	Adjustments due to Meter Reading Lag-Time (+/-)	0.0		MG
7	Metered Deliveries (add lines 5 and 6)	23,601.0		MG
8	Total Sales Meter Error and System Service Meter Error (+/-)			
	a. Residential Meter Error	455.0		MG
	b. Large Meter Error	455.0		MG
	c. Total (add lines 8a and 8b)	910.0		MG
9	Corrected Total Metered Water Deliveries (add lines 7 and 8c)		24,511.0	MG
10	Corrected Total Unmetered Water (subtract line 9 from line 4)		5,795.0	MG
Task 3	**Measure Authorized Unmetered Use**			
11	a. Firefighting and Training	303.0		MG
	b. Main Flushing	45.0		MG
	c. Storm Drain Flushing	0.0		MG
	d. Sewer Cleaning	95.0		MG
	e. Street Cleaning	7.0		MG
	f. Schools	0.0		MG
	g. Landscaping in Large Public Areas			
	1. Parks	0.0		MG
	2. Golf Courses	0.0		MG
	3. Cemeteries	0.0		MG
	4. Playgrounds	0.0		MG
	5. Highway Median Strips	0.0		MG
	6. Other Landscaping	0.0		MG
	h. Decorative Water Facilities	0.0		MG
	i. Swimming Pools	0.0		MG
	j. Construction Sites	106.0		MG
	k. Water Quality and Other Testing	0.0		MG
	l. Process Water at Treatment Plants	0.0		MG
	m. Other Unmetered Uses	38.0		MG

Figure 4.11 Boston Water and Sewer Commission Annual Water Audit in M36 format.

Line	Item	Water Volume		Unit
		Subtotal	Total Cumulative	
12	Total Authorized Unmetered Water (add lines 11a to 11m)		594.0	MG
13	Total Water Losses (subtract line 12 from line 10)		5,201.0	MG
Task 4	**Measure Water Losses**			
14	a. Accounting Procedure Errors	0.0		MG
	b. Unauthorized Connections	0.0		MG
	c. Malfunctioning Distribution System Controls	0.0		MG
	d. Reservoir Seepage and Leakage	0.0		MG
	e. Evaporation	0.0		MG
	f. Reservoir Overflow	0.0		MG
	g. Discovered Leaks	1,186.0		MG
	h. Unauthorized Use	0.0		MG
15	Total Identified Water Losses (add lines 14a to 14h)		1,186.0	MG
Task 5	**Analyze Audit Results**			
16	Potential Water System Leakage (subtract line 15 from line 13)		4,015.0	MG
17	Recoverable Leakage (estimated at 75%)		3,011.3	MG
	(24 months)		6,022.50	MG

Line	Item	Dollars		Unit
18	Cost Savings			
	a. Cost of Water Supply	$1,100.00		MG
	b. Variable O&M Costs	$0.00		MG
19	Total Costs per Unit of Recoverable Leakage	$1,100.00		MG
20	One-Year Benefit from Recoverable Leakage	$3,312,375.00		Year
21	Total Benefits from Recoverable Leakage (24 months)	$6,624,750.00		24 months
22	Total Costs of Leak Detection Project	**$600,000.00**		
23	Benefit-to-Cost Ratio		11.0 :1	

Comments:

Many of the lines reported as 0.0 in Task 4 (Measure Water
Losses) are reported as such mainly because there is no
confirmed data which can be used to calculate these amounts.
Rather than inaccurately report these figures, no water volume
is assigned to these items at this time.

The figures reported in Task 2, lines 8, are based on
the analysis of meter manufacturer standards coupled with
assumptions made regarding BWSC's meter accuracy test and
replacement programs. A more representative sample of
meters must be taken out of service and tested to obtain an
accurate measure of meter error.

Prepared by:

Name: Thomas Holder
Title: Deputy Superintendent of Operations   Date: 7/2/01

Figure 4.11 (*Continued*)

CHAPTER 5

Data Formatting and Management

Julian Thornton

5.1 INTRODUCTION

To undertake any water system audit and properly identify where losses are occurring and the magnitude of the loss it is necessary to collect data that is *accurate, organized,* and *accountable.* In Chap. 6 we discuss various equipment and methodologies for capturing data for flows and pressures accurately, using both portable and permanent field equipment. However, once we have captured the data, it is important to organize and store the data in a meaningful manner, so that we can be accountable for the subsequent decisions we will inevitably make.

In many cases good, accurate data may not be available, and the operator has to make a decision as to whether to use questionable data or estimations. In many cases it is better to do something rather than stop and do nothing. In this case one should be sure to note that the data were questionable or estimated, and the operator's assessment of what should be done to improve this in future audits and how the data should be used this time.

The following section discusses good data management techniques.

> **N**ot all water systems will have all of the data needed for a full audit, but it is better to make estimations and perform an audit than not to do one at all.

5.2 DATA COLLECTION WORKSHEET

One of the first things to do before starting to download field loggers and recorders is decide on the key factors that will be analyzed and assign relevant measurement units and decimal places to each of the parameters. For example, in most audits the operator will be measuring flow, measuring pressure, analyzing volumes, measuring levels, and accounting for time periods.

Some of the units that might be assigned are as follows.

1. Flow
   a. Metric
      ❑ Cubic meters per second
      ❑ Cubic meters per hour
      ❑ Liters per second
      ❑ Megaliters per day
   b. American Standard or Imperial
      ❑ U.S. gallons per minute
      ❑ Imperial gallons per minute
      ❑ U.S. gallons per hour
      ❑ Imperial gallons per hour

- U.S. kilogallons per day
- Imperial kilogallons per day
- U.S. millions of gallons per day (MGD)
- Imperial millions of gallons per day (MGD)
- Cubic feet per hour
- Cubic feet per day
- Acre-feet per day

2. Pressure
    a. Metric
        - Meters head of water
        - Bar
        - Kilopascals
    b. American Standard or Imperial
        - Pounds per square inch (psi)
        - Feet head of water

3. Volume
    a. Metric
        - Cubic meters
        - Liters
        - Megaliters
    b. American Standard or Imperial
        - Gallons
        - Kilogallons
        - Million gallons
        - Cubic feet
        - Acre-feet

4. Levels
    a. Metric
        - Millimeters
        - Meters
        - Millibar
        - Bar
    b. American Standard or Imperial
        - Inches column of water
        - Feet column of water

5. Time periods
    - Milliseconds (used for surge analysis and leak noise correlation)
    - Seconds
    - Minutes
    - Hours
    - Days
    - Months
    - Years

So, as we can see, there are many options for recording the various parameters. It is important to use parameters and units which are both meaningful to the country or area in which we are working and also units which are easily interchangeable. For example, we would not want to mix cubic meters per hour of flow with pounds per square inch of pressure. We might, however, use either pounds per square inch of pressure with gallons per minute of flow or cubic meters per hour of flow with meters head of water pressure.

5.2.1 Balancing Flows

When undertaking audits, which involve dynamic flows and not just volumes, it is important to balance flow inputs. To do this we usually select a unit of flow, for example, cubic meters per hour.

We then identify key points within a 24-h profile, usually minimum night flows if we are trying to identify leakage. The balance is a simple matter of adding and or subtracting individual zone flows (these might be metered areas or pressure zones) and comparing them with supply meter or production metered flows to ensure that we have all of the inflows and outflows for the system in question accounted for. (Take care if storage is located inside of the areas we are trying to balance, as filling volumes will confuse the issue.)

In situations where the system is not zoned in any way at all and is not intended to be in the future, the key points within the flow balance are

- ❑ Production meters
- ❑ Import meters or bulk supply meters
- ❑ Outlets from storage (tanks, reservoirs, and towers)
- ❑ Outlet from pumps or wells

> Top-down annual audits use volumes; bottom-up audits often use night flows.

Balancing flows may seem like a relatively simple procedure, but it can take many hours of careful analysis, especially in large systems.

It is particularly important to define one unit of measure before attempting this exercise, or the difference in one working unit and another might be confused for a missing inlet or outlet and create a lot of unnecessary work, which in turn would create unnecessary cost.

5.2.2 Balancing Pressures

It is equally important to balance pressures in a water system when attempting to identify losses, as the system pressure plays a large part in water loss, especially leakage, as discussed later in this manual.

Usually, when we want to balance pressures we work out *hydraulic grade lines* (HGLs). Hydraulic grade is the sum of the ground level plus the static pressure at that particular point, and the lines are the chosen points connected up.

Most often in water loss control situations we will need to know:

- ❑ Supply or inlet pressure
- ❑ Average zone pressure
- ❑ Critical-point pressure
- ❑ Minimum service pressure

5.2.3 Balancing Levels

In systems with large storage capacity it is also important to include the various tank or reservoir levels in the water balance, as the change in volume over time may represent significant flow and could be mistaken for loss.

In systems with a small amount of storage capacity this is not so important, but it should not be overlooked. It is always better to overanalyze than underanalyze!

5.2.4 Putting Data into a Common Format

Putting data into a common format is extremely important. Metric and imperial units should never be mixed, and even when using one or the other it is still a good idea to think about the method of data recording which has taken place in the field and the required reporting units.

> **N**ever mix incompatible units.

If working in metric units, for example, it is much easier to work in cubic meters per hour flow if you measure velocity in meters per second and calculate pipe effective area in square meters. The result will always be in meters, and then it is just a simple case of deciding the time units.

For example, we measure a velocity of 2 m/s in a pipe that has a diameter of 400 mm or 0.400 m. The area $A = \pi r^2$, or $3.142 \times 0.2 \times 0.2$ m = 0.12568 m^2. The velocity is 2 m/s so we would multiply this figure by 2, giving 0.25136 m^3/s. Now we must decide on a unit of time. Usually, when working in the field with cubic meters we use cubic meters per hour of flow. There are 60 s in a minute and 60 min in an hour, so we multiply our flow of 0.25136 m^3/s by 3600, giving 904.896 m^3/h. We know there are 1000 liters in 1 m^3, so we can also say that we have a flow of 904,896 liters/h. This number is quite large, and if it is added to other large numbers could lead to mistakes. If we want to express our flow in liters we would most likely use liters per second. If that were our desired final unit, we take our figure above of 0.25136 and multiply by 1000 to convert our flow units from cubic meters to liters. We do not need to multiply anything else, as our original measurement was in seconds. Our flow is then 251.36 liters/s. If we need to convert data from liters per second to cubic meters per hour, our common figure is 3.6. When converting liters per second to cubic meters per hour we just need to multiply our original number by 3.6 to get cubic meters per hour and vice versa.

Alternatively, if we were working in imperial units or U.S. standard units, we might measure a velocity of 2 ft/s in a pipe which has a diameter of 36 in or 3 ft. The area $A = \pi r^2 = 3.142 \times 1.5$ ft $\times 1.5$ ft = 7.0695 ft^2. The velocity is 2 ft/s so we multiply this figure by 2, giving 14.139 ft^3/s. Now we must decide on a unit of time. Often, when working in the field with cubic feet, we use cubic feet per hour of flow. There are 60 s in a minute and 60 min in an hour, so we multiply our flow of 14.139 ft^3/s by 3600, giving 50,900 ft^3/h. We know there are 7.48 gal in 1 ft^3, so we can say we have a flow of 380,734 gal/h. This number is quite large, and if added to other large numbers it could lead to mistakes. If we want to express our flow in gallons we would most likely use gallons per minute. If that is our desired final unit we divide 380,735 gal/h by 60; we do not need to divide anything else, as our figure is already in gallons. Our flow is then 6345 gal/min (sometimes abbreviated as gpm).

5.3 DATA CALIBRATION FORM

Often there is a small margin of error when measuring devices are tested. It is not always possible to recalibrate a flow meter before measuring in the field, although that option is preferable. If the flow measuring device cannot be recalibrated mechanically or electronically, it is still possible to use the data if the data are calibrated theoretically using a spreadsheet.

The spreadsheet is constructed using the calibration curves from the meter tests prior to data collection and show errors for brackets of flow. The data can then be imported into the form or spreadsheet, and automatically changed by the error attributed to that flow range. The resultant data is closer to the truth than the original. Obviously there are some cases where error will still occur, especially in the case of a particularly sensitive or unstable measurement device.

5.3.1 Equipment Calibration Form Pressure and Level

As with flow measurement devices, pressure and level sensors can have errors which cannot be corrected for before testing is undertaken. The same process as outlined above can be undertaken to ensure that pressures and levels are close to the true values.

5.4 SUMMARY

It is vitally important that data be managed properly from the start of an audit program. *Accountability* is a word that is being used very often in the water industry today. Accountability does not mean that we guarantee that all of our data is accurate. What is important is that where we have doubts as to the accuracy of the data we leave an audit trail explaining what was estimated or calculated. If we perform accountable audits with a data trail, we can always improve data accuracy over the years to come, until eventually all data will be top class.

The following checklist covers many of the aspects necessary for good data management.

- ❑ Data should be accurate.
- ❑ Data should be organized.
- ❑ Data should be accountable.
- ❑ Bad data should be clearly highlighted.
- ❑ Estimations can be made but should be clearly marked as such.
- ❑ Raw data should be kept as well as calibrated data.
- ❑ Constant measurement values should be used.
- ❑ Constant units should be used.
- ❑ A column alongside the audit sheet with relevant comments will help future auditors figure out what you did when you made your audit.

> **G**ood data management will ensure that the whole project has accountable baselines from which to judge performance and allocate new budgets.

CHAPTER

6

Equipment and Techniques

Flow Metering, Pressure Measurement, Control, and Leak Detection

Julian Thornton

6.1 INTRODUCTION

When undertaking hands-on evaluation of a water system or facility's losses, it is vitally important to use good, accurate, and valid data taken from the field. To do this, we can use portable, temporary equipment, or we can install new permanent equipment (or rehabilitate old) alongside data loggers or recorders to collect time-based trends. These allow us to analyze the actual dynamic situation. Once the situation has been identified (by an audit as discussed in Chap. 4), then detection equipment (in the case of real losses) is used to pinpoint individual problem areas.

This chapter outlines some of the equipment types and methodologies in today's market. Additional and complementary information can also be found in Chaps. 10–14, which deal more with field intervention methodologies, although there is some overlap in these chapters. Appendixes B, C, and D also contain useful and relevant data.

> There is no substitute for good, accurate, and accountable field data.

6.2 PORTABLE EQUIPMENT

Many different types of flow measurement equipment are available on the market today and, as with leak detection equipment, operators must ensure that whatever equipment is used, personnel can be trained and supported locally to use and maintain the equipment in good working order. Operators must be confident in the accuracy of the equipment being used and in its reliability in different situations. To use test equipment with uncertain credibility will only lead to confusion. Often the reason for measuring the flow is to try to identify why something is happening: why a master meter is overreading; why or if there is leakage; etc. This in itself is often difficult enough with test equipment which is familiar to the operator, without throwing in added uncertainty.

The discussion which follows is designed to highlight some of the methodologies used in field measurements, and it is not intended to promote one methodology over another or one manufacturer over another.

6.2.1 Portable Insertion Meters

One of the most commonly used portable flow meters is the insertion-type meter. Insertion meters come in various types from many different manufacturers. Each type of equipment has strong and weak points, but with good technical support, calibration, and maintenance, most of the instruments on the market today can do the job.

Some of the most common insertion meters are

- The pitot rod (see Fig. 6.1)
- The turbine insertion meter (see Fig. 6.2)
- The magnetic insertion meter (see Fig. 6.3)
- The vortex shedding insertion meter
- The heat meter

The most common forms of these meters are actually single-point velocity meters. The measurement point is selected either as a measured average point in the pipe or often the centerline is used with a calculated factor to take into account both the blockage of the meter and the difference between centerline velocity and average velocity. Once the average velocity has been either calculated or recorded, it is then multiplied by the effective area of the flowing pipe to give flow.

Some of the above-mentioned meters, in particular the differential pressure (DP) meter and the magnetic meter, have an averaging function built in. They measure the velocity at predetermined points, perform an automatic average, and give out flow units. If the purpose of the portable equipment is to make measurements on varying pipe diameters, then it is more likely that a single-point meter will be used because this will allow the user to work with virtually any pipe diameter without having to order various rod sizes.

In order to calculate either the average velocity point or the factor between the theoretic average and the centerline velocity, it is necessary to perform a velocity profile. The velocity profile is a series of velocity measurements taken across the diameter of the pipe, sometimes on both the 90° and 180° axes depending on the accuracy required—see ISO 7145-1982 E for more details on this methodology. Figure 6.4 is an example diagram and Fig. 6.5 is a sample spreadsheet used for calculating this task. As with many waterworks tasks, the use of a simple spreadsheet can save time and effort in reducing the need for repeatable calculations. The manufacturer's coefficients should be programmed into the spreadsheet for each type of meter used. Some manufacturers provide profiling programs with their equipment as part of the package.

Alternatively, many meter manufacturers supply approximate points or coefficients for uses on varying pipe sizes—see Table 6.1 for an example. It can be seen that as the pipe size increases, the coefficient increases. This is a function of the reduced effect of blockage on the pipe area by the meter.

Figure 6.1 Pitot rod installation. (*Source: BBL Ltda.*)

The decision of when to profile and when to use approximate industry standard figures really comes down to the accuracy the operator is expecting from the meter. This will be very task-specific. For example, if the operator is testing a master meter which is supposed to be accurate to ±0.5 percent, then he or she must at a very minimum perform a velocity profile. However, if the operator is looking for order-of-magnitude variation in flow to ascertain which zones have night flow for a leakage ranking exercise or using the meter to size a valve which is quite forgiving in flow pattern, then obviously it may not be deemed necessary to

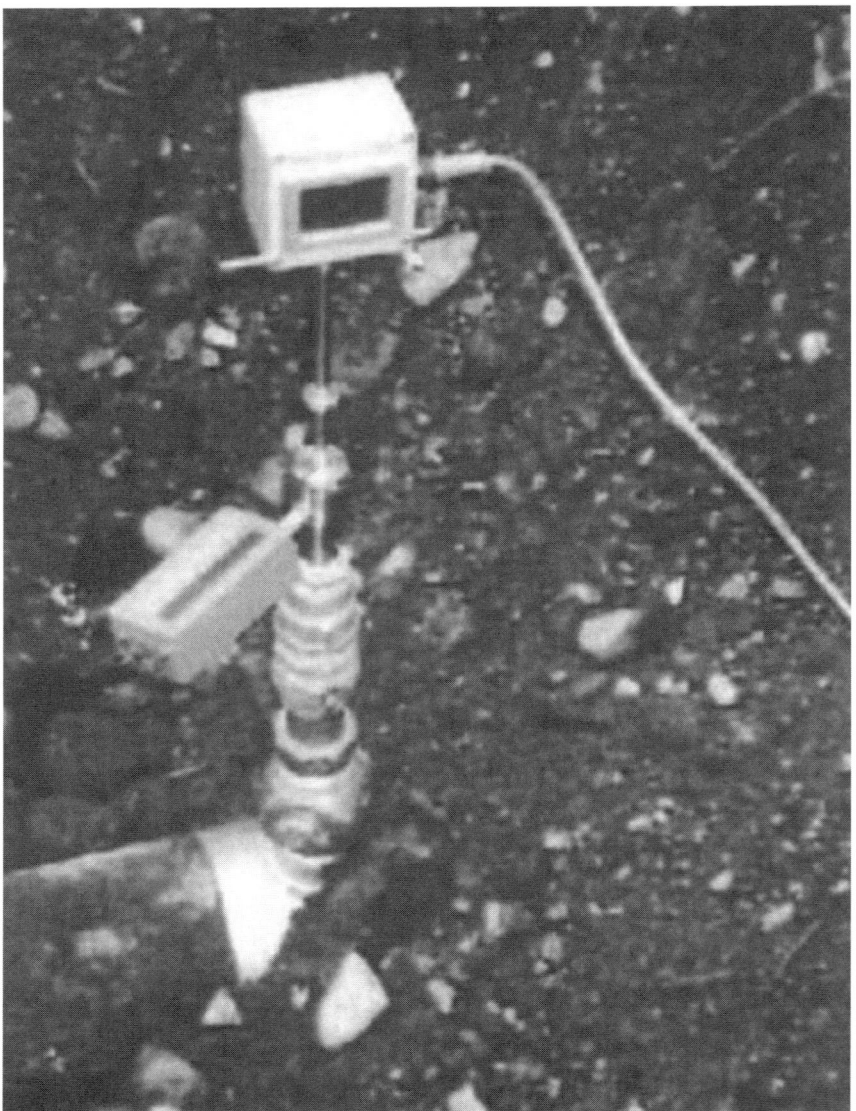

Figure 6.2 Turbine insertion meter and data logger. (*Source: Julian Thornton.*)

spend the time and effort on a velocity profile but rather use the industry standard figures. (These sorts of decisions should be noted in the comments column of the audit sheet and the data collection sheet, as discussed in Chap. 5.)

Hot Tapping

Insertion meters are usually fitted to the pipe through a hot tap or tee connection on the pipe (see Fig. 6.4). Fitting a hot tap is a relatively simple procedure and is usually done under pressure using a tapping machine (see Fig. 6.6). Care should be taken to ensure that the tap is of sufficient diameter that the rod of the meter can pass through the valve without damage. For example, if the meter rod external diameter is 1 in

Figure 6.3 Insertion mag meter installation. (*Source: Restor Ltda.*)

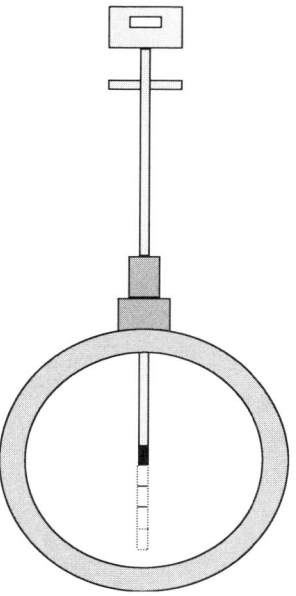

Figure 6.4 Velocity profiling.
(*Source: Julian Thornton.*)

70 CHAPTER 6 EQUIPMENT AND TECHNIQUES

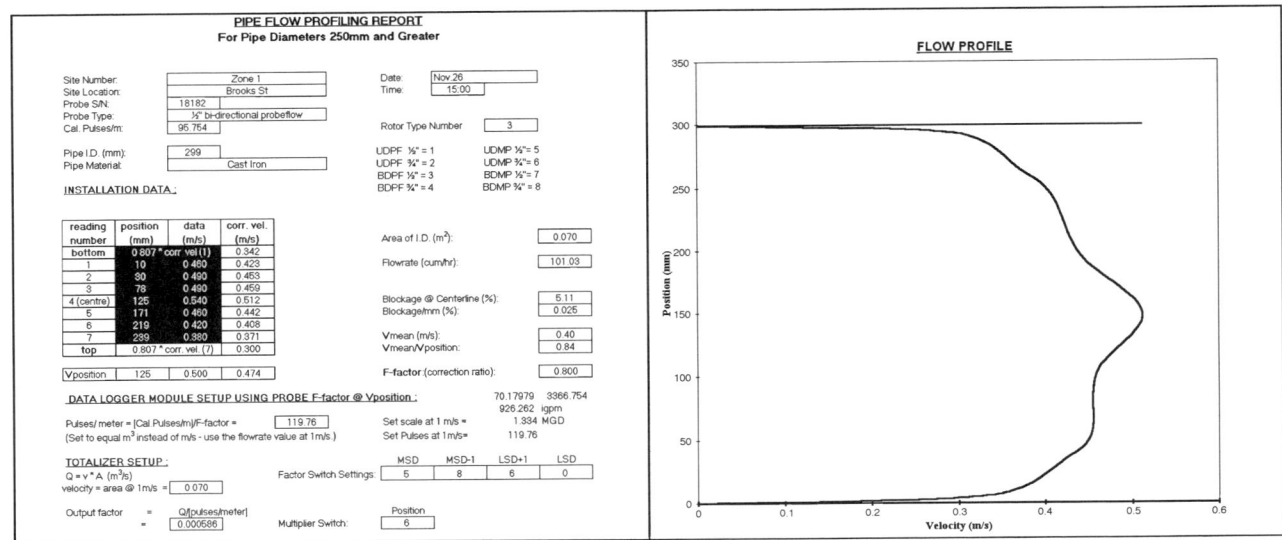

Figure 6.5 Velocity profiling spreadsheet. (*Source: Julian Thornton.*)

TABLE 6.1 Coefficients for Velocity Profiling (adapted from Quadrina data)

Pipe ID (mm)	Centerline Correction
150	0.658
200	0.753
250	0.798
300	0.823
400	0.847
600	0.863
800	0.867
1000	0.869

Source: Julian Thornton.

or 25 mm, then care must be taken to select a valve or tap which has at least a 1-in or 25-mm internal bore. Note that some valves are not identified by their internal diameter.

Consideration must also be given to the pipe material. Hot taps are usually tapped directly into the pipe if the pipe has a metallic wall, although some utilities prefer to use a weldolet (see Fig. 6.7) or a tapping sleeve (see Fig. 6.8). Whenever the pipe is not metallic, a tapping sleeve should be used. In addition, care should be taken to ensure that the cutting bit is sharp, so as not to crack the pipe [especially in the case of asbestos cement (AC) or cement pipes].

Installation Process

The first step, and one of the most important, is to select a suitable metering location, away from other fittings and disturbances in the flow, such as gate valves, meters, pressure-release valves, etc. Most manufac-

Figure 6.6 Tapping under pressure. (*Source: Restor Ltda.*)

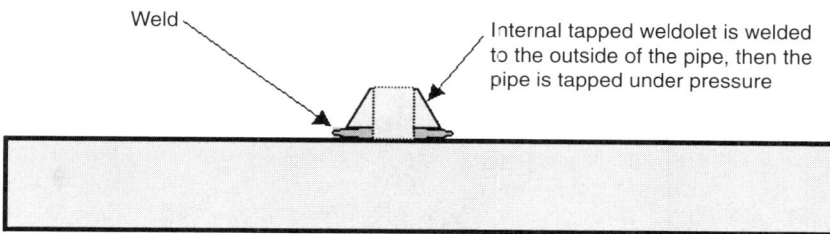

Figure 6.7 A weldolet can be used on metallic pipe. (*Source: Julian Thornton.*)

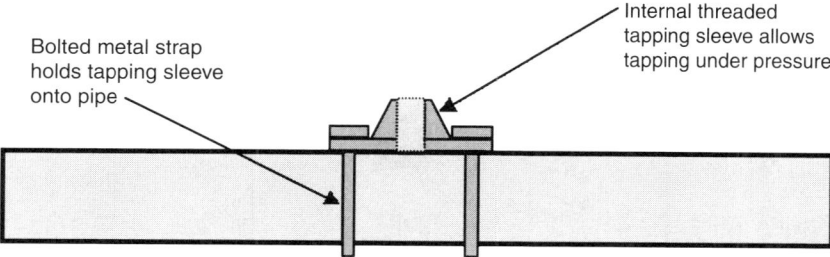

Figure 6.8 Tapping sleeve. (*Source: Julian Thornton.*)

turers indicate the number of upstream and downstream diameters of pipe needed for accurate measurement. If this information is not available, a good rule of thumb is 30 diameters from an upstream disturbance and 20 from downstream disturbances (see Fig. 6.9). Measurements can be made closer to disturbances but may result in either unstable velocity recordings

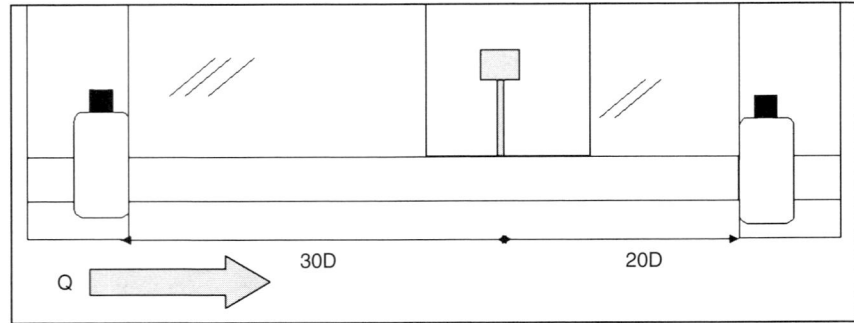

Figure 6.9 Measurements should be made away from disturbances. (*Source: Julian Thornton.*)

or unstable velocity profiles, which will result in error. (If this is the only option, then note the disturbance on the data sheet together with an estimated error, so that the notation becomes part of the audit trail.)

Once the hot tap has been installed, the exact pipe internal diameter must be measured to ensure accuracy. Figure 6.10 shows a caliper, which can be used for this process.

Profiling

After the pipe diameter has been measured, a decision must be made as to whether to profile the velocity. If a velocity profile is to be undertaken, it should be done during stable flow conditions. Depending on the pipe diameter, the operator may select a number of positions and install the meter, taking care to tighten the pressure fittings before opening the valve. Then the operator measures and records the velocity at the predetermined points and enters them into the program or spreadsheet. It is a good idea to take several readings at each point, as velocities tend to vary even during stable conditions. For additional accuracy, three profiles should be taken and an average used. The program or spreadsheet will give either a factored value to be used at the centerline, or it will predict the point at which the average velocity can be found.

Once the profile has been taken or a decision made not to profile, the operator must decide whether to fit the meter at the centerpoint and factor the data as discussed above, or locate the average point and use raw data. Usually the operator will fit the meter at the centerpoint when low velocities occur, to allow the meter to work within its minimum velocity limits.

Positioning and Fitting the Data Logger

Once the metering location has been determined, the operator should secure the locking nut to ensure that the meter is positioned correctly along the axis of the pipe (see Fig. 6.11). Most meters are quite sensitive to this positioning, and measurement errors can occur if care is not taken.

At this point the operator must fit the data logger if one is to be used. There will be further discussion on data logging in Sec. 6.4.

Key things to remember when using a single-point velocity probe to measure flows are

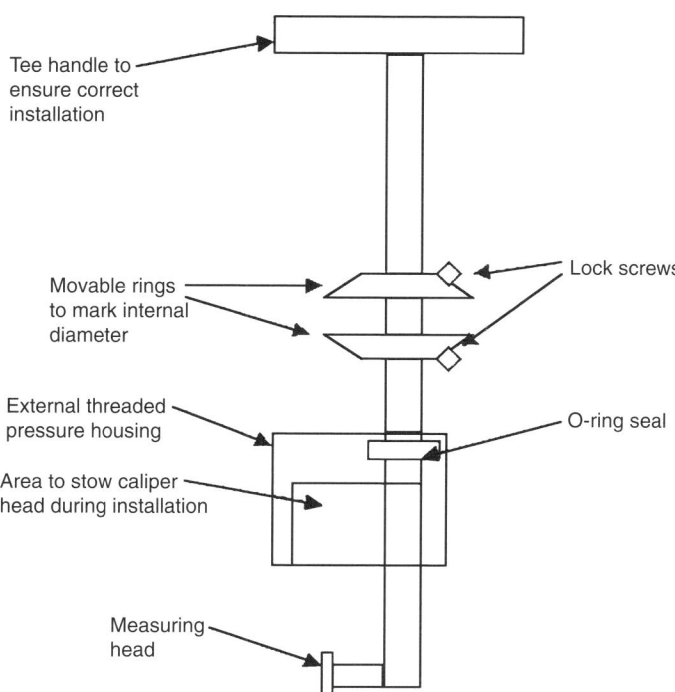

Figure 6.10 Caliper for measuring internal pipe diameter. (*Source: Julian Thornton.*)

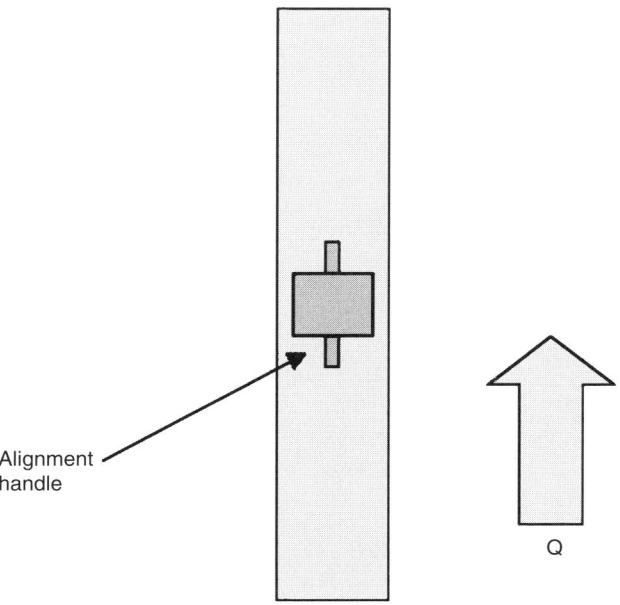

Figure 6.11 Proper alignment is important for accurate measurement. (*Source: Julian Thornton.*)

- Always record all data during setup. Clearly mark any assumptions that were made.
- The accuracy stated by the manufacturer is the accuracy and repeatability of the probe. This figure does not allow for human error during the setup process.
- An incorrect diameter measurement will seriously affect flow accuracy.
- Incorrect positioning of the probe within the velocity profile will affect the accuracy of the flow calculation.
- Incorrect factoring of a centerline velocity will also seriously affect the accuracy of a flow calculation.
- Incorrect axial positioning will affect the accuracy.
- When installing the meter, be sure to hold the meter head to ensure that it does not inflict damage when opening the valve to insert the meter.
- Be sure to withdraw the meter completely from the pipe before closing off the valve when a metering exercise is complete. Failure to do so will seriously damage the probe and also require a total system shutdown to remove the damaged probe.
- Ensure that the rod length and the available space in a chamber are compatible.
- Identify battery life and monitoring period required. Ensure that backup batteries can be fitted if required.
- Always store both raw and manipulated data in case of error during setup.

The inexperienced operator may feel that insertion metering is a lot of work with many potential sources of error. However, with care and a little practice, insertion meters can give very good performance. While they are intrusive and an entry point does have to be provided, which can sometimes be a stumbling block, this methodology does have the benefit of being able to measure the internal diameter of the pipe accurately. This is not always the case with nonintrusive methodologies and can be critical.

6.2.2 Portable Ultrasonic Meters

Portable ultrasonic meters have been around for about 20 years and have recently become very sophisticated and accurate. Some operators do not like ultrasonic meters, but with the right care during installation they can provide very accurate information. In some cases the operator may just want to get an idea about the flow variance, and in this situation ultrasonic meters are perfect because they are completely nonintrusive and do not require a hot tap or entry into the pipe.

There are various types of ultrasonic meters on the market, but the most common fall into two categories:

- Ultrasonic Doppler meters
- Ultrasonic transit time (sometimes referred to as time of flight) meters

Ultrasonic Doppler meters are normally used to measure liquids which include either particles or entrained air. The Doppler principle works on reflection, as seen in Fig. 6.12. For this reason Doppler meters are usually not used in clean-water applications.

Ultrasonic transit time meters work by sending and receiving signals from one sensor to another as seen in Figs. 6.13 and 6.14. Figure 6.13 shows a typical installation in reflex mode, where the sensors are strapped to the same side of the pipe. In this case the signal bounces off the pipe wall and back to the second sensor. Figure 6.14 shows an alternative method of installation in which the signal is transmitted directly at the second sensor.

The time taken for the signal to travel from one sensor to the other sensor changes as the velocity in the pipe changes, according to Faraday's law.

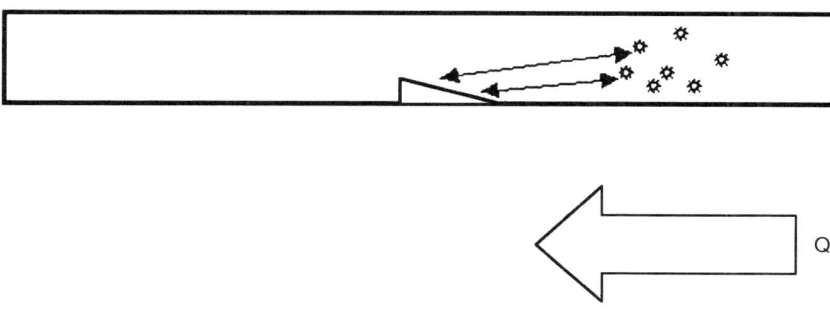

Figure 6.12 Doppler-effect sensors. (*Source: Julian Thornton.*)

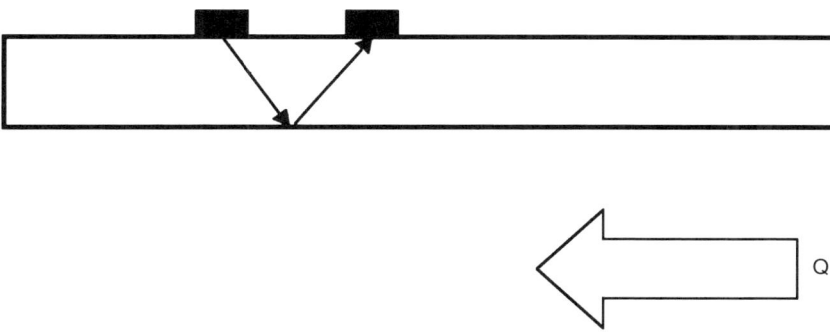

Figure 6.13 Transit time reflex pattern. (*Source: Julian Thornton.*)

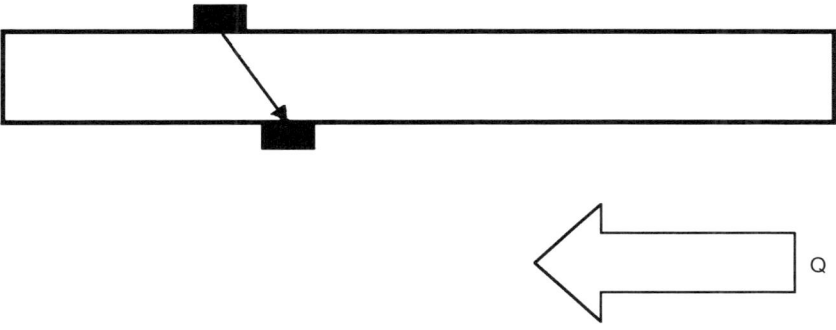

Figure 6.14 Transit time measurement. (*Source: Julian Thornton.*)

CHAPTER 6 EQUIPMENT AND TECHNIQUES

> **It is often difficult to measure flow accurately on old, corroded pipes.**

Pipe materials affect the transmission of the signal. Most experienced users of ultrasonic meters know that sometimes it can be hard to get a good signal on old, corroded cast-iron pipe, as the corrosion tends to deflect the ultrasonic beam. As the pipe material and diameter change, so does the angle of the beam as it travels through the pipe material. It is therefore very important to know the pipe material before programming the unit, as this will dictate the sensor separation. If an incorrect separation is used, then the signal received by the second sensor may be weak or nonexistent. It is also important to measure the pipe wall diameter, which is hard to do even with an ultrasonic thickness gauge because the pipe may be lined or corroded; however, best estimations and measurements need to be made.

Installing the Meter

As with the insertion meter, a location should be chosen away from other pipefittings. The pipe wall material should be identified and, where possible, internal and external measurements made. (Sometimes there may be a tap or even a piece of old pipe lying around from installation.) If not, ultrasonic thickness gauges can be used in some cases to measure the pipe wall thickness. In many cases these gauges do not measure layers of corrosion or pipe linings, which have to be estimated. It is a good idea to have a set of pipe tables with the equipment, so that in case internal measurements cannot be made, at least a good estimate from a pipe table can be entered. A complete example pipe table is given in App. E.

> **The 30-diameter, 20-diameter rule also works well for ultrasonic meters if no other rule is specified.**

The outside of the pipe should be cleaned. Corrosion or grease and dirt will affect the strength of the signal and therefore the effectiveness of the meter at this point.

The pipe information should be programmed into the unit according to the manufacturer's instructions, and the unit will in most cases tell the operator to install the sensor in either reflex or direct mode.

Before installation, the pipe should be clearly marked for sensor installation. It is important that the sensors are positioned exactly right; otherwise a poor signal may result.

Before applying the sensors to the pipe wall, conductive grease should be applied to the sensor to allow it to bond better to the pipe wall (see Fig. 6.15). The grease helps the signal pass directly into the pipe and not reflect off the outside of the pipe. Care should be taken not to overgrease the sensors, as excessive grease on the pipe—particularly small-diameter pipes, can create a situation in which the signal passes directly from one sensor to the other. (On smaller pipes the sensors tend to be close together.)

Once the sensors are in location, it is a good idea to check the signal strength by gently moving the sensor around the location. Once the highest signal strength is located, the sensors should be secured to the pipe wall. Most manufacturers supply suitable strapping for this purpose; in the worst-case scenario, duct tape works quite well.

Flow Measurement

Once the sensors are securely strapped in place, flow measurement can begin. At this point it is often recommended to undertake a zero-flow

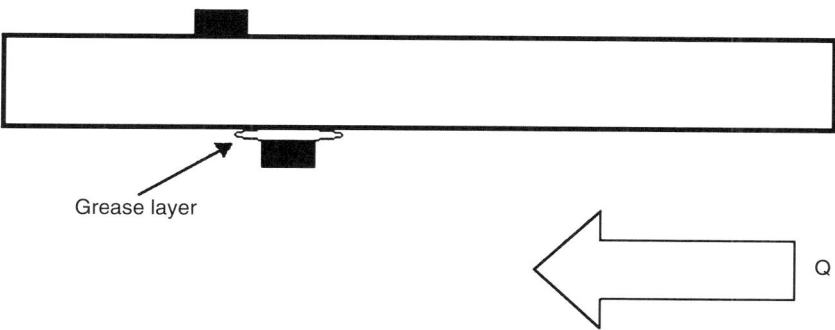

Figure 6.15 Grease helps the signal pass through the pipe wall. (*Source: Julian Thornton.*)

calibration if possible. This allows an in-place calibration of the unit. Flow should be closed off momentarily and the unit informed that the flow is zero. After this the flow may be carefully turned back on again. When turning flow on and off in a system, take care to do it very slowly, as careless operation of valves can create water hammer, which can create additional leakage and damage to the system.

At this point a data logger may be fitted to record flow and time ratios. Some units will come equipped with an internal data logger, and some may even produce real-time graphs on LCD displays.

Ultrasonic Transmit Time Meter Readings

Key things to remember when using an ultrasonic transit time meter to measure flow are

- Bad data in, bad data out. Incorrect pipe material or wall thickness data will result in error.
- Disturbances to the flow will cause errors or erratic readings.
- Old, corroded pipes are hard to measure with these devices, although not impossible.
- Always carry a tube of grease.
- Identify battery life and required monitoring period. Ensure that backup batteries are available and can be fitted if necessary.
- Always carry pipe tables.

As with the insertion meter, an inexperienced operator may feel insecure about the number of potential errors. However, experienced operators will confirm that ultrasonic meters can be used effectively in most situations as long as the limitations at each particular installation point are respected.

Selecting Portable Equipment

Some points to consider when selecting portable equipment are the following:

- The two types of portable equipment we have discussed are very flexible, and with certain limitations they can be used to gain a good idea of flow or velocity in a pipeline where no existing permanent equipment is available.

- Both types of equipment are available in battery-operated or main-power formats. Some equipment can run for longer than others can on internal batteries, and some can have backup batteries fitted with more ease than others.
- Some project situations require bi-directional flow recording, and most of the equipment available on the market today can be equipped to do this; however, care should be taken to request this function if required.
- The equipment can be used in many situations to check the accuracy of permanent metering equipment and will do so reasonably accurately as long as the data input into the setup criteria are correct. However, when checking the accuracy of permanent metering equipment, a calibrated volumetric test is preferable although not always possible.
- Local support is invaluable.

Typical Applications

Some typical applications for this type of equipment are

- Zone flow analysis
- Leak volume monitoring
- Comparative accuracy testing of permanent metering equipment
- C–factor testing
- Demand analysis
- Hydraulic model data collection
- Valve sizing
- Pump testing

Figure 6.16 shows a portable ultrasonic transit time meter being used to locate and quantify leak volume. Figure 6.17 shows an insertion meter being used to measure flow for pressure-reducing valve sizing. This picture also shows the data logger, which in this case is recording flow and pressure.

There are of course other project-specific applications for which this equipment can be used.

6.2.3 Portable Hydrant Meters

Portable hydrant meters are used to check fire flows, C factors (numbers that relate to the internal roughness of pipes; see Chap. 17), and in general when the operator needs to put a known flow onto the system, perhaps to calibrate an existing meter. Portable hydrant meters come in two main types, turbine and differential pressure. Figure 6.18 shows a turbine-type meter in action.

Differential Pressure Hydrant Meter

The differential meter is simple to use and is normally either handheld or screws on to the hydrant port. The unit is fitted to the port and the flow slowly turned on. If a hand-held device is to be used, the sensing area should be held steadily in the flow stream. The screw-on ver-

Figure 6.16 Using a portable ultrasonic meter to locate and quantify leak volume. (*Source: Julian Thornton.*)

sion is automatically positioned correctly. Then it is a simple matter of reading the gauge pressure and relating the flowing pressure to a volumetric flow, usually by means of a chart, which will be provided by the manufacturer. Normally this type of meter has no moving parts or electronic components and is therefore robust, easy to use, and easy to maintain.

When testing areas with old mains, in particular unlined old cast iron, hydrants to be tested should be flushed well before and after testing. This ensures that any debris in the line does not enter and damage the meter and that there are no dirty-water complaints after the testing.

Care should be taken when using any hydrant meter to ensure that no damage is caused to the surrounding area. The force of the flowing water from the hydrant is often sufficient to dig large holes in grass verges and often digs up and damages asphalt road or pavement coverings.

To avoid damage to the surrounding area, the use of a hydrant diffuser is recommended. The hydrant diffuser is a cone with various baffles inside, usually made of strong mesh. The diffuser either threads straight onto the hydrant in the case of the hand-held differential-type meter, or onto the downstream end of the turbine-type meter. Diffusers are simple to manufacturer and can be made by most sheet metal shops. The number of baffles required will vary with the potential flow and pressure from the hydrant.

80   CHAPTER 6   EQUIPMENT AND TECHNIQUES

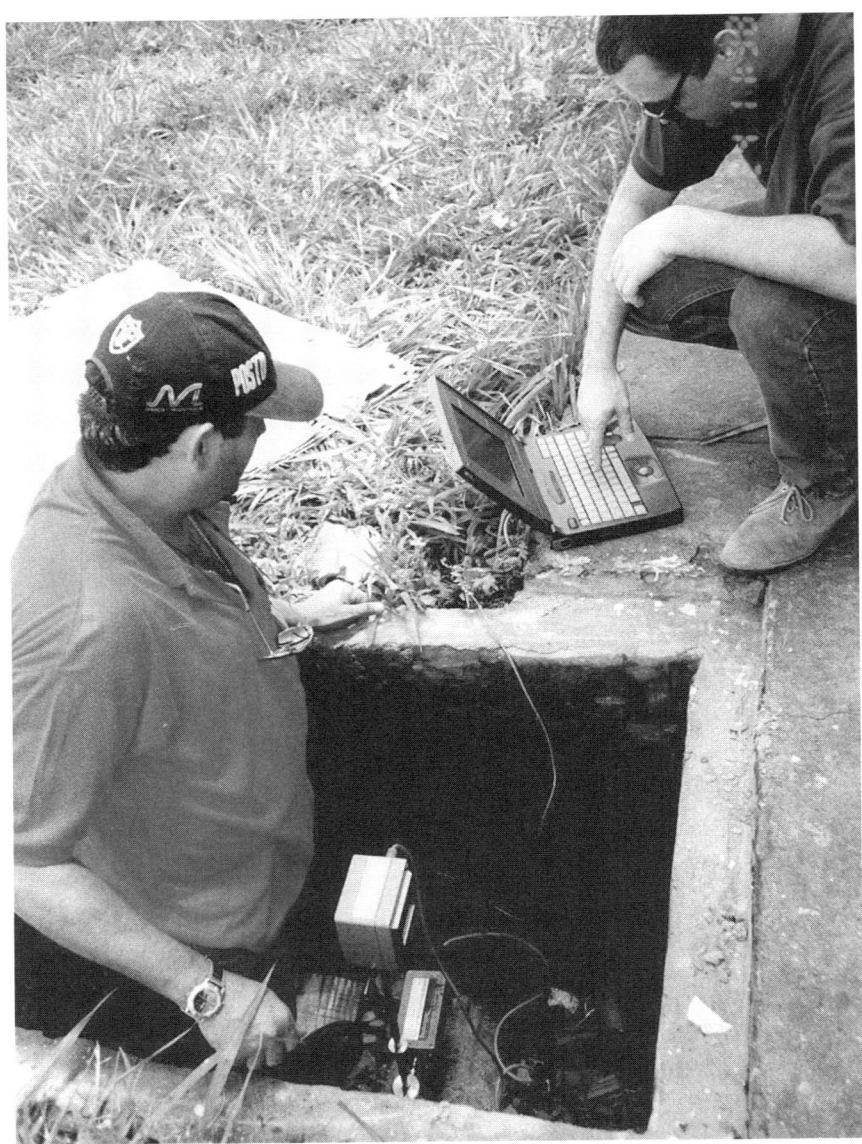

Figure 6.17 Measuring flow using an insertion turbine meter, logger, and laptop. (*Source: Julian Thornton.*)

Checklist

❏ Ensure that the area around the hydrant to be tested is barricaded off with the necessary cones and signs to warn traffic and pedestrians as to the testing, as there will be significant discharge of water.

❏ If the flows and pressures are potentially high, protect the ground where the water will flow with plastic sheeting to reduce impact and damage.

❏ Use a flow diffuser where potential damage or high flows may occur.

Figure 6.18 Turbine hydrant meter. (*Source: Julian Thornton.*)

- If using a hand-held differential pressure meter, be sure to hold the unit in the center of the flow.
- Operate hydrants slowly, to reduce negative hydraulic impact.
- Ensure that there is sufficient drainage to take away the water; there will be a significant amount of discharge.
- When testing in areas with basements, ensure that water cannot back up into basements.
- Flush hydrants well before installing the meter, to ensure that damage does not occur through rust and corrosion passing through the meter.
- Flush hydrants well after testing, to ensure that there are no dirty-water complaints.

6.2.4 Flow Loggers

Flow loggers or recorders are special types of data loggers which can take a pulse directly from the magnetic drive of virtually any meter, convert it into an electronic signal, and record rates of flow (see Fig. 6.19). Flow loggers are used for

- Comparative accuracy tests
- Demand analysis
- Rate-of-flow recording from volumetric meters
- Leak detection by zone flow analysis
- Hydraulic model data collection
- And various other projects which require flow data collection

Flow loggers are relatively easy to set up, following the manufacturer's instructions and inputting data such as flow units, pulse significance, and desired recording time.

Figure 6.19 Meter-Master flow logger. (*Source: F.S. Brainard & Co.*)

The secret to good flow logging lies in the positioning of the sensor to pick up optimum pulse strength. Obviously, missed pulses lead to missed or inaccurate flow recordings. Figure 6.20 shows examples of sensor locations for different meter types.

Some meters have a protective ring around the meter head magnets. The idea is to stop theft of water by placing a magnet around or near the drive magnets to interfere with the normal operation of the meter. This, of course, also interferes with the ability of the flow recorder to pick up the pulses emitted from the magnets. High-sensitivity units are available which can pick up pulses even in these situations, although these are usually specialty items. If using high-sensitivity units, special care should be taken to ensure that all pulses picked up are actually from the meter rotation and are not caused by external factors. If the operator feels that external factors could be interfering with the recordings, this problem can often be resolved by wrapping the unit in aluminum cooking foil. The foil has the effect of blocking the external signals and allowing the logger to pick up only the meter pulses.

Checklist

- ❏ Identify meter types to be logged and obtain pulse factors from the meter manufacturers.
- ❏ Ensure that a good pulse signal can be obtained from the manufacturer's recommended sensor location. If not, try to locate the best spot manually.
- ❏ Check pulse significance by recording a sample of flow and volume while on site and comparing this to the volume that passes through the meter according to the meter register. Perform a volume and time calculation to estimate approximate flow rates and compare it to the flow logger's calculated flow.

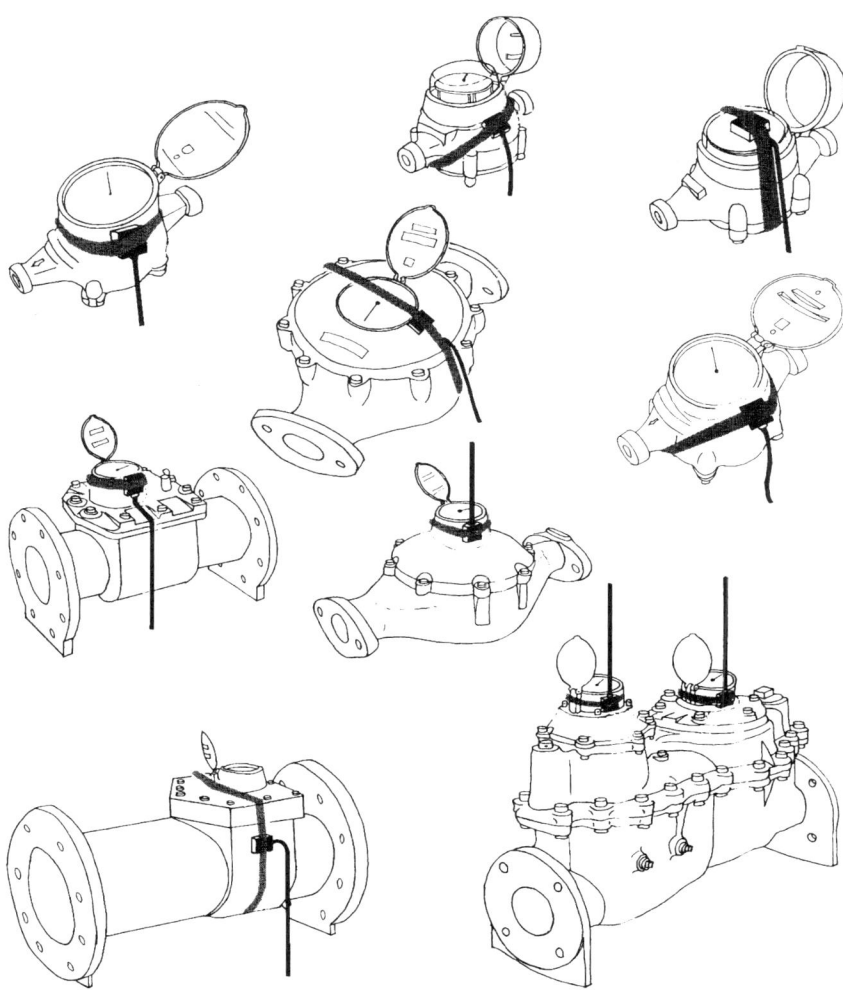

Figure 6.20 Sensor locations for different meter types. (*Source: F.S. Brainard & Co.*)

- If external signals are picked up, try shielding the unit with aluminum foil.
- Ensure that both raw pulse files and calculated flow rate files are saved. In the case of an error, the raw data files can be used to recalculate the proper flow rate.
- If the meters to be tested have security magnetic shielding, be sure to use a high-sensitivity unit.

6.2.5 Flow Charts

Sometimes a utility will have older-style flow charts. There are various types of flow charts. Some take a differential pressure from a DP meter and some take a 4–20-mA signal, which represents flow rate. Flow charts are relatively easy to set up by following the individual manufacturer's instructions, but care should be taken not to allow water to be spilled on the chart and not to allow pen ink to dry up.

Some charts have a mechanical clock and some an electronic drive to allow flow rate to be recorded over time. For a mechanical drive unit, care should be taken to select either a 30-day, 7-day, or 24-h clock. Charts should not be confused between the various options of maximum flow rate and time period.

Checklist

- ❑ Select correct clock mechanism for recording period required.
- ❑ Select correct chart for maximum flow rate required and clock mechanism installed.
- ❑ Ensure that pen is clean and charged with ink.
- ❑ Do not allow water to be spilled on the chart.
- ❑ Ensure that the mechanism is free to rotate and not jammed.
- ❑ Be sure to change the chart at the end of the recording period.
- ❑ Clearly record the date and the nature of the test on the chart after removing it for analysis.
- ❑ Ensure that the pen arm is properly calibrated at zero flow. Often the pen arm is bent when changing charts.

6.2.6 Step Testers

In Chap. 10 we will discuss step testing as a means of isolating sections of leaky main, identifying volume of loss, and ranking and pinpointing repair programs.

Step testers have been around for many years. Older step testers work on a swing-gate principle. The gate swings farther when flow is higher, which in turn moves a pen arm on a chart. The chart has a clock, which moves the chart in a circular motion as a function of time, and a graph is recorded. As the flow drops in response to a shut-in leak, the graph displays the difference in flow.

Nowadays most operators use a simple data logger and flow meter. Some step testing data loggers have a function which automatically analyzes the leak volume, and some have a radio function which allows the operator to work in a remote vehicle.

> It is important to ensure local manufacturer support when considering the first-time purchase of high-tech equipment.

6.3 PERMANENT EQUIPMENT

Previous sections of this chapter have discussed various types of temporary equipment which is often used during field testing. Now we will discuss some permanent options.

Before undertaking fieldwork where measurements must be taken, it is a good idea to inspect any permanent meters which may be used for measuring purposes. Meters should be tested to ensure proper accuracy and should be sized correctly for the flows in the field. Chapter 14 and AWWA Manuals M6 (excerpted in App. B) and M22 (excerpted in App. D) discuss these issues in greater detail.

In addition to proper calibration and sizing, it is important to understand what type of meter is being used, its principles of operation, and how it transmits data. This is necessary so that a compatible recording or logging device or mode can be selected to collect the data, or in the absence of these options manual reads can be scheduled. The following section shows some of the industry standard types of meter and shows some figures, installation recommendations, and flow limitation charts.

Each manufacturer has its own specifications for its meters, so it is necessary to have data sheets for the different manufacturers' equipment. However, in the United States, all meters are manufactured to meet or exceed AWWA standards. In other areas of the world, meters are manufactured to meet or exceed ISO specifications, which are different from those of the AWWA. AWWA and ISO specifications are readily available from the respective organizations and can usually be ordered over the Internet.

6.3.1 Meter Types

Meter types include:

- Turbine and turbo meters (see Figs. 6.21 and 6.22, and Table 6.2)
- Propeller meters (see Fig. 6.23 and Table 6.3)
- Compound meters (see Figs. 6.24 and 6.25, and Table 6.4)
- Piston residential meters (see Fig. 6.26)
- Velocity residential meters (see Fig. 6.27)
- Magnetic meters (see Fig. 6.28)
- Ultrasonic meters
- Differential pressure meters
- Vortex shedding meters

Each meter type has benefits and negative points, and it is not the intention of this book to promote one methodology or another. It is important, however, that the operator become familiar with the meter types in the area in which they will be used.

Data Transmission

Most of the meters mentioned above transmit data in one of several ways. The most common are

- Manually read, indexed, and stored information
- 4–20-mA analog output (see Fig. 6.28)
- Pulse output (see Fig. 6.29)
- Frequency output (see Fig. 6.30)

Output Correlation

In general, the various types of meters fall into three categories of output:

1. Manual read option: All of the above meter types.
2. 4–20-mA option: This is usually reserved for the electronic-type meters such as the magnetic and ultrasonic meters, the electronic vortex shedding meters, and differential pressure meters, which have electronic converters.

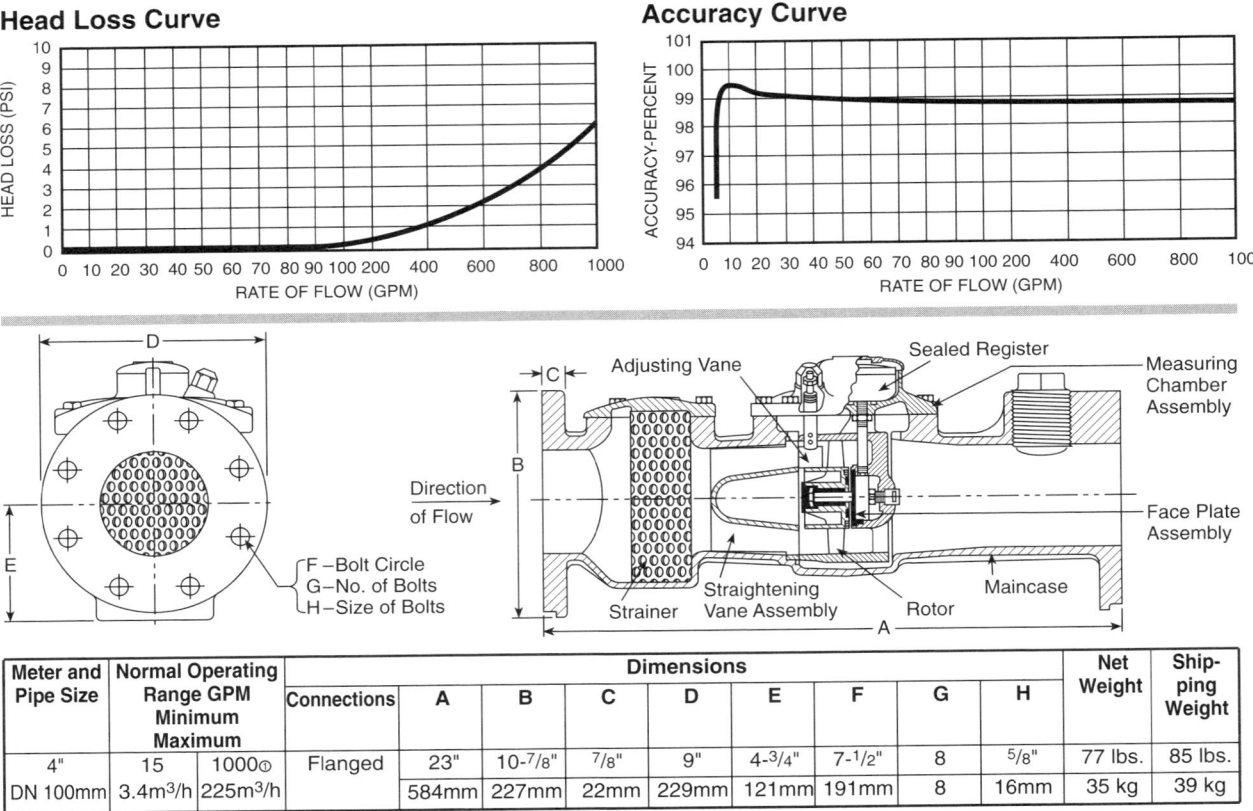

Figure 6.21 Turbine and turbo meter specifications. [*Source: Invensys Metering Systems (formerly Sensus Technologies).*]

3. Pulse and frequency output option: Pulse output is usually available from most of the electronic meters such as the magnetic and ultrasonic meters. It is also often available for the turbine, propeller meters, and residential meters (positive displacement and velocity jet types), although sometimes this option needs to be requested.

Obviously, before we can select equipment to record flows from permanently installed equipment, we need to understand the type of output the meter has, if it has one. Then we need to identify the correct recording equipment for the job. Doing this is easier when the permanent equipment is to be installed as part of the project, as the meter and the recording device can be selected ahead of time to ensure compatibility.

6.3.2 Metering Types and Characteristics

Meters can be fitted with various types of pulse output, giving faster or slower responses to changes in flow. The operator must learn to select the right type of output for the job at hand, always ensuring that there are sufficient pulses to record the flow or volume accurately, while also ensuring that there are not too many pulses to either confuse the equipment or use up the data storage allocation in the logger before the test period is complete.

> **A**lways check that the output is compatible with the recording device and the ranges are similar.

6.3 PERMANENT EQUIPMENT

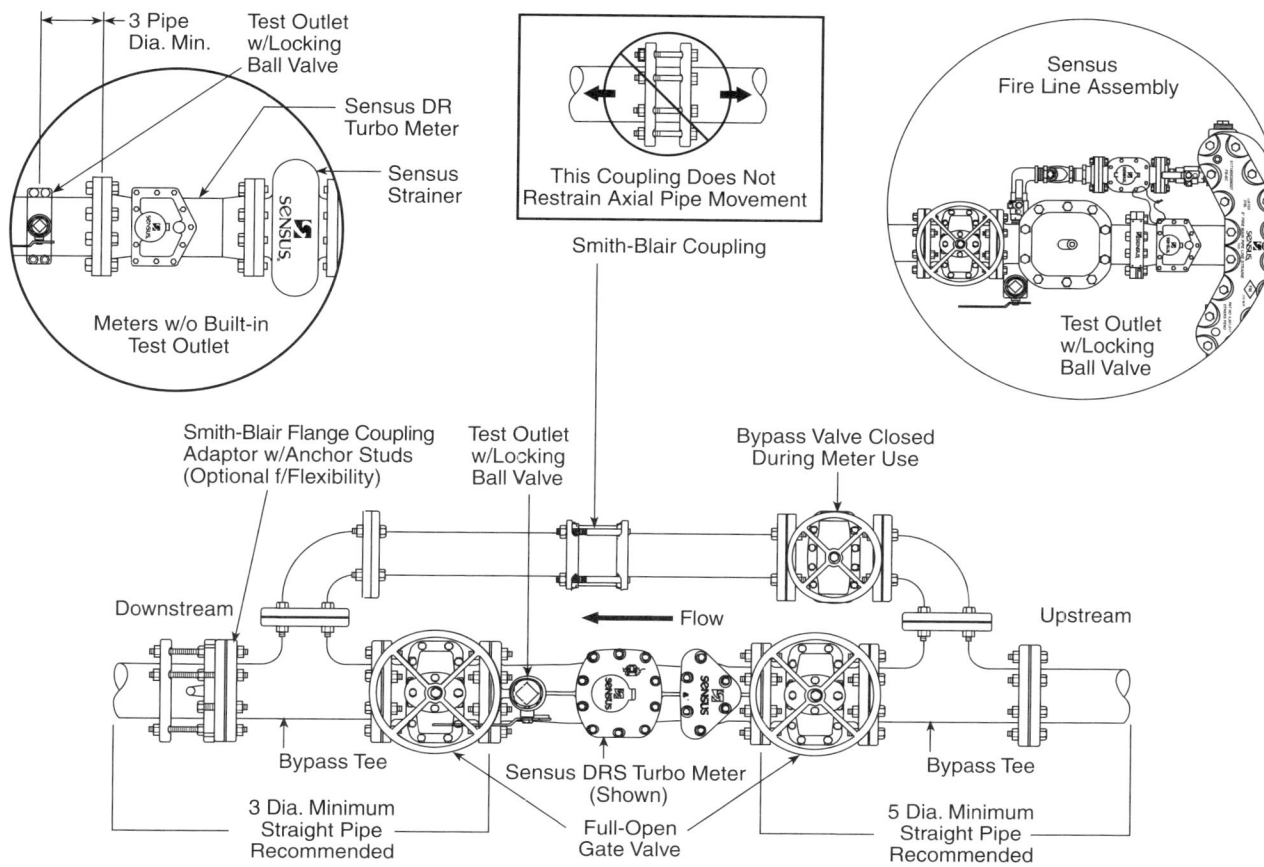

Figure 6.22 Sample turbo meter installation. [*Source: Invensys Metering Systems (formerly Sensus Technologies).*]

Electronic Meters

Electronic meters usually allow selection of the output pulse value. This is done via either mechanical switches or through programming the meter with a computer or hand-held device. This is relatively simple to do in most cases, with minimal training from the equipment supplier or by reading the operating manual.

Mechanical Meters

Mechanical meters have different types of pulse output. Most are one of the following two types.

- ❏ Reed switch sensors tend to give a lower pulse output than optical switches. The reed switch functions in conjunction with the drive magnets in the meter. See Fig. 6.29 for configurations which might be encountered. Others are also available.
- ❏ Optical switch sensors usually functions in conjunction with a special dial. See Fig. 6.30 for configurations which might be encountered. Other configurations are also available.

If the metering site is going to be underground in a chamber, it is often a good idea to ensure that the cables and meter heads are waterproof

TABLE 6.2 Flow Data for Turbo Meters

Size (in)	Model	Maincase	Normal Flow Limits (100.0% + 1.5%) (gal/min)	Extended Flow Unit (Intermittent Flows)	Low Flow (95%) (gal/min)	Headloss at Maximum Flow	End Connections
1½	W-120DR	Bronze	4–120	160	3	13.5	1½-in size, two-bolt oval, AWWA 125-lb class
1½	W-120DRS	Bronze	4–120	160	3	15.4	1½-in size, two-bolt oval, AWWA 123-lb class
2	W-160DR	Bronze	4–160	200	3	5.6	2-in size, bolt-slot oval, AWWA 125-lb class optional 2–11½-in NPT, internal threads
3	125WDR	Cast aluminum	10–350	400	10	8	2½–7½-in NST (National Standard Fire Hose Coupling Thread) furnished unless otherwise specified
3	W-350DR	Bronze	5–350	450	4	5.9	3-in size, round ANSI 125-lb class
4	W-1000DR	Bronze	15–1,000	1,250	10	3.6	4-in size, round ANSI 125-lb class
4	W-1000DRFS	Bronze	15–1,000	1,250	10	6.3	4-in size, round ANSI 125-lb class
6	W-2000DR	Bronze	30–2,000	2,500	20	6.2	6-in size, round ANSI 125-lb class
6	W-2000DRSL	Bronze	30–2,000	2,500	20	3.8	6-in size, flat face, 125-lb class
6	W-2000DRFS	Bronze	30–2,000	2,500	20	6.7	6-in size, round ANSI 125-lb class
8	W-3500DR	Bronze	35–3,500	4,400	30	8.3	8-in size, round ANSI 125-lb class
8	W-3500DRFS	Bronze	35–3,500	4,400	30	8.5	8-in size, round ANSI 125-lb class
10	W-5500DRF	Bronze	55–5,500	7,000	35	6.1	10-in size, round ANSI 125-lb class
10	W-5500DRFS	Bronze	55–5,500	7,000	35	6.9	10-in size, round ANSI 125-lb class
16	W-10,000DR	Cast iron	250–10,000	12,500	200	5.3	16-in size, round ANSI 125-lb class

Source: Invensys Metering Systems (formerly Sensus Technologies).

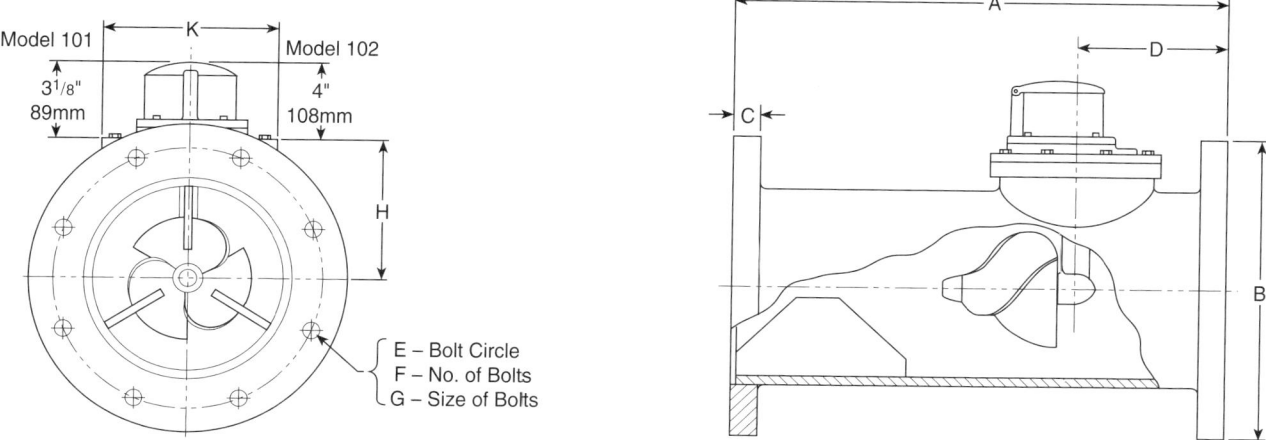

Figure 6.23 Propeller meter. [*Source: Invensys Metering Systems (formerly Sensus Technologies).*]

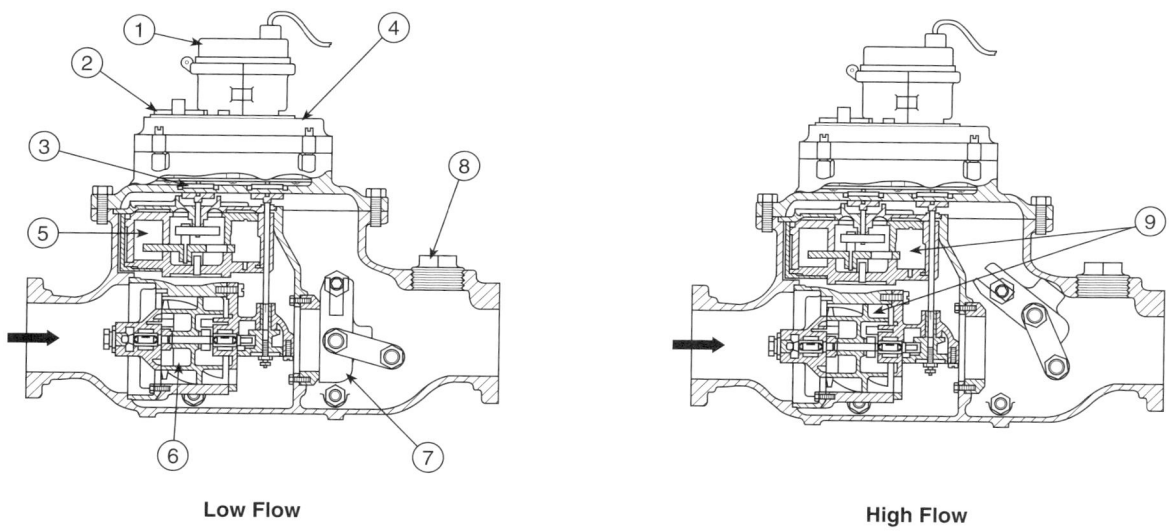

Figure 6.24 Compound meter. [*Source: Invensys Metering Systems (formerly Sensus Technologies).*]

(usually NEMA 6 rating in the United States or IP68 in the United Kingdom; other areas of the world have other classification ratings), as this tends to be an area where readings can fail. Most manufacturers have waterproof dials and sensors available, but it may be an option and not the norm.

6.4 OUTPUT READINGS

6.4.1 Understanding Pulse Recording

The preferred method for recording flows in the field with data loggers is a pulse output. The reason for this is that the logger counts a determined

TABLE 6.3 Flow Data for Propeller Meters

Meter Size	Low Flow (gal/min, m³/h)	Normal Range (gal/min, m³/h)	Dimensions										Shipping Weight
			A	B	C	D	E	F #	G	H	K	
3 in DN 80 mm	80 gal/min 18.2 m³/h	100–250 gal/min 23–57 m³/h	16 in 406 mm	7$^1/_2$ in 190 mm	$^3/_4$ in 19 mm	6$^1/_2$ in 165 mm	6 in 152 mm	4	$^5/_8$ in 16 mm	3$^3/_8$ in 86 mm	5 in 127 mm	70 lb 32 kg
4 in DN 100 mm	82 18.6	125–500 28–114	18 in 457 mm	9 in 229 mm	$^5/_8$ in 16 mm	7$^1/_2$ in 190 mm	7$^1/_2$ in 190 mm	8	$^5/_8$ in 16 mm	3$^7/_8$ in 99 mm	7$^1/_2$ in 190 mm	85 lb 39 kg
6 in DN 150 mm	160 36.3	220–1,200 50–273	22 in 559 mm	11 in 279 mm	$^{11}/_{16}$ in 17 mm	9 in 229 mm	9$^1/_2$ in 241 mm	8	$^3/_4$ in 19 mm	5 in 127 mm	9 in 229 mm	115 lb 52 kg
8 in DN 200 mm	190 43.2	250–1,650 57–375	24 in 610 mm	13$^1/_2$ in 343 mm	$^{11}/_{16}$ in 17 mm	9 in 229 mm	11$^3/_4$ in 298 mm	8	$^3/_4$ in 19 mm	6 in 152 mm	9 in 229 mm	150 lb 68 kg
10 in DN 250 mm	260 59.0	330–2,500 75–568	26 in 660 mm	16 in 406 mm	$^{11}/_{16}$ in 17 mm	10 in 254 mm	14$^1/_4$ in 362 mm	12	$^7/_8$ in 22 mm	7$^3/_8$ in 187 mm	11 in 279 mm	200 lb 91 kg
12 in DN 300 mm	275 62.4	350–3,500 80–795	28 in 711 mm	19 in 483 mm	13/16 in 21 mm	10 in 254 mm	17 in 432 mm	12	$^7/_8$ in 22 mm	8$^3/_8$ in 213 mm	11 in 279 mm	290 lb 132 kg
14 in DN 350 mm	350 79.5	450–4,500 102–1,022	42 in 1,067 mm	21 in 533 mm	1$^3/_8$ in 35 mm	12 in 305 mm	18$^3/_4$ in 476 mm	12	1 in 25 mm	9$^1/_4$ in 235 mm	13$^1/_2$ in 343 mm	450 lb 204 kg
16 in DN 400 mm	450 102.2	550–5,500 125–1,249	48 in 1,219 mm	23$^1/_2$ in 597 mm	1$^7/_{16}$ in 37 mm	12 in 305 mm	21$^1/_4$ in 504 mm	16	1 in 25 mm	10$^1/_4$ in 260 mm	13$^1/_2$ in 343 mm	550 lb 249 kg
18 in DN 450 mm	550 124.9	752–7,250 165–1,647	54 in 1,372 mm	25 in 635 mm	1$^9/_{16}$ in 40 mm	15 in 381 mm	22$^3/_4$ in 578 mm	16	1$^1/_8$ in 29 mm	11$^5/_8$ in 295 mm	13$^1/_2$ in 343 mm	620 lb 281 kg
20 in DN 500 mm	700 159.0	850–9,000 193–2,044	60 in 1,524 mm	27$^1/_2$ in 699 mm	1$^{11}/_{16}$ in 43 mm	15 in 381 mm	25 in 635 mm	20	1$^1/_{18}$ in 29 mm	12$^5/_8$ in 321 mm	13$^1/_2$ in 343 mm	820 lb 372 kg
24 in DN 600 mm	1,000 227.1	1,300–13,000 259–2,592	72 in 1,829 mm	32 in 813 mm	1$^7/_8$ in 48 mm	18 in 457 mm	29$^1/_4$ in 749 mm	20	1$^1/_4$ in 32 mm	12$^5/_8$ in 321 mm	13$^1/_2$ in 343 mm	1,000 lb 454 kg
30 in DN 750 mm	1,600 363.4	2,100–18,600 477–4,224	84 in 2,123 mm	38$^3/_4$ in 984 mm	2$^1/_8$ in 54 mm	18 in 457 mm	36 in 914 mm	28	1$^1/_4$ in 32 mm	12$^5/_8$ in 321 mm	13$^1/_2$ in 343 mm	1,150 lb 522 kg
36 in DN 900 mm	2,400 545.0	3,000–24,000 681–5,450	96 in 2,438 mm	46 in 1,168 mm	2$^5/_8$ in 67 mm	20 in 508 mm	42$^3/_4$ in 1,086 mm	32	1$^1/_2$ in 38 mm	12$^5/_8$ in 321 mm	13$^1/_2$ in 343 mm	1,350 lb 613 kg

Source: Invensys Metering Systems (formerly Sensus Technologies).

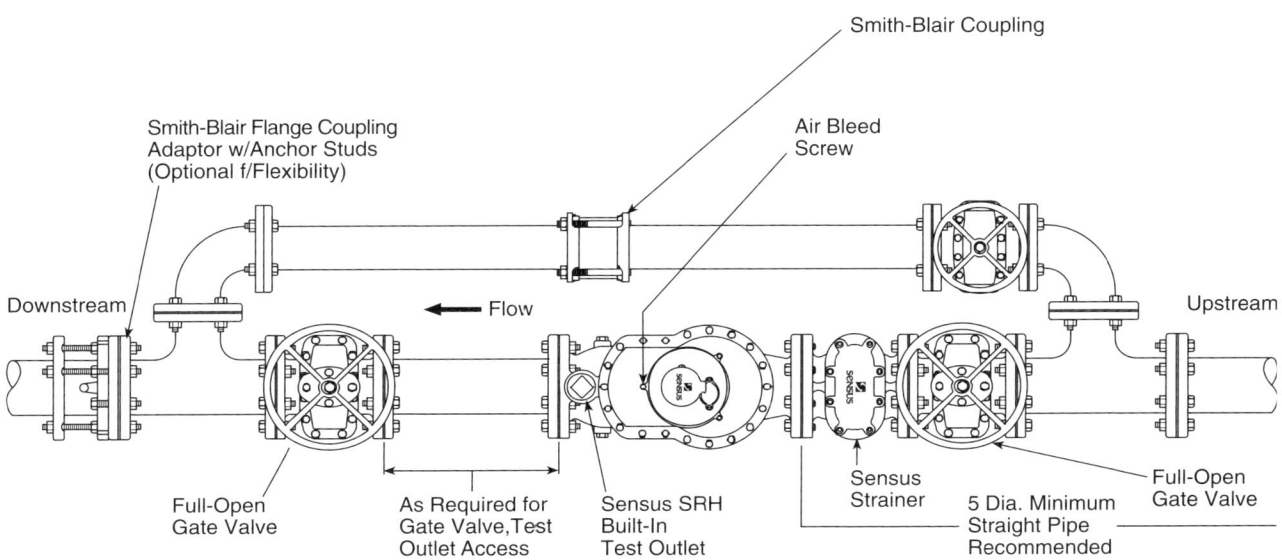

Figure 6.25 Sample compound meter installation. [*Source: Invensys Metering Systems (formerly Sensus Technologies).*]

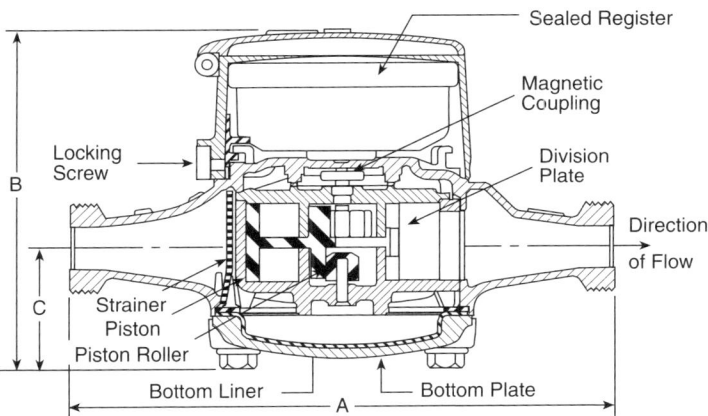

Figure 6.26 Piston residential meter. [*Source: Invensys Metering Systems (formerly Sensus Technologies).*]

number of pulses over the complete time period. Each pulse equals a determined volume of water passing through the meter. Accuracy is high and accountable if the recording equipment is set up correctly. To set up the equipment properly, operators need to understand several things about the way they will record the pulse.

Frequency Recording

Q. What is the difference between pulse and frequency?

A. A frequency is a fast pulse.

Some manufacturers' equipment have a frequency output stated in hertz. One hertz equals one pulse per second, so a frequency output is

92 CHAPTER 6 EQUIPMENT AND TECHNIQUES

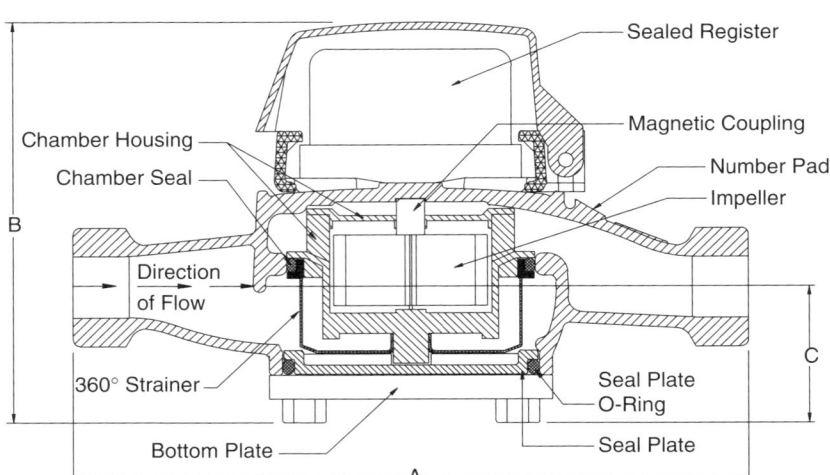

Figure 6.27 Velocity residential meter. [*Source: Invensys Metering Systems (formerly Sensus Technologies).*]

Figure 6.28 Magnetic meter. (*Source: Chris Bold, Invensys Metering RSA.*)

6.4 OUTPUT READINGS

TABLE 6.4 Flow Data for Compound Meters

Nominal Size/Model	Normal Operating Range (gal/min)	Low-Flow Accuracy @ 95% (gal/min)	AWWA Maximum Continuous Flow (gal/min)	Maximum Intermittent Flow (gal/min)	Minimum Accuracy Crossover	Headloss at Maximum Intermittent Flow (psi)
2 in SRH	2–160	$1/4$	80	160	95%	5.0
3 in SRH	4–320	$1/2$	160	320	95%	5.3
4 in SRH	6–500	$3/4$	250	500	95%	3.2
6 in SRH	10–1000	$1 1/2$	500	1000	95%	13.0
8 in manifold	16–1600	2	800	1600	95%	13.2

Source: Invensys Metering Systems (formerly Sensus Technologies).

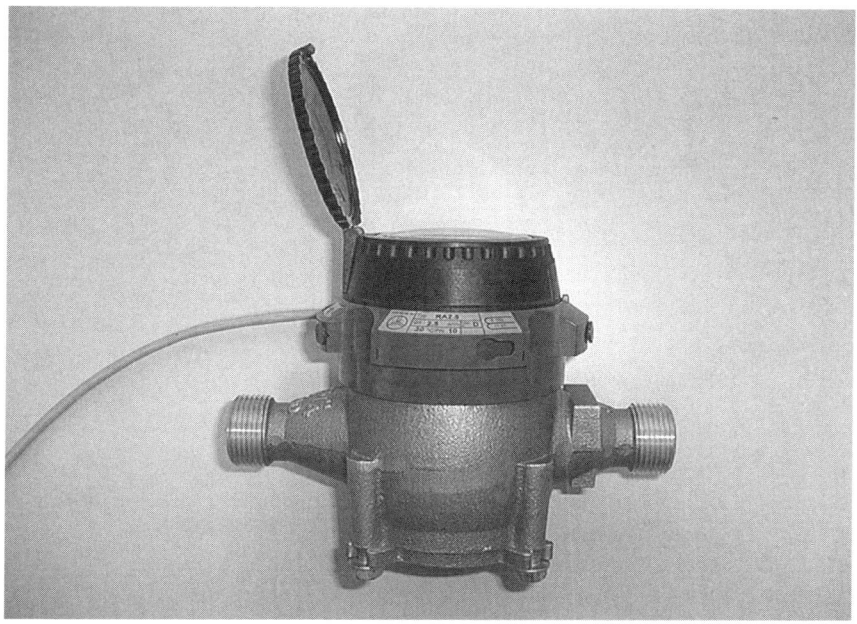

Figure 6.29 Pulse output. [*Source: (left) Chris Bold, Invensys Metering RSA; (right) Invensys Metering Systems (formerly Sensus Technologies).*]

also a pulse output. Most operators think of a pulse as a fairly slow occurrence—for example, 10 pulses per minute is considered a pulse output, whereas 10 pulses per second would probably be stated as a frequency output of 10 Hz.

Most data loggers can record pulse and frequency and can count it. In most cases, however, there is a limit to what they can count. Above the limit the logger may revert to a sampling system whereby the logger opens a window of time and counts the number of pulses recorded, then shuts down and opens up again after another predetermined period of time. If this is the case, valuable changes in flow rate may be lost between the time windows when the logger is actually recording. The effect will be a severely averaged flow profile, which may be a problem for some applications such as meter sizing.

If frequency sampling is to be used, it is recommended that the operator calculate the average frequency and ensure that the time window is

94 CHAPTER 6 EQUIPMENT AND TECHNIQUES

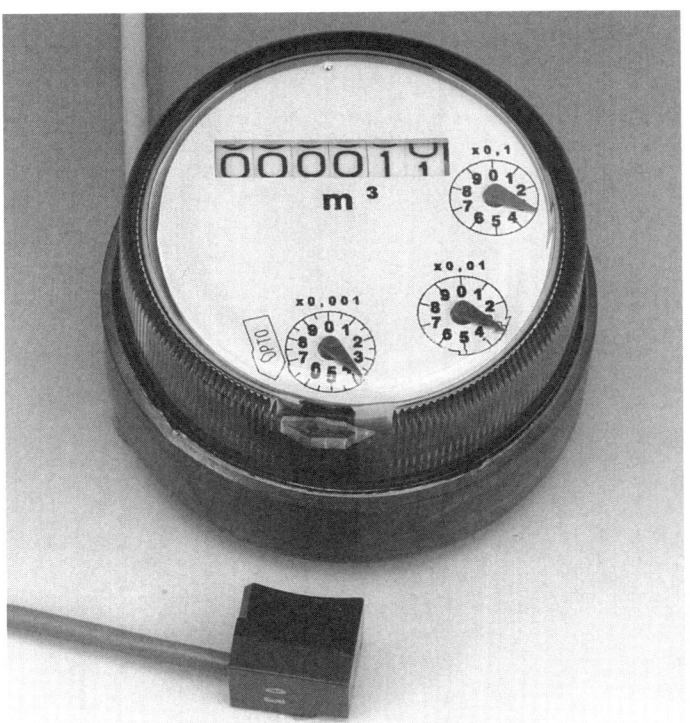

Figure 6.30 Frequency output. [*Source: (left) Chris Bold, Invensys Metering RSA; (right) Invensys Metering Systems (formerly Sensus Technologies).*]

sufficient to allow 10 pulses or more to be recorded. The effect of a missed pulse in the time window would then give an error of 10 percent or less. If better resolution is required, the operator should open the time window until the possibility of one lost pulse has a significance of less than the allowable error. See the following section for an example.

Pulse Count Recording

By far the best way of recording flow, especially in situations where portions of a flow profile may be used for volumetric analysis, is to use a pulse count option. Once the operator has decided to use a pulse count option as opposed to frequency sampling or analog sampling, then he or she must decide on one of two modes of storing the data:

- ❑ Averaging mode
- ❑ Event mode

Pulse Counting Using Averaging Mode

Most data recorders or loggers allow the user to define which type of data storage will be used. The averaging option operates by counting pulses over a certain period and then storing the average value at defined intervals. In this way the logger is not storing every single pulse (which might be a significant number, depending on the flow rate and the output unit). By not storing every single pulse, the operator can prolong the

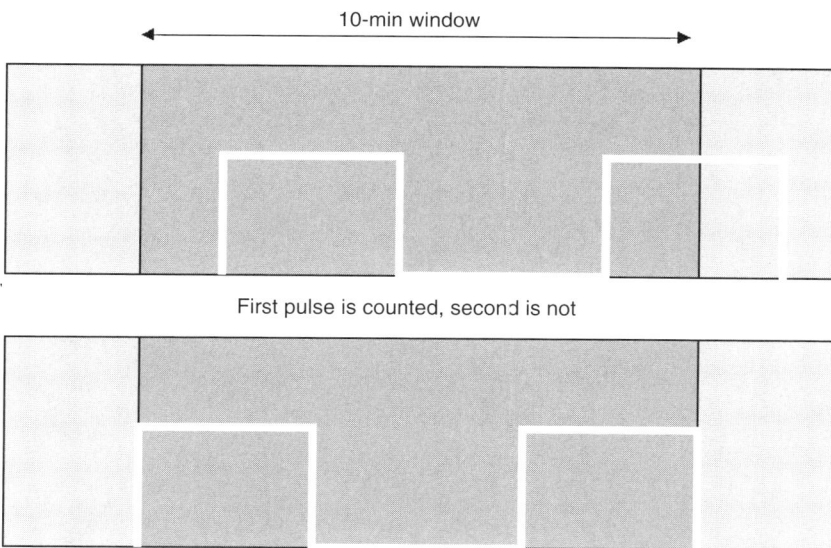

Figure 6.31 Fifty percent potential error caused by incorrect selection of pulse frequency and time window. (*Source: Julian Thornton.*)

memory of the logger to allow longer storage periods between downloads. The averaging option is useful for recording flow profiles where the important factor is the change in flow profile, as in zone flow measurement for leakage detection.

When selecting an averaging mode it is important that the operator realize that the results will be an averaged accumulation of many data points. It is important that the time window chosen reflects the type of resolution the operator is expecting to get from the data. For example, if a meter has a very slow pulse output—say, one pulse per 5 min—and the operator sets a time window of 10 min, then the logger may only record one or two pulses within that time frame. Obviously, that leaves quite a lot of room for error (1 in 2) in the assumed flow rate, depending on whether the second pulse falls just inside or outside of the time window limit (see Fig. 6.31). A better time window for this pulse would be every hour, which would mean that the logger would record 12 readings within the time window (see Fig. 6.32). In this case the potential for error is 1 in 12.

In many cases the operator wants a faster response than one reading or value every hour, so in this case the operator would change the type of pulse output.

Event Recording

Event recording means that the logger or recorder reads and stores every pulse and records the time between each pulse to infer flow rate. This is the most accurate way of recording a flow profile and should be used when sizing meters, for example, as this method shows the true peaks and spikes of usage. However, care should be taken when using this method not to fill the memory of the logger or recorder before the end of the test period.

> **E**vent recording is the best means of data collection for direct pressure demands.

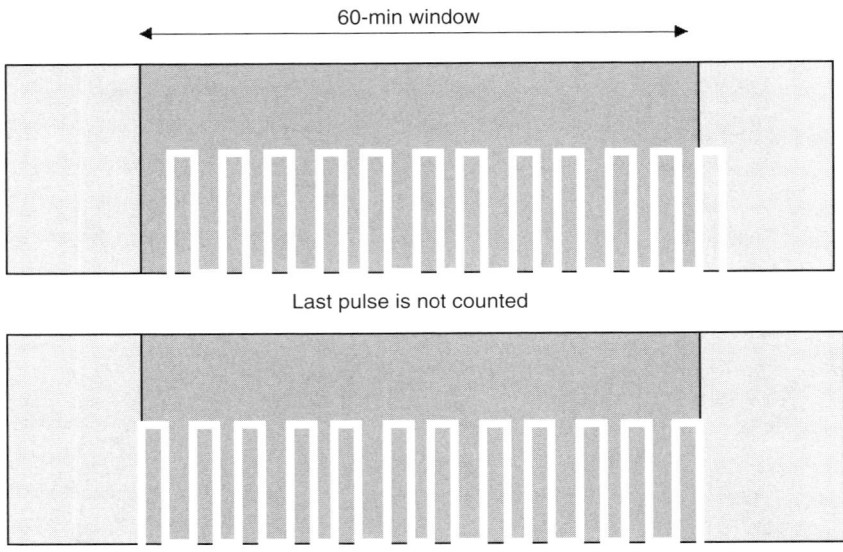

Figure 6.32 Error is reduced to 8.4 percent by better selection of time pulse frequency and time window. (*Source: Julian Thornton.*)

State Recording

State recording involves the logging of the status usually of a switch. The switch is usually on or off. This type of logging is particularly useful for recording the status of pumps over a period of time. It may be that the operator wishes to optimize the pumping routines but is unsure as to when each pump kicks on and off. By recording the pump state in a bank of pumps the operator can fine-tune the sequence and in many cases improve energy consumption efficiency.

This may seem like a lot of information for the inexperienced operator, but meter and logger manufacturers are usually eager to help out and will in many cases assist the operator with applications in the field. It is therefore recommended that inexperienced operators use equipment with local support options.

6.4.2 Understanding Analog Values

As discussed above, many meters have a 4–20-mA output. In most cases this will be a scalable output, where 4 mA equals zero flow and 20 mA equals maximum flow. In the case of bi-directional outputs, some will have a second channel and some will divide the output, for example, 4–8 mA available for negative flows and 8–20 mA available for positive flows. The manufacturer's instructions will identify the methodology used.

In most cases the data logger will wake up at a predetermined interval—say, every 5 min—take a sample of the value and store it, and go back to sleep. Again, this method works well for tasks which do not require every single peak in flow. It is possible to record continuously and register every peak using this methodology, but with most portable data loggers this would use up the available memory very quickly and therefore in most cases is not an option. This type of analog recording is often used in supervisory control and data acquisition (SCADA) systems, where the data are sampled very fast—say, every second or tenth of a second—and then transmitted to a central base. In this way the data are stored in a large computer.

Rotating Store or Store till Full

Most data loggers and recording devices have two modes for data storage.

- ❏ The *rotating store mode* functions in a cycle. Memory is used from the front to the back. As the memory begins to become full the earliest data are erased, allowing the latest data always to be on file. This type of memory allocation is used in long-term field logging exercises. Care must be taken to ensure that the number of data points to be recorded per time interval and the whole memory are sufficient that when the unit starts to over write earlier data the historical data bank covers a long enough period for the analysis required.
- ❏ *Store till full mode* is usually used when a specific event or number of events over a defined time period are to be recorded. This mode does not cycle, but rather stores data until the memory is full and then stops recording. Care should be taken when using this mode that the number of pulses and time allocation are correctly related to allow all of the events in a test to be recorded before the logger turns off.

6.4.3 Logging Summary

The best way of recording data in the field using recorders and loggers is very site-specific and depends on the resolution of data required. The operator will soon learn to match the best combination of recording device with the best combination of output options available for the measuring devices. Until the operator learns these skills it is best to rely on skilled local support to avoid unnecessary loss of data and costly rerecording.

Checklist

- ❏ Check output types.
- ❏ Check recorder options.
- ❏ When using pulse mode, select a suitable pulse generator for field application.
- ❏ Identify correct number of pulses per unit of time to ensure maximum resolution.
- ❏ Choose the averaging method when recording data over long periods of time, when lots of pulses may interfere with the recorder memory.
- ❏ Choose event logging when real demand profiles are required, as in the case of meter sizing.

- ❏ When using frequency recording, select a suitable sample rate and time window.
- ❏ When using analog recording, remember to set offset to 4 mA if using 4–20-mA output.
- ❏ Select a suitable sample rate.
- ❏ Identify either store till full mode or rotating store mode, depending on the type of test.
- ❏ Ensure that outputs and input cables are waterproof where necessary.

6.5 CALIBRATION, TESTING, AND DEAD-WEIGHT TESTS

This section discusses calibrating flow-measuring equipment, comparative flow versus volumetric testing, and dead-weight tests.

Any portable or permanent metering equipment needs to be tested on a periodic basis. The time between tests depends very much on economic factors for permanent meters (a combination of time, volume and water quality, environmental conditions, etc., see Chap. 14) and on the type of use, transportation, etc., for portable equipment. Portable equipment should be tested before and after any major field data collection exercise, and sometimes periodically in between.

Either portable or permanent equipment is usually tested in one of two ways:

- ❏ By performing comparative flows against a calibrated high-resolution permanent meter in a test rig
- ❏ By performing volumetric tests against a calibrated tank volume or weight (see Fig. 6.33)

> **A**ll permanent and temporary equipment should be tested periodically to ensure accuracy.

Figure 6.33 Volumetric meter testing. (*Source: Chris Bold, Invensys Metering RSA.*)

Sometimes permanent meters can be field calibrated by removing the metering chamber and substituting it for a recently calibrated one and comparing before and after flows to a hydrant or other controlled flow or volume source such as a reservoir or tank with known volume. Alternatively, some meter manufacturers supply filters or in-line strainers which are the same size as the meter body. (Use of in-line filters or a strainer is recommended in any case, to protect the internal parts of the meter and preserve the good working life of the meter.) The strainer can be temporarily removed and a calibrated meter chamber inserted. The two meters can then both be manually read for volume comparison or data-logged for flow time profile comparison. Once the testing is complete, the reference meter chamber goes back to the shop for retesting and calibration to ensure that the reference is indeed valid. In-the-field testing is by far the best means of testing equipment, as it is not always the equipment that fails and causes error but incorrect installation that causes error. By field testing, the equipment can be tested in situ. If an error is found in the field but not when the equipment is retested on a calibrated flow rig, then the error is obviously in the field in either temporary or permanent installation and can be rectified before further work is undertaken.

Checklist

- ❏ Always test volumetrically where possible.
- ❏ Always test in the field where possible.
- ❏ Ensure that the reference meter chamber has been recently calibrated in the case of comparison.
- ❏ Ensure that portable equipment is not damaged in transit.
- ❏ Ensure that portable equipment is calibrated frequently, at least before and after each major project.
- ❏ Ensure that permanent equipment is calibrated according to a predetermined modeled economic frequency.

6.5.1 Flow Meter Testing Summary

Flow measuring and data logging equipment, whether permanent or temporary, is often a big investment for a utility, operator, or contractor, and should be well maintained and calibrated to ensure continued accountable results. Equipment can never be calibrated too often and is all too frequently neglected and then blamed for malfunction or error. In many cases error is attributed to either lack of or incorrect operator training. However, a well-trained, confidant operator with well-maintained and calibrated equipment will in most cases come up with good, traceable, and accountable results, which are imperative for a successful water loss control program. After all, how can water loss be calculated if the results of the flow and volumetric measurements are in doubt?

6.6 PRESSURE MEASUREMENT EQUIPMENT

In addition to good flow recording, most water loss control programs require good pressure measurements. Pressure equipment may also be level-measuring equipment because pressure is the driving force in any water system, whether the pressure is provided by pumps or gravity. Pressure dictates the nature, frequency, and volume of leakage and physical losses and therefore must be taken very seriously.

6.6.1 Portable Loggers

Portable pressure or level data loggers are by far the easiest means of collecting field pressure data and transferring it to digital media. Data loggers come in all shapes and sizes, with various configurations. Some have internal pressure sensors (see Fig. 6.34) and some have external sensors. One of the most important things to understand before using a pressure logger is the type of sensor it has and its limitations.

Selecting a Sensor

Pressure sensors are usually calibrated for a maximum pressure and will output either a 4–20-mA signal or a frequency in relation to the pressure sensed. In loggers with internal sensors this is transparent to the user, as the interface is automatic; however, with external sensors the difference is vitally important—see Sec. 6.6.4, which discusses portable pressure sensors.

It is important to know the pressure limitations of the sensor to be used and ensure that it is not subjected to pressure higher than this value; otherwise, damage, in many cases irreparable, will occur. Pressure sensors are expensive, so be careful! In addition to selecting the right pressure rating for the job at hand, it is also important to select a sensor which will give the required resolution of measurement. The resolution of the sensor dictates the minimum pressure step that the logger can record. For example, a sensor may have a maximum rating of 100 m with a resolution of 1 percent. This means that the logger will record up to 100 m of

> **P**ressure sensors are expensive, and can be easily damaged by excess pressure or transients.

Figure 6.34 Data logger with internal pressure sensor. (*Source: Julian Thornton.*)

pressure in increments of plus or minus 1 m. In some cases, such as C-factor testing or level recording, this may not be sufficient, so a sensor with a higher resolution may be used—for example, maximum pressure 100 m, resolution 0.1 percent will give steps of plus or minus 10 cm. In the case of level recording specifically, usually the pressures to be measured are much lower and the requirement for resolution much greater, as 10 cm in a large reservoir could mean a huge volume of water. In this case an example might be maximum pressure 10 m, resolution 0.1 percent, which would give pressure steps of 10 mm, or resolution 0.01 percent, which would give steps of 1 mm.

Obviously, care must be taken not to use these lower-pressure sensors in distribution situations, where pressures will in many cases be much higher!

When selecting the correct pressure sensor or pressure logger for the job, remember that the system pressure changes throughout the day. In many cases, when headlosses are great during peak demand, pressure can be much higher at night, even when a standard fixed-outlet pressure-reducing station is controlling the system. So if sizing equipment for use during the day, remember to allow ample additional scale for higher nighttime pressures. If the operator desires to equalize pressures, then he should read details of modulating pressure control in the pressure management section of Chap. 12.

Testing Pressure and Level Sensors

In addition to selecting the right sensor, it is also important to perform frequent calibration of the pressure sensors, as they do tend to drift with time and use and also sometimes with temperature change.

Testing is done at two or more points, the first zero to ensure that the bottom end of the scale is secured and no offset is incurred, then, as a minimum, a pressure close to the high end of the scale should be tested. It is advised to test several other points between the top and bottom of the scale, although this is not always done because of time constraints.

6.6.2 Dead-Weight Testers

Pressures are usually induced with a dead-weight tester. Dead-weight testers come in two basic formats, a mechanical version with weights, and a hydraulic version, which is usually digital. Either model provides a very efficient and accurate way of testing sensors.

If a dead-weight tester is not available it is possible to test zero and a high pressure against a calibrated column of water such as a water tower with a known and static (at the time) level. Alternatively, a simple column can be built out of pipe in the workshop. Remember, it is not the diameter which counts, but the height, so a small-diameter tube can be used.

Once comparisons have been made, loggers and sensors are usually calibrated either electronically within the logger software or the data are downloaded in raw format into a spreadsheet and reformatted there (see Fig. 6.35).

102 CHAPTER 6 EQUIPMENT AND TECHNIQUES

c:\123r3\dist1e\D2P1.CSV
Water Mains Pressure Recording
Site No.: 0002   390   Site Name: No 350 CENTENNIAL Pressure range: 100.0 m

Sample Pressure Profile At Critical Node

MACRO COMMANDS ALT F3
NOTE: USE SELECT TO PULL IN DATA
NOTE: USE VIEW TO CHECK DATA AND RETURN
NOTE: USE CAL TO REGRESS DATA WITH CALIBRATION CURVE

Tabular Data

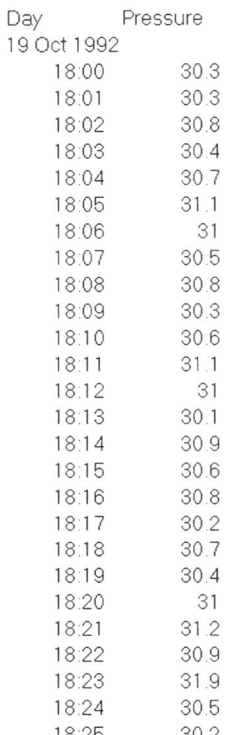

Day	Pressure
19 Oct 1992	
18:00	30.3
18:01	30.3
18:02	30.8
18:03	30.4
18:04	30.7
18:05	31.1
18:06	31
18:07	30.5
18:08	30.8
18:09	30.3
18:10	30.6
18:11	31.1
18:12	31
18:13	30.1
18:14	30.9
18:15	30.6
18:16	30.8
18:17	30.2
18:18	30.7
18:19	30.4
18:20	31
18:21	31.2
18:22	30.9
18:23	31.9
18:24	30.5
18:25	30.2
18:26	30.5

Figure 6.35 Pressure data calibration sheet. (*Source: Julian Thornton.*)

6.6.3 Portable Charts

Portable charts are an older method of testing pressures and work perfectly well. The only drawback is that in most cases data points have to be taken from the chart manually and input into a spreadsheet for analysis. This can be time-consuming for large projects where many pressures are measured and resolution requirements may be high.

When setting up a pressure chart, care should be taken to select the right chart and clock mechanism for the job. Calibration of pressure charts is done in much the same way as for the loggers and sensors described above.

6.6.4 Portable Pressure Sensors

In some situations it is preferable to use portable pressure sensors, which can either be used with a number of different logger types for different applications or can be used directly with telemetry or SCADA applications.

Care must be taken to ensure that the output type is compatible with the logger input types and that the sensor can either be powered by an internal logger battery without causing flat batteries due to overly high drawdown or that a portable battery pack or main power source is available.

Testing of pressure sensors is undertaken in exactly the same ways as stated above.

Checklist

- When using portable sensors, be sure that the output matches the input of the device that will receive the signal.
- Ensure that sufficient and suitable power sources are available.
- In all cases, check that the maximum pressure rating and resolution are suitable for the job.
- Ensure that all pressure-measuring devices are properly calibrated to zero and maximum pressure, with other intermediate check points if possible.
- Handle sensors with care, as they are sometimes fragile and should not be dropped or mishandled.
- Ensure that the sensor, logger, or chart to be used is waterproof if it will be used in a potentially wet environment.
- Ensure that suitable test equipment is on hand or available locally for periodic checks.

6.6.5 Traditional Acoustic Leak Detection Equipment

Acoustic leak detection equipment has to be among the simplest leak detection devices used in the field. However, the success of work undertaken with this type of equipment depends very much on the training of the operator. It is very important that operators gain as much hands-on training with another skilled operator as possible before undertaking a survey alone. Success depends very much on the operator's decisions based on signals received. This section discusses some of the methodology of leak detection. More detail can be found in Chap. 10.

> Successful leak detection depends very much on operator skills. Proper training is invaluable.

Mechanical and Electronic Listening Sticks

Listening sticks, or sounding rods as they are sometimes known, work by making contact with a fitting which is within a distance of the leak where the vibration sound of the leak leaving the pipe can be heard. Listening sticks come in two basic formats: mechanical and electronic.

Mechanical listening sticks were probably among the first types of leak-pinpointing equipment to be manufactured and used widely. The very first ones were solid wooden bars (dense wood) and can still be found in use. Later, stainless steel rods with diaphragms were housed in sounding cavities to amplify the noise.

As operators demanded better performance, manufacturers started to improve technology by amplifying signals and providing filters to allow the operator to filter out ambient noise, which is one of the main problems when using this type of acoustic equipment. Ambient noise, which can interfere with a listening stick survey, can come from:

- Traffic
- Demand
- Air traffic
- Gas and steam pipes

The use of any listening stick is simple. The operator identifies the area to be surveyed and then makes contact with pipefittings, hydrants, and service connections until he hears a suspected leak sound. At this point it is a simple matter of identifying where the loudest sound can be heard.

In urban areas, listening sticks are often deployed at night, when traffic noise is lower. This allows the operator more freedom to identify leak sounds without interference from other sources. Before beginning a listening stick survey, the operator should identify the pipe route and material to best estimate the required distance between contact points. This is necessary, as leak sound travels different distances in different pipe materials because of varying pipe attenuation. More details on leak surveying can be found in Chap. 10.

Checklist

- Identify pipe route before survey.
- Identify pipe material and test distances before survey.
- If traffic is going to be a problem, schedule the survey at night.
- Prepare documentation to show to residents identifying the operator and the reason that he is listening on their services.
- If using mechanical diaphragm-type units, be sure the diaphragm is in good condition, as listening sticks are often used as bor-ing bars. This destroys the contact between the rod and the diaphragm.
- If using electronic units, always check the batteries before starting a survey.

Mechanical and Electronic Geophones

Geophones are used to listen for leaks from the surface and come in two basic forms, mechanical and electronic.

Mechanical geophones have been in use for many years and are still very valid and robust pieces of equipment. The mechanical variety usually consists of brass disks with vibrating diaphragms inside. The two disks are connected by a hollow tube to a headset arrangement and are not unlike a doctor's stethoscope in appearance.

Electronic geophones were developed as the need for better performance pushed manufacturers of geophones into amplification and filtering techniques. More on geophones can be found in Chap. 10.

Geophones are used by listening above the pipe route for the sound of water escaping from the pipe. They are usually used in conjunction with a listening stick in a sonic survey. Care must be taken to listen directly

Figure 6.36 Listening for leaks using an electronic geophone. (*Source: Health Consultants, Inc.*)

above the pipe, otherwise failure may result (see Fig. 6.36). There is a distinct difference in leak sound when listening on hard or soft surfaces. The operator must learn to identify the difference and use a sounding bar where ground surfaces do not transmit sound.

Checklist

- ❑ Identify pipe route prior to survey.
- ❑ Identify ground surfaces prior to survey.
- ❑ Carry sounding bars where surfaces may be soft.
- ❑ For mechanical equipment, check diaphragms and tubes frequently to be sure they are not damaged.
- ❑ For electronic equipment, check batteries.

Tracer Gas Equipment

Tracer gas equipment consists of an injection manifold which connects a gas tank to the water main to be tested, usually with gauges to test bottle pressure and main pressure and a receiver unit which takes samples and compares them to air density.

More details on tracer gas testing can be found in Chap. 10 where Article One reviews procedures and techniques.

Tracer gas survey is a very effective way of locating either very small leaks such as those found during hydrostatic testing or hard-to-find leaks, often on PVC or large-diameter pipes where the leak sound is not transmitted far, making the use of traditional acoustic equipment ineffective.

An easy way of testing this equipment to make sure it is functioning is to hold an unlit lighter (with gas escaping) in front of the sensor and ensure that there is a noticeable deflection of the needle. If there is no deflection, it may be that the pump or filter system is either clogged with dirt or has water inside.

6.6.6 New-Technology Leak Detection Equipment

Leak Noise Correlators

Leak noise correlators were first introduced commercially into the marketplace in the late 1970s, but technology has advanced by leaps and bounds over the last few years, putting this type of equipment in the realm of new technology with changes every year. In essence, a leak noise correlator consists of a receiver unit and two radio transmitters with sensors and/or hydrophones. The sensors pick up the leak sound from the water main or pipe being tested at two points, the idea being that the leak is bracketed between the two. Using the calculation $D = 2l + Td \times V$ (as shown in more detail in Chap. 10), the correlator identifies like signals, measures the time delay between one signal and the other, and, using a known or calculated velocity along with a measured distance between sensors, calculates the position of the leak.

Most modern correlators have a number of functions built in to assist the operator in making good and accurate leak location. Some of the functions available are:

- Automatic filter selection
- Distance measurement
- Velocity calculation
- Multiple-pipe features
- Autocorrelation for single-sensor use at pipe ends
- Linear regression
- Leak location memory
- Printer
- Sensor assortment for different field situations

Leak noise correlators can be used over quite long lengths of pipe, depending on the material and diameter of the pipe and the lack of ambient noise, which could interfere with leak sounds. Most correlators can manage in excess of 500 m, and some of the newer digital versions can handle up to 3000 m in ideal conditions.

Care must be taken when using correlators to properly:

- Measure the exact distance between sensors
- Properly identify the pipe material
- Properly identify the pipe diameter
- Keep the leak close to the center of the two sensors
- Measure the velocity in the pipe section(s)
- Identify peaks which are not leakage and eliminate

More detail on leak noise correlation can be found in Sec. 10.3.2.

Many operators record leak sounds on tape or digitally on computer. It is a good idea to record a few reference leak sounds, which can be used to test the equipment on a periodic basis. Most faults which occur with this equipment are caused by misuse of sensors, which are quite sensitive to being dropped. Also, cable connections and battery life need to be constantly checked and kept in good order.

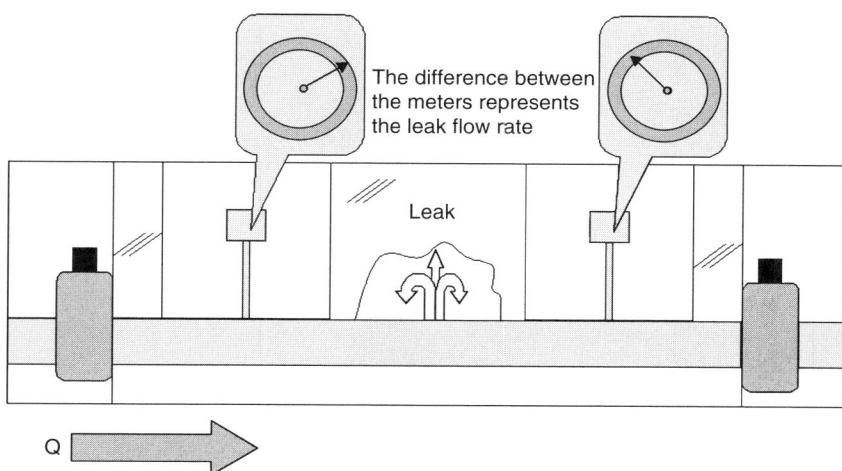

Figure 6.37 Volumetric comparison between a pair of meters can be used to find leakage. (*Source: Julian Thornton.*)

Leak Noise Loggers

Leak noise loggers have been on the market since the 1980s in various formats but have recently taken a major leap in technology as they can now not only identify areas with potential leak sound by analyzing and recording noise at night when ambient noise is at a minimum, but also correlate between sensors using a Geographic Information System (GIS) base as a reference to pinpoint the leak site.

Comparative Pairs of Meters

A common method of identifying leakage, particularly in transmission mains, is by volumetric comparison between a pair of meters, one fitted downstream of the other (see Fig. 6.37). If the meters in question are high-resolution permanent meters, it is often possible to undertake a volumetric balance. However, if the meters used for the test are temporary meters, care should be taken to identify the potential error on each installation and identify a reasonable resolution for identification. If the latter is the case, it may be preferable to undertake a night flow analysis comparison as opposed to a volumetric balance. The choice will depend on the nature of the piping and the hydraulics of the system.

Obviously, this method can only be used for identifying reasonably large leaks when used on large-diameter pipelines, but it does in many cases offer an accountable solution to testing long lengths of transmission mains.

6.6.7 Meter Testing Equipment

Meter testing equipment comes in various forms, including permanent bench testing equipment and portable field testing equipment. As discussed earlier, meter testing can be undertaken in a number of ways,

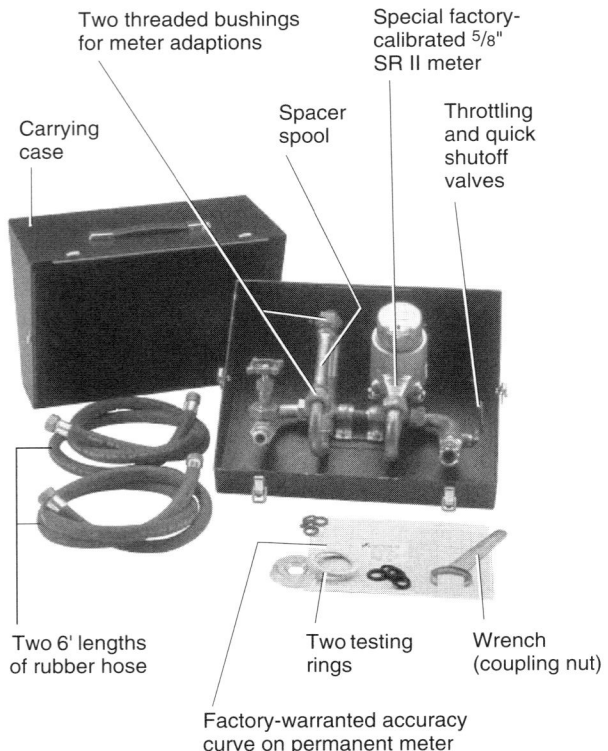

Figure 6.38 Small portable meter tester. [*Source: Invensys Metering Systems (formerly Sensus Technologies).*]

either comparatively or by means of a calibrated volume, the latter being preferable. Where possible, meters should be tested in the field to enable a complete test of the meter installation as well as unit accuracy.

Residential Meters

Residential meters are usually field-tested with a calibrated tank, which can be filled from an outside tap (while no other uses are occurring). Alternatively, a small portable tester can be fitted in line as per the kit shown in Fig. 6.38.

ICI Meters

Industrial, commercial, and institutional (ICI) accounts usually utilize large-diameter meters. If the meter is to be removed for testing, it should be subjected to an approved volumetric test. If not, it may be tested using either a portable temporary meter such as an ultrasonic or insertion meter (watch for errors in the test meter) or by comparing volumes against a set of previously calibrated meters either on a large meter tester as shown in Fig. 6.39 or on a test trailer.

Bulk Supply, Source, and Master Meters

Bulk supply, source, and master meters are also tested by one of the two methods discussed above for ICI accounts. Alternatively, insertion or

Figure 6.39 Large meter tester. [*Source: Invensys Metering Systems (formerly Sensus Technologies).*]

ultrasonic meters can be used if care is taken with inaccuracy in the test equipment due to potentially incorrect location. With proper care, the latter method is valid.

When testing these meters it is important to identify not only the accuracy of the measuring unit itself but also any telemetry equipment which may be transmitting signals back to a base unit. More information on this topic can be found in Chap. 14 and in App. C.

6.6.8 Maintaining Equipment

In the last few sections we have talked in a lot of detail about the use of permanent and portable equipment for field measurements and testing. Without data from this type of equipment it is extremely hard to assess water system condition and improve water loss figures. However, this equipment needs to be properly maintained to ensure that it gives repeatable accurate results. Bad data in produces bad data out!

Good Use and Practice

When installing permanent equipment or purchasing portable equipment for the first time, it is a good idea to identify a good practice list which should be adhered to by all operators using the equipment. Some things which might appear on the list are

- Regular maintenance
- Regular third-party testing
- Maintenance of the housing environment in the case of permanent equipment
- User log

> When budgeting for equipment purchases, also consider maintenance costs, as without proper maintenance any equipment will become worthless.

Cables and Fittings

A weak point in most equipment is the cables and fittings. People misuse cables, using them to carry equipment and to lower equipment into holes. This puts unnecessary strain on the connections and can sometimes cause irreparable damage.

Fittings should be greased and cleaned regularly with a light suitable oil to ensure that they do not get dirt into them. Waterproof seals should be checked regularly and changed when in doubt. Water ingress into fittings is one of the most common reasons for failure.

Storage and Carriage

One of the areas of most likely damage to portable equipment is during shipping and transportation of the equipment, particularly if it is to be sent via a third party over long distances.

Equipment should be purchased from the manufacturer together with a suitable durable hard carry case. If the manufacturer cannot supply this type of case, then one should be procured from a third party. Cases can be expensive, but they will always pay off in the long run.

6.7 PRESSURE CONTROL EQUIPMENT

Pressure control equipment comes in various shapes, forms, and sizes, depending on the nature of the control required. In this section we will touch on some typical applications related to water loss control. Obviously, there are many other types of valves and controllers, with other functions. More information on pressure management can be found in Chap. 12.

6.7.1 Types of Valves

The most common applications in pressure management schemes where water loss is the key factor are:

- Pressure-reducing valves (see Fig. 6.40)
- Pressure-sustaining valves (see Fig. 6.41)
- Altitude valves (see Fig. 6.42)
- Float control valves (see Fig. 6.43)
- Flow control valves (see Fig. 6.44)

More information on the use of these valves can be found in Chap. 12. Pressure valves are usually one of three types although others do exist:

- Diaphragm valves (see Fig. 6.45)
- Piston valves (see Fig. 6.46)
- Sleeve valves

Diaphragm valves usually come in two formats:

1. Globe style
   - Straight-through
   - Angle
2. Y pattern

6.7 PRESSURE CONTROL EQUIPMENT

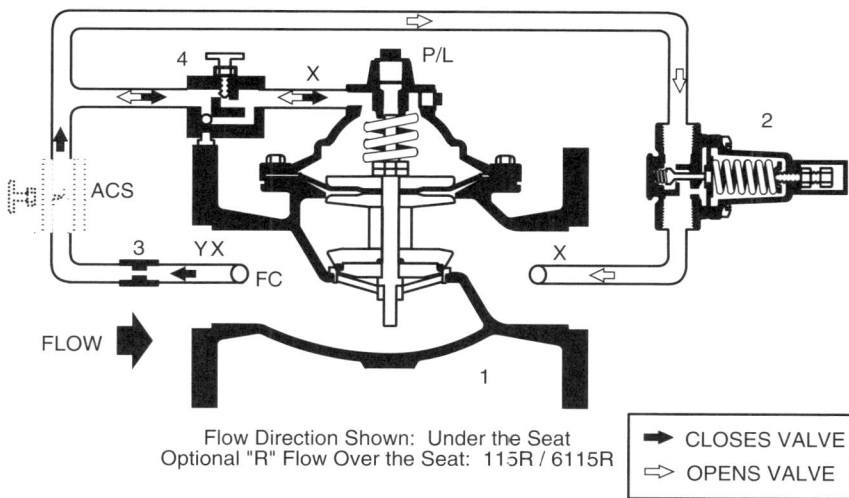

Figure 6.40 Pressure-reducing valve. (*Source: Watts ACV, Houston, Texas.*)

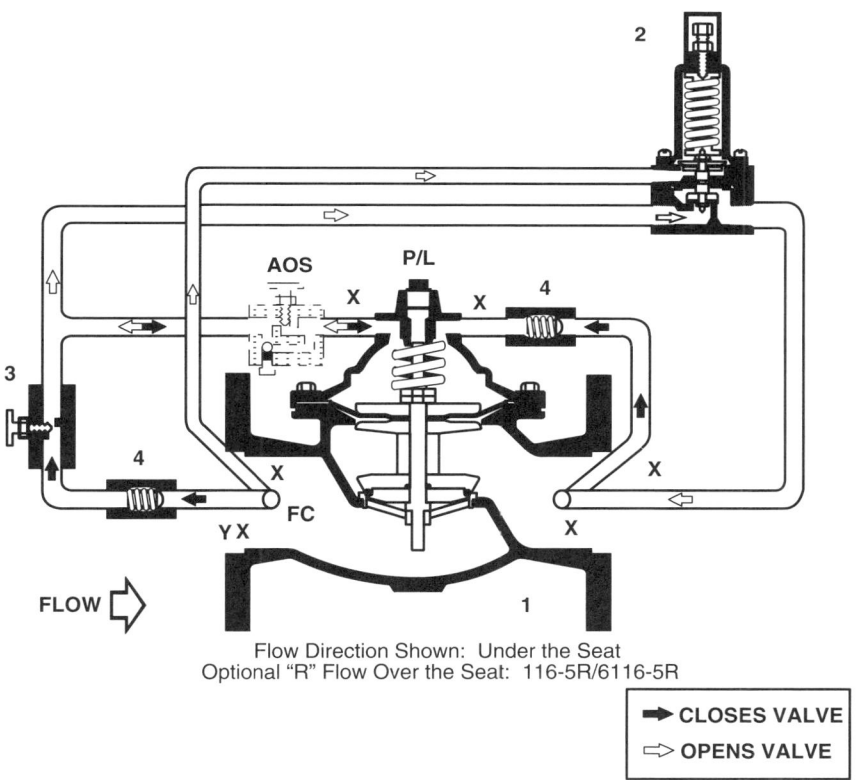

Figure 6.41 Pressure-sustaining valve. (*Source: Watts ACV, Houston, Texas.*)

The make of valve or configuration of the valve assembly chosen for the job will depend on the nature of the installation and the availability of local support. Most manufacturers have excellent installation information, and most will provide start-up and operational support at very little extra cost.

112 CHAPTER 6 EQUIPMENT AND TECHNIQUES

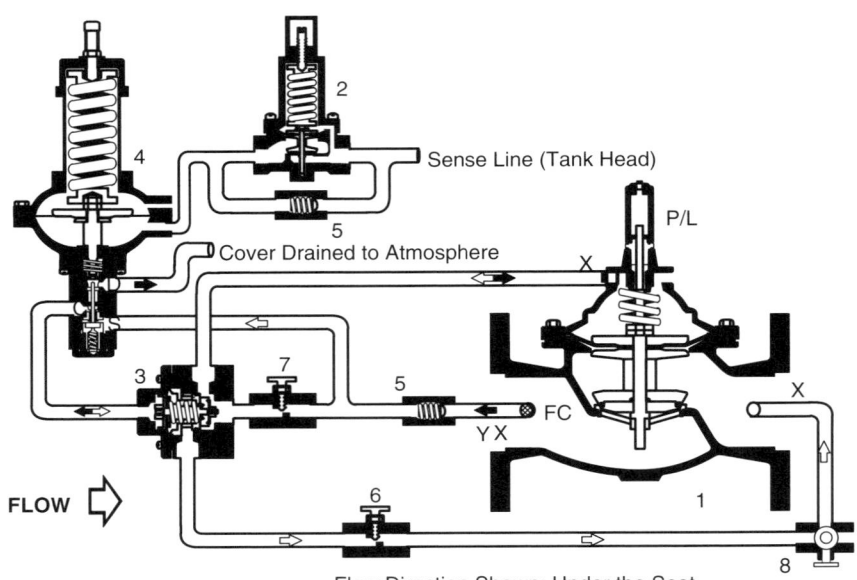

Figure 6.42 Altitude valve. (*Source: Watts ACV, Houston, Texas.*)

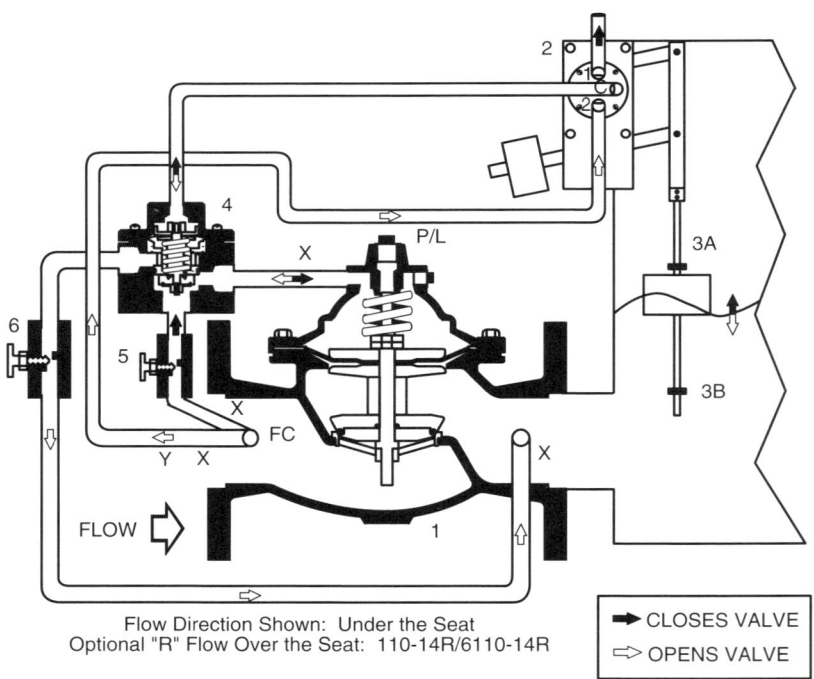

Figure 6.43 Float control valve. (*Source: Watts ACV, Houston, Texas.*)

6.7 PRESSURE CONTROL EQUIPMENT

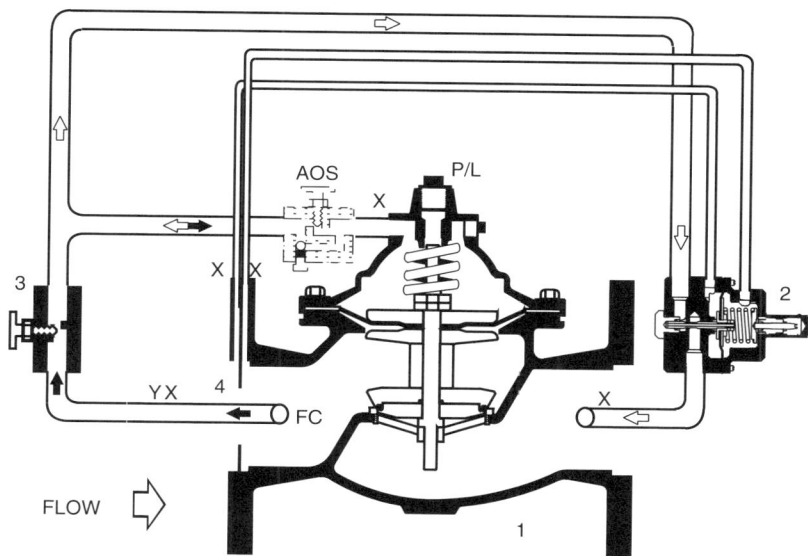

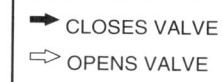

Figure 6.44 Flow control valve. (*Source: Watts ACV, Houston, Texas.*)

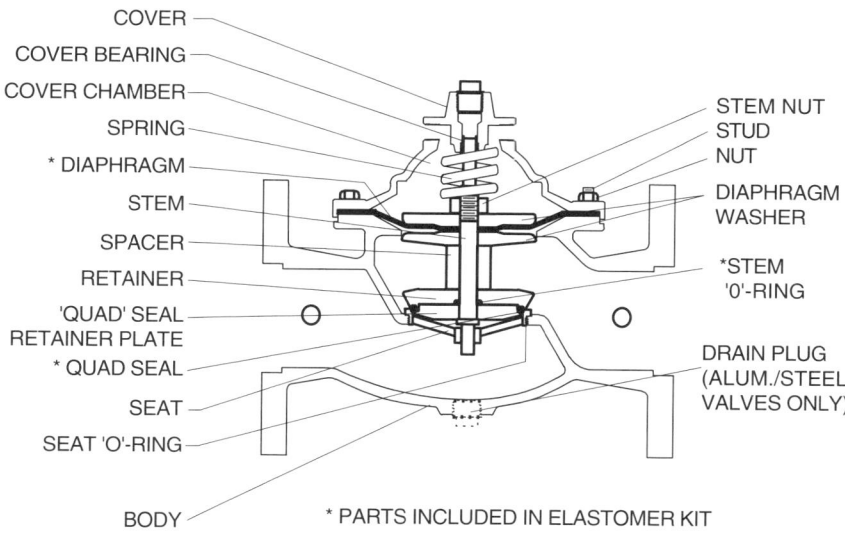

Figure 6.45 Diaphragm-type valve. (*Source: Watts ACV, Houston, Texas.*)

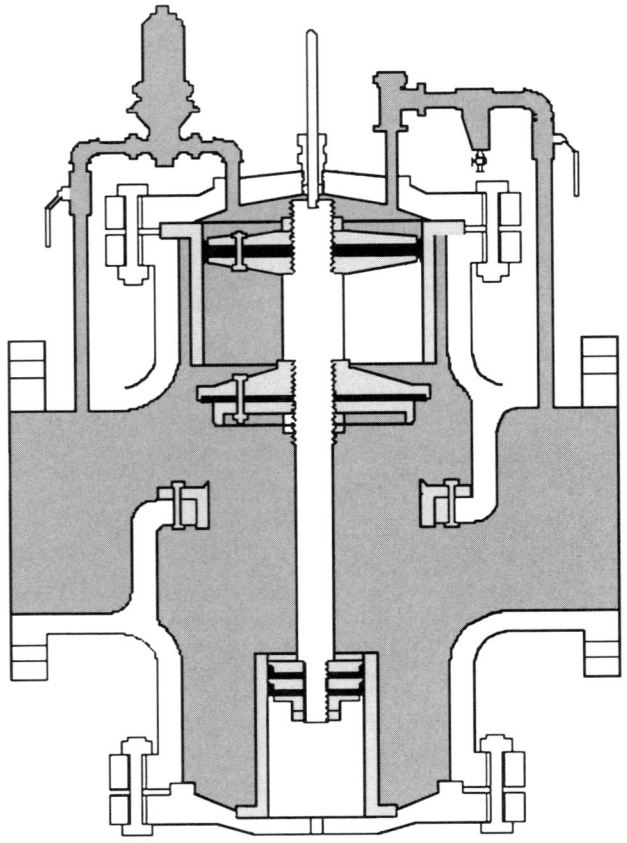

Figure 6.46 Piston-type valve. (*Source: Ross Valve Mfg. Co., Inc., Troy, N.Y.*)

All valves need regular maintenance to function properly over long periods of time. Most utilities like to settle on a particular make of valve and stock parts for those valves. This cuts down on having to stock identically sized parts for valves of various manufacturers if maintenance is to be done in-house.

6.7.2 Types of Controllers

Three basic types of controllers are available for specialized pressure leakage management. Other generic controllers may be adapted for use in this field. The three types in common uses are

- Time-based controllers
- Demand-based controllers
- Remote-node controllers

Each type of controller can be furnished with a data logger in certain configurations.

Time-Based Controllers

Time-based controllers work on an internal timer. The timer is set either to manipulate outlet pressure to various levels using an interface with a predetermined profile and pilot adaptor or to switch from one preset pilot to another by means of solenoid valves. Either of these scenarios works well but should be used in areas with fairly constant demand patterns, little seasonal and or weekend variation.

When using time-based controllers, care should be taken that the lowest pressures set can still meet emergency fire-fighting requirements.

Demand-Based Controllers

Demand-based controllers work by setting an outlet pressure by means of a pilot adaptor to a preset relationship between flow and pressure. Demand-based controllers combat headloss in water systems, ensuring that when demands are low pressure is at a minimum to reduce the effect of pressure on the leakage which is running in the system.

Demand-based controllers have the added benefit that they can be used to control pressure down below minimum fire-fighting requirements, as the controller will automatically adjust itself back to the required pressure when the hydrant is operated and the flow demand goes up. If a demand-based controller breaks at the lower pressure position, the controller is programmed to default the valve back to the higher set point, ensuring that water is available for emergency or peak demands.

Remote-Node Controllers

Remote-node controllers work by relaying a signal back to the valve and controller assembly from a remote node. The remote node selected is usually a critical node. The critical node may be a node which is at the highest elevation and therefore has the least pressure. Alternatively, it may be selected as an area with a special consumer or large consumer or an area with particularly high localized headlosses.

The remote node is fitted with a pressure logger and the logger is programmed to communicate often by way of low-power radio or cell phone with the controller on a predetermined basis. The remote logger orders the controller to allow more or less pressure into the system by opening or closing the valve, in order to maintain a stable target pressure at the remote point.

This type of control, like demand-based control, is suitable for areas with changing profiles and a need for emergency response.

Checklist

- ❏ Make sure the valves to be used can be maintained and supported locally.
- ❏ Determine what type of area is to be controlled.
- ❏ Undertake a detailed demand analysis before installation.
- ❏ Select the right type of controller to meet the requirements of the area.
- ❏ If using radio or cell phone communications, be sure the equipment is operating on an authorized wavelength.
- ❏ If using demand-based control, be sure that the meter which is fitted to provide the pulse is suitably sized and the pulse generator is suitable for the controller.

6.8 SUMMARY

In this chapter we have discussed various types of portable and permanent field equipment, which is often required or encountered in a water loss control program. While it is impossible to show all of the equipment types and configurations in this book, operators are urged to familiarize themselves with their particular equipment prior to starting fieldwork. Manufacturers will be more than willing in most cases to supply the necessary engineering manuals and often on-site support to familiarize operators with their equipment.

More details on the use of some of the equipment covered in this chapter can be found in Chaps. 10–14, and supporting information on the metering topics in Apps. B, C, and D.

> Most equipment manufacturers will provide on-site training.

CHAPTER

7

Modeling Water Losses

Julian Thornton

7.1 INTRODUCTION

Models are an excellent tool to assist the operator with water audits and in water loss management planning, but they should be used with care and diligence. Models are not magic: They are only as good as the concepts they employ, the data that are put into them, and the skill and experience of the user. So care should be taken to ensure that field data captured and coefficients used represent real conditions as closely as may be necessary for a result of required accuracy. If accountable data are not available, estimated data may be used, but the models should include comments columns reflecting the estimated inaccuracy for each component and calculating the final weighted potential inaccuracy. This chapter will discuss some simple examples of such models.

> **G**ood data in means good data out!

Modeling components of consumption has been part of network analysis models for over 30 years. However, modeling of customer meter accuracy, components of real losses (leakage and overflows), and pressure–leakage relationships developed rapidly during the 1990s, to a state of reliability where such models should be considered a standard part of the loss management practitioner's tool kit.

At this stage it should be emphasized that water loss management models are not the same things as network analysis models. Most operators, consultants, and contractors have at some time or another seen or used a network analysis model, which seeks to reproduce the flows and pressures in a distribution network subject to specific inputs and consumption patterns. Network analysis models are an extremely powerful tool for system analysis, allowing the operator to simulate almost any possible combination of scenarios within his or her systems. However, the concepts used for simulating water loss management in most network analysis models are often oversimplified, to the point where the estimated current leakage is nominally distributed globally around the nodes of the model, and then assumed to be fixed over time, and pressure-invariant. While such simplified assumptions may be valid for modeling flows and pressures in pipe systems, they are not valid for models which seek to answer the water loss management questions.

In this chapter, we will look briefly at:

- ❑ Audit spreadsheet models
- ❑ Modeling customer meter underregistration (part of apparent losses; see Chap. 14)
- ❑ Consumption analysis modeling

❏ The fixed and variable area discharge (FAVAD) concept for modeling pressure–leakage rate relationships
❏ The background and breaks estimate concept (BABE) for modeling components of real losses
❏ Application of BABE and FAVAD concepts to night flow analysis

7.2 AUDIT SPREADSHEET MODELS

This section deals with a basic audit. The model for the audit in this example has been designed on a spreadsheet following the guidelines set out in the American Water Works Association (AWWA) Manual M36.[1] [Other models are available for the International Water Association (IWA) audit and can easily be programmed in a simple spreadsheet format.] The M36 audit is relatively simple to program in a spreadsheet, as all of the calculations are clearly stated in the manual. For more details on how to carry out an M36 audit, refer to AWWA Manual M36, which explains in detail all of the steps required to perform a full audit. (An excerpt from this manual is provided in App. A of this book, which discusses the various stages of the M36 audit.)

Figure 7.1 shows entry of background data regarding metering and potential metering errors. Figure 7.2 shows input of background data regarding authorized unmetered use, then goes on to break out total losses and then identifies the leakage portion. After this it shows an estimated figure for recoverable leakage, which in this example is 75 percent. This figure can vary widely, depending, for example, on how long it has been since a leak detection and repair intervention, to locate unreported leaks, last took place. This could be modeled differently depending on local system conditions and the operator's knowledge of real and apparent losses within the system.

Figure 7.3 identifies cost per unit of recoverable leakage and the cost of the leak survey and repair to recover the leakage. At the end of the audit is a benefit-to-cost ratio which can be used to rank payback if various areas are to be compared or the operator wishes to compare payback on, for example, leak detection versus meter testing and repair.

Obviously, any scenario may be modeled using spreadsheet variables which have been programmed in. Once the initial construction work has been done, the modeling is actually very quick.

If we were to change one of the variables, for example, if a reservoir in the system were found to be leaking seriously, then we might find that the modeled results changed dramatically and the leak detection exercise becomes much less favorable, and reservoir relining or repair becomes a much better option. Figure 7.4 shows the reservoir leakage and seepage section being changed to include 50,000 kgal of loss in this category. Figure 7.5 shows the result on the benefit-to-cost ratio for leak detection, reducing from a previous 7.4:1 to a new 2.7:1. Obviously, a payback in this case of 2.7:1 is still good, but not as attractive as 7.4:1. Once the auditing process is automated into a simple model, the modeler can very quickly change parameters and numbers to see where the best cost-to-benefit ratios lie and therefore program intervention work very effectively.

> **F**or most formats, it is easy to program a spreadsheet to perform audit calculations automatically. However, the operator must fully understand the concepts being modeled.

7.2 AUDIT SPREADSHEET MODELS

	M36 WATER AUDIT WORKSHEET	(Adapted from AWWA M36)		
For:	**XYZ Allied Utility District**	**Study period**	Jan 1 1997–Jan 1 1998	
	5000 metered connections — av. flow 496 gpm Gravity feed system		Water Volume	
Line	Item	Subtotal	Total Cumulative	Unit
Task 1	**Measure the Supply**			kgal
1	Total Water Supply to Distribution System			
	Uncorrected Total Water Supply to the Distribution System (total of master meters)		257,544.0	kgal
2	Adjustments to Total Water Supply			
	a. Source Meter Error (+/-)	6,300.0		kgal
	b. Change in Reservoir and Tank Storage	0.0		kgal
	c. Other Contributions or Losses (+/-)	0.0		kgal
3	Total Adjustments to Total Water Supply (2a, 2b and 2c)		6,300.0	kgal
4	Adjusted Total Water Supply to the Distribution System (add lines 1 and 3)		263,844.0	kgal
Task 2	**Measure Authorized Metered Use**			
5	Uncorrected Total Metered Water Use (sales)	184,691.0		kgal
6	Adjustments due to Meter Reading Lag-Time (+/-)	0.0		kgal
7	Metered Deliveries (add lines 5 and 6)	184,691.0		kgal
8	Total Sales Meter Error and System Service Meter Error (+/-)			
	a. Residential Meter Error	0.0		kgal
	b. Large Meter Error	0.0		kgal
	c. Total (add lines 8a and 8b)	0.0		kgal
9	Corrected Total Metered Water Deliveries (add lines 7 and 8c)		184,691.0	kgal
10	Corrected Total Unmetered Water (subtract line 9 from line 4)		79,153.0	kgal

Figure 7.1 Background data entry. (*Source: Julian Thornton.*)

120 CHAPTER 7 MODELING WATER LOSSES

Line	Item	Water Volume Subtotal	Total Cumulative	Unit
Task 3	**Measure Authorized Unmetered Use**			
11	a. Firefighting and Training	0.0		kgal
	b. Main Flushing	0.0		kgal
	c. Storm Drain Flushing	0.0		kgal
	d. Sewer Cleaning	0.0		kgal
	e. Street Cleaning	0.0		kgal
	f. Schools	0.0		kgal
	g. Landscaping in Large Public Areas			
	1. Parks	80.0		kgal
	2. Golf Courses	0.0		kgal
	3. Cemeteries	0.0		kgal
	4. Playgrounds	0.0		kgal
	5. Highway Median Strips	0.0		kgal
	6. Other Landscaping	0.0		kgal
	h. Decorative Water Facilities	0.0		kgal
	i. Swimming Pools	0.0		kgal
	j. Construction Sites	0.0		kgal
	k. Water Quality and Other Testing	0.0		kgal
	l. Process Water at Treatment Plants	79.0		kgal
	m. Other Unmetered Uses	0.0		kgal
12	Total Authorized Unmetered Water (add lines 11a to 11m)		159.0	kgal
13	Total Water Losses (subtract line 12 from line 10)		78,994.0	kgal
Task 4	**Measure Water Losses**			
14	a. Accounting Procedure Errors	40.0		kgal
	b. Unauthorized Connections	0.0		kgal
	c. Malfunctioning Distribution System Controls	0.0		kgal
	d. Reservoir Seepage and Leakage	0.0		kgal
	e. Evaporation	0.0		kgal
	f. Reservoir Overflow	65.0		kgal
	g. Discovered Leaks	0.0		kgal
	h. Unauthorized Use	0.0		kgal
15	Total Identified Water Losses (add lines 14a to 14h)		105.0	kgal
Task 5	**Analyze Audit Results**			
16	Potential Water System Leakage (subtract line 15 from line 13)		78,889.0	kgal
17	Recoverable Leakage (estimated at 75%)		59,166.8	kgal
		(24 months)	118,333.50	kgal

Figure 7.2 Background data entry. (*Source: Julian Thornton.*)

Line	Item	Dollars	Unit
18	Cost Savings		
	a. Cost of Water Supply	$2.00	kgal
	b. Variable O&M Costs	$0.20	kgal
19	Total Costs per Unit of Recoverable Leakage	$2.20	kgal
20	One-Year Benefit from Recoverable Leakage	$130,166.85	Year
21	Total Benefits from Recoverable Leakage (24 months)	$260,333.70	24 months
22	Total Costs of Leak Detection Project	**$35,000.00**	
23	Benefit-to-Cost Ratio	7.4 :1	

Comments:

Prepared by:

Name: Julian Thornton
Title: Leak Detection Technician **Date:** 11/27/98

Figure 7.3 Cost benefit for leak detection. (*Source: Julian Thornton.*)

122 CHAPTER 7 MODELING WATER LOSSES

Line		Item	Water Volume Subtotal	Water Volume Total Cumulative	Unit
Task 3		**Measure Authorized Unmetered Use**			
11	a.	Firefighting and Training	0.0		kgal
	b.	Main Flushing	0.0		kgal
	c.	Storm Drain Flushing	0.0		kgal
	d.	Sewer Cleaning	0.0		kgal
	e.	Street Cleaning	0.0		kgal
	f.	Schools	0.0		kgal
	g.	Landscaping in Large Public Areas			
		1. Parks	80.0		kgal
		2. Golf Courses	0.0		kgal
		3. Cemeteries	0.0		kgal
		4. Playgrounds	0.0		kgal
		5. Highway Median Strips	0.0		kgal
		6. Other Landscaping	0.0		kgal
	h.	Decorative Water Facilities	0.0		kgal
	i.	Swimming Pools	0.0		kgal
	j.	Construction Sites	0.0		kgal
	k.	Water Quality and Other Testing	0.0		kgal
	l.	Process Water at Treatment Plants	79.0		kgal
	m.	Other Unmetered Uses	0.0		kgal
12		Total Authorized Unmetered Water (add lines 11a to 11m)		159.0	kgal
13		Total Water Losses (subtract line 12 from line 10)		78,994.0	kgal
Task 4		**Measure Water Losses**			
14	a.	Accounting Procedure Errors	40.0		kgal
	b.	Unauthorized Connections	0.0		kgal
	c.	Malfunctioning Distribution System Controls	0.0		kgal
	d.	Reservoir Seepage and Leakage	50,000.0		kgal
	e.	Evaporation	0.0		kgal
	f.	Reservoir Overflow	65.0		kgal
	g.	Discovered Leaks	0.0		kgal
	h.	Unauthorized Use	0.0		kgal
15		Total Identified Water Losses (add lines 14a to 14h)		50,105.0	kgal
Task 5		**Analyze Audit Results**			
16		Potential Water System Leakage (subtract line 15 from line 13)		28,889.0	kgal
17		Recoverable Leakage (estimated at 75%)		21,666.8	kgal
			(24 months)	43,333.50	kgal

Figure 7.4 Changing the reservoir leakage section. (*Source: Julian Thornton.*)

Line	Item	Dollars	Unit
18	Cost Savings		
	a. Cost of Water Supply	$2.00	kgal
	b. Variable O&M Costs	$0.20	kgal
19	Total Costs per Unit of Recoverable Leakage	$2.20	kgal
20	One-Year Benefit from Recoverable Leakage	$47,666.85	Year
21	Total Benefits from Recoverable Leakage (24 months)	$95,333.70	24 months
22	Total Costs of Leak Detection Project	**$35,000.00**	
23	Benefit-to-Cost Ratio	2.7 :1	

Comments:

Prepared by:

Name: Julian Thornton
Title: Leak Detection Technician
Date: 11/27/98

Figure 7.5 Reduced cost–benefit for leak detection. (*Source: Julian Thornton.*)

For performance comparisons, the volume data from an M36 audit can easily be reorganized into a standard international format (IWA audit) and the most appropriate performance indicators for water loss management calculated and compared with a national or international data set which takes into account key local system parameters—length of mains, number of service connections, location of customer meters, and average operating pressure.

7.3 CUSTOMER METER SIZING MODEL

For direct-feed pressure systems, customer meters need to be selected and sized to record a wide range of flow rates. Any underregistration of metered consumption appears as apparent losses in the water audit, as the "lost" water is reaching the customer but not being registered or explicitly charged for. There are many models for this type of loss analysis. The following is a simple model for a spreadsheet to analyze the potential benefit of correct sizing of existing customer meters, and is particularly appropriate for larger commercial customers.

Figure 7.6 shows the field data input screen. It can be seen here that 24 h of logged data are required for this type of model. In addition, a seasonal profile should be recorded to take into account any changes in consumption patterns between the times of field testing and maximum or minimum consumption periods. This will ensure that the meter is not overly downsized.

Figure 7.7 shows a graphical analysis. The graphical analysis in this case allows the user to make decisions about where the demand profile lies within the minimum transitional and maximum flow ranges. The model will make a decision automatically, based on economics, but sometimes the operator will prefer to override this decision because of operational system knowledge. For example, the user may be an industry which the operator knows will need additional water supply in coming months.

Figure 7.8 shows the economics page, where information such as water values per volume and meter costs are input. This is where the volumetric and financial parts of the spreadsheet come together to model the best economic scenario.

Figure 7.9 shows the final decision, with the original and new meter selections, the old and new headlosses, and a summary of savings by meter replacement.

Figure 7.10 shows the original meter as being a 4-in turbine, not a 3-in turbine as in the first scenario. Figure 7.11 shows the financial and hydraulic differences for the second scenario.

7.4 CONSUMPTION ANALYSIS MODEL

Analysis of components of consumption forms an important part of any loss reduction or conservation program. The following is a simple model for a spreadsheet to predict industrial restroom usage for industries with many employees. This type of model may be used to see the poten-

7.4 CONSUMPTION ANALYSIS MODEL

Test location	XYZ Inc.	Mandatory	Profile	Test x
Location/type of building	XYZ Street, Atlanta, GA	Date/Time	Test Profile	Seasonal
	Manufacturer	70.00	GPM	Profile
Existing meter size	3"	5/15/97	Total	
Existing meter designation	T3"	12:00	142	142.00
Month of meter test	May	13:00	133	133.00
Questionnaire filled out (y/n)	Yes	14:00	111	111.00
Tester name	Julian Thornton	15:00	111	111.00
Client name	Waterworks Dept	16:00	99	99.00
Instructions:		17:00	99	99.00
		18:00	76	76.00
1. Enter all information in the above box.		19:00	76	76.00
		20:00	65	65.00
2. Paste logger data into the table to the right or manually type the data.		21:00	65	65.00
		22:00	76	76.00
3. If available, fill in the seasonal use information from the client billing information. If not, copy average profile number, cell D3, into seasonal boxes.		23:00	65	65.00
		0:00	22	22.00
		1:00	22	22.00
4. Use graphical analysis to view the tabular and graphical representations of the test profile in different meter sizes.		2:00	22	22.00
		3:00	17	17.00
		4:00	14	14.00
5. Use meter sizing to select correct size meter.		5:00	12	12.00
		6:00	13	13.00
6. Use financial analysis to perform cost-to-benefit analysis for change-out.		7:00	10	10.00
		8:00	33	33.00
7. Internet help: e-mail		9:00	132	132.00
		10:00	133	133.00
		11:00	132	132.00

Optional	Profile		
Seasonal	Average	Normalized	Annual
Profile	Consumption	Curve % of	Average
	GPM	Annual Average	Consumption
January	23.69	1.00	23.69
February	23.69	1.00	23.69
March	23.69	1.00	23.69
April	23.69	1.00	23.69
May	23.69	1.00	23.69
June	23.69	1.00	23.69
July	23.69	1.00	23.69
August	23.69	1.00	23.69
September	23.69	1.00	23.69
October	23.69	1.00	23.69
November	23.69	1.00	23.69
December	23.69	1.00	23.69

Figure 7.6 Field data input screen. (*Source: Julian Thornton.*)

tial benefit of toilet change-out for water conservation, or it may be used to predict the amount of water needed for sanitary purposes so that it may be broken out of a measured flow profile.

Figure 7.12 shows an input table with estimations for volumetric use and frequency of use for men and women within various different buildings in a fictitious industry. Figure 7.13 shows a resultant modeled estimation of sanitary use per shift.

Figure 7.14 shows the input table again, with the volume per flush changed to reflect change out to a lower-volume toilet, in this case with

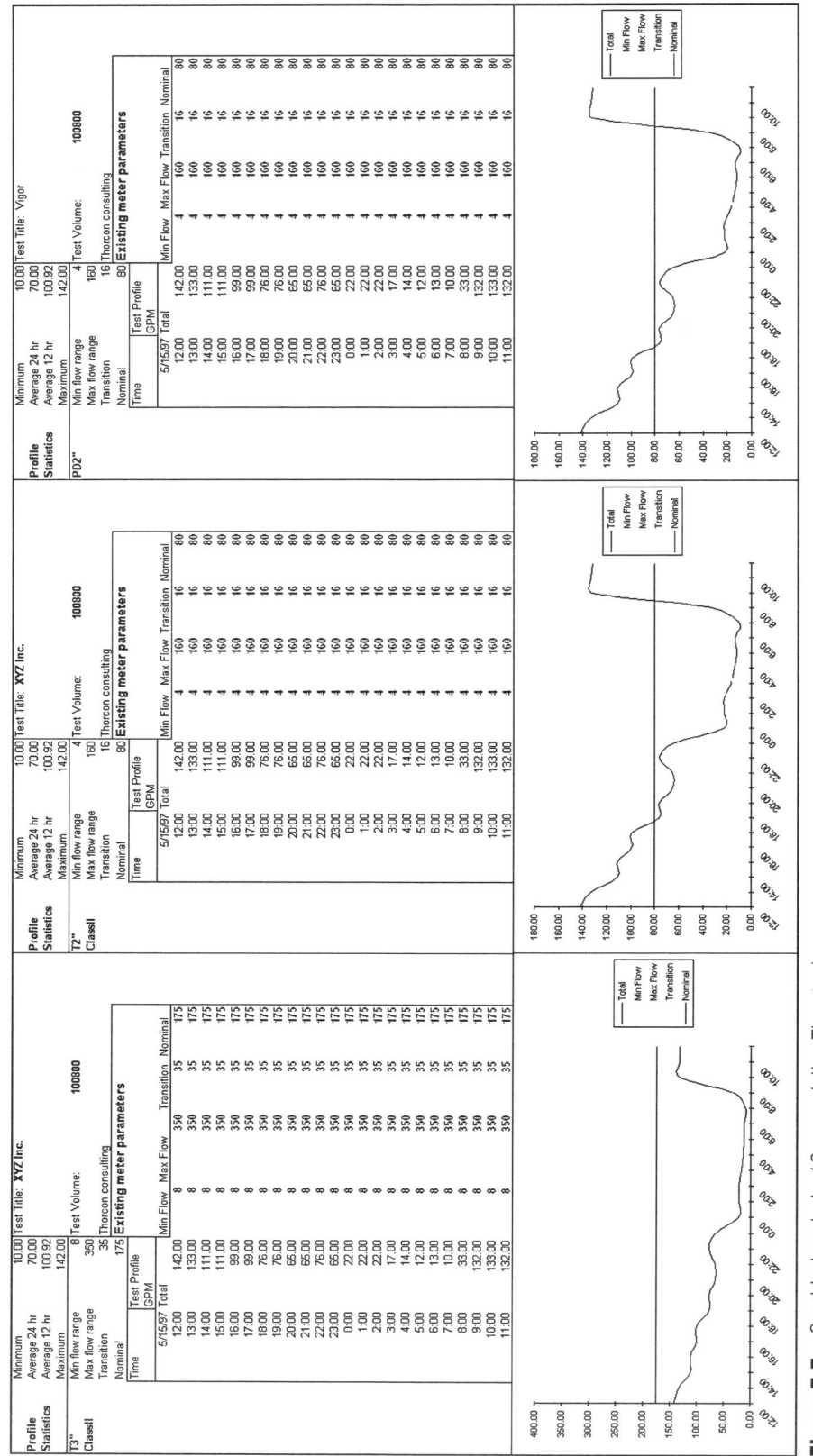

Figure 7.7 Graphical analysis. (*Source: Julian Thornton.*)

Test Title	XYZ Inc.	Client	Waterworks Dept
Test Volume	100800	Retail Water $	0.003

Notes:

Enter the retail cost of the water into cell D2.

Enter the costs of new meters per size into cells E52..E63.

Enter the cost of a meter changeout (labor) into cell P36.

Go to sheet "Meter Sizing"

Designation	Row	Min Flow	Max Flow	Transition	Nominal
T12"	Percentage	0.6	0	0.4	0
	Volume	60480	0	40320	0
	Parameter	<min	>max	<tran>min	<max>tran
	Limits	120	5000	500	2500
T10"	Percentage	0.214286	0	0.785714286	0
	Volume	21600	0	79200	0
	Limits	75	3800	380	1900
T8"	Percentage	0.0982143	0	0.901786	0
	Volume	9900	0	90900	0
	Limits	50	2400	240	1200
T6"	Percentage	0.078571429	0	0.8369048	0.0845238
	Volume	7920	0	84360	8520
	Limits	30	1400	140	700
T4"	Percentage	0.029167	0	0.069047619	0.9017857
	Volume	2940	0	6960	90900
	Limits	15	630	63	315
T3"	Percentage	0	0	0.098214	0.901786
	Volume	0	0	9900	90900
	Limits	8	350	35	175
T2"	Percentage	0	0	0.0291667	0.9708333
	Volume	0	0	2940	97860
	Parameter				
	Limits	4	160	16	80
PD2"	Percentage	0	0	0.029166667	0.970833333
	Volume	0	0	2940	97860
	Limits	4	160	16	80
PD1.5"	Percentage	0	0.5321429	0.005952	0.461904762
	Volume	0	53640	600	46560
	Limits	2.4	100	10	50
PD1"	Percentage	0	0.9017857	0	0.0982143
	Volume	0	90900	0	9900
	Limits	1.2	50	5	25
PD0.75"	Percentage	0	0.92142857	0	0.0785714
	Volume	0	92880	0	7920
	Limits	0.72	30	3	15
PD0.625"	Percentage	0	0.9607143	0	0.039286
	Volume	0	96840	0	3960
	Limits	0.25	20	2	10

Test	
Volume Gallons	100800
Test value $	302.4
Thorcon consulting	

Summary of volumes between Transition and maximum flow — Accuracy range 2%

Designation	% test volume	Perfect Volume Gal	Worst case Volume 98%	Worst case Revenue $
T12"	0	0	0	0
T10"	0	0	0	0
T8"	0	0	0	0
T6"	0.08452381	8520	8349.6	25.0488
T4"	0.901785714	90900	89082	267.246
T3"	0.901785714	90900	89082	267.246
T2"	0.970833333	97860	95902.8	287.7084
PD2"	0.970833333	97860	95902.8	287.7084
PD1.5"	0.461904762	46560	45628.8	136.8864
PD1"	0.098214286	9900	9702	29.106
PD0.75"	0.078571429	7920	7761.6	23.2848
PD0.625"	0.039285714	3960	3880.8	11.6424

Summary of volumes between Transition & minimum flow — Accuracy range 5%

Designation	% test volume	Perfect Volume Gal	Worst case Volume 95%	Worst case Revenue $
T12"	0.4	40320	38304	114.912
T10"	0.785714286	79200	75240	225.72
T8"	0.901785714	90900	86355	259.065
T6"	0.836904762	84360	80142	240.426
T4"	0.069047619	6960	6612	19.836
T3"	0.098214286	9900	9405	28.215
T2"	0.029166667	2940	2793	8.379
PD2"	0.029166667	2940	2793	8.379
PD1.5"	0.005952381	600	570	1.71
PD1"	0	0	0	0
PD0.75"	0	0	0	0
PD0.625"	0	0	0	0

Summary of volumes below minimum flow — Accuracy range 15%

Designation	% test volume	Perfect Volume Gal	Worst case Volume 85%	Worst case Revenue $
T12"	0.6	60480	51408	154.224
T10"	0.214285714	21600	18360	55.08
T8"	0.098214286	9900	8415	25.245
T6"	0.078571429	7920	6732	20.196
T4"	0.029166667	2940	2499	7.497
T3"	0	0	0	0
T2"	0	0	0	0
PD2"	0	0	0	0
PD1.5"	0	0	0	0
PD1"	0	0	0	0
PD0.75"	0	0	0	0
PD0.625"	0	0	0	0

Summary of volumes above maximum flow (Realistically these would not be used) — Accuracy range 30%

Designation	% test volume	Perfect Volume Gal	Worst case Volume 70%	Worst case Revenue $
T12"	0	0	0	0
T10"	0	0	0	0
T8"	0	0	0	0
T6"	0	0	0	0
T4"	0	0	0	0
T3"	0	0	0	0
T2"	0	0	0	0
PD2"	0	0	0	0
PD1.5"	0.532142857	53640	37548	112.644
PD1"	0.901785714	90900	63630	190.89
PD0.75"	0.921428571	92880	65016	195.048
PD0.625"	0.960714286	96840	67788	203.364

Summary most efficient size	Worst Case Total Revenue	Worst Case Accuracy	Revenue Loss Per Annum
T12"	269.136	89.00%	12141.36
T10"	280.8	92.86%	7884
T8"	284.31	94.02%	6602.85
T6"	285.6708	94.47%	6106.158
T4"	294.579	97.41%	2854.665
T3"	295.461	97.71%	2532.735
T2"	296.0874	97.91%	2304.099
PD2"	296.0874	97.91%	2304.099
PD1.5"	251.2404	83.08%	18673.25
PD1"	219.996	72.75%	30077.46
PD0.75"	218.3328	72.20%	30684.53
PD0.625"	215.0064	71.10%	31898.66

Summary

Test Location:	XYZ Inc.	Existing Meter Size:	3"	Billed/day							
Location/type of building:	XYZ Street, Atlanta, GA	Existing Meter Designation:	T3"	295.46		Months	(Change in volume and fixed income)			Cost of meter	103.4
	Manufacturer	Most efficient Designation	PD2"	296.09	Payback	18.9744706		Difference Billed $/Yr	128.64	Cost of labor	100

Figure 7.8 Economics page. (*Source: Julian Thornton.*)

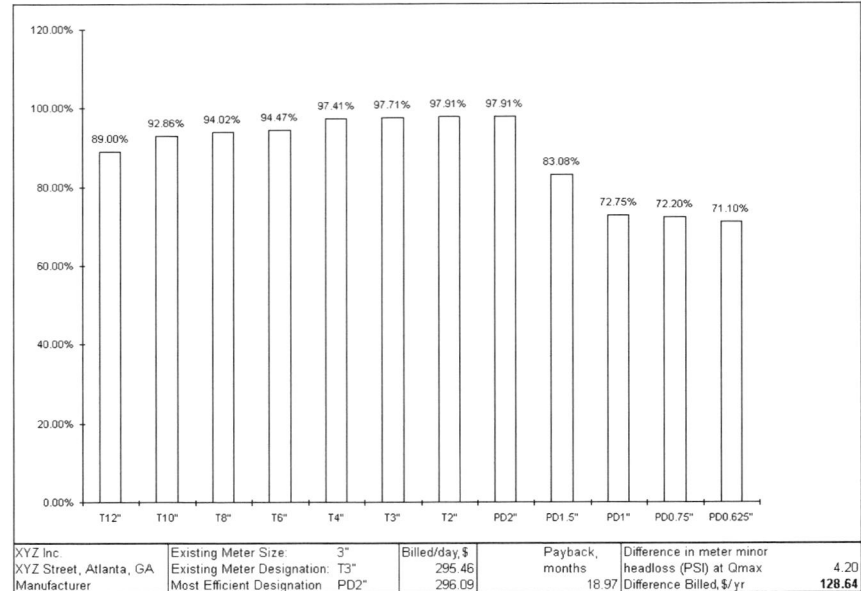

Figure 7.9 Original and new meter selections, old and new headlosses, a summary of savings. (*Source: Julian Thornton.*)

the volume reduced from 3.5 gal per flush to 1.6 gal per flush. (Excellent base information on usage can be found in the AWWA end-user survey[2] and in the U.K. managing leakage series.[3])

Figure 7.15 shows the resultant modeled reduction in use per shift.

7.5 PRESSURE ANALYSIS AND FAVAD CONCEPTS

The first example in this chapter was a simple audit model, designed to obtain an overview of the probable benefit–cost ratio of a one-off intervention for leak detection and repair.

Pressure management can also be used to mitigate the adverse effects of excess pressure. Later, in Chap. 12, we address pressure management as a means of controlling leak volumes, reducing leak frequency, and reducing wasteful consumption, as part of a water conservation strategy. The following simple spreadsheet model shows how we can add on to the audit model to look at the potential benefits of pressure management as a means of intervention against leakage and maintenance costs, as well as a brief look at how we can model reduced consumption.

The model assumes that the leakage rate from the distribution system (L; gal/min) varies with the average pressure (P) according to the relationship $L \propto P^{N_1}$.

Traditionally in North America, practitioners have used an N_1 figure of 0.5 (the square-root relationship), for example, to calculate the velocity and discharge rate through a hole of measured dimensions in a pipe at some particular pressure. However, an N_1 value of 0.5 applies only to an individual leak which has a fixed area (independent of pressure), such as occur on metal pipes. Other leaks, such as leaks on nonmetal pipes, or

Test location	XYZ Inc.	Mandatory	Profile	Test x
Location/type of building	XYZ Street, Atlanta, GA	Date/Time	Test Profile	Seasonal
	Manufacturer	70.00	GPM	Profile
Existing meter size	4"	5/15/97	Total	
Existing meter designation	T4"	12:00	142	142.00
Month of meter test	May	13:00	133	133.00
Questionnaire filled out (y/n)	Yes	14:00	111	111.00
Tester name	Julian Thornton	15:00	111	111.00
Client name	Waterworks Dept	16:00	99	99.00
Instructions:		17:00	99	99.00
		18:00	76	76.00
1. Enter all information in the above box.		19:00	76	76.00
2. Paste logger data into the table to the right or manually type the data.		20:00	65	65.00
		21:00	65	65.00
		22:00	76	76.00
3. If available, fill in the seasonal use information from the client billing information. If not, copy average profile number, cell D3, into seasonal boxes.		23:00	65	65.00
		0:00	22	22.00
		1:00	22	22.00
4. Use graphical analysis to view the tabular and graphical representations of the test profile in different meter sizes.		2:00	22	22.00
		3:00	17	17.00
		4:00	14	14.00
		5:00	12	12.00
5. Use meter sizing to select correct size meter.		6:00	13	13.00
		7:00	10	10.00
6. Use financial analysis to perform cost-to-benefit analysis for change-out.		8:00	33	33.00
		9:00	132	132.00
7. Internet help:e-mail		10:00	133	133.00
		11:00	132	132.00

Optional	Profile		
Seasonal	Average	Normalized	Annual
Profile	Consumption	Curve % of	Average
	GPM	Annual Average	Consumption
January	23.69	1.00	23.69
February	23.69	1.00	23.69
March	23.69	1.00	23.69
April	23.69	1.00	23.69
May	23.69	1.00	23.69
June	23.69	1.00	23.69
July	23.69	1.00	23.69
August	23.69	1.00	23.69
September	23.69	1.00	23.69
October	23.69	1.00	23.69
November	23.69	1.00	23.69
December	23.69	1.00	23.69

Figure 7.10 Original meter 4 in, not 3 in. (*Source: Julian Thornton.*)

small "background" leaks at joints and fittings, have N_1 values greater than 0.5, because the effective area of each leak varies with pressure. (In some very small leaks such as background leaks it is thought that the discharge coefficient may also change.) Analysis of a wide range of international data using the N_1 approach shows that N_1 for individual leakage paths or small sectors of distribution systems can vary significantly, from 0.5 at the lowest level to 2.5 at the highest level. Leakage rates from distribution systems are therefore much more sensitive to operating pressure than most North American leakage practitioners realized in the past.

130 CHAPTER 7 MODELING WATER LOSSES

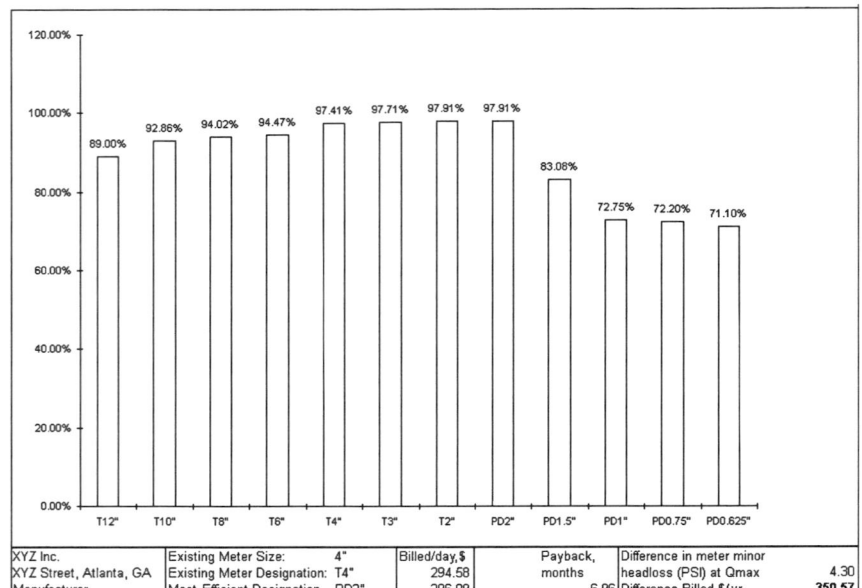

Figure 7.11 Financial and hydraulic differences in the second scenario. (*Source: Julian Thornton.*)

Figure 7.12 Input table with estimations for volumetric use. (*Source: Julian Thornton.*)

> N_1 estimates for first-pass calculations can be stated as 0.5 for holes in metallic systems, 1.5 for plastic pipe leaks, and 1.5 for small background leaks. N_1 values of up to 2.5 have been recorded.

The average N_1 value for a number of distribution sectors in Japan, Great Britain, and Brazil was close to 1.15. This is close enough to 1.0 for a linear relationship to be assumed between pressure and leakage (in relatively large areas of the distribution system), for most first-pass calculations, if the changes in pressure are not more than ±20 percent.

Operators may wish to model a band of potential results by, for example,

❑ Identifying the pipe materials in the system
❑ Identifying the predominant type of leak
❑ Deciding on a low and a high N_1 value, understanding that the true figure lies somewhere between these values

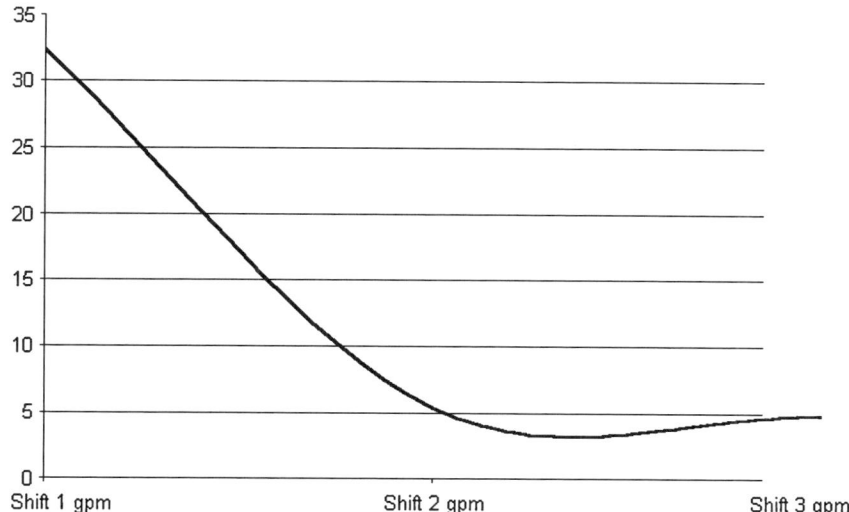

Figure 7.13 Resultant modeled estimation of sanitary use per shift. (*Source: Julian Thornton.*)

Building Total Population Distribution		Men flush Women flush Volume flush Volume wash	2 4 1.6 1	Urinal flush Urinal volume	2 1	Key Blue user enter Red calculated Black description					
		No. flushes	Volume/flush	Vol/person	No. flushes	Vol/flush	Vol/person	No. washes	Vol/wash	Vol/person	Total volume per person
Bldg. 1	414										
Men	207	2	1.6	3.2	2	1	2	4	1	4	9.2
Women	207	4	1.6	6.4			0	4	1	4	10.4
50/50											
Bldg. 2	65										
Men	36	2	1.6	3.2	2	1	2	4	1	4	9.2
Women	29	4	1.6	6.4			0	4	1	4	10.4
55/45											
Bldg. 3	40										
Men	20	2	1.6	3.2	2	1	2	4	1	4	9.2
Women	20	4	1.6	6.4			0	4	1	4	10.4
50/50											
Bldg. 4	200										
Men	120	2	1.6	3.2	2	1	2	4	1	4	9.2
Women	80	4	1.6	6.4			0	4	1	4	10.4
60/40											
Bldg. 5	270										
Men	162	2	1.6	3.2	2	1	2	4	1	4	9.2
Women	108	4	1.6	6.4			0	4	1	4	10.4
60/40											
Bldg. 6	33										
Men	20	2	1.6	3.2	2	1	2	4	1	4	9.2
Women	13	4	1.6	6.4			0	4	1	4	10.4
60/40											

Figure 7.14 Flush volume changed to reflect conservation. (*Source: Julian Thornton.*)

For example, in a nonmetal system, a lower limit of N_1 of 1.25 and an upper limit of 1.75 would be a reasonable assumption, whatever the level of leakage. For a metal system, a reasonable range would be 1.0–1.5 if leakage is low, or 0.5–1.0 for high-leakage (unreported or unfixed breaks) systems.

Alternatively, for any particular sector, the operator might go into the field, measure night flow while reducing pressure, and calculate an N_1 figure after allowing for pressure-invariant night consumption (using the BABE components approach; see Sec. 7.6).

Figure 7.16 shows an example of a spreadsheet model which pulls forward data from the audit model and calculates potential savings from reducing pressure from 80 to 45 psi. This model is extremely simplified and does not take into account changing frictional headloss in small-diameter pipe systems.

132 CHAPTER 7 MODELING WATER LOSSES

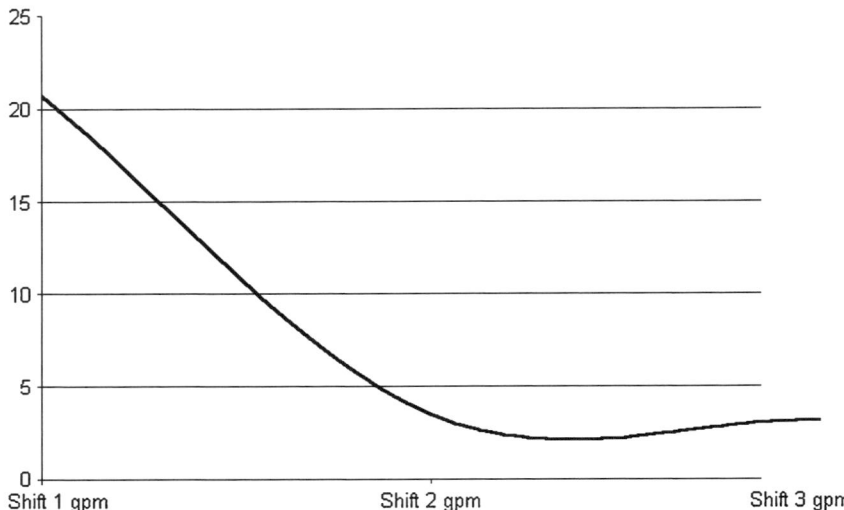

Figure 7.15 Resultant modeled reduction in use per shift. (*Source: Julian Thornton.*)

	Basic analysis of savings using a 24-month period, assuming no headloss in the system and fixed outlet control		
A.	Cost of water	$/kgal	$2.20
B.	Metered volume to sector	kgal	527688
C.	Volume of discovered leaks, if not repaired (M36 audit, line 14g)		0
D.	Volume of potential water system leakage (M36 audit, line 16)		78889
E.	Total system leakage	kgal	157778
F.	Existing average daily pressure	psi	80
G.	Desired control pressure	psi	45
H.	Power law		1.15
I.	New leakage volume	kgal	81412
J.	Total savings (E – I)	kgal	76366
K.	Total savings (J*A)	$	168006

Figure 7.16 Changing pressure from 80 to 45 psi. (*Source: Julian Thornton.*)

Line H shows where the coefficient (power law) is 1.15, and estimated total savings have been identified of 76,366 kgal or $168,006. If we were to change the power law and use the bandwidth approach, we might say that the N_1 factor can vary from 0.85 to 1.35 (depending on our knowledge of our system). Figure 7.17 shows the reduced estimated savings of 61,028 kgal or $134,262, and Figure 7.18 shows the increased estimated savings of 85,216 kgal or $187,474.

We could now use this modeled bandwidth to start to identify a budget and potential savings for our pressure management project, with reasonable certainty that the real recovered leakage figure lies somewhere inside the band.

7.5 PRESSURE ANALYSIS AND FAVAD CONCEPTS 133

	Basic analysis of savings using a 24-month period, assuming no headloss in the system and fixed outlet control		
A.	Cost of water	$/kgal	$2.20
B.	Metered volume to sector	kgal	527688
C.	Volume of discovered leaks, if not repaired (M36 audit, line 14g)		0
D.	Volume of potential water system leakage (M36 audit, line 16)		78889
E.	Total system leakage	kgal	157778
F.	Existing average daily pressure	psi	80
G.	Desired control pressure	psi	45
H.	Power law		0.85
I.	New leakage volume	kgal	96750
J.	Total savings (E − I)	kgal	61028
K.	Total savings (J*A)	$	134262

Figure 7.17 Reduced estimated savings. (*Source: Julian Thornton.*)

	Basic analysis of savings using a 24-month period, assuming no headloss in the system and fixed outlet control		
A.	Cost of water	$/kgal	$2.20
B.	Metered volume to sector	kgal	527688
C.	Volume of discovered leaks, if not repaired (M36 audit, line 14g)		0
D.	Volume of potential water system leakage (M36 audit, line 16)		78889
E.	Total system leakage	kgal	157778
F.	Existing average daily pressure	psi	80
G.	Desired control pressure	psi	45
H.	Power law		1.35
I.	New leakage volume	kgal	72562
J.	Total savings (E − I)	kgal	85216
K.	Total savings (J*A)	$	187474

Figure 7.18 Increased estimated savings. (*Source: Julian Thornton.*)

Once the savings have been identified, it is necessary to identify the cost. Chapter 8 goes into more detail about cost-to-benefit calculations. Figure 7.19 shows a continuation of the simple model, which now calculates the change in metered consumption due to the intended change in pressure. In this sheet we must identify how much of the usage is from direct pressure and how much of the usage is fixed volume. (Remember that even in countries which use direct pressure as opposed to roof tanks, there still is quite a large component of fixed volume use such as toilets, washing machines, etc.).

In most cases the change in usage will correspond to the traditional square-root relationship of $N_1 = 0.5$. In this case the impact of reduction

134 CHAPTER 7 MODELING WATER LOSSES

	Basic analysis of consumption changes using a 24-month period, assuming no headloss in the system and fixed outlet control		
A.	Retail cost of water	$/kgal	3
B.	Water sold in sector	kgal	446760
C.	% residential use		50
D.	% agricultural use		10
E.	% ICI use		40
F.	Volume sold for residential use (B*C)	kgal	223380
G.	Volume sold for agricultural use (B*D)	kgal	44676
H.	Volume sold for ICI use (B*E)	kgal	178704
I.	% of residential use from direct pressure		40
J.	% of agricultural use from direct pressure		5
K.	% of ICI use from direct pressure		5
L.	Residential volume affected (F*I)	kgal	89352
M.	Agricultural volume affected (G*J)	kgal	2234
N.	ICI volume affected (H*K)	kgal	8935
	Total		100521
O.	Existing system pressure	psi	80
P.	Desired control pressure	psi	45
Q.	New pressure-dependent residential consumption [Q = L*(P/O)$^{0.5}$]		67014
R.	New pressure-dependent agricultural consumption [R = M*(P/O)$^{0.5}$]		1675
S.	New pressure-dependent ICI consumption [S = N*(P/O)$^{0.5}$]		6701
	Total		75391
T.	Potential lost sales volume	kgal	25130
U.	Potential lost revenue	$	75391

Figure 7.19 Change in metered consumption. (*Source: Julian Thornton.*)

> The AWWA end-use study[2] has excellent information on demand volumes and types.

in leakage will mostly be proportionally greater than the impact of reduced demand. In some cases reduction of demand might actually be desired, as in water conservation projects. Figure 7.20 shows a continuation of the simple model to bring in other parameters and decisions such as whether demand reduction is considered a benefit.

Figure 7.20 also shows the final result with the assumption that conservation is not desired, and Fig. 7.21 shows the opposite case, where water conservation is desired. It can be clearly seen that the decision to take water conservation as a benefit or a cost makes a big difference in the payback period of the project. In this case, if conservation is a benefit the payback is 0.7 month, but if it is seen as a cost to the project, the payback increases to 11.5 months. A payback of 11.5 months is still pretty good, but this particular example would move considerably in a ranking list depending on the way demand reduction was viewed.

The two examples shown in Figs. 7.20 and 7.21 used an average N_1 factor of 1.15. Obviously, if the N_1 factor were either of the other limits selected, then this would also have a big impact on the final ranked location of this area.

The final figure in this series, Fig. 7.22, shows the modeled difference in payback if demand reduction is desired to defer capital construction cost for an additional supply or water source. Here the cost of water per unit for the additional supply is modeled at the same cost per unit as the existing cost. This may or may not be true in all cases and should be investigated in detail on a case-by-case basis.

	Cost-to-benefit analysis, assuming no headloss in the system and fixed outlet control	
	Costs	
A.	Cost of valve assembly	800
B.	Cost of bypass assembly	500
C.	Cost of field measurements	800
D.	Cost of installation	3000
F.	Potential reduction in revenue through pressure control	75391
	If water conservation is being implemented, mark "y" for yes here ...	n
	Benefit	
G.	Total savings through leak reduction	168006
H.	Cost to produce 1 kgal of water from a new source	0
I.	No. of kgallons reduced through leak reduction	76366
J.	Benefit from deferring installation of new source	0
	(Include only H, I, and J if water supply needs to be increased)	
	Analysis	
K.	Total cost	80491
L.	Total benefit	168006
N.	Payback on investment in months	11.5

Figure 7.20 Deciding whether demand reduction is a benefit. (*Source: Julian Thornton.*)

	Cost-to-benefit analysis, assuming no headloss in the system and fixed outlet control	
	Costs	
A.	Cost of valve assembly	800
B.	Cost of bypass assembly	500
C.	Cost of field measurements	800
D.	Cost of installation	3000
F.	Potential reduction in revenue through pressure control	75391
	If water conservation is being implemented, mark "y" for yes here ...	y
	Benefit	
G.	Total savings through leak reduction	168006
H.	Cost to produce 1 kgal of water from a new source	0
I.	No. of kgalllons reduced through leak reduction	76366
J.	Benefit from deferring installation of new source	0
	(Include only H, I, and J if water supply needs to be increased)	
	Analysis	
K.	Total cost	5100
L.	Total benefit	168006
N.	Payback on investment in months	0.7

Figure 7.21 In this case conservation is desired. (*Source: Julian Thornton.*)

Once the project has been approved on this first-pass basis, the best thing to do is to go into the field and confirm actual N_1 values for individual zones where pressure management may be installed, then perform a detailed cost-to-benefit calculation on each installation using real figures to ensure that each installation is effective.

	Cost-to-benefit analysis, assuming no headloss in the system and fixed outlet control	
	Costs	
A.	Cost of valve assembly	800
B.	Cost of bypass assembly	500
C.	Cost of field measurements	800
D.	Cost of installation	3000
F.	Potential reduction in revenue through pressure control	75391
	If water conservation is being implemented, mark "y" for yes here ...	y
	Benefit	
G.	Total savings through leak reduction	168006
H.	Cost to produce 1 kgal of water from a new source	2.2
I.	No. of kgallons reduced through leak reduction	76366
J.	Benefit from deferring installation of new source	168006
	(Include only H, I, and J if water supply needs to be increased)	
	Analysis	
K.	Total cost	5100
L.	Total benefit	336011
N.	Payback on investment in months	0.4

Figure 7.22 In this case demand reduction is desired, to defer the cost of a new source. (*Source: Julian Thornton.*)

7.6 MODELING COMPONENTS OF REAL LOSSES USING BREAKS AND BACKGROUND ESTIMATES (BABE) CONCEPTS[4]

In the early 1990s, during the U.K. National Leakage Control Initiative, a systematic approach to modeling components of real losses (leakage and overflows) was developed by Allan Lambert. Recognizing that the annual volume of real losses is the result of numerous leakage events, each individual volume loss being influenced by flow rate and duration, Lambert considered leakage events in three categories:

1. Background (undetectable) leakage: small flow rate, run continuously
2. Reported breaks: high flow rate, relatively short duration
3. Unreported breaks: moderate flow rates, duration depends on intervention policy

For each separate component of the distribution system—mains, service reservoirs, service connections in the street, service connections after the edge of the street—a value for each component of annual losses can be calculated using the parameters in Table 7.1 for some given standard pressure. The effect of operating at different pressures can then be modeled by applying FAVAD principles to each of the individual components of real losses, using appropriate specific N_1 values.

The BABE annual model was first calibrated and tested successfully using British data in 1993. It was rapidly extended to cover economic analysis to assess the economic frequency of active leakage control interventions, and since then has been used in many countries.

7.6 MODELING COMPONENTS OF REAL LOSSES USING BABE CONCEPTS

TABLE 7.1 Parameters Required for Calculation of Components of Annual Real Losses

Component of Infrastructure	Background (Undetectable) Losses	Reported Breaks	Unreported Breaks
Mains	Length Pressure Min loss rate/mi*	Number/year Pressure Average flow rate* Average duration	Number/year Pressure Average flow rate* Average duration
Service reservoirs	Leakage through structure	Reported overflows: flow rates, duration	Unreported overflows: flow rates, duration
Service connections, main to edge of street	Number Pressure Min loss rate/conn*	Number/year Pressure Average flow rate* Average duration	Number/year Pressure Average flow rate* Average duration
Service connections after edge of street	Length Pressure Min loss rate/mi*	Number/year Pressure Average flow rate* Average duration	Number/year Pressure Average flow rate* Average duration

*At some standard pressure.
Source: IWDC Ltd.

The BABE annual model can be considered a statistical model, in that it does not seek to identify every individual leakage event and calculate an annual loss volume; rather, it groups together similar events and does simplified calculations. The larger the number of events, the better the accuracy of the calculated values, so BABE annual models work quite well with systems of more than 5000 service connections.

The powerful combination of BABE and FAVAD concepts meant that, in the late 1990s, a range of simple spreadsheet models could be developed to approach a number of leakage management problems for individual systems, or a rational and systematic basis. Figure 7.23 shows the range of problems which has been modeled successfully.

7.6.1 Using BABE Modeling Concepts to Prioritize Activities

In many cases we can use a BABE-style statistical model to approximate water loss situations and then rank the losses by type and severity for intervention. In this way we can ensure that we are applying time, effort, and money to the zones or areas of a system which warrant intervention, ensuring a faster, more efficient payback on investment.

There are several commercial versions of these statistical models (and many home-made ones), most of them being extremely user-friendly and flexible. When using a commercial model, operators must first fully understand what it is they wish to perform and ensure that the commercial model has been suitably customized to the local situation. When constructing a model for a spreadsheet it is vitally important that operators fully understand the concepts being used and their limitations.

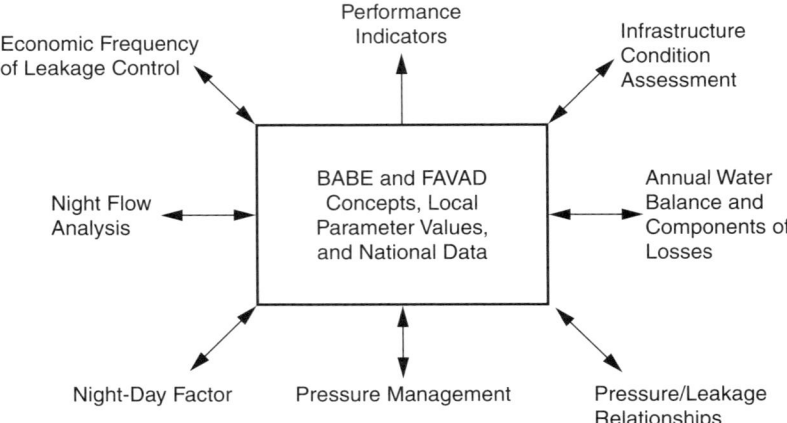

Figure 7.23 Range of problems that have been successfully modeled using BABE and FAVAD concepts. (*Source: IWDC Ltd.*)

In order to arrive at an estimation of the loss situation, most statistical models require:

- Infrastructure and system data
- Coefficients

Infrastructure and System Data

In most cases the field data required for a water loss model are flow data and pressure data. More information can be found about the collection of field data in Chaps. 5 and 6. The BABE and FAVAD approaches to modeling ensure that only a limited amount of specific data needs to be obtained, and it is obviously important to collect the data as accurately as possible to ensure that the estimation of loss is as close to the real situation as possible.

Typical infrastructure and system data needed for BABE and FAVAD models are

- Length of mains
- Volume of service reservoirs/storage tanks
- Number of service connections
- Location of customer meters
- Number of households, population, and consumption
- Number of nonhouseholds and consumption
- Average operating pressure (at night, and 24-h average)
- Numbers or frequencies of different categories of leaks and breaks
- Average duration of each category of leaks and breaks (depending on utility policies for leak detection and repair)

Coefficients

Most statistical models use coefficients and approximations developed by field testing. The coefficients are numbers which are applied to the field data to handle the assumptions on losses. These coefficients are an average of many data sets and usually provide the kind of precision required for loss management. It is important, however, that operators understand the nature of the coefficients, how and why they were applied to the calculation, so that they make any necessary changes for local conditions.

Coefficients may include:

- Typical flow rates of each category of leaks and breaks at some standard pressure (normally 70 psi or 50 m)
- Typical background leakage for mains in good condition (per mile/hour, at some standard pressure; this can be measured in an area where all locatable leaks have just been repaired)
- Typical background leakage for service connections in good condition (per mile/hour, at some standard pressure; this can also be measured as above)
- Typical numbers of residents using toilets at between 3:00 and 4:00 a.m. (or other relevant minimum night flow period)
- Typical toilet flush volume (toilet use is one of the largest residential individual uses, and the most common use of water at night other than in areas where irrigation is being undertaken)
- FAVAD N_1 values for different types of leaks and pipe materials

All the data sets are analyzed together to come up with a conclusion.

As a simple example, a night flow analysis model is used to estimate the amount of leakage in a zone. (This model is often referred to as a *bottom-up model*.) The zone consists of residential properties and no commerce or industry (*infrastructure and system data*).

One of the key factors in this model is to identify estimated legitimate night consumption and subtract it from the night flow. To do this the model makes some assumptions based on preprogrammed coefficients. In our example the model was designed in the United Kingdom but is being applied in the United States.

The model assumes that most of the use at night in a residential zone is from toilet flushes. In our example the model toilet flush volume is 1.5 gal per flush (*coefficient*). However, in the zone in which the model is being applied, the toilets have not been retrofitted and the flush is really 4 gal (*coefficient*).

So the model will ask for the population in the zone and multiply this by the estimated number of people active at night—let's say 6 percent (*coefficient*) during our analysis window of 3:00 to 4:00 a.m.

If the population in our zone is 6000 (*infrastructure and system data*), then the model assumes that 6 percent are active at some time during that period, which means 360 active flushes.

The model then identifies the flush volume from the coefficient and multiplies this by the number of active flushes. In our example this is 360 flushes × 1.5 gal per flush = 540 gal used between 3:00 and 4:00 a.m., which is 540 gal/h or 25 gal/min. However, a better estimate using the correct flush volume is 360 flushes multiplied × 4 gal per flush = 1440 gal used between 3:00 and 4:00 a.m., which is 1440 gal/h or 24 gal/min.

If the measured night flow is 50 gal/min (*field data*), the model then subtracts the estimated legitimate usage and identifies the rest as leakage. If the coefficients were applied incorrectly as shown above, the model would identify the example zone as having 41 gal/min of leakage, whereas really it would have only 26 gal/min of leakage. (It is important to ensure that the model being used reflects the conditions for which it is being used. Technology is usually transferable from one place to another, but some changes or adaptations are often necessary.

The 26 gal/min of assessed leakage could then be compared with the assessed background (undetectable) leakage; this calculation would require length of mains (30 mi), number of service connections (2500), and average operating pressure (70 psi), together with appropriate coefficients, as follows:

- Mains: 30 mi @ 8.5 gal/mi/h = 255 gal/h = 4.2 gal/min
- Service connections: 2500 @ 0.46 gal/mi/h = 1150 gal/h = 19.1 gal/min

The background (undetectable leakage) is therefore likely to be around 23 gal/min if the infrastructure is in good condition, so there may be around 3 gal/min of detectable leakage in this system. The difference in the coefficients applied make a big difference in this example as to the final ranking of that zone for intervention.

In addition to using commercial models, many operators develop their own task-specific models in spreadsheets. Spreadsheets are an excellent medium for developing models, with very little programming skill being necessary. The following model is a simple spreadsheet model developed to identify recoverable leakage in any particular zone. The model uses coefficients and data from the BABE principles.

Figure 7.24 shows an ABC zone with a measured minimum night flow of 40 m^3/h. The other columns show the breakdown of both background data and calculated consumption and background leakage of 21.84 m^3/h. The remaining 18.15 m^3/h in this case is being estimated as potentially recoverable leakage.

Figure 7.25 shows the same zone but with an entirely different assumption for the residents' active coefficient. In this case the potentially recoverable leakage is reduced to 3.15 m^3/h, making the ABC zone much less desirable for leak detection intervention.

> **W**hen using a model from another region or country, always ensure that the concepts and coefficients are applicable to your system.

7.7 MORE DETAILED COMMERCIAL MODELS

The example in the previous section was quite a simplified model, which could be copied and designed for use by most reasonably competent spreadsheet users. The following examples come from a much more in-depth model, also programmed in a spreadsheet but sold and supported on a commercial basis.

This particular model was designed to break down the components of leakage into fixed areas and variable areas, and then analyze the impact of pressure management on these two segments. In addition, the model takes into account the effects of headloss in the distribution system and the effects of pressure on consumption. The last part of the model shows a cost-to-benefit scenario. Figure 7.26 shows the concept on which the model is based.

We will show three examples for which the model could be used; obviously, many other scenarios could also be studied.

7.7 MORE DETAILED COMMERCIAL MODELS 141

Night flow analysis model

Parameter values used in the spreadsheet are as follows:
Background losses at 50-m AZNP are: Mains L/km/h 20   Assessed customer night use
                                       Underground service L/service/h 1.75   Residents % active 6.00 10.00
N1 1.15   Plumbing in property L/prop/h 0.25   Nonh'hold % active 33.00 24.00

Sector	Average zone night pressure (AZNP), m	Length of mains (L), m	Number of underground service pipes (S)	Number of billed properties (N)	Number of persons resident in DMA (R)	Percentage of non-households	Average length of mains per property (L/N), m	Assessed customer night use (NFCUA), m³/h	Sum of exceptional customer night users (>500L/h) (NFCUE), m³/h	Estimated background minimum night flow (BMNF), m³/h	Estimated leakage running presently in sector, m³/h	Measured minimum night flow, m³/h
ABC	60	33000	3345	3345	12500	0	10	7.50	5	21.84	18.15825	40.00

Figure 7.24 Night flow analysis model. (*Source: Julian Thornton.*)

Night flow analysis model

Parameter values used in the spreadsheet are as follows:
Background losses at 50-m AZNP are: Mains L/km/h 20   Assessed customer night use
                                       Underground service L/service/h 1.75   Residents % active 6.00 30.00
N1 1.15   Plumbing in property L/prop/h 0.25   Nonh'hold % active 33.00 24.00

Sector	Average zone night pressure (AZNP), m	Length of mains (L), m	Number of underground service pipes (S)	Number of billed properties (N)	Number of persons resident in DMA (R)	Percentage of non-households	Average length of mains per property (L/N), m	Assessed customer night use (NFCUA), m³/h	Sum of exceptional customer night users (>500L/h) (NFCUE), m³/h	Estimated background minimum night flow (BMNF), m³/h	Estimated leakage running presently in sector, m³/h	Measured minimum night flow, m³/h
ABC	60	33000	3345	3345	12500	0	10	22.50	5	36.84	3.15825	40.00

Figure 7.25 Different assumption for residents active. (*Source: Julian Thornton.*)

❑ Scenario One shows demand-based control to 25-m constant pressure at the most critical node in the system, in this case the highest point with the least pressure. Here, some water demand conservation can be tolerated.

❑ Scenario Two shows a more finely tuned control, which could be used for systems where conservation is not desired.

❑ Scenario Three shows an extreme control scenario, which may be used during drought conditions or in systems where rotational supply is in effect due to lack of water resources.

It should be noted that all data are fictitious and the relationships between fixed and variable areas are not indicative of relationships the operator may encounter in the field.

7.7.1 Scenario One

The base data for all of the scenarios are shown in Fig. 7.27. It can be seen that in the more detailed model 24 h logged field data are input as the basis for modeled control scenarios. This zone has two large consumers who consume water 24 h per day. These are broken out, and data for these consumers are measured in the field and input into the model

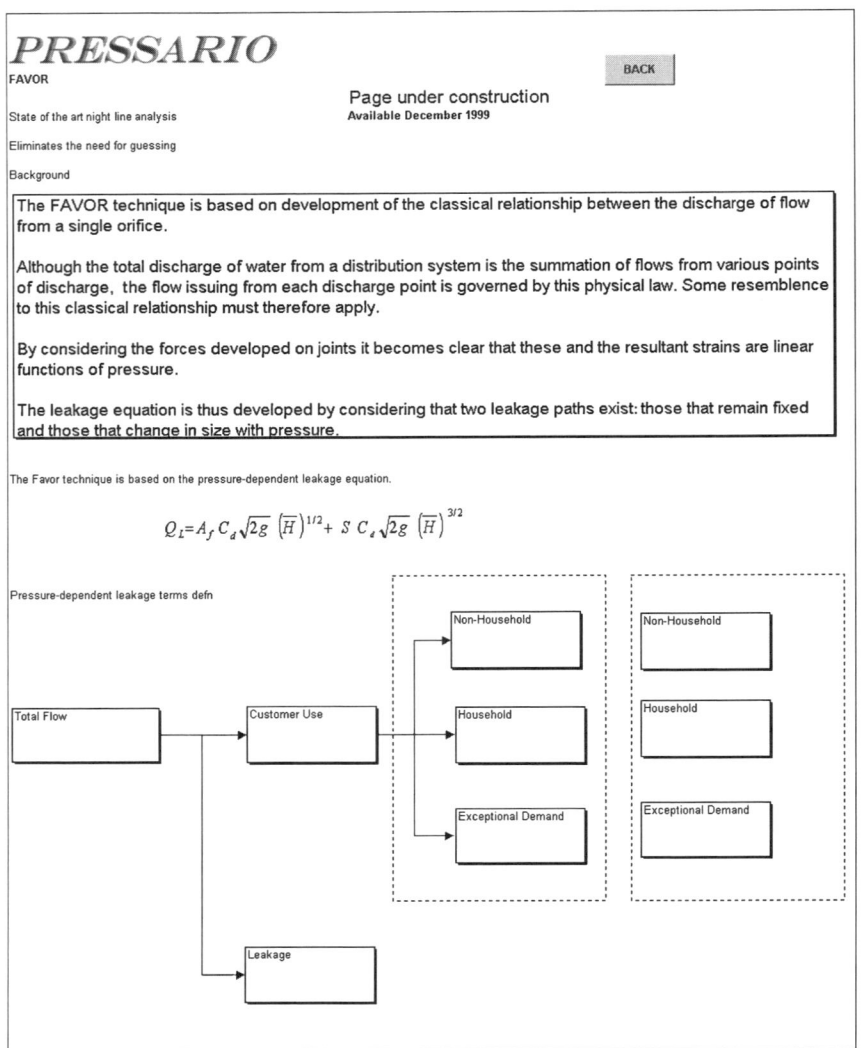

Figure 7.26 Model concept. (*Source: Pressario by John H. May.*)

separately. Figure 7.28 shows where these data are input. Next the components of usage are estimated and input as seen in Fig. 7.29. This gives estimated night consumption for other consumer groups who have less individual impact on the minimum night flow at which leakage is assessed.

In order to identify the pressure leakage relationships, a field test is undertaken to identify the change in flows as pressure is dropped. Data from this field test can be seen in Fig. 7.30. The model predicts the overall areas of leakage in the left-hand corner.

The model shows a graphical breakdown of consumption and fixed and variable area leakage in both area and flow terms. In this case the model predicts a high probability of bursts or breaks on plastic mains. Figure 7.31 shows details.

The next screen is where the modeler will input desired control scenarios to see the impact on both leakage and consumption within the area. Various types of potential control mechanisms are available to the modeler,

7.7 MORE DETAILED COMMERCIAL MODELS

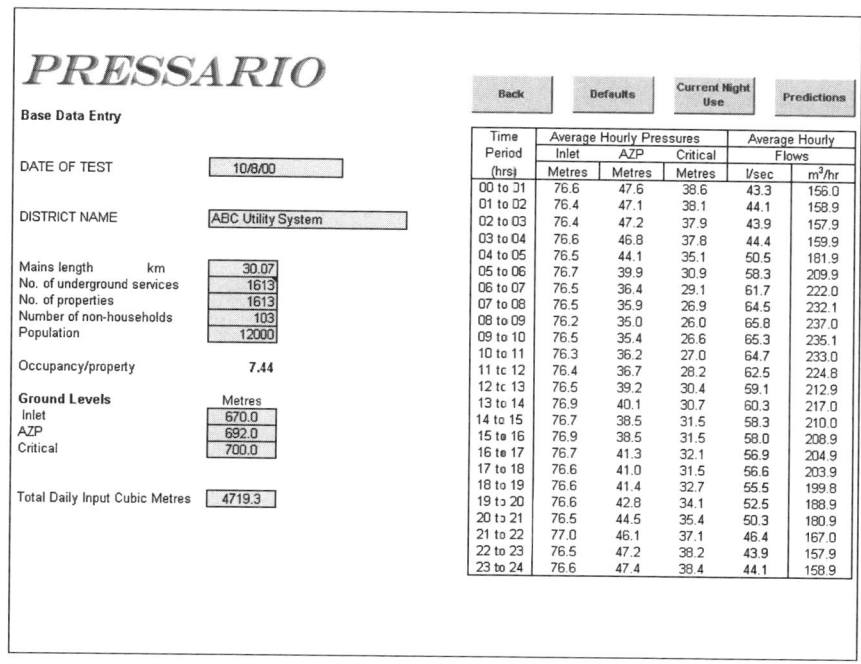

Figure 7.27 Base data. (*Source: Pressario by John H. May.*)

Figure 7.28 Large consumer data. (*Source: Pressario by John H. May.*)

including fixed outlets, time, and various types of demand based modulation. Figure 7.32 shows the input screen. It can be seen that in this scenario headloss has been taken into account and also pressure-dependent demand.

Figure 7.33 shows the pressure-reduction valve control required to affect the control scenario requested. This screen may be used to identify the parameters for controller programming. Figure 7.34 shows the difference before and after control. In this case there is more impact on night flows than on day flows, although some reduction in peak consumption can be seen.

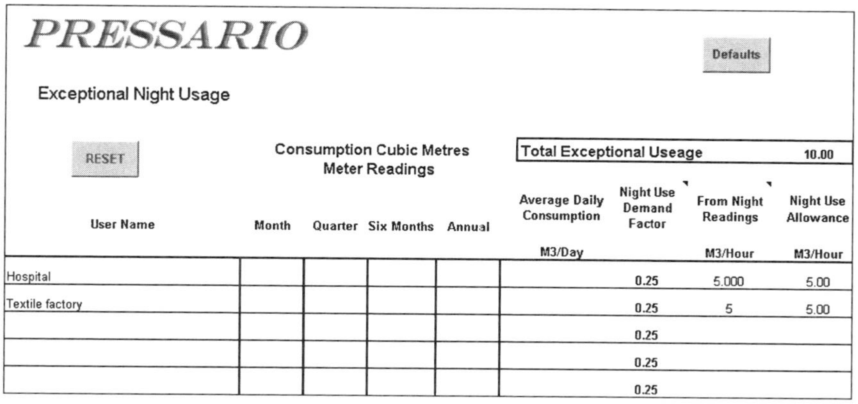

Figure 7.35 shows the screen in which the pressure-dependent demand is defined. This screen is extremely important, as this will define the predicted reduction in metered consumption.

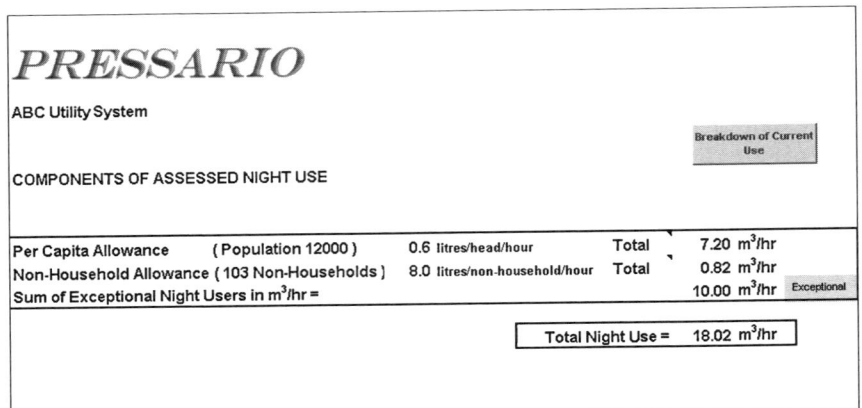

Figure 7.29 Components of usage. (*Source: Pressario by John H. May.*)

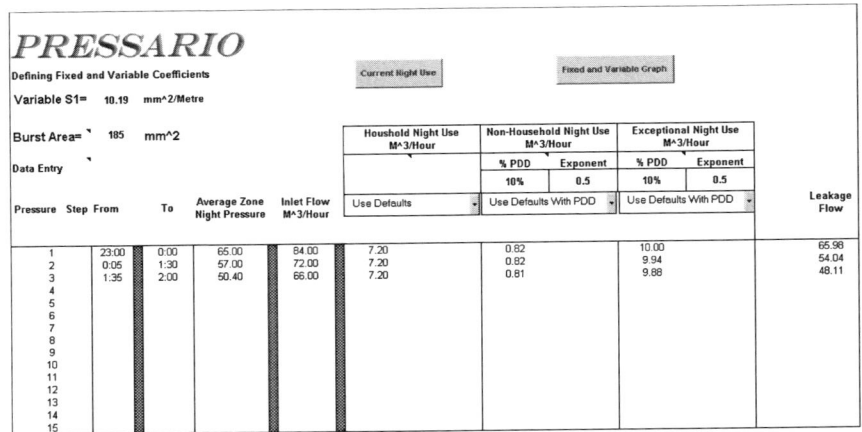

Figure 7.30 Field test data are input to determine pressure–leakage relationship. (*Source: Pressario by John H. May.*)

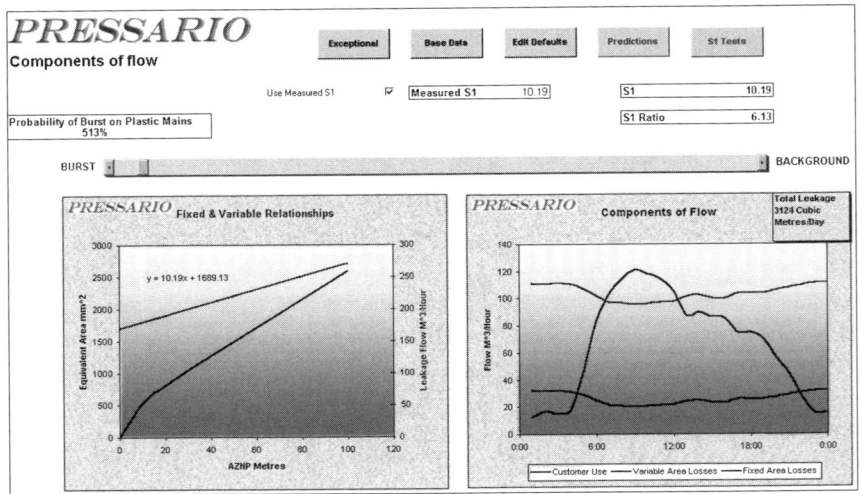

Figure 7.31 Graphical breakdown of consumption and fixed and variable area leakage in both area and flow terms. (*Source: Pressario by John H. May.*)

7.7 MORE DETAILED COMMERCIAL MODELS

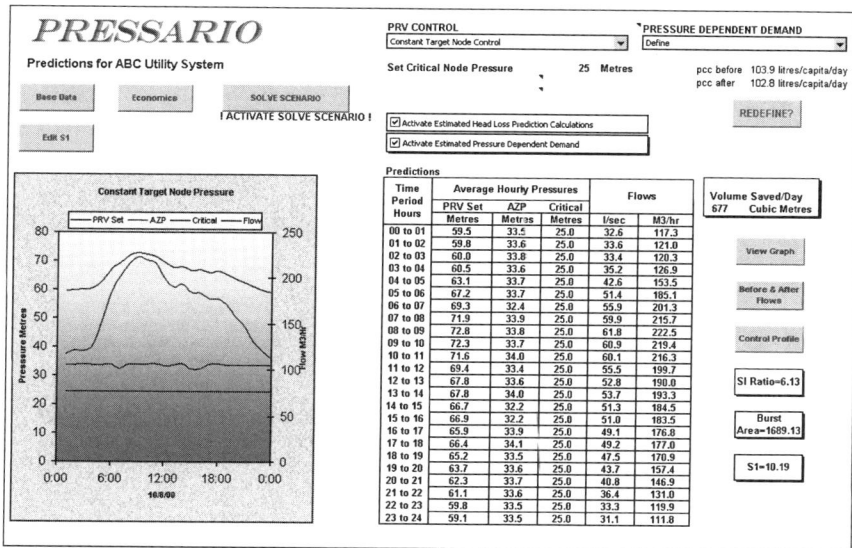

Figure 7.32 Input screen for desired control scenarios. (*Source: Pressario by John H. May.*)

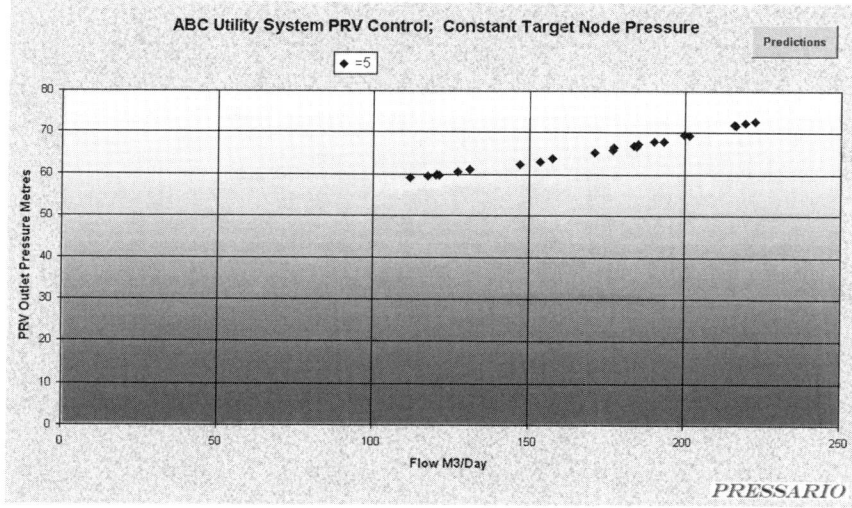

Figure 7.33 PRV control required to effect the control scenario requested. (*Source: Pressario by John H. May.*)

Finally, Fig. 7.36 shows the economics of the control scenario modeled. The base data are shown on the left and the results on the right. It can be seen that total daily input has been reduced by 15.7 percent, mainly by leakage, although some consumption reduction has occurred. The economic breakdown shows the impact of each reduction, including estimates for reduction in new break repairs due to better pressure management. The annual leakage savings are shown as a positive benefit, and loss of revenue due to reduced consumption is shown as a project cost. The payback in this case is 3.5 months, although the cost of water in this system is high.

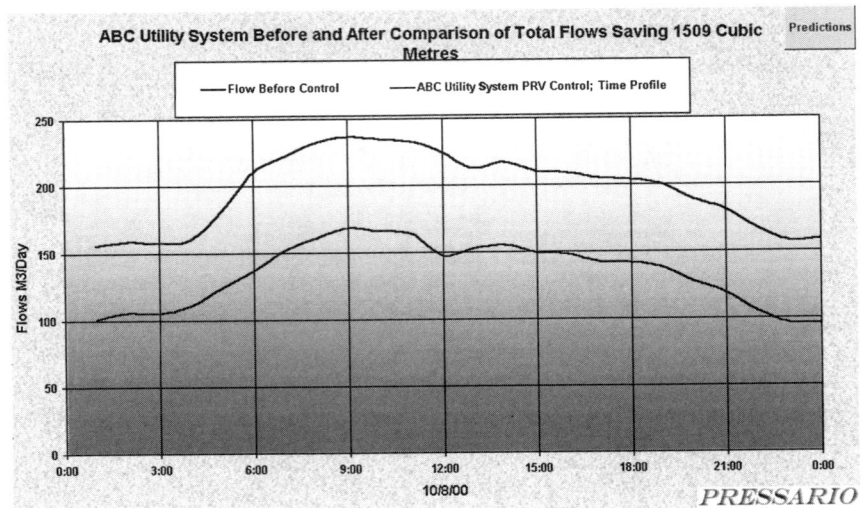

Figure 7.34 Resultant difference before and after control. (*Source: Pressario by John H. May.*)

PRESSARIO

Accounting for pressure dependent demand

Activate After Data Entry

User Guide

Time Period				Household Use			Non Household Use			Exceptional Use		
(hrs)	Inlet Pressure	Total Flow	Customer Use	Total	Pressure Dependent Percentage	Pressure dependent component	Total	Pressure Dependent Percentage	Pressure dependent component	Total	Pressure Dependent Percentage	Pressure dependent component
	Metres	M3/hr	M3/hr		Exponent	**0.5**		Exponent	**0.5**		Exponent	**0.5**
00 to 01	76.642	156.0	12.6	2.57	15%	0.38	0.00	15%	0.00	10.00	15%	1.50
01 to 02	76.435	158.9	16.5	6.47	15%	0.97	0.00	15%	0.00	10.00	15%	1.50
02 to 03	76.435	157.9	15.2	5.16	15%	0.77	0.00	15%	0.00	10.00	15%	1.50
03 to 04	76.59	159.9	18.0	**7.20**	15%	1.08	**0.82**	15%	0.12	**10.00**	15%	1.50
04 to 05	76.487	181.9	45.9	30.89	15%	4.63	0.00	15%	0.00	15.00	15%	2.25
05 to 06	76.694	209.9	83.2	63.22	15%	9.48	0.00	15%	0.00	20.00	15%	3.00
06 to 07	76.538	222.0	103.2	83.17	15%	12.48	0.00	15%	0.00	20.00	15%	3.00
07 to 08	76.487	232.1	114.2	94.23	15%	14.14	0.00	15%	0.00	20.00	15%	3.00
08 to 09	76.176	237.0	121.2	101.20	15%	15.18	0.00	15%	0.00	20.00	15%	3.00
09 to 10	76.487	235.1	118.3	98.31	15%	14.75	0.00	15%	0.00	20.00	15%	3.00
10 to 11	76.28	233.0	114.4	94.45	15%	14.17	0.00	15%	0.00	20.00	15%	3.00
11 to 12	76.435	224.8	105.1	85.14	15%	12.77	0.00	15%	0.00	20.00	15%	3.00
12 to 13	76.487	212.9	87.7	74.68	15%	11.20	0.00	15%	0.00	13.00	15%	1.95
13 to 14	76.9	217.0	90.0	75.96	15%	11.39	0.00	15%	0.00	14.00	15%	2.10
14 to 15	76.694	210.0	86.4	71.43	15%	10.71	0.00	15%	0.00	15.00	15%	2.25
15 to 16	76.9	208.9	85.3	70.32	15%	10.55	0.00	15%	0.00	15.00	15%	2.25
16 to 17	76.694	204.9	75.0	60.01	15%	9.00	0.00	15%	0.00	15.00	15%	2.25
17 to 18	76.59	203.9	74.8	60.76	15%	9.11	0.00	15%	0.00	14.00	15%	2.10
18 to 19	76.642	199.8	69.9	56.89	15%	8.53	0.00	15%	0.00	13.00	15%	1.95
19 to 20	76.59	188.9	55.8	43.79	15%	6.57	0.00	15%	0.00	12.00	15%	1.80
20 to 21	76.487	180.9	44.3	33.29	15%	4.99	0.00	15%	0.00	11.00	15%	1.65
21 to 22	77.004	167.0	26.8	16.85	15%	2.53	0.00	15%	0.00	10.00	15%	1.50
22 to 23	76.487	157.9	15.3	5.27	15%	0.79	0.00	15%	0.00	10.00	15%	1.50
23 to 24	76.642	158.9	15.8	5.79	15%	0.87	0.00	15%	0.00	10.00	15%	1.50
		Totals cubic metres		1247		187	1			347		
		Litres/capita/day		103.9								

Figure 7.35 Defining pressure-dependent demand. (*Source: Pressario by John H. May.*)

7.7.2 Scenario Two

In Scenario Two we simulate a model for a system which does not consider conservation of demand a benefit, but rather just wishes to reduce leakage. Figure 7.37 shows a control scenario which has been entered to attempt to bring pressure down at night and return it to normal levels during peak demand, to ensure that consumption stays at normal levels.

Figure 7.38 shows the flows before and after this scenario. It can be seen that peak flows are maintained while night flows are cut drastically, with a larger impact on leakage and a lesser impact on consumption.

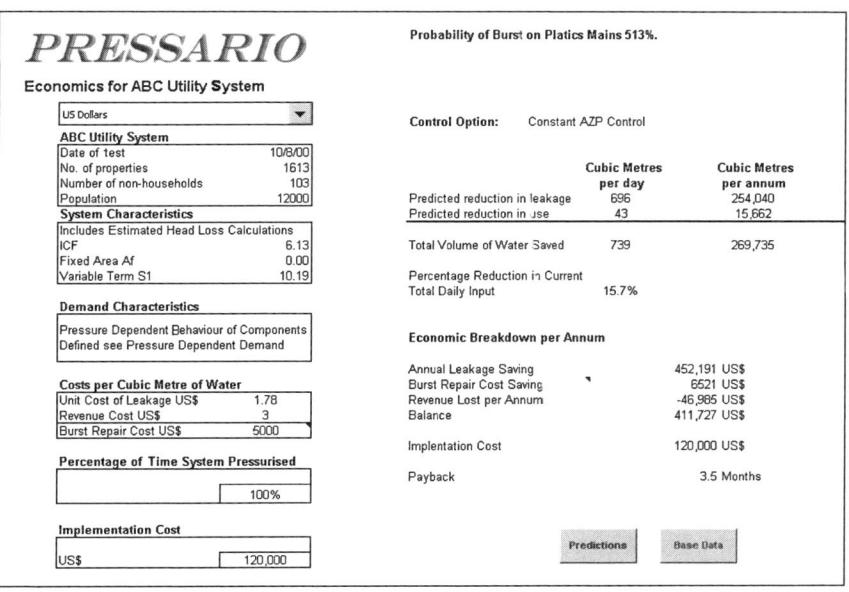

Figure 7.36 Economics for scenario modeled. (*Source: Pressario by John H. May.*)

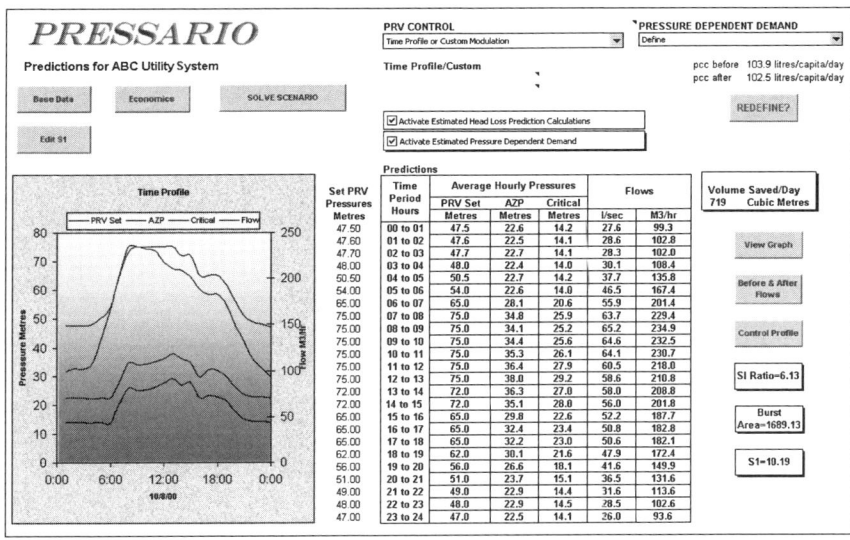

Figure 7.37 New control scenario entered. (*Source: Pressario by John H. May.*)

The economics for this scenario as shown in Fig. 7.39 change significantly, with leakage being reduced slightly less (as pressures are higher during the day) and consumption being reduced by significantly less. The payback on this scenario is slightly better than for Scenario One, or 3.4 months.

7.7.3 Scenario Three

Scenario Three reflects either an emergency drought situation or a system where there is not enough supply to meet demand and the system has to be put into rotational supply conditions.

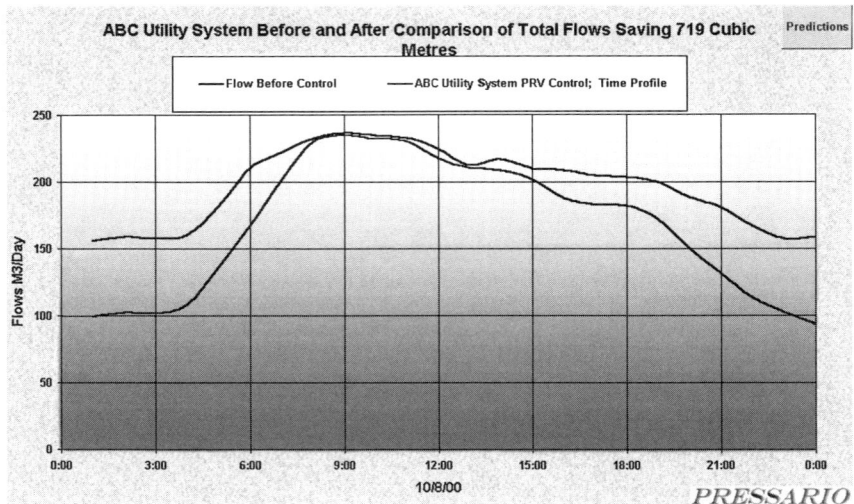

Figure 7.38 Flows before and after this new scenario. (*Source: Pressario by John H. May.*)

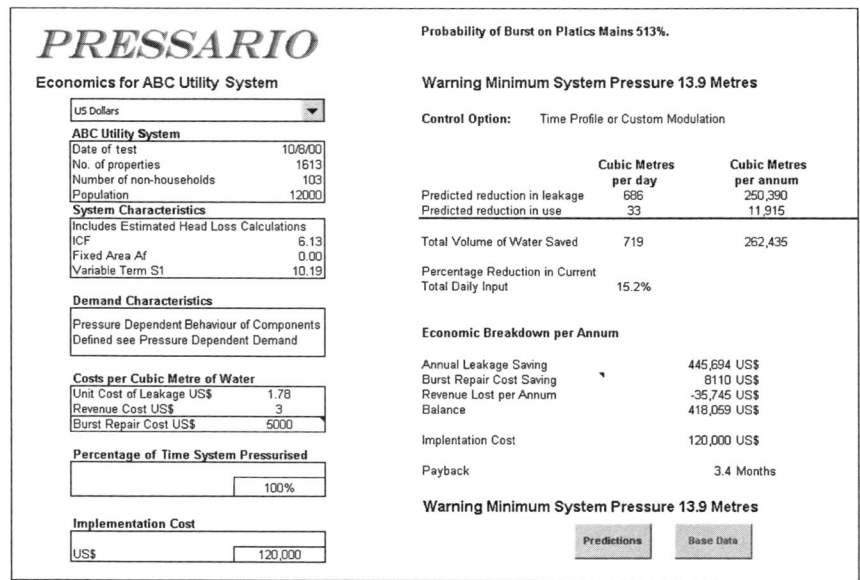

Figure 7.39 New economic output. (*Source: Pressario by John H. May.*)

Rotational supply conditions can be devastating on pipe condition, as the pipe is constantly charging and discharging, causing high levels of new leakage. In addition to problems with high maintenance costs, rotational supply is also dangerous, as water quality can suffer greatly, especially in situations where sewage is leaking close to water pipes. As the pipe is taken out of service and pressure drops, certain sections can enter vacuum conditions and suck in unwanted contamination, which can even be fatal to consumers. By rotating the system but maintaining minimum pressure in the system, unwanted contaminants cannot enter the system.

Figure 7.40 shows an example of a modeled control scenario where pressures are built up at night, perhaps to allow storage to fill. During

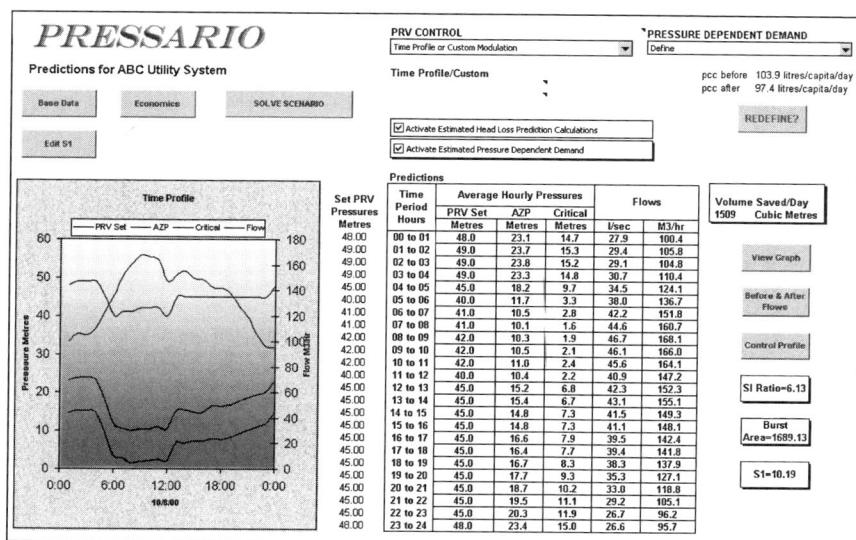

Figure 7.40 This example shows pressure building at night to fill storage. (*Source: Pressario by John H. May.*)

peak conditions pressures are seriously curtailed into this zone, to allow water to be transported to other zones during peak demand.

Figure 7.41 shows more savings during the peak than at night, which would be expected from this type of control profile.

Figure 7.42 shows the economics of this control mode. The system demand has been reduced by 32 percent, with most of the impact still being on leakage.

The three scenarios shown are fictitious but serve to show some of the ways a detailed model can assist the utility in predicting benefits from various potential types of control, before implementation. Benefits are not only economic, but can also accrue in improved service and water quality. Models can also be used to predict the best way to control a system under emergency conditions such as drought or after an earthquake, when severe loss of water results in low pressures or empty pipes.

7.8 SUMMARY

In this chapter we have shown models covering a variety of different tasks which make up a water loss control program:

- ❑ Auditing and cost-to-benefit analysis (top-down analysis)
- ❑ Meter testing and sizing
- ❑ Demand analysis
- ❑ Leakage analysis
- ❑ Pressure management

Some of the models shown were simple home-made models, and others were professionally supported models. In all cases the important factor is the validity of the data being used and the knowledge of the person doing the modeling. In all cases it is necessary that operators understand

150 CHAPTER 7 MODELING WATER LOSSES

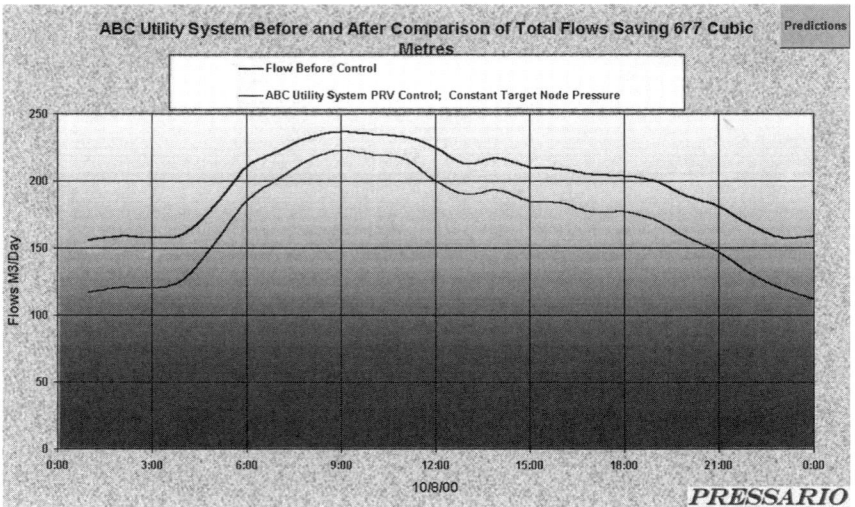

Figure 7.41 More savings during peak than at night. (*Source: Pressario by John H. May.*)

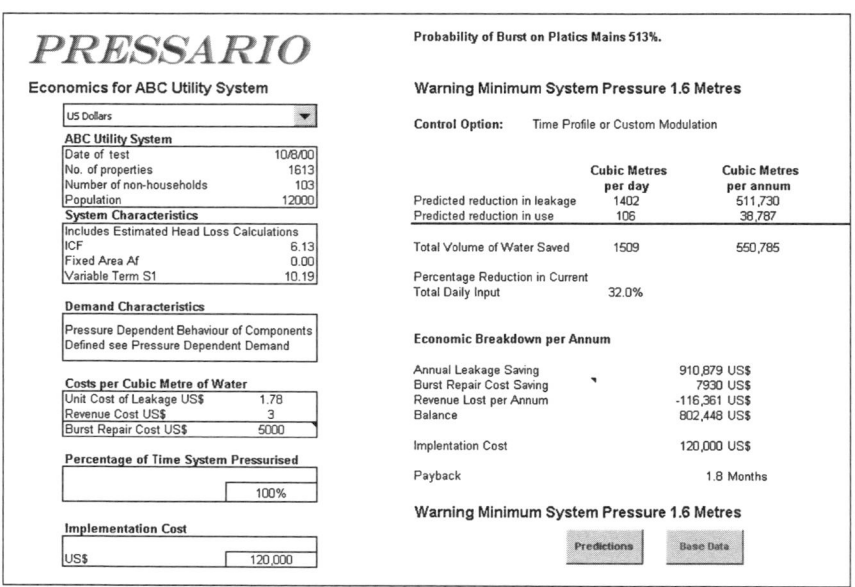

Figure 7.42 New economic output. (*Source: Pressario by John H. May.*)

> When modeling, it is important to be accountable. Always mark clearly all assumptions and estimations along with the model goals and outputs.

the limitations of the models and the data that they are using and what impact that may have on final decisions for intervention, budget allocation, and team resource.

As noted in Chap. 4, accountability is a key factor. If good data are not available, then estimates can be used when modeling, but it is important to note any estimates carefully so that others can interpret the results properly.

7.9 REFERENCES

1. American Water Works Association, Manual M36, *Water Audits and Leak Detection,* 2d ed., AWWA, 1999.
2. *Residential End Uses of Water,* AWWA Research Foundation, 1999.
3. *Managing Leakage* (full set of reports), Water Research Centre, England, 1994.
4. Lambert, A. O., Myers, S., and Trow. S., *Managing Water Leakage: Economic and Technical Issues,* Financial Times Energy Publications, 1998.

CHAPTER

8

Completing the Audit and Determining Cost to Benefit

Julian Thornton

8.1 INTRODUCTION

Once the field measurements have been made, the data recorded and checked, downloaded, the entries made into our selected audit sheet or model, and the draft results output, we need to consider the validity of the results.

As with any audit in any industry, we need to make a certain number of estimations. These estimations may be about the accuracy of field data, missing system data, or the coefficients we use in our models. Estimations are usually made using some or all of the following:

- ❏ Modeled estimations
- ❏ Estimations made from good operator knowledge
- ❏ Estimations made from common sense
- ❏ Estimations made from data sets recorded in other areas or systems

As we go through the auditing, modeling, and estimating process we must record all assumptions and estimations made and the reason for making them. When we have finished the audit we must select the key criteria (the ones that will make the most substantial difference to the results) to check.

8.2 CONFIDENCE FACTORS

Confidence factors can be assigned to measurements made and to estimated data. These confidence factors can then be used to assign a bandwidth of possible real results on either side of the estimated result. Confidence factors can be assigned either as a result of detailed statistical calculations or based on operator gut feel. Either is valid as long as they are recorded with the reasons for assignment.

Original model run				
Stopped meters 0 flow bracket	41 %		Volume increase	
Meter underread low flow	9.5 %	Stopped meters	15.47% %	
Meter underread low/med flow	23 %		9384350	gallons
Meter underread from USDIST	2 %	Wear on high volume	2.06% %	
			1251300	gallons
		Underread from USDIST	2.35% %	
Meter wear medium flow	4 %		1426233	gallons
Meter wear medium flow	4 %	Total increase	19.88% %	
Meter wear medium flow	3 %		12061883	gallons
Meter wear high flow	2 %			
Meter wear high flow	2 %			

High-end estimate	High	Original			
Stopped meters 0 flow bracket	90	41 %		Volume increase	
Meter underread low flow	40	9.5 %	Stopped meters	15.47% %	
Meter underread low/med flow	10	23 %		9384350	gallons
Meter underread from USDIST	4	2 %	Wear on high volume	2.06% %	
				1251300	gallons
			Underread from USDIST	2.35% %	
Meter wear medium flow	7.5	4 %		1426233	gallons
Meter wear medium flow	10	4 %	Total increase	19.88% %	
Meter wear medium flow	12.5	3 %		12061883	gallons
Meter wear high flow	15	2 %			
Meter wear high flow	20	2 %			

Low-end estimate	Low	Original			
Stopped meters 0 flow bracket	40	41 %		Volume increase	
Meter underread low flow	20	9.5 %	Stopped meters	15.47% %	
Meter underread low/med flow	3	23 %		9384350	gallons
Meter underread from USDIST	0	2 %	Wear on high volume	2.06% %	
				1251300	gallons
			Underread from USDIST	2.35% %	
Meter wear medium flow	2	4 %		1426233	gallons
Meter wear medium flow	5	4 %	Total increase	19.88% %	
Meter wear medium flow	7.5	3 %		12061883	gallons
Meter wear high flow	10	2 %			
Meter wear high flow	12.5	2 %			

Addidtional comments: This sensitivity analysis was undertaken for XYZ Utility using data taken from national sources. The high and low levels were selected using operator knowledge and gut feel.

Figure 8.1 Sample sensitivity calculation. (*Source: Julian Thornton.*)

8.3 SENSITIVITY ANALYSIS

Once we have decided which are the key elements in our process and the confidence that we have in the estimations or measurements that we made during the data compilation, we may undertake a sensitivity analysis on these key elements.

A sensitivity analysis need not be a complicated procedure and really is just a matter of taking the best- and worst-case scenario for each of the key elements and seeing what the effect is on the overall audit results.

The compound result of all of the best- and worst-case scenarios is the final bandwidth of results which can be used to make the cost-to-benefit analysis, either on a particular task or on a complete project. Figure 8.1 shows a sample sensitivity calculation.

> Sensitivity analysis gives us an idea of the best- and worst-case scenarios. Our real answer will lie somewhere between these.

8.4 THE ECONOMICS OF LOST WATER

8.4.1 Water Prices and Usage Figures

Water prices vary greatly throughout the world and usually are a function of

- ❏ Availability
- ❏ Cost of production or import

TABLE 8.1 Water Prices and Usage around the World

Region	Access to Water within <200 m	Average Consumption (liters per capita per day)	Average Cost (US$/m^3; 1993 data)
Africa	69.08%	53.56	1.30
Arab countries	88.23%	157.86	0.65
Asia	87.46%	160.7	0.54
Industrialized countries	99.60%	262.34	2.24
Latin America	86.87%	182.79	0.91
Transitional countries	99.08%	306.59	0.41
All	84.38%	161.33	1.08

❑ Infrastructure cost
❑ The strength of the demand for water
❑ Environmental impact of withdrawal
❑ Social makeup
❑ Political incentives
❑ Awareness of future requirements

Table 8.1 shows a sample span of prices throughout the world.

Water usage figures also vary widely, as can also be seen in Table 8.1. Water usage varies with similar characteristics as those listed above. Obviously, the availability of water, the cost to transport it and deliver it, and the level of demand affect the cost of the water, and the cost of the water will affect the final cost-to-benefit analysis for any given scenario.

8.4.2 Calculating Costs and Benefits for Economic Analysis

In order to identify appropriate intervention methods and rank both the methods of intervention to be used and zones or areas of a water system for intervention, we must understand not only the physical volumes of loss and the potential recovery in each of the categories of intervention identified but also their economic impact on the bottom line. In order to do this we must consider a number of factors, including:

❑ Whether the system is pressurized 365 days a year, 24 hours a day.
❑ Whether the system is subjected to rotational or uncontrolled water shutoff. If so, then real losses are often valued at the resale value of water until such time as the rotational supply situation is resolved.
❑ Whether the system has sufficient water resources and infrastructure to meet system growth requirements or whether new sources of water will be required. If so, what will be the cost of the new sources?
❑ Whether the system has available water resources close by in the case of expansion and whether the system can draw more water from those resources without negative legal or environmental impact.
❑ The true marginal cost of the production or purchase of water. (In the case of a production plant, the fixed costs will remain current even if the water losses are reduced. In the case of water

purchase, the supplier may have minimum operating levels of finance and be forced to increase the unit cost of water provided if volumes purchased drop.)
❑ The true sales cost by category of consumer for water sold.
❑ The impact of sewerage charges, which are a function of the metered water.
❑ The cost of the intervention.

As an example, consider a water supply system that is not able to supply water 24 hours a day 365 days per year, in all supply sectors. (Believe it or not, this is quite common and most definitely not just a Third World problem.) Some supply sectors are therefore subjected to rotational supply. The rotational supply is put into effect on a manual basis as and when local reservoirs need to be refilled. There are three main types of negative impact from this situation:

❑ Water quality and public health problems
❑ Poor public image
❑ Reduced revenue, increased costs for maintenance and new construction

It is hard to quantify the dollar value of increasing the water quality or improving public image, although both have definite value. In some cases lives have been lost due to poor water quality. However, we can quantify the benefits of resolving the third impact in the following areas:

❑ Reduced leakage volume and frequency at marginal or variable cost
❑ Increased revenue at sales cost
❑ Reduced maintenance costs for leak repair as the leak frequency is reduced and reinstatement costs as the number of excavations is reduced
❑ Deferred expenditure for capital expansion such as new reservoirs, treatment facilities, transmission pipes, etc., to eventually meet the needs of the people who are affected by the rotation

If the system has a mixture of sectors which are 100% supplied and partially supplied by a rotation system, it is necessary to develop and understand a formula to calculate the real financial benefits to undertaking intervention in either type of zone.

The following discussions itemize some of the considerations to be taken into account when calculating a cost-to-benefit ratio for recovery of water losses.

8.4.3 Considerations for Cost-to-Benefit Calculations in a 100 Percent Supplied System

Cost-to-intervene calculations usually take into account the following:

❑ Engineering costs, EC
❑ Construction costs, CC
❑ Product costs, PC

- Increased maintenance costs, IMC, to maintain the new equipment and software
- Reduced revenue, RR (in the case that pressure management will be used), from direct pressure supply activities such as garden watering, customer-side leaks, etc. (This is a very small percentage in some countries, as most of the use is commercial and residential storage; even in countries where direct use is common, only a portion of that use is actually direct pressure—the rest is filling volumes such as toilet tanks or washing machines.)

Benefit calculations usually take into account the following:

- Reduced volume of losses at marginal cost, RL (marginal cost is the cost to supply the water less the fixed overhead)
- Reduced maintenance costs, RM, from reduced leak frequency (calculated as the cost of a leak repair at various points of the system multiplied by the amount reduced)

The benefits can be considered over any time period, often 12 or 24 months.

Considering these factors, we would use the following formula to derive a cost-to-benefit ratio for potential sectors in a 100 percent supply area: Cost-to-benefit ratio = cost/benefit. Therefore, EC + CC + PC + IMC + RR/(RL per month + RM per month) × 24.

> The American Water Works M36 audit uses a 24-month time period, although actual economic intervention periods can be calculated as discussed later in this chapter.

8.4.4 Considerations for Cost-to-Benefit Calculations in a System with Rotational Supply

Cost calculations in a system with rotational supply usually take into account the following:

- Engineering costs, EC
- Construction costs, CC
- Product costs, PC
- Increased maintenance costs, IMC, to maintain the new equipment and software
- Reduced revenue, RR (when pressure management may be used), from direct pressure supply activities such as garden watering, customer-side leaks, etc.

The benefit calculations usually take into account the following:

- Reduced losses at marginal cost, RL (marginal cost is the cost to supply the water less fixed overhead)
- Reduced maintenance costs, RM, from reduced leak frequency (calculated as the cost of a leak repair at various points of the system multiplied by the amount reduced)
- Increased revenue at sales cost, IR (the overhead to supply the water is fixed, so the increased revenue should be calculated at retail prices)
- Deferred expenditure, DE, for supplying more water by increasing reservoir or treatment capacity. (This is calculated using the loan cost to borrow the money to construct, multiplied

by the percentage rate applied to the loan, multiplied by the time during which the construction can be deferred before natural increase in demand by demographic growth requires expansion.)

Considering these factors, we would use the following formula to derive a cost-to-benefit ratio for potential sectors in a rotational supply area: Cost-to-benefit ratio = cost/benefit. Therefore, EC + CC + PC + IMC + RR/[(RL per month + RM per month + IR per month + DE per month) × 24] (assuming in this example that the DE portion can be deferred for the complete 24-month period).

By performing simple field and model analysis, we can grade the potential sites to ensure maximum benefit.

8.5 COST-TO-BENEFIT ANALYSIS SHEET

As with data calibration, we need to identify a standard spreadsheet, which can be used to rank and grade our data. Sample sheets can be seen in Chap. 7.

8.5.1 The Economics of Intervention

A nice way of showing cost-to-benefit options and indeed the final bandwidth of results is by using graphs. Figure 8.2 shows a simple graph for estimating the optimal frequency of intervention. In this example the options of intervention over different periods of time were considered. Obviously, the more frequent the intervention, the more costly the program will be, but the unreported leak run time will be reduced. There will be a point at which the program will cost more than the benefit and the trend line will reverse. The optimal point is the point where the total cost curve, representing intervention and cost of loss, is at a minimum.

The total cost method is ideal for utilities that have a regular leak detection program and therefore do not have a backlog of leaks. For those utilities that do not have a regular program it is a good idea to undertake an annual program until the number of unreported leaks reaches a steady state.

Most systems will find that in the first few years of aggressive intervention this point is very hard to reach. The two key parameters in determining an economic intervention frequency are the marginal cost of water and the average number of new unreported leaks per year.

An efficient way of identifying when to intervene is not to preset an acceptable payback period but rather to monitor the rate of rise of new loss. This could, for example, entail monitoring night flow trends in a sector after all locatable leaks have been found. The utility identifies the cost to survey and locate leakage throughout the whole sector; this is the "do something limit" or the intervention level. The utility also identifies the value per volume of water lost, which may change depending on the season or source or may be constant. Then the utility monitors the increasing night flow trend, turning flow data into weekly or monthly volumes of loss.

> The old saying that a picture is worth a thousand words is most certainly still relevant.

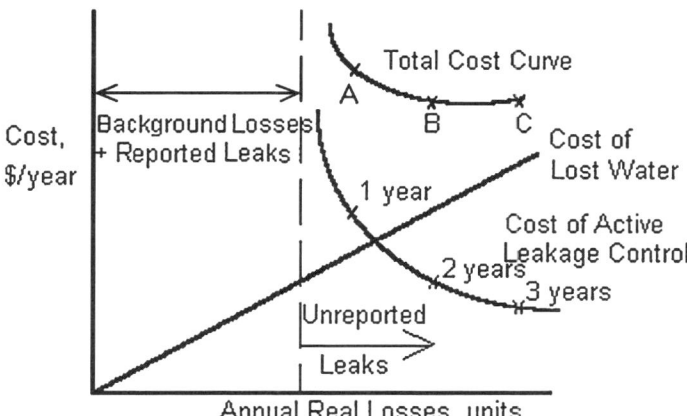

Figure 8.2 Simple cost-to-benefit graph for frequency of intervention. (*Source: Julian Thornton.*)

Once the aggregated value of the loss equals the cost to intervene, this is the most economic time for an intervention. The intervention will be undertaken and the whole process will start again, providing a track record of loss management. As the utility employs different technologies such as pressure reduction, surge control, or service replacement, it will often see a noted reduction in the climb rate of new leakage. This will in turn set a new frequency for intervention.

By constantly monitoring performance and noting reasons for change in rate of rise, the utility will quickly see which programs are the most beneficial and economic for the system.

Readers who wish to research this practice in more detail should consult *Managing Water Leakage: Economic and Technical Issues*, by Lambert, Myers, and Trow.[1]

Checklist

1. Clearly mark an estimation so that other operators can understand the nature of the assumptions in the audit.
2. Identify all elements of project cost.
3. Identify all elements of project benefit:
   - ❏ Monetary
   - ❏ Soft cost
   - ❏ Political
   - ❏ Social
   - ❏ Public health
   - ❏ Public awareness
4. Perform a sensitivity analysis on key elements of the calculations.
5. Display results using an acceptable margin.

8.6 SUMMARY

By this point in the manual we have discussed the need to be aware of the problem of losses, the nature of water loss, and the need for progressive measures of loss management. We have identified how to undertake an audit to define where the problems are and to what extent the problems affect the efficient operation of the system. We have covered data collection in the field and how to store and manipulate it properly. We have discussed using computer models to assist in decision making, and how to check the sensitivity and accuracy of those models. Finally, in this chapter we have discussed cost-to-benefit calculations, which can be attached to the various stages of auditing and modeling.

It is worth mentioning again briefly at this point the need to use the proper performance indicators to track the performance of a utility or technical program within the utility. Chapter 4 discusses this subject, and also the part of Chap. 18 by Lambert, Huntington, and Brown, and the case study by Kunkel discuss this issue in detail.

For further information on the proper use of performance indicators, the reader may wish to consult the IWA's *Performance Indicators for Water Supply Services*.[2] This publication provides performance indicators for all levels of water system management, not just the loss management portion we have been discussing in this manual.

8.7 REFERENCES

1. Lambert, A. O., Myers, S., and Trow, S. *Managing Water Leakage: Economic and Technical Issues*, Financial Times Energy Publications, 1998.
2. IWA, *Performance Indicators for Water Supply Services*, Manual of Best Practice, International Water Association, 2000.

CHAPTER

9

Potential Solutions for Controlling Water Losses

Julian Thornton

9.1 INTRODUCTION

As discussed in Chap. 2, water losses come in two basic forms:

- ❑ Real losses (sometimes referred to as physical losses)
- ❑ Apparent losses (sometimes referred to as nonphysical losses)

In addition to reducing water loss during an optimization project, water system operators often look to reduce energy losses caused by inefficient pumping. Pump scheduling and pump efficiency programs often form part of a water loss control or system optimization project. More information on the various methodologies for reducing pumping-related energy costs can be found in Chap. 18, Article Four. This chapter will briefly review some of the most common methods of reducing losses and leads into in-depth discussions in Chaps. 10–14.

Lambert and Herner[1] published one of the most recent definitions of real and apparent losses in the IWA *Blue Pages,* as follows:

"'Real losses' are physical water losses from the pressurized system, up to the point of customer metering. The volume lost through all types of leaks, bursts [often referred to as breaks], and overflows depends on frequencies, [system pressures,] flow rates, and average durations of individual leaks.

"'Apparent losses' consist of unauthorized consumption (theft or illegal use), and all types of inaccuracies associated with production metering and customer metering. Underregistration of production meters, and overregistration of customer meters, leads to underestimation of real losses. Over-registration of production meters, and underregistration of customer meters, leads to overestimation of real losses."

Each water system will have different types and degrees of loss, and each has a potential solution with an associated cost. Cost to benefit is discussed in more detail in Chap. 8. However, before the cost-to-benefit ratio can be calculated, the potential solutions have to be identified and graded.

In addition to having a good return or cost to benefit, it is also important when considering intervention to consider the local conditions and the sustainability of the method or solution adopted. Water losses don't go away: they keep coming back. Water loss control is not a one-time project; it is a continuous and changing solution to an ever-changing problem.

9.2 SOLUTIONS FOR REAL LOSSES

Some potential solutions for real losses are

- Leak detection to locate nonvisible leakage
- Increased response to visible reported leakage to reduce annual loss volumes
- Zoning to identify volumes of loss in a continuing and efficient manner
- Pressure management to reduce volumes of loss and frequency of new leaks
- Level control to reduce overflows from storage
- Corrosion control to reduce frequency of new leaks
- Mains replacement
- Mains rehabilitation
- Service replacement

There are many reasons for real losses in a water system. They may result from

- Pressure
- Corrosion
- Vibration from traffic loading
- Incorrect backfill
- Poor materials or workmanship
- Lack of periodic maintenance
- Environmentally related, such as cold weather

The intervention methods chosen will depend very much on which factors are contributing to real losses in any particular system. Real loss optimization and methods of intervention are discussed in more detail in Chaps. 10–14. Each of these chapters discusses one of the topics on the four arrows shown in Fig. 9.1.

9.3 SOLUTIONS FOR APPARENT LOSSES

Some potential solutions for apparent losses are

- Production meter testing
- Sales meter testing
- Correct meter sizing
- Correct meter specification (the best meter is not always chosen for the job, in particular in low-flow situations)
- Meter replacement
- Improved meter reading
- Improvements in billing
- Location of illegal or unregistered connections
- Revenue recovery or prepaid systems in areas of low payment

There are also many reasons for apparent losses in a water system. These may include:

9.3 SOLUTIONS FOR APPARENT LOSSES

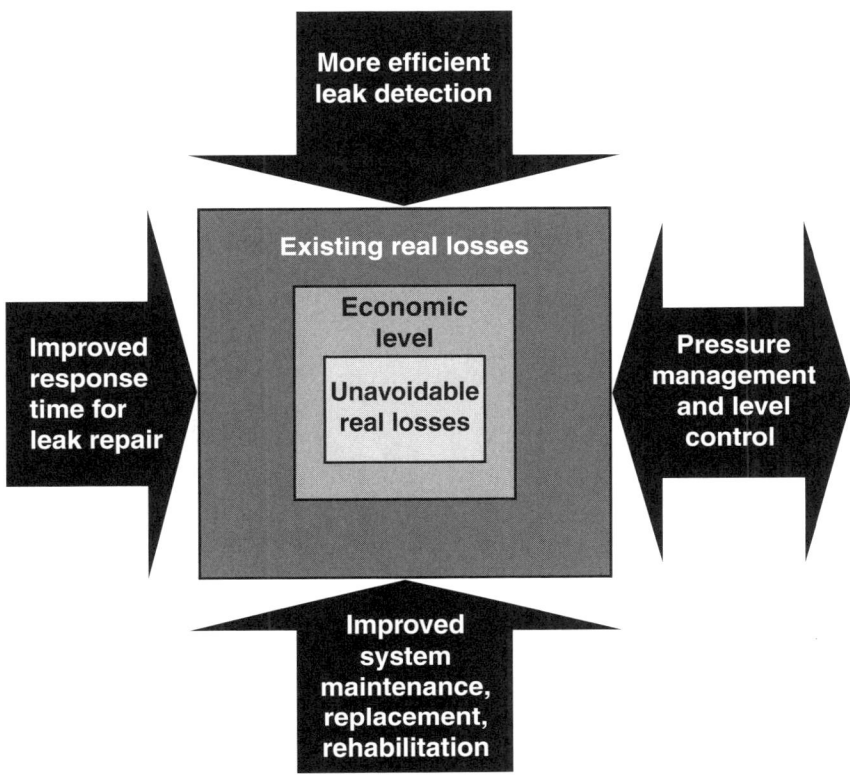

Figure 9.1 Four components of an active real loss management program. (*Source: Julian Thornton.*)

- Water quality affecting water meters
- Environmental conditions such as extreme heat or cold affecting water meters
- Lack of periodic testing and maintenance
- Lack of proper sizing considerations during original installation
- Incorrect installation of water meters
- Out-of-date meter population database
- Uncontrolled population growth
- Theft
- Inefficient reading and billing methods

Again, the methods of resolving the losses will vary greatly according to the reasons for the apparent losses in any given system. The four arrows in Fig. 9.2 show a component breakdown for apparent losses like that for real losses in Fig. 9.1. More details on apparent losses intervention are given in Chap. 14, and in-depth detail on metering subjects is provided in Apps. B and C, where excerpts from American Water Works Association Manuals M6 and M33 can be found.

The following discussion provides a brief overview of some of the most common field intervention methods in use today. Most of the following tasks can be grouped together to form total solutions or can be undertaken as individual, stand-alone programs. (More detailed discussions of field intervention can be found in the following chapters.)

164 CHAPTER 9 POTENTIAL SOLUTIONS FOR CONTROLLING WATER LOSSES

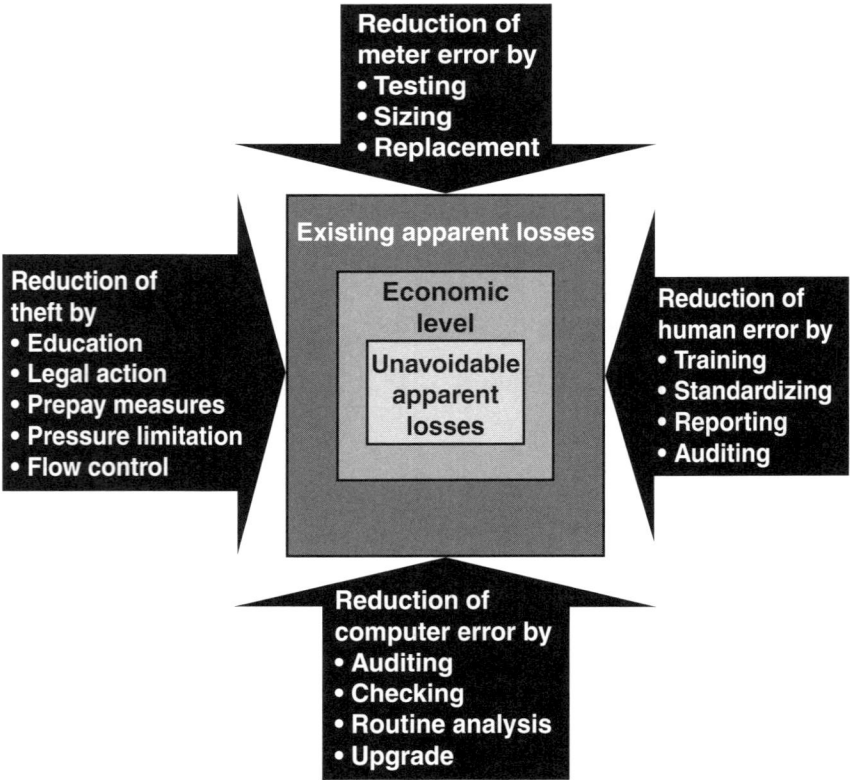

Figure 9.2 Four components of an active apparent loss management program. (*Source: Julian Thornton.*)

9.4 REAL LOSS INTERVENTION

9.4.1 Leak Detection to Locate Nonvisible Leakage

Leak detection surveys can be undertaken on a periodic basis to identify and repair system leakage that does not normally come to the surface and get repaired as part of the routine maintenance tasks performed during the year. This type of service will be repeated frequently, the frequency depending on the economic breakdown of the cost of intervention versus the benefit. Often at the beginning of an aggressive leak reduction program it is necessary to repeat this exercise various times, to take out what are known as backlog leaks. Once the leakage is brought under control, then the correct frequency for intervention can be modeled.

As technology advances, the size of leaks locatable and the speed of survey improve, making previously unavoidable (background) losses economically viable to reduce.

9.4.2 Increased Response to Visible Reported Leakage

In addition to leak survey and repair, it is a good idea to identify the benefit of increasing the levels of service of repair of leaks which do come to the surface and form part of the annual maintenance program. Annual real losses comprise the leak flows multiplied by the time the leaks run. Obviously, if normal repair time can be improved, then the annual losses will also come down. See the case study in Chap. 18 involving SABESP, which discusses some of the benefits of increasing speed of repair.

9.4.3 Zoning to Identify Volumes of Loss in a Continuing Manner

Many water systems run on an open basis and do not have natural zones. In this case it is often beneficial to zone the system, either on a permanent basis or a temporary basis, and measure flows into the zones. Night flows can be analyzed and give a good indication of the level of leakage in the zone. More details on analyzing night flows can be found in Chap. 10. Simple flow measurements on a regular basis can direct the leak location and repair programs mentioned above and ensure that leak-pinpointing crews are used efficiently.

9.4.4 Pressure Management to Reduce Volumes of Loss and Frequency of New Leaks

In addition to zoning for flow analysis, it is a good idea to zone to control pressures. This ensures that system pressures do not get too high, putting unnecessary strain on old joints and connections. Many systems are already controlling pressures with fixed-outlet systems, or geographically controlled zones. However, additional benefit can often be gained by modulating the outlet pressure to ensure good system pressures during the day, when demand is on the system, and lower pressures at night, when in many cases a large percentage of the flow is leakage. In many cases fixed-outlet systems are running "upside down" due to headlosses in the system. By upside down we mean that the pressures within the zone are lower during peak demand due to higher headlosses, and pressures are higher at night due to lower headlosses. This can be rearranged by flow modulation of the outlet pressures and is explained in detail in the pressure management section of Chap. 12.

9.4.5 Level Control to Reduce Overflows from Storage

Many storage reservoirs and tanks have faulty level controls, which can lead to high levels of losses, especially at night when the demand on the system is reduced and the storage naturally starts to fill. Measuring the flows and levels around a tank can easily rectify this situation, and fitting the right valve—usually an altitude or ball valve—will control the levels to the desired shutoff.

9.4.6 Corrosion Control to Reduce Frequency of New Leaks

Another factor which plays an important part in reducing the frequency of new leakage, and therefore maintaining lower annual levels of loss, is corrosion control. Corrosion control is discussed briefly along with some methods of rehabilitation in Chap. 13. However, these subjects are very involved, and other literature should be consulted for more detail.

9.4.7 Mains Replacement

In many cases, because of the cost, mains replacement is one of the last resorts to resolving real losses. Mains replacement is usually expensive and can cause disruption, especially in dense urban environments. However, when leakage reaches a certain frequency it is often the only realistic method of reducing real losses. Mains replacement usually forms part of the annual maintenance program for a water system and is usually scheduled not only because of leak frequency but also to increase carrying capacity.

One point should be made about mains replacement, and this is that great care should be taken when handling and storing the pipe sections prior to installation. Plastic-based pipes should be carefully stored in a cool, dry place, away from sunlight, which can destroy the molecular structure of the pipe, making the pipe brittle and more prone to breaks in the future. Metal-based pipes should be treated with care, ensuring that they are not scratched or dented, as these scratches and dents can form locations for new corrosion.

New methods of trenchless mains replacement can be used to lower the levels of disruption caused by open trench replacement. Some of these methods are discussed briefly in Chap. 13.

9.4.8 Service Replacement

Most operators know that a large part of system real losses occurs at service connections (more so in dense urban environments than in rural systems). Service leakage is often due to corrosion, stray current problems, and vibration or poor materials quality. In most cases service replacement can be undertaken without major disruption to traffic, even in dense urban situations.

9.5 APPARENT LOSS INTERVENTION

9.5.1 Production Meter Testing

The production meter is the first meter which should be tested in the system, as this is the equipment which gives us the bulk supply figures for audits and routine balances. Production meter calibration can have an effect on both real and apparent losses.

9.5.2 Sales Meter Testing

The water meter has often been referred to as the cash register of the water utility, as it is usually the first line of measurement in the billing cycle. To ensure that the utility maximizes revenues and stays accountable for charges made to customers, it is necessary to undertake periodic testing of all meters in the system, both production and supply meters and revenue meters.

Meter testing is often scheduled on the basis of time passed since installation or last test or volume passed since installation or last test. American Water Works Association Manual M6 shows some recommended (Public Service Commission) test frequencies by state. (A brief overview of relevant sections of this manual can be found in App. B.)

- Age
  - Maximum 20 years (California)
  - Minimum 5 years (Colorado, Kentucky, and New York)
  - Average of 22 states listed was 9.36 years
  - The most common period was 10 years (14 of 22 cases)
- Volume
  - Maximum 750,000 (New York, New Jersey)
  - Minimum 100,000 (Alabama, Illinois, Indiana, Maine, Michigan, Pennsylvania)
  - Average 250,000 (of 10 states cited)
  - Most common volume was 100,000 (6 of 10 states cited)

In most cases a testing program will start with testing older meters and those with the heaviest volume, although operators should learn to create financial models to dictate on an economical basis when is the best time to test and change meters for their particular system. Meter testing is discussed in more detail in Chap. 14.

9.5.3 Correct Meter Sizing

Many older utilities have in the past used the rule-of-thumb meter sizing method of either downsizing one size from pipe size or just giving the customer the size of meter he requests. This and downsizing of large plants and facilities has resulted in many meters being oversized.

Meters have greatest difficulty reading low flows, and oversized meters tend to work in the low-flow range most of the time, meaning that they will in most cases underread.

In most cases, meter downsizing programs provide very fast paybacks as well as aiding the utility in updating meter population databases and identifying unmetered connections. A good example of this is shown in the Boston correct meter sizing case study in Chap. 14.

9.5.4 Correct Meter Specification

In addition to testing meters and sizing them correctly, it is important to ensure that the meter is correctly specified for the job it will do. Various recent industry reports and case studies have cited typical residential flows as having between 10 and 20 percent of the flow occurring near the minimum flow capacity of the meter. If this is really the case we should consider using lower-flow meters. Low-flow meters are readily available on the market today. Readers with an interest in this area should consult the ISO standards and compare Class C and Class D meters with what they currently have. Some recent cases have shown huge volumes of water being recouped by changing from standard meters to these types of meters.

9.5.5 Correct Meter Replacement

If testing has been undertaken and meters have been specified for a change-out program, it is important to consider appropriate installation to ensure that the meter will operate properly in a sustainable manner. Most manufacturers provide installation layout diagrams showing the necessary upstream and downstream configurations. Some diagrams can be found in Chap. 14.

When installing new meters it is also important to consider environmental conditions such as extreme heat or cold and ease of reading. If automatic meter reading (AMR) or external readers are desired, it is important to consider electronic registers.

9.5.6 Improved Meter Reading

It is surprising in this day and age of computers and information technology to see how many utilities still use a manual read system. Manual read systems are acceptable if the readings are taken on an efficient and systematic basis; however, many utilities suffer from "estimated read syndrome," at surprisingly high rates. This plays havoc with accountability and loss control and also makes it hard to derive a baseline from which to draw benefit from a loss control program.

If a utility decides to continue to use a manual read system, then controls must be put into place to ensure that readings are never estimated, or at least only in very rare cases. Controls may be external registers.

Alternatively, a utility may decide to automate the reading by installing AMR. AMR has often been hard to justify purely on an improved meter-reading basis, but for a utility that also wants to reduce real losses, an AMR system can bring many other benefits such as block use analysis and pressure monitoring for real loss control exercises such as leak detection and pressure management.

9.5.7 Improvements in Billing

As well as improving meter reading, it is important to make sure that the billing data are properly analyzed and efficiently processed. Meter database and billing database management is as essential as meter testing. It is very important to separate metered volume from billed volume.

9.5.8 Location of Illegal or Unregistered Connections

It is surprising how many connections go unbilled for one reason or another. In some areas customers may actively bypass or steal water through illegal connections or bypassed meters. In other cases installations get overlooked, especially in the case of commercial or industrial properties with multiple inlets. These types of connections are often located during leak detection exercises when operators sound each service connection. These operators can be trained to also check the legitimacy of metered connections at the same time.

9.5.9 Revenue Recovery or Prepaid Systems in Areas of Low Payment

Some utilities suffer from poor payment histories in certain areas. In many cases the utility is not allowed to shut off the water completely and is mandated to provide a minimum to sustain life. However, the utility can instigate either prepayment schemes by which the client purchases a card much the same as a telephone calling card, or the utility can install restriction devices either on the customers' service or by installing control valves on a zonal basis to reduce the impact of lost water.

9.6 SUMMARY

Obviously, each system has its own individual problems, some of which may have been covered by the brief overview in this chapter. The following chapters cover in more detail some of the intervention methods and describe how to set up field intervention methods. Field intervention is a very hands-on activity and can be equipment-intensive. If the operator is unfamiliar with this type of work, we suggest a careful review of Chap. 6 before selecting equipment.

9.7 REFERENCE

1. Lambert, A., and Hirner, W., *Blue Pages* (the IWA information source on drinking water issues), final draft, August 21, 2000.

CHAPTER

10

Reducing Real Losses in the Field
Active Leakage Management

Julian Thornton

> *Case Study One*
> Conservation Project Saves $24M for Utilities
> *Tim Brown*
>
> *Case Study Two*
> Locating Leaks with Acoustic Technology
> *Anthony N. Tafuri*
>
> *Case Study Three*
> Severn Trent Leakage Management Process
> *Martin Kane*
>
> *Article One*
> Tracer Gas Helium Testing Conduits and Closed Systems Procedures and Methodology
> *Dave Southern*
>
> *Article Two*
> Water Main Leakage Detection by Means of Ground Penetrating Radar
> *Rodney Briar*
>
> *Case Study Four*
> Systematic Water Loss Reduction through Technology Application
> *Kenneth J. Brothers, P.Eng.*

10.1 INTRODUCTION

In Chap. 9 we reviewed the difference between real and apparent losses, and we identified how to calculate the volume and value of the individual losses in Chap. 4. Once the value and nature of the losses have been identified and quantified and economic sustainable limits identified, realistic targets can be set. Once the targets and budgets for intervention have been identified, the most suitable methodologies for reducing and controlling the losses can be identified.

This chapter presents some of the most common practical methods of water loss reduction by active leakage detection. (Other common techniques, using pressure management, are reviewed in Chap. 12.) The diagram at the right shows how improved leakage detection fits in our four-component approach to reducing real losses.

In order to schedule field activities properly, it is first necessary to prioritize the losses. Most utilities have limited budgets, so the methods of intervention with the shortest paybacks are usually the ones that are put into place at the start of the program. In this way the programs start to self-fund out of savings after a certain period of time, depending on the utility's ability to create budgets for capital and maintenance activities out of other budget figures (not all utilities can be this creative with their books).

In order to implement and track physical intervention in the field it is necessary to prioritize activities and create a methodology list. (The following discusses a possible plan for a complete field intervention, not just leakage management. The chapters that discuss the task in more detail are given in parentheses.)

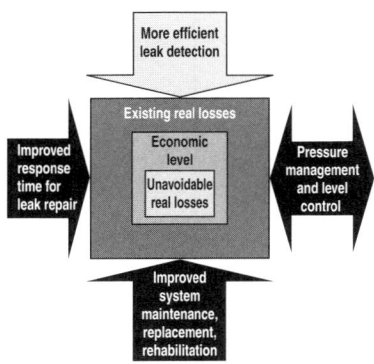

1. Update system plans to ensure efficient leakage control program and a sustainable maintenance program (this chapter).
2. Check the master and or supply meters for accuracy to ensure that real figures are used in the audit and definition of real losses (Chap. 14 and App. B).
3. Identify existing programs such as sales meter testing or leak control (meter testing is discussed in Chap. 14).
4. Identify existing reporting for water purchased, delivered, and loss accounting, and ascertain if a suitable baseline can be identified from which to match performance (Chaps. 4 and 7).

5. Identify hydraulics of the system, pumped, gravity, wells, or storage (Chap. 12).
6. Identify whether zone flow analysis has or can be undertaken (this chapter).
7. Undertake zone flow analysis if possible.
8. Rank zones for leak detection.
9. Identify the best means of leak detection for the type of pipe and ground conditions.
10. Undertake leak detection.
11. Report and rank leaks located.
12. Repair leaks using priority as above (Chap. 11).
13. Identify whether reservoirs or storage might be a point of water loss, through either leakage or overflow.
14. If possible losses are occurring, equate losses to a value and undertake remedial measures to reduce losses.
15. If sales meter testing has not been done, select suitable samples and tests (Chap. 14).
16. Identify economic frequency for meter change-out (App. B).
17. Change out meters (Chap. 14 and App. B).
18. Identify economically sustainable level of leakage.
19. Maintain both leakage and meter change-out programs at economical frequency.
20. Identify a reporting procedure and time frame to ensure that the above programs are sustainable.
21. Identify a responsibility structure to ensure that somebody takes final responsibility for the performance of the loss control program.

The above list identifies many of the tasks required for a water system loss management program. However, the order in which items 3–16 are undertaken will vary from utility to utility depending on the results of the initial audit (see more on auditing in Chap. 4 and App. A). Obviously, if a utility is experiencing higher losses from the metering side than the leakage side, the utility will want to attack the apparent losses first (and vice versa).

This chapter has been prepared by various contributors, and provides technical advice, comments by specialists, and various real case studies, which may be used to compare situations and results.

10.1.1 Mapping

The first thing which must be done when considering tackling losses in the field is to ensure that the maps and plans of the system and its components are accurate and as up to date as possible.

The media on which water company plans are kept vary widely, from distribution systems with the latest software (GIS), to systems with up-to-date paper plans, to systems with an up-to-date picture in someone's head, to systems in which no one has any idea where anything is! Obviously, the cost of updating such systems will vary greatly.

Systems with good plans and organized, structured background data tend to be more profitable, as the managers responsible for day-to-day

> In many cases the volume of real losses may be higher, but the monetary value of the apparent losses may be higher, as they are recovered at sales cost. Depending on the reason for the intervention—loss of revenue or loss of volume or resource—the utility will prioritize accordingly.

decisions have tools at their fingertips with which to make decisions about the performance of their organization. Systems with very little background data find it very hard to set a realistic baseline for performance and therefore tend to get out of control and make panic-based decisions. Even after instigating a project to improve performance, these systems find it hard to justify the results, as they do not have reliable baselines from which to measure. A system in this position should consider putting the data in order and plans in place prior to beginning any other intervention.

GIS (Geographic Information System) is becoming a very popular way of managing system plans and provides a very user-friendly graphical interface with the system plans. GIS also brings other benefits, as it allows direct interaction with other system tools such as financial and billing databases, telemetry and SCADA systems, complaints logs, emergency logs, and routine maintenance activities records such as leak location and repair data. A full GIS system can also be interlinked with a hydraulic model, which is a decision-making tool used by many water utilities. As the GIS is linked with the model, there is less need for costly model updates—the model is automatically changed as the GIS is worked on.

The Global Positioning System (GPS), which is now being used by many utilities to automatically register or locate system components and major features within the system through the use of satellite positioning, is often used alongside GIS. The GPS sounds like it might be rather high tech and difficult to operate, but it is quite the opposite. GPS data downloads automatically into most GIS databases. Costs of GPS systems vary widely with the resolution required, but costs in most cases in most countries are not prohibitive. Figure 10.1 shows GPS being implemented in the field in Pietermaritzburg, South Africa, as part of an overall upgrading of plans and loss management program funded by the federal government.

If a utility has no system plans or intends to make a major upgrade, GIS is a very well recommended route. There are many packages available on the market today, although it is important to use a package suited to the level of the local operators, whether in-house or contract. The software should be well supported in the region, and easily upgraded. Figure 10.2 shows a layer from a GIS system put into place in SABESP, São Paulo, Brazil, during implementation of a leakage management program. This particular figure shows municipal blocks, roads, pipes, and number of connections per block. This plan was used to determine areas where pressure management might be implemented. Figure 10.3, from the same company, shows another zone where pressure management and leak detection and repair is being carried out. The GIS system is being used to map reported and unreported leak locations for repair and monitoring of leak frequency.

After making the decision as to whether a software package or paper plans will be used, it is necessary to decide what is required of the plans in whatever format, to allow programming of an updating exercise. In addition to the requirements listed below, the operator should consider the size and scale of the plans to be generated or updated. In the case of GIS the operator can select areas and manipulate them to whatever size

CHAPTER 10 REDUCING REAL LOSSES IN THE FIELD

Figure 10.1 Using GPS to locate fittings. (*Source: Restor.*)

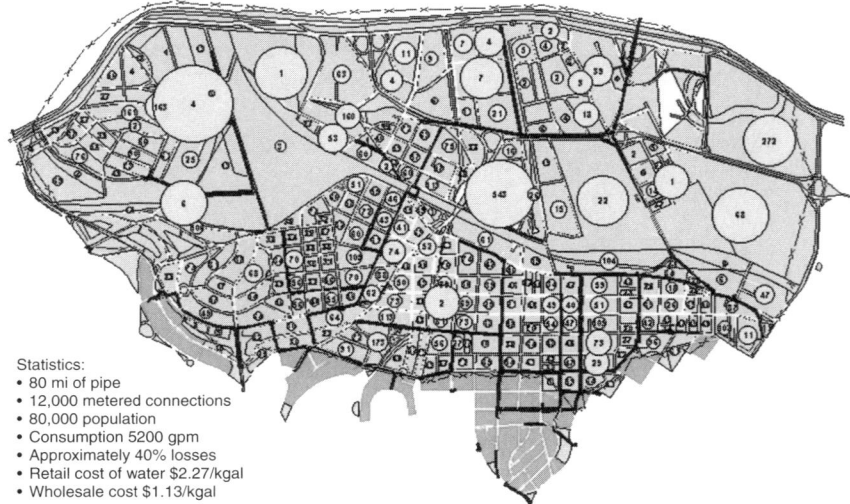

Statistics:
- 80 mi of pipe
- 12,000 metered connections
- 80,000 population
- Consumption 5200 gpm
- Approximately 40% losses
- Retail cost of water $2.27/kgal
- Wholesale cost $1.13/kgal

Figure 10.2 Geographic information system (GIS) used to determine potential areas for pressure management. (*Source: SABESP/BBL Ltda, Contract 66.593/96.*)

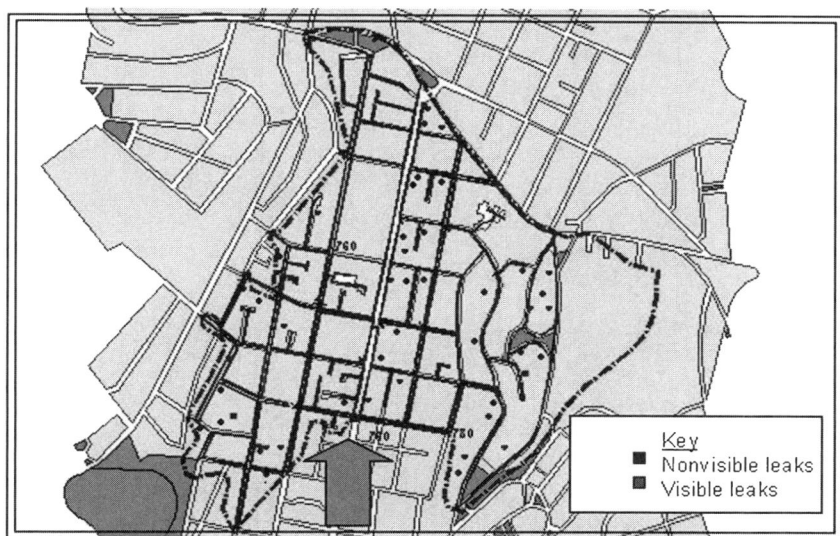

Figure 10.3 Using GIS to map reported and unreported leaks. Contract 4.134/97. (*Source: SABESP/BBL Ltda.*)

is required through use of zoom tools. This is not true for paper plans, however. Obviously, there needs to be enough detail to be able to make accurate decisions, but we also need the scale to be small enough that we can review an area in its entirety. Many utilities use a scale of 1:2000 for urban areas where a lot of detail and dense interconnections are required. Rural systems or areas often use plans to a scale of 1:5000, as there is not so much crowding and it is preferable to see a larger area at one time.

In general, to allow a thorough loss management strategy to be implemented, plans must be available with the following basic information:

- Roads with road names
- City or municipal blocks
- Meter book routes
- Water reticulation system diameters, pipe material, and, where possible, age (entire system transmission included)
- Clear identification of major consumers
- Ground levels and contours to at least 5-m intervals
- All water sources, wells, treatment stations, transfer points, storage
- All valves, control valves, master meters
- Clear identification of any zones within the system and their function (pressure control, zone flow analysis, step testing, billing, municipal land use, etc.)

Once a list has been created as to what features are required on the plans, a careful desktop review must be undertaken to decide what is present and usable, what is present but out of date, and what is missing altogether. Once this information is known, teams can be put together to collect the necessary data.

> It is often difficult to cost-justify expensive updating of plans, but this is a vitally important component of water loss management and efficient system operation.

Most fieldwork is done with pipe and cable locators, metal detectors, and recording devices such as ground-probing radar (GPR) instrumentation or even a simple clipboard and paper, depending on the level of technology required for the project and the project budget. Figure 10.4a shows a picture of pipe location being undertaken in the field using a locator. Location work is also often done with GPR, as shown in Fig. 10.4b.

All data collected must be stored in a manner which is easily accessible to all members of the team.

(a)

(b)

Figure 10.4 (a) Using a pipe locator. (*Source: Heath Consultants, Inc.*) (b) Using GPR. (*Source: R.V.M. Surveys C.C., RSA.*)

When instigating an update of systems plans, much thought must be given as to how to implement a system to ensure sustainable good plans. It is important to have buy-in from all sectors of the water utility, such as finance and maintenance.

10.2 LEAK DETECTION

Leak detection in various forms has been around for many years and is a process which has recently become very high tech. However, it is important to note that the higher technology only enables us to do things faster, more efficiently, or to a higher standard; many of the older technologies are still relevant.

Each piece of equipment forms part of a tool kit and is not usually the only solution to leak problems. Most important, whatever level of technology the leak detection team is using, they must be confident in their use of it and in the ability of the equipment to find and pinpoint leakage accurately.

10.2.1 Visual Survey

The most basic form of leak location is the visual survey. A visual survey consists of walking the lines looking for either leaks which appear above the ground or, in very dry countries or regions, areas that have suspicious green growth patches above the water lines. Figure 10.5 shows a leak that could easily be located by visual survey. This particular leak is on an above-ground air valve. Other leaks, which are not quite so obvious, are also often picked up.

While the visual survey is not very high tech, it should not be underestimated, particularly by utilities which have suffered from lack of good and frequent maintenance. The visual survey is also a very quick method of locating a large break or burst.

10.2.2 Acoustic Survey

The acoustic survey has been around for years and is probably the most familiar kind of leak detection survey. Operators use a variety of instruments with two distinctly different methods. The first method identifies an area where leak sound can be heard on fittings, and the second method identifies where under the ground is the exact point of the leak.

The first method tests fittings such as hydrants, meters, and valves and is often referred to as hydrant survey.

10.2.3 Hydrant Survey

Although many fittings are tested, such surveys are often referred to simply as hydrant surveys in countries like the United States and Canada; fire hydrants are above ground and easily accessible. They are also found on virtually every street corner, meaning that merely testing at hydrants can provide good coverage of most areas. A hydrant survey can thus be

Figure 10.5 Visible leakage from an air valve. (*Source: Mark Loveday.*)

a time-saving quick-run-through method. (Care should be taken when surveying plastic-based mains in this way, as leak sounds do not transmit as far as on metallic-based mains.)

This type of survey can be done with a number of instruments, ranging from mechanical listening sticks (wooden or steel, with a mechanical diaphragm) to electronic listening sticks which amplify the leak sound. Whatever equipment is used, it is important to understand the leak frequencies which may be encountered and the types of pipe material which are being tested. Leak frequencies vary depending on the type of leak, the type of pipe, the back fill, and whether the hole around the leak is water logged.

The three following types of situation generate leak frequencies.

- *Friction sound* is the sound created by water forcing its way through the pipe wall and making vibrations along the pipe. This tends to be a higher-frequency leak sound and can range anywhere from 300 to 3000 Hz (see Fig. 10.6). In general, high-frequency leak sounds are easy to recognize but do not travel very far along the pipe.
- *Fountain sound* is the sound of water circulating around the leak site and tends to be lower frequency, in the range of 10–1500 Hz (see Fig. 10.7).
- *Impact sound* is the sound of a leak impacting on the walls of the hole around the leak and the sound of the impact of rocks, which often are thrown around the leak. This sound also occurs in the range of 10–1500 Hz (see Fig. 10.8).

For operators using electronic listening sticks with frequency filters, it is important to open up the filters as wide as possible at the beginning of the search, so as to allow all leak sounds to enter the equipment. As the

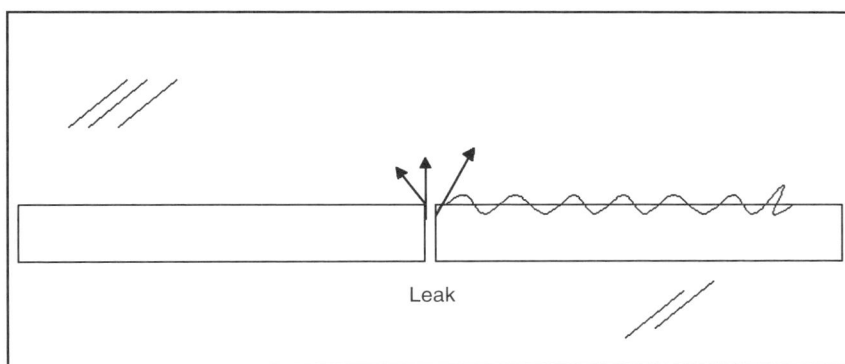

Figure 10.6 Friction sound. (*Source: Julian Thornton.*)

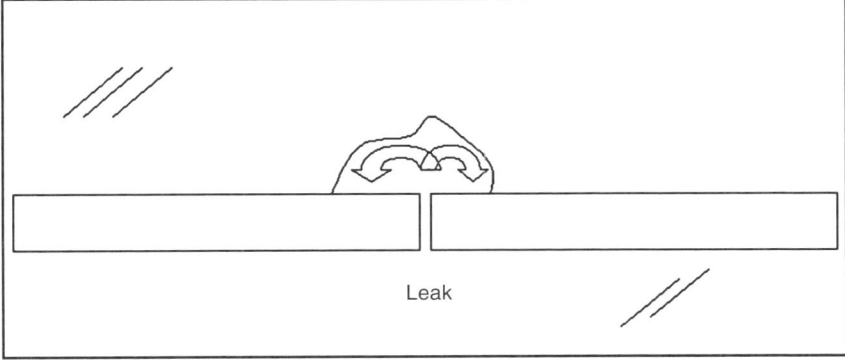

Figure 10.7 Fountain sound. (*Source: Julian Thornton.*)

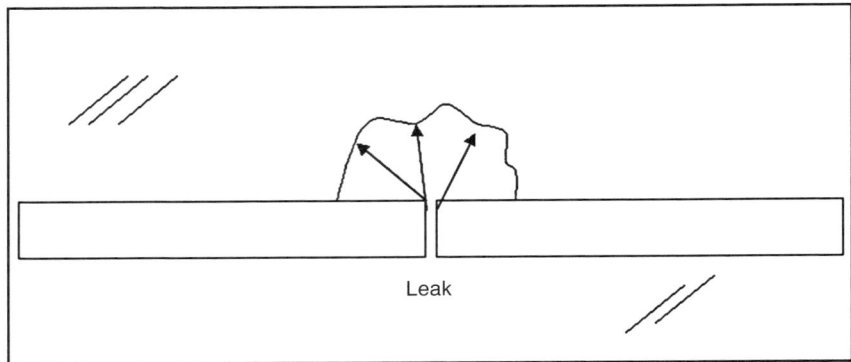

Figure 10.8 Impact sound. (*Source: Julian Thornton.*)

search narrows, the filters can be narrowed down to allow other unwanted sounds, such as traffic or consumption, to be filtered out.

Different pipe materials and pipe diameters convey leak sounds in different ways, so it is important to understand the system to ensure that the sounding locations selected are within hearing distance of any particular leak. In general, metallic pipes allow leak signals to travel farther and plastic pipes lesser distances. Asbestos cement and cement pipes tend to be in the middle. In very good conditions the author has heard leak sounds travel up to 1 km, but a more conservative high end should be taken to be around 250 m. The lower end, particularly on plastic pipe, may be as low as 10 or 15 m. Leak sounds also travel through the water. These tend to be at lower frequencies, which travel farther and are usually picked up when using correlators with hydrophones.

The operator must be trained to understand not only how to use the acoustic sounding equipment but also its limitations, so that he can vary the survey criteria in terms of the limitations of his equipment and the likely distance that leak sound will travel. After all, if sounding is undertaken every 200 m but the pipes are plastic, there is a very good chance that unless the leak just happens to be next to a hydrant or fitting which is being sounded, the leak could very easily be missed. Every new system and every different part of the system must be treated on its own merit. Sounding may be a simple technique, but the planning must be done by someone who understands the limitations and factors involved in each part of the system.

10.2.4 Geophone (Ground) Survey

The second type of acoustic survey involves ground microphoning. Ground microphoning (also referred to as geophoning and ground miking) can also be done with two specific types of equipment; either mechanical or electronic geophones. Figure 10.9 shows a mechanical geophone.

Geophoning consists of walking above the pipe and listening for the sound of a leak through the ground. In this case we are relying on the transmission of sound through a different medium, the soil, as opposed to the pipe work as in the first methodology.

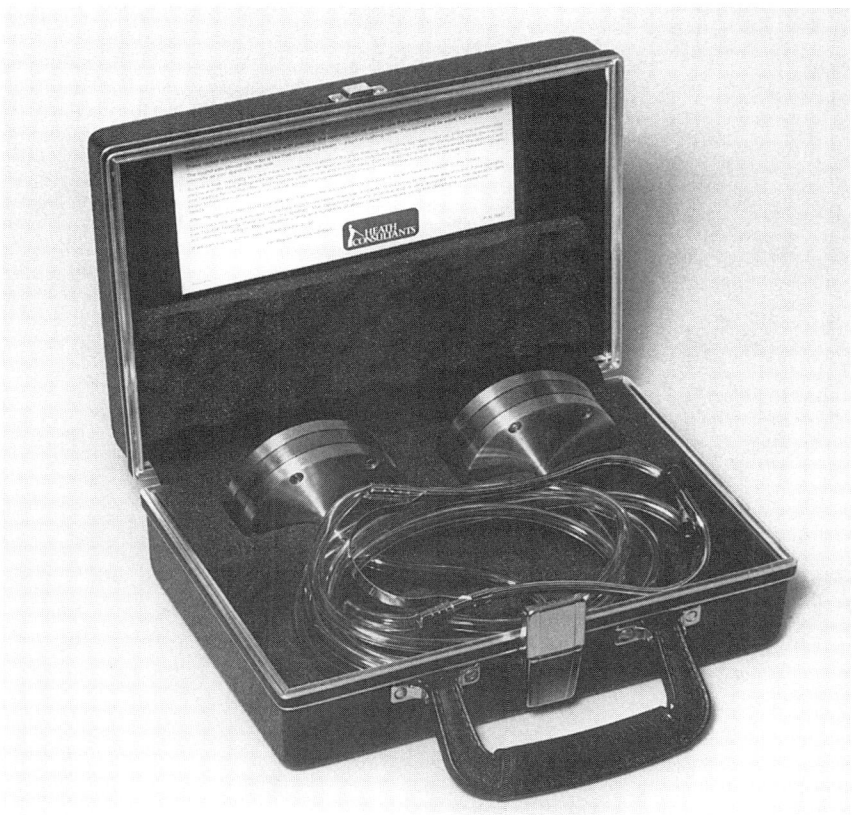

Figure 10.9 Mechanical geophones. (*Source: Heath Consultants, Inc.*)

Different soil types also transmit the sound in different ways, and operators learn to differentiate between heavy clay soils and light sandy soils. Again, operators must consider a suitable distance between test points to be sure not to miss a leak location.

Geophoning is usually carried out in conjunction with the first techniques identified above, although it can be used alone, especially in areas with few fittings and predominantly plastic pipe.

Leak sounds located by geophone surveys tend to be of lower frequency and tend to be from the last two kinds of leak sound mentioned above. (In some cases experienced operators can use a geophone sensor to perform the hydrant and fittings survey mentioned above, when they know that the leak sounds are predominantly low-frequency).

Again, inexperienced operators should consider opening up the filters to their maximum at the start of the survey to ensure that all potential frequencies are located. More experienced operators learn to use different filters in different situations.

The only way to become a good acoustic leak detection person is to gain experience, and the only way to do that is to get out in the field and spend as much hands-on time as possible. Usually, in the beginning we are afraid to make judgment calls and harder-to-find leaks tend to get overlooked, because the operator is afraid that a hole may be dug for no reason. However, if the situation allows, it is better to dig and find some other reason for the noise than not to dig. The operator gains valuable knowledge even if a leak is not located. This is not always possible, of course, in a commercial leak detection project.

In all acoustic surveys it is important to ensure that the operator is listening directly above the pipe, which reinforces the need for good system plans. In addition, we need to understand some of the other ambient system noises, which may throw us off the trail or mislead us.

The surface on which we are listening also plays a big part in allowing us to pick up leak sound easily. In general, it is easier to pick up leak sound when searching over concrete or asphalt or any hard surface as opposed to when listening over mud or grass. If listening over the latter it is advisable to use ground stakes and listen on the stake itself. The distance between listening points should be greatly reduced over soft surfaces. Also, be careful when listening between hard and soft surfaces, as the same leak will sound louder over concrete (or other hard surface) than over soft soil. The leak may actually be in a verge, but the ground over the leak in soft soil may not transmit noise as well as a neighboring piece of pavement. When locating under these circumstances it is advisable to use a boring bar to make test holes, to see if water can be seen surfacing. Obviously, a detailed locate must be done first, to ensure that cables or other underground utilities are not damaged during this process.

> The nature of the ground surface affects our ability to hear leaks well.

Things to Watch For

- ❏ Demand and usage
- ❏ Traffic or airplanes
- ❏ Pressure control stations
- ❏ Meters
- ❏ Dogs (they can also bite!)
- ❏ Gas lines
- ❏ Steam lines

Presurvey Checklist

- ❏ Prepare good system plans at workable scales.
- ❏ Clearly mark limits of zone(s) to be tested.
- ❏ Locate unknown pipe lengths.
- ❏ Identify a suitable distance between testing points per main type.
- ❏ Identify large users who could interfere with sounding.
- ❏ Identify where pressure valves are.
- ❏ Prepare protective clothing.
- ❏ Prepare a suitable leak location form.
- ❏ Charge batteries for electronic equipment.
- ❏ Check sensors against a reference sound such as a tap running to ensure sensitivity.
- ❏ Take I.D., as you will be entering private property from time to time.
- ❏ Take the necessary signs and cones to warn traffic.

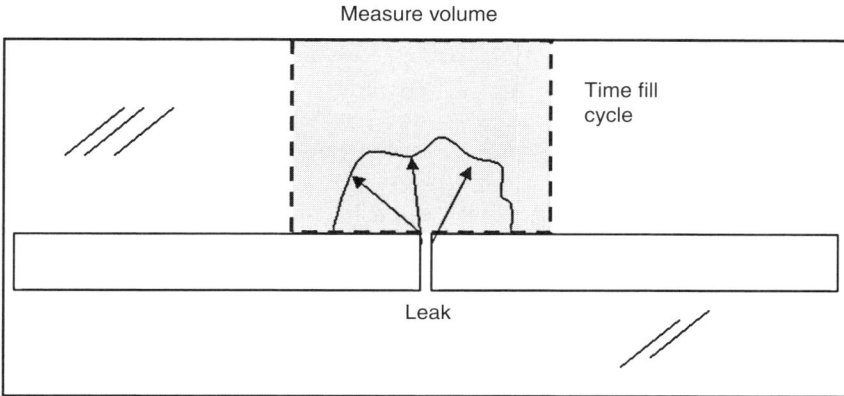

Figure 10.10 Volumetric measurement of leaks. (*Source: Julian Thornton.*)

In many instances operators prefer to undertake any kind of acoustic or sonic leak detection at night, when demand is at its lowest and traffic is less of a problem.

Postsurvey Checklist

- Clearly record all suspected leak points on prepared sheets.
- Clearly identify the points on the maps.
- Attempt to rank the leaks by severity of loss and potential damage to life or property.
- Prepare a repair work order.
- Identify a realistic time frame for repairs to be undertaken, ensuring that the worst leaks are repaired first.
- Where possible, visit the leak site during repair to make a photographic record of the leak.
- Attempt to make volumetric measurements for larger leaks, to assist in preparing the annual balance (see Fig. 10.10 for methodology).

The following case study identifies what is probably one of the larger success stories in leak detection history in the United States to date.

10.3 CASE STUDY ONE

Conservation Project Saves $24M for Utilities

Tim Brown, Heath Consultants, Inc.

One of the largest energy and water conservation projects carried out in the United States saved $24.4 million per year at a cost of $2.7 million for 278 water utility companies in the state of Tennessee as of January 1991. This is a benefit-cost ratio of 9.5:1 and represents a payback period of just 38 days. The average system savings was $91,398 per year.

The Tennessee department of Economic and Community Development, Energy Division, provided the funding for and implemented the water accountability project. More than 400 water utilities were eligible for participation in the program.

The Tennessee Energy and Water Conservation Program was submitted to the State of Tennessee and was nominated by Governor Ned McWherter for national award consideration by the U.S. Department of Energy. Government officials, scientists, engineers, and others then evaluated the project. Upon completion of the evaluations, the program won honors in the National Awards Program for Energy Innovation.

The Tennessee Association of Utility Districts oversaw and administered the project on behalf of the State of Tennessee, and the water system audits, meter accuracy testing, and leak detection/pinpointing surveys were performed by a consultant. The project was divided into two phases: (I) to identify energy and water loss and to make recommendations for corrective action; (II) to conduct a leakage detection/pinpointing survey of the distribution system. Phase II was scheduled to be initiated if the benefit-to-cost ratio was favorable.

10.3.1 A Description of the Program

In January 1988, Heath Consultants, Inc., contracted with the Energy Division of Tennessee's Department of Economic and Community Development to conduct a two-phase program to identify energy and water loss and to make recommendations for corrective action.

Phase I of the program included a detailed audit of the water produced and purchased, operational costs, electric consumption (pumping costs), and the daily operations of the utility. All testable master and commercial/industrial water meters 2 in and larger were tested to determine their accuracy, since inaccurate meters figure significantly in determining the water system's product accountability.

The purpose of Phase I was to accurately determine the amount of unaccounted-for water, and the cost of that water based on the cost to produce and/or purchase and distribute it. This figure is known as the avoidable cost. The total avoidable energy loss (Btu's) and total dollar loss to the utility due to unaccounted-for water was used to determine the benefit-to-cost ratio for corrective action.

Before the program began, there was confidence that the program would save a great deal of energy, water, and money, but it appears that the magnitude of savings was grossly underestimated!

As of November 9, a total of 119 audits had been completed, yielding a cumulative total of $9,010,224 of avoidable cost within 1 year at a cost to the State of Tennessee of only $409,132. This represents a payback period of 16.6 days. Included in the avoidable cost is 72,698,052,000 avoidable Btu's, representing $1,496,860 of energy savings. In addition, a total of $342,909 of avoidable revenue loss per year was identified, due mostly to inaccurate meters.

It was also reported that a total of 77 systems had completed the leakage detection phase, pinpointing a total of 4,175,118,600 gal/year of water loss due to system leakage. This represents $4,793,863 of lost water per year. The cost of pinpointing this lost water was only $511,944, with a payback period of just 39 days. The utilities needed only to excavate the pinpointed leakage locations and repair the leakage to realize their savings.

The innovation, transferability, energy savings, and economic impact encompassed by this program made it a winner. Innovation was demonstrated in the program's ability to assist water systems throughout Tennessee to identify and correct deficiencies in order to operate more efficiently. Utilization of funds from the state "oil overcharge fund" to finance this extensive program was a breakthrough. Many of the managers and operators of the system had never been exposed to such in-depth study of virtually all aspects of an operating water system. An objective third-party review and a full explanation of all activities that were being performed throughout the program, coupled with the reduction of energy and water loss identification, have made this a very popular program.

Transferability comes into play when experts "transfer" their knowledge to system operators and their own in-house staff. For example, the utility operator observes the meter testing process and learns how and why accuracy testing of meters is an important aspect of controlling loses. Having observed this procedure, the operator may elect to conduct meter accuracy tests in-house, utilizing existing personnel.

When a leakage control survey is found to be necessary, the consultant often requests that a utility company employee accompany the leakage technician while the survey is in process, so that the utility employee can be trained to detect and pinpoint sources of leakage in the distribution system. The first thing the operator learns is that the vast majority of leaks, for a variety of reasons, will not come to the surface. Many water operators have now learned that they cannot wait for leaks in the system to find them; they must go out and find out where the leaks are. This knowledge alone goes a long way toward controlling leakage in the system. When the leakage control survey is completed, each utility is left with a program to control its leakage in the future and to respond to leakage complaints with a logical and systematic plan.

Energy savings generated by the program are very easy to demonstrate. The program identified the amount of energy loss which can be saved by corrective action. Most loss is due to system leakage, which can be pinpointed and repaired, resulting in immediate energy savings due to the decrease in pumping required.

The obvious economic impact of this program comes from savings which were realized by the individual water systems. Many will now be able to operate on a more economically sound basis. They may now be able to upgrade their systems to operate more efficiently to supply higher-quality water to consumers at a reasonable cost, with end savings realized by the consumers. Many systems will capitalize on their potential return on investment for this type of service. Future budgets within these water systems should include such services to allow the systems to be maintained and operated under sound economic practices. This will conserve energy and drinking water, which is beneficial to the industrial and residential growth of every modern state.

10.3.2 Correlator Survey

The late 1970s and early 1980s brought the advent of a new technology called leak noise correlation. Leak noise correlators come in many

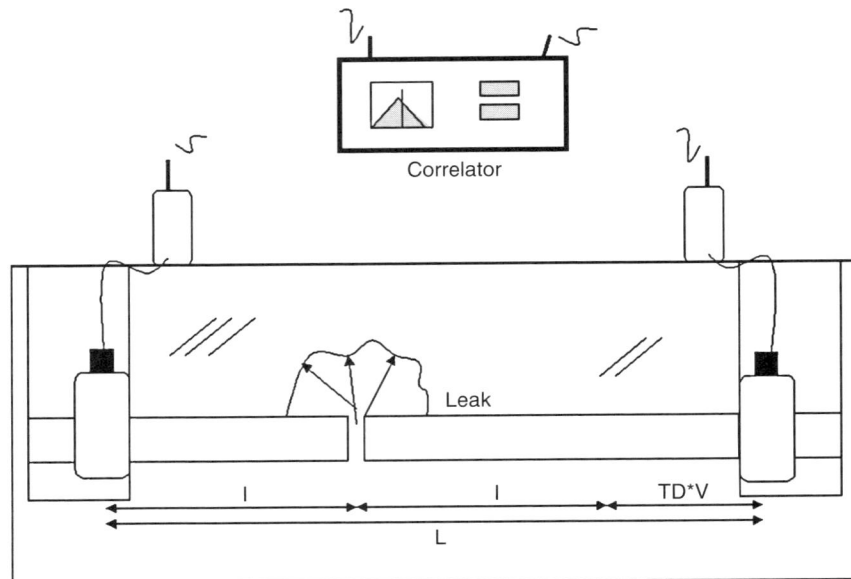

Figure 10.11 Principles of leak noise correlation. (*Source: Julian Thornton.*)

shapes and forms and have become very sophisticated devices, but they all operate using the same principle of locating a leak sound across a determined distance of pipe spanned by two sensors and analyzing the time taken for the same portion of sound to reach each sensor.

The calculation principle is set out in the formula $L = TD \times V + 2I$, where L is length, TD is the time delay for the signal to reach the farthest sensor after reaching the first sensor, V is the speed at which the leak sound can travel either in the pipe wall or in the water, and I is the leak position from one sensor and fitting. This is shown more clearly in Fig. 10.11. The calculation is then reworked to give us a way to calculate the leak position, which is $I = L - (TD \times V)/2$.

Correlators have become very sophisticated devices, and each manufacturer has its own special way of improving leak pinpointing efficiency. The manufacturer of each model should give detailed training. Some basic concepts, however, are true for any model. Weak points in the theory usually occur when correlating over long distances when the leak is quite close to one end. This creates a high time delay, TD. Because in many cases the velocity is estimated, a longer time delay multiplied by an incorrect velocity will throw the pinpointing way off the actual leak position.

Many correlators have a velocity calculation feature, which should be used; however, a simple little trick to ensure reasonably accurate location is always to try to keep the leak area fairly close to the middle of the range. This can be achieved by running a quick correlation to locate the leak roughly, then move the sensors to centralize the suspected leak point. As discussed in the last section, no one piece of equipment can do everything. It is often a good idea, after locating a suspected leak location with a correlator, to get out the ground microphone or geophone and listen

> **K**eep the leak close to the center of the sensors and the potential error in location will be less.

over the suspected point. If a positive leak sound is also heard through the geophone, this will boost the confidence of less experienced operators and make them feel more comfortable about digging a hole. In the case of unskilled operators, this can often save the day.

Obviously, any estimated or miscalculated data will throw off the calculation, so it is important to measure the distance between the two points accurately. This is usually done with an electronic measuring wheel. The measuring wheel must be driven over the exact pipe route, not necessarily straight between the two sensor fittings. The velocity used for the calculation can be taken from internal or external tables. However, it should be checked in the field using the velocity measurement function. Remember, the velocity will change every time the pipe material or diameter changes. Most correlators allow the operator to input various pipe materials and diameters to provide a likeness to what is in the field. The time delay is measured by the correlator and is out of the operator's control; however, as mentioned previously, the operator can try to keep the time delay to a minimum by centering the leak as near as possible between the sensors. As with acoustic surveys, it is a good idea to open up filter settings during the first few runs to see which are the predominant frequencies. Filtering can be done by more experienced operators after the initial frequencies have been analyzed.

The more you check and double-check the data, the more likely you are to achieve success.

One other situation which can throw off inexperienced operators occurs when correlating on a length of main where the leak is actually not on the main but rather on a tee main or service. The correlator does not know that the leak is off on the tee and will show the leak on the tee itself. It will be up to the operator to know the system well enough to check the line going off from the tee. See Fig. 10.12.

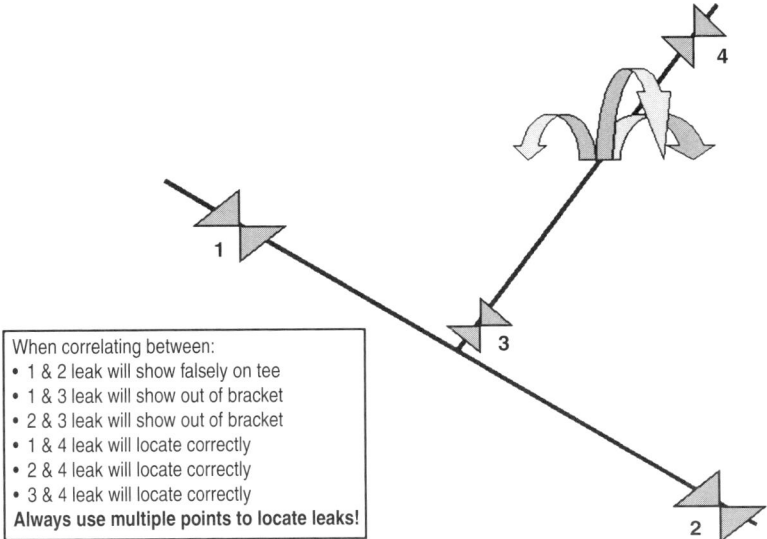

Figure 10.12 Tee connection rule. (*Source: Julian Thornton.*)

188 CHAPTER 10 REDUCING REAL LOSSES IN THE FIELD

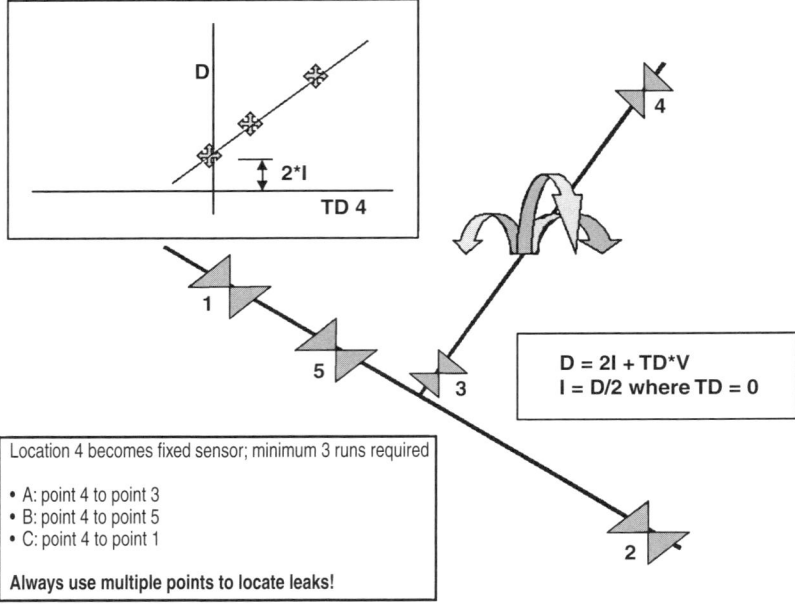

Figure 10.13 Linear regression of results. (*Source: Julian Thornton.*)

A good way to avoid errors is to perform three or more correlation runs with varying lengths of pipe between the sensors, then do a linear regression of the data as shown in Fig. 10.13. This has the effect of averaging out the errors in each velocity calculation and giving a closer answer. Some correlators have this facility built in, but any experienced operator can use this method by manually plotting the results of varying length and changing time delay.

Some utilities specify leak noise correlator-only surveys, in which each and every section is subjected to a test with the correlator. While this is often thought as an all-encompassing way of testing the system, it is often best to mix acoustic surveys with correlator surveys, as a correlator will only pinpoint a leak if it can pick up leak sound at two points. With plastic pipe, in which leak sound does not travel very far, forcing a leak noise correlator-only survey could mean that some leaks are not pinpointed.

Whether performing an acoustic survey or a correlator survey or indeed a mixture of the two, it is always a good idea to check system pressures. This is particularly important in areas which have trouble supplying water 24 h a day, areas with extremely high headlosses, areas with aggressive pressure control, and areas where there is an emergency burst or break and the pressure has been brought abnormally low. It is usually hard to generate enough sound to locate a leak when pressures are below 15 m or 35 psi, although in some cases sound is generated.

Leak pinpointing is still very much a personal art as opposed to a science. Operators with experience will do things in their own particular way and achieve excellent results, whereas others with less experience may suffer periods of disappointment or indeed puzzlement. However, the more time an operator spends in the field and tackles the

> **L**eak sound does not travel well in plastic pipe. It is often better to use hydrophones when testing plastic pipe, as sound travels farther in water.

problems at hand even when apparently unsolvable, the more proficient he will get.

10.3.3 Summary

It is very important that good records are kept, of both successful leak pinpointing and also unsuccessful attempts. There is always something to be learned by all attempts. Figure 10.14 shows a sample leak report form, which can be used during pinpointing exercises.

Leak pinpointing is an art and a science, and every day there is a major leap in technology. Case Study Two, which is adapted from *Journal AWWA,* shows some of the recent advances in technology (from the petroleum industry) as well as discussing in detail the water system infrastructure in the United States today.

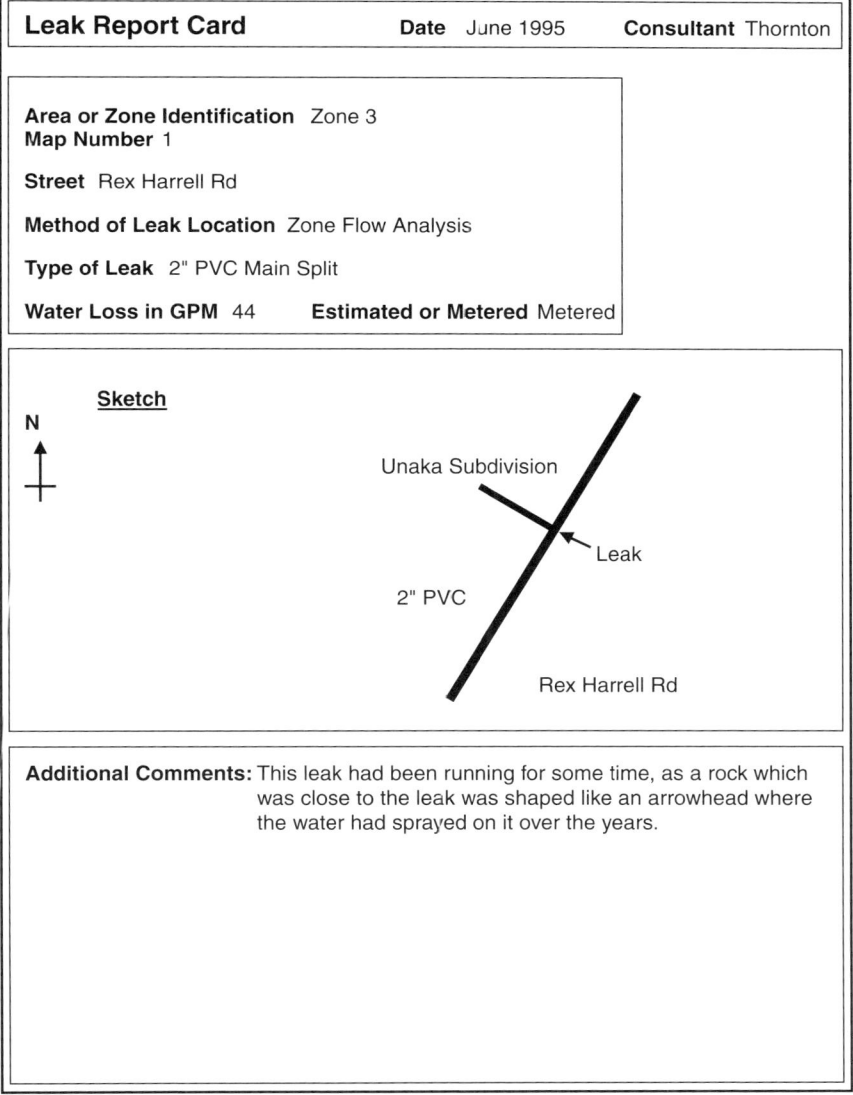

Figure 10.14 Sample leak report form. (*Source: Julian Thornton.*)

> Researchers were able to pinpoint leaks as small as 0.1 gph using acoustic emission techniques.

10.4 CASE STUDY TWO
*Locating Leaks with Acoustic Technology**

Anthony N. Tafuri

10.4.1 Background

Many water distribution systems in this country are almost 100 years old. About 26 percent of piping in these systems are made of unlined cast iron or steel and are in poor condition. Many methods that locate leaks in these pipes are time-consuming, costly, disruptive to operations, and unreliable in finding small leaks. This case study presents the results of research conducted at the U.S. Environmental Protection Agency's Urban Watershed Research Facility in Edison, New Jersey. The project sought ways to use acoustic technology to pinpoint leaks as small as 0.1 gph (0.1 mL/s) in petroleum pipelines, a regulatory requirement for those lines. Because all experiments were conducted using water and on pipelines of size and material similar to those found in many water distribution systems, results also apply to these pipelines. Although leaks of 0.1 gph (0.1 mL/s) are unusually small to search for in water distribution systems, researchers were able to locate small leaks within 1 ft (0.3 m), which is compatible to the best practice of commercially available leak pinpointing technology for water distribution systems.

Numerous aging drinking water systems have difficulty complying with the increasingly stringent requirements of the Safe Drinking Water Act, which indicates that distribution system water-quality problems are widespread and here to stay. Water main breaks occurred at a rate of about 75,000 per year, according to data from 1993 to 1996.[1] At current-day replacement rates, pipes will have to last 200 years—a period that far exceeds the design service life of the pipe. The potential may increase for structural failure leading to serious health risks caused by intruding pathogens or water-borne diseases.

In the United States, 24 percent of water-borne disease outbreaks reported in community water systems during the past decade were caused by contaminants that entered the distributions systems—not by poorly treated water.[2] Of approximately 200,000 public water systems in the United States, about 30 percent serve 90 percent of the population. Potable water is conveyed within these 60,000 community systems in an estimated 880,000 mi (1.4 million km) of pipe. Much of this pipe was installed after World War II, is more than 30 years old, and is judged to be in good condition. However, 26 percent is made of unlined cast iron or steel and is judged to be in fair or poor condition at best, from a structural and hydraulic viewpoint.[3] These systems require accelerated repair and replacement to avoid harming public health and to reduce the loss of valuable water. Such losses are estimated to represent on a national basis about 30 percent of the cost of water distribution.[4] Improved technology for detecting leaks more quickly and accurately, and ultimately for predicting where failure is likely to occur, will help address this problem.

*Adapted from *Journal AWWA*, vol. 92, no. 7 (July 2000), by permission. Copyright © 2000, American Water Works Association.

Evaluation technologies should be nondestructive; the ideal technology for assessing distribution system integrity should be nonintrusive, should not interrupt operations, and should be adaptable to the complex array of materials and conditions present. These requirements are not easy to accommodate. Water distribution pipes are typically constructed of many materials (e.g., cast iron, ductile iron, or concrete), are located under roads and sidewalks at various depths, and have many bends, taps, valves, and connections. These factors increase the difficulty of applying nondestructive evaluation methods.

Acoustic leak detection is the nondestructive evaluation method most commonly used by the water industry. When a pressurized pipeline develops a leak, the escape of water creates a sound that can be targeted mechanically. Locating the leak generally involves two phases. An initial survey detects sound by placing listening rods or aquaphones in direct contact with pipes or their appurtenances (e.g., fire hydrants or control valves). In the second phase, suspect leaks are pinpointed by ground microphones that listen for leak sounds on the pavement or soil directly above the pipe. The process is fairly simple, but it may not be successful because of problems with pipe location, background noise, and poorly compacted soil. Small leaks commonly cannot be detected at all.

An alternative tool for phase II pinpointing is a leak correlator, in which magnetic sensors are placed directly on a pipe on either side of the suspected leak. The sensors can be attached to valves, hydrants, services, or an exposed main. The sensors are then connected to an amplifier that uses wireless radio transmitters to send the leak sound to the correlator. The correlator determines the time lag between the measured leak signals by calculating the cross-correlation function. The location of the leak relative to each sensor is then calculated using an algebraic expression relating the time lag, the distance between sensors, and the velocity of sound propagation in the pipe. Leak correlators are generally considered to be state of the art in the water industry, and they are satisfactory for most professional users who are looking for leak rates in the gallon-per-minute range. Problems such as background noise interference and attenuation of leak signals along the pipe, which are normally encountered when locating leaks with acoustic equipment, are less obvious in this range of operation.

The U.S. Environmental Protection Agency's (USEPA's) National Risk Management Laboratory is the lead laboratory in a multiagency research program that studies the use of acoustic principles to detect and locate leaks in underground pressurized pipelines. The program, which is sponsored by the U.S. Department of Defense, involves the Army, the Navy, the Department of Energy (DOE), and the New Jersey Institute of Technology through a grant from the National Science Foundation. The program's objective is to develop a technology that will satisfy federal regulations that require petroleum pipelines to be tested regularly for leaks by a device that is able to detect 3.0, 0.2, or 0.1 gph (3.0, 0.2, or 0.1 mL/s) for continuous, monthly, or yearly monitoring, respectively.[5] When the project began in 1995, no single method was able to simultaneously detect and locate leaks at these rates without disrupting service.

10.4.2 Acoustic Emission Technology

At the program's inception, it was recognized that acoustic emission (AE) technology could satisfy these requirements for a variety of liquid-filled buried pipelines. Partners in the program were interested in pressurized pipelines associated with fueling facilities, retail gasoline service stations, and low-level liquid waste management systems. Access to these underground pipelines is commonly obtained through access covers, inspection pits, hydrants, and valve pits spaced about 500 ft (152.2 m) apart. Thus, the objective became to detect and locate a leak of 0.1 gph (0.1 mL/s) by use of a 500-ft (152.4-m) maximum sensor spacing. Achieving this goal would allow problems to be identified faster and more accurately and repairs to be made in a more timely and cost-efficient manner. Product loss would also be minimized.

Three Classes of Leaks Identified

Most nondestructive testing methods require some type of reference so that equipment can be calibrated and signals interpreted. Fabricated leak plugs were developed and used to generate controlled and repeatable leaks, both in the laboratory and in the field. Three classes of leak sources were defined for the project.[6]

Class 1 Sources Typified by Leak Plug

Class 1 sources are typified by a wall-penetrating leak plug designed by the DOE's Oak Ridge National Laboratory to address the characteristics of its double-wall stainless steel pipeline. The plug is relatively large, allowing experiments with various orifice geometries; however, the outside dimensions constrain its use on 2-in (50-mm) pipe to sections with end caps. Laboratory experiments on other test sections indicated that those plugs did not work well on single-wall pipes.

Class 2 Sources Resulting from Cracks

Class 2 sources are produced through a very long, through-the-wall crack. To simulate this source in the laboratory, a threaded cap was installed on the ends of pipe sections. Leaks were introduced through the threads by adjusting the cap; the leak rate was determined by measuring the volume of water captured in a beaker during a fixed period of time. Low leak rates, <0.1 gph (0.1 mL/s), were achieved by slightly loosening the threaded end cap under low water pressure.

Class 3 Sources from Small Leaks

Class 3 sources represent small, wall-penetrating leaks. Plugs were fabricated from socket head bolts with the center drilled out to precise specifications. Several were fabricated with orifice sizes varying from 0.07 to 0.04 in (0.117 to 1.02 mm) by putting pressurized water behind the plug; a leak could be established with a well-defined repeatable rate. With various plug and pressure combinations, the leak rate could be varied from 0.25 to 20 gph (0.2 to 20 mL/s).

Laboratory-Scale Test Chambers

Two test chambers were constructed to evaluate sections of pipe to be tested in the field, as well as to simulate the geometry of various leaks. For most studies, a 2-in-diameter (50-mm) steel pipe test section was fitted on one end with the class 1 leak plug and in the middle with the class 3 leak plug, to generate leaks of various rates. The other end of the pipe was equipped with a hose fitting that allowed the pressure of the liquid or gas to be regulated and an access hole for a pressure transducer. A second test chamber was created with a section of 4-in-diameter (100-mm) steel pipe. Each test section was housed in a clear acrylic tank equipped with a drain to regulate the accumulation of liquid in the backfill material. Soil could be changed to study the effects of backfill; in addition, the pipes could be rotated to simulate the leaks in different positions. By filling the pipe with liquid and internally pressurizing the cavity, a leak could be generated, and the resultant AE signal could be measured and characterized. The test chambers enabled visual observation of each experiment and provided a tool for transferring laboratory results to the field and for investigating leak behavior and leak detection capabilities in general.

Field-Scale Controlled-Condition Pipe Loops

A field-scale experimental loop system was designed and installed at the U.S. EPA's Urban Watershed Research Facility (UWRF) in Edison, New Jersey, to represent operational facilities. The buried systems included the following:

- A 500-ft-per-leg (152.4-m) 2-in-diameter (50-mm), Schedule 40, galvanized steel loop with interconnections [2-in-diameter (50-mm) pipe loop]
- A 500-ft (152.4-m), 12-in-diameter (300-mm) coated conduit with an insulated annulus space and a 4-in-diameter (100-mm) steel carrier pipe connected to a 500-ft (152.4-m), 4-in-diameter (100-mm) steel pipeline to form a 1000-ft (304.8-m) loop [4-in inside-diameter (ID)/12-in outside-diameter (OD) (100-mm/300-mm) insulated pipe loop]
- A 1000-ft (304.8-m), 12-in-diameter (300-mm), Schedule 40, coated steel loop composed of two 500-ft (152.4-m) legs and a "test-pit" area [12-in-diameter (300-mm) pipe loop]
- A 500-ft (152.4-m), double-wall, 2-in-ID/4-in-OD pipeline connected to a 115-ft (35-m), 2-in-diameter (50-mm) steel pipeline [2-in-ID/4-in-OD (50-mm/100-mm) double-wall pipe loop]

Access tubes [4-in-diameter (100-mm) polyvinyl chloride pipes] installed along the pipe loops provide for easy sensor mounting without excavation. The test-pit area provides access for replacing 20-ft (6-m) spool sections so that various types of pipe can be studied, as well as changing leak rates and backfill materials, orienting calibrated leaks and spools containing actual failures, and controlling backfill moisture content.

Research Divided into Phases

The research program was divided into three phases: (1) laboratory studies on small lengths of pipe in the test chambers, (2) controlled-condition experiments on the large-scale pipe loops at the UWRF, and (3) preliminary field assessments at operational facilities. One facility was operated by the U.S. Army at Fort Drum, New York, and another was operated by the DOE at the Oak Ridge Reservation Y-2 Weapons Plant in Tennessee.

Data were acquired and processed using two AE systems: a high-frequency (10–50-kHz) system and a low-frequency (3–8-kHz) system. Both systems used a two-channel AE instrument and resonant transducers (Physical Acoustics Corp., Princeton, New Jersey). The leak location was displayed automatically in real time using the tuned linear location approach developed under this program and an algorithm featured in the equipment used.

Initial Experiments

Initial experiments in the 2-in (50-mm) test chamber were performed without any backfill. Under these test conditions, leak orientation appeared to have no effect on the AE signal. The signal was considered to be extremely weak for a given leak rate of about 0.1 gph (0.1 mL/s), which was produced by a class 3 plug and an internal test pressure of 10 psi (70.31 kg/m^2). A signal strength of <0.12 V, expressed as the root-mean-square voltage, between the 500th and 520th second, correlated with laminar flow of water through the leak plug (see Fig. 10.15). To strengthen the signal, nitrogen gas was introduced into the flow. Between the 530th and 560th second, when the gas began to mix with the leaking water, flow changed from laminar to turbulent two-phase flow. Finally,

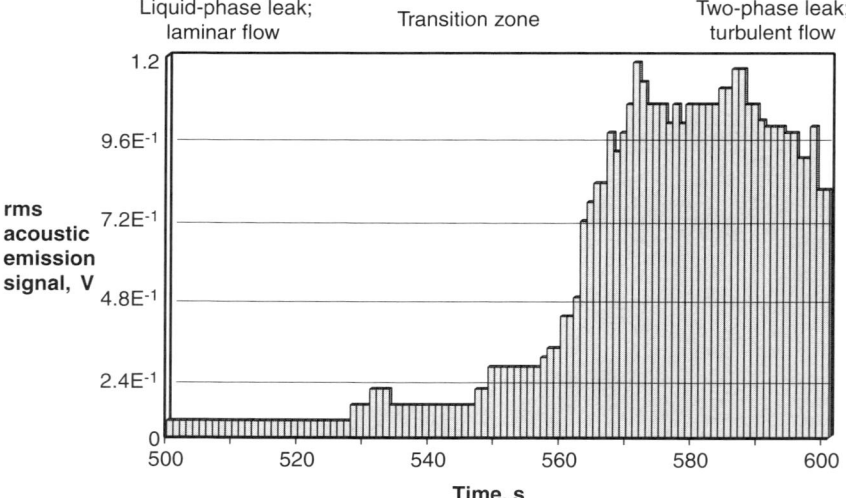

Figure 10.15 Root-mean-square (rms) signal transition relative to composition of leak. [*Source:* Journal AWWA, *vol. 92, no. 7 (July 2000), by permission. Copyright © 2000, American Water Works Association.*]

after the 560th second, the signal increased an order of magnitude to a maximum of about 1.2 V; the highly turbulent two-phase flow developed into a fine mist as it exited the leak plug.

The same procedures were applied to the 4-in-ID/12-in-OD (100-mm/300-mm) test chamber with similar results. In addition, the effects of leak orientation and backfill were studied using dry sand for backfill. With a class 3 leak plug installed, the test section was pressurized (without a gas blanket) to 60 psi (421.86 kg/m^2). Signal strength measurements were made using a 15-kHz resonant transducer with a plug oriented upward, sideways, and downward (see Fig. 10.16). The strongest signals were detected when the plug was oriented upward or downward, whereas the sideways orientation always generated a much weaker signal. The AE responses were similar for both the 2-in and 4-in (50-mm and 100-mm) sections. However, as discussed later, this leak orientation influence was not substantiated by the data collected during the larger-scale controlled-condition experiments or preliminary field evaluations.

Laboratory studies also investigated a 21-ft (6.4-m) length of the 2-in-ID/4-in-OD (50-mm/100-mm) double-wall stainless steel pipe section. Class 1 leak plugs, with orifice diameters of 0.004–0.04 in (0.1–1.0 mm), were installed and pressurized from 13 to 100 psi (91.4 to 703.1 kg/m^2). The resulting leaks ranged from 5 to 1400 gph (5 to 1400 mL/s). AE signal strengths were measured from the output of two sensors (mounted on the pipe section) having peak resonances of 30–60 kHz, respectively. The AE signal strength was linearly proportional to the rate of leakage through the leak plugs. Various experiments allowed calculation of the acoustic power radiated into the pipe by the leak. It was additionally determined that the

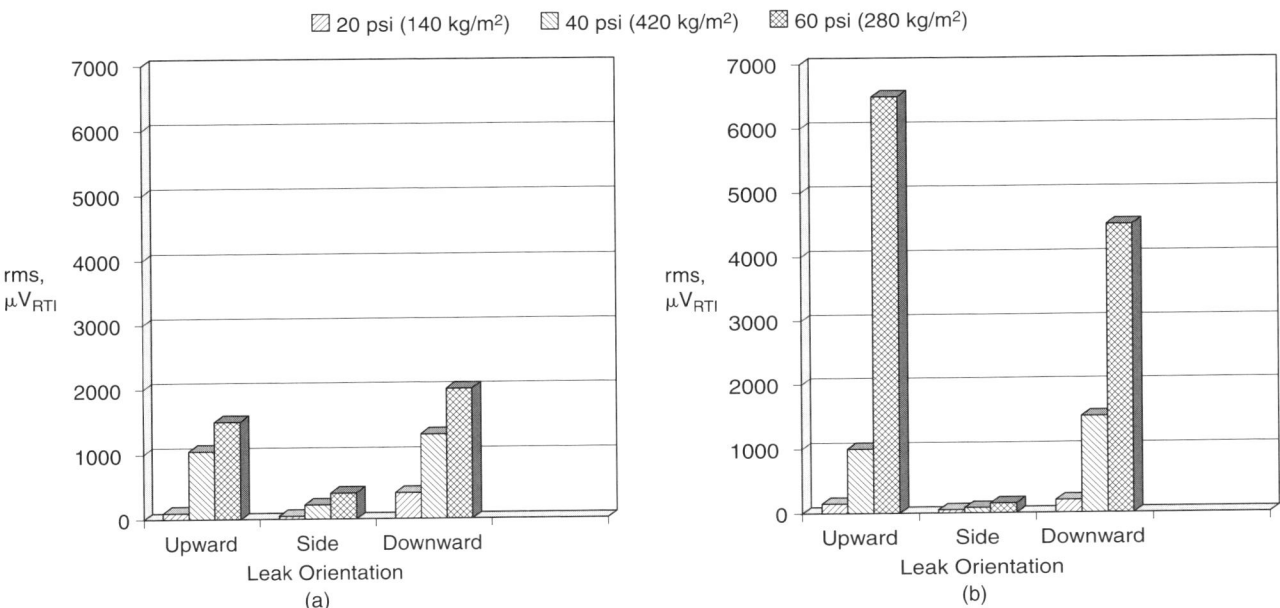

Figure 10.16 Root-mean-square (rms) signal versus orientation of leak plug and pressure, 2-in-diameter (50-mm) (*a*) and 4-in-diameter (100-mm) (*b*) pipe loop. μV_{RTI} = microvolts referred to the preamplifier input. [*Source:* Journal AWWA, *vol. 92, no. 7 (July 2000)*, by permission. Copyright © 2000, American Water Works Association.]

leaks could be modeled with quadruple radiation as the dominant source, thus providing for the first time a quantitative basis for applying AE technology to the detection and location of two-phase (gas–liquid) leaks.

The laboratory studies provided data for the development of procedures that were used in the controlled-condition experiments and in preliminary field evaluations. The test chambers were used to develop methods for increasing the leak signal above the value produced by the normal method of pressurizing the pipe and using its contents to test the pipe hydrostatically. In one method, a pressurized gas blanket was applied above the liquid contents of the pipe. This produced a two-phase flow (for a gas–liquid mixture) through the leak under the same test pressures, resulting in a one- to two-orders-of-magnitude increase in the AE signal at the leak source. However, in larger pipes greater than 2 in (50 mm) in diameter, this technique did not significantly increase the AE signal.

Field-Scale Controlled-Condition Experiments

Experiments were conducted at the UWRF. Leaks were introduced into the 2-in-diameter (50-mm) pipe loop in the same way as they were simulated in the laboratory. A class 3 leak plug was installed, the pipe was filled with water, and the system was pressurized with nitrogen to a pressure that would generate a given leak rate. During a tightness test to check for leaks, several leaks were discovered in the inlet and outlet fittings. One was accidentally created at the location of a "blank" plug (no leaking orifice) by a small grain of sand that had become lodged between the pipe surface and the O-ring that was used to seal the head of the cap screw. This small leak of 0.014 gph (0.014 mL/s) was measured by filling a graduated beaker during a fixed period of time. The leak rate was accurately detected and successfully located using AE and a 25-ft (7.6-m) sensor spacing.

Attenuation was measured and two series of leak tests were performed on the 4-in-ID/12-in-OD (100-mm/300-mm) insulated pipe loop. In the first test, a class 3 leak plug was installed in the conduit, and the annulus space was pressurized to 10 psi (70.31 kg/m^2). The resultant leak signal detected on the coated conduit attenuated rapidly. The rubber protective coating was removed in spots to place the sensor on the bare metal. This change improved the signal very little.

In the second test, sensors were coupled to the carrier pipe at the beginning and end of the pipe loop at a sensor spacing of 500 ft (152.4 m); a class 3 leak plug was installed in the first leg of the expansion loop 159 ft (48.5 m) from the sensor. The pipe loop was pressurized to 70 psi (4921.7 kg/m^2), which produced a 17.4-gph (17.4-mL/s) leak through the leak plug. Data were collected using the tuned linear location results shown in Fig. 10.17. The AE system detected and located the leak at about 150 ft (47.7 m); the precise leak location was 159 ft (48.5 m). A second signal was detected near the 300-ft (91.4-m) position before the transient signal ended during the first 10 min of data collection. The location of this transient signal corresponded to the actual location of the tee for a 30-ft (9.1-m) branch in the pipe loop. The data show that items such as elbows and tees disrupt the wave propagation of leak noise, making it more dif-

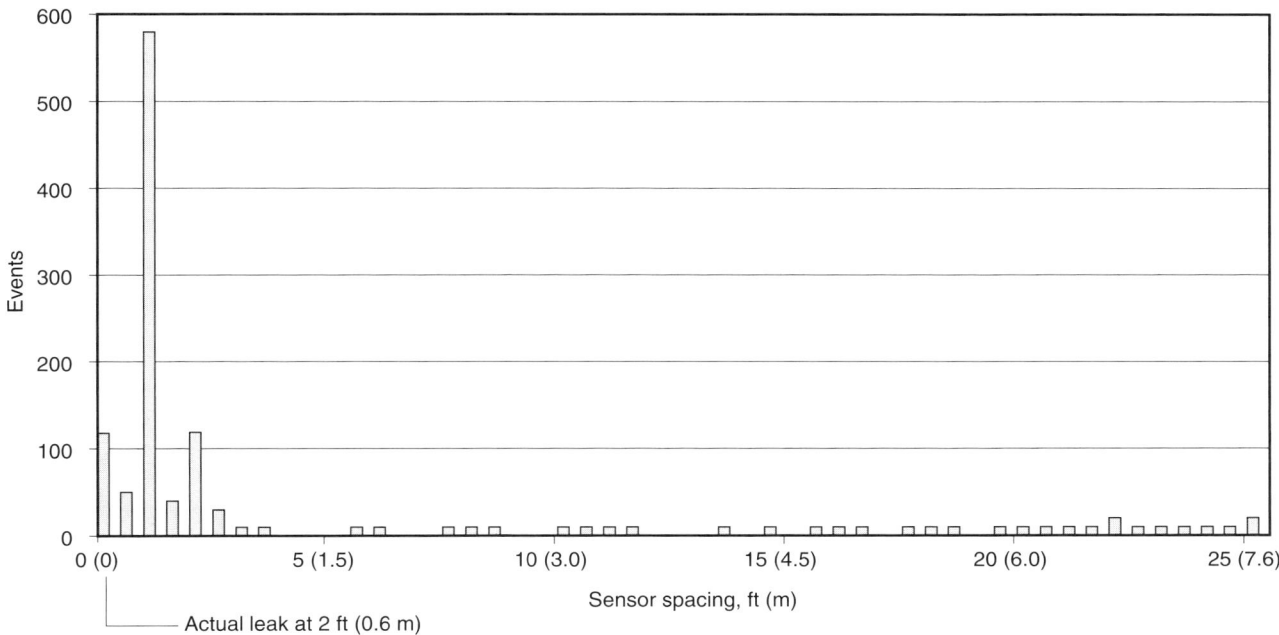

Figure 10.17 Tuned linear location of 0.014-gph (0.14-mL/s) leak rate from 2-in-diameter (50-mm) pipe loop. [*Source:* Journal AWWA, *vol. 92, no. 7 (July 2000), by permission. Copyright © 2000, American Water Works Association.*]

ficult to locate leaks. Nonetheless, the AE system successfully located a 17.4-gph (17.4-mL/s) leak rate at a sensor spacing of 500 ft (152.4 m), despite the tee, branch, and expansion loop.

An additional test on the 4-in-ID/12-in-OD (10.2-mm/30.48-mm) insulated pipe loop confirmed the previous results. The sensor was relocated to the end of the 30-ft (9.1-m) branch (tee section) at a sensor spacing of 363 ft (110.4 m). The objective was to evaluate the ability of the AE system to detect and locate a leak where the wave propagation from the leak had to traverse the expansion loop, make a 90° turn at the tee, and travel another 30 ft (9.1 m) within the branch before being detected by the second sensor. Despite these wave propagation obstacles, the system located the leak at 164 ft (49.8 m), only 5 ft (1.5 m) from the actual leak source.

Experiments on the 12-in-diameter (300-mm) pipe loop demonstrated good leak location accuracy for a 1.2-gph (1.2-mL/s) leak rate with sensor spacing of 100 ft (30.4 m) and for a leak rate range of 10–20 gph (10–20 mL/s) with sensor spacing of 200 ft (60.8 m). The effects of soil loading and leak orientation were also studied. The buried 12-in-diameter (300-mm) pipe was equipped with two sensors (each with a peak resonant frequency of 5 kHz) mounted 154.5 ft (47.1 m) apart, with a class 3 leak plug located 58.5 ft (17.8 m) from one sensor position. A 17.4-gph (17.4-mL/s) leak rate was generated by pressurizing the pipe loop to 70 psi (4921.7 kg/m^2). Data were collected and an estimate of the leak location was derived by combining three leak location methods: complex coherence, cross-correlation, and signal difference. The leak was determined to be 60.8 ± 4 ft (18.5 ± 1.2 m) from the sensor position, only 2.3 ft (0.7 m) from the leak source.

Another test was conducted with the sensors located 100 ft (30.4 m) apart and the leak plug installed at 40 ft (12.0 m). A 1.2-gph (1.2-mL/s) leak rate was generated into a sand backfill. In this test, the tuned linear location technique was used to process the data and to estimate the location of the leak. The leak was detected to be at the 37-ft (11.2-m) position, 3 ft (0.9 m) from the actual leak.

Preliminary Evaluations on Operational Pipelines

Early in the research program, an opportunity to study an operational system was presented by the U.S. Army at Fort Drum. A high-temperature water distribution system was made available for evaluation of the AE instrument, software, and sensor-transducer arrangements that were being developed. The field evaluation showed that the AE system using the tuned linear location method could successfully locate several leaks on an operational insulated pipeline that distributed 1400 gpm (84,000 mL/s) of 300°F (148.9°C) water at 300 psi (2109 kg/m^2) through pipe diameters that ranged from 5 to 6 in (125 to 150 mm). Several sections of insulated supply and return lines were assessed by both low- (3–8-kHz) and high-frequency (10–50-kHz) AE systems and sensor spacing that ranged from 232 to 874 ft (70.7 to 266.4 m).

Three evaluations were conducted while the pipeline was in full operation. The first evaluation focused on the return and supply lines between two access covers. For both sections, the carrier was 5 in (125 mm) in diameter, and the length of pipe tested was 602 ft (183.5 m). The second evaluation included the same sections of line with the annulus pressurized to 10 psi (70.31 kg/m^2). The third evaluation was conducted on a supply line with a 16-in-diameter (400-mm) carrier and sensor spacing of 874 ft (266.4 m). Several leaks were detected on the operating supply line.

Field evaluations were also conducted at the DOE's Oak Ridge Reservation Y-12 Weapons Plant on an operational 400-ft (121.9-m) section of a 1-in-ID/3-in-OD (25-mm/80-mm) double-wall pipeline that transferred gaseous hydrogen fluoride. The study investigated the feasibility of installing a continuous AE leak monitoring system. Data collected indicated that the AE technology was applicable to an operational double-wall pipeline. In addition, the data allowed researchers to specify the type of AE instrument, sensor, sensor spacing, and mounting technique, and to identify the leak location method that would be best suited for continuously monitoring this particular DOE process.

It was initially expected that only baseline and pressurization data would be collected at the DOE site. The line was pressurized to 30.3 psi (213 kg/m^2) against an initial target pressure of 30 psi (211 kg/m^2). The AE system immediately detected a leak. Monitoring was continued to determine whether the system stabilized. After 20 min, the DOE plant operator reported a pressure drop of 0.1 psi (0.7 kg/m^2). Operations personnel indicated that a loss so small over such a long period of time would have been previously dismissed as instrument error.

Four Methods Useful for Leak Detection

Four signal processing methods were evaluated and found to be useful for leak testing: (1) tuned linear location, (2) signal difference, (3) cross-correlation, and (4) complex coherence analysis. Similar approaches are described in the *Non-destructive Testing Handbook* (see Ref. 6). As a result of the research, the tuned linear location method was developed, and the signal difference method was modified.

Conventional Approach

Researchers tried unsuccessfully to locate the 0.014-gph (0.014-mL/s) leak accurately by use of a conventional linear location approach in real time. By varying the detection threshold of the AE channel with the highest "hit" rate (which corresponds to the sensor closest to the leak), the system was tuned to produce the linear location plot shown in Fig. 10.17. The location of the leak shown in Fig. 10.17 is within 1 ft (0.3 m) of the location of the actual leak on the 2-in-diameter (50-mm) pipe loop with a sensor spacing of 25 ft (7.6 m). The amount of threshold adjustment was determined by matching the hit rate on the two sensors that straddled the leak. Initially, each channel was tuned manually by adjusting the detection threshold to achieve a desired hit rate, until it was realized that this could be done automatically by use of the algorithm that was already available in the AE instrument.

Signal Difference to Estimate Leak Location

Previous studies of leak location report on how leaks can be located when leak orifice produces a continuous signal.[7] By measuring the signal strength at each sensor location and taking the difference, the leak location can be estimated by comparing the differential measurements with the signal difference-versus-position curve generated from an attenuation plot. This concept was extended by applying the same principle to the peak amplitude measurements taken from transient leak signals. The method should not be applied without an attenuation curve to reference the differential peak amplitude measurements. Additionally, the attenuation curve must be developed for a distance equal to or greater than the sensor spacing used. In this study, an attenuation curve was developed for a distance of 150 ft (45.72 m) using a 0.012-in (0.3-mm), 2H lead pencil break source; a 0.02-in (0.5-mm), 2H pencil break source; and a spring-loaded center punch. The data were fitted to produce the attenuation curve for the 2-in-diameter (50-mm) pipe loop. The differential peak amplitude measurement placed the "detected leak" to within 1 ft (0.3 m) of the "actual leak."

Overall, the tuned linear location technique provided the best results for all cases. The modified signal difference method provided an additional approach and served as a means for verifying the tuned linear method. The cross-correlation and coherence analyses worked in some cases but not all. Because the latter two methods rely on collecting and processing waveform data, they collect fewer data during a fixed period of time compared with the collection and processing of conventional AE hit data. Additionally, these methods were more susceptible to dispersion than the other methods. In summary, all three methods worked to various degrees, as long as an adequate data set was collected. About an hour was required to collect data for all of the methods studied.

Precautions must be taken in deciding whether to use the high- or low-frequency system and the source location method for data analysis. During the field evaluation at Fort Drum, the data collected using the low-frequency cross-correlation system provided good results on the 602-ft (183.5-m) section of the 5-in-diameter (125-mm) carrier pipe when both artificial and natural signals were used. However, the same system arrangement did not work during tests on the 874-ft (266.4-m) section of the 16-in-diameter (400-mm) pipeline. Conversely, the data collected using the high-frequency, tuned linear location system provided good results on the 874-ft (266.4-m) section of the 16-in-diameter (400-mm) carrier pipeline using both artificial and natural signals. The same system arrangement was less effective during tests conducted on the 602-ft (183.5-m) section of the 5-in-diameter (125-mm) pipeline.

Close Sensor Spacing (see Fig. 10.17)

As sensors are moved closer to the leak source, the metal-borne signal (which travels at higher velocities than the liquid-borne signal) becomes detectable. Conversely, as the sensor is moved farther away from the source, the metal-borne signal attenuates until it can no longer be detected. Concurrently, the liquid-borne signal, which tends to travel at lower speed and to attenuate less, dominates the "detected signal."

The following illustrates two findings regarding sensor spacing. First, the characteristic velocity that was used in the test was 3225 fps (983 m/s). When sensor spacing exceeded 100 ft (30.5 m), calibration tests were conducted using a spring-loaded center punch to launch a stress wave that propagated from one sensor to the other. The time of flight was measured and compared with the sensor spacing to derive a wave velocity that could be used by one of the source location techniques. The speed of sound in water, 4790 fps (1460 m/s), produced more accurate results when sensors were spaced at 100 ft (30.5 m) or greater. However, it is possible that the spring-loaded center-punch calibration technique may not reproduce the same wave propagation behavior as a leak. Thus, it is recommended that in situations in which the calibration signal velocity differs substantially from the water velocity, the leak location analysis should be performed twice, using each velocity, and the results should be compared. The second point illustrated is the potential for different wave modes traveling at different velocities to trigger the AE system: one sensor was placed 10 ft (3.0 m) from the leak plug and the other 40 ft (12.2 m) from the leak plug.

Tuned Linear Location Method to Locate Multiple Leaks

Static pressure tests produced the best results on all pipe loops at the UWRF. However, the tuned linear location method successfully and accurately located several simultaneous leaks on operational pipelines, as was demonstrated during the field tests on the Fort Drum and Oak Ridge systems. For each case, 1 h of data collection was sufficient to locate leaks accurately. Simultaneous leaks were detected and located on the operational system at Fort Drum using a sensor spacing of 874 ft (266.4 m).

Strongest Signals from Vertical Leaks

Laboratory studies on leak orientation indicated that vertically oriented (up or down) leaks produce the strongest AE signals and thus are more likely to be detected and located. However, they had little influence in the field on larger-diameter pipelines. Without the flow restriction caused by soil loading, there would be no leak to detect regardless of leak orientation.

Effect of Soil Loading on Leak Detection

Soil was removed from above a leaking class 3 leak plug, and a sensor was mounted next to this leak. The soil was then replaced over the plug in 1-ft (0.3-m) intervals up to a depth of 3 ft (0.9 m). The AE signal strength measurements were taken at each 1-ft (0.3-m) interval above the plug, resulting in data establishing that adding more than 3 ft (0.9 m) of backfill had no effect on the AE signal strength.

Two-Phase Flow Does Not Enhance Detection

Maximum sensor spacing and leak location were studied for various leak plugs and leak rates. These tests were initially performed to strengthen the signal using pressurized water and then with nitrogen gas to generate two-phase flow. Although the addition of nitrogen gas enhanced the laboratory tests and the experiments on the 2-in-diameter (50-mm) pipe loop, no improvement in leak location was detected during any of the other controlled-condition tests.

Effects of pipe diameters were evaluated. Leak detection and location were accurately pinpointed on operational pipelines up to 16 in (400 mm) in diameter and with sensor spacing of more than 870 ft (265 m). The effects of pipe diameter were examined during all phases of the program. Developed procedures enabled the characterization of dispersion effects, the selection of an appropriate leak location method, and the determination of the characteristic wave propagation velocity.

Coatings on Pipes

Coatings were studied relative to their effects on acoustic coupling (sensor to pipe) and on AE leak location performance. Experiments were conducted on the 4-in-ID/12-in-OD (100-mm/300-mm) pipe loop and in the field studies. Attenuation was measured on both empty and filled pipe loops and on the operational pipelines during and before operation. Coal tar, one of the most typical coatings, did not change signal attenuation. Evaluations of sensor coupling techniques determined that direct sensor coupling worked well and that removing or disturbing the protective coatings was not warranted. Velocity was measured using the lead break and the spring-loaded center-punch calibration methods.

Elbows and tees may lead to false positives. Elbows and tees were compared with straight pipe. If such features are not taken into consideration, their effects may lead to false positives, i.e., erroneous detection of a leak. However, these problems can be overcome by using elbows and tees as locations for mounting sensors.

10.4.3 Summary

The overall objective of this program was to develop a portable AE system that could detect and locate a 0.1-gph (0.1-mL/s) leak on an operational system with maximum sensor spacing of 500 ft (152.5 m). A secondary objective was to develop a continuous monitoring system that could detect and locate a leak rate of at least 3.0 gph (3.0 mL/s). Ongoing research is needed to satisfy these objectives completely; however, several accomplishments have shown the capabilities of the AE system developed under the program. These include the following.

- The ability to detect and locate a 0.014-gph (0.014-mL/s) leak to within 1 ft (0.3 m)—with 25-ft (7.6-m) sensor spacing—of its source showed that the system could detect and locate the mandated leak rates.
- The system accurately detected and located multiple leaks in an operational insulated pipeline with flows up to 1200 gpm (1200 mL/s) at a sensor spacing of 874 ft (266.4 m).
- The system accurately located a 1.2-gph (1.2-mL/s) leak at 100-ft (30.5-m) sensor spacing on the 12-in-diameter (300-mm) pipe loop.
- The system showed continuous monitoring capabilities in its application to the operational DOE double-wall pipeline.

Findings from ancillary studies supported the field application and further development of the AE system. These findings include the following.

- The optimal system frequency range to locate the leak in the signal processing methodology was determined to be 5–15 kHz.
- The effects of pipe diameter, leak orientation, soil loading, pipe coatings, elbows, and tees were identified and accounted for in the field application of the system. Vertically (up–down) oriented leaks produced the strongest AE signals; without soil cover, no AE signal, regardless of leak orientation, was observed. Elbows and tees interfere with leak detection, whereas mounting sensors on the elbows and/or tees reduces or eliminates this interference. Although coatings on pipes are prime concerns in nondestructive testing, coal tar, one of the most typical coatings, did not interfere with signal processing.
- Several new transducers were designed, fabricated, and evaluated to accommodate the characteristics of the pipelines under study and the conditions experienced in the field.

Current research is investigating the capabilities of acoustic technology for continuous remote monitoring of the integrity of distribution systems, e.g., incorporating sensors to monitor for leaks and/or defects, wall thinning, and biofilm growth.

10.4.4 Acknowledgments

The U.S. Environmental Protection Agency, through its Office of Research and Development, collaborated in the research described with the Department of Defense, Department of Energy, and National Science Foundation, through a grant to the New Jersey Institute of Technology (NJIT). The Physical Acoustics Corporation of Princeton, New Jersey, was also involved as a subcontractor to NJIT. The work was supported by a grant from the Department of Defense. Mention of trade names or commercial products does not constitute endorsement or recommendation for use.

It is obvious that some of the techniques here are not directly transferable on a commercial scale today; however, it is research like this which is responsible for the new technology which is constantly emerging in the marketplace, which allows us to keep lowering the economic and technical minimum levels of leakage in our systems.

10.5 FLOW ANALYSIS, STEP TESTING, AND ZONING

Flow analysis, step testing, and zoning analysis has been practiced for many years. The analysis involves setting up either permanent or temporary zones (see Figs. 10.18 and 10.19) and then analyzing the relationship of flows within the zones. In North America the zones often tend to be fairly large and temporary, as operators do not like their systems shut in, for water quality and fire-fighting reasons. In Europe, operators tend to operate small permanent zones on a continual basis.

10.5.1 Flow Analysis

A simple way to compare flows to determine if an area has high leakage is to determine the nighttime flows as a percentage of the average

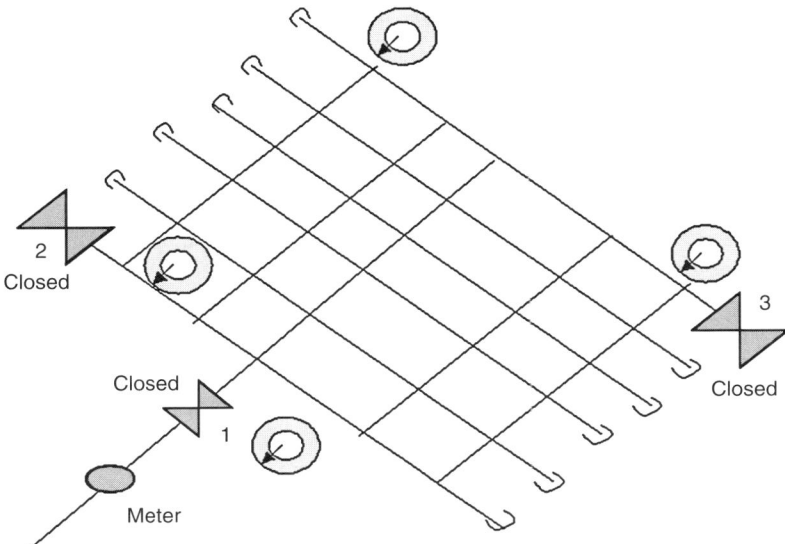

Figure 10.18 Zone flow analysis. (*Source: Julian Thornton.*)

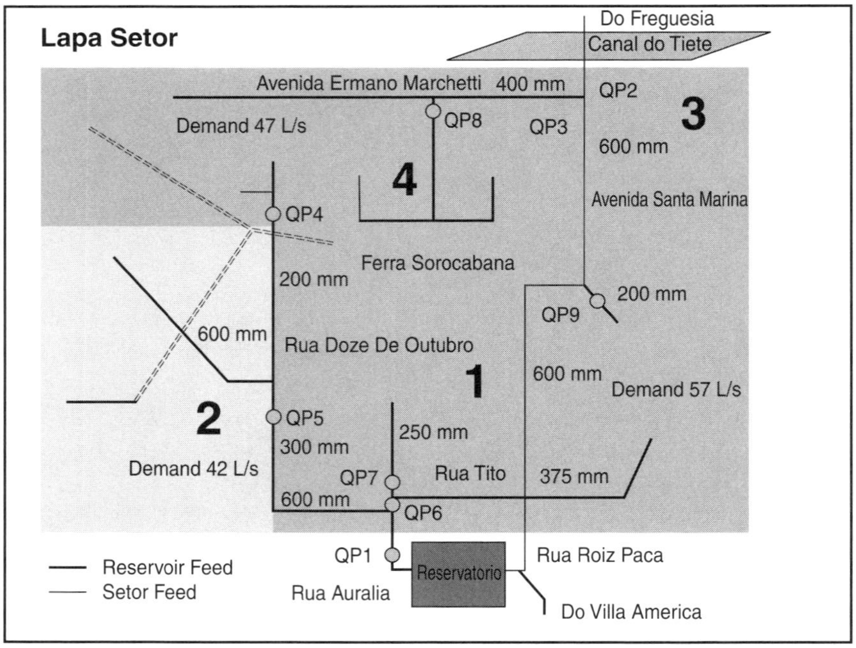

204 CHAPTER 10 REDUCING REAL LOSSES IN THE FIELD

Figure 10.19 Zoning Contract 66.593/96. (*Source: SABESP/BBL Ltda.*)

> If nighttime flow is a large percentage of average daytime flow, this could indicate a potential for leakage if there are no significant night users.

daytime flows. If the percentage is high, the potential for leakage is high. See Figs. 10.20 and 10.21. This method is very system- and zone-dependent, as obviously some areas (in particular, industrial areas) could have 24-h consumers who form a large part of the daily flow pattern. In this case the operator would identify all of the zones with high night/day flow relationships and then program them for leak pinpointing. Once leaks have been repaired and the flows returned to what would be considered a normal ratio, the zones are monitored to see if and when the leakage returns. Then the next intervention can be scheduled. Another way to use night flows is to perform a component-based analysis of the zone and its flows and break out what is estimated to be legitimate flow for consumer use, the rest being leakage. Chapter 7 shows some simple models which can be used in this type of approach.

Before any kind of zone analysis can be undertaken, it is necessary to do

- ❑ Desktop study
- ❑ Demand analysis
- ❑ Field checks to ensure that the zone is isolated

Desktop Study

The desktop study is undertaken to identify suitable areas on paper or on the GIS system. The areas should be chosen by ease of isolation, using as many natural boundaries as possible. Fewer valves closed in will ensure fewer problems in the future. Particular care should be taken when initially selecting the zones, not to shut in feeder or transmission mains, which are necessary to feed other parts of the system. Often the differ-

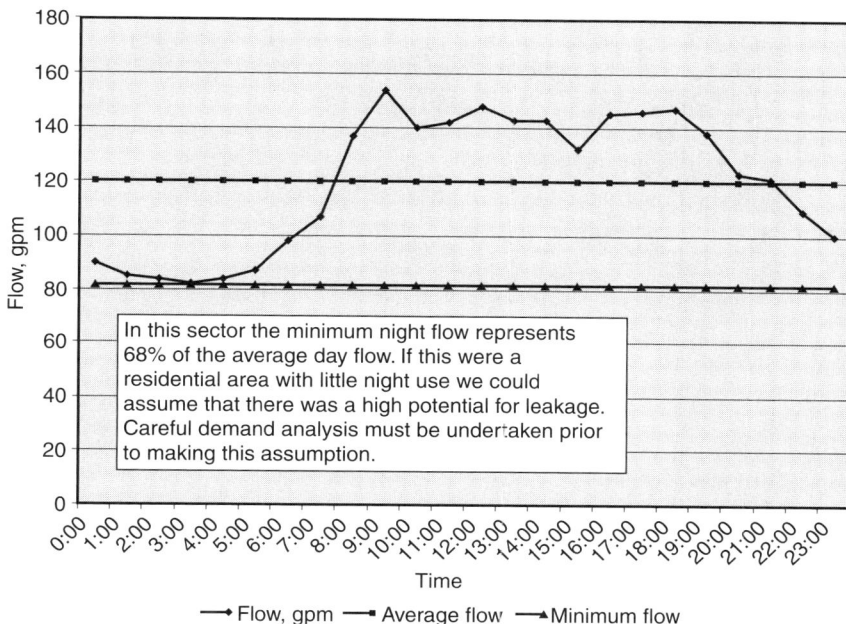

Figure 10.20 Comparison of nighttime flow and average day flow. (*Source: Julian Thornton.*)

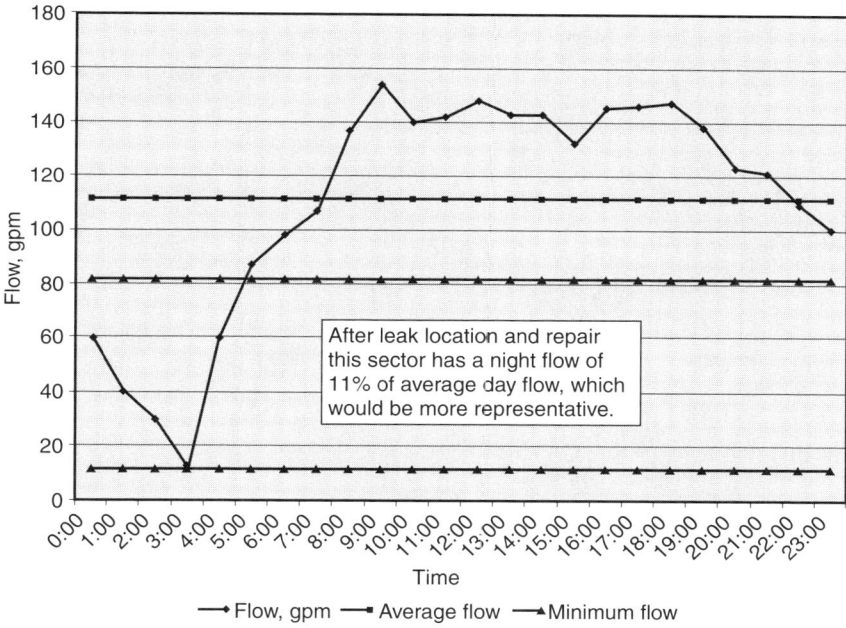

Figure 10.21 Comparison of nighttime flow and average day flow after leak repair. (*Source: Julian Thornton.*)

ence between a successful, viable, and sustainable zone flow program depends on the proper selection of zones at the desktop stage. This task is often underemphasized, as operators are either keen to get out into the field or are not confident in the quality of the plans.

Demand Analysis

As with many other forms of system intervention or analysis such as pressure management or hydraulic modeling, it is vitally important to undertake a demand analysis to understand the kind of consumers and the individual needs of those consumers.

Due to the nature of the estimations of legitimate demand at night and therefore the recoverable or controllable leakage, it is common practice to identify large 24-h consumers such as industrial, commercial, and institutional users. These consumers should either be recorded as to actual consumption at night during the analysis period (usually between 2 and 4 a.m.) or should be valved out of the test so as not to influence the predictions. Particular care should be taken when considering zone flow analysis near hospitals and kidney dialysis patients.

Field Checks to Ensure That the Zone Is Isolated

Before relational analysis can be undertaken with field measurements, it is necessary to ensure that the zone to be tested is properly isolated from neighboring zones. The zone does not have to be a single feed, although this is simpler to control. If multiple feeds are to be used, then it is important to ensure that the measuring devices and recorders or data loggers are properly synchronized before testing.

Many operators like to undertake a zero shut test to ensure that the zone is properly isolated from neighboring zones. This entails shutting down the supply or supplies and measuring the pressures throughout the system to ensure that there are no feeds or valves holding the pressure up. Figure 10.18 shows a sample layout for a zero shut test. This should be the preferred methodology for all zone isolation testing, although it is not always possible to undertake given the nature of demands on the system.

The use of pressure loggers for this exercise can greatly benefit the operator if something is found to be entering the system. In this case the operator merely compares ground levels to the static pressure read and can identify from this information where the potential source is; see Fig. 10.22.

In some systems the operator may prefer not to undertake a zero shut test. This may not be possible for a number of reasons:

- ❏ Large consumers requiring constant supply
- ❏ Kidney patients undergoing dialysis
- ❏ Social or political influence

If this is the case the operator may perform one of two other types of testing to ensure that the system is tight (although they may not be as accurate):

- ❏ Dead-end testing
- ❏ Boundary valve sounding

Dead-End Testing

Dead-end testing involves identifying all of the potential boundary valves which should be closed. Then the operator identifies a gate valve back into the zone (it can be on either side of the boundary valve), which

> **T**he demand analysis phase is one of the most important parts of the loss management program and should not be overlooked.

10.5 FLOW ANALYSIS, STEP TESTING, AND ZONING 207

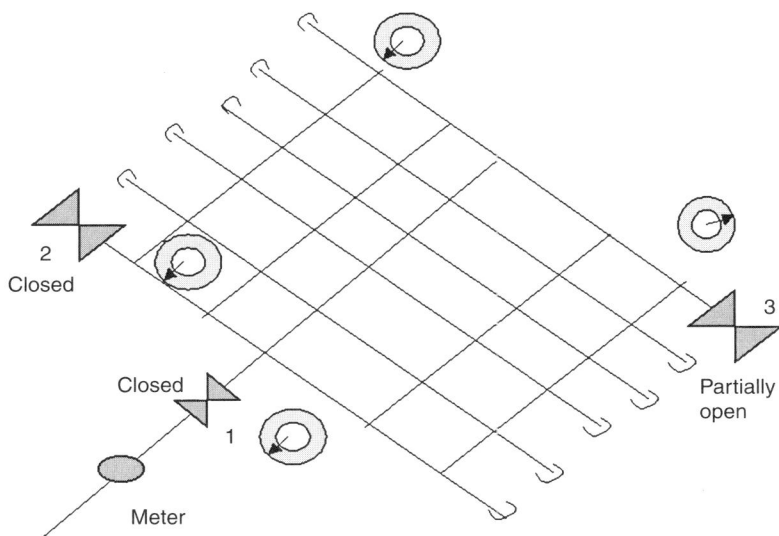

Figure 10.22 Identifying a potential zone breach. (*Source: Julian Thornton.*)

has either a hydrant or a pressure fitting between the chosen gate and the boundary valve. This short piece of main is valved off and the pressure checked at the test point. See Fig. 10.23. If the pressure can be bled off the short section of dead-end main, then the boundary valve is closed. If the pressure cannot be bled off, then either the gate valve or the boundary valve is passing by. The valves can be sounded using sonic testing equipment to see which valve is not functioning properly, and the problem can be rectified. This methodology is then applied to all boundary valves.

The only negative to this approach is that there may be unknown supplies into the zone which may not be picked up by this method.

Boundary Valve Sounding

All boundary valves are tested for the sound of water passing under a partially closed or defective gate. This method is effective only where there is a significant difference in pressure between one zone and its neighbor and should be undertaken only during periods of high demand (otherwise water may not be being drawn from one side to the other). The sounding technique is not very effective for finding boundary valves which are completely open, as these valves create very little friction sound, which is created by the water squeezing through the partially shut gate.

The sounding-only technique should be used as a last resort. It should generally be undertaken in conjunction with a physical valve condition test. The valve condition test is undertaken with a valve wrench. The valve is checked to see if it is completely open or completely closed.

Some systems have valves which close anticlockwise and others clockwise. Unfortunately, some systems have a mixture of the two. If uncertain, the operator must check.

> **S**ounding of boundary valves to ensure a closed zone is effective only if there are different pressures between the zones and a demand.

208 CHAPTER 10 REDUCING REAL LOSSES IN THE FIELD

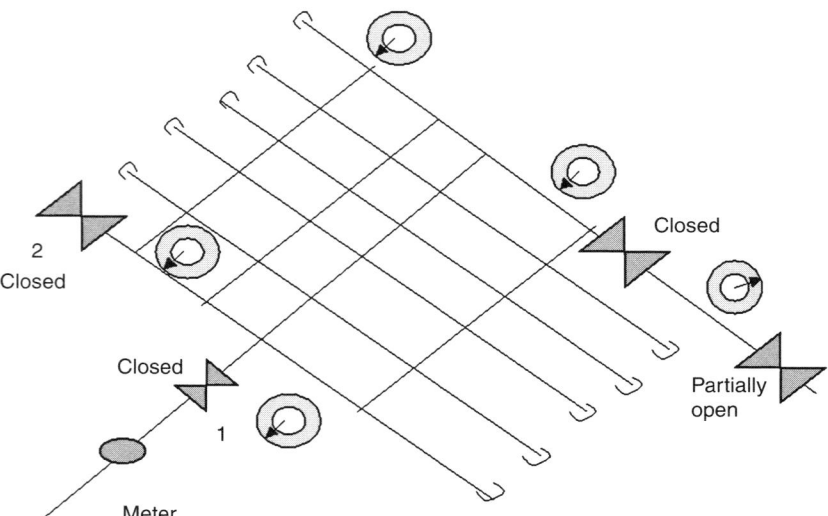

Figure 10.23 Dead-end testing. (*Source: Julian Thornton.*)

> Some valves close clockwise and some anticlockwise.

To check whether a valve is completely open or closed it is necessary to rotate the valve completely in one direction and squeeze shut. If the valve is actually shut, as the operator believes, it will take two clicks to open the valve. If the valve is actually open, it will take only one click. By clicks we mean that as the tension is taken off the valve spindle in the closed position, the stem will feel loose. Then, when the gate is lifted from the seat, it will again feel loose. These are known as clicks. The intermediate positions make it harder to turn the spindle. The opposite is true when the valve is completely open, as the operator will only need to release the tension from the spindle and then it will be loose or free all the way to the closed position, as there is no need to release the gate from the seat.

Once the zone has been proven to be isolated, field measurements can be taken. Chapter 6 deals with the equipment used for field measurements in more detail.

Many systems are taking zone flow analysis very seriously and are using complex commercial models to determine efficient levels of leakage. By efficient we mean not spending more on intervention than the value of the recovered loss, but also spending a realistic amount to keep leakage to a minimum. A more detailed discussion of auditing and modeling can be found in Chaps. 4 and 7.

Part of the data input for these models requires a certain degree of demand analysis, therefore forming a fail-safe system of data collection and entry and saving time.

Once zones have been ranked, intervention can begin. Some utilities just send out pinpointing crews, but others perform a step test first.

10.5.2 Step Testing

Step testing involves isolating sections of main from the zone and the zone meter, which is recording the flow on a data logger or loggers. Every time a section with a leak is isolated, a marked drop will be seen on the flow graph (see Fig. 10.24). This drop represents the leak volume, which

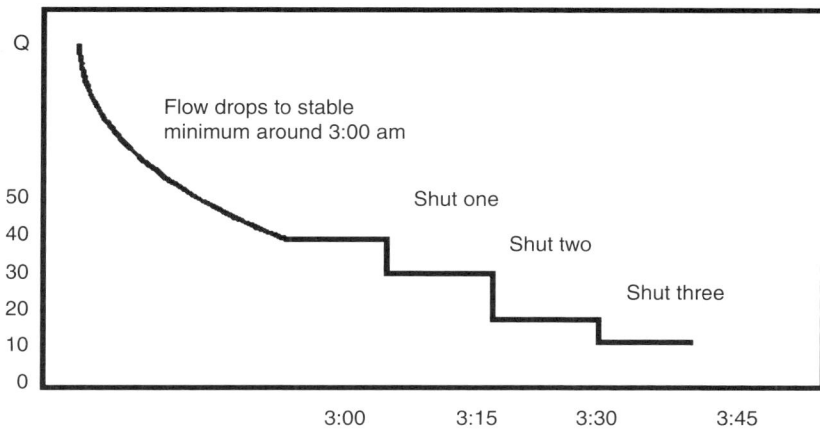

Figure 10.24 Step test. (*Source: Julian Thornton.*)

is nice to have recorded for cost-to-benefit calculations and program tracking and also saves time by directing leak pinpointing crews only to those sections of main where leakage has been proven to be occurring. Again, when undertaking step testing it is very important to know who the customers are in the area and who cannot be isolated. Step testing is usually carried out at night.

Step testing can be carried out in other forms too. Many rural water systems do not have zones, but they do have sections of plastic pipe with not very many fittings, in which it is often hard to find leaks. One interesting way of setting these systems up for sustainable leakage management is to install measuring points in the system either permanently or temporarily. These points are then analyzed at night, when there is very little use. (See Fig. 10.25.) Any sections showing significant flow are marked as having leakage and intervention is scheduled. This method saves costly surveys over miles and miles of main that may not have leakage and helps the operator home in on the sections with significant leakage. When the volume of leakage is known, the operator can justify higher levels of effort to locate the leak than he would be able to do if he were unsure as to the existence of leakage. In some cases the cost of excavating additional test holes can be justified to track a leak when its volume is known. If the operator were unsure that the leak is actually on that section, he would not wish to spend the additional time and money.

Every time a new test hole is excavated it is a good idea to build a simple chamber around it. It will then be one more test hole for the next time the utility needs to survey.

Valve Squealing

A simpler form of the above methodology can also be performed at night by squeezing down certain gate valves in the system and listening on the gate valves with a listening device as described earlier in this chapter. This methodology is similar to that described for testing boundary valves.

The valves which remain quiet are the ones with downstream sections of main that have little or no leakage. The ones that squeal have either

CHAPTER 10 REDUCING REAL LOSSES IN THE FIELD

Figure 10.25 Using an ultrasonic meter to measure leakage. (*Source: Julian Thornton.*)

usage or leakage downstream, and those sections can be justified for more detailed intervention.

10.5.3 Noise Logger Surveys

Noise loggers are a relatively new technology, having been around in various forms for the last 10 years or so. In general, the loggers are placed in strategic positions within either permanent or temporary zones, which have been identified as having either a high night flow-to-day flow ratio or a level of leakage which is above economic, depending on the analysis used. The loggers record the sounds in the system over time and can be interrogated by a software package. Areas where suspected leakage is occurring (areas with continued high noise at night) can be marked for further interrogation.

Recent advances in technology now allow loggers to correlate with one another, in some cases saving the need for a separate acoustic or correlation intervention.

The following case study shows how Severn Trent Water, in the United Kingdom, uses noise loggers *en masse* to reduce leakage to very low levels. These noise loggers are not correlating noise loggers, but they are permanently installed devices which transmit data via radio to a receptor unit. Some utilities may find it cost-effective to employ permanent loggers, while others may find it more cost-effective for teams to deploy temporary loggers. Each system should study the options and costs related to those options and choose the best methodology for the situation.

10.6 CASE STUDY THREE
Severn Trent Leakage Management Process
Martin Kane

10.6.1 Overview

This case study describes how Severn Trent (ST) has harnessed new technology as part of its leakage strategy and achieved a step change in leakage performance. Critical to ST's success has been the deployment of a device called Permalog, which has allowed the end-to-end process of leakage detection and repair to be reengineered. Permalog is not a panacea in and of itself. It is the way it is deployed and the overall leakage process within which it sits that delivers real business benefit.

10.6.2 Severn Trent

The U.K. operation of Severn Trent provides water services to 8 million people across an area of 8000 mi^2. The catchment includes the U.K.'s Second City of Birmingham and 10 other major industrial cities in central England that hold a large proportion of the engineering and industrial base. Leakage is given a special focus within the company's distribution system operations, with all activity and management reporting to a Leakage Process Manager.

10.6.3 Background

Severn Trent was originally formed in 1974 as part of a major reorganization of the U.K. water industry. Leakage levels gradually rose from then until the early 1980s under a generally passive approach to detection. Leakage was not one of the more visible targets for the water industry, and the government was generally relaxed about the figures. In 1983, the 3-year average leakage level was 574 ML/day or 30 percent of water supplied. From this time on an active policy was followed, and by 1986 ST had established an extensive coverage of district meter areas (DMAs), with 2200 DMAs now in play. The impact of DMAs and pressure management initiatives was immediate, and leakage levels had been reduced to 525 ML/day by 1989. Following U.K. water industry privatization in 1989, leakage levels rose across the industry. By 1994, ST's leakage level had reached 665 ML/day, or 32 percent of water supplied.

In 1995 the United Kingdom experienced a drought, with a number of water companies having to issue restrictions to customers to conserve supplies. The media embarked on a campaign to raise the profile of leakage and water resources as national issues. The company declared a program of mains reinforcements and source development to ensure that we were capable of providing sufficient water under worst-case scenarios. In parallel with this activity, the company committed to reducing leakage by 50 percent by the year 2000. The leakage target agreed to with the government at that time was 342 ML/day to be achieved by March 2000.

10.6.4 Severn Trent Leakage Management Strategy

The leakage management strategy addressed the following issues:

- Accurate measurement of the company's district meter areas (DMAs)
- Implementation of valve control procedures to track all strategic and tactical valve operations to ensure DMA integrity
- Pressure management and optimization of existing pressure-release valves
- Meter reading/repair frequencies set and achieved
- Information technology (IT) systems enhanced to manage leakage data and provide effective targeting of the detection effort
- Introduction of response-time targets to fix leaks once detected
- Investment in the latest leakage detection equipment, including leak noise correlators and Aqualogs
- Development of an in-house leak detection training facility, with all appropriate staff trained and assessed in the latest methods and best practice

The new process showed significant reductions in the first 3 years, from 665 to 399 ML/day, albeit rather expensively, as the initiative was labor-intensive. While the cost of fixing leaks remains fairly static, the cost of detection follows an exponential curve, as the lower the leakage level, the greater is the cost of reducing leaks.

The Economic Level of Leakage

Given mounting criticism from the U.K. government, national media, and consumer groups on the levels of leakage being reported by the industry, the debate turned to what the economic level of leakage for any water company should be. Each water company undertook this work for its own operation, with the outputs differing to reflect the diverse nature of the mains systems, supply headroom, and cost of water among other factors.

Leakage targets were then defined which, while meeting government and regulatory requirements, resulted in the lowest cost to the company of supply and demand over a 30-year planning horizon.

The outcome of the work was a least-cost curve, which sets out a target for ST of 330 ML/day for the year 2002/2003, equating to a leakage rate of about 18 percent of total supply. The curve shows that lower levels of leakage rapidly become significantly more expensive, driven by the increased costs of detection.

The study also concluded that the company's approach to leakage control represented water industry best practice. However, in the light of the U.K. water regulator's (OFWAT) determination of a 14 percent reduction in income from 1 April 2000, the cost of achieving 342 ML/day during the period 1999/2000 together with the predicted costs of making any further inroads into the target of 330 ML/day were becoming prohibitive using traditional leak detection methods.

Traditional Leakage Detection

Over the past 20 years, leak localization (finding the general area in which leakage is located) has been carried out in the United Kingdom using one of three methods:

1. "Stop tap bashing" still accounts for the vast majority of leak localization activity, although it is a slow and highly repetitive task.
2. "Step testing" is still used to localize significant mains bursts, although it is less popular nowadays due to the necessity for night work and the potential for water-quality problems.
3. "Noise logging" has been introduced effectively in recent years, with an overall improvement in cost efficiency.

All these methods are labor-based and therefore incur significant operational expenditures. In the current climate, water companies face the challenges of achieving and maintaining lower leakage levels at lower cost, as well as improving customer service by maintaining continuity of supply, reducing the overall amount of water lost through leakage, and responding more quickly to incidents when they occur.

The noise logging devices currently on the market, such as Aqualog, showed the potential for the technology to pinpoint where to look. However, the units were quite "intelligent" and required programming each time they were used. They were also expensive. Retrieval of data required a visit with a laptop PC to download the readings. These were then taken back to base for analysis of what leaks might exist and where. In short, they did not produce the reduction in process time needed to match future cost-reduction requirements.

10.6.5 Permalog

In late 1998, Severn Trent entered discussions with Palmer Environmental about an opportunity to become involved in the development of an exciting new concept called Permalog, which addressed this fundamental issue. As the name suggests, this is an extension of the Aqualog concept.

In order to overcome the costs of data retrieval, the new device had a built-in processor. It would still listen for noise in the pipe, but it would be able to track a stable pattern and send an alarm when noises were heard that were outside the anticipated profile.

Rather than downloading data from the logger, the area in which loggers have been deployed is patrolled with a remote hand-held receiver, known as a Patroller. This is usually deployed from a moving vehicle that drives round the area under investigation and automatically receives, decodes, and analyzes transmissions from the loggers, communicating its findings to the operator.

Each logger transmits to the Patroller whether it hears a noise or not. Clearly, any two adjacent loggers hearing a noise indicate that a leak exists on the main between those points. Once a potential leak has been exposed, a two-person team goes out and pinpoints the exact location using standard correlation techniques.

Early 1999: ST Pilot Trials

During the first half of 1999, ST piloted the Permalog system. The aims were to prove the performance of the new technology in a realistic environment and to demonstrate the operational and economic benefits that could be obtained. Several DMAs, representative of ST's distribution network, were selected as suitable locations.

One of these DMAs was Castle Donington, located in the east of ST's region. A total of 173 Permalog units were deployed across the whole DMA. Immediately following deployment, an initial Permalog patrol was carried out which identified many areas of interest for follow-up, and 13 leaks, including a significant mains burst, were confirmed by correlation and/or surface sounding. Once these had been repaired, a second patrol identified a further five leaks. Each patrol took only 2–3 h to complete.

July 1999: Full-Scale Trials in Worcestershire

In July 1999, ST purchased 14,500 Permalogs in order to carry out full-scale deployment in four of its nine operational areas. Of these loggers, 10,000 were assigned to just one of the company's operational areas, Worcestershire:

- Area: 1981 km^2
- Population: 616,900 people
- Properties: 250,000
- Average daily consumption: 170,000 m^3/day
- Length of water mains: 4610 km
- Service reservoirs: 59
- District meter areas: 300

Over the course of eight months, ST developed world-leading leakage processes in the application of this new technology, enabling it to reduce leakage levels in Worcestershire from 27 to 15 percent of water supplied. This equates to water saving of 21,000 m^3/day.

The process has now moved on to cover Birmingham, a city of over 1 million people, where an innovative "lift-and-shift" strategy, using Permalogs, has delivered significant reductions in leakage.

This "lift-and-shift" strategy has further built on the process management innovations developed for Worcestershire, provided for enhanced cooperation between distribution network controllers and field service staff, and allowed for more directed reward schemes for field staff. The process has been closely coordinated with pressure reduction and optimization schemes and is now being rolled out across the whole of the company's distribution operation.

Benefits of Technology

Permalog challenges the traditional concepts of leakage detection. Traditional methods relied on targeting an area and then deploying any number of field teams to sound between fittings to find out whether leaks existed. As leakage levels fell, the cost of both maintaining the position and bettering it became proportionally higher.

New technology has allowed the end-to-end process of leakage detection and repair to be totally reengineered. The shift from manual sounding to a reliable IT-based alternative that sits as a constant monitor on the network enables leaks breaking out on the network to be detected much sooner. Repairs can be scheduled to follow the detailed detection phase, as it is known, with more certainty that leaks will be found.

As total leakage is a function of the number of leaks breaking out and the time they run before detection and repair, the optimized detection process using Permalog can generate significant benefits in leakage management. Once areas have stabilized, the deployment density can be reduced and/or the patrolling frequency reduced, depending on the desires of the client, the money available, and the targets to be met. The benefits from Permalog lie not just in the application of the technology, but also in the ability to reengineer the whole leak detection and repair process.

10.6.6 Achievements to Date

Leakage levels in ST have nearly halved since the introduction of its Leakage Control Strategy in 1996. Unaccounted-for water (UFW) in 2000 stands at 340 ML/day, compared with 665 ML/day in 1994. The U.K. regulator (OFWAT) has set mandatory industry targets for 2001–2002, and STW's target was set at 333 ML/day.

The company currently has the second lowest unit leakage level in the U.K. industry. More significantly, it is now firmly in the lead position in the "virtuous quadrant" of low leakage and low water consumption by its unmeasured domestic customers.

Demand has correspondingly reduced from a level of 2100 ML/day in 1995 to a current level of 1880 ML/day. This reduction obviously has significant implications for the company's resource strategy. The additional costs of active leakage control are currently estimated to be of the order of £16 million per year. Clearly, there has to be an economic level of leakage below which further investment cannot be justified.

OFWAT has assessed ST's own calculation of the economic level of leakage as robust and recommended that the company should be measured against its progress toward this level. They do require, however, that the economic level of leakage be recalculated on a biannual basis. Given the achievements to date, it is likely that the company's success in improving pressure reduction, Permalog, and other leakage detection and repair techniques will become widespread, leading to an ongoing downward direction for the economic level of leakage for the water industry.

10.6.7 Conclusion

Harnessing new technology such as Permalog has enabled ST to make a step change in leakage performance and maintain its industry-leading position. It must be emphasized, however, that new technology is not in itself a panacea in the war against leakage. At the end of the day, Permalog is simply one strategic tool that sits firmly alongside a number of tried and tested methods and practices. It is the strategic leakage management process within which it is used that delivers the real benefits.

Tracer gas leak detection is often used for hard to find leaks, hydrostatic test failures, and demand-side small leaks and slab leaks. The following article is a very detailed review of the methodology and equipment needed for this kind of testing and has been written by Dave Southern of Hetek Solutions Inc. of Canada (formerly known as Heath Consultants Limited).

10.7 ARTICLE ONE

Tracer Gas Helium Testing Conduits and Closed Systems: Procedures and Methodology

Dave Southern, Technical Service Operations, Hetek Solutions Inc., London, Ontario, Canada

10.7.1 Summary

The Gasophon was developed as a self-contained portable leak detector to locate leaks in buried conduits from the ground surface, using helium as a tracer gas.

The tracer gas process for locating leaks in buried conduits from the ground surface involves inserting a known amount of tracer gas—helium gas or a helium/air mixture—into the pipeline or closed system after the existence of leakage has first been determined by a hydrostatic or other test. A failure to hold static water pressure at a predetermined pressure for a predetermined time is considered a hydrostatic test failure.

The helium-type tracer gas procedure is generally utilized after other traditional methods of leak detection have been exhausted. Preparation for the test requires removing product from the pipeline system, placing test holes over the pipeline or system at predetermined intervals, securing a supply of helium, and conducting background gravity tests in the test holes. All other methods of leak detection should be explored, other than excavation, prior to attempting the helium tracer gas procedure.

10.7.2 Equipment and Materials

Gasophon: General Description

The procedure is carried out using a portable, highly sensitive, battery-operated gas detector. The instrument can be utilized to detect a wide variety of gases with specific gravities different from atmospheric air. The instrument has the ability to locate gases of many types, light or heavy, flammable or nonflammable. The expansion of the scale through three measuring ranges enables even the minor gas concentrations to be detected and differentiated. The instrument does not indicate the tracer gas percentage of volume, and consequently, cannot be considered a quantitative instrument. The heavier or lighter the gas being detected in relation to atmospheric air, the more sensitive the instrument is to that gas. Helium is a desirable gas for locating leaks in pipeline or closed systems because it is a nontoxic, inert gas which is readily available in pressurized cylinders. It is the second-lightest gas, having a specific gravity of

0.137, and because of this, is easily detected in small concentrations by the Gasophon from 0.5-in test holes placed in the ground over buried systems.

Principle of Operation

A simplified explanation of the operation of the Gasophon is as follows. The velocity of sound in air is approximately 330 m/s. With gases, the velocity of sound will be greater or less than air depending on whether the gas is heavier or lighter than air. The instrument consists primarily of two tubes, one for sample measurement and the other containing atmospheric air as a standard for comparison. Each tube is fitted with a sonic transmitter at one end and a receiver at the other. An internal pump draws the sample to be analyzed through the measurement tube. The difference in the velocity of sound in the two tubes is directly proportional to the specific gravity and amount of tracer gas in the sample. The velocity difference is measured electronically and is displayed as a meter deflection. The heavier or lighter the gas, the more sensitive is the instrument.

Plunger Bar

The plunger bar is a manual impact tool designed to drive a 0.5-in-diameter rod to a depth of up to 3 ft.

Helium Injection Gauges and Manifold

The regulator is a standard gas regulator such as an L-Tec Trimline model R-76-150-580 with gauges for 0–5000 psi on the inlet and 0–100 psi on the outlet side. The 0.5-in-diameter outlet hose goes to the inlet of the manifold. The manifold has a 0.5-in-diameter inlet and a standard Chicago fitting for the compressor inlet. At the other end of the manifold is another Chicago fitting, which can be removed if the manifold is to be attached directly to the injection point instead of being attached to the compressor outlet.

Compressor

Any compressor with an outlet in the range of 175 ft^3/min, such as the Ingersoll-Rand 175, can be used provided that it has a Chicago fitting on the outlet and it has no oiler.

Pneumatic Rock Drill

Any 30–60-lb air-operated rock drill with a 1-in chisel bit is suitable.

Helium Gas Cylinders

Industrial-grade helium is readily available from welding suppliers.

10.7.3 Prerequisites

The success rate for the type of test is relatively high but there is no guarantee that every leak can be located. In order to avoid problems and assume that the operator has a reasonable probability of success, the following background information is required.

Type, Diameter, and Length of Pipe or System to Be Tested

The type of pipe (PVC, ductile iron, other) is necessary, as different pipe materials may have different test pressure limits. Also, the pipeline or system may have to be located if as-built drawings are not available. If the pipeline is PVC, does it have a tracer wire? The diameter and length of the pipeline or system are needed to calculate the volumetric capacity, which is necessary to calculate the amount of helium required.

Results of Any Hydrostatic Testing

The test requirements for residential water mains are that the main hold a pressure of 150 psi for a period of 1 h to be accepted. There is also a provision for allowable loss based on AWWA Standard C600-82. In the event of a test failure, it is important to know the amount of water required to raise the pressure back to 150 psi, as this will indicate the relative size of the leak or leaks. Whether the pressure drops to that of an adjacent tied-in main to another pressure, or to atmosphere pressure, during the test may indicate that a tie-in valve is bypassing or that a joint leak is reseating itself at a certain pressure. If this occurs, the pressure during the helium test must be greater than the lowest pressure during the hydrostatic test, to ensure leakage of the tracer gas from the joint. At the same time, care should be taken not to pressurize the pipe above design limits.

Depth of Main and Type of Fill

Information about the depth of the main and the type of fill is used to determine the appropriate venting time and spread pattern into the soil atmosphere of the tracer gas from the leak location. If the fill material is not uniform along the entire length of the main, test hole spacing may have to be adjusted. If native backfill is used, it is possible that organics may be present which could release methane gas into the soil atmosphere during the decomposition process. Since methane gas is significantly lighter than air, it will register on the device as a light gas similar to helium. For this reason, test holes are tested and a baseline established prior to helium injection.

As-Built Drawings of Depth and Alignment of Main

Plans should be provided, if possible, to determine whether the conduit is a straight run of pipe or if mechanical fittings, such as elbows or tees, are present. If so, more than one blow-off point may be necessary. Any change of depth in the main which produces a low area where water could sit in the pipe must be known, because if this water cannot be removed by purging or use of a pig, the pipe should be excavated and exposed at the low point and cut or drilled to release the water. Failure to do this may result in a "no leak" result if the leak happens to be in this section. In order for the gas (helium) to escape, the water must first be pushed out of the leak. This can sometimes take considerable time, particularly with larger-diameter mains.

Location of Other Underground Utilities

Utility locates should be obtained by the client prior to placement of test holes. It is very embarrassing/hazardous to punch or drill a hole in one pipe (gas) while trying to find a leak in another. It can also be very expensive.

Access to an Air Compressor

For large-diameter mains, or long runs of pipe, a compressor is required to blow the main down and during injection, to provide air as a carrier for the tracer gas. Also, in the event that some or the entire main is under asphalt, concrete, or a heavy frost cap, the test holes may have to be placed using a pneumatic rock drill.

Personnel Requirements

The client must provide sufficient personnel to place the test holes, to open and close blow-off valves, and to provide traffic control where required.

Weather Conditions

Tracer gas testing is somewhat dependent on weather conditions, particularly rain. Rain which is heavy enough to saturate the ground and/or fill in the test holes makes testing impractical. A heavy frost cap, which extends below test hole depth, may cause tracer gas to spread under the cap and not be detectable at the test hole directly above the leak.

10.7.4 Field Operations

Based on the information received, the operator will proceed with field operations in the following order.

Preparation of Main for Helium or Helium/Air Injection

1. The main should be emptied prior to the arrival of the operator. Valves at the high and low ends of the main should be opened, and the product allowed to vent or drain out. This includes opening the valve at the end of any branches if it is not a straight run of pipe.

Caution: From the time that the conduit or closed system is dewatered or emptied and put under pneumatic, rather than hydraulic, pressure, it is important to understand that we are no longer dealing with a liquid system, but with a high-pressure gas system, with all the additional hazards that this involves.

Since liquids are not compressible, an event such as an explosive decompression of the pipe, or a joint/fitting failure while under hydraulic pressure, even at the hydrostatic test pressure of 150 psi, would only result in the loss of a small amount of water in a short time with little risk of injury. However, since gases are compressible, the result of an explosive decompression under pneumatic pressure is very different. Enough force may be generated to blow the pipe apart and send many razor-sharp fragments of the main in all directions. In one incident, a piece of 8-in-diameter main was dug out of a tree 50 ft away from the excavation. All main-line valve caps and curb box tops should be removed, to provide a venting point for the helium.

2. The compressor is then connected to the fitting at the high end of the main, making sure that the Chicago fittings are wired together. The blow-off valve(s) are closed, and the main is pressurized with air to a reasonable pressure. With the compressor still running, the blow-off valves are then opened, beginning with the blow-off closest to the compressor, and the main is depressurized. This will help to purge the main of any residual product left after draining. This procedure can be repeated as many times as necessary to ensure that as much product as possible is removed from the main.

Test Hole Placement and Marking

Test holes are made in the ground, as opposed to surface sampling, for the following reasons.

- Unlike natural gas pipelines, which are relatively shallow, other conduits are generally at depths of 5 ft or greater, which means that the helium exiting from the leak is going to be diluted to a much greater extent as it vents through the soil atmosphere.
- Most natural gas pipeline leaks have been venting gas into the soil for an extended period of time, as opposed to the relatively short venting time of a helium leak.
- The placing of a test hole creates a miniature well in the ground, 0.5 inch in diameter and 18 in deep, to which the helium can migrate without being diluted by atmospheric air.

Advance testing of the test holes, prior to helium injection, provides a baseline background for comparative purposes for subsequent tests. Locations of all underground utilities should be noted before test holes are placed. Where the conduit or system to be tested is under fill material with no hard surfaces, the standard method is to place 0.5-in-diameter plunger bar holes at uniform intervals and depths along the conduit and at known fitting locations prior to inserting the tracer gas. Bar hole spacing is determined by depth of main and type of cover from available information. Generally, a bar hole interval of about 10 ft is most practical; however, if joint and fitting locations are known, this can be expanded. Conversely, if the fill material is granular, such as sand or gravel, the bar hole spacing may be shortened. The depth of the bar holes should be about 18 in except in frost conditions, when it is necessary to go below the frost cap.

Where the main to be tested is under concrete, asphalt, or frost, all of the above applies, except that the holes are placed with a compressor and rock drill. It would also be prudent prior to drilling to arrange for filling of these holes after testing.

After placement, the test holes are numbered with paint. Care must be taken at this point to spray the marking paint adjacent to, and not near the test hole, as the solvents in the paint well affect the test. This marking identifies the test hole if records are kept, and makes each test hole easy to locate should a subsequent test run be required.

Testing the Existing Soil Atmosphere prior to Injection

The soil atmosphere in each test hole is analyzed to detect lighter-than-air readings on the "10" scale prior to injection of helium. The "10" scale is the most sensitive setting. This is done to predetermine the density of the soil atmosphere relative to atmospheric air in the ground cover over the buried conduit system. A Lexan test probe approximately 3 ft long by 0.25 inch in diameter with side holes in the bottom 6 in is ideal. It is reasonably strong and clear. If any water is sucked up by the equipment from a test hole, it is readily visible before it enters the equipment. This pretest procedure is necessary to detect and either eliminate or compensate for the following conditions.

1. Neutral reading: no needle movement. This is the ideal situation, where the specific gravity of the soil atmosphere is the same as atmospheric air. This generally occurs on new installations where clean-engineered fill is placed over the conduit. No remedial work is required in this case, and helium injection can proceed without site complications.
2. Upscale deflection: lighter-than-air reading. Since the instrument is designed to detect any gas with a specific gravity difference to air, upscale readings prior to helium injection indicate the presence of other lighter-than-air gases in the soil atmosphere. The most common gases are
   a. Naturally occurring methane gas: Methane gas is sometimes present when native material containing organics is used for backfill over the main. This gas should be removed from the soil atmosphere by purging if possible. However, if this is not practical, and the readings are relatively small (full scale on 10 or less), it may be possible to use the 100 scale by enriching the helium/air mix being injected into the main.
   b. Natural (pipeline) gas: This gas may be in the test area if there is a leak in an adjacent natural gas distribution pipeline. Natural gas (pipeline gas) consists largely of methane and displays similar specific gravities as naturally occurring methane (CH_4). Unlike naturally occurring methane gas, pipeline gas is introduced into the soil atmosphere under pressure, and can travel a considerable distance through the soil. When natural gas leakage is confirmed using appropriate equipment, the gas company should be notified *immediately,* in the interest of public safety. Helium testing is not recommended until the gas leak is repaired and the natural gas has been purged from the soil atmosphere.
3. Downscale deflection: heavier-than-air reading. Since the instrument is also designed to detect gases with a higher specific gravity than air, downscale readings prior to injection of helium indicate the presence of heavier-than-air gases in the soil atmosphere. The most common contaminants are
   a. Heavy hydrocarbons (petroleum products): These products are generally found in the soil around or adjacent to fuel storage and distribution areas. After confirmation with a CGI and

charcoal filter, the appropriate authorities (fuel safety branch) should be notified immediately. Helium testing is not recommended until remedial work, such as removal or treatment of the contaminated soil, is completed.

b. Carbon dioxide: This is generally present when clean fill, native or engineered, is used over the conduit system. It is generally caused by the aerobic decomposition of organics in the fill. The presence of carbon dioxide or other heavier-than-air gases in the soil atmosphere over the conduit being tested is important when using lighter-than-air tracer gas. The heavy gas and the light tracer gas may mix, resulting in a neutral gravity and a corresponding readout on the instrument (no reading). Carbon dioxide can be filtered out by passing the sample through a filtering medium of calcium oxide, or the helium/air ratio can be increased to compensate.

c. Water vapor: Helium testing can proceed in this case, provided there is not enough water vapor present to condense inside the equipment. A water trap or hydrophobic filter placed in the inlet sample line can usually reduce or eliminate this problem.

Injection of the Helium/Air Mixture into the Water Main

1. An air compressor has been connected to the injection point. If the injection point is a hydrant, or a service line with a curb valve, the hydrant or curb valve must remain open for the duration of the test. Closing either of these after injection of helium will allow the gas to escape through the hydrant barrel or the curb valve through the self-draining mechanism. The conduit pressure is allowed to return to zero by shutting off the compressor feed valve on the manifold and opening the blow-off(s). The gauges are attached to the helium bottles, and the helium feed line is connected to the manifold. It is recommended at this stage to inject a certain amount of 100 percent helium into the main prior to opening the delivery valve of the compressor. This is accomplished by opening the tank valve on one of the helium bottles, and bleeding off about 500 lb of the 2500 lb pressure in the tank. This slug of helium will mix with the air already present in the conduit and will ensure that a detectable amount of helium will appear at the blow-off(s).

2. The mixed gases are then injected into the system. It is important to differentiate between "delivery pressure" and "maximum output pressure" of the compressor.

    a. Maximum output pressure: Most compressors, which have a delivery volume of 175 cfm, have a "maximum output pressure" of 100–120 psi before they automatically shut down to idle. This pressure can be determined by running the compressor, shutting the outlet valve, and noting the pressure at which the compressor goes idle. For our purposes, this pressure reading is not important unless the situation calls for a test pressure on the conduit above the maximum output pressure of the compressor. If the maximum output pressure is 110 psi and for some reason the main has to be pressurized to 120 psi, the compressor outlet valve can be closed, and the additional pressure can be made up by feeding helium from the tanks by turning the regulator up to 120 psi.

    b. Delivery pressure: A compressor running unrestricted with the outlet valve open generally has a delivery pressure of about 30 psi. When this compressor is attached to a main with the blow-off(s) open, it is still not working against a great deal of backpressure. The amount of backpressure depends on the diameter of the conduit, the length of the run, and how much of a restriction is caused by the blow-off(s). When the blow-off(s) are closed, the pressure in the conduit begins to rise, as does the delivery pressure of the compressor. When equilibrium is reached at the maximum output pressure, the compressor will go to idle.

3. Compressed air from the compressor delivery point and the tracer gas are fed into the main at a ratio of approximately one unit of tracer gas to 10 units of air. This is accomplished by noting the delivery pressure of the compressor, which will remain constant as long as the blow-off(s) remain open, and adjusting the regulator on the helium gauges attached to the tanks to read 5 psi above this pressure. The delivery pressure can be determined at any time during the injection process by shutting off the helium feed valve.

4. The blow-off points at the ends of the system are not closed until a sample taken with the Gasophon indicates the presence of the tracer gas. Testing should be done in sequence, starting with the blow-off nearest to the injection point. Each blow-off point is closed when the proper concentration of tracer gas is indicated. The time required for the helium to reach the blow-off point will, of course, vary directly according to the total volume of the conduit and the delivery volume from the compressor and helium tanks.

5. When all blow-offs are closed, the delivery pressure of the compressor will start to rise as backpressure is built up in the conduit system. By constantly monitoring this delivery pressure (see par. 3 above) and adjusting the regulator on the helium tank upward to maintain 5 psi higher difference, a readily detectable concentration (±10 percent) of helium will be injected into the conduit system.

6. When the appropriate test pressure is reached, the compressor delivery valve and the helium tank valve are closed. The test pressure can be observed on the helium injection gauge. To ensure that there is a detectable spread of the tracer gas into the soil atmosphere prior to testing the bar holes, a time period of at least 1 h should be allowed after the conduit is pressurized. The only exception to this procedure would occur if a significant pressure loss (2–5 psi, depending on the total volume of gas in the main) were observed before the 1-h interval had passed.

Testing for Tracer Gas in the Soil Atmosphere

Bar holes are tested with the Gasophon in order, starting with the injection point and working toward the blow-off(s). A sample of the soil atmosphere is drawn into the Gasophon and analyzed. The result of this analysis is noted and compared with the initial test at the same location. If no readings show up on the initial test, the procedure can be repeated at regular intervals until the tracer gas is detected. When repeating the testing, be certain always to start at the injection point and work downstream. This equalizes the time that the gas has been in each section of the conduit. This is particularly important when dealing with long runs of pipe involving numerous bar holes. Assuming a quarter-mile section of conduit with test holes spaced at approximately every 10 ft, we are dealing with at least 132 bar holes. If the operator finishes the run at the blow-off and then works back to the injection point, a leak which is near the injection point may have introduced so much helium into the soil that pinpointing becomes difficult due to the spread.

Pinpointing Helium Locations

"Pinpointing" is defined as the process of determining the exact location of the excavation needed to effect leakage repair. Whether the process is successful is determined by a number of factors. Helium is a nontoxic, noncombustible, and inert gas having a specific gravity approximately one-tenth that of air. Because the molecule is so small, it has the ability to leak in detectable quantities from even the smallest of failures. It spreads from a leak by displacing the natural soil atmosphere that normally occupies the space between soil particles. The shape and size of the leakage pattern is determined largely by the resistance of the soil atmosphere to gas venting from the leak. Because helium is considerably lighter than the soil atmosphere, it tends to rise rapidly through the soil to the surface. However, if restricted, the leaking helium will seek the path of least resistance in developing a spread pattern. Some factors influencing a helium-spread pattern are as follows.

1. Line pressure: Generally, the higher the pressure in the main, the faster the helium will vent to the surface.
2. Leak sizes: The volume of gas entering the soil in a given time period directly influences the size of the leak pattern. In general, the larger the leak, the wider is the spread pattern.
3. Depth of cover: If all other factors are equal, an increase in the depth of cover will result in a wider spread pattern at the surface and an increase in venting time to the surface. Leak patterns are normally in the shape of an inverted cone.
4. Type of cover
    a. Light soils: Engineered fill (sand and gravel) and light, porous soils offer little resistance to the flow of gas. The spread pattern, if unrestricted by frost or a hard surface, is generally a small circular pattern with very little lateral spread. It may be necessary to space the bar holes somewhat closer together under these conditions. Also, particularly in sand, surface sam-

pling with a bellows-type probe may be successful if there is a pressure drop in the main indicating leakage. The helium may be venting between the bar holes.
   b. Medium soils: Because loamy soils are less porous, there is more resistance to the flow of gas. Therefore, a somewhat larger pattern normally occurs at the surface. A bar hole spacing of 10 ft should be sufficient.
   c. Dense soils: Heavy clay soils greatly restrict the flow of gas, and the spread pattern can be large. There is also the possibility that the helium may not appear in the test hole directly above the leak or in any adjacent test holes, because the natural sealing qualities of the clay can force the gas to follow cracks, fissures, and voids in the soil and vent elsewhere. Extreme care must be taken when locating leakage in this type of material. Additional bar holes may be necessary.
5. High groundwater level. Displacement of the soil atmosphere by a high water table which covers the conduit causes resistance to the flow of gas. However, this is generally not a problem. In some cases, where the groundwater level is at or near the bottom of the test holes, when the conduit is pressurized with air prior to injecting the helium, bubbles will appear, indicating a leak. This area should be noted, and the bubbles can be tested for helium after injection. It is best to proceed with the helium test in any event, since it is never certain that the leak which is bubbling is the only leak in the conduit.
6. Other underground utilities: Other underground utilities in the area of the test may act as well and collect helium, particularly in the case of dense clay soils. Such things as natural gas curb boxes, Bell and hydro risers, and sanitary, storm, and Bell manholes create wells in the ground to which the helium can migrate. It is a good policy to locate and test these collection points as well as the test holes. Curb boxes and main-line valve boxes on water mains can act as collection points. Care must be taken when pinpointing helium leaks on water mains under test. For example, a helium reading at a water curb box may indicate two situations, particularly if the curb box is on the water-main side of the street. The curb box itself may be leaking, or helium may be migrating along the service trench from the main connection or an adjacent point. This can sometimes be confirmed by placing a bar hole next to the curb box. Since the curb valve is directly under the curb box, a leak on this valve should vent directly up. Consequently, readings in the adjacent bar hole should be zero or very minor.
7. Frost: A layer of frost over the water main has a pavement-like effect and causes a larger spread pattern. Frost penetration is dependent on soil moisture content, soil type, temperature, and the insulation properties of the surface. Snow cover and sod act as insulators and retard the penetration of frost. Pavement has poor insulating qualities and allows deeper penetration.

Variations in soil moisture and surface insulation cause irregular frost penetration. The frozen areas restrict venting, while the non-frozen areas permit venting. As is the case with dense soils (clay), the helium will sometimes vent through cracks and fissures in the frost cap rather than into the test holes. Any frost pattern which is deeper than the test holes makes the helium test impossible, as the helium will spread under the frost cap and not be detectable in the test holes.

When the tracer gas is detected, the maximum point of concentration is determined by comparing the individual readings in each test hole quantitatively using the Gasophon. If two or more bar holes show a similar concentration or if the initial bar hole spacing is greater than the size of the desired excavation needed to effect the repair, additional bar holes can be placed between the existing ones to determine the maximum concentration of the tracer gas in the soil atmosphere. Where two bar holes give the same reading, the Gasophon pump can be used to purge each hole. The time required for each reading to go to zero will indicate which hole is closer to the leak. It often happens that one of the test holes can be purged, while the other cannot.

Never assume that there is only one leak present in the system under test. It is important that approximately 1 h after the first leak has been pinpointed, another test of all bar holes and venting points be conducted before leaving the site. This will normally ensure that no more leaks are present. In some cases, the client may request that the operator stand by during excavation and repairs, as he may wish to have you present when the system is retested. There are several tests that the consultant can make to help the client pinpoint the leak after excavation.

Excavation, Leak Location, and Repair

Before excavation of the helium indication begins, two operations must be performed. First, the helium injection manifold and gauges must be removed. Before the regulator is put away, turn the regulator adjustment knob counterclockwise until it is loose. This will relieve the pressure on the spring and diaphragm. Second, all pressure remaining in the conduit should be relieved for safety reasons prior to repair.

It is possible that the client may excavate the helium location and not find the source of helium in the excavation. Since there will still be residual helium in the soil on each side of the excavation, helium readings taken in bar holes placed horizontally each way may indicate the direction of the leak. Once exposed, the leak location can be reconfirmed with the Gasophon. After repairs are completed, the client may recharge the conduit and conduct a hydrostatic test to confirm integrity.

Ground-probing radar (GPR) is being used more and more as part of the arsenal of tools in the field. GPR is most commonly used for locating underground utilities, but Rodney Briar, of Johannesburg, South Africa, has had many years of success using GPR for leak location and provided the following article.

10.8 ARTICLE TWO
Water Main Leakage Detection by Means of Ground-Penetrating Radar
Rodney Briar

10.8.1 History

As with much innovative technology, ground-penetrating radar was initially developed by the U.S. military during the Vietnam War to assist in finding the enemy in their underground passages and bunkers. The technology eventually found its way into the commercial field and, initially, Geophysical Survey Systems, Inc., developed ground-penetrating radar into a viable tool for shallow penetration up to about 5 m and named their product "subsurface interface radar." This name relates to the fact that ground-penetrating radar is able to produce an image of what is below the ground, by reflecting radar frequency waves, emitted by a transmitter, from any interface in the ground, such as earth/water, earth/rock, rock/air, etc., back to the receiving antenna. Usually this antenna is built into the same box as the transmitter, and drawn over the ground, producing data which can be processed and converted into a vertical cross section or slice of ground below where the transmitter/receiver, henceforth referred to as the "antenna," has been drawn.

Dr. Hylton White, a South African physicist, was working in the United States in the 1980s and became involved with ground-penetrating radar. Upon returning to South Africa he continued his involvement, working first for the South African Chamber of Mines and then for a company that obtained the sales rights for Geophysical Survey Systems, Inc., in South Africa, during 1990. At this same time, a large contract for the replacement of water mains in the Central Business District (CBD) of Johannesburg was under way, covering half the water mains in the CBD, around 170 city blocks, and this contract was widely publicized in the press, attracting the attention of Dr. White. The rest, as the saying goes, is history.

The water mains in Johannesburg were principally steel, and in particular, reticulation pipes of 150 mm diameter and below had been laid under the footpaths. For ease of laying the precast concrete slabs forming the footpath, they had been laid on a fine sand available locally, residue from the processing of gold, which apart from still containing a tiny fraction of gold, now being processed out of the remaining stockpiles of the "mine sand," also contained residual acids from the processing.

During the heavy thunderstorms which occur in summer in Johannesburg, the thunderstorm capital of the world, the acids were slowly leached out of the "mine sand" into the subsoils and corroded the steel water mains.

All the city blocks had basements several floors deep, and water leakage into the basements was a huge problem. Conventional leakage detection methods were being used to trace the leaks, but with a plethora of other municipal services also in the ground, all of them also in a parlous state due to the action of the "mine sand" acids, the success rate in finding leaks was lower than normal with listening methods. The leaks rarely surfaced, as Johannesburg is built on a high, rocky ridge, with excess water entering into the ground, disappearing, and finding its way to the older, abandoned, small mine workings close by. The only time conventional sounding methods could be used, due to other noises affecting results, was during very restricted times, say 1 to 5 a.m. These other noises included the many other leaks which confused the correlators, the longest straight length without a tee being 70 m and then another tee across the street at a further 15 m, traffic, airplane noise (Johannesburg is under the flight path into the International Airport), noisy electrical cables and transformers close to both internal and external water pipes, etc., problems which face the successful detection of leaks in most major cities.

Ground-penetrating radar is affected only by one external influence, high-voltage overhead cables, which do not exist in city centers, and even then they only leave a particular pattern on the screen if on-site detection is being used, or on the computer monitor if additional later processing is being used to enhance the data.

10.8.2 Growth in Use of Ground-Penetrating Radar

The introduction of ground-penetrating radar meant that reported leaks into basements could be searched for at any time, and if the leak were not found in the adjacent small reticulation pipe, the bulk main or mains in the roadway could also be checked. This checking of the bulk mains could not be done by listening methods because the size of the mains restricted the effective length over which correlators could work and the access points for listening were already much farther apart than on the smaller reticulation pipes.

The use of radar grew slowly and even spread to suburban conditions, where "nuisance water," water causing a nuisance, was present and could not be traced to an adjacent leak by other methods already owned by the municipal authorities and in place with trained operators. Radar was the last resort. As an example, there had been a leak into the basement of the Johannesburg Stock Exchange Building which had defied all attempts at detection for over a year. Radar was finally called in, and the mains within the area were scanned without result. A scan around the building revealed that water was making its way to the basement wall from an adjacent telephone cable trench. This water was traced back by radar to a leak within the CBD over 1 km away. The water was traveling in the cable trench under the ducts, where it is always difficult to compact the earth during the laying of any pipe, and exiting at a low point in the cable duct run into the Stock Exchange Building.

The successes of radar were such that in 1993 the Johannesburg Municipality was beginning to think about trying to reduce the leakage losses from the 35 percent rate of the late 1980s and early 1990s to a more acceptable figure. A leakage detection contract was envisaged, and in order to choose the methods for this attack on the leakage situation, head-to-head trials between radar and an expert imported correlator team from a European manufacturer of correlators was held in a suburb of Johannesburg, finding leaks on reticulation pipes. Each of the two methods found exactly 11 leaks, and it was decided to draw up a contract for leakage detection on 600 km of the water pipes in Johannesburg. A further test was done, using three different correlators, and it was found that the maximum length and diameter of pipe over which the correlators at the time available in South Africa could give reliable results was 300 m and 300 mm, respectively, on steel pipes with deteriorating results, fiber cement, and PVC pipes. The contract was let using correlators on pipes up to 150 mm diameter and radar on pipes above that size.

The writer was the contract manager, and the project was a huge success, with correlators finding 30 leaks in 450 km of pipe tested and radar finding 5 leaks in 150 km of pipe scanned. Although the success rate of the radar appears to be inferior to that of the correlators, the bulk mains are usually constructed to a higher standard than reticulations and with better supervision during construction. There are also no consumer connections, which accounted for half the reticulation leaks found by the correlator.

With each leak found being assessed by the municipality in terms of the volume of water lost, the contract was also a financial success, with a payback ratio over the value of water being lost over one year being 5.5:1.

The 1980s had been a time of drought in South Africa, but the 1990s proved to be the opposite, and in addition, with the impending supply of additional water from the Lesotho Highlands Water Scheme, planned in times of water shortages, due in 1997, the will of the municipality to reduce the leakage losses was reduced and the radar was put to other uses where there was a need to "see" into the ground, which is in fact what the radar does.

The finding of the source of nuisance water and suspected leaks as opposed to scanning, however, continued, and in addition, several smaller contracts were entered into for scanning with the radar for leaks in Johannesburg and Cape Town for distances up to 100 km at a time. By this time, an alternative method had been developed for finding leaks on reticulation pipes, which involved merely listening with listening sticks at available access points on the reticulation and then correlating only where a leak noise was heard. This increased the price differential between acoustic methods and radar, and effectively restricted radar to distribution and bulk mains as opposed to reticulation pipes. However, whenever there was a difficult leak to find, or when several noises confused the geophones and correlators, for example, radar was always called in. In addition, the radar was used for confirming the presence of leaks found by other methods where the signal might have been doubtful and the cost of excavation was high, such as in a street in the Central Business District.

One of the smaller contracts of about 70 km was carried out for the South African Bureau of Standards, and the bureau was impressed that radar was used in place of listening sticks, which it was using on the reticulation in the township of Soweto, due to meters being inside the properties, the confined nature of the area, increasing transient noise, the inherent dangers of night listening, and the practice of all rubbish being dumped at street corners on top of the valves, where the rubbish awaited picking up by loading shovels into lorries that also picked up all surface signs of the presence of a valve. The bureau was so impressed with the ground-penetrating method of leakage detection that the method was included in the next edition of the Code of Practice on Water Loss Control, S.A.B.S. 0306 (1999).

By the end of the 1990s, then, ground-penetrating radar was well established as a leakage detection tool in South Africa, as well as all the other uses to which it can be put.

10.8.3 The Present

Despite the good rains of the 1990s and the additional water from the Lesotho Highlands Water Scheme, the change in the political scene in South Africa saw a tremendous effort made to provide treated and reticulated water to many millions of people who previously had to fetch water from local boreholes and wells. This reticulation has brought relief to the previously disadvantaged peoples of South Africa but, in part because there is little skill and money available to maintain the domestic side of the new systems, and they often terminate at an outside standpipe, which is often left running, there is a tremendous new load on water resources to the extent that the Lesotho Highlands Scheme will be augmented and extended, and municipalities now have to report and account to the national government for unaccounted-for water, or losses from pipelines.

This pressure on the municipalities has forced them once again to consider leakage detection seriously, and the result in Johannesburg alone is a 3400-km leakage detection contract, ongoing in the year 2000. That will cover about 50 percent of the water mains in the city. Of the 3400 km mentioned, 800 km are bulk mains, for which ground-penetrating radar was chosen.

Due to the current pressure on water resources, other relatively large municipalities in South Africa will require leakage detection throughout the early years of the new millennium.

10.8.4 The Method

The radar unit that has been used for leakage detection in South Africa since 1990 is a model S.I.R. 3, sold by Geophysical Survey Systems, Inc., of New Hampshire, with a model 39 visual display interface, which allows instant color monitoring of the scan results instead of the standard paper roll black-and-white printout. A high-quality cassette tape recorder is used to record interesting data for later processing and selective printout of results. Clients like to see a "picture" of a leak! The antenna is drawn along the ground and is connected to the radar unit by a 30-m-long cable. The whole system is carried in a pickup truck. A minimum of three persons is required to scan for leaks: a driver, who, due to the walking pace at which the antenna is drawn, can also watch the moving image on the screen for signs of leakage; a person to handle the antenna; and a third person to look after the cable. The whole system runs at 12 V and is powered from a 12-V, 50–100-A/h vehicle battery, which will last for a day before recharging is necessary. It is not connected directly to the vehicle battery due to the voltage fluctuations during charging.

More modern and compact systems allow one person to be able to carry the radar unit, look at a small monitor, and draw the antenna, but in South Africa, for several reasons (security, the parlous exchange rate to the dollar, etc.), investment in a new system has been resisted until a continuous year's work ahead can be seen. The present radar unit has been very reliable, only once in 10 years needing outside help, when it was sent to Allied Associates Geophysical in the United Kingdom for repair.

The radar unit has an almost infinite combination of color palettes which can be used, all of which are useful under different circumstances and tasks, but of course, once one has used the radar with some success one tends to use generally the same palette, which in our case has a background of black and gray until a strong reflector to the radar shows up, one of which is water, but pieces of steel, concrete, and rock are also strong reflectors. Strong reflectors then show up as white shapes, which are very visible when one is trying to watch the monitor while simultaneously driving the vehicle.

It has to be said that, in order to eliminate images which are similar to leaks but aren't, a digital geophone is used, information on which can be obtained elsewhere. The geophone is used to listen above the potential leakage site, and because a modern digital one is used, it is usually possible to ascertain whether a leak is present. The readout gives nine comparable readings which can be used to build up a "curve" of results, which indicate a leak if maximum noise is coincident with the radar image, and the geophone also listens for constant noises, which a leak is likely to be, as opposed to variable transient noises, which have been a problem for geophone operators for years.

It is not possible here to give a detailed description of how to set up ground-penetrating radar for leakage detection, because there are now so many different makes and models available. So the most useful remaining information which can be given are the advantages and disadvantages of the use of radar. The reader will have gathered by now that one of the most significant developments in leakage detection is the use of a combination of techniques to achieve reliable results. Because of the nature of the leak detection companies (most of them have an agency for some form of leak detection equipment), the predominant form of attack is single-technique. This approach is now becoming discredited in all branches of engineering geophysics, and the philosophy is spreading to leakage detection. However, the two-pronged attack is taking longer to manifest itself in leakage detection because most leakage detection is carried out by small or one-person companies with few financial resources to maintain a stock of different methods in their armory.

10.8.5 Pros and Cons of Radar Leakage Detection

Radar is to all intents and purposes unaffected by the day-to-day interference suffered by acoustic methods, and scans or investigations can be carried out almost anywhere quite quickly. Results can be inspected on the site, either during scanning or immediately after scanning by rerunning the recorded data, or at base on a larger, clearer monitor, before or after processing to enhance the images. The only geographical terrain where radar is difficult to use is where the antenna cannot easily be drawn smoothly across the ground, i.e., through very long grass, shrubbery, boulder-strewn areas, and steep cross falls. However, if there were many kilometers of this type of terrain to cover during the execution of a leakage detection exercise on a major cross-country pipe route, it is possible to use aerial means, such as a helicopter, to carry the equipment, and from which to run the survey. Careful cost–benefit calculation prior to undertaking such a means of leakage detection would, of course, have to be carried out.

The combination of radar and geophone is a powerful tool, but using listening sticks and correlators on smaller-sized pipes is cheaper and just as effective. Radar comes into its own when the pipe sizes get above 150 mm, the pipe materials change from steel to fiber cement, or PVC, and access points are only available at more than 200-m centers. In effect, when the conditions make the use of listening methods either unreliable or too expensive, then radar should be used. In addition, there are other circumstances when radar needs to be brought into play. The presence of certain noise generators within a reticulation system, or close thereto, can make the use of listening methods impossible. In-line pressure-reduction valves, large district water meters, and all forms of electrical substations close to the pipes transmit high noise levels which can either drown out leak noises to a geophone, and/or confuse a correlator, as does, in particular, overhead electrical district reticulation mounted on steel poles. It is then necessary to check with radar for reliable results.

Sometimes other fittings in the reticulation will cause a "leak noise," such as a series of severe bends, and although a "dry" hole is not too much of a problem in the grass verge of a quiet suburb, it is a disaster in the middle of a busy city street junction. Under those circumstances, a scan with radar will confirm whether it is worth excavating or whether the listening methods have picked up a false signal. As stated, confirmation with two methods is always better, particularly when the alternative is the needless generation of a traffic snarl-up or worse. When a complaint is received by the manager of a water network that there is a leak on a main which is, for instance, getting into the adjacent ducts of the local telephone system and causing faults and cable damage, and listening methods do not produce results, it is possible for radar to ascertain whether groundwater or overirrigation, for example, are the culprits, by scanning the ducts and the possibly innocent main to compare relative water contents in the trenches. While its use is not strictly related to leakage detection, radar can also assist managers of water networks in the location of underground pipes and chambers which have been "lost" because of inadequate drawings, or construction work and road improvements which have either unknowingly—or sometimes knowingly—been carried out over valve chambers and other access points to mains. Unfettered access to chambers assisted by ground-penetrating radar is in itself an undoubted useful tool in the fight against unaccounted-for water.

10.8.6 The Future of Ground-Penetrating Radar in Water Main Leakage Detection

As with all electronic devices to date, it is certain that radar will become more portable, faster, specialized, and easier to use. Taking the above four adjectives in order, portability is already high, but will probably be improved by manufacturers producing radar which can be carried above ground in probably the same manner as a briefcase, with the images being observed on one eyepiece of a pair of goggles similar to virtual reality viewers, the whole being run for a day by a small pack of lithium-ion batteries.

Speed of leakage detection could increase to the point already reached by radar when used for carrying out road condition surveys, up to 50 km/h or more, urban traffic speed, with the antennas rear-mounted on the vehicle just above road level. Unmetalled surfaces above reticulation systems could be covered in the same way by quad bikes.

While the ground-penetrating radars produced today are normally for general-purpose applications, anywhere a need to "see" into the ground is of advantage, future radars will be produced for specific applications. A major simplification of the setup and controls could be made if only one function, such as leakage detection, were envisaged. For instance, mine-worthy, flameproof radar is now the norm down mines, but the data-recording capacity is not up to that needed for other purposes. Production of radar for specialized purposes will make it easier for specialists in those fields to use it.

10.9 INFRARED TESTING AND TESTING FOR RESERVOIR LEAKAGE

10.9.1 Infrared Testing

Infrared thermography can be used as a method of testing for leaks which do not surface. The method is quite expensive and in many cases is undertaken by flying over the areas to be tested. The method will work only if the temperature of the water escaping from the leak is different from that of the ground into which the leak is leaking. This method has been used successfully for testing transmission mains in rural areas, but would probably not be the method of choice for dense urban areas, where there would be too many confusing traces from other sources, such as sewer lines, for example. The author has also heard of some operators using this method to detect reservoir leakage.

10.9.2 Testing for Reservoir Leakage

Large amounts of leakage can be lost through either leakage from the structure of the reservoir or from reservoir overflow. Leakage from the structure itself is probably more common in older underground brick or block-built reservoirs which have not been lined, but leakage can occur in other forms of storage too.

The easiest way to check for leakage is to isolate the reservoir from the system by closing the inlet and outlet valves. This is usually done at night. Once the reservoir is isolated, a depth test over time can be performed either by simply measuring carefully the drop in level over time or by installing a high-resolution level data logger to measure the drop over time. It is then just a matter of calculating the area of the reservoir, calculating the volume per area times the drop measurement, and calculating the volume of loss. Care must be taken to ensure that the outlet valve is not letting by. Calculations can be more difficult when the structure is not uniform, as the area changes as the level drops. Most utilities should have accurate as-built drawings showing exact measurements.

> **A** level drop test can be performed to check whether reservoir leakage is present.

If the reservoir is found to be leaking, then one way of finding the actual leaks is to send in a diver with fine sand. The fine sand is sprinkled along the walls of the reservoir and the base and is drawn into patterns where the suction of the leak takes effect. In many cases, though, if the leakage is significant, the reservoir should be programmed to be lined, as long as the basic structure is still structurally sound.

Storage overflow losses are more common where storage is in a remote location and water is not easy to see running down a street, as it would be in an urban situation. Overflow usually happens at off-peak times when headlosses and demand are low in the system. This is most often at night.

Overflow pipes should be inspected to see if there are obvious markings on the ground or wet patches where water has been ejected. See Fig. 10.26. Another simple method is to wedge a ball or object into the pipe during the day. If the object moves, it is likely that there has been an overflow situation. A more detailed analysis can be undertaken by using a high-resolution level logger. When the level of the overflow pipe is reached, loss starts to occur. Coupled with a temporary meter at the inlet to the tank, it is easy to calculate the volume and value of the loss.

Once the value of the loss is calculated, a suitable and cost-effective method of intervention may be installed. The simplest forms of level control are mechanical float valves or altitude valves, which are discussed in Chap. 12. However, utilities often use remote control systems and SCADA to control tank levels. In some cases the problem occurs because these more sophisticated methods are prone to lightning strikes. Sometimes the problem is no control or inefficient manual control, and sometimes the problem is lack of maintenance on simple mechanical controls. In all cases the loss should be resolved in a cost-effective fashion.

Figure 10.26 Evidence of storage overflow. (*Source: Julian Thornton.*)

Utilities with SCADA or telemetry systems in place can utilize these systems to periodically read zonal meters and analyze the condition of the losses through periodic modeling and assessment. The following case study identifies the efforts of the water utility in Halifax, Nova Scotia, Canada (Halifax Regional Water Commission) to combat losses through the use of advanced technology and SCADA.

10.10 CASE STUDY FOUR

Systematic Water Loss Reduction through Technology Application

Kenneth J. Brothers, P.Eng., Halifax Regional Water Commission, Halifax, Nova Scotia, Canada

This presentation summarizes the results of a systematic water loss reduction through a phased approach of acoustic leak detection programs, SCADA and zone metering implementation, and the introduction of noise correlator technology. Several system case studies will be reviewed to track the reduction of water leakage recaptured from the application of these technologies over periods of time.

This case study occurred in the Halifax Regional Municipality, in the Province of Nova Scotia, Canada. Nova Scotia is located on the eastern edge of Canada's Maritime Provinces. The Halifax Regional Municipality is the capital region of Nova Scotia and the cultural, commercial, and financial center of the province with a population of approximately 325,000 people.

The Halifax Regional Water Commission is a newly formed regional water utility comprising three former water utilities in the Halifax metropolitan area. In 1996, the provincial government legislated the amalgamation of four municipalities into the Halifax Regional Municipality. The three former utilities, Halifax Water Commission, Dartmouth Water Utility, and Halifax County Water Utility, were amalgamated into the Halifax Regional Water Commission. These former water utilities operated independently and had varying degrees of leak detection activity, SCADA application, and metering.

Before one can evaluate any water loss reduction, it is interesting to review the infrastructure and history of the water systems under evaluation. The City of Halifax was established in 1749. It was originally serviced with water by the Halifax Water Company, which was incorporated in 1844, to provide central water services to the peninsular Halifax area. Today, the Halifax Regional Water Commission comprises 1100 km of distribution water main, 6500 fire hydrants, 12 storage reservoirs, 11,000 valves, 120 pressure control/boosting facilities, and approximately 69,000 service laterals throughout the system. The water system has operational pressures varying from 20 to 135 psi. The infrastructure pipe material is primarily metallic, consisting of approximately 58 percent ductile iron, 38 percent cast iron, 2 percent PVC, and 2 percent asbestos cement and concrete cylinder pipe.

In addition to the system makeup, the ground conditions in the Halifax Regional Municipality vary widely, with a high percentage of fractured bedrock in the older portions of the city and some of the newer outlying areas. In terms of percentages, 60 percent of the system is installed in rock or bedrock conditions, 20 percent in red clay, 15 percent in glacial till, and the balance in a blend of rock and till conditions. The soil conditions present a further challenge to leakage control in the Halifax Regional Water Commission. The majority of the older system, consisting of cast iron, is located in fractured bedrock and is located near the Halifax harbor. It has been our experience that a high percentage of water mains breaks do not surface in these areas.

10.10.1 Unaccounted-for Water—Metered Ratios versus Textbook Reductions

In any discussion of unaccounted-for water, it is important to establish what, in fact, is being referred to in terms of water loss or water loss parameters. Unaccounted-for water is currently expressed in terms of percentage unaccounted for. This percentage is determined on an annual basis as a difference between the annual production (production master meter) and annual metered sales to customers, yielding the metered ratio or gross unaccounted for. In the Province of Nova Scotia, the Provincial Regulatory Board requires utilities to achieve a minimum 30 percent metered ratio difference.

In 1996, when the utilities were amalgamated to form the new commission, it was interesting to note how each utility accounted for water and rationalized the unaccounted-for figure. The following reductions were applied in some of the former utilities:

Water loss accounted for	1%
Flushing, fires, public works	3%
Meter error	2%
Allowable leakage	3%
Unmetered water/theft	1%
Subtotal:	10%

Many textbook values have been applied in the previous example, and, in fact, had been used by our utility to rationalize unaccounted-for water losses with these factors. It is now our practice not to apply these factors to artificially reduce the unaccounted-for water losses. The commission utilizes metered ratio and establishes a goal of 15 percent gross unaccounted for (metered ratio) as a target guideline for our older master metered systems. Our newer systems are held to a higher standard, reflecting the construction practice, pipe material, and testing requirements, which have resulted in tighter water systems. The Regional Water Commission also operates seven small systems and, as a comment for comparison, our small systems operate under 5 percent unaccounted-for metered ratio.

10.10.2 Case Study

The Halifax Regional Water Commission, as previously noted, is comprised of three former water utilities. The commission segregates the system into three discrete areas, referred to as the East, West, and Central regional systems. The Central system case review illustrates a 17-year history of unaccounted-for (metered ratio) reductions from a figure of 44 percent in 1982 to a systematic reduction in water leakage to a figure of 10.8 percent in 1998. How were these results achieved, and why have they been sustainable over the past 6 years?

The first phase of the water leakage reduction program focused on acoustic detection techniques and annual programs of comprehensive system surveys. In our example, this was clearly the most important activity, resulting in significant water leakage reductions, and is the most cost-effective in terms of personnel and equipment resources. In the Central region system example, biannual acoustic leak detection surveys were conducted in the first three years. By 1985, the gross unaccounted-for water figure was reduced from 44 to 24 percent. This is a 20 percent reduction in water supplied to the system over a 3-year period. It is clear that system knowledge and leak detection expertise evolved in the 3 years and contributed significantly to the 20 percent reduction in water production.

In 1985, the Central system initiated a SCADA (Supervisory Control and Data Acquisition) system in the water utility. Zone metering was an integral part of the SCADA system as well as reservoir inflow/outflow, and pressure-reducing valve flow rate and discharge pressure monitoring. The SCADA implementation program has continued since 1985 in various parts of the system. Each year additional facilities are added to the system as well; all new systems constructed by developers are required to install utility-approved SCADA equipment. The SCADA system is a powerful tool in the identification of water loss and leakage flow rates. The Central system achieved a high degree of zone metering by 1992. Our technical staff became expert at undertaking diagnostic review of system flow rates and determined the base nighttime flow rates within each master metered zoned area. These nighttime flow rates, typically measured between 2 and 4 a.m., were instrumental in identifying differential flow rates within the master metered areas. The identification of flow rates was correlated to typical water main sizes (100 gpm, equivalent to 6-in-diameter circumferential shear) for the scheduled dispatch of leak detection crews for pinpointing survey.

During the period from 1985 to 1992, the Central region experienced a continual reduction in unaccounted-for water, from a value of 24 percent in 1985 to 13.8 percent in 1992. This is a further 10.2 percent reduction in water supplied to the system over this period of time. In 1992, the utility purchased a noise correlator to augment the acoustic leak detection programs, which were fully developed at this time. The leak detection crews applied noise correlator technology only after acoustic leak detection procedures did not isolate the leak location quickly. It was felt that staff became competent in noise correlator programs after a 12-month learning curve, which we feel subsequently achieved further reductions in the gross unaccounted-for water, from 13.8 percent in 1992 to 10.8 percent in 1998.

A review of the Central region unaccounted-for water profile from 1982 to 1998 indicates a successive, sustainable reduction in unaccounted-for water leakage of over 33 percent in the system.

Does application of leak detection surveys, SCADA monitoring, zone master metering, and noise correlator technology yield a predictable reduction in unaccounted-for water? It is our view that these results are predictable and, given similar water system conditions and operating parameters, the results are not only predictable, but are repeatable.

Since 1996, the application of technology for water leakage reduction has been applied to the two other regional systems forming the new Halifax Regional Water Commission; the East and West systems already had acoustic leak detection programs in effect. However, the full application of SCADA systems and zone metering was not achieved. The commission has undertaken an aggressive campaign of installation of new zone metering (standard metered installation) facilities throughout the system. This has resulted in immediate decreases in water loss in the East region. The application of noise correlator technology has been a valuable asset in the identification of leakage during winter conditions and reductions in pinpointing exercises in more difficult conditions and terrain.

Since 1996 the East region has experienced reductions in unaccounted-for water leakage from 27 percent in 1996 to 16.7 percent in 1999. This 10 percent reduction in water loss has occurred from increased activity in existing leak detection programs, SCADA application, and master metering. It is also consistent with the similar reductions in the Central region of approximately 10 percent from the initiation of SCADA and zone metering. Although this program is continuing in the East region, we can anticipate further reductions in water loss upon completion of the zone metering plan and further diagnostics of nighttime flow rates in the East region. We anticipate upwards of a 3 percent reduction in water leakage over the next 5 years in this region.

The scope of this presentation cannot accommodate the many attributes of SCADA monitoring systems and the diagnostic power of a comprehensive monitoring system. However, base nighttime flow rates within defined master metered areas provide valuable information on leak occurrence, volumetric flow rate, and in some situations, approximate locations of water leakage. Priority response by leak detection and repair crews based on this information significantly reduces the "run leak time" of water breaks, which considerably affects the total unaccounted-for water leakage.

The commission's plan for water accountability includes the establishment of revised meter reading routes to coincide with zone master metering areas. This final stage of a comprehensive water metering and zone master metering plan for the commission will facilitate quarterly water audits for unaccounted-for water, which will assist the commission in priority leak detection surveys and enhance our water main renewal priority evaluation program as well. It is the final stage of our comprehensive plan to account for water, not only on a system-wide basis, but also on a micro-zone-metered basis to extract maximum information from SCADA, zone, and premises metering. It is through the technology application that utilities can reduce water leakage and maximize delivery of production water to our customers.

In summary, unaccounted-for water may be greatly reduced through leak detection surveys triggered through zone and master metering technology, which will result in significant and sustainable long-term reductions in water leakage through appropriate application of new technologies in our water works profession.

Interestingly, since Ken Brothers wrote this case study, his company has adopted the new techniques of ILI analysis in accordance with IWA recommendations discussed in Chap. 4 and is also undertaking leak noise mapping. The following details his latest achievements.

> We have, since this case study was written, developed and implemented what we feel is a best-practice method for water leak identification, specific for utilities that have an existing leak detection program.
>
> We changed our approach to leak detection, and now conduct a noise mapping program, using in-house standards method for equipment, techniques for sounding, recording noise, noise validation process, and corresponding spreadsheet documentation of follow-up and results, using a mix of staff (technical and utility workers). We used the IWA ILI index (Lambert/IWA) as a measurement of results.
>
> In summary, we achieved the following this past summer. Immediately after undertaking the standard leak detection program in our three operational regions (1100 km of water main), and after repairing all the identified leaks, we found 700 system noises for validation and repaired an additional 197 water leaks of various types and sizes. This is about two-thirds of the total number of annual water leaks we would repair in a typical year!
>
> We will be preparing a comprehensive paper on this project in the near future. The beauty of our program is that we can demonstrate the actual results through our well-developed SCADA master metering system.
>
> From an ILI measurement before the program, each of the regions was calculated, with results varying from about 1.6, 6.6, and 11 ILI for each region, to a go-forward ILI rate of 1.2, 2.2, and 8. The 1.2 and 2.2 are excellent results, and we are now focusing on the third region to reduce the ILI to 3 this year. We expect to sustain these results over the long haul through state-of-the-art SCADA and zone master metering, and using the best technologies in noise correlation available today.
>
> Our next step is to apply pressure control techniques, where applicable, to further reduce unavoidable and real leakage.
>
> Staff is very motivated in this program, which has been ongoing for about 5–6 months. Our approach includes IWA/ILI measurements, standard practice in noise detection, a validation process, residual noise mapping, leak pinpointing, water recapture calculations, baseline nighttime master metering measurement for leak quantification, and new technology applications. We are striving for world-class results, and feel we have the systems, techniques, staff, specialties, and motivation to achieve this lofty goal.

10.11 SUMMARY

In this chapter we have discussed some of the more traditional methods of leakage management and location. We have also detailed some methodologies such as tracer gas testing and testing by radar, which have not to the authors' knowledge been widely publicized. The intent of this chapter is to give some guidance as to possible methods of intervention against leakage, in particular underground, nonvisible leakage. There may be other methods which should be given consideration, and obviously each situation merits a careful analysis of the options before commitment is made to one methodology or group of methodologies. However, if we don't intervene, the leakage situation will only worsen!

Note: Various manufacturers of equipment and consultants were mentioned during the course of the narrative. It is not the intention of the author, the publisher, or any other associated person or body to promote the use of one technology or company over another.

10.12 REFERENCES

1. American Water Works Association, *Stats on Tap,* 1996, www.awwa.org/pressroom/statswp5.htm.
2. Clark, R. M., et al., "Urban Drinking Water Distribution Systems: A US Perspective," *Proc. Conf. Water Conservation, Water Supply, and System Integration,* Valencia, Spain, 1998.
3. American Water Works Association, *Stats on Tap,* 1998, www.awwa.org/pressroom/statswp5.htm.
4. Kirmeyer, G. J., Richards, W., and Dery-Smith, C., *An Assessment of Water Distribution Systems and Associated Research Needs* (90658), AWWA Research Foundation, Denver, Colo., 1994.
5. *Federal Register,* Underground Storage Tanks; Technical Requirements and State Program Approval; Final Rules, 53:185, Part II EPA, 40 CFR Parts 280 and 281, September 23, 1988.
6. McIntire, P. (ed.), *Acoustic Emission Testing,* vol. 5 of *Nondestructive Testing Handbook,* 2d ed., American Society for Nondestructive Testing, Columbus, Ohio, 1987.
7. Miller, R. K., et al., "A Reference Standard for the Development of Acoustic Emission Leak Detection Techniques," *Nondestr. Testing Eval. Int.,* vol. 32, no. 1, 1999.

CHAPTER

11

Speed and Quality of Leak Repair

Julian Thornton

> **Case Study One**
> The Economics of Leak Detection and Repair
> *Ellen E. Moyer, James W. Male, I. Christina Moore, and John G. Hock*
>
> **Case Study Two**
> Water Temperature Predicts Maintenance Peaks
> *Scott Potter*

11.1 INTRODUCTION

The diagram at the right shows where the speed of leak repair comes into the overall four-component picture of reduction of real losses.

We have discussed various methods of locating leaks and their respective volumes in Chap. 10. It is very important to rank those leaks for severity of loss or danger to life or property and schedule them to be repaired as soon as possible. Figure 11.1 (adapted from BABE work by Allan Lambert; see Sec. 7.6) identifies components of loss. (More on component analysis can be found in Chap. 7.) Some leaks are reported and others unreported.

It can be seen that all leaks are pressure-dependent: More pressure equals more loss. However, annual loss is also based not only on the time until you are aware that leakage is occurring, i.e., the time between zone flow analyses or leak surveys, but also the time needed to pinpoint the leakage and then the time needed to repair the leakage. The latter is true for both reported and unreported leaks. There is no point in having a costly program in place to become aware of and pinpoint new unreported leakage in hours or days if it then takes weeks or months to fix the leaks!

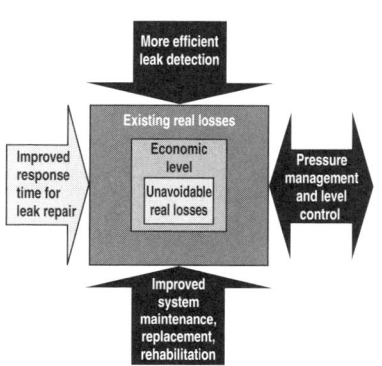

Case Study One is an adaptation of an article which was published back in the 1980s regarding the leak detection and repair program put in place by Westchester Joint Waterworks. Although the article is 20 years old and the monetary values are outdated, the concepts it put forward are still valid.

11.2 CASE STUDY ONE

*The Economics of Leak Detection and Repair**

Ellen E. Moyer, James W. Male, I. Christina Moore, John G. Hock[†]

This article analyzes the costs and benefits of a leak detection and repair program for a publicly owned water utility. The characteristics of different types of leaks and the accuracy of the detection methods used are also discussed. The results of the analysis demonstrate that benefits outweigh costs and that the program has led to a substantial decrease in unaccountable-for water without incurring greater repair costs for the system.

---

*Adapted from *Journal AWWA,* vol. 75, no. 1 (January 1983), by permission. Copyright © 1983, American Water Works Association.

[†]Ellen E. Moyer is a research assistant, James W. Male is an associate professor, and I. Christina Moore is a research engineer, all with the Department of Civil Engineering, University of Massachusetts, Amherst, MA 01003. John G. Hock is manager of the Westchester Joint Water Works, Mamaroneck, NY 10543.

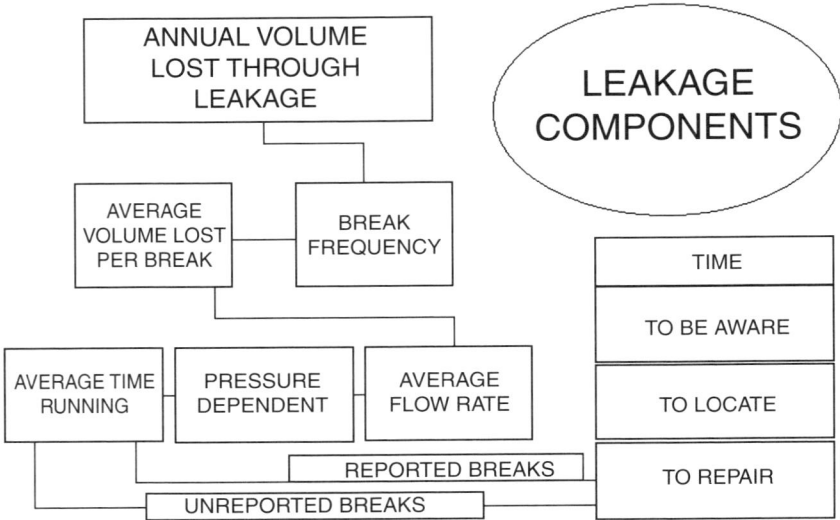

Figure 11.1 Components of loss. (*Source: Julian Thornton.*)

Systematic leak detection and repair (LD&R) of water distribution systems is an effective conservation technique that is increasingly economical as both the costs of supplying water and the demand for water continue to rise. The favorable economics of LD&R are illustrated through a case study of the Westchester Joint Water Works (WJWW) in Mamaroneck, New York. Some insights into leak characteristics and the accuracy of sonic leak detection were also obtained during the study and are discussed.

With the growing interest in and the success of LD&R programs, there is a large body of recent literature about LD&R. Good descriptions of leak detection methods are provided by Cole[1,2] and Heim.[3] Beckwith,[4] Brainard,[5] and Kingston[6,7] discuss both direct and indirect benefits.

The published results of economic case studies of LD&R programs in Louisville, Kentucky,[8,9] Los Angeles, California,[10] Gary, Indiana,[11] and Arlington, Massachusetts[12] summarize overall costs and benefits of these programs. All of these case studies provide evidence of the advantages of LD&R. This case study evaluates the economic aspects of different types of leaks and their repair.

11.2.1 Background

WJWW is a well-managed public water utility serving a population of 50,000 people located mainly in three suburban communities of New York City. The distribution system consists of 348 km (188 mi) of pipe (primarily tar-lined cast-iron pipe that is 10–80 years old), 1334 hydrants, and two pumping stations. The system delivers approximately 37.8 ML (10 million gallons) per day. Much of the water is gravity-fed and comes to WJWW in a nearly finished state. Therefore, pumping and treatment costs are relatively low.

The utility purchases water from New York City at a price of $103 per million gallons as long as WJWW's per-capita consumption does not exceed the consumption in New York City for the same period. Water used in excess of this amount is billed at a rate of $702 per million gallons. This pricing policy provides considerable incentive to conserve water. WJWW's efforts to do so are aimed at reducing unaccounted-for water, which is defined as the difference between water entering the distribution system and water sold to customers as registered on consumer meters.

Short-term conservation efforts have also been implemented during drought emergencies in the past. In addition, WJWW reduces unaccounted-for water by testing and repairing or replacing water meters and systematically rehabilitating the distribution system. WJWW has also considered initiating public education programs and an increasing block rate structure.

Two factors made WJWW an excellent choice for a study of the costs and benefits of LD&R. First, WJWW has surveyed its entire distribution system for leaks three times in the past 6 years. Second, the manager of the utility has put emphasis on keeping records. Thus all the data necessary for a detailed economic study were readily available.

The repair of detected leaks has led to a substantial decrease in unaccounted-for water at WJWW, as shown in Fig. 11.2. Before institution of the LD&R program, the amount of unaccounted-for water rose steadily, to a high of 29.5 percent in 1974. By the end of the three surveys in 1980, unaccounted-for water had been reduced to 18.8 percent, and it continued to decline to a low of 16.3 percent for the 12-month period ending June 30, 1981. No doubt other factors such as meter maintenance and management changes are also responsible, but the leak detection and repair program appears to be the primary cause of the decline of WJWW's unaccounted-for water.

11.2.2 Objectives of the Study

The primary objective of this study was to assess the benefits and costs of WJWW's LD&R program. This included an analysis of the characteristics

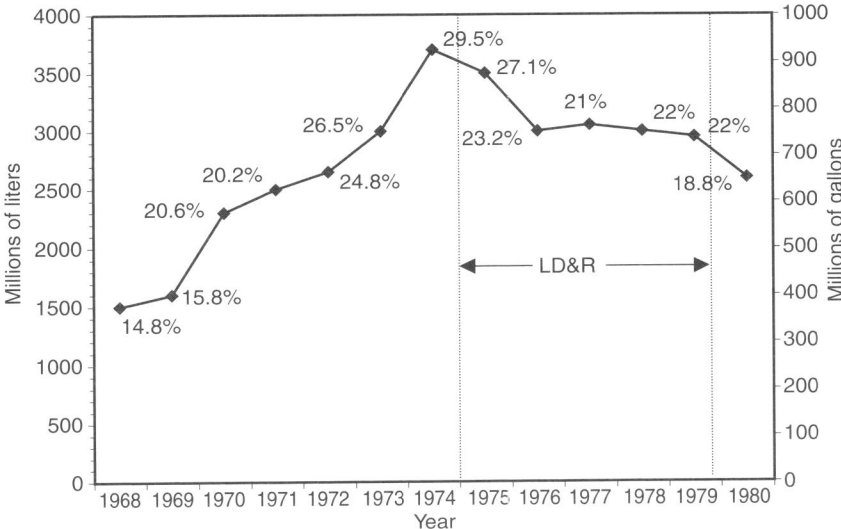

Figure 11.2 WJWW unaccounted-for water at year-end, 1968–1980. (*Source: "The Economics of Leak Detection and Repair."*)

of different types of leaks to gain insight into how water loss and repair costs varied among them. A secondary objective was to evaluate the accuracy of the sonic detection methods used in the surveys by comparing detection and repair reports.

11.2.3 Methodology

Data Collection

WJWW has kept detailed records of its leak detection and repair activities including (1) reports completed by the surveyor at the time of leak detection; (2) reports completed by the foreman of the repair crew during repair; and (3) general records concerning water usage, costs, and unaccounted-for water. These records were reviewed to obtain the information listed below. From the leak surveyor's report, the following information was recorded for each leak: (1) leak identification number; (2) date of detection; (3) street, address, and, where applicable, hydrant or valve number; (4) locations of detection soundings—hydrant, main valve, curb valve, or ground surface overlaying a main; (5) probable leak location—hydrant, service, main, joint, valve; (6) ground cover—asphalt, soil, other; (7) leak class—(A) 0–0.315 L/s (0–5 gpm), (B) 0.315–0.630 L/s (5–10 gpm), or (C) more than 0.630 L/s (more than 10 gpm); (8) surveyor's initials; and (9) diagram of the immediate area marked with the suspected leak location.

From the repair crew foreman's report, the following information on each leak was recorded: (1) leak identification number; (2) date of repair; (3) leak location—hydrant, service, main, other, or no leak found; (4) comments regarding the nature of and the possible cause of the leak and type and size of the leak, where applicable; (5) estimate (in gpm) of rate of flow from the leak; (6) duration—estimate of whether the leak had existed for a long or a short time; (7) labor hours spent repairing the leak; (8) type and area to be repaved; and (9) material used for the repair.

The following information was obtained for each year of leak detection in order to calculate costs and benefits: (1) wholesale purchase price of water; (2) chemical costs; (3) power costs; (4) cost of the leak detection survey; (5) repair crew pay rate; (6) unit costs of various types of pavement; and (7) cost of materials.

Finally, the following information was recorded for each year: (1) total amount of water purchased and sold; (2) total repair costs for hydrants, mains, and services; (3) total revenues, operating expenses, and profits; (4) expenditures for meter maintenance, insurance, legal services and liability, and overtime.

Once compiled, the data were coded and stored in a computer file for analysis.

Analysis

Simple statistical methods were used to determine the benefits and costs associated with LD&R and to study the accuracy of leak detection.

To facilitate economic comparisons, all economic values were converted to 1980 dollars. Costs for material and water were converted by using the U.S. Bureau of Labor Statistics' (BLS) *Consumer Price Index for*

All Urban Consumers for all items for the northeastern region of the United States; labor costs were converted by using the average of *Engineering News-Record*'s (ENR) *Common Labor Index and Skilled Labor Index*. Both the BLS and ENR indexes cite an average rate of change of 7.1 percent per year for 1975–1980, although values were calculated by using each year's specified rate.

Benefits were defined as the value of the amount of water that would have been lost from the leak had it continued to leak for 1 year from the time of discovery at its discovered leakage rate. The choice of 1 year is somewhat arbitrary. The water saved by detecting and repairing a leak in a given year was equal to the volume that would have been lost had the leak repair been delayed. Since the WJWW surveys are conducted every 2 years, a conservative assumption is that leaks detected now would have been left undetected for a year, on average, without the current survey. Lost water was valued at the lower wholesale purchase price ($103 per million gallons) plus power and chemical costs ($11–$20 per million gallons), which were obtained for each year by dividing the total cost of power and chemicals by the total amount of water produced.

LD&R costs included the costs for labor (wages plus 55 percent for fringe benefits), pavement, materials, detection (total cost of survey for the year divided by the number of leaks detected that year), and 20 percent overhead on all costs.

Net benefits were calculated simply as benefits minus costs.

Results: Analysis of Benefits and Costs

In Table 11.1, benefits, costs, and net benefits have been summed for each of the three surveys of the system and for the three surveys combined. The total yearly amount of leakage stopped by LD&R over the 6-year period was estimated to be 10,469 ML, representing total benefits of $401,413. Total LD&R costs were $239,052, resulting in total net benefits of $162,361 for the 6-year period.

As might be expected, survey 1 differed considerably from surveys 2 and 3. Because it was the first survey of the system, a greater number of serious leaks, many of which had probably been running for some time, were detected. Nearly as much leakage was stopped in survey 1 as in surveys 2 and 3 combined, and survey 1's resulting net benefits were more than twice as great as those of the other two combined.

TABLE 11.1 Costs and Benefits of WJWW Leak Detection and Repair (1980 Dollars)

Survey Number	Survey Years	Number of Leaks Detected	Water Saved (ML)	Value of Water Saved ($)	Total LD&R Costs ($)	Net Benefits ($)
1	1975–1976	182	5,092.706	212,689	98,519	114,170
2	1977–1978	179	2,711.468	99,350	75,650	23,699
3	1979–1980	137	2,664.663	89,375	64,883	24,492
Total	1975–1980	498	10,468.837	410,413	239,052	162,361

Source: "The Economics of Leak Detection and Repair."

The largest expense of the LD&R program was for detection services, which accounted for 46.3 percent of total LD&R costs. The remaining costs were for labor (21.8 percent), overhead (16.7 percent), pavement (10.4 percent), and materials (4.8 percent).

Figure 11.3 shows the total repair and replacement costs for the years 1968–1980. In addition, a breakdown of the costs for repair and replacement of mains, services, and hydrants is shown. Other costs include overtime and leak detection survey costs. The curves labeled mains, services, hydrants, survey, and overtime sum to yield the total distribution system repair and replacement costs. The repair and replacement costs shown in Fig. 11.3 do not differentiate between repair costs resulting from leak detection and other repair and replacement costs. As a comparison, the dashed line shows the total leak detection and repair costs.

It is interesting to look at WJWW's total repair and replacement costs for the distribution system before and after the LD&R program began. Even with leak detection and repair costs included, the overall costs did

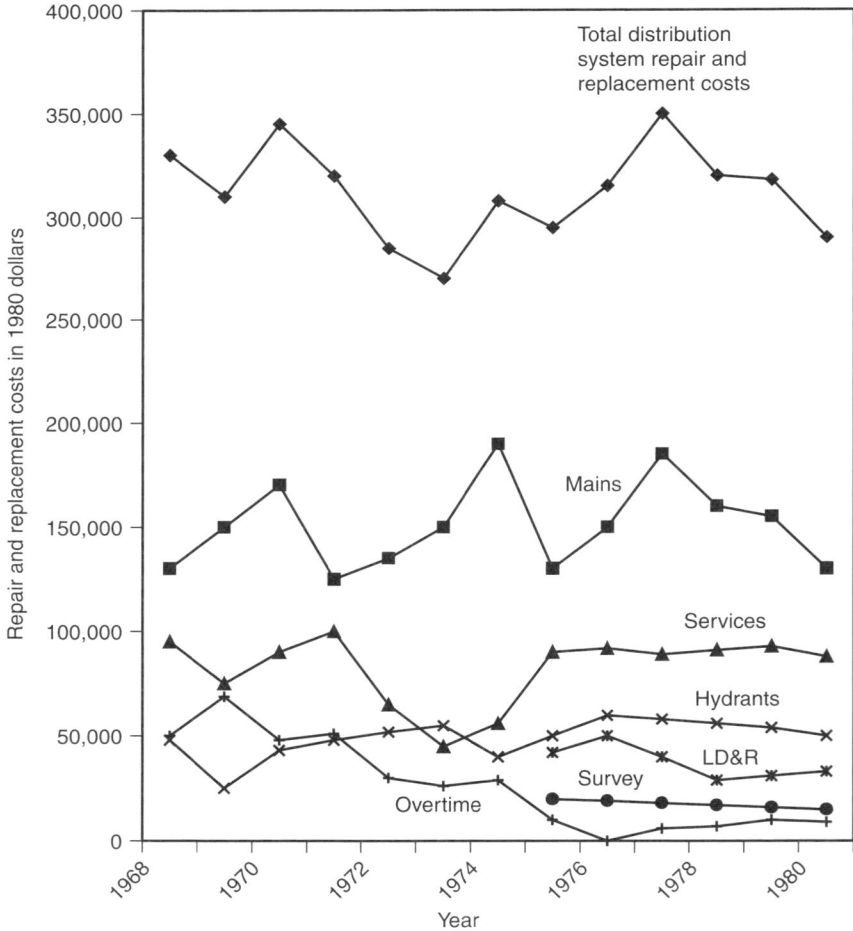

Figure 11.3 Distribution system repair and replacement costs, 1968–1980. Overhead and foreman's pay are included in LD&R costs but not in the other categories. (*Source: "The Economics of Leak Detection and Repair."*)

not rise by an inordinate amount. In fact, when costs are compared on an equal basis (1980 dollars), the total repair costs decreased considerably after 1977.

Between 1975 and 1980, fluctuations in the total system repair costs curve are due predominantly to changes in expenditures on mains. The curves are almost parallel. Main expenditures increased during the first survey period (1975–1976), when many large leaks were being discovered and repaired. The subsequent decrease in the cost of main repair and replacement may also be linked to the LD&R program. The detection and repair of small leaks in the early survey may have prevented development of costly main breaks later on. There is some evidence[13] that main breaks sometimes occur because the supporting soil is washed away by leaked water, making the pipe less able to resist forces such as water hammer and traffic.

The high total repair cost for 1977 is partially explained by very severe winter weather causing a high incidence of main breaks.

11.2.4 Analysis by Type of Leak

Costs and benefits of different types of leaks varied significantly. Leaks were divided into six categories (1) hydrant; (2) service—customer service leaks found on the customer's side of the corporation stop (repairing these leaks is the responsibility of the customer); (3) service—WJWW service leaks found on WJWW's side of the corporation stop; (4) main leaks; (5) no leak found—a dry hole where a leak was originally thought to exist but not found, even after multiple excavations; and (6) other—leaks found in the distribution system of other utilities, mainly gas utilities and water utilities of neighboring towns.

For purposes of analysis, dry holes and other leaks were assigned leakage rates of zero, since no WJWW water was being lost from them. However, costs were incurred in their discovery. Customer-service leak repair costs are paid by the customer and so are not included in LD&R costs.

Table 11.2 tabulates the number of leaks for the six leak categories. In general, hydrant leaks were the most numerous and the smallest of the leaks, whereas main leaks were the largest and fewest in number. As shown in Table 11.3, the mean leakage rate of all 498 leaks was 0.67 L/s.

TABLE 11.2 Number of Leaks Detected for Different Types of Leaks

Type of Leak	All Surveys	Survey 1	Survey 2	Survey 3
Hydrants	298	96	116	86
Service—customer	38	16	13	9
Service—WJWW	70	26	29	15
Main	75	39	16	20
No leak found	10	4	2	4
Other	7	1	3	3
All types	498	182	179	137

Source: "The Economics of Leak Detection and Repair."

TABLE 11.3 Discovered Leakage Rates for Different Types of Leaks (L/s)

Type of Leak	All Three Surveys		Survey 1		Survey 2		Survey 3	
	Mean	Range	Mean	Range	Mean	Range	Mean	Range
Hydrants	0.19	0–1.58	0.31	0–1.58	0.12	0–0.82	0.14	0–0.88
Service—customer	0.91	0.44–1.64	0.86	0.44–1.64	0.85	0.63–1.14	1.07	0.76–1.2
Service—WJWW	1.02	0.13–3.15	1.04	0.19–3.15	0.85	0.19–1.89	1.29	0.13–3.15
Main	2.28	0.06–9.46	2.33	0.06–9.46	2.30	0.19–6.31	2.14	0.13–7.89
No leak found	0.00		0.00		0.00		0.00	
Other	0.00		0.00		0.00		0.00	
All types	0.67	0–9.46	0.89	0–9.46	0.48	0–6.31	0.62	0–7.89

Source: "The Economics of Leak Detection and Repair."

The mean leakage rate for survey 1 also had the highest mean leakage rate for hydrant leaks. Service leaks were largest in survey 2. The mean leakage rate of main leaks was fairly constant for all surveys.

The accuracy of the repair foreman's estimates is important because they are the basis for calculation of economic benefits. These estimates appear to be fairly accurate, based on comments from WJWW personnel and also on preliminary results from an additional case study in which leakage rates were actually measured.

Table 11.4 presents a more detailed analysis of the economic aspects of LD&R. The average LD&R cost of all leaks was $480 per leak with the highest mean for survey 1 and the lowest for survey 2. Mean LD&R costs for WJWW services and mains were $800 to $1000 per leak, two or three times the cost of hydrant LD&R. Since other leaks and customer service leaks are repaired by others, their LD&R costs are essentially detection costs plus overhead.

The overall average net benefit of LD&R was $326 per leak but was about four times higher in survey 1 than in survey 2 or 3. The variation of net benefits with type of leak depended primarily on leakage rate and secondarily on LD&R costs, since these costs varied much less than leakage rates. Main LD&R yielded net benefits of about $1900 per leak. At the other extreme, mean net benefits for the usual small hydrant leaks were −$105; service leaks were intermediate. All dry and other leaks had negative net benefits.

It should be pointed out that the cost of the survey for a given year was apportioned equally among the leaks found in that year. This is a somewhat gross approximation. For instance, hydrant leaks are generally much easier to detect than large main leaks and should probably be assigned a smaller portion of the survey cost. Since any apportioning method would be arbitrary, it is more useful to consider repair costs exclusive of survey costs to gain a clearer idea of the actual expense of different types of leak repairs and the resulting net benefits. The costs and net benefits of leak repair (excluding detection costs) are presented in Table 11.5.

TABLE 11.4 Average Costs, Benefits, and Net Benefits of LD&R for Three Surveys (Dollars per Leak)

	All Surveys			Survey 1			Survey 2			Survey 3		
Type of Leak	Cost	Benefit	Net Benefit	Cost	Benefit	Net Benefit	Cost	Benefit	Net Benefit	Cost	Benefit	Net Benefit
Hydrants	332	227	−105	363	404	41	301	136	−165	339	153	−186
Service—customer	274	1081	807	287	1149	862	245	950	705	290	1147	857
Service—WJWW	836	1202	366	880	1369	489	742	969	227	942	1361	419
Main	839	2781	1942	822	3074	2252	899	2699	1880	823	2274	1451
No leak found	645	0	−645	970	0	−970	464	0	−464	410	0	−410
Other	267	0	−267	232	0	−232	242	0	−242	305	0	−305
All types	480	806	326	541	1168	627	423	555	132	473	652	179

Source: The Economics of Leak Detection and Repair."

TABLE 11.5 Average LD&R Costs, Benefits, and Net Benefits for Different Types of Leaks Exclusive of Detection Costs (1980 Dollars)

Type of Leak	Cost	Benefit	Net Benefit
Hydrants	64	227	163
Service—customer	2	1081	1078
Service—WJWW	571	1202	631
Main	568	2781	2213
No leak found	393	0	−393
Other	8	0	−8
All types	212	806	594

Source: "The Economics of Leak Detection and Repair."

11.2.5 Accuracy of Sonic Leak Detection

By statistically comparing the survey and repair reports, indications of the accuracy of sonic leak detection in determining leak size and location were obtained. In general, surveying was found to be highly accurate. The fact that only 10 leaks, or 0.5 percent of the total, were dry holes attests to the effectiveness of sonic leak detection and the skill of the surveyors. Dry holes, which involved multiple excavations and ended in eventual repair, were not included.

Another indication of the accuracy of surveying is the leakage rate estimate made by the surveyor during detection compared with the assessment by the repair foreman upon examination of the leak. The estimates of the leakage rate were reported in two ways: the estimate at the time of detection was reported as a class A, class B, or class C leak, whereas the estimate at the time of repair was reported as an actual leakage rate. Table 11.6 shows the close correlation between the two sets of estimates. From the data shown in Table 11.6, it is apparent that surveyors had the most trouble classifying intermediate-size leaks. They often put leaks that should have been in the C category in the B category.

A third indication of the accuracy of leak detection is obtained by comparing the actual leak location that was discovered by the repair crew and the suspected location of the leak as predicted by the surveyor. Table 11.7 is a simplified presentation of the results of a correlation analysis of these two variables. In many cases, the surveyor predicted more than one leak location; these cases are combined in the last column regardless of whether the correct location was included in the predicted location. The percentages in each row sum to 100.

On the whole, actual and predicted leak locations correlate closely. Of all hydrant leaks, 96.6 percent were correctly located by the surveyor; the locations of 86.8 percent of the customer service leaks, 55.9 percent of the WJWW service leaks, and 84.5 percent of the main leaks were also correctly predicted. All dry holes were originally thought to be main leaks.

TABLE 11.6 Correlation of Estimates of Leakage Determined by the Surveyor (Leak Class) and by the Repair Foreman (Leakage Rate)

Leak Class*	All Three Surveys			Survey 1			Survey 2			Survey 3		
	Number of Leaks	Mean	Range	Number of Leaks	Mean	Range	Number of Leaks	Mean	Range	Number of Leaks	Mean	Range
A 0–0.315 L/s (0–5 gpm)	276	0.23	0–4.73	96	0.34	0.03–4.73	104	0.16	0–3.15	76	0.18	0–1.26
B 0.315–0.63 L/s (5–10 gpm)	157	0.81	0–4.73	61	0.93	0–4.73	49	0.62	0–4.73	47	0.86	0–4.73
C More than 0.63 L/s (more than 10 gpm)	59	2.37	0–7.46	24	2.97	0–9.46	24	1.62	0–6.31	11	2.66	0–7.89
Total	492	0.67	0–9.46	181	0.89	0–9.46	177	0.49	0–6.31	134	0.62	0–7.89

*Data for leak class were missing for six of the leaks.
Source: "The Economics of Leak Detection and Repair."

TABLE 11.7 Correlation of Leak Location Predicted by the Surveyor with the Actual Leak Location Found by the Repair Crew

Actual Leak Location	Predicted Leak Locations Corresponding to Categories of Actual Leak Locations (%)				
	Hydrant	Service	Main	Multiple Predicted Locations	Total
Hydrant	96.6	0	2.7	0.7	100
Service—customer	0	86.8	13.2	0	100
Service—WJWW	0	55.9	16.2	27.9	100
Main	2.8	4.2	84.5	8.5	100
No leak found	0	0	100.0	0	100
Other	0	16.7	83.3	0	100

Source: "The Ecnomics of Leak Detection and Repair."

11.2.6 Discussion

In order to perform the preceding analyses, it was necessary to make several assumptions concerning the characteristics of leaks and the value of the water itself. In most cases, these assumptions underestimate benefits and overestimate costs. Consequently, net benefits of LD&R are likely to be greater than those presented here.

Net benefits of LD&R were narrowly defined for this analysis as WJWW's direct monetary savings that resulted from not having to purchase, pump, and treat water that would have been lost to leakage over a 1-year period. This is probably the minimal reasonable value that could be assigned to this water, which is initially inexpensive. Numerous other benefits, some tangible and some not, deserve mention.

Briefly, they are (1) reduced property damage as a result of fewer breaks, which has beneficial implications in terms of inconvenience, energy, material, public relations, insurance claims, and lawsuits; (2) less overall wear and tear on the distribution system and on pumping and treatment facilities; (3) lower total repair costs since, by using LD&R, leaks are repaired systematically and efficiently, correcting many problems before they become serious; (4) deferred construction of new facilities because of decreased total demand; (5) reduced risk of contamination; (6) valuable information derived from LD&R about the state of the distribution system; (7) improved public relations and increasing awareness by the public of the importance of conserving water and other resources; (8) detected leaks on other utilities' distribution systems; and (9) decreased expenditures on leak-related expenses. (WJWW's costs for insurance and legal services have decreased since LD&R began, although it is impossible to determine by how much the LD&R program contributed to the decrease.)

11.2.7 Conclusions

Analysis of the LD&R program showed significant net benefits even when conservative estimates were used to determine benefits. The actual net benefits of WJWW's LD&R program could be far greater than the results of the analysis indicate.

Sonic leak detection, probably the least expensive method currently available, was found to be an effective tool for predicting locations and sizes of leaks.

WJWW has benefited from its LD&R program. The first survey of the system was particularly beneficial, since it effected a dramatic reduction in unaccounted-for water. By 1981, unaccounted-for water was down to 16.3 percent at WJWW. Less frequent surveying is now required to keep unaccounted-for water at a reasonable level. Since 1981 WJWW has been using a leak detection van equipped with leak correlators recently acquired by the county.

Several general recommendations can be made as a result of this case study:

- Sonic LD&R is an effective and economical way to conserve water. In general, the more expensive the water, the greater the net benefits of LD&R will be. In the case of WJWW, each survey had positive net benefits.
- The frequency with which a distribution system should be surveyed to yield optimal benefits would vary from utility to utility depending on water cost, LD&R costs, age of the system, climate, and other factors. This aspect of LD&R needs further study, as does the related area of leak formation and development.
- The leak classes used by surveyors to categorize suspected leaks by size should be reexamined. The B class (0.315–0.630 L/s [5–10 gpm]) appears to be too narrow to distinguish between intermediate and large leaks. Expanding the B class to range from 0.315 to 1.260 L/s (5 to 20 gpm) could result in a more meaningful determination of repair priorities.
- The importance of record keeping became increasingly apparent during this study. Because WJWW had kept complete, detailed records, it was possible to assess the value of the LD&R program and, furthermore, to do so in the context of related factors such as total repair costs and other water conservation efforts.
- If a water utility has limited resources for repairing leaks after completion of a leak detection survey, it should concentrate on those leaks that would yield the greatest return for the money spent on repair. If this were the case for WJWW, the utility should concentrate on main leaks first and then on leaks whose repair would yield lower net benefits.
- Highly skilled and experienced personnel are required for sonic leak detection to give the best results. Many factors, such as soil type, traffic noise, other nearby utility systems, and type of pipe, influence the sound of the leak the surveyor hears and must be considered in order to locate the leak correctly. Incorrect location estimates result in costly multiple excavations or dry holes.
- In summary, leak detection and repair is an effective means of water conservation. Of the many options available for conserving water, LD&R is a logical first step. If the utility does what it can to conserve water, customers will be more cooperative in other water conservation programs, many of which require individual

effort. A leak detection and repair program can be highly visible, encouraging people to think about water conservation before they are asked to take action to reduce their own water use.

11.2.8 Acknowledgments

The research reported in this case study was supported in part by the U.S. EPA Office of Water Research and Technology, grant 14-34-0001-0497.

The case study above discussed in detail the planning, execution, and detailed record keeping and analysis of a leak detection study (for the location of nonreported leaks) and the subsequent repair of the leaks. However, another important factor in reducing annual volumes of real losses is to optimize the speed and quality with which both reported and nonreported leaks are repaired, as this too will have a large impact on annual volumes of loss. The case study by SABESP, which is presented in Chap. 18, discusses the huge benefits gained, at its utility by this type of action.

Case Study Two shows how one utility predicts when leakage is going to occur by tracking water temperature. In this way the utility is able to prepare itself for new leak situations and ensure that leak run time is kept to a minimum.

11.3 CASE STUDY TWO
*Water Temperature Predicts Maintenance Peaks**

Scott Potter

The Louisville Water Co., a Kentucky utility chartered in 1854 as a municipal corporation, is a nationally recognized utility with demonstrated technical competence in all areas of water utility management. LWC is a member of the Partnership for Safe Water and one of the first utilities to be evaluated by the AWWA QualServe program. As such, LWC is proactive in dealing with legislation and regulations under the Safe Drinking Water Act and continuously maintains a rigorous research program to deal effectively with possible future requirements of state and federal regulations. As part of its proactive program, LWC replaces or rehabilitates 45 mi of water main each year—approximately 1.5 percent of the system—for an annual capital expenditure of $10 million.

Evaluating main breaks is an important part of the replacement and rehabilitation program. When looking at the entire transmission and distribution system, one factor stood out more than others as a contributor to main breaks—finished water temperature (FWT). The age of the cast-iron pipes varies throughout the LWC distribution system, from older than 130 years to brand-new. The pressure also varies significantly, from a minimum of 40 psi to 100 psi, as do the soil conditions, from clay to sand. These variations do not appear to affect the number of breaks

*Adapted from *Opflow*, vol. 26, no. 7 (July 2000), by permission. Copyright © 2000, American Water Works Association.

throughout the system as much as FWT and, except for the temperature of and drought effect on the soil, were not considered in the following discussion.

11.3.1 Verifying Operational Observations

LWT decided to examine FWT closely because several people within the operational group used it as an informal indicator for probable increased break activity. Experienced operations staff knew that as finished water temperatures dropped toward 40°F (4.4°C), break activity would increase. A detailed survey of temperature trend data confirmed this informal observation.

LWC experiences dramatic water temperature changes because raw water temperatures from the Ohio River vary from 33°F (0.5°C) to 85°F (29.4°C) over the course of a year. One riverbank infiltration well has a slight moderating influence on FWT in LWC's elevated service area, but the breaks appear consistently throughout the system, and the FWT discussed here represents the main plant's discharge temperature. Analysis demonstrates that extreme temperatures (either low or high) produce above-normal break activity.

Data collected from December 25, 1998, through March 8, 2000, demonstrate a strong correlation between the FWT and the propensity for main breaks. When the FWT reached 39°F (3.9°C) on two separate occasions, the number of main breaks increased dramatically, and when the FWT approached 90°F (32.2°C), main break activity increased as well (Fig. 11.4).

The first interval of extremely low FWT was during the first 20 days of January 1999, when the water temperature was below 39°F (3.9°C). Workers from Local 1683 of the American Federation of State, County and Municipal Employees repaired 163 main breaks over this 3-week period, for an average of 7.76 breaks repaired per day. On January 21, the finished

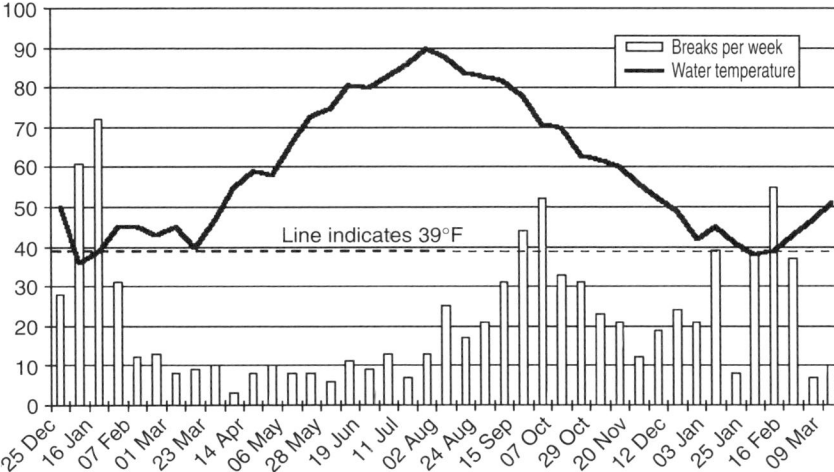

Figure 11.4 Finished water temperature and break correlation. (*Source:* "*Water Temperature Predicts Maintenance Peaks.*")

water temperature reached 40°F (4.4°C), and continued to increase to 44°F (6.6°C) through the next month, and main break activity leveled off. It is important to note that the FWT never fell below 40°F (4.4°C) after January 21. From January 24 through March 13, 56 main breaks were repaired—an average of 1.17 per day. This represents an 85 percent decrease from the early January main break rate.

On March 18, the FWT began to increase for the spring and summer period. Break activity also began to increase. Figure 11.4 shows that, while not exactly parallel, summer break activity also increases with higher FWT.

The colder the FWT, however, the stronger is the correlation in increased break activity. The second survey period when the finished water temperature fell below 39°F (3.9°C) was between January 21, 2000, and February 11, 2000. During that interval, the union repaired 134 main breaks, averaging 6.38 break repairs per day. The peak, 99 breaks from December 23, 1999, through January 14, 2000, is presumed to have been caused by a combination of rapidly declining finished water temperature and persistent drought.

While the FWT did not drop as low as 39°F (3.9°C) the rapid decrease in temperature, increase in water density, and severely dehydrated soil conditions caused by drought-generated break activity almost equal that when the FWT actually reached 39°F (3.9°C).

11.3.2 Year-Long Analysis

LWC experienced a total of 967 main breaks in 1999, an average of 2.65 breaks per day. This was a record year for the company: the two periods when FWT fell below 39°F (3.9°C), coupled with the prolonged drought (August 2–November 9, 1999), contributed to the extraordinary number of breaks (the shifting, cracked soil conditions and high water demand during the drought are also considered to be factors in the breaks during those periods). The number of breaks per day when the FWT was 39°F or lower was 70 percent higher than the number of breaks per day when averaged over the entire year.

The data strongly support the conclusion that a FWT of 39°F (3.9°C) or lower will result in a dramatic increase in the number of main breaks to be repaired. The reasons for this phenomenon have not been specifically researched by LWC. There is general consensus within LWC that the density of water maximizing in this temperature range plays a large role. The cast iron within our system appears to be more susceptible to a rapid decrease in FWT: a rapid transition to 39°F (3.9°C) in this material produces even higher break activity.

Data from the LWC Distribution Operations ground temperature measurement system, which provides constant soil temperature measurements at 1-ft intervals, from 1 to 6 ft, demonstrated that the soil temperature at 3 ft and below never fell below 45°F (7.2°C) over the 1999–2000 winter season. This indicates that there may be a slight heating effect on water within the buried infrastructure of the distribution system at temperatures below 39°F (3.9°C). Also, if finished water temperature trends are at extremely high levels, break activity may increase, too, especially if soil conditions are poor.

11.3.3 Conclusions

Finished water temperature is a great advance warning system. LWC Distribution Operations uses this information for advance planning and the identification of the need to initiate the winter emergency plan. For instance, if long-range weather forecasts indicate extreme low temperatures over a sustained period and FWT is dropping quickly toward or is already below 39°F (3.9°C), it is reasonable to assume that the unusual break activity is going to begin to persist. This may require contractual assistance in main break repair, notification to authorities of the possibility of longer-than-normal repair completion rates, and other activities.

LWC is also gathering data to identify the effects, if any, of mixing the demonstration Riverbank Infiltration Well discharge water with water from our normal Ohio River source on finished water temperature. An unexplored possible benefit to the Riverbank Infiltration Well water is that water's temperature stability when compared with Ohio River water. Other research, including the continuing examination of soil temperatures, is also planned. In observing and analyzing the patterns that contribute to a problem such as main breaks, LWC can continue to be proactive in its efforts to supply safe drinking water to its customers.

11.4 LEAK TYPES

Figures 11.5–11.8 show various leaks. Figure 11.5 shows a cast-iron ring fracture. Figure 11.6 shows a ferrule or corporation stop leak, Fig. 11.7 shows a pinhole leak, and Fig. 11.8 shows a leak caused by rusted hydraulic couplings.

11.5 SAFETY

When fixing leakage, proper signing should be used to warn traffic that excavation is being undertaken and to protect passers-by. In addition to signs, proper care should be taken to protect workers in the excavation. Each area will have its own rules and regulations, which should be adhered to.

11.6 SUMMARY

Leaks must be repaired to ensure that loss volumes are kept to a minimum. Surprisingly, many utilities do not always repair known leaks! This may in some cases be an economic decision or one based on distribution logistics, but in some cases it is just a lack of awareness of the impact on annual loss volumes. Chapter 7 discusses in detail some of the methods available to model annual losses and the impact of various interventions on those losses. The impact of an improved repair program can be easily modeled.

Not only must leakage be repaired, it must be done in a manner which will ensure that this particular leak will not recur in the short

Figure 11.5 Cast iron ring fracture. (*Source: BBL Ltda.*)

Figure 11.6 Ferrule or corporation stop leak. (*Source: Julian Thornton.*)

term. Unfortunately, quality of repair is an area which is sometimes overlooked.

The time until leak repair is carried out will almost always have a large effect on the annual volume of real losses, whether it is leak repair from surfacing reported leaks or unreported leakage which is located during a routine leak survey. Many small leak volumes soon add up to one large leak volume!

Figure 11.7 Pinhole leak. (*Source: BBL Ltda.*)

Figure 11.8 Leak caused by rusted hydraulic couplings. (*Source: BBL Ltda.*)

11.7 REFERENCES

1. Cole, E. S., "Methods of Leak Detection: An Overview," *J. AWWA,* vol. 7, no. 2, p. 73, 1979.
2. Cole, G. B., "Leak Detection: Two Methods That Work. Part I," *OpFlow,* vol. 6, no. 5, p. 3, 1980; "Part II," *OpFlow,* vol. 6, no. 6, p. 3, 1980.
3. Heim, P. M., "Conducting Leak Detection Search," *J. AWWA,* vol. 71, no. 2, p. 66, 1979.
4. Beckwith, H. E, "Economics of Leak Surveys," *J. AWWA,* vol. 56, no. 5, p. 56, 1964.
5. Brainard, F. S., Jr., "Leakage Problems and the Benefits of Leak Detection Programs," *J. AWWA,* vol. 71, no. 2, p. 64, 1979.
6. Kingston, W. L., "Leak Detection Is Worth the Effort," *Am. City & County,* vol. 9, no. 11, p. 59, 1978.
7. Kingston, W. L., "A Do-It-Yourself Leak Survey Benefit-Cost Study," *J. AWWA,* vol. 71, no. 2, p. 70, 1979.
8. Campbell, F. C., "Distribution Systems Leakage Survey," *J. AWWA,* vol. 62, no. 7, p. 400, 1970.
9. Payne, B. E., "Shrinking Non-revenue Water," Reuben Donnelly, 1966.
10. Laverty, G. L., "Leak Detection: Modern Methods, Costs and Benefits," *J. AWWA,* vol. 71, no. 2, p. 61, 1979.
11. Pilzer, J. E., "Leak Detection—Case Histories," *J. AWWA,* vol. 73, no. 11, p. 565, 1981.
12. New England River Basins Commission, *Before the Well Runs Dry: A Handbook for Designing a Local Water Conservation Plan,* 1980.
13. Betz, Converse, Murdoch, Inc., *New York City Water Supply Infrastructure Study: Manhattan,* Vol.1, report for the U.S. Army Corps of Engineers, May 1980.

CHAPTER

12

Pressure Management

Julian Thornton

> **Case Study One**
> Advanced Water Pressure Management in the Berea–Alexander Park Supply District, Johannesburg, South Africa
> *Allen Young*
>
> **Case Study Two**
> Ramallah Case Study for Reducing Leakage from Al Jalazon Refugee Camp Water Network
> *Nidal Khalil*

12.1 INTRODUCTION

System optimization is in many cases far more cost-effective than system expansion and most always has a more positive environmental impact. Many water systems are designed considering the minimum level of pressure required for the demand types, but in many cases no consideration is made for maximum pressure levels. If no consideration or only basic consideration was made at the time of installation, there may very well be room for optimizing the pressures within a system. Pressure management is one of the most basic and cost-effective forms of optimizing a system and can in many instances provide fast paybacks on large investments. The diagram at the right shows where pressure management fits into the four-component scenario of real losses management.

Pressure management has been around for many years in various forms, but only in the last few years has the use of advanced pressure control been used on a wide basis in system optimization and in loss reduction and management programs.

This chapter follows all the planning stages for a pressure control scheme, from deciding whether it is necessary and to what extent, to cost justification and practical field installations.

This chapter is not designed to replace a complete valve manual. Valve manuals are available from most manufacturers, and a variety of manuals and standards is available from the Instrumentation, Systems, and Automation Society (ISA). This chapter is rather a very practical, "hands-on" guide to using pressure management (pressure reduction, level control, flow control, and pressure-sustaining valves) as one of the many tools to reduce losses and operate water distribution systems in a more efficient manner.

> **M**any systems are designed with minimum pressure requirements in mind but not maximum pressure limitations, so many systems have areas which are grossly over-pressured.

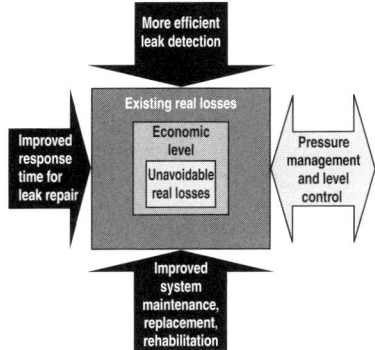

12.2 WHY UNDERTAKE PRESSURE MANAGEMENT SCHEMES?[1]

12.2.1 Positive Reasons

Leakage Reduction

The reduction of leakage is a subject which is on the minds of most water utility engineers and managers throughout the world. In other chapters of the book we have discussed various types of leakage reduction programs, of which pressure reduction is one. As with all of the other techniques for reducing leakage, pressure management is just one tool,

> **P**ressure management is one of the most basic tools available for total loss management.

which should be used where applicable in conjunction with other technologies and methodologies.

Recent studies and research have shown that both leakage volume and new leakage frequency are reduced greatly by the reduction and stabilization of pressure within a distribution system. Obviously, not all systems can tolerate pressure reduction and indeed many systems suffer from lack of pressure; however, there are still many, utilities that are operating at pressures in excess of those required, which would benefit greatly from a pressure management scheme. When considering reduction of leakage, practitioners usually think of pressure-reducing functions, but in many cases, particularly pumped systems, leakage can be reduced greatly by surge anticipation.

Water Conservation

In direct pressure use situations (see Fig. 12.1), pressure reduction can be an effective way of controlling unwanted demand. A simple example is whether someone cleans his teeth for 5 min at a high pressure or for 5 min at a low pressure. If the tap is left on for the duration, much less water will be consumed at the lower pressure. This is not the case in tank-fed residential situations (see Fig. 12.2), as the head controlling the demand is a function of the height above the equipment being used, not the incoming pressure. (Work is being undertaken by practitioners in areas with residential tanks to better understand the role that pressure management may play in the reduction of ball valve leakage, which often goes undetected because meters have trouble reading these low flows. It has been noted that the ball valves stop leaking below certain pressures, with no further intervention needed. This may in effect mean that pressure management can also have a positive effect on apparent losses.)

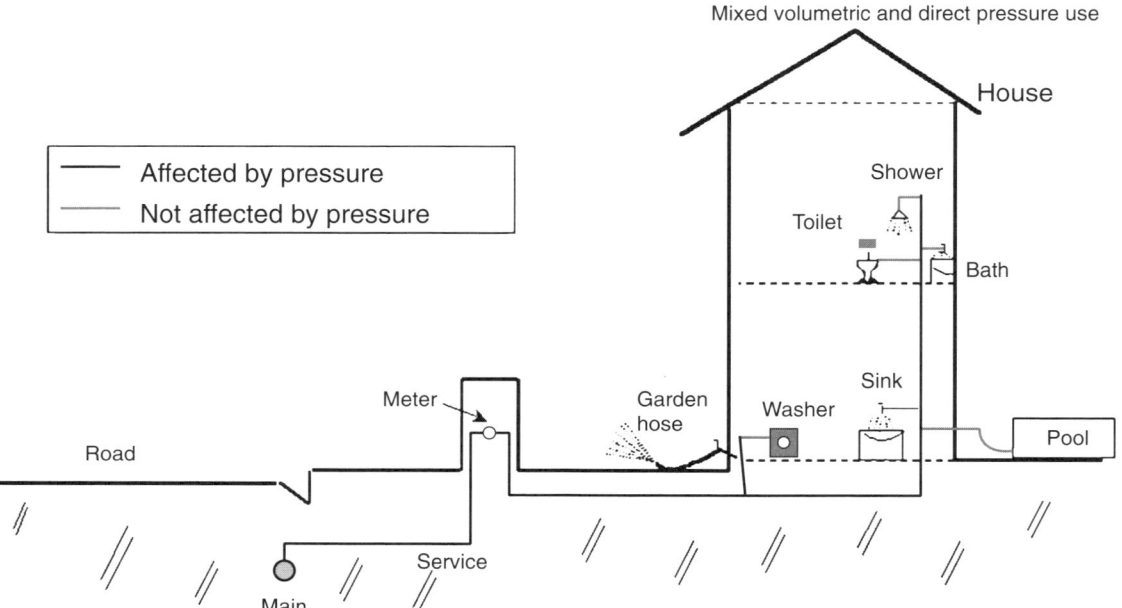

Figure 12.1 Residential demand, direct pressure. (*Source: Julian Thornton.*)

12.2 WHY UNDERTAKE PRESSURE MANAGEMENT SCHEMES? 263

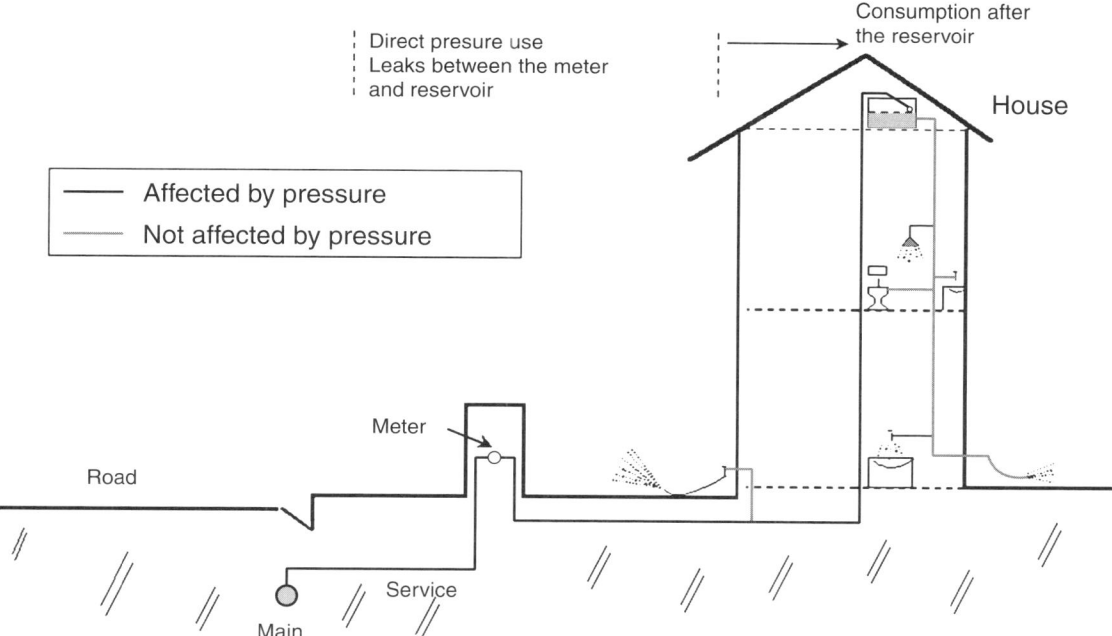

Figure 12.2 Residential demand, tank fed. (*Source: Julian Thornton.*)

While many utilities may not want to reduce demand, because it will have a negative impact on billing, many other utilities have found that it is much more cost-effective to reduce demand than to implement costly capital expansion programs to increase supply or meet excessive demand peaks. Utilities with direct feed systems should carefully analyze the demand types within residences and commercial industrial customers, as many demands are volumetric and will not be affected by pressure reduction (except in terms of fill time).

12.2.2 Nonpayment

Some utilities are faced with a nonpayment situation which is difficult to resolve due to political or social pressure: they have to continue to supply water even if the customer is not paying. In these situations, pressure management to reduce consumption while maintaining a minimum level of supply is of the utmost importance to optimize losses and conserve resources.

Some utilities are simply not permitted to increase their supply because of environmental restrictions. Instead, pressure reduction can be performed on a zonal basis or indeed on an individual customer basis, as the situation requires.

Pressure reduction can also be used as an emergency measure for drought control: levels of demand and leakage can be drastically cut until reserves return to normal.

12.2.3 Efficient Distribution of Water

Many water distribution systems have problems supplying some customers, while others enjoy a constant source of water. The reasons may

> Pressure management is not only about pressure reduction but also in some cases about increasing pressure, sustaining pressure, surge control and level management).

be aging infrastructure, poor design, geographic constraints, or demographic layout. Pressure management using not only pressure-reducing techniques but also pressure-sustaining techniques, boosters, or flow control can ensure that the system distributes its resource as evenly as possible, providing required volumes for a majority of the customers.

Guaranteed Storage

The implementation of pressure management schemes can assist utility operators in ensuring that reservoirs and storage tanks remain at realistic levels to meet demands. This may be done by using a mixture of pressure-reduction, pressure-sustaining, and flow control valves. See Fig. 12.3. Level control also ensures that storage is not allowed to overflow during off-peak hours when system demand and headloss is low and pressures are highest. Reservoir overflows can form a large part of a utility's leakage if not properly controlled.

Reduced Hydraulic Impact

Hydraulic impact, surge, and transient waves are caused by quick changes in system conditions. Unfortunately, most systems have situations where an operator closes a valve too quickly or the opposite. Maybe a hydrant is operated quickly in an emergency, or a large consumer suddenly stops drawing water. Without valve control in the system, transient waves are allowed to travel backwards and forwards within the system, causing damage at any weak point. While pressure-relief valves and surge-arrestor valves are the tools for this type of situation (see Fig. 12.4), simple pressure management schemes limiting pressure to those required are also effective in reducing the negative impact of transient waves. Simple pressure-reducing valves installed to maintain lower pressures will also damp the potential negative effects.

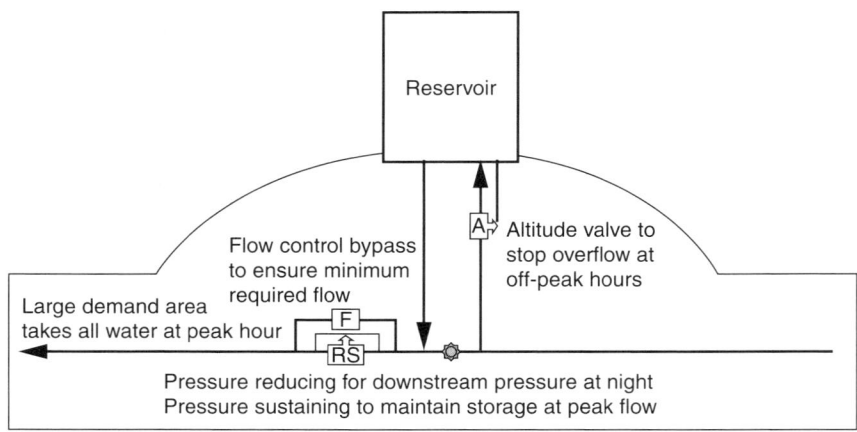

Figure 12.3 Pressure management often uses a mixture of valve types. (*Source: Julian Thornton.*)

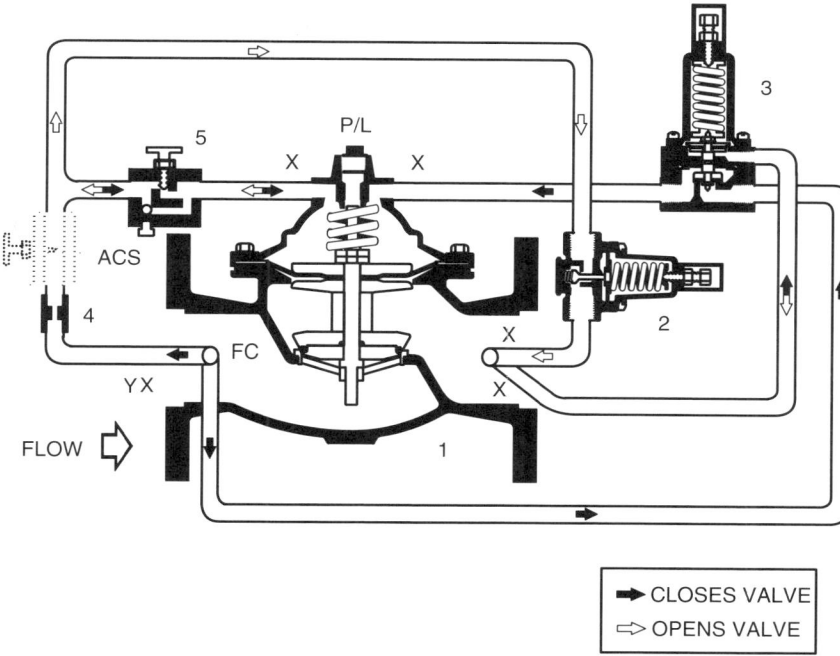

Figure 12.4 Surge-arrestor valve diagram. (*Source: Watts ACV, Houston, Texas.*)

Reduced Customer Complaints

Pressure management schemes are designed not only to reduce pressure but also to provide a constant supply of both water pressure and volume. Some customers experience periods of the day with low pressures caused by high headlosses in the system. High velocities, some of which may be due to uncontrolled demand downstream of the customer, cause high headlosses. Other customers complain of pressures which are too high and cause either discomfort or damage to equipment. Uncontrolled leakage can also cause lack of supply for customers.

Contrary to belief, pressure management can increase customer satisfaction.

12.3 POTENTIAL CONCERNS

It may seem from the preceding pages that pressure management is the answer to all a utility's problems! However, a poorly implemented program may also cause problems of its own. When discussing a pressure management scheme for a utility that does not currently have control or is intending to increase the level of control, the usual concerns are

- Fire flow concerns
- Loss of revenue
- Reservoirs not filling at night

12.3.1 Fire Flow Concerns

Where fire flows are a concern, sectors can have multiple feeds, controlled by pressure-reduction valves with flow-modulated capability. Therefore, if there is a fire, the system has sufficient hydraulic capacity to maintain pressures and flows for fire fighting, as required, for example, by National Fire Protection Association (NFPA) regulations in the United States and Canada. The valves automatically regulate pressure as determined by demand requirements plus the minimum safe operating limit at residual conditions.

Systems which do not have the benefit of the more efficient flow-modulated valves often have a large sleeper valve, either in parallel with the operational valve or at a strategic entrance to the sector. This valve opens when the system pressure drops due to additional headloss created by fire flow. This large valve generally remains closed unless an emergency situation occurs. The use of a large, nonfunctioning valve may in many cases not be as cost-effective as the more modern and efficient demand-modulated options, but sometimes the range of demand dictates that a second parallel valve be used.

The NFPA regulations require that systems have an available residual pressure of 20 psi while a hydrant is flowing and 40 psi static head. Hydrants are coded as discussed in Chap. 17 as to their flow capacity at these standard reference pressures. When setting up potential pressure-controlled sectors, these limits, along with insurance regulations for the types of property in the sector, should be taken into account. Most countries have some kind of fire code, which should be followed when planning a pressure management scheme.

> **W**hen setting up pressure management zones, fire codes must be respected.

12.3.2 Loss of Revenue

As far as the loss of revenue is concerned, systems with high leakage will almost always see a positive benefit from pressure management, even when stacked against the potential loss of revenue, due to reduction of pressure in residences or industry. Any lost revenue is included in the cost-to-benefit calculations as a cost against the project, just as installation and product costs are. This is also true for systems with lower losses and high costs to produce or purchase water. In situations where a loss of revenue cannot be tolerated, pressure management can be limited to nighttime hours, when legitimate consumption is at its lowest and system pressures are at their highest.

Remember also that many systems are enforcing water conservation programs. Pressure reduction is also a water conservation program.

A large portion of water use within a household is caused by toilets; tank-type toilets use a fixed volume of water for each flush, regardless of the pressure. There are also many other fixed-volume uses within a residence which will not vary significantly with pressure (see Fig. 12.1).

When considering pressure management for a sector, consider the per-capita use, and whether it is excessive. Sample per-capita uses are given in Table 12.1. If use is excessive, pressure management will become a natural part of a conservation program. If overall usage is not excessive, the utility should determine the components of consumption within the

TABLE 12.1 Estimated Per-Capita Use of Water in the United States, 1990

State	Liters per Capita per Day	Gallons per Capita per Day
Alabama	379	100
Alaska	299	79
Arizona	568	150
Arkansas	401	106
California	556	147
Colorado	549	145
Connecticut	265	70
Delaware	295	78
District of Columbia	678	179
Florida	420	111
Georgia	435	115
Hawaii	450	119
Idaho	704	186
Illinois	341	90
Indiana	288	76
Iowa	250	66
Kansas	326	86
Kentucky	265	70
Louisiana	469	124
Maine	220	58
Maryland	397	105
Massachusetts	250	66
Michigan	291	77
Minnesota	560	148
Mississippi	466	123
Missouri	326	86
Montana	488	129
Nebraska	435	115
Nevada	806	213
New Hampshire	269	71
New Jersey	284	75
New Mexico	511	135
New York	450	119
North Carolina	254	67
North Dakota	326	86
Ohio	189	50
Oklahoma	322	85
Oregon	420	111
Pennsylvania	235	62
Rhode Island	254	67
South Carolina	288	76
South Dakota	307	81
Tennessee	322	85
Texas	541	143
Utah	825	218
Vermont	303	80
Virginia	284	75
Washington	522	138
West Virginia	280	74
Wisconsin	197	52
Wyoming	617	163
Puerto Rico	182	48
Virgin Islands	87	23
United States Total	397	105

Source: Soley et al., in L. W. Mays (ed.), *Water Distribution Systems Handbook*, McGraw-Hill, New York, 2000.

268 CHAPTER 12 PRESSURE MANAGEMENT

> For a detailed breakdown of water usage in the United States, the AWWA has an excellent residential end-use study, which can be purchased through the AWWA Internet site, www.awwa.org.

sector (residential, commercial, industrial), the volumetric consumption, and the consumption tied directly to pressure. The potential benefits of loss reduction versus reductions in revenue can then be analyzed. Cost-to-benefit calculations are discussed in more detail in Chap. 8.

12.3.3 Reservoir Filling

Regarding reservoirs not filling at night because of reduced system pressure, many pressure reduction programs concentrate on the smaller mains, therefore allowing reduction of losses in selected areas, while allowing normal system pressure in the larger trunk or transmission lines. (As in the example in Fig. 12.3, a complete pressure management project can in some cases actually improve reservoir filling characteristics). This is particularly important in pumped systems, where the storage tanks balance on the system pressure; gravity systems are less affected.

Reservoirs are usually connected with larger pipes, so there should not be a problem in many cases. Most utilities find that nonvisible leakage tends to occur in smaller pipes and service connections, so the effectiveness of a potential pressure management program should not be reduced significantly by the exclusion of larger pipes in the control area. See Fig. 12.5.

12.4 TYPES OF PRESSURE MANAGEMENT

Pressure management comes in various forms, from basic sectorization of a gravity system to dynamically controlled automatic control valves (ACVs). Different distribution systems may have different requirements or indeed multiple requirements. Some of the most common forms of pressure management are discussed below.

12.4.1 Sectorization

Sectorization is one of the most basic forms of pressure management, but it is very effective. Subsectors are divided either naturally or by phys-

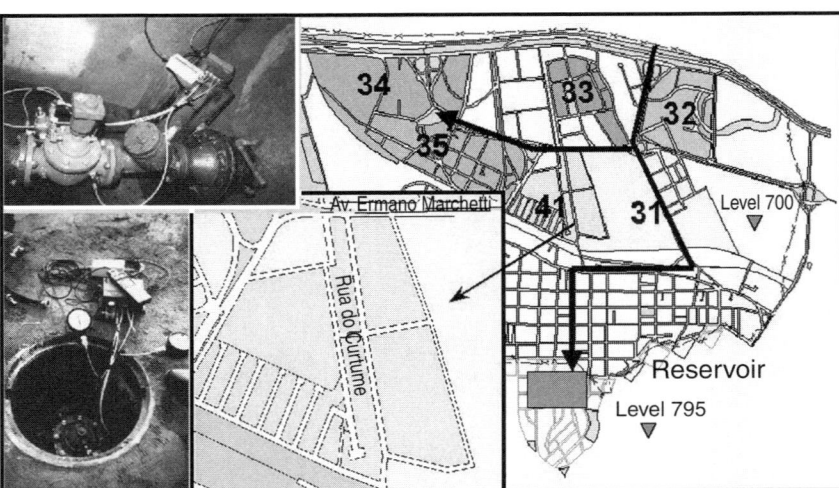

Figure 12.5 Pressure management in subsectors. (*Source: Julian Thornton.*)

ical valving. The sectors are usually quite large and often have multiple feeds, so they do not usually develop localized hydraulic problems because of valve closures. Systems with gravity feeds are usually sectorized by ground level, and systems with pumped feeds are usually sectorized depending on the level of elevated tanks or storage.

One of the hardest parts about controlling pressure solely by using sectorization is enforcing boundary valve control. Nowadays telemetry devices are available which transmit valve status to a central control every time the valve is operated, allowing managers to control the integrity of the sectors and ensuring that they are returned to normal after either an emergency or a maintenance procedure.

Sectorization in its simplest form does not require the implementation of costly ACVs and controllers, but it is often not completely efficient without them. Many systems that have had sectorization in place for many years are finding that it is cost-effective to implement more advanced controls in addition to the basic controls already in place.

12.4.2 Pump Control

Many utilities use pump control as a method of controlling system pressure. Pumps are activated or deactivated depending on system demand. This method is effective if the reduced level of pumping (usually at night) can still maintain reservoir levels. With recent energy conservation concerns this methodology should be carefully reviewed as to the efficiency of energy use. The pump(s) may operate outside the designed profile if they are subjected to upstream valve throttling or demands outside the design limits. Inefficient pumps can cause huge increases in electricity consumption and sometimes even expensive fines for overuse during peak times.

Properly controlled pumps, particularly with variable-speed drives, can provide very effective system pressure control.

12.4.3 Throttled Line Valves

Many system operators recognize the need to reduce system pressure and partially close a gate or butterfly valve to create a headloss and reduce pressure. This method is least effective, however, as the headloss created will change as system demand changes. At night, when a distribution system needs the least pressure, the pressure will be higher; and during the day, when the distribution system needs the most pressure to supply demand, the pressure will be lower. This creates a classic case of an upside-down zone.

> Throttled system valves are the least effective way of controlling pressure.

12.4.4 Automatic Control Valves—Fixed Outlet

Automatic control valves (ACVs) are a traditional method of control and use a basic hydraulically operated control valve (see Figs. 12.6 and 12.7). In Sec. 12.8.5 we will be discussing controllers and varying profiles. The fixed-outlet control valve method is effective for areas with low headlosses, demands which do not vary greatly with the seasons, and uniform supply characteristics. Fixed outlet control in other areas may be inefficient, as outlet pressures have to be set high enough to meet minimum

Figure 12.6 Pressure-reducing valves working in parallel. (*Source: Julian Thornton.*)

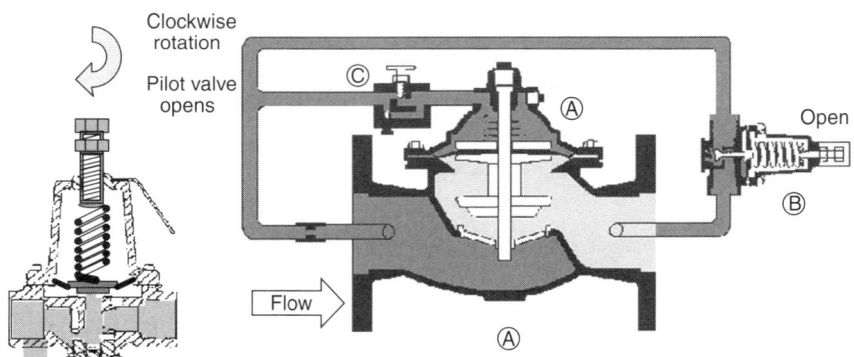

Figure 12.7 Pressure-reducing valve diagram. When pilot valve B is open, the pressure in the control circuit does not exert any force on the membrane of main valve A; therefore, the main valve will open. (*Source: BBL Ltda.*)

pressures during peak demand. As system demand falls, usually at night, the headlosses in the system fall and system pressure returns toward the static pressure, which in many cases is far in excess of that required to meet nighttime demand plus fire demand. See Fig. 12.8.

12.4.5 Leakage Control—Pressure Leakage Theories

It has recently been shown that the relationship between leakage and pressure is not related merely to the square root of the pressures in question, but rather to an expanding power law. As well as PVC pipe leaks,

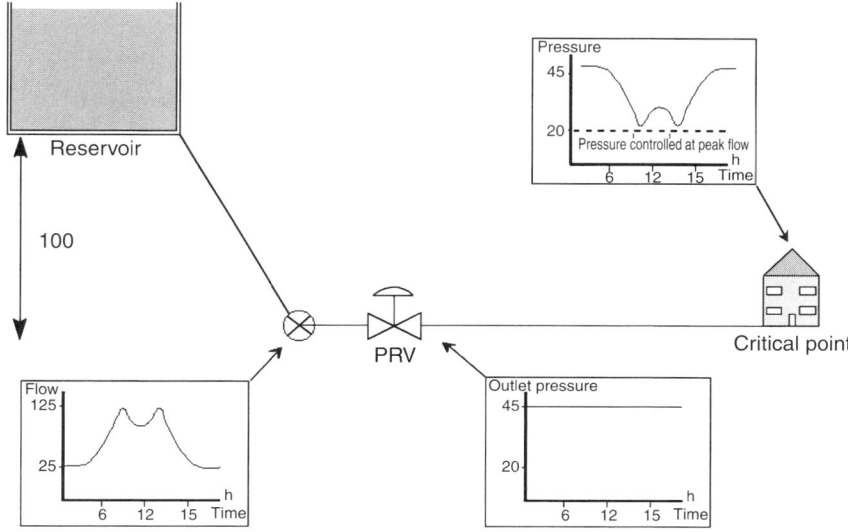

Figure 12.8 Effects of fixed outlet control. (*Source: BBL Ltda.*)

many other types of leaks, particularly at joints, are subject to a change in area as pressure changes. This means that potential benefit in pressure reduction on the volume of these leaks has a much greater impact, as not only the velocity of leak flow changes, but also the leak area.

Traditional Calculations

Traditional calculations for reduction in leakage through reduction in pressure assumed a fixed-area leak. The calculation for this type of situation was and still is as follows:

When the pressure is changed from P_0 to P_1, the leakage rate changes from L_0 to L_1. Therefore, $L_1 = L_0 (P_1/P_0)^{0.5}$.

Example A zone with fixed-area leakage has a leak rate of 500 gpm at 80 psi. If the pressure were reduced to 50 psi, what will be the savings in leakage rate?

Answer $L_1 = L_0 (P_1/P_0)^{0.5}$ = new leakage = $500(50/80)^{0.5}$ = 500 − 395 = 105 gpm.

Fixed and Variable Area Discharge (FAVAD)

Leakage can be described in terms of either fixed or variable paths. Fixed-area leakage could be pinholes in galvanized service line or a hole in a cast-iron pipe. This type of leakage follows the traditional calculation shown in the last paragraph. Savings through reduction in fixed-area leakage are usually more conservative than in areas with variable-area leakage.

Variable-area leakage normally occurs in systems with some kind of PVC or plastic pipe, systems with joint leaks (often found in systems with AC piping or old hydraulic couplings), and systems with background leakage (commonly referred to as unavoidable leakage).

Variable-area leakage is not calculated using the traditional square-root formula but rather with a power which is very much system-dependent.

CHAPTER 12 PRESSURE MANAGEMENT

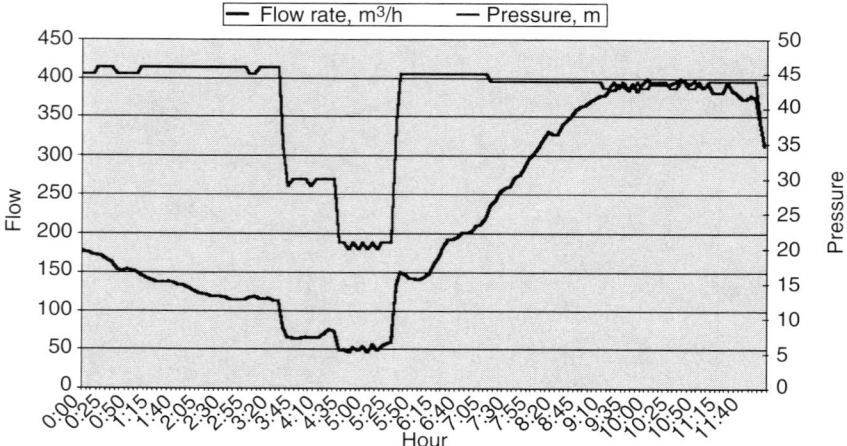

Figure 12.9 Reduction in pressure provides a reduction in leak flow rate. (*Source: SABESP/BBL Ltda.*)

Powers range from 0.6 to 2.5 and should be calculated on a zone-by-zone basis. Data collected internationally throughout the 1990s from many systems, including those of the United Kingdom, Japan, Germany, and Brazil, show that a power factor of 1.15 is representative of large zones with varied materials.

Calculating the power factor is quite simple and can be undertaken in the field with either data loggers or manually by using flow and pressure readings. This type of testing is commonly referred to as step testing.

To calculate the correct power factor, the pressures and flow should be read at night during stable demand conditions. The pressure should be lowered by either reducing the pressure on an existing pressure-reducing valve or by throttling a gate valve. The corresponding drop in flow will dictate the power factor. Usually the power factor used is an average of three or more drops or steps. See Figure 12.9 for a sample step test result.

Using the internationally accepted average power of 1.15, our sample calculation of reduced leakage changes to $L_1 = L_0 (P_1/P_0)^{1.15}$ = new leakage = $500(50/80)^{1.15}$ = 500 − 291= 209 gpm. It can be seen that we have an additional saving of 104 gpm. This additional saving is a function of the changing area of the leak(s) in this example. We must therefore conclude that, other than for systems with 100 percent fixed-area leakage (which is very hard to find), the traditional method of calculating potential savings from reduction of pressure is, to say the least, very conservative and misleading.

Background Leakage

Although many utilities are undertaking very efficient leak assessment, detection, and repair programs, there always remains an element of leakage which is undetectable. This is often referred to as background leakage. This leakage is made up of many small pinhole leaks, joint leaks and drips, etc., which cannot be detected by traditional means. The only efficient way of reducing the impact of background leakage (other than a complete mains and service replacement program, as discussed in Chap. 13) is to control pressure efficiently.

High background leakage will often be found in systems with high service density, high hydrant density, or where maintenance is difficult because of a heavy urban development.

Reduction of New Leak Frequency

Pressure management helps to reduce not only the volume of leakage and background leakage, but also the frequency of new leaks. Pressure is not the only factor in new leakage, but it is often a significant one. Other factors may include ground conditions, traffic conditions, pipe material, stray currents, temperature, and backfill.

12.4.6 Overflow Control

When discussing pressure management and its impact on water loss, it is important also to discuss level management in reservoirs, tanks, and storage.

Water loss from overflows in storage facilities is too often overlooked, as it is deemed not to be significant and often tanks are in out-of-the-way locations so overflow is not always evident.

Overflows usually occur at night (when pressure conditions are often at their highest, due to lack of demand and headloss on the system) and are caused by either lack of level controls or malfunctioning controls. Level control can be performed manually by pump control, by SCADA, which involves automatic control by computer-linked software, or by simple hydraulic control, using either altitude valves or ball valves. Sometimes a utility will use a sophisticated series of automatic controls; however, external forces such as lightning may affect them. A simple hydraulic backup is often cost-effective.

Most tanks and reservoirs have an overflow pipe. If a utility wishes to discover if overflow is occurring, it is a simple task to inspect the point where the overflow pipe dumps water. If there is recent evidence of water being discharged, then either the level should be data logged and compared with the overflow level, or if data logging technology is not available a simple solution is to locate a ball in the overflow pipe and inspect the position of the ball each day. If the ball has come out of the pipe, then there has been an overflow.

Pressures and levels should be monitored and the level of loss analyzed. A simple cost-to-benefit exercise will identify whether a new system of control is warranted.

> **P**ressure management includes the management of reservoir and tank levels, which can often be the source of considerable annual losses.

12.4.7 Monitoring Points

For any pressure management project, it is necessary to monitor as a minimum the following points:

- ❑ Supply nodes
- ❑ Storage nodes
- ❑ Critical nodes
- ❑ Estimated average nodes

Supply nodes are any points which supply a system or subsector of a system. A supply node may also be an outlet point from one zone to another. In some cases it may be necessary to monitor bi-directional flows.

Storage nodes include any reservoir, tank, standpipe, or location where water is stored.

A critical node is a point where supply may be at its weakest—for example, a high level within the system or a point where there is high headloss in the supply pipe. Alternatively, it may be a point where a user cannot be left without water—for example, a manufacturing facility or a hospital.

An *estimated average node* is a location that is chosen to be representative of average conditions (ground level, pressure, headloss, etc.) within the system or zone.

12.4.8 Flow Measurements

In general, flow measurements should be taken at any supply or exit point. A supply point may be a pumping station, treatment plant, storage facility, well, or bulk transfer point to the system or zone. It may be deemed necessary during a demand analysis to measure demands from large consumers, if they are considered to be large nighttime water users.

Measurements should be taken for a minimum of 24 hours, but preferably for 7 days or more. The decision on how long to measure usually comes down to cost.

Care should be taken when measuring flows to ensure that these flows are easily related to changing seasonal trends. The ideal situation in areas of changing demands is to monitor for 1 year, but this is very rarely possible. The next best thing is to normalize annual demands and relate a week of flow monitoring to the normalized curve.

Flows should be measured accurately, with calibrated equipment, although an accuracy of ±10 percent is usually acceptable because the valves to be installed have quite a wide range.

> The longer the measurement period the better; however, measurement periods are usually limited by the cost.

12.4.9 Pressure Measurements

Pressure measurements should be taken at all of the node points mentioned above. Pressure should be measured with a reasonably high-resolution logger (±0.1 percent full-scale), which should be calibrated for accuracy and drift before and after the field installation. Further information on the measuring process can be found in Chap. 6.

12.4.10 Using Hydraulic Models to Identify Locations for Installation

It is not necessary to use a computerized hydraulic model to select areas for pressure control, but if one is available it can be used to identify areas with high pressures and also to identify areas of high headloss, where the more advanced dynamic controllers could be used beneficially. See Fig. 12.10. In general, the model should be reasonably calibrated and include any extreme demands necessary, such as fire flows or seasonal adjustments. A model calibrated to ±15 percent is generally acceptable for this type of work.

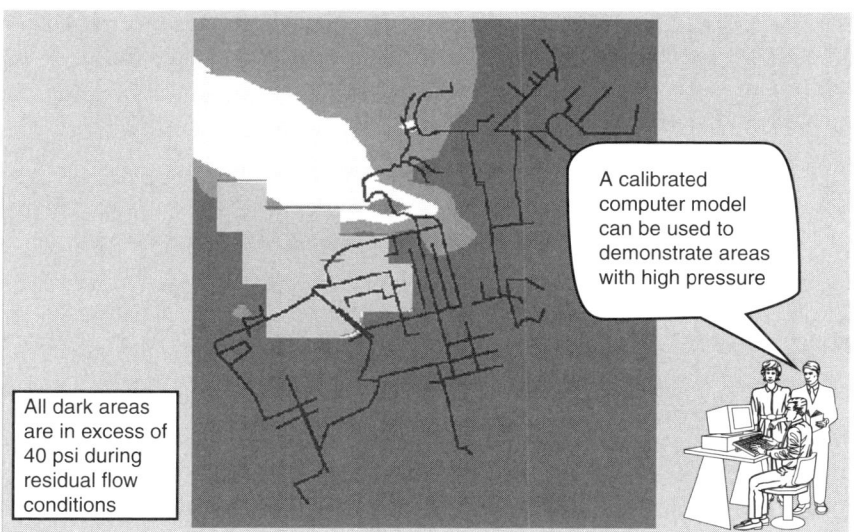

Figure 12.10 Hydraulic models can be used to show areas of high pressure. (*Source: Julian Thornton.*)

Using a model is a very nice way of quickly identifying potential areas, although it is still necessary to go into the field and make field measurements, as often the situation in the field changes, valves are left closed, new leaks occur, etc.

Understanding the Hydraulics of Your System before Implementation

In addition to using hydraulic models to locate pressure control stations and field measurements to acquire critical data, it is also very important to understand exactly how the sector functions hydraulically. This analysis is normally undertaken in the demand analysis phase of the project and should identify:

- Percentage of direct pressure consumption
- Percentage of consumption from individual storage tanks
- Distribution feeds, by pump or by gravity
- Breakdown of consumer usage as to residential, commercial, and industrial
- Level controls for elevated storage
- Pump shutoff controls

The results of this research will form the basis of the control scheme, providing limits of control and cost-to-benefit assumptions.

12.4.11 Using Statistical Models to Calculate the Potential Benefit of a Scheme

Once we have identified an area, made field measurements, and identified how the water is used within the sector, we can proceed to the decision-making stage. During this phase we identify how much control we can effect without disrupting normal supply and what benefit this

276 CHAPTER 12 PRESSURE MANAGEMENT

control will have on reduction of leakage volume, reduction of new leak frequency, deferral of new source schemes, and in some cases water conservation.

The diagrams in Figs. 12.11 to 12.15 show a simple example of a statistical model which was created in Excel to solve certain sets of equations, analyze field data, and produce a simple cost-to-benefit analysis. (Similar examples can also be found in Chap. 7.) This particular model was made to interface with results from the AWWA M36 audit sheet. The model data shown are fictitious and not necessarily representative. They assume a 24-month period and also that there is virtually no headloss in the system (often the case in North American water systems). The model assumes a fixed outlet control and shows the difference in payback given various scenarios.

Figure 12.11 shows the data captured from the AWWA M36 audit. Figure 12.12 shows a calculation for identifying potential loss of revenue through pressure reduction. These first two figures are basic data that is required to set up the calculation. The following three figures identify different scenarios of a cost-to-benefit calculation. Figure 12.13 shows a scenario with conservation not being an issue, so the cost of lost revenue is included in the project cost. The scenario in Fig. 12.14 assumes that conservation is a relevant issue, so the cost of lost revenue is discarded.

Figure 12.15 illustrates a scenario in which conservation is an issue and also includes the cost to provide additional source water. If a utility is facing the need to conserve, it is quite likely that it is also facing the need to increase source capacity. If it can defer that capital expenditure, there is usually a huge benefit. It can be seen that there is a vast difference in the paybacks as the decision making changes.

The cost of the equipment in these spreadsheets is very low compared to a real-life situation, but it can be seen as the process progresses that the cost of the equipment is a very small part of the equation.

Deferral of capital expenditure is often realized by reducing losses. The value of the lost water in these cases usually exceeds the marginal production cost.

	Basic analysis of savings using a 24-month period, assuming no headloss in the system and fixed outlet control		
A.	Cost of water	$/kgal	$2.20
B.	Metered volume to sector	kgal	527688
C.	Volume of discovered leaks, if not repaired (M36 audit, line 14g)		0
D.	Volume of potential water system leakage (M36 audit, line 16)		78889
E.	Total system leakage	kgal	157778
F.	Existing average daily pressure	psi	80
G.	Desired control pressure	psi	45
H.	Power law		1.15
I.	New leakage volume	kgal	81412
J.	Total savings (E – I)	kgal	76366
K.	Total savings (J*A)	$	168006

Figure 12.11 Estimated reduction in leakage. (*Source: Julian Thornton.*)

12.4 TYPES OF PRESSURE MANAGEMENT 277

	Basic analysis of consumption changes using a 24-month period, assuming no headloss in the system and fixed outlet control		
A.	Retail cost of water	$/kgal	3
B.	Water sold in sector	kgal	446760
C.	% residential use		50
D.	% agricultural use		10
E.	% ICI use		40
F.	Volume sold for residential use (B*C)	kgal	223380
G.	Volume sold for agricultural use (B*D)	kgal	44676
H.	Volume sold for ICI use (B*E)	kgal	178704
I.	% of residential use from direct pressure		40
J.	% of agricultural use from direct pressure		5
K.	% of ICI use from direct pressure		5
L.	Residential volume affected (F*I)	kgal	89352
M.	Agricultural volume affected (G*J)	kgal	2234
N.	ICI volume affected (H*K)	kgal	8935
	Total		100521
O.	Existing system pressure	psi	80
P.	Desired control pressure	psi	45
Q.	New pressure-dependent residential consumption [$Q = L*(P/O)^{0.5}$]		67014
R.	New pressure-dependent agricultural consumption [$R = M*(P/O)^{0.5}$]		1675
S.	New pressure-dependent ICI consumption [$S = N*(P/O)^{0.5}$]		6701
	Total		75391
T.	Potential lost sales volume	kgal	25130
U.	Potential lost revenue	$	75391

Figure 12.12 Estimated reduction in consumption. (*Source: Julian Thornton.*)

	Cost-to-benefit analysis, assuming no headloss in the system and fixed outlet control	
	Costs	
A.	Cost of valve assembly	800
B.	Cost of bypass assembly	500
C.	Cost of field measurements	800
D.	Cost of installation	3000
F.	Potential reduction in revenue through pressure control	75391
	If water conservation is being implemented, mark "y" for yes here . . .	n
	Benefit	
G.	Total savings through leak reduction	168006
H.	Cost to produce 1 kgal of water from a new source	0
I.	No. of kgallons reduced through leak reduction	76366
J.	Benefit from deferring installation of new source	0
	(Include only H, I, and J if water supply needs to be increased)	
	Analysis	
K.	Total cost	80491
L.	Total benefit	168006
N.	Payback on investment in months	11.5

Figure 12.13 Estimated cost to benefit where conservation is not desired. (*Source: Julian Thornton.*)

278 CHAPTER 12 PRESSURE MANAGEMENT

	Cost-to-benefit analysis, assuming no headloss in the system and fixed outlet control	
	Costs	
A.	Cost of valve assembly	800
B.	Cost of bypass assembly	500
C.	Cost of field measurements	800
D.	Cost of installation	3000
F.	Potential reduction in revenue through pressure control	75391
	If water conservation is being implemented, mark "y" for yes here ...	y
	Benefit	
G.	Total savings through leak reduction	168006
H.	Cost to produce 1 kgal of water from a new source	0
I.	No. of kgallons reduced through leak reduction	76366
J.	Benefit from deferring installation of new source	0
	(Include only H, I, and J if water supply needs to be increased)	
	Analysis	
K.	Total cost	5100
L.	Total benefit	168006
N.	Payback on investment in months	0.7

Figure 12.14 Estimated cost to benefit where conservation is desired. (*Source: Julian Thornton.*)

	Cost-to-benefit analysis, assuming no headloss in the system and fixed outlet control	
	Costs	
A.	Cost of valve assembly	800
B.	Cost of bypass assembly	500
C.	Cost of field measurements	800
D.	Cost of installation	3000
F.	Potential reduction in revenue through pressure control	75391
	If water conservation is being implemented, mark "y" for yes here ...	y
	Benefit	
G.	Total savings through leak reduction	168006
H.	Cost to produce 1 kgal of water from a new source	1.2
I.	No. of kgallons reduced through leak reduction	76366
J.	Benefit from deferring installation of new source	91639
	(Include only H, I, and J if water supply needs to be increased)	
	Analysis	
K.	Total cost	5100
L.	Total benefit	259645
N.	Payback on investment in months	0.5

Figure 12.15 Estimated cost to benefit where a new source is required. (*Source: Julian Thornton.*)

Obviously, every utility faces its own unique challenges. This simple spreadsheet example does not cover all of those challenges but does serve to demonstrate some of the more frequently faced decisions.

A simple model can be constructed by most users in Excel by following these guidelines, but various commercial models are also available for purchase. Extracts from some of the more detailed models and analysis can be found in the Chap. 7, which deals in more detail with data analysis and decision making using computerized models.

The decision to purchase a model or construct one should really lie with the type of staff a utility has and the time that is available. While the calculations are not really complex, it can in some cases be false economy to try to build a model when a small investment will buy a tried and tested version.

Most commercial models are flexible, but care should be taken to assure that the model purchased takes into account the hydraulic characteristics of the utility system in question. As discussed earlier, there are significant differences between hydraulic characteristics of demand for a system which uses residential storage and a system which has direct pressure feed.

The following case study, by Allen Young of BKS in South Africa, details the efforts of Johannesburg City Council to reduce leakage by pressure management. Detailed statistical models were used during the process selection process.

12.5 CASE STUDY ONE
*Advanced Water Pressure Management in the Berea–Alexander Park Supply District, Johannesburg, South Africa**

Allen Young, BKS Pty Ltd.

12.5.1 Background

The Eastern Local Council of the Greater Johannesburg Metropolitan Council (GJMC) calculated in 1999 that 18.6 percent of the volume of all water acquired from its bulk water supplier was being lost in the storage and distribution system, over and above an allowance of 12 percent for expected losses.

No data are available on what proportion of the unaccounted-for water may be due to leakage from pipes, but a rough indication based on comparison of observed night flows in the Berea–Alexander Park district suggests that as much as 50 percent of this might be due to leakage.

In November 1998 the concept of advanced pressure management using electronic controllers fitted to pressure-reducing valves to achieve modulation of pressures was presented to engineers of the GJMC by Julian Thornton of the Brazilian company BBL/Restor. Case studies based on his

*By permission of the Greater Johannesburg Metropolitan Council.

experience in applied pressure management in São Paulo, Brazil, were presented, which provided the motivation for the council to include pressure management as one of its strategic initiatives, to reduce unaccounted-for water.

The ambit of the GJMC project was to locate existing pressure-controlled districts that would benefit from modulated pressure control, or alternatively, to identify suitable districts for introduction of pressure management. The latter approach proved to be the most fruitful, and the Berea–Alexander Park supply district was selected as one of the districts that presented good potential for successful implementation of advanced pressure management.

12.5.2 District Selection Criteria

The criteria that favored selection of this district were

- Its large size and potential isolation from neighboring supply areas
- Lack of known low-water-pressure problems
- Adequate static pressures and topography that lends itself to an overall reduction of the hydraulic grade across the district
- Suitable positions for PRVs that would allow adequate working head for pressure control.

12.5.3 Description of the District and Condition of the Pipes

The district covers an area of some 1370 hectares and is a predominantly residential area with a range of lot sizes from 0.06 to 0.15 hectare. The residential area has commercial centers consisting mainly of clustered shops and small shopping complexes, with the Bruma office park and hotels on the eastern side of the district. The area houses four sports clubs and the Kensington golf course, which are potentially large users of water for irrigation.

Other vital statistics of the supply area are set out below:

- Number of consumer connections: 8577
- Population: 30,230
- Types of consumers
  - Residential: 95 percent
  - Commercial: 5 percent
- Mains lengths
  - Primary mains: 200–750 mm in diameter, 34 km
  - Secondary mains: 20–160 mm in diameter, 136 km
- Pipe materials, proportion of total length, average age
  - Steel, 71 percent, 45 years
  - HDPE, 10 percent, 13 years
  - UPVC, 18 percent, 14 years
  - Fiber cement, 1 percent, 52 years

The district is supplied with water from two linked reservoirs situated on the western side of the district. These reservoirs are constructed with approximately the same top water levels, and supply a maximum static pressure in the lower parts of the district of 12 bar, while a minimum static pressure of 4.9 bar is provided at the highest point in the district. There was no existing pressure-reducing valve sites within the zone.

The older supply mains in this zone are of rolled steel with a 6-mm wall thickness and caulked spigot and socket joints. Pipes are coated internally and externally with bitumen. Although external bitumen coatings on older pipes are generally intact, internal linings exhibited loss of binding, with entrapped pockets of water between the lining and pipe wall with resultant pockets of advanced corrosion under the lining. Spalling of the bitumen lining was also observed. Exposed caulked joints were found to be weeping, which together with the general internal condition of the pipes, indicated that leakage from the older primary mains, which comprise about 15 percent of the distribution system, is a likely occurrence.

Little is known of the condition of the smaller-diameter piping and consumer connections, but the age of the older sections of the district (>45 years) indicates that corroded house connections could be a cause of leakage.

12.5.4 Preinstallation Investigation and Initial Pressure Management Plan

Following a desktop study in which the district was identified as a possible candidate for pressure management, preinstallation investigations were undertaken in order to develop a pressure management plan for the district. Field investigations included the following:

- Gathering of infrastructure and demographic data for the district
- Field inspection of proposed PRV sites and consumer types in the critical high areas
- Checking of normally closed valves that isolate the district from adjacent supply areas
- Measurement of flows and pressures at the feed points, and logging of pressures at critical high points and other selected points in the reticulation (Use was made of portable electromagnetic and turbine insertion meters as an economical method to measure transient flows.)
- Analysis of data and the estimation of leakage reduction for a proposed diurnal modulated pressure profile using a statistical model
- Estimation of costs for installation of PRVs and performance of a cost-to-benefit analysis to test the viability of the proposal

The pressure management plan was to install two new PRV stations on the two reservoir feeds into the district. Pressures would be modulated to give an average reduction of 1–2 bar throughout the zone, with a maximum reduction of pressure of 2 bar during off-peak times based on a target pressure at the critical high point of 3 bar.

Use would be made of Technolog Autowat PRV control equipment to modulate pressures. Theoretical pressure modulation profiles for the two PRVs were designed as a starting point. The interaction between the two PRVs is a function of the headloss in the reticulation and required observation in the field. The pressure control profiles would have to be set empirically in the field once the effect of installation of PRVs on the pressures and flows at the feed points and critical points had been observed. It was anticipated that one of the new PRVs (the one at a lower elevation) would be set at a fixed outlet and would probably remain closed except during peak draw-off periods. Pressure modulation in the zone would then take place by controlling only the other PRV.

The practical location of pressure control stations took the following into account:

- ❑ The PRV should have the maximum upstream working head possible, taking into account the expected headloss through the PRV.
- ❑ The points at which the least number of PRVs would be required to effect control on the district.
- ❑ The ease with which the site would fit into the existing pipe layout.
- ❑ The ease of access to the future station.
- ❑ The environmental acceptability of the site.

Field measurements of flow and pressure were carried out at the proposed PRV sites and at the critical high point and average pressure points in the district. Portable electromagnetic insertion flow meters were used to obtain temporary measurements of the flows.

The field data were used in a statistical software model that estimated the effect of modulating pressures on reduction of background and burst leakage in the pipe system, taking into account estimated reduction in pressure-related consumption.

The field data were further used to size PRVs and meters for the proposed pressure control stations.

A calculation was done of expected saving of water. The forecasted monetary saving through reduction of leakage using the statistical model was R795, 816 per annum. A cost-to-benefit calculation yielded a payback period of some 8 months, confirming the economic viability of the pressure management plan, and it was decided to proceed with implementation of the pressure management plan.

12.5.5 Final Design

Each PRV station comprised a 250-mm Claval PRV with a pot strainer mounted upstream and a Meinecke meter positioned 5 diameters downstream of the PRV. A 250-mm bypass was constructed around the meter and PRV to facilitate future maintenance.

Reinforced concrete chambers were constructed, one of which was partially positioned under a roadway. The chamber was enlarged and partially repositioned to create access from the sidewalk—an important consideration for safety and ease of access for regular data downloads and checking of equipment.

12.5.6 Results

After commissioning of the PRVs, they were both set to maximum fixed outlets. It was found that the PRV at Montague Street remained shut (inlet 7.5 bar/outlet 6 bar), while the PRV in Berea Road continued to feed the zone without a problem (inlet 5 bar/outlet 4 bar).

Night flows were observed to have diminished from 350 to 110 m^3/h.

The outlet pressure of the PRV in Berea Road was seen to drop from 3.2 bar during the daytime to 2.2 bar at night. The pressure at the critical high point became more even and fell only by 0.5 bar on average. (This may indicate a smaller uncontrolled feed into this area that is able to sustain pressures at the high point of the zone, which will require further investigation.) The pressure at the average zone pressure point was reduced on average by 1 bar and showed less fluctuation, indicating that leakage was being reduced throughout the district and there was less stress on the pipe network.

In addition to reduction in night flows, there was a clear reduction in consumption during peak periods. Part of this was due to reduction in leakage and part due to reduced pressure related consumption, e.g., garden sprinkling.

Total reduction in consumption (leakage and usage) was calculated:

- Volume: 2259 m^3
- Period: 23.75 h
- Savings/day: 2283 m^3
- Savings per annum: 833,141 m^3
- Rand value: R1,749,595

The minimum reduction in leakage was estimated for the period 21:00 to 05:00 when normal consumption is minimal:

- Volume: 1110 m^3
- Period: 6 h
- Savings/day: 1110 m^3
- Savings per annum: 405,223 m^3
- Rand value: R850,968

The total cost for the two sites including professional and construction costs was R850,000. A realistic payback period of 6–9 months was therefore achieved.

12.6 CALCULATING COST-TO-BENEFIT RATIOS

Once the data have been entered into the model and calibrated to a certain degree of confidence, the model can be used to analyze the cost of a potential project and its estimated benefit. The components and the diameters to be installed, the type of bypass and chamber, the ground type, the type of control to be effected, and the type of maintenance program to be actioned after installation determine the cost. An additional cost can be a small reduction in revenue from the direct-pressure user component. This cost should only be used if there is no need for water conservation within the utility. If the utility is trying to get its consumers to reduce consumption, then pressure control will form a very efficient part of this program and reduction of consumption will become a benefit and not a cost.

The benefit is calculated from the reduction in leakage volume, reduction in maintenance costs, deferral of costs to build a new source of water if water is scarce, reduction of supply to nonpaying customers, and increased storage management.

The cost to benefit is calculated as a function of the cost divided by the benefit and is usually displayed as a ratio and also as the number of months required to pay back the initial investment. In most utility cases a good payback is somewhere within 24 months. In many cases advanced pressure control provides paybacks of less than 12 months, due to the huge impact on leakage and the simplicity of installation.

More detailed information on calculating cost to benefit can be found in Chap. 8.

12.7 AUTOMATIC CONTROL VALVES

Various types of automatic control valves are available on the market. Some use a diaphragm, some a piston, and some have a collapsible sleeve arrangement. However, most of the valves bodies of each manufacturer are interchangeable for type of control. For example, a valve which was designed originally as a pressure-reducing valve can easily be changed into a level control valve, a pump control valve, a flow control valve, or any number of other functions, by changing the way that it is tubed and the type of pilot control fitted.

Hydraulic ACVs basically work by using the upstream force of the pressure to either open or close the valve by entering or leaving the head of the valve, in function to the pilot setting. This can be seen in Fig. 12.7.

When considering a pressure management scheme for the first time, it is a good idea to talk to various valve manufacturers to ensure that their valves can be altered for various functions, by altering the tubing and piloting. This will ensure the optimization of your investment as systems change character.

For example, a utility may install a 6-in valve for pressure control in a zone this year, correctly sizing all parts. Then, 2 years later, a construction company constructs a large condominium, changing the demand conditions. The 6-in valve may now be undersized, but it is a simple job to replace this valve with a more representative one and reuse the 6-in valve in another location. Care should be taken when sizing the bypass assembly to allow for flexibility.

Another important point is maintenance. For ease of maintenance, most utilities use valves from only one or maybe two different manufacturers. This saves on having to stock the same size parts for different makes of valve. Obviously, price is a concern for most projects, but local support and product and spares stocking should also be taken into consideration. After all, in any ongoing project the initial investment is often only a small part of the overall investment over the life of the project.

Pressure management, like all of the other loss control concepts discussed in this manual, is not a static concept but rather a constantly changing project, which follows the ever-changing needs of the utility.

12.7.1 Pressure Reduction

Pressure reduction is probably one of the most common forms of pressure management being practiced today, with very positive impacts on

> **M**ost valve bodies can be piloted for a number of different pressure management activities.

leakage. A valve not dissimilar to the diagram shown in Fig. 12.7 performs pressure reduction hydraulically. Placing more or less tension on the control spring changes the pilot stem position and the pilot valve opens or closes. As the available orifice size in the pilot changes, more or less water is forced into or out of the head of the valve, making it modulate toward either the open or the closed position. Pilots can be adapted to be fitted with a controller, as explained later in this section.

In most cases pressure control will be undertaken in a zone which has excess pressure throughout the zone. However, in some larger zones, where cost-to-benefit ratios are good, it may be necessary to boost water to certain high critical locations. While this may seem ridiculous, it is a simple matter of performing the cost-to-benefit calculations to see if the ratio is good.

It is not uncommon to find that in addition to smaller sectors which allow a large amount of pressure reduction, i.e., a valley, other, larger zones with the potential for only a very small amount of pressure control will also give very good paybacks. Figures 12.16 and 12.17 show an example of this situation from one of the SABESP installations in São Paulo, Brazil.

> **L**arge areas with only a small amount of potential for pressure reduction may still give good results.

12.7.2 Determining Installation Points

Once the potential sector has been chosen and control points identified either on paper or in a computer model, it is very important to go into the field and locate the exact spots where the valve assemblies will be installed. Other underground utilities should be carefully located prior to excavation.

Care should also be taken when siting a valve on an inclined piece of road, to ensure that inlet pressure always exceeds the required maximum outlet pressure, plus a few extra pounds which are required headloss across the valve to make it function.

Once the spot has been located, it is a good idea to make a location diagram, ensuring that the valve housing is constructed exactly in the right

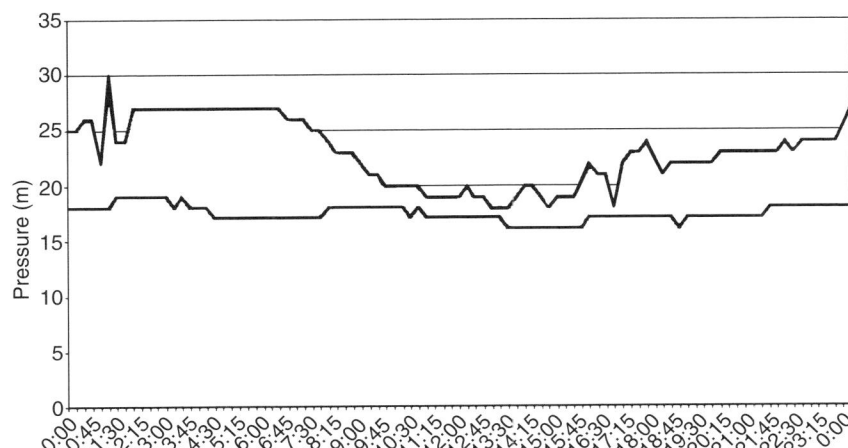

Figure 12.16 Pressures before and after management, contract no. 69.502/96. (*Source: SABESP/BBL Ltda.*)

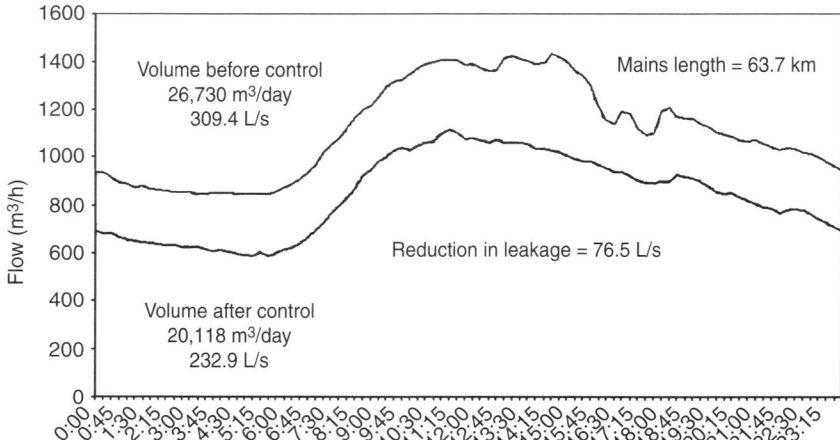

Figure 12.17 Volumes and flows before and after pressure management, contract no. 69.502/96. (*Source: SABESP/BBL Ltda.*)

location and that this location can be easily located at a later date—for example, if it has been asphalted over. See Fig. 12.18 for a sample location diagram.

12.7.3 Multiple Valve Sectors

Some sectors cannot be hydraulically fed only from one point. This may be due to a fire flow volume requirement, or high headlosses during peak demand periods, or any number of other reasons, such as water quality concerns. This does not necessarily mean that the zone is not viable.

Zones with several feeds are quite viable and reasonably easy to set up, as long as careful thought is given to the hydraulic reactions of one valve against its counterpart. It is important to rank valves in order of importance and ensure that the control set points reflect the order of ranking. For example, some valves may be required to function only during periods of high headloss, i.e., during peak or emergency demands. These valves would then remain closed during the rest of the day. Hydraulic grade lines (HGLs) can be used to ensure that the valves are balanced out.

Normally the valves on larger feeds will be set to respond more quickly to changes in demand, while the other valves feeding the system may be set with a slightly longer response time.

12.7.4 Reservoir and Tank Control

As discussed previously, there are various ways of controlling tank levels and many utilities may already have this under control. We are therefore only going to discuss two simple hydraulic solutions to level control, ball valve control and altitude valve control. These two methodologies are probably the simplest and most maintenance-free solutions to reducing water loss through overflow. Some utilities with SCADA systems, which have lightning problems, may also wish to consider this as a back-up system, which will operate hydraulically and independently of the automated system they may have.

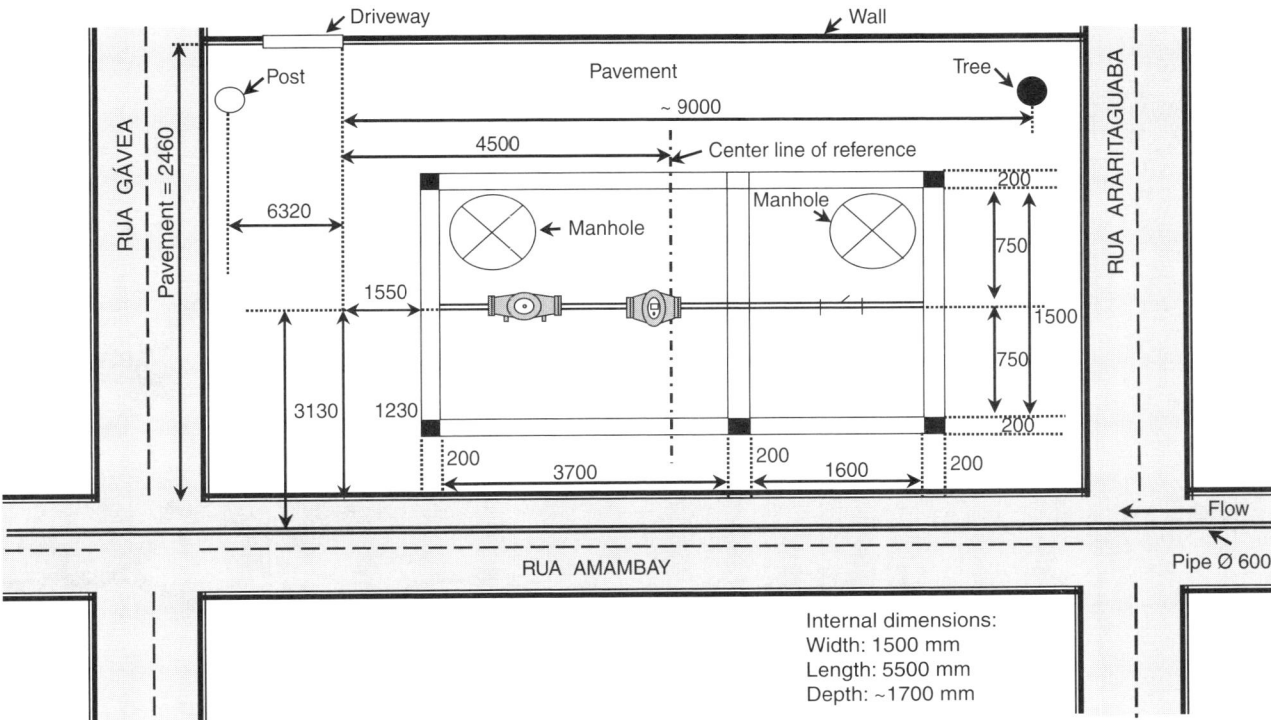

Figure 12.18 Valve location diagram. (*Source: BBL Ltda.*)

Ball Valve Control

Ball valves operate very simply by means of a floating ball on the surface of the water. The newer units have a ball connected to a pilot system, which in turn operates the main valve, as per the diagram in Fig. 12.19. In reservoirs with turbulence, it is important to make sure that the ball assembly is installed in a stilling well or a calm location as per the picture in Fig. 12.20. This ensures that the turbulent surface does not affect the control, making the valve open and close inappropriately. The ball valve assembly is ideal for storage facilities, which fill from the top, as opposed to bottom-filling tanks and storage.

Altitude Valve Control

The altitude valve uses a column of water which equals the level of the tank to control a pilot valve, which in turn opens and closes the main valve as per the diagram in Fig. 12.21. Altitude valves are usually installed on bottom-filling tanks and storage, as can be seen in Fig. 12.22, but they can be installed on top-filling tanks if the sense line is connected to the outlet pipe.

Care should be taken not to install the altitude valve too far away from the tank, as this will create delayed reactions and poor control. Manufacturers supply good installation diagrams, which should be adhered to. Altitude valves can be installed on unidirectional pipes and on bi-directional pipes and can be set for on/off response or relational response.

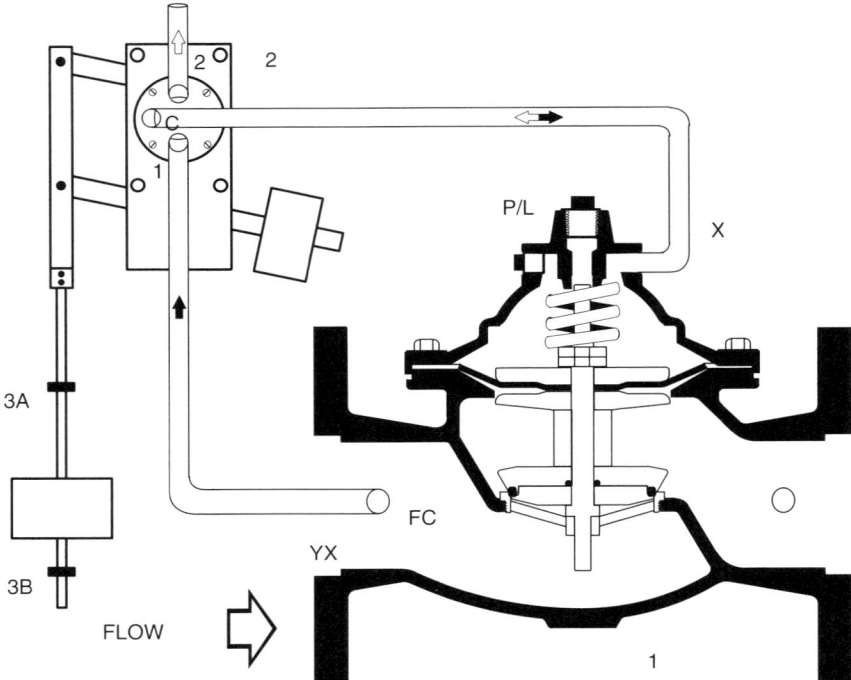

Figure 12.19 Ball valve control. (*Source: Watts ACV, Houston, Texas.*)

Figure 12.20 Calibrating a ball valve. (*Source: Watts ACV, Houston, Texas.*)

Demand Control Using Flow and Sustaining Valves

Some systems find that during peak flow conditions certain parts of the system are hydraulically deficient. When this occurs, in many cases certain consumers will use most of the water while leaving only a small volume for others. This is often so in the case of very large consumers, who create large, localized headlosses. It is also often seen in the case of irregular settlements in developing countries.

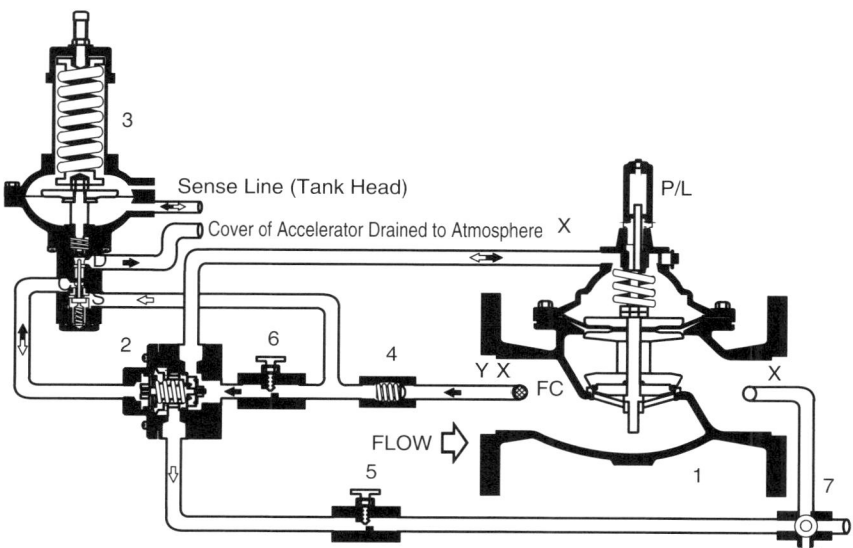

Figure 12.21 Altitude valve diagram. (*Source: Watts ACV, Houston, Texas.*)

Figure 12.22 Altitude valve installation. (*Source: Watts ACV, Houston, Texas.*)

Flow control valves and pressure-sustaining valves can be used to reduce the impact of these situations and ensure a constant supply of water for all consumers. These valves, in conjunction with pressure-reducing valves in areas of excess pressure, help to ensure that an even supply pressure occurs in all parts of the system.

Sustaining and flow control features can also be added to pressure-reducing valves (see Figs. 12.23 and 12.24), making the valve an efficient tool for directing water around the system and ensuring a good turnover of reservoir water, which in turn ensures good water quality. This type of control is often necessary to protect reservoir volumes during times of high demand, as some reservoirs will empty more quickly than others. In such a situation, without control, some reservoirs may always be virtually empty, while others may never be empty.

Figure 12.23 Sustaining valves can be used to protect supply. (*Source: Watts ACV, Houston, Texas.*)

Figure 12.24 Sustaining valve installation. (*Source: Watts ACV, Houston, Texas.*)

Sectors with Large Industrial Customers

In sectors with very large consumers, care should be taken to make sure that flow and pressure profiles used for sizing valves are representative of the highest and lowest demand periods. In some cases it may be advantageous for both the utility and the customer to install an onsite storage facility, if one is not already installed. Once the large consumer has some storage, either flow or pressure-sustaining valves can be used to control the demand of the large consumer and minimize the peak impact caused through high localized headlosses.

Site visits for all large consumers are a must to ensure that their demand needs and emergency needs are taken care of properly. The cost of these surveys and in some cases retrofitting fire sprinkler systems or providing storage may sometimes be included in the project cost. This would be so in the case of a customer who did not want to change his system and was holding up the whole project. Obviously, this would only be done at the utility's expense in favorable-payback situations.

Figure 12.25 Parallel installation can be used to extend flow range. (*Source: Watts ACV, Houston, Texas.*)

Sectors with Large Seasonal Variations

In sectors with large seasonal variations it may be necessary to install multiple feeds, or valve installations in parallel. The parallel installation would consist of a large valve, which will provide high flows during peak conditions, usually at weekends or holiday periods in the case of tourist areas. The smaller valve would then function most of the time. In many cases where a controller is installed, the controller need only be installed on the valve which is most active. In certain cases the larger valve will function most of the time, with the small valve operating just at night, during periods of minimum night flow. See Fig. 12.25 for an example setup.

Sectors with Weak Hydraulic Capacity

In sectors with weak hydraulic capacity it may not be uncommon to find potential for pressure reduction at night, whereas during the day there is not sufficient pressure. Pressure reduction at night, however, can often still be justified and would depend on the cost-to-benefit analysis for final decision. As mentioned earlier, often pressure-sustaining valves in conjunction with pressure-reducing valves are used. An alternative if possible is to install a small tank, which will pick up lost pressure only during the peak hours when the system is most stressed and pressures are uncontrollable.

> **P**ressure management can be undertaken at off-peak hours only if the system is weak during peak demand times.

12.8 VALVE SELECTION AND SIZING[2]

Valve selection and sizing is often done using average values of flow and pressure; because valves are fairly forgiving, in most cases the valves work. This is not recommended practice, however!

In the case of pressure reduction for leakage control, it is strongly recommended that flows and pressures be measured in the field. As discussed previously, the impact of reduced pressure on leakage is often critical to the operation of the valve, and without accurate data, valves may be installed incorrectly and operate erratically.

Field measurements are also beneficial when seasonal corrections need to be made and ensure that the valves can cope with top-end flows without creating too much headloss. This is also true when calculating the effects of emergency water use such as fire fighting.

All valve manufacturers provide valve sizing charts. Table 12.2 is an example of such a chart; some additional information follows (*Source:* Watts ACV, Houston, Texas).

1. Pressure reducing valves: Selection of the correct size pressure reducing valve is a relatively simple process. Criteria for selection are minimum flow, maximum flow, and pressure drop across the valve. Following are explanations of the three types of PRV installations. These also apply to any functions combined with the reducing function such as reducing/check and reducing/solenoid valves.

TABLE 12.2 Quick Sizing Chart

System Flow Range (gpm)	Size Range (gpm)	Size Range (gpm)	Size Range (gpm)
Single-Valve Installations			
1–100	1¼ in		
1–150	1½ in		
1–200	2 in		
20–300	2½ in		
30–450	3 in		4 in (30–680)
50–800	4 in	ACV 6000	6 in (50–1,025)
115–1,800	6 in	series	8 in (115–2,300)
200–3,100	8 in	valves	10 in (200–4,100)
300–4,900	10 in		
400–7,000	12 in		
500–8,500	14 in		
650–11,000	16 in		
Parallel Installations			
1–400	1¼ in (1–100)	2½ in (20–300)	
1–800	1¼ in (1–100)	3 in (30–500)	
1–1,000	1½ in (1–150)	4 in (50–850)	
1–2,000	2 in (1–200)	6 in (115–1,800)	
1–3,800	1¼ in (1–100)	3 in (30–500)	8 in (200–3,100)
30–3,800	3 in (30–500)	8 in (300–1,800)	
1–5,400	1¼ in (1–100)	3 in (30–500)	10 in (300–4,900)
30–5,400	3 in (30–500)	10 in (300–4,900)	
1–8,000	1½ in (1–150)	4 in (50–850)	12 in (400–7,000)
50–8,000	4 in (50–850)	12 in (400–7,000)	
1–9,500	1½ in (50–850)	4 in (50–850)	14 in (500–8,500)
50–9,500	4 in (50–850)	14 in (500–8,500)	
1–13,000	2 in (1–200)	6 in (115–1,800)	6 in (650–11,000)
115–13,000	6 in (115–1,800)	6 in (650–11,000)	

Select the flow range that meets your system requirements and correct valve(s) size(s). *Note:* Maximum flow rates in this table allow for continuous flows at velocities of 20–22 ft/s.
Source: Watts ACV, Houston, Texas.

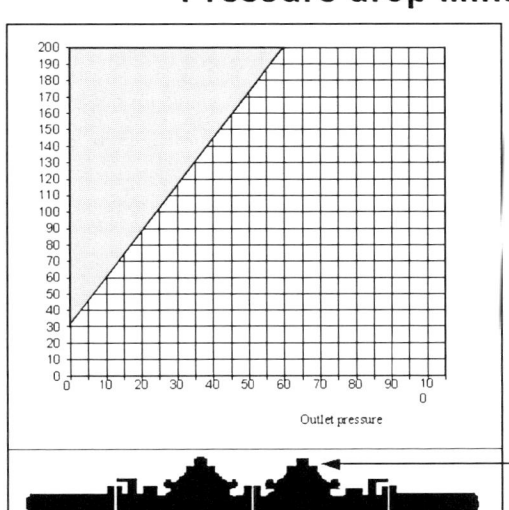

Figure 12.26 Cavitation chart. (*Source: Watts ACV, Houston, Texas.*)

a. Single valve installation: A single reducing valve can be applied if operating flow requirements are within the capacity of one valve size, and pressure drop is outside the cavitation zone.
  (1) Select the valve size from a sizing chart that is within the range of flow to high flow. (Consider requirements of lowest-demand equipment.)
  (2) Check pressure drop (inlet–outlet) to confirm that desired outlet pressure is above the recommended lowest outlet setting, to avoid cavitation conditions. (Check a cavitation chart, for example, Fig. 12.26).
b. Parallel installation: If flow requirements fall outside the capacity of a single valve, an additional smaller valve installed in parallel may be required. In parallel installations, the larger valve handles the requirements for maximum flow down to its low-flow capacity. The small valve extends the low-flow range. Total capacity of this installation is equal to the sum of the maximum flow of both valves.
  (1) Select valve size combinations from the sizing chart that are within low to high flow system range.
  (2) Check pressure drop (inlet–outlet) to confirm that desired outlet pressure is above index psig, or check the cavitation chart (Fig. 12.26).
c. Series installation: If pressure drop requirements cause the outlet pressure to be below the index psig, or fall in the cavitation zone, then two valves in series may be required. Each valve will function outside the cavitation zone to safely drop the high inlet pressure, in two steps, to the desired outlet pressure. Valve size is based on the minimum–maximum flow ranges explained previously.

2. Isolation shutoff valves: Butterfly or similar valves should be installed in the line upstream and downstream of the automatic control valve to allow for maintenance service without draining the system or exposing service personnel to the pressure.
3. Installation recommendations and requirements: Avoid mounting valves 6 in and larger in a vertical discharge position (valve stem horizontal or cover pointing sideways). If your installation requires this mounting position, consult the manufacturer or specify at time of order.

12.8.1 ACV Types: Diaphragm, Piston, Rolling Diaphragm, Sleeve

All pilot-operated hydraulic ACVs operate using similar principles, although the mode of control may be very different. Each manufacturer will cite benefits and will attempt to justify why his valve is better. Full-bore valves such as the rolling diaphragm and sleeve will cite low head-losses at high flows, while globe-style diaphragm manufacturers will cite stable modulation and control. At the end of the day it is important for the utility engineer to understand what it is he or she wishes to achieve and then to select the best valve for the job.

In addition to technical benefits, the engineer should also consider two other very important points, local support and ongoing maintenance costs. One of the biggest problems of utilities that have many valves installed is the cost and availability of quick maintenance. A utility should try to avoid installing many different makes of valve, as the cost of stocking parts increases significantly and the availability of local support drops drastically.

12.8.2 Valve Sizing and Limits: Qpmax and Qpmin, Cavitation, Headloss

In many cases control limits are set as a function of the maximum pressure controllable at the valve, while providing a constant minimum pressure at the critical node(s). If a substantial amount of pressure is to be controlled, the manufacturer's cavitation chart should be consulted to ensure that the valve is operating within its limits. See an example in Fig. 12.26.

If the valve is found to be potentially operating within the cavitation zone, then installation of two valves in series should be considered as discussed in Sec. 12.8.4. See Fig. 12.26.

When sizing valves, care should be taken to check the potential headloss through the total valve assembly (gate valves, filter, meter, control valve, and pipe fittings), especially when the pressure during peak hours is already low and modulated control is desired only during off-peak times. If care is not taken, supply may be reduced during peak hours, resulting in complaints of no water. See Fig. 12.27 for a reference spreadsheet, which was made by BBL Limitada out of various manufacturer's information. This spreadsheet can be used to check headlosses across all of the fittings within the bypass assembly.

As a consequence of the pressure control, the existing flow profiles will be reduced, in particular when a high level of leakage is present. Leakage

**ESTIMATED HEADLOSS
BYPASS**

Company: Date:
Site: Site Identity:
Sector: Zone:

DATA TO BE SUBMITTED IN: [   ]

Ø inline piping: [   ] m   Q: [   ] m³/h
                    A = [   ] m²   v: m/s
Ø bypass piping: [   ] m   Length: [   ] m
                    A = [   ] m²

HEADLOSS DUE TO FRICTION CAUSED BY PIPING (see Moody diagram):

$h_f = f \dfrac{L \times V^2}{D \times 2g}$ $R_e =$ $\therefore f = $ [   ] $h_f = $ m
Roughness coef.=

HEADLOSS DUE TO FRICTION CAUSED BY COMPONENTS:

$h_f = K \dfrac{V^2}{2g}$ *From manufacturers' catalogs:*

[   ] Reducer   0.00 m   [   ] Y filter   [   ] m
[   ] Amplifier   0.00 m
[   ] Tee (lat. exit) (K=1.30) 0.00 m [   ] Meter   [   ] m
[   ] Bends 90° (K=1.20)   0.00 m
[   ] Bends 45° (K=0.40)   0.00 m   [   ] PRV   [   ] CV
[   ] Gate valve   0.00 m $\Delta p = \left(\dfrac{Q}{3.6\,CV}\right)^2 0.702:\ 0.00\ \text{m}$

TOTAL HEADLOSS :   _____ m

Observations:

Calculations made using liquid = H_2O at 20°C

Formulas used: Reynolds, Darcy, Moody diagram

Mechanical joint included in piping

Figure 12.27 Calculating headloss in the installation. (*Source: BBL Ltda.*)

is reduced as a function of a power of the pressure before and after control. This power can vary depending on the type of pipe material and types of materials present in the system.

Care should be taken when selecting valve sizes that the flow cannot fall below the minimum acceptable flow for the valve, after leakage has been reduced. If this happens the valve may control erratically, as it is controlling at the almost-closed position, so any small modulation affects the flow and pressure more than when it is modulating at the nominal position. This may result in either higher maintenance costs or increased leakage.

Where a valve has to deal with both high- and low-flow conditions it is often common to install a small bypass valve around the main control valve to ensure smooth hydraulic control.

12.8.3 Parallel Installations: Fire Control, Large Flows, Variations in Flow Pattern

Where flow patterns vary greatly or there is a requirement to meet safety or periodic emergency demands, it is normal practice to install valves in parallel. These types of installations usually take the form of one large valve and one small valve working together to increase the effective flow range. Figure 12.28 shows an example of two valves of different diameters working together. In this situation the large valve takes account of most of the daily flow and the small valve deals just with the relatively low nighttime flow. In this case, if a controller were fitted it would be fitted to the large valve and the smaller valve would have a fixed outlet pressure just slightly higher than the minimum modulated pressure of the larger valve. As the larger valve modulates toward the closed position, reducing the outlet pressure, the smaller valve kicks in. A difference in outlet settings of around 5 psi gives a smooth changeover.

Where the larger valve is fitted to meet fire flow conditions and stays closed most of the time, the opposite is true. If a controller were fitted, it would be fitted to the smaller valve and the outlet of the larger valve would be set just below that of the smaller valve. As the demand increases and the small valve starts to create excessive headloss, therefore forcing outlet pressure down, the larger valve will modulate open and feed the system with a fixed outlet pressure and large volume.

Parallel installations can also be undertaken where a large flow dictates that two large valves are installed in parallel. Figure 12.29 shows an

Figure 12.28 Parallel installation, unequal diameters; contract no. 3.066/98. (*Source: SABESP/BBL Ltda.*)

Figure 12.29 Parallel installation, equal diameters. (*Source: Julian Thornton.*)

example of a parallel, equal-diameter installation. It also shows how a controller can be hooked up to one pilot, which then controls two valve head chambers. This is normal practice and works well as long as valve opening and closing speed is calibrated equally for both valves. In some cases (usually dual, large-diameter valves), a pilot with a larger-diameter nozzle should be used to allow a larger volume of water to pass through while still maintaining control and not being effected by localized headloss.

12.8.4 Series Installations: Large Pressure Drops

Where a large amount of pressure needs to be cut and a single valve would enter into the cavitation zone, two equal-diameter valves operating in series can be fitted. In the case of a controller being fitted, it would be fitted to the second or downstream valve, allowing the first valve to cut pressure from upstream to required maximum and the second flow-modulated valve from required maximum to required minimum. Figure 12.30 illustrates a complex example of two valves in parallel and series with a controller.

12.8.5 Using Controllers to Make Valves More Efficient

The advent of intelligent and cost-effective controllers now allows the use of conventional fixed outlet valves in a more efficient manner. The controller effectively allows multiple set points for downstream pressure, depending on either time-based or system demand-based requirements. In many cases, just altering the set point by the amount of headloss difference between daytime and nighttime conditions can effect huge savings in leakage.

In addition to standard leakage control, controllers can be used to make valves function for emergency situations, for example, earthquake control. When an area is hit by an earthquake, it is very possible for a

Figure 12.30 Parallel and series installation with controller. (*Source: Julian Thornton.*)

major transmission line to rupture, leading to possible depletion or complete loss of storage. By calibrating controller profiles, the controller can reduce the amount of reservoir loss or shut down the line completely, saving valuable resource and the headache of trying to refill storage under emergency conditions.

Time-Based Control

Time-based control can be effected by using a controller with an internal timer. Control is affected in time bands in accordance with demand profiles. This methodology is very effective for areas with stable demand profiles and headlosses and is usually used where cost is an issue but advanced pressure management is desired. Time-based modulation controllers can be supplied with or without data loggers and/or remote links. Some manufacturers connect the controller to the pilot valve and alter the set point of the pilot valve by introducing a force against the existing force of the pilot spring, as shown in Fig. 12.31. Other manufacturers use a timer and a solenoid valve to reroute control through preset pilots.

Demand-Based Control

Demand-based control is the best type of control for areas with changing conditions, headloss, and fire flow requirements and the need for advanced control. This type of control is affected by controlling outlet pressure in relation to demand by connecting the controller to a metered signal output. Modulation of outlet pressure is achieved by altering the force against the pilot spring. The controller is normally supplied with a local data logger and optional remote communications; see Fig. 12.32.

12.8 VALVE SELECTION AND SIZING

Figure 12.31 Time-based controller fitted to valve pilot. (*Source: Julian Thornton.*)

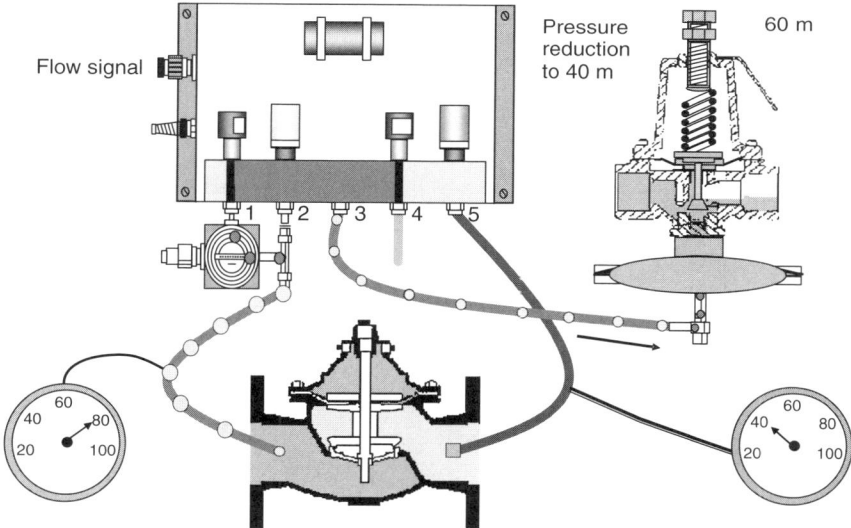

Figure 12.32 Demand-based control diagram. (*Source: BBL Ltda.*)

Control can be effected with a preset profile, which shows the changing relationship of demand and headloss in the sector. Alternatively, a direct communications link can be created between the controller and the critical point. Obviously, the second option involves communications and therefore higher costs, which are not always necessary.

In general, installation costs are higher for this type of control, but additional savings and guaranteed fire flows due to more intelligent control usually make this type of control more desirable.

SCADA

Some utilities have SCADA systems. Many SCADA systems are designed to run the transmission-level system and are not designed for the distribution level, due to the cost involved. However, utilities are starting to cost-justify system optimization, and SCADA should not be ruled out as an excellent, though costly, means of managing pressure within the distribution system. (The author has seen installations and presentations from Canada, Australia, and Japan, where this type of control is effected.) Pressure control sites which are hooked up to SCADA usually function with butterfly or gate-style valves with actuators. The valves are opened and closed automatically as a result of either pressure or flow requirements. Generally speaking, this type of control is the most expensive to install but is by far the most efficient. It may not always be necessary, however, and cost-to-benefit calculations may not justify the installation of a full SCADA system just for loss control. If a utility already has distribution-level SCADA in place, then it may be very cost-effective to add modules.

12.9 VALVE INSTALLATION

Once the valve has been sized and control limits identified, it is time to decide how to install the valve. Valves can be installed in a number of ways at varying costs. We will discuss various options.

12.9.1 Where to Dig

Where to dig may seem obvious: the easiest location! Sometimes, however, decisions may be made quickly and on paper, without the proper site investigations. On-site investigation is probably one of the most important tasks and should not be taken lightly.

Before selecting a location it is necessary to locate all other underground utilities (see Chap. 10 on locating underground utilities). Identify an area where traffic will be easiest to contend with in the case of an underground chamber, and consider properties which may be affected by the excavation.

12.9.2 Mainline or Bypass

Once a reasonable location has been located, it is necessary to decide if the valve assembly will be installed on the mainline, allowing for a smaller-diameter bypass which is cheaper to install, or if the valve assembly will be installed on the bypass, allowing the valve chamber and access manhole to be installed in the verge for easy access when the mainline is under the road. See Figs. 12.33 and 12.34 for these options.

12.9.3 Headloss Concerns

When sizing a bypass and fittings for valve installation, it is important to consider the headloss which will be created at peak flow, plus emergency demand not only through the valve but also through all of the fit-

Figure 12.33 Valve being installed on mainline, contract no. 69.502/96. (*Source: SABESP/BBL Ltda.*)

tings. Remember that if the valve will be modulated to various set points with a controller, it may be desirable to open the valve almost completely, making it transparent during high demands, and control only during low-demand periods. If the bypass or adjoining fittings is downsized to reduce cost, headlosses may be created in excess of the minimum control desired. Careful consideration of fittings sizing should be given at this stage. In some cases economy is justified and in others it is not; each case is site-specific and should be analyzed during the cost-to-benefit stage.

Figure 12.34 Valve being installed on bypass, contract no. 3.066/98. (*Source: SABESP/BBL Ltda.*)

12.9.4 Hydraulic Connections

When considering the type of connections to use, it is important to consider the type of existing pipework. If the existing pipework is flanged cast iron or ductile, then anchorage for horizontal movement is not so critical a problem. If the existing pipework is made up of asbestos cement or to a lesser degree bell and spigot PVC-type pipe, then consideration should be given to potential horizontal movement during periods of control. In all cases vertical movement should be considered.

12.9.5 Anchorage

Depending on the size of pipework and valves, each case should be considered independently and calculations should be made to ensure that no vertical or horizontal movement of the installation is possible. It is normal to put thrust blocks on the 90° bends of the bypass, but additional thrust restraint should be calculated if the valve itself will be installed on the bypass.

12.9.6 Chamber or Above-Ground Installation

In some cases it is desirable to install the valve assembly above ground. See Fig. 12.35 for an example. Above-ground installations are very practical and allow the operator to avoid confined-space entry requirements, potential flooding problems (especially in areas with high groundwater conditions), and generally working in cramped conditions. Obviously, there are negatives to above-ground installations, in that the installation takes up a lot of space, particularly in the case of a large-diameter installation. Above-ground installations also attract more attention and may become the focus of vandalism in certain environments.

Figure 12.35 Above-ground assembly, contract no. 3.066/98. (*Source: SABESP/BBL Ltda.*)

Each site should be considered on an individual basis and the merits of each type of installation considered.

12.9.7 Valve Commissioning

It is always better for a skilled and experienced operator to perform valve start-up. Sometimes, for any number of reasons, the valve will not function as it should, and without the necessary knowledge an unskilled operator could create serious problems in the system if the valve controls erratically; at the very least, a lot of time and effort may be spent attempting to resolve an essentially simple problem.

12.9.8 Start-up Procedures

Start-up procedures are of course site-specific, and it is a good idea to make a checklist before beginning. A sample start-up procedure for a pressure-reducing valve, which also has rate-of-flow control and a sustaining function, follows (*Source:* Watts ACV, Houston, Texas).

Installation/Start-up

Start-up of an automatic control valve requires that proper procedure be followed. Time must be allowed for the valve to react to adjustments and the system to stabilize. The objective is to bring the valve into service in a controlled manner, to protect the system from damaging overpressure.

- ❏ Clear the line free of slag and other debris.
- ❏ Check to ensure that the orifice plate is installed in the valve inlet flange and that the inlet sensing port is not covered by the retainer ring. If necessary, rotate until the space aligns with the port.
- ❏ Install the valve so that the flow arrow marked on the valve body/tag corresponds to flow through the line.

> It is best to have a skilled and experienced operator perform valve start-up.

❑ Close upstream and downstream isolation valves.
❑ Open ball valves or isolation cocks in the control tubing, if the main valve is so equipped. Failure to open these will prevent the valve from functioning properly.

Step 1. Preset pilots.
 a. Rate of flow: Adjust *out*, counterclockwise, to start valve at a lower flow rate.
 b. Pressure sustaining: Turn sustaining control adjustment screw *out*, counterclockwise, backing pressure off the spring, to allow it to stay open while adjusting other controls.
 c. Pressure reducing: Adjust *out*, counterclockwise, backing pressure off the spring, preventing possible overpressuring of the system.

Step 2. Turn the adjustment screws on the closing speed and opening speed controls, if the main valve is so equipped, *out*, counterclockwise 1½ to 2½ turns from full closed position.

Step 3. Loosen a tube fitting or the cover plug at the main valve to allow air to vent during start-up.

Step 4. Pressure the line, opening the upstream isolation valve slowly. Air is vented through the loosened fitting. Tighten the fitting when liquid begins to vent.

Setting the Rate of Flow Control

Step 5. Slowly open the downstream isolation valve until the valve is fully open.

Step 6. With a demand for flow on the system, the valve can now be adjusted for the proper flow rate. This requires a meter to read the flow that the valve is providing.

Step 7. While reading the meter register, adjust the rate-of-flow control:
❑ Turn the adjustment screw *in*, clockwise, to increase the flow rate regulated.
❑ Turn the adjustment screw *out*, counterclockwise, to reduce or lower the flow rate regulated.

Setting the Pressure-Reducing Control

Note: Reducing control is set higher than sustaining control.

Step 8. Fine-tune the pressure-reducing control to the desired pressure set point by turning the adjustment screw *in*, clockwise, to increase or *out*, counterclockwise, to decrease downstream pressure.

Step 9. Opening speed flow control adjustment: The opening speed flow control allows free flow into the cover and restricted flow out of the cover of the main valve.
 a. If recovery of pressure is slow upon increase downstream demand, turn the adjustment screw *out*, counter clockwise, increasing the rate of opening.

b. If recovery of downstream pressure is too quick, as indicated in a rapid increase in pressure, probably higher than the desired set point, turn the adjustment screw *in*, clockwise, decreasing the rate of opening.

Step 10. Closing speed control adjustment: The closing speed needle valve regulates fluid pressure into the main valve cover chamber, controlling the valve closing speed. If the downstream pressure fluctuates slightly above the desired set point, turn the adjustment screw *out* counterclockwide, increasing the rate of closing.

Setting the Sustaining Control

Step 11. Setting the sustaining control requires lowering the upstream pressure to the desired minimum sustained pressure.

Step 12. Leave the downstream isolation valve full open and close the upstream isolation valve until the inlet pressure drops to the desired setting.

Step 13. Adjust the sustaining control screw *in*, clockwise, until the inlet pressure begins to increase, or *out*, counterclockwise, to decrease, stopping at the desired pressure.

Step 14. Allow the pressure to stabilize.

Step 15. Fine-tune the sustaining setting as required, as detailed in step 13.

Step 16. Open the upstream isolation valve to return to normal operation.

12.9.9 Air

When starting up a new installation or restarting one that has been subjected to zero pressure, it is common for air to be entrapped in the head of the valve. The result is that the valve will not control properly. Usually in this case the valve will not close or modulate toward the closed position. It is common practice to install a small air-release valve on the head of the valve, as can be seen in Fig. 12.36. Alternatively, one can be installed on the line. The latter is usually done when the utility wishes to have an indicator stem installed on the head of the valve to show the valve position. A line air valve will also handle a larger capacity and can be used as part of the systems air-release feature.

12.9.10 Modulation Speed

Valve modulation speed is always an issue, whether a controller is fitted or not. The hydraulic speed controls should be set to allow smooth, controlled modulation. In addition, when changing the outlet pressure in the case of advanced modulation, the controller reaction speed should be matched with the system needs. As a rule of thumb, larger valves need to be controlled faster, as their hydraulic reaction time is longer due to the larger amount of volume in the head of the valve.

> Modulation speed is a critical issue for all types of control valve and should be addressed according to independent system conditions.

Figure 12.36 Air valve installed on PRV head, contract no. 3.066/98. (*Source: SABESP/BBL Ltda.*)

Smaller valves should be modulated slower. A reasonable band for modulation by the controller is between 10 and 25 s per control pulse. Where the head volume is very large, as in the case of two valves running from one controller in parallel, it may be desirable to change the pulse volume, or the time the solenoids are open, to allow more forcible control. Most manufacturers provide detailed manuals explaining how to do this with their equipment. However, the operator should have a feel for the type of reaction needed for the system. System needs may include fire fighting response or large consumer draws. Obviously, a valve should not be modulated too quickly, or it will set up a very negative hydraulic reaction. If unsure how the system will react to control, it is a good idea to put out pressure data loggers in the system, logging very fast, then experiment with the pulse size and frequency on the controller(s) to see which combination gives the smoothest control.

12.9.11 Stability

Before fitting a controller it is a good idea to log the system pressures to see if the valve(s) are controlling stably without the controller. If speed controls are set incorrectly or if the system has multiple feeds and the outlets are set incorrectly, the valves will hunt. This should be corrected before trying to establish control through a more advanced regime.

12.10 MAINTENANCE CONCERNS

After the valve has been installed and properly commissioned and calibrated, it is very important to put into place a periodic maintenance schedule to ensure ongoing efficient operation of the valve. The time between maintenance visits is usually determined by water quality, the location of the installation (if it may be vandalized), or the variability in demand requiring changed modulation profiles. Maintenance should include but not be limited to the following.

1. Valve maintenance
   - ❏ Clean principal filter and secondary filter
   - ❏ Check tubing for leaks or kinks
   - ❏ Check operation of control isolation valves
   - ❏ Check pressure gauges
   - ❏ Check smooth modulation of valve
2. Controller maintenance
   - ❏ Check battery
   - ❏ Check input cables
   - ❏ Check logger functionality
   - ❏ Check modulation speed
3. Subsector maintenance
   - ❏ Check boundary valves
   - ❏ Check night flows
   - ❏ Check critical node pressures
   - ❏ Check critical node validity
   - ❏ Repair new leakage

> **V**alves, like any other equipment, need regular maintenance to ensure ongoing efficient operation.

In addition to the above, the chamber should be periodically checked for leakage and seepage, air quality, and general usability. Chamber manhole covers should also be checked periodically and greased to allow easy lifting.

The following case study shows benefits derived from pressure management in Palestine, which is an extremely water-stressed country in one of the most arid areas in the world.

12.11 CASE STUDY TWO

Ramallah Case Study for Reducing Leakage from Al Jalazon Refugee Camp Water Network

Nidal Khalil, Jerusalem Water Undertaking

12.11.1 Background

Jerusalem Water Undertaking (JWU) is located in the central part of the West Bank, 16 km north of Jerusalem. JWU provides drinking water to most of the population centers in Ramallah and Al Bireh Governourate. The Governourate includes one major urban area, the twin cities of Ramallah/Al Bireh, and about 100 villages, municipalities, and refugee camps.

The region that obtains its water services from the JWU is a densely populated area with approximately 205,000 people in 1999. It consists of the two largest municipal areas of Ramallah and Al Bireh cities, which form the political, economical, and cultural heart of the Governourate, and four other towns, Betunia, Beit Hanina, Bir Zeit Dier Dibwan, and

Silwad, which fall within 15 km radius to the northeast and west of the heart of the district. It also includes some 40 villages and four refugee camps. JWU is a self-sufficient, autonomous, nonprofit, national utility established in 1966. JWU, governed by a board of directors, has a well-deserved reputation for efficiency, and responsiveness, to community needs. JWU has the responsibility for the planning, design, maintenance, and overall management of the water supply schemes in its service area.

12.11.2 Introduction

Al Jalazon refugee camp was established after the Arab–Israeli war of 1948. It is situated on approximately 0.85 km^2 of land located 6 km to the north of Ramallah City and is inhabited by 6400 inhabitants. Until 1980 the camp lacked a water distribution network and a decent sewage collection system.

In 1980 a water network was installed in the camp. It was laid using galvanized pipes for diameters 2 in and less and steel pipes with internal cement lining and external asphalt coating for diameters greater than 2 in. The bad condition of the open sewage collection system resulted in increasing the corossivity of the soil and accelerated the deterioration of the network.

12.11.3 Definition of the Problem

The crowded conditions in the camp, the large variations in the topography, the large inlet pressure, and the condition of the sewage collection system all contributed to the following.

1. Over 50 percent of the network is subjected to pressures ranging from 10 to 16 bar.
2. Over 60 percent of the galvanized network and house connections are badly corroded.
3. A large difference in the average daily water billed (322 m^3) and what is registered at the inlet connection (520 m^3) indicated the existence of a serious problem, which needed to be solved.
4. A large number of reported bursts due to corroded pipes (20 in 1998) compared to the length of the network (6.56 km).

12.11.4 Investigations Made

Using the BABE software (see Sec. 7.6), JWU conducted an investigation of the amount of leakage from the network. It consisted of measuring the inflow and pressure at the inlet connection and the pressure at the point of average zone night pressure (AZNP) and target point over 24 hours. All the data were incorporated into the BABE software, and results for the expected amount of leakage, usage, and total inflow were obtained as shown in Table 12.3. Figure 12.37 shows the relation between pressure and flow before pressure reduction for the different flows (use, losses, and total flow). In addition, JWU investigated samples of pipes obtained from the reported bursts caused by corrosion or other causes and found that most of the samples showed that the network was badly

TABLE 12.3 Pressure and Flow before Control

Time Period (hours)	Average Hourly Pressure at			Average Inflows			
	Inflow Point (m)	AZP Point (m)	Target (m)	Measured (L/s)	Measured (m³/h)	Losses (m³/h)	Use (m³/h)
00 to 01	158.0	135.0	108.0	3.9	14.0	5.46	8.54
01 to 02	160.0	139.0	111.0	2.8	10.0	5.63	4.38
02 to 03	162.0	139.0	110.0	2.2	8.0	5.63	2.38
03 to 04	162.0	139.0	111.0	2.2	8.0	5.63	2.38
04 to 05	163.0	139.0	111.0	1.9	7.0	5.63	1.38
05 to 06	157.0	130.0	100.0	2.5	9.0	5.26	3.74
06 to 07	145.0	124.0	92.0	3.6	13.0	5.02	7.98
07 to 08	131.0	102.0	75.0	4.4	16.0	4.13	11.87
08 to 09	115.0	86.0	58.0	6.4	23.0	3.48	19.52
09 to 10	112.0	81.0	51.0	7.5	27.0	3.28	23.72
10 to 11	108.0	78.0	44.0	11.4	41.0	3.16	37.84
11 to 12	110.0	80.0	48.0	9.7	35.0	3.24	31.76
12 to 13	116.0	88.0	50.0	8.6	31.0	3.56	27.44
13 to 14	120.0	85.0	55.0	8.3	30.0	3.44	26.56
14 to 15	125.0	95.0	65.0	8.6	31.0	3.84	27.16
15 to 16	133.0	102.0	73.0	6.9	25.0	4.13	20.87
16 to 17	126.0	98.0	68.0	7.5	27.0	3.97	23.03
17 to 18	115.0	82.0	55.0	7.5	27.0	3.32	23.68
18 to 19	104.0	75.0	46.0	6.4	23.0	3.04	19.96
19 to 20	115.0	82.0	50.0	8.6	31.0	3.32	27.68
20 to 21	130.0	102.0	61.0	7.5	27.0	4.13	22.87
21 to 22	140.0	115.0	85.0	6.7	24.0	4.65	19.35
22 to 23	149.0	124.0	95.0	4.7	17.0	5.02	11.98
23 to 24	155.0	129.0	100.0	4.4	16.0	5.22	10.78
Averages	133.79	106.21	75.92	6.02	21.67	4.30	17.37
Maximum	163.00	139.00	111.00	11.39	41.00	5.63	37.84
Minimum	104.00	75.00	44.00	1.94	7.00	3.04	1.38
Daily Totals in m³/day					520.0	103.2	416.8
Losses						19.8%	

Source: Jerusalem Water Undertaking.

corroded and needed replacement. Figure 12.38 shows a picture of a sample of old pipe against a sample of new pipe.

12.11.5 Results of Investigation

The results of the investigation showed that the difference between the total amount of water which entered the system and the estimated use was 103.2 m³. This amount was associated mainly with leakage from the network and was purely theoretical.

In order to deal with the problem, the pressure at the inlet was reduced (using pressure-control valve) by 23 m and another set of readings is taken at the same points for the pressure and flow. This reduction resulted in reducing the leakage from 103.2 to 85.8 m³/day, thus realizing a saving of 17.4 m³/day while maintaining the same amount of usage. Table 12.4 and Fig. 12.39 show the results of this reduction.

310 CHAPTER 12 PRESSURE MANAGEMENT

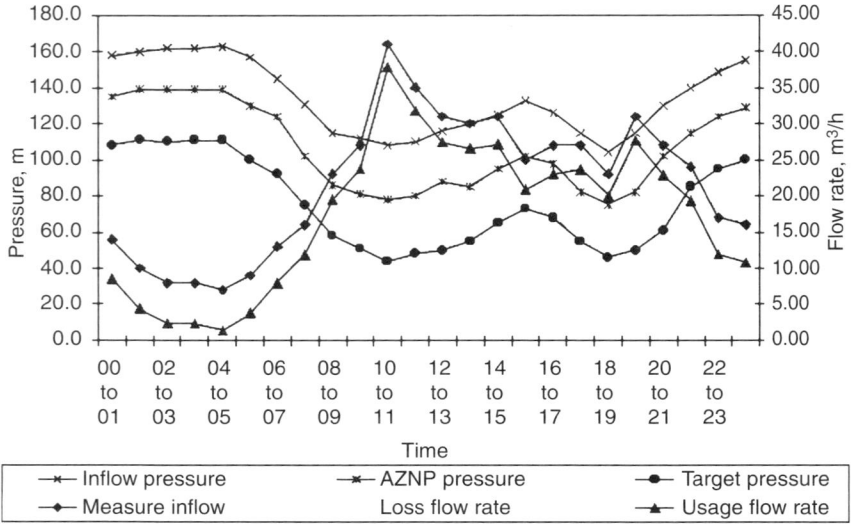

Figure 12.37 Pressure and flow profiles for Al Jalazone Camp before pressure reduction. (*Source: Jerusalem Water Undertaking.*)

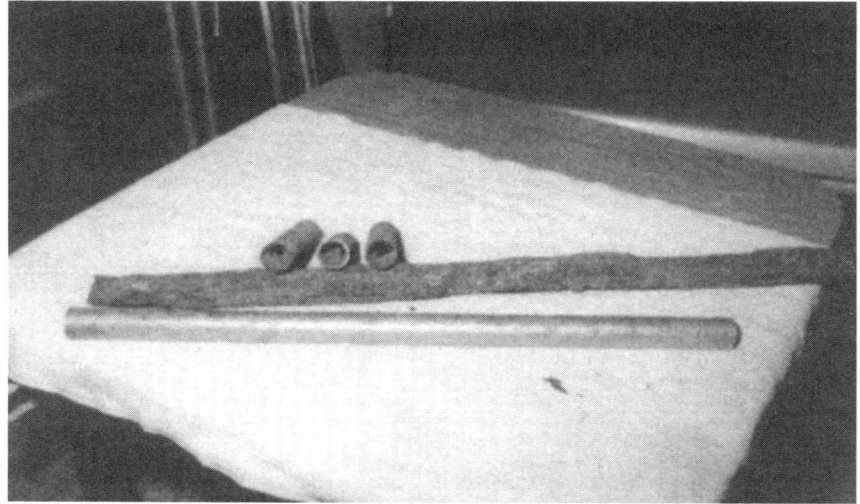

Figure 12.38 Corroded piping. (*Source: Jerusalem Water Undertaking.*)

12.11.6 Solutions Proposed

In light of the water savings achieved at a reasonably low reduction of pressure at the inlet connection and without actual pressure management of the system, it was proposed to do the following.

1. Install a permanent pressure-control valve at the inlet connection in order to reduce the inlet pressure to 130 m (otherwise water will not reach the target point). The total cost of the proposed system was $US6200.

TABLE 12.4 Pressure and Flow after Control

Time Period (hours)	Average Hourly Pressure at			Average Inflows			
	Inflow Point (m)	AZP Point (m)	Target (m)	Measured (L/s)	Measured (m³/h)	Losses (m³/h)	Use (m³/h)
00 to 01	135.0	115.0	84.0	3.9	14.0	5.44	8.56
01 to 02	135.0	115.0	87.0	2.8	10.0	5.44	4.56
02 to 03	142.0	120.0	91.0	1.9	7.0	5.67	1.33
03 to 04	140.0	119.0	89.0	2.2	8.0	5.63	2.38
04 to 05	139.0	119.0	90.0	1.4	5.0	5.63	−0.63
05 to 06	131.0	106.0	79.0	1.9	7.0	5.01	1.99
06 to 07	124.0	90.0	64.0	3.1	11.0	4.25	6.75
07 to 08	103.0	70.0	46.0	4.7	17.0	3.31	13.69
08 to 09	90.0	60.0	32.0	5.3	19.0	2.84	16.16
09 to 10	81.0	50.0	20.0	6.7	24.0	2.36	21.64
10 to 11	75.0	40.0	7.0	9.4	34.0	1.89	32.11
11 to 12	85.0	42.0	14.0	8.3	30.0	1.99	28.01
12 to 13	87.0	45.0	20.0	10.0	36.0	2.13	33.87
13 to 14	95.0	57.0	27.0	8.6	31.0	2.69	28.31
14 to 15	92.0	55.0	28.0	8.1	29.0	2.60	26.40
15 to 16	100.0	70.0	41.0	8.1	29.0	3.31	25.69
16 to 17	90.0	54.0	33.0	6.7	24.0	2.55	21.45
17 to 18	87.0	51.0	25.0	6.7	24.0	2.41	21.59
18 to 19	76.0	49.0	17.0	7.2	26.0	2.32	23.68
19 to 20	86.0	49.0	16.0	6.1	22.0	2.32	19.68
20 to 21	95.0	68.0	35.0	8.3	30.0	3.21	26.79
21 to 22	111.0	80.0	51.0	6.4	23.0	3.78	19.22
22 to 23	117.0	91.0	63.0	5.8	21.0	4.30	16.70
23 to 24	125.0	100.0	72.0	5.8	21.0	4.73	16.27
Averages	105.88	75.63	47.13	5.81	20.92	3.57	17.34
Maximum	142.00	119.00	91.00	10.00	36.00	5.67	33.87
Minimum	75.00	40.00	7.00	1.39	5.00	1.89	−0.63
Daily Totals in m³/day					502.0	85.8	416.2
Losses						17.1%	

Source: Jerusalem Water Undertaking.

2. Sector the camp into two zones in order to reduce the difference in elevation between low points and high points.
3. Replace the water network taking into consideration the existing sewage collection system. A new design for the whole system was developed; and the estimated cost of replacing the network was US$336,336.

12.11.7 Financing of Solutions

JWU, being a nonprofit public utility, does not have the necessary funds to finance large replacement projects. It depends on external donors to finance such projects. The Al Jalazon replacement network project (among other projects) was proposed to a number of donors. Unfortunately, so far we have not been successful in mustering the necessary funds. As for

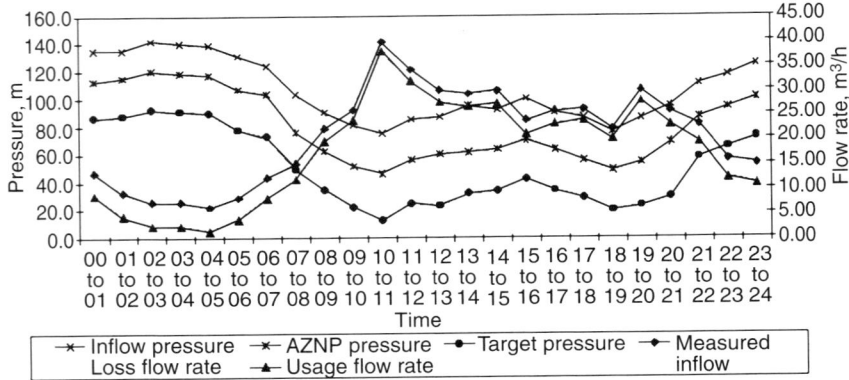

Figure 12.39 Pressure and flow profiles for Al Jalazone Camp after pressure reduction. (*Source: Jerusalem Water Undertaking.*)

installing the pressure control system, the cost was reasonable and within the capacity of JWU.

Sectoring the network was not possible due to the many interconnections within the network.

12.11.8 Solutions Implemented

The pressure control system was implemented and is functional as planned. Some replacement work was done on the network. The open sewage system was not corrected.

12.11.9 Results of Implementation and Loss Reduction

Following are the results of installing the pressure control system at the inlet connection:

- Saving in water in 1 year: $17.4 \times 365 = 6351$ m^3
- Marginal cost of 1 m^3 of water: $0.684
- Total value of water saved per year: $6351 \times 0.684 = \$4344$
- Costs of material and labor: $6200
- Payback period: 1.4 years

This saving will be realized without doing major replacement work on the network.

12.12 SUMMARY

Pressure management is one of many tools which may be used by leakage management practitioners to combat either leakage volumes or increasing leak frequencies. It may also be used in conjunction with demand reduction programs. Water efficiency programs are discussed in more detail in Chap. 15.

Pressure management is a suitable means of controlling water losses in all areas of the world, from highly industrialized nations to developing countries.

12.13 REFERENCES

1. Thornton, J., "Pressure Management," *Opflow,* October 1999.
2. Thornton, J., "Correct Selection Sizing and Advanced Operation of PRV's," ABES National Congress, Salvador, Brazil, 1998.

CHAPTER

13

Pipe Maintenance, Rehabilitation, and Replacement

Julian Thornton

> ***Case Study One***
> Ductile Iron Pipe in Stray Current Environments
> *Richard W. Bonds, P.E.*
>
> ***Case Study Two***
> Leakage: How Low Can You Go? Cheadle Water Works Project: A Unique Opportunity to Minimize Leakage
> *Ian Elliott*
> *John Foster*

13.1 INTRODUCTION

Underground piping is one of the largest investments a utility can have and the cost to maintain and or replace old piping is often prohibitive, due not only to the physical costs of the pipework itself but also to the excavation and reinstatement in often dense urban situations. Unfortunately, maintenance is often overlooked, as the problem is out of sight and therefore can be out of mind until an emergency situation occurs. However, any good proactive loss management program should address ongoing maintenance as one of the key issues. The diagram at the right shows where maintenance, rehabilitation, and replacement figures in our "four-arrows" concept of real losses control.

Pipe maintenance can come in many forms and can be undertaken at varying time frequencies, depending on the nature of the problem, the attitude of the operator, and the seriousness of the situation. However, some of the more frequent maintenance programs encountered to counteract losses are corrosion control, and pipe lining and replacement. In the case of pipe replacement, new technologies are being used to undertake trenchless replacement.

This chapter touches on some of the problems and methodologies encountered in the market today, although both subjects are sciences in their own right and have been widely discussed and published in other documents and publications.

13.2 PIPELINE CORROSION

There are many forms of corrosion, including:

- ❑ Galvanic
- ❑ Oxygen concentration cell attack
- ❑ Bacteriological
- ❑ Stray current
- ❑ Pitting
- ❑ Crevice
- ❑ Selective dissolution

- Stress related
- Erosion
- Fatigue related
- Impingement
- High temperature

However, as water operators we usually have to deal with the following types:

- Galvanic
- Bacteriological soil borne
- Bacteriological water borne
- Stray currents

As with any of the other components which can cause a systems water loss, the reasons for corrosion in any one particular water system are varied and often complex and should be studied on an individual basis. Some of the standard methods of corrosion control are

- Protective external coatings
- Pipe relining
- Insulation of pipe joints
- Water treatment using corrosion inhibitors
- Cathodic protection
- Stray current drainage bonds

Corrosion control can be a complex issue involving a very proactive approach to system maintenance, but in most cases paybacks on this type of service are very fast, with significant reduction in the number of new breaks and leaks in very short time periods.

13.3 CASE STUDY ONE

Ductile Iron Pipe in Stray Current Environments

Richard W. Bonds, P.E., Ductile Iron Pipe Research Association, Birmingham, Alabama

13.3.1 Introduction

Stray currents pertaining to underground pipelines are direct currents flowing through the earth from a source not related to the pipeline being affected. When these stray direct currents accumulate on a metallic pipeline or structure, they can induce electrolytic corrosion of the metal or alloy. Sources of stray current include cathodic protection systems, direct-power trains or streetcars, arc welding equipment, direct-current transmission systems, and electrical grounding systems.

To cause corrosion, stray currents must flow onto the pipeline in one area and travel along the pipeline to some other area or areas, where they then leave the pipe (with resulting corrosion) to reenter the earth and complete the circuit. The amount of metal lost from corrosion is directly proportional to the amount of current discharged from the affected pipeline.[1]

Fortunately, in most cases, corrosion currents on pipelines are only thousandths of an ampere (milliamps). With galvanic corrosion, current discharge is distributed over wide areas, dramatically decreasing the localized rate of corrosion. Stray current corrosion, on the other hand, is restricted to a few small points of discharge and, in some cases, penetration can occur in a relatively short time.

Considering the amount of buried iron pipe in service in the United States, stray current corrosion problems for electrically discontinuous gray iron and ductile iron pipe are very infrequent. When they are encountered, however, there are two main techniques for controlling stray current electrolysis on underground pipelines. One technique involves insulating or shielding the pipeline from the stray current source; the other involves draining the collected current by either electrically bonding the pipeline to the negative side of the stray current source or by installing grounding cell(s).[2]

Inquiries to the Ductile Iron Pipe Research Association (DIPRA) show that, of the different sources of stray current mentioned previously, impressed current cathodic protection systems on nearby structures have been the major concern of water utilities. As a result, DIPRA has conducted research for many years on the effects of stray currents from cathodic protection systems on both bare and polyethylene encased iron pipe. The cause, investigation, and mitigation of this source of stray current on iron pipe are the focus of this case study.

13.3.2 Ductile Iron Pipe Is Electrically Discontinuous

Ductile iron pipe is manufactured in 18- and 20-ft lengths and employs a rubber-gasketed jointing system. Although several types of joints are available for ductile iron pipe, the push-on joint and, to a lesser degree, the mechanical joint, are the most prevalent.

These rubber-gasketed joints offer electrical resistance that can vary from a fraction of an ohm to several ohms, which is sufficient for ductile iron pipelines to be considered electrically discontinuous. A ductile iron pipeline thus comprises 18- to 20-ft-long conductors that are electrically independent of each other. Because the joints are electrically discontinuous, the pipeline exhibits increased longitudinal resistance and does not readily attract stray direct current. Any accumulation, which is typically insignificant, is limited to short electrical units.

Joint resistance has been measured at numerous test sites as well as in operating water systems. Forty-five joints were tested at a DIPRA stray current test site in an operating system in New Braunfels, Texas. In 830 ft of 12-in-diameter push-on-joint ductile iron pipe, nine joints were found to be shorted. Such shorts sometimes result from metal-to-metal contact between the spigot end and bell socket, due to the joint being deflected to its maximum. Due to oxidation of the contact surfaces, however, shorted joints can develop sufficient resistance over time to be considered electrically discontinuous with regard to stray currents.

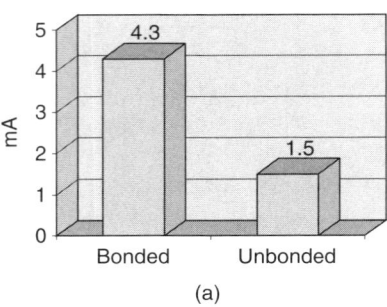 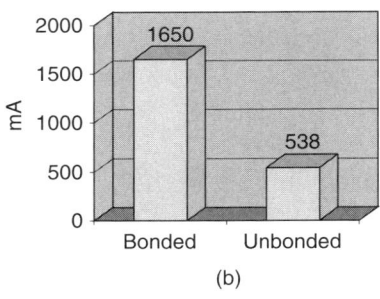

Figure 13.1 Effects of joint bonding. Laboratory installation rectifier output: 8 A. (*a*) With polyethylene. (*b*) Without polyethylene. (*Source: Ductile Iron Pipe Research Association.*)

The ability of electrically discontinuous ductile iron pipe to deter stray current was demonstrated in an operating system in Kansas City, Missouri, where a 16-in ductile iron pipeline was installed approximately 100 ft from an impressed current anode bed. A 481-ft section of the pipeline was installed so that researchers could bond all the joints or only every other joint. When current measurements were made on this section of pipeline, it collected more than 5½ times the current when all the joints were bonded than when every other joint was bonded.

The effect of joint bonding on stray current accumulation has also been demonstrated in the laboratory. The pipe was installed so that researchers could test combinations of bonded joints, unbonded joints, polyethylene-encased pipe, and bare pipe. It was found that pipe with bonded joints collected three times more current than pipe with unbonded joints (Fig. 13.1). Also, when exposed to the same environment, the bare pipe collected more than 1100 times the current collected by the pipe encased in 8-mil polyethylene.[3]

13.3.3 Cathodic Protection Systems

Cathodic protection, which is a system of corrosion prevention that turns the entire pipeline into the cathode of a corrosion cell, is used extensively on steel pipelines in the oil and gas industries. The two types of cathodic protection systems are galvanic and impressed current.

Galvanic cathodic protection systems utilize galvanic anodes, also called sacrificial anodes, that are electronically more active than the structure to be protected. These anodes are installed relatively close to the structure, and current is generated by metallically connecting the structure to the anodes. Current is discharged from the anodes through the electrolyte (soil in most cases) and onto the structure to be protected. This system establishes a dissimilar metallic corrosion cell strong enough to counteract normally existing corrosion currents (Fig. 13.2). Galvanic cathodic protection systems normally consist of highly localized currents, which are low in magnitude. Therefore, they are generally not a stray current concern for other underground structures.[4]

Stray current corrosion damage is most commonly associated with impressed current cathodic protection systems utilizing a rectifier and

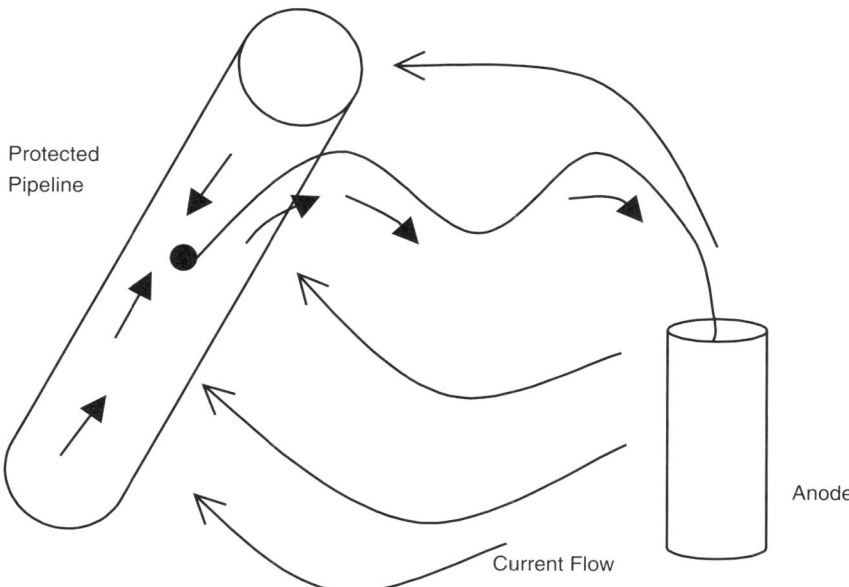

Figure 13.2 Galvanic cathodic protection system. (*Source: Ductile Iron Pipe Research Association.*)

anode bed. The rectifier converts alternating current to direct current, which is then impressed in the cathodic protection circuit through the anode bed. The rectifier's output can be less than 10 V or more than 100 V, and from less than 10 A to several hundred amperes. The impressed current discharge from the ground bed travels through the earth to the pipeline it is designed to protect and returns to the rectifier by a metallic connection (Fig. 13.3). Unlike galvanic cathodic protection systems, one impressed current ground bed normally protects miles of pipeline.

13.3.4 Ductile Iron Pipelines in Close Proximity to Impressed Current Anode Beds

Whether an impressed current cathodic protection system might create a problem on a ductile iron pipeline system depends largely on the impressed voltage on the anode bed and its proximity to the ductile iron pipeline. In general, the greater the distance between the anode bed and the ductile iron pipeline, the less is the possibility of stray current interference.

If a ductile iron pipeline is in close proximity to an impressed current cathodic protection anode bed, a potential stray current problem might exist. Around the anode bed (the area of influence), the current density in the soil is high, and the positive earth potentials might force the ductile iron pipeline to pick up current at points within the area of influence. For this current to complete its electrical circuit and return to the negative terminal of the rectifier, it must leave the ductile iron pipeline at one or more locations, resulting in stray current corrosion.

Figure 13.4 shows a ductile iron pipeline passing close to the impressed current ground bed and then crossing the protected pipeline at

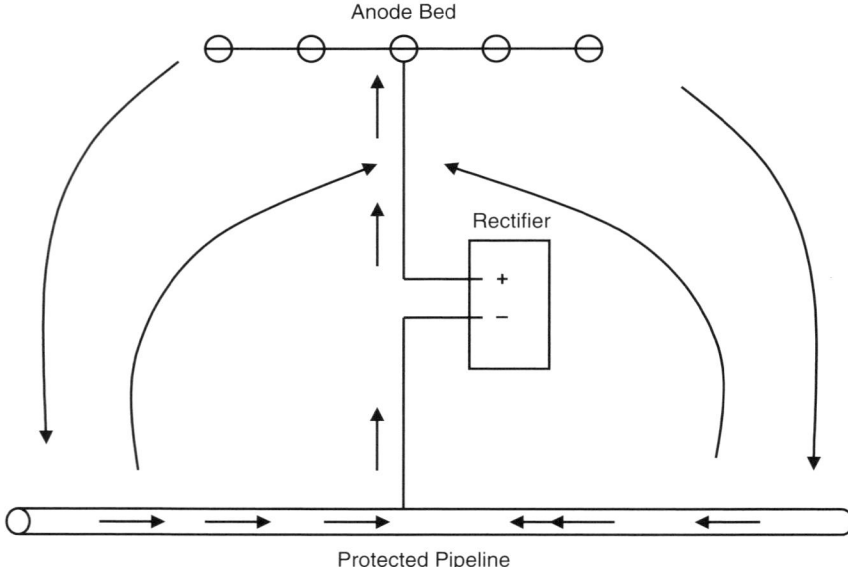

Figure 13.3 Impressed current cathodic protection system. (*Source: Ductile Iron Pipe Research Association.*)

a more remote location. Here, if the current density is high enough, current is picked up by the ductile iron pipeline in the vicinity of the anode bed. The current then travels down the ductile iron pipeline, jumping the joints, toward the crossing. It then leaves the ductile iron pipeline and is picked up by the protected pipeline to complete its electrical circuit and return to the negative terminal of the rectifier. At the locations where the current leaves the ductile iron pipeline, usually in the vicinity of the crossing and/or in areas of low soil resistivity, stray current corrosion results.

Figure 13.5 shows a ductile iron pipeline paralleling a cathodically protected pipeline and passing close to its impressed current anode bed. Again, if the current density is high enough, the ductile iron pipeline may pick up current in the vicinity of the anode bed, after which the current flows along the ductile iron pipeline in both directions and leaves to return to the protected pipeline in more remote areas. This may result in current discharging from the ductile iron pipeline in many areas, usually in low-soil-resistivity areas, rather than concentrated at the crossing as in the previous example.

Normally, electrically discontinuous ductile iron pipeline will not pick up stray current unless it comes close to an anode bed where the current density is high.

13.3.5 Pipeline Crossings Remote to Impressed Current Anode Beds

Usually, a stray current problem will not exist where a ductile iron pipeline crosses a cathodically protected pipeline whose anode bed is not in the general vicinity. A potential gradient area surrounds a cathodically

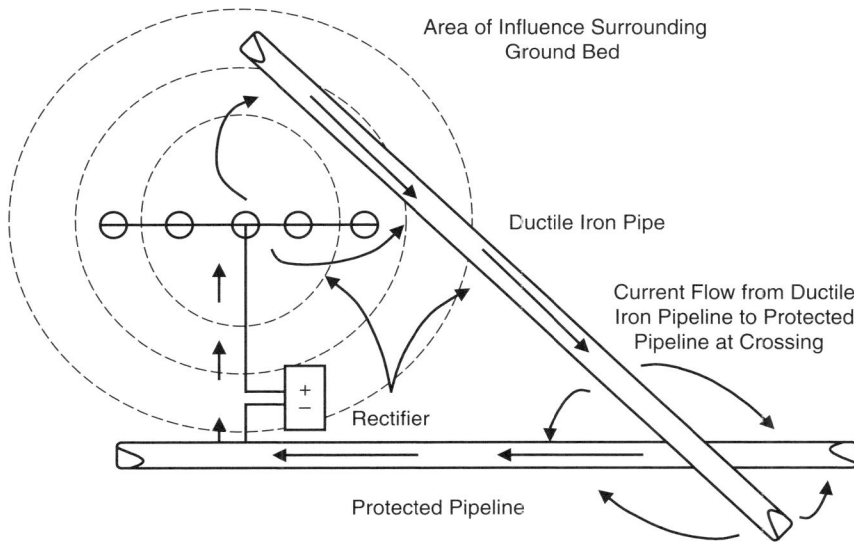

Figure 13.4 Stray current from a cathodic protection installation. (*Source: Ductile Iron Pipe Research Association.*)

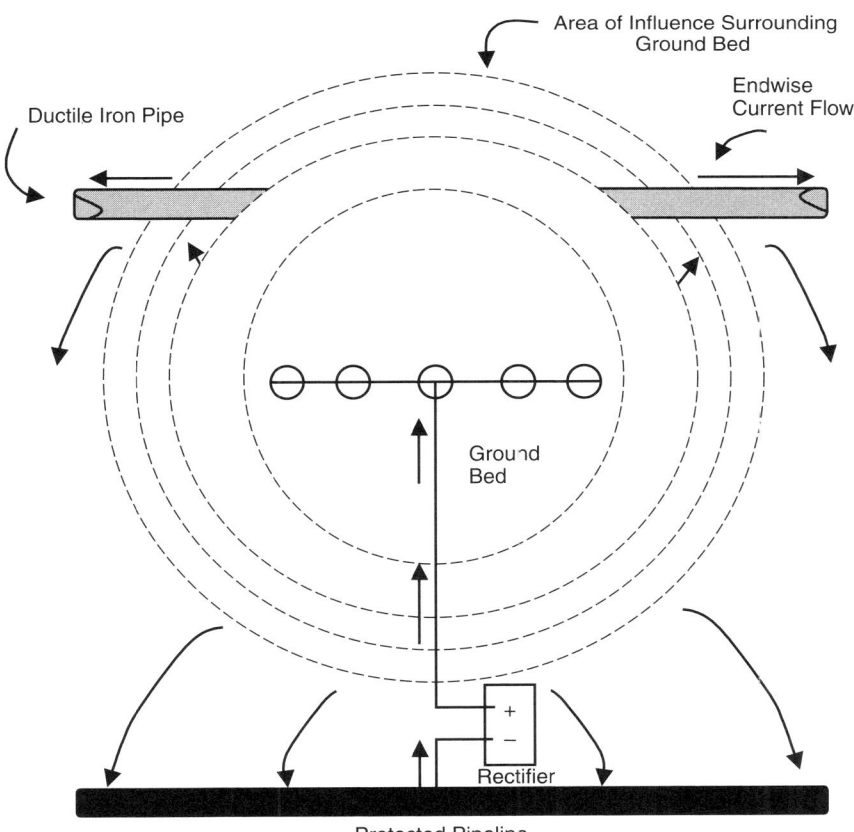

Figure 13.5 Ductile iron pipeline paralleling a cathodically protected pipeline and passing close to its impressed current anode bed. (*Source: Ductile Iron Pipe Research Association.*)

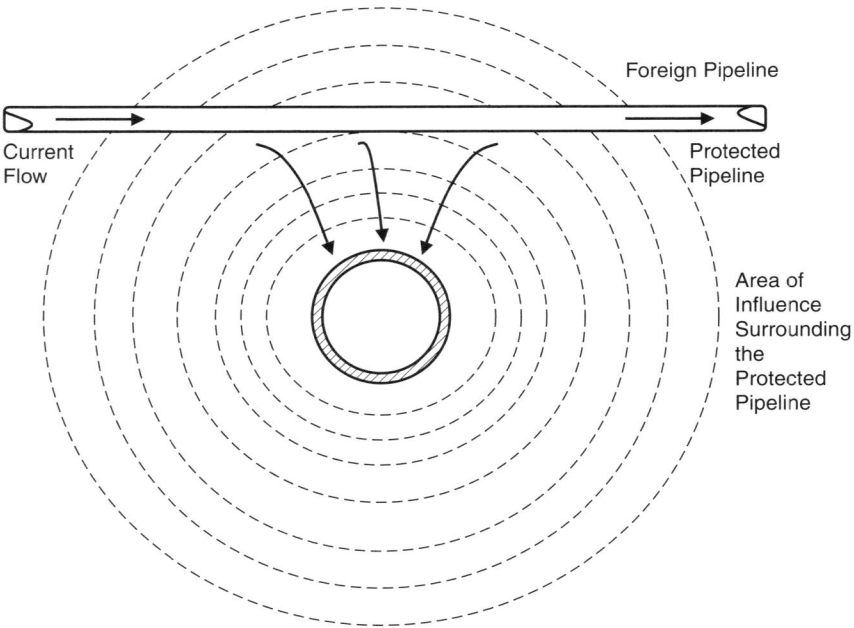

Figure 13.6 Foreign pipeline passing through potential gradients around cathodically protected bare pipeline. (*Source: Ductile Iron Pipe Research Association.*)

protected pipeline due to current flowing to the pipeline from remote earth. The intensity of the area of influence around a protected pipeline is a function of the amount of current flowing to the pipeline per unit area. If a foreign pipeline crosses a cathodically protected pipeline and passes through this potential gradient, it tends to become positive with respect to the adjacent earth. Theoretically, the voltage difference between pipe and earth can force the foreign pipeline to pick up cathodic protection current in remote sections and discharge it to the protected pipeline at the crossing, causing stray current corrosion on the foreign pipeline (Fig. 13.6). Because the intensity of the potential gradient around the protected pipeline is small, negligible for well-coated pipelines, and because ductile iron pipelines are electrically discontinuous, stray current corrosion is rarely a problem for ductile iron pipe systems crossing cathodically protected pipelines if the impressed current anode bed is remote. At these locations, the ductile iron pipeline can be encased with polyethylene per ANSI/AWWA C105/A21.5 for 20 ft on either side of the crossing for precautionary purposes.

13.3.6 Investigation of the Pipeline Route Prior to Installation

It is important to inspect the pipeline route during the design phase for possible stray current sources. If stray current problems are suspected, mitigation measures can be designed into the system, the pipeline can be rerouted, or the anode bed can be relocated.

If, during the visual inspection, an impressed current cathodic protection rectified anode bed is encountered in the general vicinity of the proposed pipeline, one method of investigating the possibility of potential stray current problems is to measure the potential difference in the soil along the proposed pipeline route in the area of the anode bed. This can be done by conducting a surface potential gradient survey using two matched half-cell electrodes (usually copper–copper sulfate half-cells) in conjunction with a high-resistance voltmeter. When the half-cells are spaced several feet apart in contact with the earth and in series with the high-resistance voltmeter, earth current can be detected by recording any potential difference. The potential gradient in the soil, which is linearly proportional to the current density, can be evaluated by dividing the recorded potential difference by the distance separating the two matched half-cells.

When conducting a surface potential gradient survey, one-half-cell can be designated as "stationary" and placed directly above the proposed pipe alignment while the other half cell is designated as "roving" (Fig. 13.7). Potential difference readings are then recorded as the roving half-cell is moved in intervals along the proposed route. A graph of potential versus distance along the proposed pipeline can then be constructed. Normally, depending on the geometry of the ground bed and the cathodically protected pipeline and foreign pipeline locations, the highest current density will be found closest to the anode bed. Usually, the higher the current density, the greater is the possibility of encountering a stray current corrosion problem on the proposed pipeline.

The installation of a ductile iron pipeline typically will not change the potential profile appreciably. This allows the engineer to make recommendations based on the surface potential gradient survey conducted prior to pipeline installation. Figures 13.8 and 13.9 are surface potential gradient survey graphs of stray current test sites located in New Braunfels, Texas, and in San Antonio, Texas, respectively, which compare the current density profile before and after installation of the ductile iron pipeline. As can be seen, there is very little difference in the current densities of the two profiles

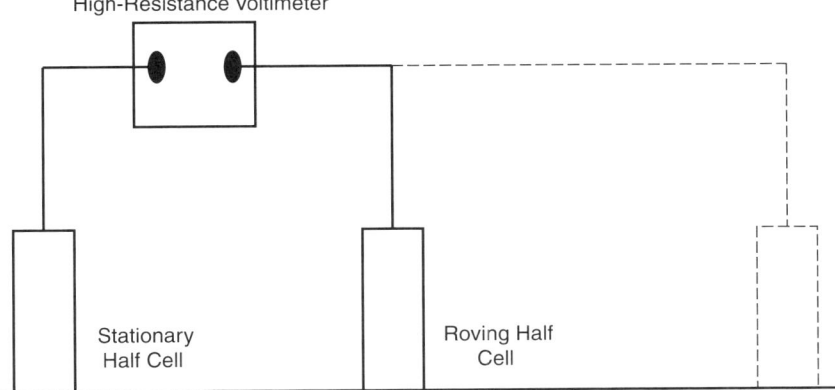

Figure 13.7 Surface potential gradient survey. (*Source: Ductile Iron Pipe Research Association.*)

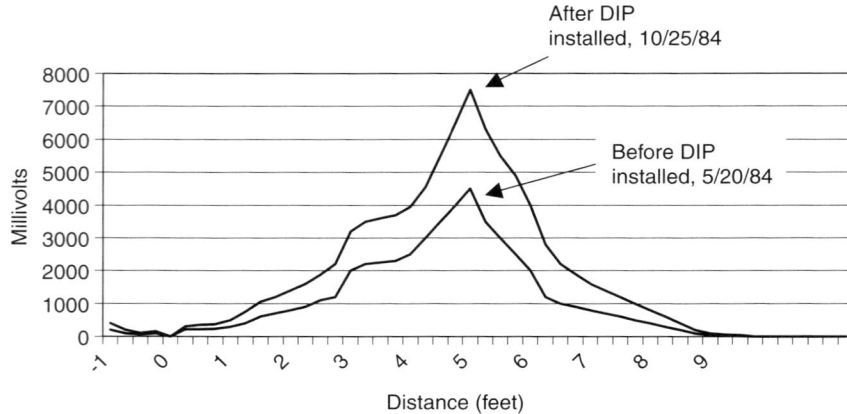

Figure 13.8 Potential profile comparison: New Braunfels, Texas, May 20 and October 25, 1984. (*Source: Ductile Iron Pipe Research Association.*)

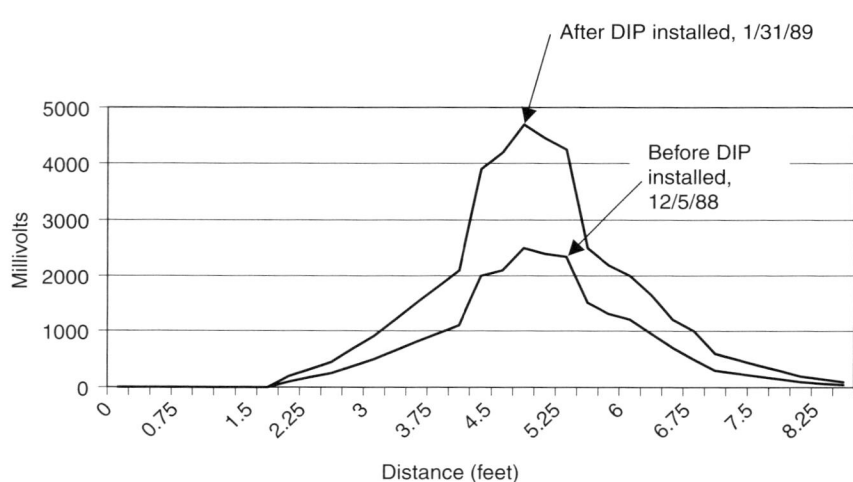

Figure 13.9 Potential profile comparison: San Antonio, Texas, December 5, 1998 and January 31, 1989. (*Source: Ductile Iron Pipe Research Association.*)

regarding their slope and their boundaries—a fact evidenced in numerous other installations and test sites.

Pipeline installations can vary by geometry, soil resistivity, water table, pipe sizes, pipeline coating, rectifier output, etc. However, by knowing the potential gradient prior to installation, the engineer can predict—using conservative values—whether the proposed pipeline will be subjected to stray current corrosion.

13.3.7 Mitigation of Stray Current

Electrical currents in the earth follow paths of least resistance. Therefore, the greater the electrical resistance of a foreign pipeline, the less it is susceptible to stray currents. Ductile iron pipelines offer electrical resistance at a minimum of every 18–20 ft, due to their rubber-gasketed joint systems. This, in itself, is a big deterrent to stray current accumulation. The effect of joint electrical discontinuity can be greatly enhanced by encasing the pipe in loose dielectric polyethylene encasement in accordance with ANSI/AWWA C105/A21.5.

The electrical discontinuity of ductile iron pipelines and the shielding effect of polyethylene are effective deterrents to stray current accumulation and are all that is required in the vast majority of stray current environments. This would include any crossing of cathodically protected pipelines and/or where the ductile iron pipeline parallels a cathodically protected pipeline. At these locations the potential gradient is created by the protective current flowing to the protected pipeline and is normally small.

There are isolated occasions where electrical discontinuous joints and polyethylene encasement would not be adequate to protect the pipe, e.g., the ductile iron pipeline passing through, or very close to, an impressed current cathodic protection anode bed. When this is encountered, consideration should be given to rerouting the pipeline or relocating the anode bed. If neither of these options is feasible, the potential area of high-density stray current should be defined (this can be accomplished by concluding a surface potential gradient survey), the ductile iron pipe in this area should be electrically bonded together and electrically isolated from adjacent pipe, polyethylene encasement should be installed in accordance with ANSI/AWWA C105/21.5 through the defined area and extended for a minimum of 40 ft on either side of the said area, and appropriate test leads and "current drains" should be installed. A typical installation is shown in Fig. 13.10.

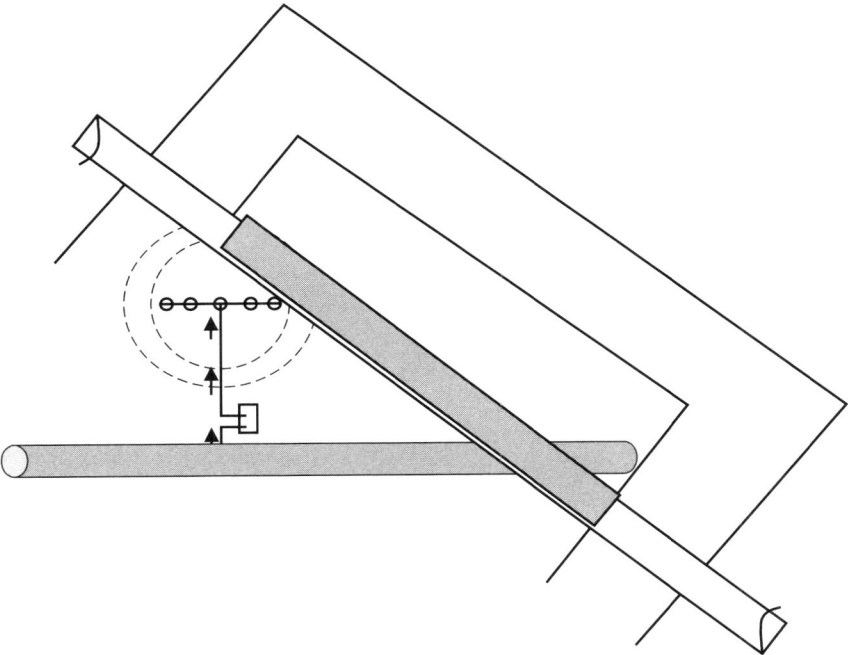

Figure 13.10 Typical installation. (*Source: Ductile Iron Pipe Research Association.*)

In the defined area, the ductile iron pipe most probably will collect stray current. This area needs to be electrically isolated from adjacent piping that will not be collecting stray current. One method of achieving this is to install insulating couplings. Bonding of joints in this area ensures that corrosion will not occur at the joints.

Polyethylene encasement of the pipe in the defined area dramatically reduces the amount of stray current collected. This helps to contain the area of influence and reduces the power consumption of the cathodic protection system. The polyethylene encasement extending on either side of the area shields the pipe from collecting stray current.

Test leads for monitoring are normally installed on each side of the insulators and in the location of the crossing, if one exists. By having test leads on each side of the insulators, their effective electrical isolation can be ascertained. The test leads on the insides of the insulators can also be used to check whether the bonded section is, in effect, electrically continuous.

The collected current then will need to be effectively drained back to the cathodic protection system. This can be accomplished by installing a resistance bond from the affected area of the ductile iron pipeline to the protected pipeline or to the negative terminal of the rectifier. Resistance can then be regulated to achieve a desired potential on the ductile iron pipeline and reduce the current consumption from the cathodic protection system. Another method of draining the collected current is to design and install grounding cells. These grounding cells normally consist of anodes located in areas of current discharge.

13.3.8 Conclusions

DIPRA has conducted numerous investigations in major operating water systems where ductile iron pipelines crossed cathodically protected gas and petroleum pipelines. These investigations involved rectifiers and anodes located in the immediate vicinity (within several hundred feet of the crossing), as well as those located at remote distances.

When the anode beds were remote to the crossings, all investigations indicated that the amount of influence on the ductile iron pipe was negligible and would not be considered detrimental to the expected life of the system. In installations where the anode bed was located in the immediate vicinity, the findings were influenced by factors such as rectifier output, soil resistivity, diameter of the respective pipelines, condition of the coating on the protected line, etc. Despite these variables, several observations confirmed the findings of laboratory tests. The most significant was the efficacy of rubber-gasketed joints and polyethylene encasement in deterring stray current from ductile iron pipelines.

Throughout the United States, thousands of ductile iron and gray iron pipelines cross cathodically protected pipelines, yet very few actual failures from stray current interference have been reported. This is additional strong evidence that stray current corrosion will seldom be a significant problem for electrically discontinuous ductile iron pipelines. The bonding of joints and the use of galvanic anodes or drainage bonds may well be a solution to stray current interference in high-current-density areas, but these systems must be carefully maintained and monitored. If the anode-grounding cell becomes depleted or the drainage connection broken, the bonded ductile iron pipeline will be more vulnerable to stray current damage than if the pipe had been installed without joint bonds. Therefore, such measures should be taken only where stray current interference is inevitable. In most cases, passive protective measures such as polyethylene encasement are more desirable.

13.4 PIPE REPLACEMENT AND REHABILITATION

Underground pipelines have a limited life and may need to be rehabilitated or replaced for a number of reasons, including:

- High break or leakage rate
- High occurrence of joint leaks
- Encrustation or corrosion (internal or external)
- Hydraulic carrying capacity
- Structural reinforcement
- Threat to life or property

In this manual we are focusing on water loss management. Rehabilitation and periodic maintenance can effectively add years to the life of a pipeline, although different methods will be more or less effective and costly in different situations.

In general, pipe replacement methods such as the first options mentioned below will be more effective for reduction of leakage, particularly if the pipe is seriously damaged structurally. However, there are many case studies which do show good results in water loss reduction from spray linings.

Some of the methods of replacement and rehabilitation are discussed below.

13.4.1 Pipe Replacement Methods

Main and Service Replacement

Obviously pipes can be replaced by laying new pipe and discarding or removing the old one, but this is often extremely costly and in some cases completely impractical, especially in dense urban environments. In many utilities a service replacement program will resolve large volumes of loss, as in many cases the largest annual volumes of real loss lie in the smaller leaks on service lines which run for longer periods undetected or reported. Careful analysis of the components of annual loss should be undertaken prior to any replacement or rehabilitation.

Trenchless Technologies

Other methods of pipe replacement can be undertaken using "no dig" or trenchless technologies, which are usually cheaper and almost always less disruptive.

> Trenchless technologies for pipe replacement repair and maintenance are often more cost effective, especially in dense urban environments.

Some of the methods of trenchless pipe replacement are discussed below.

Slip Lining

Slip lining is probably one of the simplest of "no dig" replacement techniques. In this case the old pipe is cleaned out and a new, smaller-diameter pipe is drawn through or pushed through the old one. The new pipe is of a smaller diameter and usually made of polyethylene (PE). Once the new pipe is in place, the service connections are usually excavated and reconnected.

Slip lining does reduce the original diameter of the pipe, and care should be taken that enough hydraulic carrying capacity remains for the job at hand. However, in many cases, particularly with old cast iron pipe, the old pipe, although a larger diameter, may have corroded to give a much smaller effective diameter.

Close-fit lining is another type of slip lining where a deformed liner is inserted into the pipe and then restored to its original size once in place.

Pipe Cracking or Pipe Bursting

In situations where the hydraulic carrying capacity needs to be maintained or indeed increased, pipe cracking can be undertaken. The old pipe is prepared and then a conical wedge is drawn through ahead of the new pipe. In this way it is possible to use the old pipe as a guide for the new pipes; however, the new pipe is actually larger than the old one.

The methods mentioned above will in all cases assist in reducing leakage as well as providing other benefits such as increased hydraulic capacity and clean, safe water supply conditions.

13.4.2 Rehabilitation Methods

In most cases where structural integrity is not found to be a problem, pipes can be cleaned and lined. The liners tend to be ether cement or epoxy and are not in most cases designed to be structural or to reduce leakage, but rather, provide a clean smooth environment, to ensure a healthy water supply and a lower friction factor.

Pipe Cleaning

Before any kind of relining intervention can be undertaken, it is important to clean the pipe properly to ensure that the lining can bond with the pipe wall, without pockets of debris or corrosion, which could later form problem areas.

Pipes with internal corrosion can be cleaned in a number of ways; some of the most common are

- ❑ Air scouring
- ❑ Rotating chains and rods, and scraper trowels
- ❑ Pigging

Air Scouring

Air scouring is undertaken using an air compressor to inject air pressure at a slightly higher pressure than the water pressure. As the air is introduced into the line and then let out downstream, a surge is set up which has the effect of ripping the corrosion off the pipe walls. Air is usually injected and purged through selected fire hydrants while the main is under pressure, although it is a good idea to close surrounding distribution valves to limit discoloration of the water in surrounding areas. Air scouring should be properly supervised, and after any program a mains flushing exercise should be undertaken to ensure that there are no health hazards, dirty water, or entrained air complaints. One of the benefits of air scouring is that it is not necessary to excavate to undertake the cleaning work. This methodology is often used when lining is not going to be undertaken but an improved hydraulic capacity is required.

Rotating Chains and Rods, and Scraper Trowels

When a section of main has been identified for relining, access pits have to be excavated prior to application of the lining. These same pits are used to pull rotating rods or chains through. Alternatively, scraper trowels can be pulled through the line. After the main has been scraped, it should be flushed and often is pigged prior to lining.

Pigging

Pigs come in various shapes and sizes and can be used for the initial cleaning or to clean up after a rodding or scraping exercise as discussed above. The pigs are inserted into the main through the pits which will be used for the relining process and are retrieved at the end of the section. Some pipelines have "pig traps" which allow regular pigging of the line even when a relining exercise is not warranted.

Spray Linings

Epoxy

Epoxy linings have been approved by many environmental agencies throughout the world, but not all are approved and care should be taken when considering their application. Epoxy lining is sprayed onto the pipe wall through the use of a towed centrifugal pump gun. Epoxy linings are usually quite thin and therefore have less negative impact on internal pipe diameter and effective hydraulic capacity. Epoxy linings also tend to dry quickly, allowing the mains to be put back into service quickly.

Cement

Cement linings are also widely approved and provide an excellent way of improving internal pipe condition and improving "C" factors. Cement lining is often the lining of choice for new pipelines, too.

Cement linings are applied by centrifugal spraying and also by trowelling, depending on the pipe diameter. Cement linings tend to take a little longer to dry than epoxy, and as the lining is thicker, care should be taken to ensure that effective hydraulic capacity is not reduced below acceptable limits.

13.4.3 When to Replace or Rehabilitate

From a water loss reduction perspective, the decision to replace or rehabilitate a pipeline is often made on a cost-to-benefit basis, although other factors such as those shown below will often influence the decision:

- Environmental considerations
- Health concerns
- Structural problems
- Emergency hazards
- Demand growth
- Reduced hydraulic capacity
- Lack of alternative supplies

The cost of not replacing or rehabilitating the pipe can be evaluated using the following factors:

- Average historic break frequency
- Cost of volume of lost water per incident
- Cost of damage caused by blow-out
- Cost to repair the main
- Cost to reinstate the surrounding area

This cost should then be compared to the cost to replace or rehabilitate the main in question and the lifespan of the proposed intervention.

Case Study Two shows the efforts of Severn Trent in the United Kingdom to apply the total replacement methodology (not just mains and services but meters and a reservoir too) in a test case in a small water system. (See Sec. 13.4.1.) By replacing everything, Severn Trent had a unique opportunity to start afresh with a completely new system and monitor minimum leakage levels and true rate of rise. This data are invaluable as base data.

13.5 CASE STUDY TWO

Leakage: How Low Can You Go? Cheadle Water Works Project: A Unique Opportunity to Minimize Leakage

Ian Elliott, Director of Engineering, Severn Trent Water Ltd., 2297 Coventry Road, Birmingham B26 3PU United Kingdom

John Foster, Principal Engineer, Severn Trent Water Ltd., 2297 Coventry Road, Birmingham B26 3PU United Kingdom

The case study covers rehabilitation of water distribution infrastructure to a small town of circa 8000 people in the United Kingdom. The work involved the replacement of 32 km of distribution mains with diameters ranging from 50 to 350 mm with new MDPE pipe utilizing a number of "no dig" techniques. All customer services were renewed, including in many cases customer-owned service pipes. Radio-read meters were installed in all properties.

The result will be a unique opportunity to determine "lowest practical leakage level" and gain a detailed understanding of degradation of systems integrity with time.

13.5.1 Background

Severn Trent Water in the United Kingdom

Severn Trent Water is a leading provider of water supply and wastewater services in the United Kingdom. Severn Trent Water Limited is part of the Severn Trent Plc group and has a market capitalization of $6.5 billion. The utility provides water and sewage services to 8 million people across the heart of Britain and to communities in fifteen of the United States. Since privatization in 1989, more than $6.4 billion has been spent on mains upgrading, replacing the distribution system and services.

Severn Trent Water has the lowest average water service charges in England and Wales, and the best overall quality in the United kingdom. It achieves 99.9 percent compliance with UK and European drinking water standards—the most stringent in the world. In addition, the level of treatment provided by its sewage treatment works is the highest in the country.

Water Charges

UK domestic water supplies have historically been charged for through a "rateable value" system based on the value of the property, regardless of the number of occupants or the amount of water usage. The privatization of the water industry in 1989 introduced compulsory metering of all new properties to ensure realistic charges based on water usage. Generally, existing properties remain unaffected but customers are able to opt for a meter to be installed and pay charges based on usage if they choose to do so.

Severn Trent Water (STW) introduced compulsory metering of high-usage properties, e.g., those with swimming pools or garden sprinkler systems, and provided free meter installation for those customers who wished to switch to a different method of charging.

Service Pipes

Pipes linking properties to the mains network have usually had shared ownership, the water company being responsible for the pipe up to the property boundary, whereas thereafter the pipe becomes the sole responsibility of the owner. This has raised concerns in two specific areas—leakage on the customer side and potential problems where properties have been fed via a shared pipe.

13.5.2 Cheadle Water Works Company

The private Cheadle Water Works Company had been established in the early nineteenth century and fed the small market town of Cheadle, Staffordshire, in the North Midlands of England. The town is located entirely within the Severn Trent Water supply area and consists of approximately 3800 properties of varying age, with a mixture of rural properties on the town fringes and a major industrial user.

In 1997 average demand was recorded as 2.5–3 Mliters per day, when accepted per-capita demand figures suggested the figure should be of the order of 1.25 Mliters/day.

The company's assets were in very poor condition, consisting of 20 km of very old, unlined iron mains, and two very old, leaking service reservoirs (ca. 1830 and 1935). The limited resources of the Cheadle Water Works Company and restrictions of the old distribution system had already resulted in the new development on the fringes of town being supplied with water by STW.

The CWW was unable to meet current water demand effectively and did not have sufficient financial resources to fund the investment required to meet increasing demands on the fringes of the town for new development areas. Massive rate increases to begin the process of improving its assets were being considered, but instead an approach was made to STW to take over the company for a nominal sum.

Severn Trent recognized that a significant program of rectification was required and set up a Project Board to manage the assimilation of the CWW.

13.5.3 Project Organization

A Project Board was established early in 1997 and included representatives from all disciplines to be involved. The overall task of the Project Board was to coordinate the activities of the individual aspects of the project, which included Mains & Service Renewal, Metering, Borehole, and Reservoir. Project Engineers for each one of these activities regularly reported to the Project Board to allow an overall Coordination Program to be controlled.

The overall project was planned in three stages:

1. Establish the true factual situation at takeover.
2. Gather data and implement short-term solutions to safeguard supplies and improve levels of service.
3. Define and implement a longer-term strategy for the implementation of the old Cheadle system into STW.

Problems

The CWW had suffered from a serious lack of investment for many years, which had resulted in the following problems:

- A continuous gardening watering ban which had been in place for some years
- Variable water quality
- Inconsistent and unfair charging policy

- Inadequate distribution network
- Leakage estimated to be of the order of 50 percent of the water into supply
- Inadequate pressures in some key areas
- Few operable valves and no method of zoning the distribution system
- No network meters to monitor flows
- Only marginal chlorination, with inadequate safety measures
- Leaking unstable service reservoirs with inadequate security measures
- Inoperable stop taps or in many cases none at all

Data and Information

Investigations into the mains network was complicated by the fact that there were no adequate record plans. The only mains layout plan dated from the 1920s.

Sample sections of the most critical mains were taken and confirmed the system to be undersized and in very poor condition, showing significant problems of leakage and encrustation. No guarantee could be given for any of the existing pipes, and it was concluded that it would be impractical and uneconomical to try and repair the system piecemeal. The decision was taken to renew the complete distribution network.

Network design was commissioned and a Stoner model was established to determine the main distribution system required. This work was completed in November 1997 and indicated that a minimum of 27 km of new mains was required.

Initial Priorities

The immediate concern was to secure supplies into the area and enable the garden watering ban to be lifted. This was effected by providing a short new link main into a neighboring STW supply zone and upgrading an existing booster. Work was completed in the summer of 1997.

The Cheadle reservoirs were then taken out of normal service and the whole town was fed from the adjacent system, bypassing the old works. This had a dramatic effect on the water lost through leakage and meant that the long-standing garden watering ban could be lifted.

With the short-term objectives secured, the process of developing the project in more detail commenced.

13.5.4 The Cheadle Project

Outline Proposals

The base project included the complete replacement of the distribution system and the construction of a new reservoir together with the refurbishment of the borehole source and provision of a treatment system.

In addition, the company decided to extend the project by offering to provide every property with a new separate service pipe and meter, where practical. This would create a discrete area in which every property would be supplied through new pipework and a meter, which could be remotely read by computer from the office base at Leicester some 60 m away.

This would provide the company with valuable information with respect to:

- Leakage levels in a newly refurbished distribution system
- Leakage detection and localization
- Water usage patterns
- A complete new distribution system, which could be monitored over a period of years to provide information on developing leakage patterns to assist with defining an economic level of leakage

Construction and installation work was broken down into three main elements:

1. Mains and service pipes
2. Reservoir and borehole reconstruction
3. Meter and radio-read installation

Mains and Services

The work was to include the replacement of the whole of the mains network and individual service pipes to provide separate pipes to each individual property where possible. The time scale was a major factor in tendering the contract on a design and construct basis—reducing the time period for design and construction by approximately 3–4 months.

The tenderers were provided with the network design indicating pipe sizes and lengths, with the documents stating that the tenderers had the choice of construction method. It was also implicit within the tender that the contractor would be responsible for specific customer care aspects, including service pipe surveys and the normal warning procedures for disruptions to supplies.

The company's intention was also to replace as many service pipes as possible, including private side services up to the property wall. This work could only be done with the consent and agreement of the owners, and because of the lack of any record information, involved a complete survey of all the properties to identify pipe runs.

During the survey, customers were asked if they would wish to have their pipes replaced. Initial indications showed a potential takeup of almost 90 percent, though this was later complicated by the issues surrounding common services, on which owners were given certain conditions which had to be met before work could proceed.

Teamwork

The unique nature of the project and the high impact on both individual customers and the town as a whole resulted in an almost unique teamwork approach to the design and construction process.

The mains and services project team included representatives from the following groups:

- Engineering: overall project management, contractual detail and customer interface, quality and specification compliance, supervision of construction.

- ❏ Contractor: surveys of the service pipes, design details, construction methods, and management and customer liaison
- ❏ Operations: confirmation of final design details and operational advice on the existing network
- ❏ Customer relations: additional contact point for general customer issues and support with specific problem areas
- ❏ Highways Authority: continued liaison for planning of roadwork and advice on quality issues
- ❏ Marketing: general public relations and media contact

Construction Methods

The restrictions within the town meant that the most cost-effective methods would be low-dig techniques (conditions permitting). The final choice of method was determined by the main contractor, D. J. Ryan. The following methods were used (see Fig. 13.11).

- ❏ Pipe sizes ranged from 25 mm for service pipes to 350 mm for distribution mains. All pipes used were MDPE.
- ❏ The contract used a very open teamwork approach. Close liaison took place at all times with the local Highways Authority, which included regular progress and program meetings to review the effects on traffic and standards of construction.

Specific Problems Addressed

Specific problems addressed included:

- ❏ Traffic management
- ❏ Density of construction operations
- ❏ Ground conditions

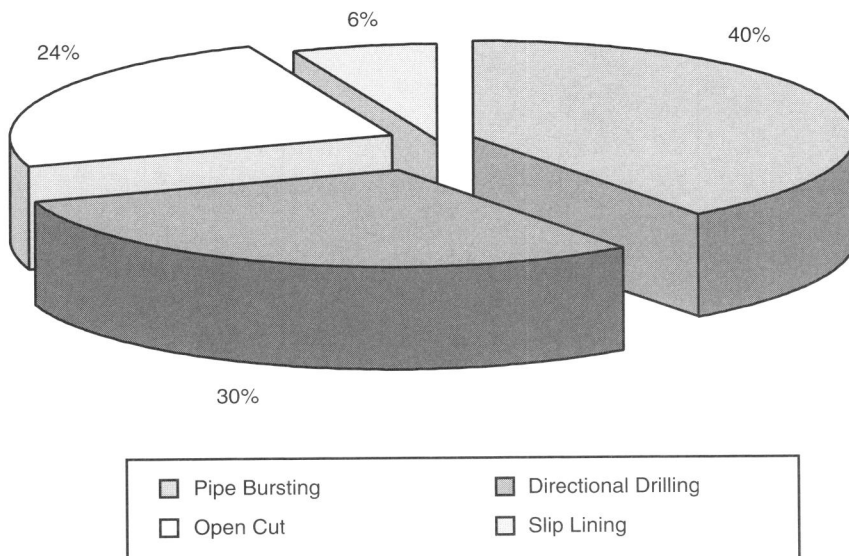

Figure 13.11 Construction methods. (*Source: Severn Trent Water, Ltd.*)

- Maintenance of supplies during works
- Vulnerability of the existing system
- Quality problems during the rehab work due to high velocities
- Resolution of customer issues
- Customer options

The unique nature of the project and the resulting cooperation required from domestic customers resulted in a hierarchy of options being developed. Customers were given the following options:

- Meter fitted either internally or externally (within the boundary stop tap box)
- Service pipes could be replaced up to the property wall by agreement with the customer
- Payment for water could be either by metered rates or through the rateable value charging system at the customer's request

This provided customers with an extensive range of choices, which would enable them to find an option most suited to their own circumstances.

Though all properties would eventually be fed through a meter, the ideal scenario of all properties having their own meter and separate service pipe was not achievable because of problems with access to certain properties and the issue of common service pipes.

Reservoir and Boreholes

The existing reservoirs were known to be in very poor condition, only 25 percent of the total capacity available for use because of structural and leakage problems.

The site itself was in a very difficult location, but in spite of this, the limitations on other water sources within the area resulted in the decision to refurbish the boreholes and provide new treatment and storage facilities rather than abandon the site.

The main construction problems revolved around the restricted nature of the site, which was located on a hill at the center of the town and included a very difficult access route, which was narrow and winding and ran very close to very old properties.

Two well/boreholes existed on the site; these were up to 70 m deep. A closed-circuit television (CCTV) survey found them to be in good overall condition, and samples showed good-quality water. Test pumping proved the No.1 borehole capable of yielding up to 2.5 Mliters per day.

Construction work commenced on site during September 1998.

Meters

Current STW Metering Policy

STW is now promoting internal metering because of the customer benefits, but as a result needs to identify a cost-effective method of remote reading because of potential access difficulties. Current STW policy is to fit Fusion System Equity water meters with remote read by a Talisman Touch Pad read system.

Metering on the Cheadle Project

The Cheadle area offered the unique opportunity of a high-density meter population within the context of a pipework system which will be almost completely new. The new network, besides giving improved customer service, could provide an opportunity to offer further customer benefits and to give the company a chance to obtain very precise information on system leakage and general usage.

The objectives were identified as follows:

- To monitor all inlet and outlets within the Cheadle district metered area (DMA) system
- To enable automated calculation of leakage based on the integrated sum of inlet and outlet totals
- To enable ad-hoc readings of individual or groups of meters and to provide customers with up-to-date information on their own water usage
- To enable automatic collection of nighttime readings when usage is at its lowest, to determine unusual flow patterns
- To support possible future automation of readings for billing purposes

Radio-Read Systems Design

The system was designed to serve meters, which could be installed either internally within the property or externally in the boundary stop tap box. System Equity electronic meters were used for internal fits and Kent encoded meters for boundary box locations.

Each of the meters is linked to a Genesis Meter Module (GMM) radio transponder supplied by Itron via a waterproof wired connection developed by Fusion Meters for use with all utility meters and remote devices.

The town area was broken down into a network of 15 radio areas, each of which included repeaters, slaves, and master receivers. These were sited on street light standards and telegraph poles with the relevant permissions. Signals were then relayed via Vodafone wide-area link to a remote system computer. The system has been developed to read and collect information from every meter within the network at 15-min intervals between 1 and 3 a.m. This includes six main distribution meters and the consumption meters, which will number in excess of 3500.

The network design included designation of overlapping areas, the siting of repeaters, slaves, and masters being crucial to allow areas to overlap and enable rerouting capability for maximum system dependability given the topographical nature of the area.

Difficulties Encountered

Logistics and Timings

The operational complexity was heightened by the need to follow closely behind the pipeline rehabilitation work. The need for operational flexibility and traffic management requirements resulted in a varying program, which made planning difficult.

High Water Levels

Groundwater levels in the lower areas of the town began to rise as abstraction from the borehole ceased.

Customers Issues

Customers were given the choice of internal or external meter installation. Initial survey figures indicated that the preference would provide an 80:20 split. As data from the first installations built up, a ration of 50:50 was observed. However, as time progressed, the ratio was raised to 70:30 as word spread that internal installations were performed with great attention to customer preferences.

13.5.5 Progress to July 1999

Mains and Services

- Over 32 km of main have been replaced, and most of the old mains have been abandoned.
- 3150 "Company" service pipes have been replaced, and some 2200 "Customer" service pipes have been renewed.
- The mains and service work was substantially completed by the end of March.

Reservoirs and Borehole

- The new reservoir has been constructed, and final testing is in progress.
- The new borehole head works has been constructed, and the new disinfecting and control equipment is being commissioned.
- Completion is expected by the end of August 1999.

Meters and Radio-Read Installation

- So far, 3537 properties have been fitted with meters.
- The meters at over 2600 properties can now be read remotely on either an ad-hoc basis or automatically and as individuals or in groups.
- Up to nine readings per meter can be taken during the nighttime period between 1 and 3 a.m.
- Installation and commissioning of the automatic meter reading (AMR) system continues and is expected to be completed by the end of August.

See Table 13.1 and Figure 13.12.

13.5.6 Scheme Benefits

Customer Benefits

The major customer benefits were identified as follows:

- Vastly improved supplies to modern-day treatment standards with a secure source

TABLE 13.1 Costs

Item	Major Contractors	Forecast Final Cost ($M)
Secure supplies		0.29
Mains and services	D. J. Ryan & Sons	6.72
Reservoirs and borehole	Mowlem Construction	2.05
Meter and radio read installation	Kennedy Iron, Ltd.	2.25
Other		0.16
Total		11.47

Source: Severn Trent Water Ltd.

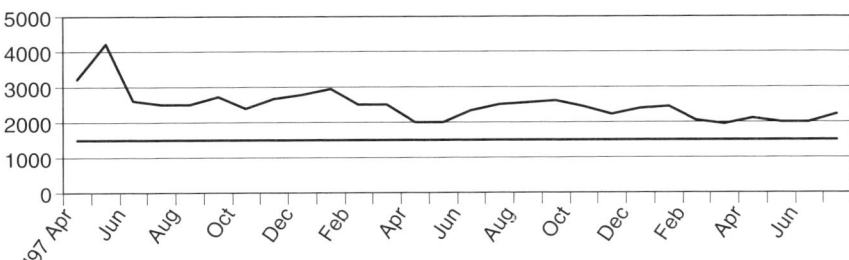

Figure 13.12 Water into supply. (*Source: Severn Trent Water, Ltd.*)

- Removal of the garden watering ban
- The option to monitor water usage (and therefore potential costs) directly through a meter, without any commitment to switch from the rateable charging system
- The opportunity for leakage on customer pipes to be detected
- A telephone contact to give the opportunity for customers to discuss their individual circumstances and options to switch to lowest charging method
- The possibility of individual meter readings remotely to verify reported leakage

Severn Trent Water

Major benefits to STW from the system can be described as follows:

- Almost 4000 new customers with completely new assets and a secure borehole source
- Significant reduction in water lost through leakage
- Future provision of accurate usage and leakage calculations
- Data to enable the monitoring of leakage changes and water usage over time (including an accurate assessment of private-side leakage)
- Information on daily and seasonal variations
- A potential long-term assessment of the new pipe network and changes over time
- An assessment of the effect of high meter concentrations on general water usage indication of "lowest practical" leakage level

13.5.7 Summary and Conclusions

- STW set out to renew infrastructure and establish a complete metered system to monitor water usage and minimize leakage using a radio-read system—the project should be fully operational by the end of August 1999 to provide on-line leakage detection and a new model system.
- It is the first known project of its kind, combining the provision of an almost totally new system, including the replacement of customer-owned pipes, with all outlets metered and measured by an intelligent system.
- The work has been carried out within a very short time scale and overcame some significant problems during the process.
- Initial benefits of the improved service to customers are already apparent: supplies are more reliable and quality more consistent, and the garden watering restriction has been lifted.
- The longer-term benefits from the leakage and water usage information cannot yet be quantified, but data gathered should provide benefits to both STW and the water industry as a whole.

13.5.8 Acknowledgments

The authors wish to thank Severn Trent Water for giving permission to publish this case study.

13.6 SUMMARY

Most utilities today have older, less effective pipes, which are coming to the end of their useful life. As technology advances, trenchless replacement and rehabilitation options are becoming very attractive options to traditional mains replacement. Careful tracking of reported leak and break frequencies along with hydraulic and camera inspections will allow the operator to quickly identify those sections of main which can no longer be maintained cost effectively in their current condition.

13.7 REFERENCES

1. Peabody, A. W., *Control of Pipeline Corrosion,* National Association of Corrosion Engineers, Houston, Tex., 1967.
2. Wagner, E. F., Loose Plastic Film Wrap as Cast-Iron Protection, *Journal AWWA,* vol. 56, no. 3, pp. 361–368, March 1964.
3. Stroud, T. F., Corrosion Control Measures for Ductile Iron Pipe, National Association of Corrosion Engineers, Conference, 1989.
4. Smith, W. Harry, *Corrosion Management in Water Supply Systems,* Van Nostrand Reinhold, New York, 1989.

CHAPTER

14

Resolving Apparent Losses

Julian Thornton

> ### *Case Study One*
> Proper Meter Sizing for Increased Accountability and Revenues
> *John P. Sullivan, Jr., P.E.*
> *Elisa M. Speranza*
>
> ### *Case Study Two*
> Putting the Water Works G-Man to Work
> *Roger W. Esty*

14.1 INTRODUCTION

Apparent losses come in various types, and each utility should properly identify the types and costs associated with the particular losses by undertaking a detailed audit as discussed in Chap. 4. The diagram at the right shows a preliminary component analysis of apparent loss management, similar to the one we used for real losses in earlier chapters. (Work is currently being undertaken by the author and other international specialists to improve this preliminary component analysis.)

In many cases a number of actions can be undertaken to combat apparent losses. This chapter discusses some of those actions.

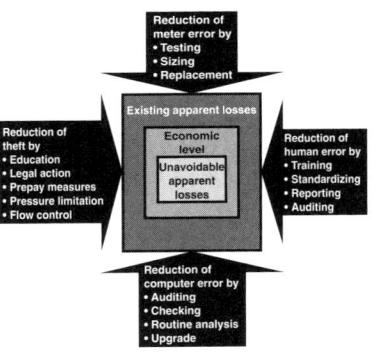

14.2 MASTER METER CALIBRATION

Master meters are often overlooked and do not get installed properly or do not receive the correct periodic maintenance once they have been installed. For any utility that wishes to embark on an optimization project, the master or supply meter is one of the first meters which should be checked and properly calibrated. We are discussing the master meter in this chapter because meter testing and sizing is being addressed here; however, the master meter actually affects both real and apparent losses.

> The master meter is one of the most important registers a utility has with which to measure performance.

14.2.1 Master Meter Types

Master meters come in various types, shapes, and sizes, some of them being

- ❑ Differential pressure meters
- ❑ Venturi meters
- ❑ Dall tube meters
- ❑ Orifice plate meters
- ❑ Proportional flow meters
- ❑ Magnetic meters
- ❑ Ultrasonic meters
- ❑ Turbine meters
- ❑ Propeller meters
- ❑ Vortex shedding meters

It does not really matter which type of meter a utility has installed. What matters are that the meter is functioning to its specification, the data being recorded are compatible with other system data, and the utility understands the limits of the meter that is installed.

14.2.2 Master Meter Testing Program Steps

There are various steps to be taken when considering a master meter testing program.

Master Meter Pretest Checklist

1. Identify and locate all meters.
2. Locate all manufacturers' specifications for those meters.
3. Determine whether the meter is installed to the manufacturer's specification.
4. Determine whether on-site testing can be undertaken.
5. Determine the type of on-site testing to be undertaken and its realistic limitations.
6. Identify an allowable band of error between the test volume and metered volume.
7. Determine whether any testing or repair work has been carried out in the past.
8. Identify a local supplier or contractor who can calibrate meters which fall outside allowable limits.
9. Identify suppliers of replacements for meters which cannot be calibrated.
10. Establish a realistic budget for the work.
11. Establish a realistic time frame for the work to be carried out.
12. Establish an accountable tracking mechanism to show clearly both baseline usage before calibration and calibrated usage after the program.

Posttest Checklist

1. Clearly identify and record any major changes in calibration, for both span and zero.
2. Identify the impact on the annual water balance.
3. Store both raw and massaged data for future reference.
4. Identify local extraction limitations and how they are affected by the new results.
5. Put in place a periodic testing program to ensure that the meters stay calibrated and the new results accountable.

14.2.3 General Meter Testing and Repair

Meter testing for any meter, whether revenue or master, should be done on a periodic basis to ensure accountable results. Some countries and states have recommended time frames or usage volumes, which dictate good testing practice.

> In App. B the reader can find Chaps. 4 and 5 of the AWWA M6, and in App. C Chap. 4 of the M33, which gives excellent detail on the workings of some of the meters mentioned here.

14.3 METERS*

Water meters are mechanical devices. As such they do wear and lose accuracy after an extended period of operation. The larger meters are fre-

---

*Sections 14.3 and 14.4 and the material from the beginning of Sec. 14.5 to the end of Sec. 14.5.1 were written by Jack Jackson of Invensys Metering Systems (formerly Sensus Technologies).

quently ignored as long as they continue to record consumption. These meters may be few in number, but they represent a significant amount of revenue for a water system. Typically, the largest 10 percent of the meters measure 40–60 percent of a system's consumption and may constitute a significant revenue contribution. If these larger meters mean so much to a water system's financial health, why are they not maintained to provide peak performance? The answer is that large meters are difficult to repair, spare parts are expensive, assemblies are sometimes complex, and knowledgeable and trained repair personnel are needed to maintain them.

In addition, maintenance is delayed because meter installations are frequently crowded or piping compromises have to be made. Many times there is no bypass piping to continue supplying water to the customer during a meter accuracy test, or it is difficult to dispose of the test water. Also, workspace around the meter may be restricted and unsafe.

Liability, safety issues, and span of control of the testing personnel can sometimes be a concern to the system's management. Because these larger meters are so important, their operating condition must be monitored on a systematic and timely basis. One alternative approach is to test these meters on-site by qualified test personnel.

14.3.1 Testing Rationale

On-site testing of the larger meters is one way to ensure their proper operation, and another is to test the meters at the water system's facilities. There are certain advantages to testing the meters the latter way, and in some cases it will be prudent to do so. In the majority of situations, on-site testing is more economical in time and resources. From a technical standpoint, the piping in a meter installation may have a definite influence on a meter's accuracy, and these deviations may frequently be detected by on-site testing.

> On-site testing of meters can often be the preferred method, as the site is tested for suitability as well as the meter for accuracy.

One method of maintaining proper performance for certain types of larger meters is to replace the operating components and assemblies while leaving the meter body in place. This is acceptable. If the measuring and registration functions are within one integral assembly, no accuracy tests are required after installation. However, if separate assemblies are involved, one cannot be certain everything is in place and operating properly. In this case, due to the potential loss of significant revenue, a final on-site test is recommended.

Due to the revenues these larger meters produce, a formal on-site meter testing program should be part of every water system's maintenance operation. Increased revenue and water accountability gains quickly offset the initial investment and continuing costs of such testing programs. Case studies show such returns to be both immediate and dramatic.

14.3.2 Requirements

A number of factors are involved in establishing an on-site meter testing program. The meter pit must have adequate and safe space in which personnel can operate safely. OSHA safety requirements should be followed, and all testing personnel should be properly and sufficiently trained.

The piping arrangements around the meter must include some method to positively isolate the meter, while still maintaining an adequate flow to the end user. Temporary or permanent bypass piping should be utilized when possible. Some larger meters have built-in test plugs; others do not. For installations requiring test outlets, these can be fabricated in a number of ways. Service saddles and reducing tees are the most frequently used approaches. These need to be installed according to the recommendations of the meter manufacturer and located so that the connecting hose to the on-site tester is correctly located downstream of the meter. To assist the testing operation, it is suggested that a short length of pipe be permanently attached to the test outlet, along with a shut-off valve, which can be locked into position. This will save considerable time and expense in site preparation.

A prerequisite to testing is that all the downstream isolation valves provide a positive shutdown. If the valve is not completely operational and for some reason there is leakage, an accurate test is not possible. The lower the test flow rate, the higher the significance of such a leak.

Finally, personnel assigned to perform the tests must be properly trained and have the appropriate test equipment. Certain techniques and procedures should be followed when using the test equipment. The consequences of discharging large volumes of water at high flow rates must be understood, appreciated, and considered. Improper use of the equipment may be harmful to testing personnel, the meter, the surrounding area, and the general public.

Techniques for performing the tests, selecting the appropriate test flow rates, determining the accuracies, and reaching conclusions must be known and carefully followed to obtain valid test results. Meter testers may be considered field specialists or technicians.

14.3.3 Testing Equipment

Residential meters, that is, those meters up to 1 in in size, may be tested on-site in several ways to determine accuracy. The test equipment and methods for determining the accuracy of these meters are not applicable to testing the larger meters. The larger meters require specialized test equipment, which can handle a wide range of flow rates and provide accurate, valid data. These devices may either be purchased as a manufactured assembly or fabricated by the water utility.

The equipment for the larger meters is available as a portable test package, installed on trailers, or mounted in a van or pickup truck. Regardless of the style, these testers all contain certain basic elements, which are required to properly test turbine, compound, and propeller meters.

Because of the wide flow ranges involved, a tester includes at least two, and sometimes three, meters of varying capacities. A shut-off valve is typically located downstream of each meter to control the flow rate for the various tests. A pressure gauge is required to check both the line pressure and the residual pressure at the tester. Sometimes resettable registers and/or flow raters are included to reduce the time required to conduct a complete test. (Pictures of the various meter testers mentioned here can be found in Chap. 6 of this book.)

Flexible hoses are required to connect the test equipment to the test connection of the meter being tested. Due to the static pressures and hydraulic forces present, all hoses must be in good condition and positioned as straight as possible between the two meters. For the larger testers, it is important that the tester itself be anchored by means of a vehicle, or a type of hold-down device, because the tester will want to move and become unstable during operation.

The master meters used on the testers should be protected and handled with care (not thrown around). They should also be tested and recalibrated periodically to ensure that accurate measurement is being maintained.

14.3.4 Safety First Checklist

1. Inspect area: Look for adequate area for discharge and run-off of tested water. It is not unusual to discharge in excess of 10,000 gal of water in a large meter test segment. Make sure water will not run back into meter pit while testing. Be aware of sidewalks and streets where pedestrian traffic may occur. A 300-gpm stream of water from a tester can be dangerous to vehicular and pedestrian traffic.
2. Close valves: Close both upstream and downstream valves to isolate the meter from line pressure. This must be done prior to removing any test plugs.
3. Bleed pressure: Bleed all residual line pressure from the meter assembly prior to removing any test plugs. Generally, this can be accomplished by loosening a bleed screw found on the meter cover. If a bleed screw is not present, a main flange or drain plug may be loosened to relieve any pressure.
4. Install tester: After ensuring there is no line pressure, remove the test plug, install test pipe, hoses, and tester. Make sure all equipment is laid out across the ground in a straight manner, with both hoses (inlet and outlet) having no sharp and/or irregular bends.
5. Check hoses and purge: Inspect setup for good layout and begin purging air from equipment by opening the small valve on the tester and slowly opening the meter supply valve. Continue until line is under full pressure and all air is bled from setup.
6. Secure tester: It may be necessary to secure the tester and hoses if high pressures/flow rates will be involved. Chaining the tester to a fixed object and/or driving large stakes through the holes in the tester may accomplish this.
7. Adhere to warning tag: In no case deviate from the warning tag. The warning tag that is affixed to the tester has been developed from extensive product testing in field situations.

14.4 LARGE METER INSTALLATIONS
14.4.1 Bypass Piping

Large meter settings are relatively expensive and require considerable preliminary planning. Large meters are heavy, and removal for service or testing can be costly and time-consuming. Provisions for fire service must also be given serious consideration.

Every large meter should be installed with a bypass line and valve that is needed to supply water to the customer while the meter is being serviced. For non-fire line meter applications, the bypass should be sized one nominal size smaller than the meter being tested, down to 2-in size. For fire service metering applications the bypass line size should be the same nominal size.

There are numerous reasons why the bypass line is necessary. Certain classes of water customers require continuous service, such as hospitals and health care facilities, for example. Without bypass lines the meters serving these customers cannot be tested or repaired, which may result in the loss of significant revenue.

Preassembled meter packages are designed to provide all necessary equipment for the complete meter installation and help provide for fast, easy installation. These packages are supplied by most meter manufacturers and are especially valuable to many utilities that may not have the tools or equipment required to handle the installation of large meters.

14.4.2 Test Outlets

Most meter companies recommend a spool piece downstream of the meter that is at least two pipe diameters in length. This is also used to eliminate any turbulent flows on the exit side of the meter's measuring element. On top of the spool piece should be a tagging saddle, brass nipple, and a ball or gate valve that is required for field testing of the meter. Most compound and fire service meters have test plugs built into the meter casings. The 4-in and 6-in sizes have 2-in test plugs, while 8-in and larger sizes may have 3-in test plugs. Prior to installing the meters, the test plugs should be removed and replaced with brass nipples and either ball or gate valves to facilitate field testing of the meters. The valve is also needed to relieve pressure from the meter before opening the main casing. There have been numerous accidents where the test plug has blown out during removal when the main gate valves were leaking and the meter was under pressure. Consider this situation in utilities where the line pressures are 150 psi, or greater. *Be careful!*

Nearly all turbine meters produced prior to 1992 offered no test outlet in the meter body and required a separate spool piece and test nipple installation. When test plugs are fitted in the meter bodies, a separate test tap becomes unnecessary. These test outlets range from 1 to 2 in in size, depending on the meter sizes. Additionally, some commercially available fire line meter assemblies and compound meters are provided with a test riser outlet assembly with a locking ball valve and fire hose coupling for proper testing. See Figs. 14.1–14.3 for installation recommendations for turbo, compound, and fire line meters.

14.4.3 Testing Procedures

Prior to testing it is necessary to know what the typical accuracy curve is for each specific brand, model, and size of meter being tested. This infor-

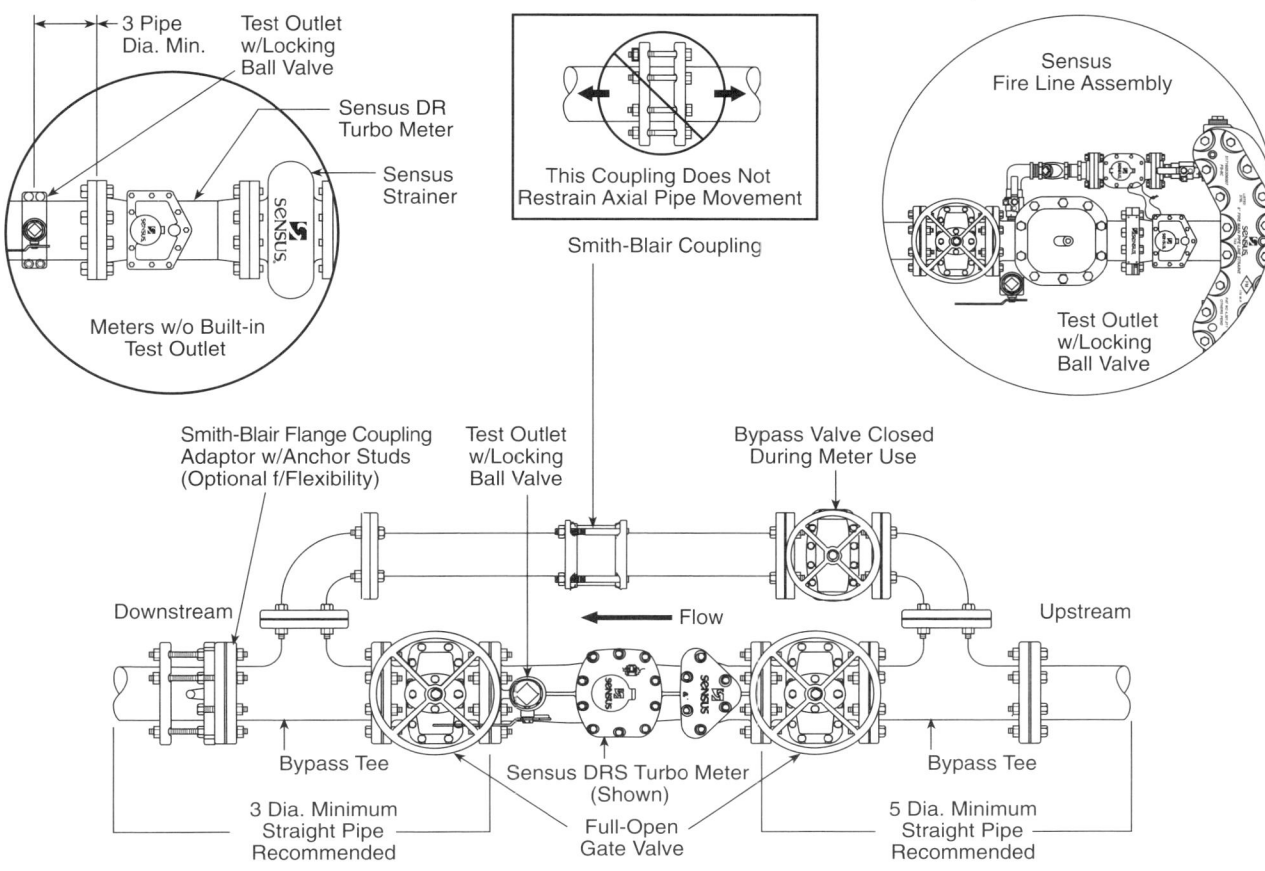

Figure 14.1 Installation recommendations for turbo meters. [*Source: Invensys Metering Systems (formerly Sensus Technologies).*]

mation may be obtained from the meter manufacturer's literature. A local chart can be made up which lists the flow rates at which each type of meter should be tested in order to properly assess its operating condition.

For positive displacement meters, AWWA Manual M6 provides the three flow rates (low, mid, and high) which apply to all meter brands. For turbine and propeller meters, either Manual M6 or the manufacturer's meter literature should be consulted (Fig. 14.4). For compound meters, it is important to know where "crossover" is located so that it can be specifically tested. The manufacturer's accuracy curve is a proper information source, as different brands of compounds offer variant crossover flow rates. This information is not currently available in AWWA Manual M6.

The various suppliers of large meter test equipment provide detailed procedures for conducting accuracy tests. In general, the tester is connected and the line flushed. The preliminary test should be at a relatively high flow to determine if there are any leaks or unknown taps in the pipeline. The flow rate should be set approximately to 50 percent of the meter's capacity and the test conducted for 10 sweeps of the dial for adequate resolution. After determining the accuracy of the meter, the test should be rerun for half the volume. The second accuracy test's results

348 CHAPTER 14 RESOLVING APPARENT LOSSES

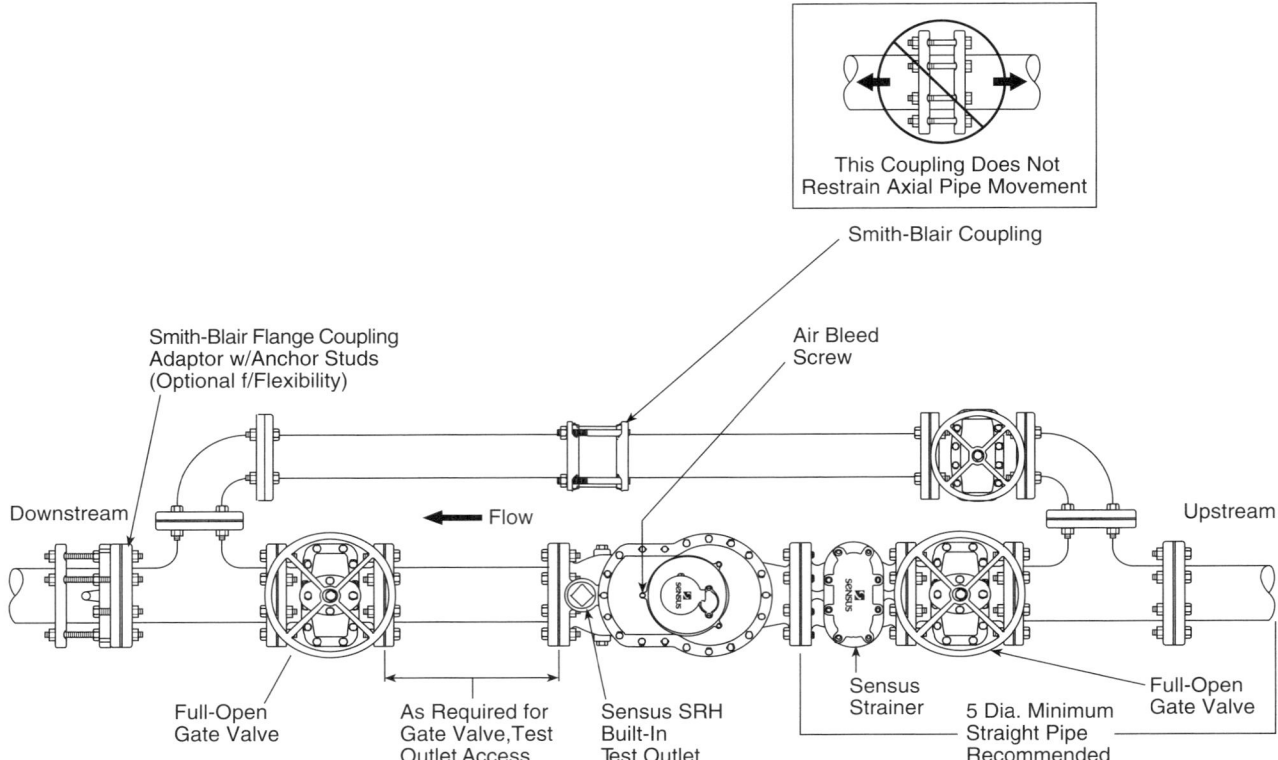

Figure 14.2 Installation recommendations for compound meters. [*Source: Invensys Metering Systems (formerly Sensus Technologies).*]

should be within one-half percent of the first test. If not, a leak or other uncontrolled flow of water should be suspected. If the meter in question has a flow indicator, it may indicate water movement as a result of downstream isolation valve leakage.

The Test Site

The test site should be safe and allow adequate dissipation for the test water being discharged. Determine how to disperse water that will be used during the testing operation. Again, it is typical to discharge 5000 to 10,000 gal of water during a test sequence.

Examine the test site for available test plugs, bypass piping, and upstream and downstream isolation valves. Make all necessary shut-offs and connect the tester according to the manufacturer's recommendations.

Prior to running any test, determine the make, model, and manufacturer of the meter in question and document these data on the meter test sheet (Fig. 14.5).

Field Testing

Field testing is usually confined to meters larger than 2 in in size and is recommended for all sizes of current and compound types. Few meter shops are equipped with tank facilities for the quantities of water needed to test these larger meters. Furthermore, the accuracy of some current and compound meters may be affected by the configuration of pipe and fit-

> **S**afety should be considered, especially when testing large meters, as large volumes of water are discharged.

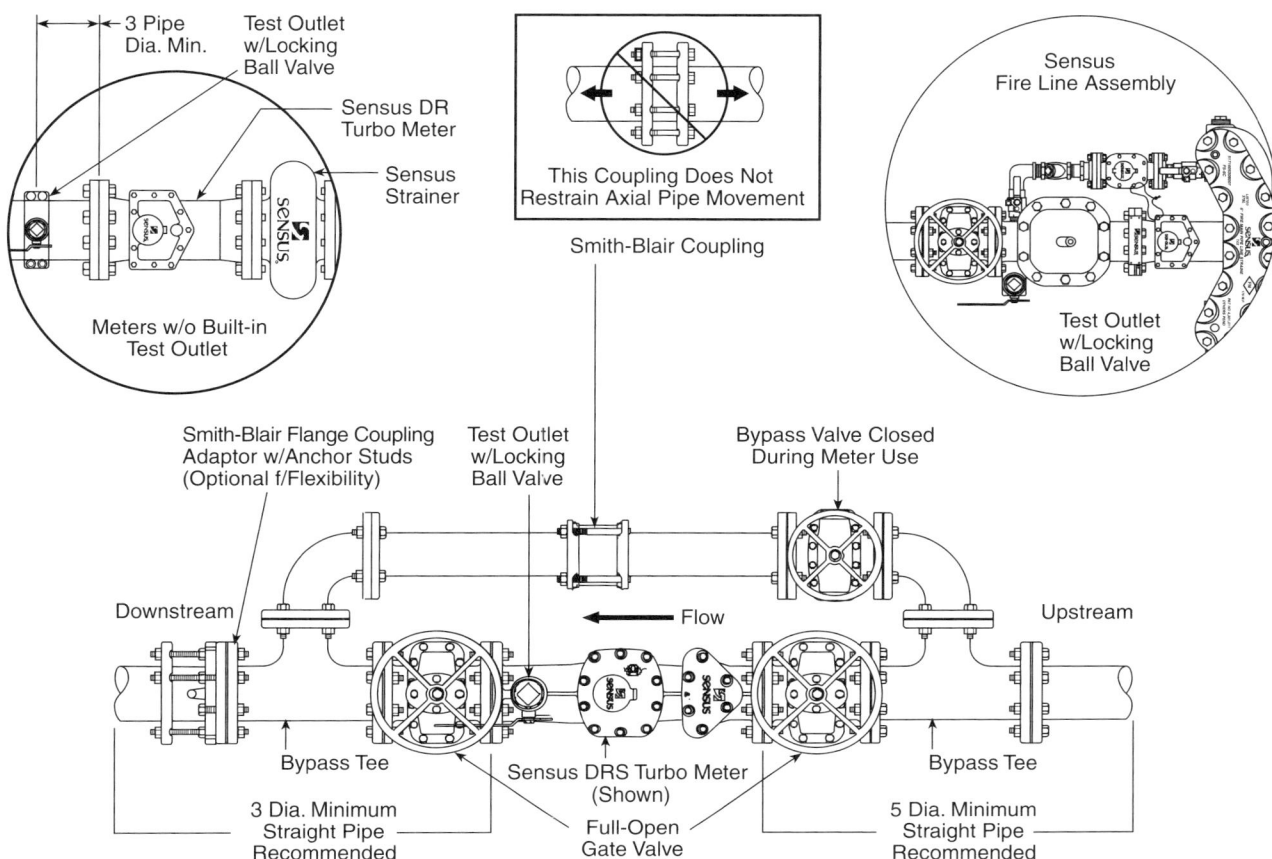

Figure 14.3 Installation recommendations for fire line meters. [*Source: Invensys Metering Systems (formerly Sensus Technologies).*]

tings directly ahead of the meter. Field testing of meters is essentially the same as shop testing, except that instead of using a tank to measure the test water, a comparison is made between the meter to be tested and one that has been previously calibrated (known to be accurate). The two meters are connected in series, and the test water is discharged to waste (as the calibrated meter is not 100 percent accurate on all flows, it may be necessary to adjust for its accuracy variance at different rates of flow, to ensure proper test results).

One very important point to remember in field testing is that both meters must be full of water and under positive pressure. The control valve for regulating flow, therefore, should always be on the discharge side of the calibrated meter. A valve on the inlet side of the meter being tested or one located between two meters for controlling rates of flow should not be used, as inaccurate results may be obtained.

The discharge of test water often presents a problem in field testing of meters. An additional problem is the selection of a test meter with sufficient capacity to deliver the high rates of flow required for maximum flow test rate. It is often necessary to use a rate lower than that set forth for the maximum flow test of larger meters, and in many instances a maximum of 500 gpm or less can be used. This rate, however, is usually

		DISPLACEMENT METERS (AWWA C700)										
Size in.	Maximum Rate (All Meters)				Intermediate Rate (All Meters)			Minimum Rate (New and Rebuilt) ①			Maximum (Repaired) ①
	Flow Rate gpm	Test Quantity		Accuracy Limits percent	Flow Rate gpm	Test Quantity		Accuracy Limits percent	Flow Rate gpm	Test Quantity		Accuracy Limits percent	Accuracy Limits percent (min.)
		gal.	ft.³			gal.	ft.³			gal.	ft.³		
5/8	15	100	10	98.5 - 101.5	2	10	1	98.5 - 101.5	1/4	10	1	95 - 101	90
5/8 × 3/4	15	100	10	98.5 - 101.5	2	10	1	98.5 - 101.5	1/4	10	1	95 - 101	90
3/4	25	100	10	98.5 - 101.5	3	10	1	98.5 - 101.5	1/2	10	1	95 - 101	90
1	40	100	10	98.5 - 101.5	4	10	1	98.5 - 101.5	3/4	10	1	95 - 101	90
1-1/2	50	100	10	98.5 - 101.5	8	100	10	98.5 - 101.5	1-1/2	100	10	95 - 101	90
2	100	100	10	98.5 - 101.5	15	100	10	98.5 - 101.5	2	100	10	95 - 101	90

	Size in.	Maximum Rate				Intermediate Rate				Minimum Rate			
		Flow Rate gpm	Test Quantity		Accuracy Limits percent	Flow Rate gpm	Test Quantity		Accuracy Limits percent	Flow Rate gpm	Test Quantity		Accuracy Limits percent
			gal.	ft.³			gal.	ft.³			gal.	ft.³	
CLASS I TURBINE METERS (AWWA C701)	1-1/2	80	200	20	98 - 102	35	100	10	98 - 102	12	100	10	98 - 102
	2	120	300	30	98 - 102	50	200	20	98 - 102	16	100	10	98 - 102
	3	250	500	50	98 - 102	75	300	30	98 - 102	24	100	10	98 - 102
	4	400	1000	100	98 - 102	125	500	50	98 - 102	40	100	10	98 - 102
	6	1000	2000	200	98 - 102	200	500	50	98 - 102	80	1000	100	98 - 102
	8	1500	3000	300	98 - 102	300	1000	100	98 - 102	140	1000	100	98 - 102
	10	2200	5000	500	98 - 102	500	1000	100	98 - 102	225	1000	100	98 - 102
	12	3300	7000	700	98 - 102	700	2000	200	98 - 102	400	1000	100	98 - 102
CLASS II TURBINE METERS (AWWA C701)	1-1/2	90	300	30	98.5 - 101.5	10	100	10	98.5 - 101.5	4	100	10	98.5 - 101.5
	2	120	300	30	98.5 - 101.5	10	100	10	98.5 - 101.5	4	100	10	98.5 - 101.5
	3	275	600	60	98.5 - 101.5	20	100	10	98.5 - 101.5	8	100	10	98.5 - 101.5
	4	500	1000	100	98.5 - 101.5	20	1000	100	98.5 - 101.5	15	100	10	98.5 - 101.5
	6	1100	2500	250	98.5 - 101.5	40	1000	100	98.5 - 101.5	30	1000	100	98.5 - 101.5
	8	1800	4000	400	98.5 - 101.5	50	1000	100	98.5 - 101.5	50	1000	100	98.5 - 101.5
	10	3000	6000	600	98.5 - 101.5	75	1000	100	98.5 - 101.5	75	1000	100	98.5 - 101.5
	12	4000	8000	800	98.5 - 101.5	100	1000	100	98.5 - 101.5	120	1000	100	98.5 - 101.5
		Maximum Rate				Intermediate Rate ②				Minimum Rate			
COMPOUND METERS (AWWA C702) (Test at intermediate rate not necessary.)	2	100	100	10	97 - 103	10 - 15	100	10	90 - 103	1/4	10	1	95 - 101
	3	150	500	50	97 - 103	10 - 15	100	10	90 - 103	1/2	10	1	95 - 101
	4	200	500	50	97 - 103	20 - 25	100	10	90 - 103	3/4	10	1	95 - 101
	6	500	1000	100	97 - 103	25 - 35	100	10	90 - 103	1-1/2	100	10	95 - 101
	8	600	2000	200	97 - 103	35 - 45	100	10	90 - 103	2	100	10	95 - 101
	10	900	2000	200	97 - 103	—	—	—	90 - 103	4	100	10	95 - 101

FIRE-SERVICE TYPE (AWWA C703) ③

	TURBINE MAIN LINE TYPE WITH BY-PASS ③							TURBINE MAIN LINE TYPE WITH BY-PASS ③						
Meter Size in.	Minimum Rate (95 percent min. accuracy limit)		Cross-Over Rate (90-103 percent accuracy limit)		Maximum Rate (98.5-101.5 percent accuracy limit)			Meter Size in.	Minimum Rate (95 percent min. accuracy limit)		Intermediate Rate (98.5-101.5 percent accuracy limit)		Maximum Rate (98.5-101.5 percent accuracy limit)	
	Flow Rate gpm	Test Quantity gal. / ft.³	Flow Rate gpm	Test Quantity gal. / ft.³	Flow Rate gpm	Test Quantity gal. / ft.³			Flow Rate gpm	Test Quantity gal. / ft.³	Flow Rate gpm	Test Quantity gal. / ft.³	Flow Rate gpm	Test Quantity gal. / ft.³
4	④	100 / 10	25 - 35	1000 / 100	750	2000 / 200		4	10	1000 / 100	20	1000 / 100	750	2000 / 200
6	④	100 / 10	50 - 60	1000 / 100	1500	5000 / 500		6	20	1000 / 100	40	1000 / 100	1500	5000 / 500
8	3	100 / 10	50 - 60	1000 / 100	2500	5000 / 500		8	30	1000 / 100	50	1000 / 100	2500	5000 / 500
10	3	100 / 10	55 - 65	1000 / 100	4000	8000 / 800		10	35	1000 / 100	75	1000 / 100	4000	8000 / 800

① A rebuilt meter is one that has had the measuring element replaced with a factory-made new unit. A repaired meter is one that has had the old measuring element cleaned and refurbished in a utility repair shop.

② Cross-over flow rates vary depending on meter model and brand. These values are for Sensus (Rockwell) Compound Meters. Consult manufacturers for other brands.

③ The values listed are for Sensus meters only.

④ Flow rate for FireLine 1-1/2 – 3" gpm depending on bypass meter. Flow rate for UL/FM Compact at 3 gpm.

Figure 14.4 Test flow rates. [*Source: Invensys Metering Systems (formerly Sensus Technologies).*]

sufficient to indicate a large meter's accuracy in the higher range of flows. Obviously, lesser flow rates should be used only as an expedient, and the established test rates should be used wherever possible. *It is safe to assume that the test curve will "flatten out" after reaching peak registration, which is approximately 10 percent of the meter's rated capacity. Stay within required limits for registration.*

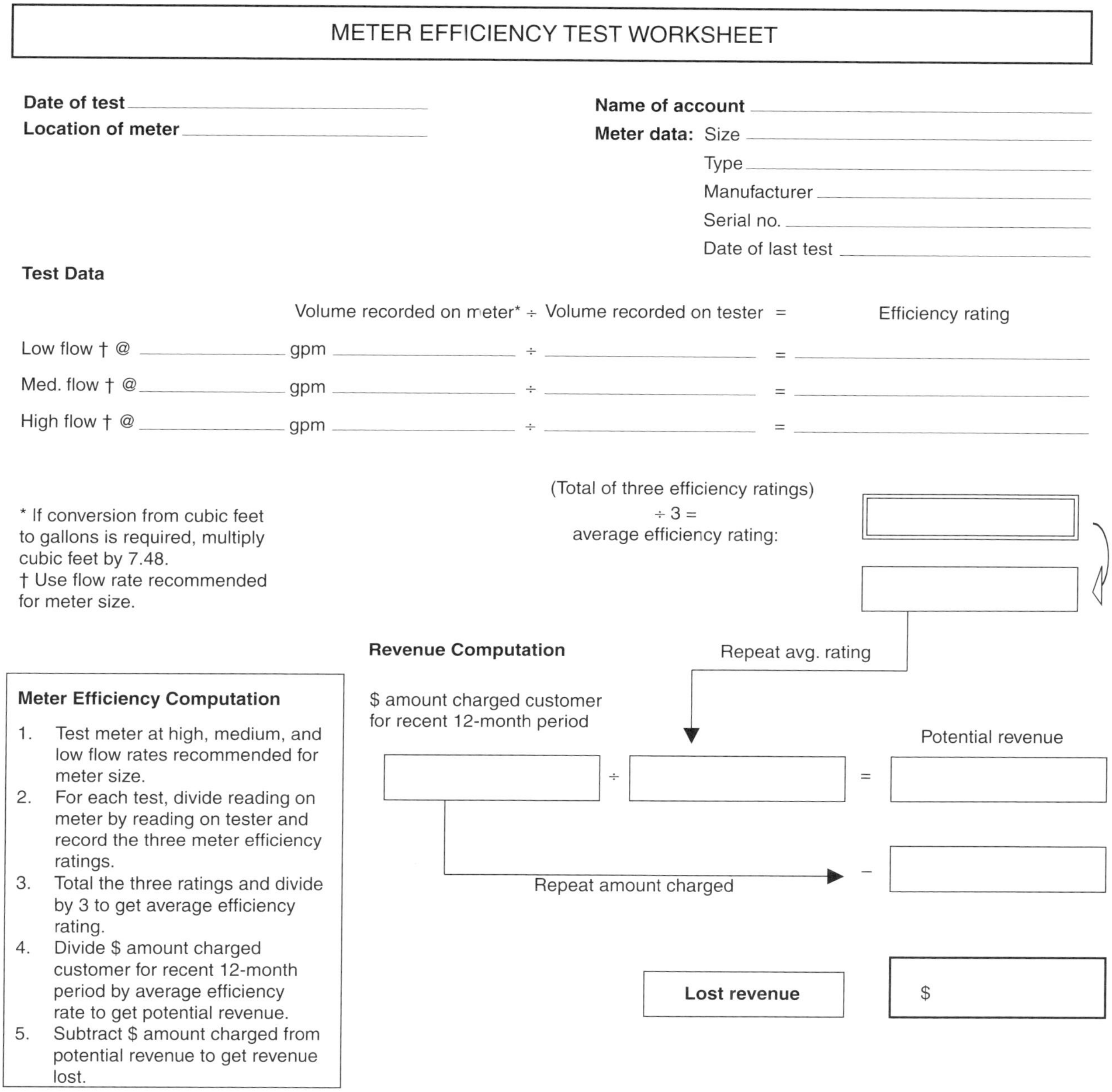

METER EFFICIENCY TEST WORKSHEET

Date of test _____
Location of meter _____

Name of account _____
Meter data: Size _____
Type _____
Manufacturer _____
Serial no. _____
Date of last test _____

Test Data

Volume recorded on meter* ÷ Volume recorded on tester = Efficiency rating

Low flow † @ _____ gpm _____ ÷ _____ = _____
Med. flow † @ _____ gpm _____ ÷ _____ = _____
High flow † @ _____ gpm _____ ÷ _____ = _____

* If conversion from cubic feet to gallons is required, multiply cubic feet by 7.48.
† Use flow rate recommended for meter size.

(Total of three efficiency ratings) ÷ 3 = average efficiency rating: _____

Repeat avg. rating

Revenue Computation

$ amount charged customer for recent 12-month period

_____ ÷ _____ = _____ Potential revenue

Repeat amount charged

_____ − _____

Lost revenue | $ _____

Meter Efficiency Computation

1. Test meter at high, medium, and low flow rates recommended for meter size.
2. For each test, divide reading on meter by reading on tester and record the three meter efficiency ratings.
3. Total the three ratings and divide by 3 to get average efficiency rating.
4. Divide $ amount charged customer for recent 12-month period by average efficiency rate to get potential revenue.
5. Subtract $ amount charged from potential revenue to get revenue lost.

Figure 14.5 Meter efficiency test worksheet. [*Source: Invensys Metering Systems (formerly Sensus Technologies).*]

Following is a recommended large-meter maintenance program.

1. Make a meter survey and inspection.
   a. Are there test plugs?
   b. Is there a bypass?
   c. Are there isolating valves on both sides of the meter?
2. Test the meter for accuracy.
   a. Begin the test from 0 gpm.

b. Repair, if necessary.
   c. Change the meter, if necessary.
3. Record the usage rate for 48 h
   a. Compare the 48-h usage rate with the test accuracy results.
   b. Review the usage rate at the changeover point with the accuracy of the meter at that flow rate.
   c. Change the meter size, if necessary.
4. Prepare a long-term surveillance program:
   a. Manual surveillance
   b. Computer surveillance
5. Inspect and test the meter whenever the performance is not consistent with past performance history accumulated after steps 1–4. When meter performance is not consistent, the meter should be inspected for:
   a. A change in customer usage rates
   b. Meter malfunction

Points to Consider

When testing a meter on-site, it should be remembered that the basic concept is to compare the accuracy of the meter in question with the calibrated meter tester. The calibrated meter has its own performance characteristics and is not 100 percent accurate across its entire flow range, so it should have an available compensation curve describing this.

The testing sequence is usually from the low flows to the higher flows. Experience has shown that when most meters begin to wear or lose accuracy, it is at the lower rather than the higher flows. If a meter is performing accurately up through the lower 25 percent of its capacity, it will normally test accurately through the rest of the range. This is especially true of the very large meters. Another item to consider is what to do with the discharge water after it has gone through the tester. One thousand gallons per minute is a lot of water. A sudden high flow rate such as this could reduce the supply pressure available or disturb any debris in the service line to the customer. It could also reduce the water supply in a nearby portion of the system, or cause considerable damage if the water is not discharged properly.

When conducting tests, it is suggested that no test be less than 1 min in duration and that the meter's sweep hand make at least one complete revolution. The residual pressure on the tester should never be less than 20 psi when running a high-flow test. Also, for safety, the tester should not be operated on lines with static pressure exceeding 80 psi unless provisions are made to secure the tester.

Detailed records are important so that trends in performance can be monitored, along with the accuracies obtained at the various flow rates. Remember to record the meter's registration before and after the testing, so the customer is not charged for the water used during the test.

The local utility should establish and determine how inaccurate a meter can become before repairs or replacement is required. A cost-versus-benefit analysis is required, based on factors important to the local water system management. Considerations should also include the sizing and selection of the replacement meter. Another factor might be the updating of the meter installation for future on-site testing.

14.5 FIELD METER TESTING—A CLOSER LOOK

The test site should be looked over for a safe run-off area for test water. Make all necessary shut-offs and hook up test meter.

- ❏ Set up equipment and begin testing.
- ❏ Slowly flush all air until maximum flow is reached. (Maximum flow is achieved when either the valve is wide open or the tester pressure gauge drops to 20 psi.)
- ❏ Read and reset the registers and run the maximum flow as previously run for a quantity of at least one sweep of the dial on the meter being tested.
- ❏ Repeat the above test, increasing the quantity by 2 times the quantity of the first test. Compare the first test percentage to the second. The difference should not be greater than ±5 percent. If it is greater, review these possible causes for the difference:
  1. Tested meter is in need of repair. Running of low flows may confirm this suspicion.
  2. Tested meter may have a badly worn register causing excessive pointer play. Tapping on the register lens and observing the amount of pointer movement can confirm this.
  3. Air trapped in line. Flush line and rerun tests.
  4. Isolation valve(s) leaking, causing inconsistent tests. Check by looking at tested meter's register low-flow indicator for movement over a 1–5-min period.
  5. Strainer clogged with debris or partially blocked.
  6. Test meter may be clogged with rocks, debris, or damaged during flushing.
- ❏ There are many other causes for inconsistency of test data. Inconsistency problems must be resolved before continuing or there will be no validity to the test results.
- ❏ Continue testing, referring to AWWA Manual M6 (Fig. 2; see App. B for more details on the M6) for test rates on specific meter types, and/or it is preferred to use the meter manufacturer's suggested test rates. Remember that one additional test rate to run is the average customer flow usage rate. This will give you a good feel for how efficiently the meter is operating at revenue-producing area flow rates.
- ❏ Compound meters require a test at a crossover flow rate. To achieve the exact crossover rate will require the use of the pressure gauge and the rate of flow display. Slowly open the rate control valve. When a rise in the pressure gauge needle is noted, the rate indicated by the register is crossover. If crossover is not detected, close the rate valve until the gauge drops back again. Repeat opening and increasing flow until crossover is obtained. This procedure will take practice but is worth the effort.

> **D**o not let other people intimidate you during your testing. Follow the instructions, and let the test results speak for themselves!

14.5.1 The Evaluation

Evaluating a water meter requires both experience and confidence in the operator's skill and training. This section is intended to address the correlation of the test data to the meter. Years of experience are necessary on some meters to properly correlate testing with a conclusive "fix" to the meter. This section provides tests and causes for Sensus meters only.

To evaluate a tested meter, one must have the utmost confidence in one's test data. Hopefully, all of the previous information in this section has been used to obtain the data to be reviewed. Sensus allows a $\pm 1\frac{1}{2}$ percent accuracy spread on turbo meters, fire line meters, and compound meters tested at normal operating ranges. At low flows and crossover, Sensus allows $\pm 1\frac{1}{2}$ percent to -5 percent accuracy, which exceed AWWA standards, prior to setting an independent standard. See AWWA C701, C702, C703, and C704 for additional guidelines.

Points to remember when reviewing meter test data include the following:

- Always run low, medium, and high flows (manufacturer's or AWWA recommendations).
- Always run a minimum of one sweep on the tested meter register.
- Review normal operating range tests. Is the minimum flow test 95 to 101.5 percent? If either is not within the range, the meter should not be geared or adjusted to meet specifications without repair. Complete replacement may be required.
- Turbo meters show loss of registration first at low flows, due to bearing wear. Make sure that a good job has been done in testing at low flow rates.
- Compound and fire line meters have a crossover flow. Take time when testing to determine this rate. Evaluate each side of crossover as a cause for failure along with valve problems. Do not attempt to isolate measuring chambers and conduct isolated tests.

Again, you, as the expert, will need to use all of your training and expertise to evaluate and diagnose tested meters, therefore, never stop looking, listening, and learning.

See Table 14.1.

The 10 Rules of Successful Testing*

1. Maintain personal and workplace safety at all times.
2. Keep hoses straight while in use.
3. Adequately purge all entrapped air.
4. After isolation valves are closed, check for register "creep" (slow advancement).
5. Always read register twice.
6. Never test when equipment is in need of repair or calibration.
7. Always check equipment before and after both internal and external damage.
8. Take your time. Do not let others pressure you into making a mistake.
9. Always rerun questionable accuracies and increase the quantity of the rerun test.
10. For additional assistance, contact your local meter supplier.

Source: Invensys Metering Systems (formerly Sensus Technologies).

TABLE 14.1 Meter Evaluation

Turbo Meter Evaluation

Meter Size	Adj. Vane	Test Data	Possible Cause
4-in W-1000	+15°	90% @ gpm 97% @ 100 gpm 99% @ 700 gpm	Broken rotor blades Rotor bearings and/or thrust bearings worn Debris caught on blade
4-in W-1000	+5°	100% @ 10 gpm 103% @ 100 gpm 105% @ 700 gpm	Jetting from debris (in strainer or caught on flow strainer) Installation effects Air entrapped in line Coating on rotor and/or chamber
4-in W-1000	+30°	94% @ 10 gpm 98% @ 100 gpm 99% @ 700 gpm	Adjusting vane moved to (−) from original test Installation effects Improper repair

This meter could be recalibrated by moving the value to 0°

Compound Meter Evaluation

Meter Size	Test Data	Possible Cause
3-in SRH	**Low-Flow Tests** 105% @ 0.5 gpm 102% @ 3.4 gpm 98% 10 gpm	Leaking downstream isolation valve
3-in SRH	**High-Flow Tests** 88% @ 25 gpm 94% @ 55 gpm 99% @ 280 gpm	Damage to propeller High-flow chamber wear Coordinator wear Vertical shaft binding and/or bushing wear
3-in SRH	95% @ 0.5 gpm 99% @ 3.4 gpm 100% @ 10 gpm 106% @ 25 gpm 108.7% @ 150 gpm 108% @ 280 gpm	High-flow side geared too high Debris causing jetting Installation effects

Fire Line Meter Evaluation

Meter Size	Test Data	Possible Cause
6-in compact fire line	192% @ 4 gpm 96% @ 45 gpm 99% @ 500 gpm	**Bypass Meter** Broken rotor blades Adjusting vane moved to (−) Rotor bearings and/or thrust bearings worn Debris caught in blade **Detector Check Valve** Worn seat Debris preventing closure
6-in compact fire line	100% @ 4 gpm 100.3% @ 45 gpm 105% @ 500 gpm	**Large Meter** Jetting from strainer and/or installation Adjusting vane moved to (+) Coating on rotor and/or chamber
6-in compact fire line	100% @ 4 gpm 103.3% @ 45 gpm 101% @ 100 gpm 100% @ 500 gpm	Leaking downstream isolation valve **Bypass Meter** Adjusting vane moved to (+) Coating on rotor and/or chamber

Source: Invensys Metering Systems (formerly Sensus Technologies).

14.5.2 Revenue Meter Replacement

The decision to replace revenue meters is often made on an age basis, with many states indicating a preferred change-out time by age or volume. However, the decision should really be made on an economic basis. Simple economic models can be made in house using the variables listed below. More detailed commercial models are available to allow more in-depth modeling.

Revenue meter data, whether from small residential or large commercial meters, will become less accurate as a function of

- Stopped meters
- Bypassed or nonmetered clandestine accounts
- Debris
- Air
- Overly high velocity
- Very low velocity
- Temperature
- Incorrect installation
- Chemical buildup
- Human error

Obviously the variables of time and volume should be applied to the above factors.

It is a common-sense decision in most cases to test meters periodically as discussed above, particularly large consumer meters which in many cases provide a substantial part of the revenue income for a utility. In addition to testing the meters, care should be taken to ensure that the meters are properly sized.

Sizing Meters

When sizing meters, it is very important not to base the decision on only 24 h of data. A customer's consumption can vary greatly throughout the day, week, and year, for various reasons. Care should be taken to locate seasonal use information and also to understand the type of consumption in each specific case.

Commercial and Industrial

For commercial properties that do not operate on weekends, the largest variations will normally be between weekday use and weekend use. Seasonal consumption changes generally depend on the type of manufacturing, but if personnel use most of the water, seasonal variations are normally fairly small, other than at holiday times. If, on the other hand, most of the water is used in process, the variation will depend very much on the type of process which is being run.

Plants often shut down for vacation, and care should be taken that profiles are not taken during these periods.

Residential Properties

Residential properties often have a big seasonal change to reflect the hot weather. Often more than 50 or 60 percent of residential consumption in industrialized nations in medium to high-class residences is for outdoor

> **S**easonal use variations should be checked carefully when sizing meters.

irrigation. Care needs to be taken when sizing the meter to understand the type of consumer/area (normal—residential high, medium, or low, or vacation property).

In cities, consumption is usually higher during the summer months, excepting holiday times when many people leave for vacation areas. In these cases, much of the additional water is used for additional showers and also use of garden hoses. This type of additional use tends to occur at peak times and can have a big impact on any potential meter downsizing decision.

When looking at consumption profiles in coastal areas, extra care must be taken to consider the season. For a large apartment block in such an area, the occupancy rate might change from 10 to 100 percent between winter and summer. However, most of the use will still occur at peak times, as people prepare for the day or evening ahead.

In general, it is essential to collect as much data as possible, in the form of data logged profiles and volumetric billing data. In addition, we can improve decision making by categorizing users and understanding end usage. A good meter sizing decision can have a big impact on revenue and nonrevenue water figures, and a poor decision can be very costly in time, money, and public opinion!

> During the demand analysis phase of any loss management program, customers should be categorized for ease of data analysis and subsequent tracking of results.

14.5.3 Tilted Meters

Another common reason for meter inaccuracy is meters which are tilted. Meters are often tilted to one side to allow the meter reader easier access to read the meter. This procedure must be avoided at all times, as this will in most cases cause inaccuracy of up to and sometimes in excess of 5 percent.

> Tilted meters cause errors; meters should be installed as per the manufacturer's recommendations.

The following case study shows the results of a major meter sizing program in Boston, Massachusetts. Kind permission to reprint the study was given by AWWA and the Boston Water and Sewer Commission (BWSC). Many figures presented in the original version of this article have not been reproduced here for size reasons. Those readers who would like to look at these figures should refer to the original article or contact AWWA.

14.6 CASE STUDY ONE

*Proper Meter Sizing for Increased Accountability and Revenues**

John P. Sullivan, Jr., P.E., Chief Engineer, Boston Water and Sewer Commission

Elisa M. Speranza, Special Project Manager, Boston Water and Sewer Commission

14.6.1 Background

Each year in the water industry, billions of gallons of water are "lost." The American Water Works Association Research Foundation (AWWARF) estimates revenue losses due to "unaccounted-for water" range from $158

*Adapted from *Proceedings of 1991 AWWA Annual Conference,* by permission. Copyright © 1991, American Water Works Association.

million to $800 million nationwide. The problem of unaccounted-for water, which has been the subject of dozens of studies, reports, and books, can basically be summarized as follows:

1. All of the water purchased does not reach its intended destination and
2. The retailer is never paid for some of the water which does reach its intended destination

The Boston Water and Sewer Commission (BWSC, the Commission) provides retail water and sewer services to over 1 million people who live and work in the City of Boston. The BWSC is the largest customer of the Massachusetts Water Resources Authority (MWRA), a regional authority that provides wholesale water and sewage treatment to 60 communities.

Because the Commission purchases about 40 percent of the total water sold by the MWRA, when Boston's water distribution system is losing water, an artificially high demand in efforts, including leak detection and repair, are aimed at reducing the amount of water lost between the time the Commission purchases it and the time it is sold to the customers. Through these efforts, the Commission makes an expensive contribution toward the goal of avoiding expensive and environmentally damaging water augmentation projects in the future

Just a few years ago, the diversion of the Connecticut River to supplement Greater Boston's water supply seemed like a certainty. Through aggressive "demand management" programs, demand on MWRA water system has been reduced from 317.2 million gallons per day (mgd) in 1976 to 290 mgd in 1990. The latter figure is well below the system "safe yield" of 300 mgd. As a result, the Connecticut River diversion project has been placed on hold indefinitely.

The second problem, unbilled water usage, is a potential untapped revenue source for the Commission. In these times of fiscal austerity, the BWSC's ratepayers must be assured that all customers are paying their fair share, and that the Commission is maximizing its income to meet the rising costs of providing water and sewer services.

As shown in Fig. 14.6, from 1976 to 1990, water consumption in Boston dropped by 26 percent, from 150 to 110.2 mgd. Between 1998 and 1990, the Boston Water and Sewer Commission brought water consumption down by 9.3 percent from the 1988 figure of 121.5 mgd, to 110.2 mgd. The dramatic drop in water usage can be credited to the Commission's aggressive leak detection, repair, and other water conservation programs. Unaccounted-for water—the difference between the amount of water purchased from the MWRA and the amount billed to BWSC customers— dropped by 18 percent, from 33 to 27 percent of the total.

14.6.2 Unaccounted-for Water Task Force

While the consumption decrease in Boston is significant, the city's unaccounted-for water percentage is still unacceptably high. According to industry studies, unaccounted-for water values of 20–30 percent are not uncommon for older systems, particularly those in the northeast. However, the Commission believes a concerted, agency-wide focus can significantly reduce unbilled water, even in an older urban system.

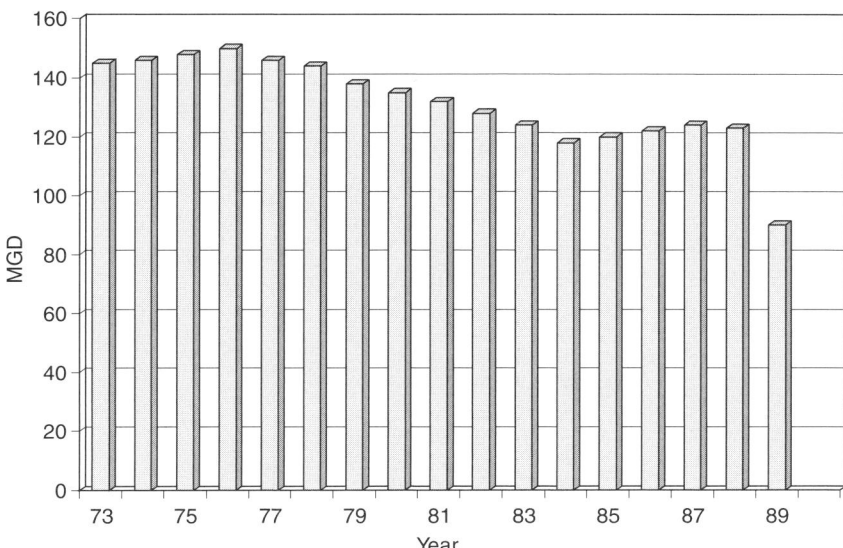

Figure 14.6 From 1975 to 1990, water consumption in Boston dropped by 26 percent, from 150 to 110.2 mgd. During the period of time shown by the graph, the minimum night flow is 44.629 mgd and the average daily flow is 79.815 mgd, a ratio of 56 percent. (*Source: Boston Water and Sewer Commission.*)

In response to the challenge of accounting for more of the water the Commission purchases, the Executive Director formed an Unaccounted-for Water Task Force in March 1990 to conduct a comprehensive review of the source of unaccounted-for water, and to investigate potential strategies to address this issue. The task force was unusual in that it included staff from various departments which had not necessarily dealt with the question of unbilled water in the past, including field services, meter installation, billing, water operations, planning, and engineering services.

Unaccounted-for water had been previously reviewed as part of BWSC's Water Distribution Study conducted by Camp, Dresser and McKee on behalf of the Commission in May 1987. Based on 1985 data, the consultants determined that there was a gap of about 32 percent between what the Commission purchased from the MWRA and what it billed its customers. The study acknowledged the downward trend in unbilled water since the Commission's creation in 1977, and described various efforts—particularly leak detection and repair—which resulted in a reduction of water purchased and an increase in water billing.

The study also identified various reasons for unbilled water, including: metering and billing problems, unmetered consumer use, unmetered public use, unavoidable leakage, and potentially recoverable leakage. Therefore, a percentage of what was termed "unaccounted-for" water was, in fact, being used for legitimate purposes, but was not being identified as such, nor was it necessarily being billed.

At that time, potentially recoverable leakage represented the largest share of unaccounted-for water at 18.5 mgd—about 49 percent of the 38.1

mgd total unaccounted-for water and 16 percent of the total average daily water purchase of 119 mgd. The task force estimated that potentially recoverable leakage still represents almost half of all unaccounted-for water.

In December 1990, the Task Force issued its first report. Many specific recommendations were made, including suggested revisions to metering, billing, and record-keeping practices, new efforts to improve water accountability, and the continuation of successful programs such as leak detection. The task force's recommendations in the metering area dealt with proper sizing, reading, slippage, repair, and replacement of meters. This case study focuses on one of these issues, proper meter sizing.

14.6.3 Past Metering Practices

The BWSC has over 86,000 meters in service, about 10 percent of which are larger than $1\frac{1}{2}$ in. Past metering practices generally required that meter size be determined by the size of the supply pipe. Thus a 1-in meter was installed on a 1-in service pipe, a 2-in meter on a 2-in pipe, etc. The pipe size was determined using applicable plumbing codes, taking into account total required volumes and maximum allowable pressure drop through the pipe.

These extremely conservative calculations, made by developers, often resulted in the installation of meters which were larger than was needed. It should be noted that many older fixtures used more water than their modern counterparts, so usage assumptions were, perhaps correctly, higher.

Because the cost of water was so low, the city was not generally concerned about missing water at low flow rates through oversized meters. The major concern was to guarantee no additional pressure loss through the meter. On pipe sizes 3 in and larger, the common practice was to install a compound meter, a complicated mechanical device capable of recording low, moderate, and high flow. The meter consisted of a small, $\frac{5}{8}$-in to 3-in metering device and a larger, 3-in to 8-in turbine-type meter, which would work together to record the total flow.

The compound meters actually worked quite well. Unfortunately, however, preventive maintenance programs were inadequate and the compound meters fell into disrepair. By 1974, most of the larger compound meters were partially or totally malfunctioning. Based on historical repair records and a cost comparison of turbine to compound meters, the city decided to replace the 3-in and larger meters with new, state-of-the-art turbine meters. The newer turbine meters were far less complicated and easier to maintain, but they could not accurately register flows less than 5 gpm for 3-in meters, 10 gpm for 4-in meters, and 20 gpm for 6-in meters. It was generally assumed that most of the flow in a building serviced by these larger meters would fall into the meter's range. The amount of water used at lower flow rates was unknown and was not a factor in determining meter type and size.

In 1976, the combined water and sewer rate was about $1.02/1000 gal. Since then, the combined water and sewer rate has increased by 429 percent to $5.40/1000 gal in 1991. In 1985, when the MWRA took over the ailing metropolitan water and sewer system, rates began to rise sharply. Massive capital projects such as the $7 billion Boston Harbor cleanup and a proposed water filtration plant will drive rates up even further over the next 10 years. By the year 2000, the BWSC predicts that the average family in Boston will be paying $14.40/1000 gal—over $1000 in annual water and sewer bills. Consequently, the cost of water has gone from being an insignificant factor in meter sizing to being an extremely important consideration.

In 1988, the Commission began to investigate the possibility of downsizing meters. All new accounts generally have been required to install a meter which is one size smaller than the nominal pipe size, and developers are required to submit forecasted water demands. Although the Commission recognizes that compound meters could accomplish the goal of accounting for water at all rates of flow, experience and judgment dictate that, in most cases, simpler devices would better serve the Commission's needs. Because there was no available methodology for evaluating whether existing meters were properly sized, the BWSC developed a pilot project to address this issue.

14.6.4 Project Approach

The problem confronting the Commission in implementing the Task Force's recommendations was how to account for more water used at low rates of flow, without violating the customer's high-end flow requirements. The project approach developed was based on the theory that (1) water lost at low flow rates was significant enough to warrant a major effort to recover it and (2) people do not use water at previously assumed rates.

Project Team

The Commission formed a meter downsizing project team as a subcommittee of its Unaccounted-for Water Task Force. The team consisted of staff from the engineering services, field services, meter installation, and meter reading divisions.

Since August 1990, the project team has met every Monday morning to coordinate its efforts. At these meetings, staff bring cases to the table from various sources and agree on the proper meter size for a particular customer. Water requirements are analyzed, potential problems are discussed, and proper meter sizes are assigned based on the operating ranges of different-sized meters.

Among the factors taken into consideration was that 2-in disc meters, in contrast to the older turbine meters, can accurately register as low as 2 gpm and will record flow as low as 1 gpm with 95 percent accuracy. The trade-off in using the smaller meters is a limit to the operating range (a 2-in meter's maximum is 160 gpm, versus 350 gpm with a 3-in meter) and the added pressure loss (10 psi loss at 160 gpm with a 2-in meter and 1 psi at 160 gpm with a 3-in meter).

Databases

The first task which faced the project team was to develop lists from which the meter testing, investigation, and installation crews could work. The project team has focused primarily on meters over $1^1/_2$ in. Although they represent only 10 percent of the meters in service, large meters account for roughly 63 percent of the water the Commission sells.

The first priority was to evaluate recently changed large meters. These meters would be easier to downsize because fittings would be new, control valves were functioning, and the Commission would have had recent contact with the customer. In addition, accurate consumption information would be more readily available.

A second list involved the generation of a database of large users with apparently too-low average daily use (ADU) records. The MIS department was asked to generate a list of meters over 3 in with ADUs of 0–300 ft^3/day.

A third category of meters investigated was derived from special cases which were brought to the attention of staff through various other sources, such as customer services, construction, or the routine large meter testing and change-out program.

Account Investigations

The project team has assigned special crews to conduct field investigations and collect detailed information on water usage, including number and type of fixtures, type of building, number of stories and units, and whether the building has central air conditioning and a pump. A current meter reading is taken, and measurements are made from flange to flange. The investigating crew is instructed to gather all the information it can when visiting the site, in order to avoid duplication of effort, and to provide enough background material for the project team to determine the proper meter size.

Flow Testing

The project team decided that more accurate rate-of-flow measurements were needed to make better-informed decisions about meter sizing. When determinations regarding proper meter sizing cannot be made using fixture unit evaluations or other methods, flow search equipment is used.

Because the computer technology available to analyze rate-of-flow information is relatively new, the Commission sought out manufacturers of equipment which would meet its needs. Two manufacturers, F. S. Brainard Co. and Schlumberger Industries, responded by supplying, respectively, the Meter Master and the Flow Search. During the Commission's pilot project, both companies have fine-tuned their equipment and related software packages using input from Boston meter tests.

Both the Meter Master and Flow Search work on a similar principle. Magnetic pulses, which vary from meter to meter, are emitted from the spin of the turbine or disc in the meter. Depending on the meter size, the pulses reflect different volumes of water, which are defined in the related software. The data are imported to a computer, where the software translates the pulses into total rates of flow over various time intervals. The Meter Master, which is designed to work with most makes of meters, uses a sensor placed on various locations on the meter, wherever the signal is the strongest.

The Flow Search was designed for use with Neptune meters. The sensor is placed directly on top of the meter, after the register is removed, and picks up pulses from there. The register head then fits back on top of the sensor to continue recording consumption.

It should be noted that the BWSC runs flow tests during expected periods of peak flow, usually for 3-day periods, sometimes during the week and sometimes over a weekend. Meter sizing decisions are often based on a combination of flow test data and best engineering judgment, because testing over longer periods of times is not always possible. So far, of over 400 large meters downsized, only one has been upsized again, due to pressure problems.

The commission has recently purchased three Meter Masters and three Flow Searchers to conduct flow testing on large meters which are candidates for downsizing. Six more Flow Searchers are on order.

Rate-of-flow testing equipment is also used to track changes in consumption.

Follow-up

In order to track the progress of the program and to obtain estimates of water recovered, a specialized database was established. The database records account number, address, old meter sizes, work orders, new meter size and number, and former average daily use. The Commission then takes meter readings 30, 60, and 90 days following the installation of the smaller meter, and keeps a running total of the change in ADU recorded.

At first, due to the condition of the customer database and the prevalence of estimated readings, it was difficult to obtain accurate estimates of actual usage. The project team decided to track both the change in water consumed and the change in billed consumption by deriving a "true" actual former usage number from the last available actual reads. When compared with the estimated usage, the net changes in billed consumption and actual consumption were surprisingly quite close.

Unusually large gains or losses in consumption are investigated in order to ensure that the data have been entered correctly, and to explain any aberrations in consumption.

The Commission plans to read accounts one year after downsizing, to obtain a truer picture of change in recorded usage, accounting for seasonal consumption and other factors.

14.6.5 Results

In analyzing "candidates" for downsizing, the project team attempted to identify trends and generalizations among various categories of customers. While downsizing is not always appropriate, and does not always result in an increase in recorded consumption, the team has identified many accounts where smaller meters would likely have an immediate impact on unaccounted-for water.

The following cases, in public housing, apartments, schools, commercial and institutional buildings, and municipal property, are representative of some of the emerging trends the team has identified, which have provided guidance for subsequent decisions.

Public Housing

The Boston Housing Authority (BHA) is the BWSC's largest customer, bringing in over $6 million in revenue annually. Because the BHA represents such a significant portion of the BWSC's customer base, the Unaccounted-for

Water Task Force decided to focus attention on a representative sample of the BHA's accounts.

In examining water consumption trends at various public housing developments, the project team took several factors into account:

- Most daily housing apartments do not have dishwashers, central air conditioning, or in-unit laundry facilities, which would drive up water consumption.
- Most developments are no more than four stories high, which would obviate most pressure considerations.
- Many developments have undergone recent renovations, which would likely include the installation of newer, water-saving fixtures.

Most meters at BHA developments were 3 or 4 in in size. Based on the factors above, and on earlier flow measurements, the Unaccounted-for Water Task Force recommended that most meters at public housing developments could be downsized to 2 in. The downsizing project team has implemented that recommendation, which has resulted in significant increases in accounted-for water at several developments.

The 2-in meters have delivered sufficient high-end flows to buildings with 50 to 124 units, and have picked up thousands of gallons of water which previously slipped by the larger meters at low rates of flow.

For example, at the Bunker Hill housing development, a Meter Master rate-of-flow recording system was used to monitor flows in one 119-unit building in the development in which a new 2-in meter was installed. Unfortunately, an analysis of the impact of downsizing at Bunker Hill was skewed by several aberrations in the Commission's billing account system. If the "problem" accounts are removed from the analysis, the ADU of 15 accounts jumped from 109,118 gpd to 127,220 gpd after downsizing, a 17 percent net increase of 18,102 gpd.

At the BHA's Fidelia Way housing development, the Commission downsized two 4-in turbine meters, with an operating range of 10–450 gpm, to two 2-in displacement meters, with a normal operating range of 2½–160 gpm. Rate-of-flow test data showed that the flow through one of the 4-in turbine meters ran below the minimum flow rate nearly all night. After downsizing, test data showed an increase in registered water from midnight to 6:00 a.m. of 13,039 to 18,477 gal, or a 42 percent increase. It also showed an increase from 21,866 to 28,667 gal from 10:00 p.m. to 6:00 a.m., or a 31 percent increase.

Apartments/Condominiums

Market-rate and luxury apartments were studied as a separate category from public housing for several reasons, including:

- Different lifestyles of market-rate tenants and condominium dwellers would indicate different rates of water consumption, particularly where fixtures such as dishwashers, washing machines, and central air conditioning are present.
- Many multiple-unit apartments and condominiums are in high-rise buildings where water is pumped to upper floors; the presence of a pump eliminates most water pressure concerns, which

should always be taken into account in making meter sizing decisions.
❑ When apartments have been recently renovated, new plumbing code requirements and the availability of water-saving fixtures may influence water usage.

At the Foundry, a newly renovated condominium complex in South Boston, a 2-in meter was installed in place of an existing 3-in turbine meter. From June 1988 through March 1990 the average daily use was recorded at 1728 gal. After the smaller meter was installed, the recorded average daily use jumped 33 percent, to 2304 gal.

At 65 Commonwealth Avenue, a 16-unit condominium complex, a 4-in meter was downsized to 2 in. The ADU before downsizing was 419 gpd. After the meter was changed, the ADU jumped to 3456 gpd—an increase of 74 percent.

Down the road at 12 Commonwealth Avenue, the 3-in meter at a 57-unit apartment building was downsized to 1½ in, resulting in an increase of 3 percent from 6006 to 6193 gpd.

Schools

The project team has analyzed fixture units and potential water demand at several public and private schools in the City and believes most meters in schools are oversized.

For example, at St. John's parochial school, an old 3-in meter was feeding 16 sinks, 24 toilets, and 5 urinals. The account was using approximately 1668 gpd. A 3-in meter was found supplying 25 toilets, 25 urinals, and one shower at the Beethoven School in West Roxbury, with an ADU of 785 gpd. At Boston High School, a 4-in meter was supplying 80 toilets, 24 sinks, and one shower, with a pumped system. That school uses approximately 1242 gpd. All three schools are scheduled for downsizing to 1½-in meters based on an analysis of the fixture units and data previously collected in buildings with similar fixture units.

At the end of the 1991 school year, the Commission plans to downsize approximately 120 meters at Boston public school buildings, mostly installing 1½-in displacement meters in place of the 3-, 4-, and 6-in turbines currently in service.

Commercial

Downsizing is also appropriate for many commercial buildings, although careful attention must be paid to the type of business and potential process-related fluctuations in water use.

Subsequent to flow testing at 109 Lincoln Street, which houses offices and a garage, a 4-in meter was downsized to 2 in, resulting in a 42 percent increase in ADU, from 3104 to 4421 gpd. After downsizing, tests showed the maximum flow at 33 gpm, well within the range of the 2-in meter, which probably could have been downsized even further to a 1-in meter.

At an office building at 40 Court Street, a 3-in turbine meter was downsized to 1½ in following a flow test which showed a maximum flow of 35 gpm.

According to the project team's analysis, commercial laundry facilities most often require larger size meters. Although large amounts of water are not used on a constant basis, flow measurements have shown peaks in consumption which require the higher maximum flow through a large meter.

Institutional

Many of Boston's largest water units are institutions such as universities, hospitals, and museums. Therefore, the project team decided to focus attention on institutional users as a separate category. Within user categories, and even within accounts, water use can vary widely. For example, university dormitories have different consumption patterns than classroom buildings and therefore might require different meter sizes.

Northeastern University facilities personnel were reluctant to allow the Commission to downsize meters. Flow testing, however, showed that many of the meters feeding the university could safely be downsized. At Ryder Hall, 139 Forsyth Street, flow testing showed a maximum flow through a 3-in meter of 25 gpm, with a minimum of 0. After downsizing, the new $1^1/_2$-in meter picks up flows of around 2 gpm which were previously missed.

At another Northeastern building at 370 Huntington Avenue, the maximum flow through a 3-in meter was about 21 gpm, with a minimum of 0. This meter was downsized to $1^1/_2$ in, and flow measurement now shows a $^1/_2$ gpm flow running all night. Thirty days after downsizing these two Northeastern meters, the ADU has increased 146 percent, from 9993 to 24,624 gpd. At Wentworth Institute, 550 Huntington Avenue, a 3-in turbine was downsized to a 2-in disc, resulting in a 126 percent increase, from 3231 to 7286 gpd.

Flow measurement at the Museum of Fine Arts indicated that the average flow rate was 25 gpm with a minimum flow of 7 gpm. The project team therefore concluded that the 4-in meter should be downsized to 2 in to capture low flow. In addition, a large constant night flow was detected.

At the Young Men's Christian Union, 48 Boylston Street, a 3-in meter was flow tested at a maximum flow of 43 gpm, with a minimum of 0. After the meter was downsized to $1^1/_2$ in, there was a 20 percent increase in recorded consumption, from 6440 to 7719 gpd.

Municipal

The City of Boston is the BWSC's second largest customer. Almost all city facilities are metered, and represent a broad spectrum of user categories, from municipal office buildings to fire and police stations to parks and other recreational facilities.

At the City's largest recreational facility, the James Michael Curley Recreational Center (known as the L Street Bathhouse), flow testing indicated that the 6-in meter could be safely downsized to 2 in. Since that meter was changed, the ADU has increased by 35 percent, from 9784 to 13,240 gpd. In addition, a constant flow was measured at night, indicating a leak at the premises. This figure also shows that the normal operating minimum of a 6-in turbine meter was far too high to pick up the water being used at low rates of flow.

Flow testing at the Boston Fire Department building at 125 High Street showed a maximum flow of only 20 gpm through a 3-in meter, with usage under 4 gpm most of the time. This meter is scheduled to be downsized to 1 in.

A 4-in turbine meter at the Curtis Hall Municipal Building was downsized to 1½ in following a flow test showing a maximum flow of 8 gpm and a minimum flow of 0. A close-up view of the low-flow testing after the meter change reveals a constant minimum flow.

14.6.6 Conclusions

Downsizing Works

Downsizing has been successful for a wide variety of the Commission's accounts. Although results have varied from case to case, as of May 17, 1991, the Commission has recovered over 57,474 ft^3 (429,905 gpd or 156,915,320 gal per year) of water by downsizing over 400 meters 1½ in and larger. When multiplied by the BWSC's current water and sewer rates, the downsizing effort could generate over $700,000 annually to offset the Commission's rate revenue requirements, in addition to cutting unaccounted-for water.

In some cases, the value of downsizing meters is immediately apparent, such as has been shown by the analysis of most public housing and schools. In all cases, previously held assumptions about meter sizing should be questioned.

To illustrate this point, the project team analyzed several cases to determine what the meter size should have been, using standard fixture unit assumptions, versus the actual rates of flow measured with flow testing equipment. At 216 Tremont Street, a nine-story office building, calculation yielded a fixture unit value of 1838, which would indicate a meter size of 3 in. Flow measurement indicated this meter could be downsized to 1½ in.

A 12-story office building at 40 Court Street had a 3-in meter, as was dictated by the fixture value of 3256. Flow testing accurately predicted that a 1½-in meter would be sufficient.

In some cases, even though the fixture value indicated a 1½-in meter would be appropriate, 3-in and even 4-in meters were installed. Such was the case at 12 and 65 Commonwealth Avenue, both of which have recently had meters downsized to 1½ in based on flow measurements.

The Commission recognizes that the meters being replaced may be performing slightly below AWWA standards, which may contribute to the increase in registered consumption after the new meter is installed. However, data from the BWSC's large meter testing program indicated that most large meters are within accuracy standards, so the Commission does not consider this a significant factor in measuring downsizing results.

Factors Affecting Downsizing Decisions

Several outside factors not previously considered have significant impact on water consumption and rates of use.

Pumps

When analyzing the proper meter size for a building over five stories, it is essential to know whether the building uses a pump to deliver the water to upper floors. If the water is not pumped, an undersized meter may have an adverse impact on water pressure. A smaller meter may be used if the building is equipped with a pump.

Air Conditioning

Air conditioning make-up water is another factor which should be taken into consideration. Central air conditioning can use from 3 to 10 gal of water per minute to make up for evaporation, depending on the size of the unit. Flow testing and water consumption evaluations performed during the winter will not be accurate during the summer months, when air conditioning increases water usage.

Flushometers

A third factor to take into account is the amount of water required by "Flushometer" fixtures. Depending on the type of fixture, Flushometers can use water at a rate of approximately 35 gpm in a 15-s burst. It is also important that sufficient pressure be maintained so the fixtures will reset properly.

Space Limitations

In some cases space limitations prevent smaller meters from being installed. For example, at a Boston University building at 632 Beacon Street, the pipe leading to a 3-in meter must be replaced and a new flange installed before a properly sized 2-in meter can be installed. At the Commission's headquarters at 425 Summer Street, a similar situation exists, making downsizing a complicated and time-consuming endeavor.

Water Conservation Devices

As mentioned previously, where water conservation devices such as low-flow showerheads, faucet aerators, toilet dams, and low-flush toilets have been installed, previous estimates of water usage should be reconsidered when making meter sizing decisions.

Additional Benefits of Downsizing

Capital Costs

An obvious additional benefit to meter downsizing is the reduction of capital costs for large meter replacement.

Leak Detection

During the course of flow investigations, many leaks have been found at customers' premises. As mentioned, flow search at the City's James Michael Curley Bathhouse revealed a constant night flow of 3 gpm. The Commission received a letter of thanks from another customer, the Roxbury Boys and Girls Club, for discovering a 6-gpm leak as a result of a flow search investigation. Although discovery of these leaks and the sharing of data can offset some of the additional revenue generated by meter downsizing, it fosters positive customer relations.

Water Conservation

The Commission has also determined that smaller meters act as flow restricters by increasing headloss. Therefore, downsizing can actually promote water conservation without the installation of new water-saving fixtures.

Data Collection

Another important benefit of the program has been the opportunity to clean the customer database through detailed meter investigations. Meters which had not been read for long periods of time have been located and are now read regularly. Illegal connections have been discovered and remedied, preventing water theft. In general, the program has enabled the Commission to gain more knowledge about its system and about customer water consumption, both of which have contributed to a significant reduction in unaccounted-for water, and an increase in revenues.

The preceding case is a recent effort in the 1990s by the BWSC to become more accountable. The next case discusses accountability issues, which were around in the 1930s. It appears that accountability has been and will be a problem faced by utilities throughout history and throughout years to come! In both cases many of the problems were resolved by good, accurate field data collection. The technology of the equipment used in each case changed significantly, from mechanical to electronic, but the methodology and reasoning stayed the same.

14.7 CASE STUDY TWO
*Putting the Water Works G-Man to Work**

Roger W. Esty, Superintendent, Danvers Water Works, Danvers, Massachusetts

One of the perplexing problems to handle in a water works plant is the customer who comes in and kicks about the large meter bill that he has just received. Of course he argues that it was impossible for him to use that amount of water. The meter must be wrong. He requests that it be taken out and another put into its place.

We try to place ourselves in his position and just imagine for a few moments that we have received a bill that perhaps amounts to $9.00. Naturally, if we had been receiving a bill for from $3.00 to $5.00 each month and then this big one came in, we would be upset. Whether he is right or wrong when he approaches the window to register his protest, he argues in a disagreeable tone of voice to the poor girl behind the counter.

In the olden days, after a lot of arguing by both parties, pro and con, the customer left in a huff, vowing his vengeance on the poor water works girl or official. Today, when we receive the same kind of customer we agree with him that the bill is high, in fact out of all proportion to

---

*Adapted from *Water Works and Sewerage* (a Gillette publication), vol. 85, no. 9 (September 1938), by permission of F. S. Brainard and Company.

what he has been accustomed to paying. Then we compare with him previous readings of his meter during a similar period, and question him carefully and as diplomatically as we can. If, then, we do not determine satisfactorily, to the customer as well as ourselves, the cause or reason for the bill, we then call in the "G-Man" of the Water Department.

14.7.1 What Is This G-Man?

What, or who, is this "G-Man"? Well, this is a nickname given to a machine that really ferrets out and records what is going on at the house of complaining customers. Its real name is "The Meter Master." It is a device, which perhaps is about three years old, and is the result of the conception and ingenuity of a real New England water works man, Frank S. Brainard, of the Water Bureau of Hartford, Connecticut.

While there have been devices that have done this work in a similar manner, it remained for Mr. Brainard to perfect an instrument that registers on a chart the total consumption and the numerous times the water is used during a given period. It is so flexible that it can be connected to any type or style of consumer meter without disturbing the recording device of the meter.

This recorder is operated by a clock mechanism, which can be so regulated that one revolution of the chart is made in 6 hours or in 24 hours. The charts are arranged with an inner and outer circle so that either the reading of a single dial meter (like an ordinary house meter) or two readings may be registered simultaneously, which is required for a compound meter.

Although the charts can be run to secure records for 6 hours or 24 hours, for general purposes it is far more desirable to use the 24-hour chart. Only in cases of extremely large flows, or in district testing, is it more satisfactory to use a 6-hour chart.

If the meter were the ordinary house type, then the first index hand would register 1 cu ft. As the meter register hand makes one complete revolution (indicating that 1 cu ft of water has been drawn), the arm of the recorder makes a stroke up and down the chart. The first index hand can also be set to register 10 cu ft when making one complete movement up and down. This, however, has to be done on a round reading register. It cannot be thus used on a straight reading register. The complete record of rates of use, as well as total consumption, is shown when the chart makes a complete revolution in the period selected—6 or 24 hours.

In making tests, the meter cover or cap is removed and a cap ("cartridge," so called) is placed over the first index hand of the meter register. Then a holder, fastened by cap screws, is placed on top of the meter. This holds in place the flexible cable, which connects the recording instrument of the "Meter Master" to the meter. If a compound meter is to be tested, then two cables are utilized, one to each register.

14.7.2 Some Experiences

In order that those who are not familiar with the machine, and its value to the water works superintendent, may get a better conception of its usefulness, I will cite several instances where it has revealed conditions on a service that the ordinary inspection and meter test would not pick up. Then the reader of this article will, perhaps, be more familiar with the advantages that this machine has over any other means of detection work that we have knowledge of.

First, a Simple Case

The customer complains about the size of his bill. He says it is impossible to use so much water, and he says that there are no leaks. We go to his place. He has five overnight camps. We inspect them and find a toilet leaking in the first camp. He says it doesn't amount to anything. We connect the Meter Master to the meter and run a 24-hour test. Total registration 20 cu ft, 16 cu ft of this proving to be leakage. This (at 20¢ per 100 cu ft) amounts to $2.88 for the three months. Results? The customer admits his error, is pleased with the recorded chart, and thanks us with a smile for conducting the test.

Another Case

Two houses on one meter, owner and wife in front house. Young couple and three old people over 70 in second house. Bill about $9.00 per quarter. We found by the chart reading that water was being used some time during 23 of the 24 hours. All during the night toilet flushings were indicated on the chart. We assumed, and since have been convinced, that this night consumption was caused by these old people. No more complaints from this customer. He had a true picture of just what is going on at this house all of the time.

Third Case

Two houses on one meter. Two families in front house and one in the rear. Bill about $12.00 per quarter. We found that on five different occasions 4 cu ft of water was used. This amounts to 30 gal. We questioned each family as to their habits and found that the family in the second house was the cause. Their hot water storage tank, back of the kitchen stove, got so hot it would rumble. The lady of the house would open the hot water faucet and let it run off, until it became cool, and then shut it off. This would cool it down for several hours. The cause? They had two small copper coils around the two oil burners in the stove. They were around the top of the burners and, consequently, received all of the heat. I suggested lowering the coils halfway down on the burner. It has eliminated the high meter bill. This wouldn't have been discovered in any ordinary fixture inspection and, possibly, by no amount of questioning.

Fourth Case

A two-family house, and a bill of $12.00 per quarter. From the meter charts we found that on occasions, from one hour to as long as three hours, a continuous stream was running. We ran this test for three days, as it was so unusual. An inspection on one of these occasions immediately located the cause. A family of two adults and three children occupied the second floor of this tenement. The lady of the house placed her washing in a set tub and then turned on the faucet and left the water running while she did the housework. In ten or fifteen minutes she would come back, work the garments around in the water and go off about her work again. This was kept up until the washing was completed.

14.7.3 Applications Are Several

The value of this testing and recording device is twofold: First, to run flow tests on services to find out how often and at what rate the water is drawn. This is, in order to determine whether the meter is too large or too small for that particular service. We have also operated at the same time a pressure-recording gauge and this will record the drop in pressure when the water is drawn. Secondly, to check large or out of the ordinary meter bills. In this case it provides visual evidence of exactly how much water is used and when, and thus gives the customer an actual picture of his consumption demands. In addition, it furnishes information regarding the presence of leaking fixtures, the use of improper amounts of water by toilets, and other possible sources of waste.

The rate recorder has been used in conjunction with flow tests on large meter installations to determine when heavy demand occurs and at what rate the water is drawn. It has a particular value in connection with night flow testing of the distribution system. In these tests the town is divided into districts and supplied through fire hose and in most instances through one 2-in meter. The test is generally started at 6:00 p.m. and continued through the night to 6:00 a.m. Valuable flow data can thus be obtained economically, and without much effort.

Our source of supply is in the Town of Middleton, 5 mi from the center of Danvers. Most of this town has public water service. It is very interesting to study the daily demands of this little town. We serve a population of about 1200 people in Middleton, through 300 active services on 7 mi of pipe. All services are metered.

The maximum demand, according to the chart, is between the hours of 9 and 10 in the morning, when they used 52 gpm. The minimum demands come between the hours of 3 and 4 in the afternoon, when they used at the rate of 12 gpm. The figures are very conclusive that the systems as well as the household fixtures are tight.

The latter is another instance of the use to which this machine can be put. This test only took one day to run, but we changed charts every 6 hours in order to study more closely the variations in the demand for water for an entire town. It only goes to show how little water a small residential town actually requires.

In expressing my personal sentiments about the value that this machine has to the water works man, if I may be permitted to express myself in slang, I should say that "it is the slickest thing that has ever come down the pike" to help the water works manager learn just what's going on in his system, and to convince dissatisfied customers that the Water Department is no longer in the guessing business.

14.7.4 The Editor's Comments: Every Water Works Needs a "G-Man"[*]

Nothing in a busy day of a water works manager is less satisfying than to have a complaining customer leave the office still with a doubt in his mind that his meter has not been overread; or, else, that it is obliged to be registering too much water. Little is more important to water utility heads in these times than the good will of the public, born of satisfied customers. On this phase of operation and management more thought is being given today than ever before.

In handling dissatisfied customers, a pleasing personality is worth a whole lot; the value of argument, which may sound worthy, sincere, and logical, cannot be discounted. But the most valuable contribution to happy public relations is an admission that there is at least a possibility that the meter may not be 100 percent correct; that it can and will be checked for accuracy. When such is done, the customer usually feels that he has had every attention due to him, and the act registers as an appreciated service from the utility.

However, it costs something to remove, check, and reset meters. Furthermore, it is but part answer to the problem—why the large water bills? Today, we hear much concerning a means of circumventing meter removals and tests, the findings from which can almost invariably be predicted in advance. Instead of cross-questioning the customers or members of his household, and presenting time-consuming argument, as convincing as possible as to what are the most likely reasons for the excessive water bill, it is now possible to say— "Mr. Customer, we feel that there is one best way to check into your complaint. It will very likely save you some money, and it will with certainty save a lot of speculation and argument. We will put our 'G-Man' to work on this case." Did you say "who or what, is this Water Works G-Man?" That is answered in the first article in this issue, "Putting the G-Man to Work," the contribution of George W. Esty, superintendent of the Water Department of Danvers, Mass.

The "Water Works G-Man," as Mr. Esty has dubbed the device, is a portable compact instrument which, when attached to a water meter in service, records the timed frequency of water operation and the rate of flow through the service line every minute during a 24-hour day. In this manner, the customer's meter is made to tell its own story, while in actual service, for the benefit of all—particularly the customer. The chart reveals clearly whether the meter is "sleeping" when the customer's family is sleeping, or whether there are leaks somewhere, which do not let the meter sleep. Perhaps the "G-Man" on watch 1440 minutes of the day taking notes in the form of red lines on the instrument chart will reveal the next day that someone flushed the toilet at 1:31 a.m. What's important about this particular recording is that the person forgot to jiggle the handle on the flush tank before going to bed. Result, the ball valve remained open part way for $6\frac{1}{2}$ hours. The wasting water deprived the meter of its rest the balance of the night, and kept the "G-Man" busy

[*These are the comments of the editor of the publication in which this case study originally appeared.]

putting down "black marks" (only they are red) on the chart against a forgetful flusher.

Now, when Mr. Customer is shown the chart revealing the day's history of water drawings, toilet flushings, how long it took Mazie to wash the dishes, how long it took and how much water was used to sprinkle the lawn or garden—and, what that running flush tank did to the meter register—he really learns "things that every customer should know." With the report of the never-sleeping "Water Works G-Man" on the table, and with a few words of interpretation by the water official, there is no room for argument or need for speculation. Mr. Customer recalls that lately he had had to go back and jiggle the flush tank handle, himself, on two occasions. Also, he may see plainly that the new spray dishwashing gadget he bought for Mazie last month took more water than the less sanitary dishpan method, which he didn't like. Another customer may have learnt that his meter couldn't get any rest at night because of what proved to be a number of hidden small leaks along his service line across the lawn. These are just samples of the things, that the "Water Works G-Man" reveals to the mutual satisfaction of Mr. Customer and the water official handling customer complaints, even if embarrassing to the first.

In the customer's mind, there's something uncanny about an instrument that reveals such irrefutable evidence. Any distinction between the recording device and the meter proper is lost sight of. Faith in the meter being restored, and respect for a management employing such methods being established to a high degree (plus a bit of customer embarrassment), results in payment of the bill on the spot—frequently a word of commendation and appreciation being added. And, as Mr. Esty points out, it is the next water bill that clinches the matter of improved relations. It brings the "Thank you" (with a smile) when customer and manager next meet.

As recorded in previous issues of this magazine, testimony as to the value of such 24-hour recording instruments, placed on complaining customers' meters, is also being heard from other managers attending meetings of water works associations. Only last month, before the Virginia Section of AWWA, D. R. Taylor, superintendent of the Water Department of Roanoke, Virginia, recited his experiences with recording instruments placed on customers' meters. The resulting record chart had proved the most effective means yet discovered for amicably and promptly settling high bill complaints and refuting claims of plumbers, employed by the customer, that all leaks had been repaired. In contrast, methods and experiences of some other managers sounded archaic. What it would take Esty's or Taylor's G-men one setting to discover, took several inspections and poking-about to uncover. And, even then, it was a matter of some luck that peculiar hidden causes of consumer water waste and high bills were detected.

In addition to its constant value in settling customer complaints satisfactorily and economically, Mr. Esty cites and describes other uses and applications of the "Water Works G-Man." Amongst these is the selection of the proper meter for 24-hour duty on important services, resulting in more accurate recordings and revenue increases such as quickly paying for the instrument. Another use is that of checking the tightness of mains through records made on bypass meters during night flows.

It is nice to see that a case from almost 70 years ago is still very much relevant today! Accountability is a question which water works operators will always have to deal with.

Additional information on proper meter sizing can be found in App. D and was supplied with permission of Brad Brainard, who is writing this for the new AWWA M22.

14.8 BILLING SYSTEM ANALYSIS AND RECTIFICATION

We have discussed the importance of meter testing, repair, setting, sizing, and replacement on apparent losses and must now discuss and consider an area which is often overlooked: the methodology of collecting the data from the meter, depositing it into a central database, correctly interpreting the data, and generating accurate billing information. While this may seem like something simple, it is often a "magical" area where data go in and come out, but nobody really knows exactly how the data are processed.

Recent advances in technology have brought about automated meter reading (AMR). In many cases when a system is automated, the billing system is also rehabilitated and updated. Automated billing systems allow the utility to track many conditions to ensure efficient collection and analysis of water consumption. However, care should be taken that the operators of the system are well trained in the use of the software, which can often be complicated. When purchasing an automated system it is often wise to purchase local support and maintenance too.

If AMR is not an option, then the utility must periodically check and implement the following tasks to ensure accountability:

- ❏ Audit meter reader reports for accuracy.
- ❏ Audit the number of estimated reads.
- ❏ Ensure that the meter book routes cover new parts of the system if the system is expanding.
- ❏ Ensure that there is no overlap.
- ❏ Ensure that the meter readings can be realigned to match with master meter readings for nonrevenue water analysis and auditing.
- ❏ Ensure that the actual raw data are accessible in addition to billed data, which are often massaged.
- ❏ Ensure that any meter type changes have been entered into the system—size, type, use, diameter, max volume, type of reading units.
- ❏ Track the number of zero-volume accounts and compare with the number of vacant properties listed, or holiday homes.
- ❏ Compare municipal land use data with billed water data to identify metered accounts, or theft of water.
- ❏ Check for serious declines in usage patterns, particularly for large consumers, to identify bypassed meters or use of internal wells without permits.

14.8.1 Audit Meter Reader Reports for Accuracy

Some meter readers are better than others are. Some will record all information required of them and additional information, but others tend to estimate more than read. The utility must ensure that actual reads are taken on all accounts where possible.

In some cases meters are hard to read, such as:

- Houses with dogs
- Meters which are at an inconvenient angle
- Meters in basements with owners not at home and no external reader
- Meters in chambers which are full of water or debris
- Meters which are misted over

Meter readers must be prompted to report these conditions immediately, so that the situation can be rectified. They must also be told not to invert the meters to an angle easier to read (as in the second example above), as this has been found to cause serious errors. More details of research on this subject can be found at the website of the South African Bureau of Standards, www.sabs.co.za.

In the case of dogs, basements, and accessible chambers, external readers can be conveniently placed outside the property boundary.

> Estimated reads are to be avoided where possible, as they can skew data and hide the real situation.

14.8.2 Audit the Number of Estimated Reads

It is important to know how many estimated reads a utility has and what effect these might have on billing and accountability. It is not usually good public relations to estimate bills on a constant basis. Also, if a utility is considering changing-out meters or moving to an automated system, estimated reads can often mask the true volumetric benefits, as no clear baseline for volumetric use and billing can be established.

14.8.3 Ensure That the Meter Book Routes Cover New Parts of the System If the System Is Expanding

Some systems are experiencing rapid growth. It is important that this growth is tracked and meter routes are adjusted to cover new areas. Where meter reading is done manually, it is also important to consider realistic workloads, to ensure that the estimated read scenario does not become prevalent.

14.8.4 Ensure That There Is No Overlap

In rapidly changing systems or during reorganization, it is possible to overlap workloads, meaning that meters may be read twice. If a utility has council boundaries or operational zones, the divisional boundaries must be clearly defined on both the operational and billing plans. These should be the same but are often quite different, as unfortunately departments do not always share information, as they should. GIS systems are beginning to reduce this type of error, as information is readily available to all departments and any changes are made at a central location and automatically updated at other workstations.

14.8.5 Ensure That the Meter Readings Can Be Realigned to Match with Master Meter Readings for Nonrevenue Water Analysis and Auditing

For the purposes of auditing the system's performance, it is necessary to be able to align the sales meter readings with the supply meter readings. Most utilities perform some kind of block analysis to ensure that this is possible.

14.8.6 Ensure That the Actual Raw Data Are Accessible in Addition to Billed Data, Which Are Often Massaged

As we mentioned in the introduction to this section, the billing system can often be a black box where data go in and come out in the form of bills. However, it is sometimes not possible to access the actual raw measured volumes per read. These data are important when auditing a system's physical performance and should always be available to supplement massaged financial data.

> Ensure that raw data are always accessible.

14.8.7 Ensure That Any Meter Type Changes Have Been Entered into the System—Size, Type, Use, Diameter, Maximum Volume, and Type of Reading Units

As we mentioned earlier, a utility should perform periodic testing and repair or replacement of its meters. It is also vitally important to record the results of this testing and any remedial measures taken in the billing system so that any severe changes in use can be accounted for. For example, if a large user has recently been downsized, it is possible that its bill will rise drastically. However, if raw data have not been kept as mentioned above and the date of intervention not recorded, it is often common that the client will call and complain. When the billing system operator pulls up the client's historical record and sees a huge difference with no reason for the difference and nothing to compare the volumes to, often they will rebate the customer without further investigation. This will obviously cause revenue loss, confusion, and a lack of accountability to the end user.

14.8.8 Track the Number of Zero-Volume Accounts and Compare with the Number of Vacant Properties Listed, or Holiday Homes

It is important to track the number of zero-volume-use accounts and compare that to the number of vacant properties listed in an area. (In the case of vacation areas, the number of registered holiday homes should also be identified.) By tracking the zero accounts that are not vacant properties, the utility will have an idea of the mechanical performance of the revenue meters. This data will be used in the economic model for dictating testing and change out periods.

14.8.9 Compare the Municipal Land Use Data with Billed Water Data to Identify Unmetered Accounts or Theft

It is important to track municipal land use data and trend it with billing records to identify where potential theft or unmetered supply is occurring. In some situations clients will make their own service connections, or a building or temporary supply may get overlooked and never be transferred to the billing system. In these cases the municipal land use data should show populated use although the billing system will not show a history of billing. In this case meter readers working that route should check out the area.

14.8.10 Check for Serious Declines in Usage Pattern, Particularly of Large Consumers, to Identify Bypassed Meters or Use of Internal Wells without Permits

As we mentioned earlier, some large consumers represent a significant portion of the overall income to a utility. At a very minimum, these customers should be tracked to see how their historic usage profiles vary. Any significant downturn in usage could represent a malfunctioning meter, incorrect reads, a bypassed meter, or maybe even an internal unauthorized source such as a well. The latter, if not properly controlled, could also cause backflow problems and lead to potential water system contamination. Significant changes in usage could also be represented by legitimate changes in usage, such as water efficiency measures, or, in the case of industrial consumers, relocation of parts of a manufacturing process to another plant.

14.9 SUMMARY

Obviously, a utility wants to maximize revenue as much as possible (by reducing apparent losses) while minimizing variable overheads such as leakage (by reducing real losses). With careful auditing and planning supported by periodic testing and intervention, most utilities should be able to maintain a reasonable level of performance, with little extra effort over and above the normal maintenance activities.

We have not discussed the effects of system pressure on meter accuracy, or apparent losses, but there is most definitely a connection. Practitioners are currently investigating the effects of very low flows on water meters and the nature of those low flows.

In many cases the low flows can be caused by toilet tank flushes or roof tank filling in cases where roof tanks are in use. As pressure is managed to lower levels, the small flows, which are often system-side leakage, can in many cases be significantly reduced or stopped. As these smaller flows are reduced, the effective unregistered water volumes are also reduced, thereby reducing apparent loss volumes. This research will be reported in greater detail in subsequent editions of this manual.

14.10 BIBLIOGRAPHY FOR CASE STUDY ONE

- American Water Works Association Research Foundation, *Water and Revenue Losses: Unaccounted-for Water,* 1987.
- Male, J. W., Noss, R. R., and Moore, I.C., *Identifying and Reducing Losses in Water Distribution Systems,* Noyes Publications, Park Ridge, N.J., 1985.
- Boston Water and Sewer Commission, *Water Distribution Study,* Camp, Dresser and McKee, May 1987.
- Schlumberger Industries, *Water Division Product Catalog,* 1989.
- Hensley, J. (ed.), *Cooling Tower Fundamentals,* Marley Cooling Tower Company, Kansas City, Mo., 1982.

CHAPTER

15

Water Efficiency Programs

Bill Gauley, P.Eng.
Principal, Veritec Consulting, Inc.

15.1 INTRODUCTION

Developing programs to improve water efficiency is fast becoming a preferred alternative for municipalities faced with a need to expand their water supply or wastewater infrastructure. Improving water efficiency is almost always more environmentally responsible and can often be considerably more cost-effective than expanding capital works.

This chapter is intended to help those planning to implement a water efficiency program (WEP) to focus on elements that will be important to the overall success of the project. Success here is defined as achieving the maximum cost-effective water savings through implementing publicly acceptable measures. The material in this chapter should help program designers to establish specific goals for water demand reduction, as well as to understand that if the goals are not specific it will be impossible to quantify program success.

This chapter will also explain that it is only after the program's overall goals have been established that it will be possible to identify which demand components should be targeted and, ultimately, which water efficiency measures will be best suited to achieve these goals.

The section dealing with water saving targets will explain the importance of knowing both the maximum potential water savings and the target water savings associated with a water efficiency plan, and why there is usually a difference between these values.

Important implementation issues are identified later in the chapter; some of these issues are often overlooked or misunderstood, and are therefore described more fully.

The final section outlines the importance of monitoring and tracking program results. It describes some of the tools often used to assess program performance as well as some of the more common monitoring misconceptions.

Understanding the material outlined in this chapter should help both program designers and implementation staff to have a better understanding of some of the more basic elements involved in implementing a successful water efficiency program.

> **C**apital expansion is extremely expensive and in some cases virtually impossible. In these cases systems will first reduce their system real losses, while maintaining current billing levels. If the reduction in real losses is not significant enough to defer the capital construction, then demand reduction is undertaken.

15.2 WHY PLAN A WATER EFFICIENCY PROGRAM

Since the 1980s it seems that an increasing number of municipalities and agencies are implementing water efficiency programs. Some even require that the potential for water demand reduction be determined and evaluated before approval to expand the water or wastewater infrastructure will be granted. Even when it is not mandated, many municipalities are showing fiscal responsibility by considering the economical and environmental benefits associated with demand-side management (water efficiency) versus supply-side management (infrastructure expansion).

Unlike the old adage, "art for art's sake," there should be very clear and well-defined reasons for undertaking a water efficiency program. Fortunately, in today's environment there are generally a myriad of reasons for doing so. Some of the more common reasons include:

- ❑ A need to expand water or wastewater treatment plants or infrastructure
- ❑ Nearing the capacity of water source (e.g., reservoir or aquifer)
- ❑ An interest in being environmentally responsible

Whatever the reason, for the program to succeed it is important that the overall goal is understood and accepted by all involved parties—politicians, works department, the public, etc. After all, it is the program goal that will dictate which demand component (as described in the next section) should be targeted, and it is the target demand component that will dictate which water efficiency measures should be included.

It is essential, therefore, for the program designer or implementation team to understand the different system demand components and how they relate to the various water efficiency measures that are commonly implemented as part of a water efficiency program.

> Some municipalities are mandating conservation before granting system expansion or extraction rights.

15.3 SYSTEM DEMAND COMPONENTS AND HOW THEY RELATE TO A WEP

Throughout the year, most water supply systems experience a range of water demand rates[1]—often changing with the season. Figure 15.1 illustrates the various demand components commonly experienced by water supply systems. These demand components tend to form a demand pyramid, with the base of the pyramid comprising the system's average daily base demands, and the top of the pyramid representing the system's peak day demand.[2] Each of these demand components is described in detail later in this section.

Because most water efficiency measures target a specific demand component, it is important that these measures are properly selected based on the program's goals. Improperly chosen measures may not only be ineffective, but even worse, may actually have a negative impact on the program (e.g., they may reduce system revenues).

---

[1] Note that this statement refers to demand rates, not billing rates.
[2] Typically, water treatment facilities are designed to meet peak day demands, while system storage is utilized to meet peak hour demands.

15.3 SYSTEM DEMAND COMPONENTS AND HOW THEY RELATE TO A WEP 383

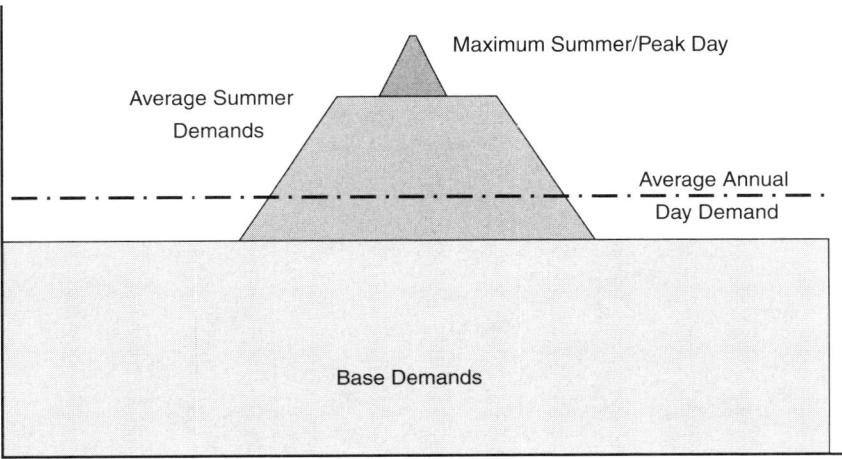

Figure 15.1 Typical demand pyramid. (*Source: Bill Gauley.*)

15.3.1 Base Demands

Generally, an assortment of demand types contribute to a system's overall base demand. Those that are related to residential indoor water use, such as toilet flushing, showering, clothes washing, etc., generally experience little variation from season to season. Distribution system leakage, and most nonirrigation and noncooling water demands in the industrial/commercial/institutional (ICI) sector, are also fairly constant throughout the year. Base demands form the largest component of average winter[3] day water demands. As seasonal temperatures rise, however, irrigation and other seasonal demands increase. In fact, in the heat of summer, as much as 50 percent or more of a system's total water supply can be related to irrigation and cooling.

Base demands are affected by changes in population size, number of employees, and demographics. However, since base demands are not generally affected by changes in the weather, they tend to be reasonably constant from year to year.

Commonly, a significant percentage of a system's base demand is discharged to the sanitary sewer system. Therefore, water efficiency programs targeting reductions in wastewater flows (i.e., to defer wastewater infrastructure expansion) should focus on reducing base demands (as well as other inflow and infiltration, explained in the next section).

Although reducing base demands does not decrease the demand volumes related specifically to irrigation and other outdoor water uses, it does lower the peak demand rate by taking a "slice" off the bottom of the pyramid (see Fig. 15.2) and lowering the entire demand pyramid (see Fig. 15.3). Note also in Fig. 15.3 that the peak demand is reduced by the same demand rate (not the same percentage) as the base demand reduction.

A sometimes overlooked, yet critical, aspect of a water efficiency measure is the sustainability of the water savings. Are the savings maintained in subsequent years, or must the reduction measure be repeated or

---

[3] The term "winter" is used here to describe any nonirrigation season.

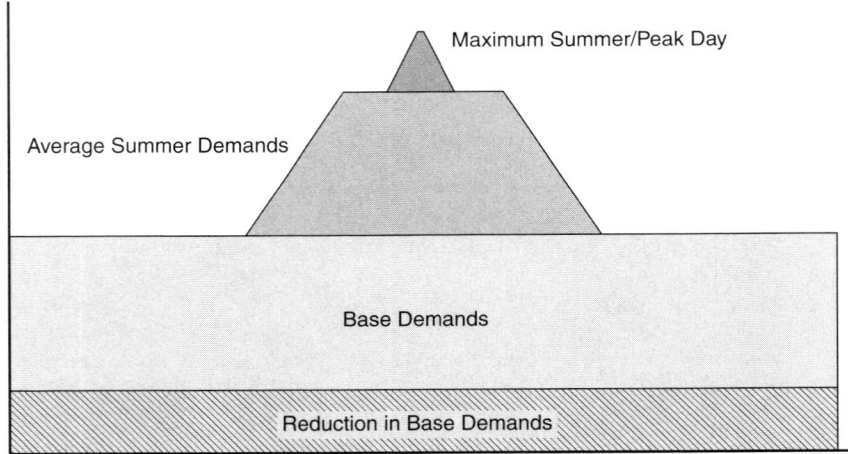

Figure 15.2 Peak demand rate lowered. (*Source: Bill Gauley.*)

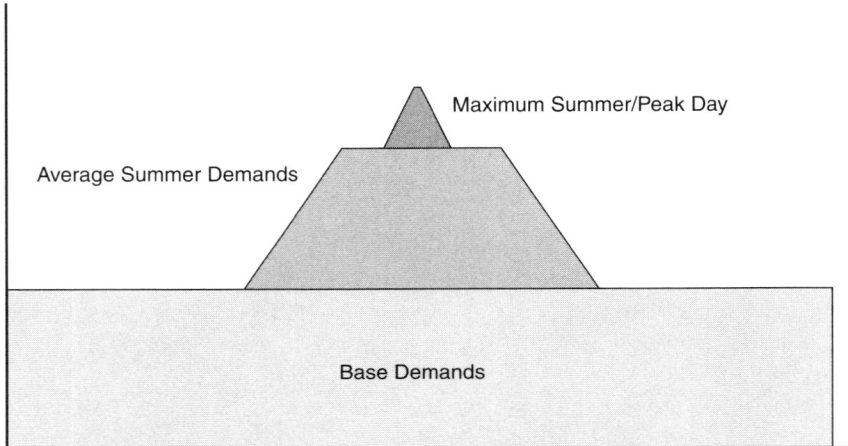

Figure 15.3 Entire demand pyramid lowered. (*Source: Bill Gauley.*)

reinforced? Generally, water savings that are not sustained are of little value to a municipality.[4]

15.3.2 Base Sewage Flows

Sanitary sewage flows are, in general, relatively constant throughout the year. Flows in systems with high levels of inflow[5] or infiltration[6] (I&I) will vary depending on changes in groundwater levels or precipitation. Generally speaking, when rainfall events are eliminated, there is relatively little variation in sanitary sewage flow rates from season to season.

---

[4] The exception to this comment is in the use of temporary emergency measures, such as watering bans during periods of drought.
[5] Infiltration: groundwater seeping into sewers through cracks and joints.
[6] Inflow: surface water being directed into the sewer.

Typical sewage reduction programs generally involve water efficiency measures targeting base demands (replacing toilets, showerheads, clothes washers, etc.), or reducing the levels of I&I.

Municipalities that maintain combined sewer systems, where both sanitary sewage and storm water are collected by the same system, often include I&I reduction measures in their water efficiency program.

15.3.3 Average Annual Day Demand

Some system operators calculate average annual day demand (AADD) by dividing the total annual water production by 365 (i.e., the number of days in a year). This value actually represents the average annual day production and includes water lost through system leakage and other unaccounted-for water demands. This volume can be divided by the total population[7] serviced by the system to determine the average daily gross[8] per-capita water demand (or, more accurately, the average daily gross per-capita water production).

It should be noted that the average daily net per-capita water demand, i.e., the average volume of water attributed specifically to personal use, is generally determined by dividing the total volume of water billed to residential customers by the total residential population. This value is usually presented as a demand rate, typically gallons per capita per day (gcd) or liters per capita per day (lcd). Demand rates can be determined for population subsets as well (single-family households, multifamily apartment buildings, industrial facilities, commercial sites, etc.) and can also reflect seasonal demand variations (average summer day single-family household water demand, average winter day commercial site water demand, etc.).

The AADD is an academic value that changes from year to year (differences in summer irrigation demands, for instance, can have a significant effect on the AADD). Since they are a blend of the various seasonal demand components, AADD values generally do not provide sufficient data to design water efficiency programs that target either base demands (affecting both water and wastewater treatment infrastructure) or peak demands (affecting only water supply infrastructure). For example, two systems could have identical AADDs and yet have completely different operational characteristics (see Fig. 15.4).

For this reason, it is not usually practical to base a water demand reduction target or a water efficiency program on AADD demands. In fact, programs that reduce AADD while not reducing either base flow demands or peak demands may accomplish nothing more than reducing system revenues.

---

[7]The term "population" generally refers to the residential population, i.e., those persons living in single-family and multifamily households within the community. Care should be exercised when evaluating municipalities where significant portions of their population work outside the community, or where a significant number of their employees actually reside outside the community.

[8]Gross per-capita water demands include residential water demands, ICI water demands, fire-fighting demands, mains flushing, as well as all unaccounted-for demands.

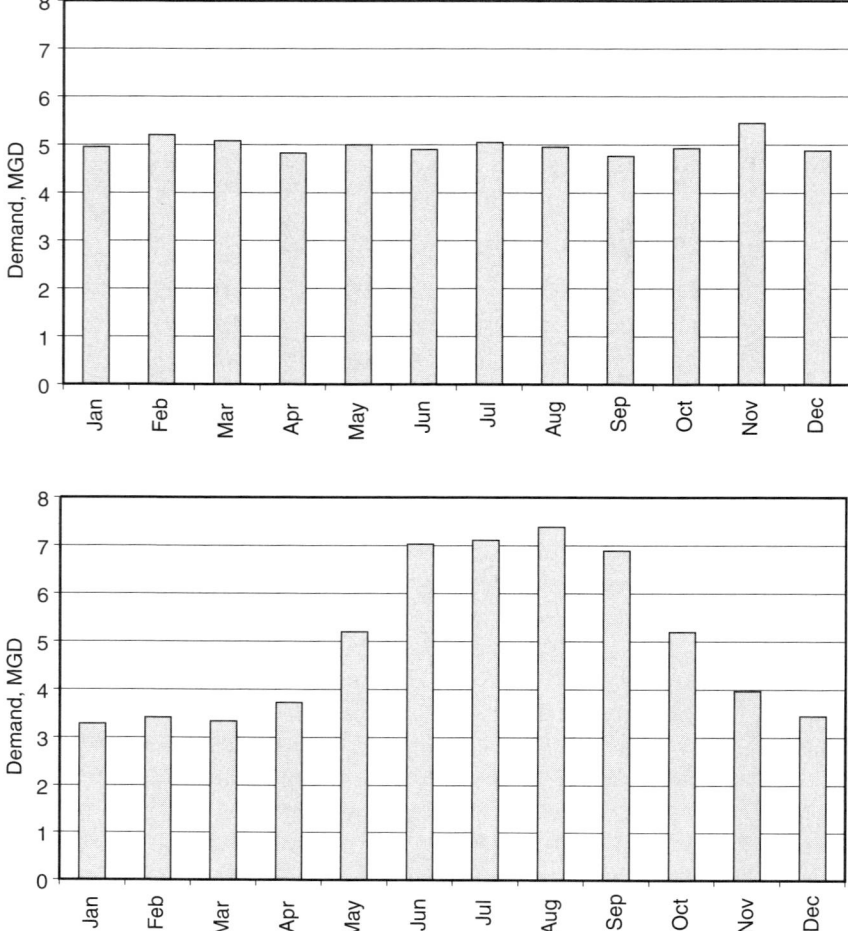

Figure 15.4 Two systems could have an identical AADD (in this case 5 mgd) and yet have completely different operational characteristics. (*Source: Bill Gauley.*)

Example 1

A community has determined that their wastewater treatment plant is nearing capacity and decides to implement a water efficiency program to extend the life of the facility. They set their goal as a 10 percent reduction in the average annual day demand.

A year later they are surprised to find that although they did achieve a 10 percent reduction in AADD, their wastewater flows were unchanged. After some investigation they determine that the reduction in AADD was related entirely to reduced summer irrigation (perhaps the result of a cooler than average summer with higher than average precipitation rates) and not to their water efficiency program.

Although reduced irrigation demands did reduce the community's AADD, it had no affect at all on the wastewater flows.

The community decides that future programs aimed at extending the life of wastewater treatment plants will focus on reducing base flows rather than AADD.

Example 2

A community has determined that their water treatment plant is nearing capacity and decides to implement a water efficiency program to extend the life (capacity) of the facility. They set their goal as a 10 percent reduction in the average annual day demand.

A year later they are surprised to find that although there was no reduction in AADD, they actually achieved their target of reducing the system's maximum water demands, thereby extending the capacity of the plant. After some investigation they determine that because of a warmer than average April and September, the total volume of summer irrigation demands was slightly higher than average. However, their landscape irrigation reduction program, which involved providing customers with informative bill stuffers, radio and TV ads, etc., had the desired affect of reducing customer irrigation demands during the hottest and driest part of the summer.

Although the program did not reduce AADD at all, it did achieve the goal of extending the capacity of their water treatment facilities.

The community decides future programs aimed at extending the life of water treatment plants will focus on peak demands rather than AADD.

15.3.4 Maximum Summer/Peak Day Demands

The peak day demand is usually defined as the highest water demand recorded during a single 24-hour period in any calendar year and, as such, it changes from year to year. Although technically the peak day demand occurs only on a single day, in reality there can be several peak-type days (maximum summer demands) within a year, and they may occur sequentially (e.g., during a hot, dry period) or at several times throughout the year.

A system's peaking factor is a mathematical value determined by dividing the peak day demand by the average annual day demand and, as such, also changes from year to year. Dissimilar systems can have the same peaking factor. The largest peaking factor experienced over a period of several years is often used as a design peaking factor[9] when planning new water supply infrastructure components.

It is important to note that although peak day demands are generally related to outdoor irrigation demands and usually occur after long periods of dry, hot weather, they may also be the result of large water main breaks, fires, or industrial demands, or a combination of any of these factors.

There can be significant benefits associated with reducing peak day water demands, e.g., deferring the need to expand water treatment facilities or distribution infrastructure, to enable the current infrastructure to service an expanding population, etc. As a result of these benefits, many municipalities implement at least some type of program targeting outdoor irrigation, e.g., watering restrictions (odd/even day watering, time-of-day, etc.), bill stuffers, radio/TV/newspaper articles or advertisements, irrigation audits, etc.

---

[9] Design peaking factors that are higher than actual historical values contain an additional margin of safety.

388 CHAPTER 15 WATER EFFICIENCY PROGRAMS

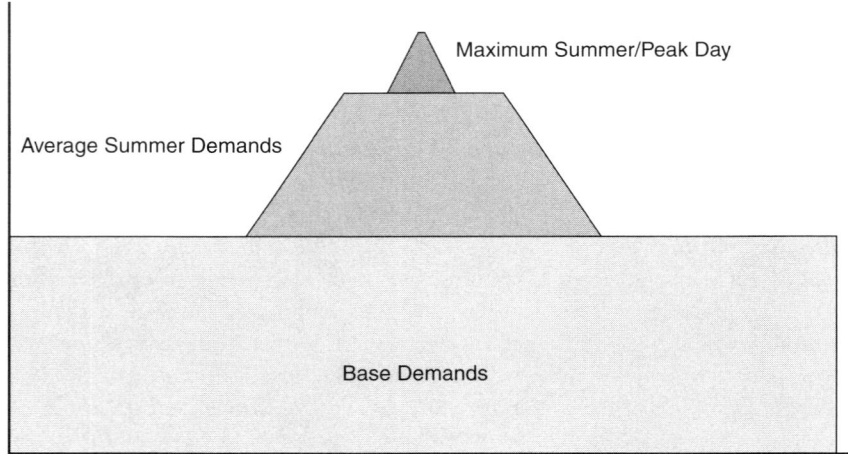

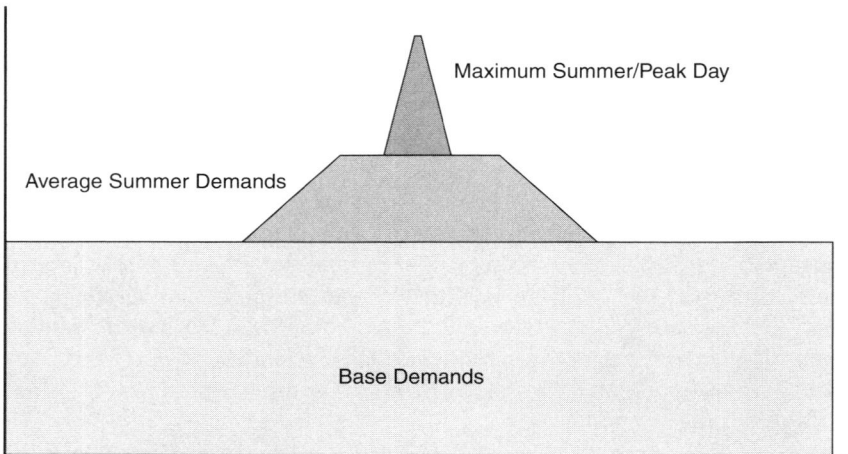

Figure 15.5 Water efficiency programs that reduce average summer day demand but do not reduce peak day demand will not achieve the program's goal. (*Source: Bill Gauley.*)

Although it can be very important to reduce peak day demand, it is important to note that water efficiency programs that reduce average summer day demand but do not reduce peak day demand will not achieve the program's goal, but will reduce water sales revenues. See Fig. 15.5.

Peak day demands are usually determined by monitoring daily water production volumes at water treatment plants or in well fields. Changes in weather patterns (e.g., hot and dry versus cool and wet summers) can cause large variations in peak day demands from year to year.

Agencies or municipalities implementing water efficiency programs targeting peak day demands must be aware that any savings achieved through their program may be either exaggerated or eclipsed by the generally significant demand changes related to weather.

Ideally, the program should reduce peak day demands alone (thus extending the service life of the water treatment or distribution infrastructure) while not affecting average summer day or base demands (thus not reducing revenues). For this reason, some municipalities are implementing pilot programs to quantify specifically the peak day water savings achieved as a result of implementing their irrigation reduction measures.

These pilot programs involve bulk monitoring[10] of a study area and a control area. The water efficiency measure (or intervention) is implemented only in the study area. Bulk monitoring must be used for this type of program because it is not possible to quantify savings using water billing data.[11]

Similar to base demand reduction programs, the sustainability of the water savings is critical. Since the target of many peak day reduction programs involves changes to customer irrigation habits rather than changes to fixtures or equipment (e.g., installing new toilets or showerheads), maintaining peak water demand savings over time may be more complicated than sustaining base demand savings.

15.3.5 Summary

Although many water efficiency measures affect more than one demand component, they are generally intended to target a specific goal such as peak or base water demand reductions, or wastewater flows reductions. It is important that the proper measures be selected to address the program's specific demand component target. Table 15.1 relates commonly implemented water efficiency measures to the type of demand component they most effect.

15.4 WATER SAVING TARGETS

Once you have determined whether your water efficiency program will focus on peak day demands or base demands (or both), and which measures you will implement, you must determine how much water your program can realistically be expected to save, i.e., how effective your

TABLE 15.1 Impact of Water Efficiency Measures

Measure	Primary Impact
Toilets	Base demand programs
Showerheads	Base demand programs
Clothes washers	Base demand programs
Landscape irrigation	Peak water demand programs
Seasonal pricing	Peak water demand programs
Watering restrictions	Peak water demand programs
Cooling water reduction	Peak water demand programs
Gray water reuse	Peak water demand and base demand programs
Public education	Peak water demand and base demand programs

Source: Bill Gauley.

[10] Bulk monitoring involves recording the water demands of a large group of customers, e.g., an entire subdivision, simultaneously, by installing water meters directly in the supply water mains. The use of bulk monitoring eliminates the Hawthorne effect, which is described later.
[11] Water billing data, even when bills are issued every month, do not provide the details necessary to identify demand parameters on individual days, nor do these data account for changes in weather conditions.

WEP will be. The first step in this process is to determine the theoretical maximum savings of your program.

15.4.1 Theoretical Maximum Savings

The theoretical maximum savings (TMS) is a calculated value; it assumes that the measure is implemented perfectly and achieves 100 percent market penetration. The TMS establishes the upper boundary of the water savings target—a program cannot save more water than the TMS.

It is important to note that the TMS does not consider any of the program delivery elements that would be required to achieve 100 percent participation. For example, the TMS does not require any knowledge of incentive amounts, installation criteria, how removed fixtures will be disposed of, how marketing will be performed, the cost-effectiveness of the measure, etc.

The TMS is later used as a tool when determining the target WEP water savings. Some examples of calculating TMS values are illustrated in the following examples.

Example 3

A municipality with a population of 50,000 has a sewage treatment plant that is nearing capacity and decides to implement a toilet replacement program to extend the life of their plant.

They decide to use the information from the AWWARF[12] Residential End Use Study (REUS) to establish the approximate TMS for this measure in their municipality. The study states that nonefficient toilets are used 4.92 times per person per day, with 4.1[13] gal (15.5 liters) per flush. The study also states that efficient toilets are used 5.06 times per person per day with 1.9 gal[14] (7.2 liters) per flush.

A sample household survey has identified that only an insignificant number of existing toilets are currently water efficient. The approximate TMS for their measure is determined as follows:

- Water demand related to toilet flushing—existing nonefficient: 50,000 population × 4.92 flushes/capita/day × 4.1 gal/flush = 1,008,600 gal/day
- Water demand related to toilet flushing—projected efficient: 50,000 population × 5.05 flushes/capita/day × 1.9 gal/flush = 479,750 gal/day
- TMS = 1,008,600 gal/day − 479,750 gal/day = 528,850 gal/day

In other words, if the entire population replaced their existing toilets with water-efficient models and achieved savings similar to those stated in the REUS, the municipality would save 528,850 gal/day (gpd).

---

[12] American Water Works Association Research Fund.
[13] The REUS states that nonefficient toilets are flushed an average of 4.92 times per person per day and account for 20.1 gal of water. Average flush volume is, therefore, 4.1 gal per flush.
[14] The REUS states that efficient toilets are flushed an average of 5.06 times per person per day and account for 9.6 gal of water. Average flush volume is, therefore, 1.9 gal per flush.

Example 4

A community with a population of 50,000 needs to reduce peak water demands to postpone a planned expansion to the water treatment plant. The peak demands in the system occur in the summer and are the result of extensive landscape irrigation. The town decides to implement a water efficiency program targeting residential irrigation to reduce the system's peak demands. There are about 14,000 single-family households in the community.

The utility has completed a billing data analysis and determined that the average household water demands during the nonirrigation season (i.e., winter) is about 200 gpd, while the average household summer day demand is 280 gal, and the peak summer day demand is 350 gal. Since they are trying to postpone expanding their treatment plant they decide to focus on reducing the peak summer day demand.

The average "additional" household demand (i.e., demand in excess of the winter day demand) on the peak summer day equals 150 gal. The town is aware that some of this additional demand is related to vehicle washing, filling swimming pools, etc., and they estimate[15] that approximately 65–70 percent of additional demand, or about 100 gal per household per day, is related to landscape irrigation.

In this case, the TMS depends on the type of measure used to address the goal.

Scenario A

Program designers know that virtually all of the irrigation demand could be eliminated if a mandatory watering ban was enforced[16]; this would mean that about 100 gpd per household could be saved. The TMS is 14,000 households × 100 gal/household/day = 1,400,000 gpd.

This type of restriction, however, is generally not popular with customers or politicians and is usually used only in an emergency situation, such as a severe drought.

Scenario B

Program designers review the results achieved in other jurisdictions and estimate that about 25 percent of the irrigation demand could be saved through voluntary restrictions or through the free distribution of hose timers, rain gauges, brochures, etc., to residents by temporary employees, or by utilizing detailed and informative bill inserts, etc.[17] Although this type of program will have a smaller TMS, it is expected to be much more acceptable to the residents. Assuming that 100 percent of the population

---

[15] Although it is almost always necessary to estimate certain values and make certain assumptions when determining the TMS, they should be based on sound engineering judgment and properly referenced.

[16] Seattle, Washington, issued a mandatory ban on all summer lawn watering during a 1992 drought and, essentially, eliminated the normal peak water demand. The ban was, however, unpopular with both customers and elected officials.

[17] Savings estimates are for illustration only and are not intended to reflect actual probable savings.

participated and each home saved 25 percent of its irrigation, the TMS is 25 percent × 100 gpd × 14,000 households = 350,000 gal.

15.4.2 Realistically Achievable Savings Target

A program's realistically achievable savings target (RAST) is usually somewhat less than the TMS value for several reasons, such as:

- Actual customer participation rates will be less than 100 percent, especially if participation is voluntary.
- Not all measures will achieve 100 percent of their potential water savings, especially when the measure requires changes to customer water using habits.
- Not all measures can be implemented cost-effectively; for example, a water efficiency measure would not be considered cost-effective if the cost per unit of water was greater to implement it than it is to expand the water supply.
- Not all measures can be implemented on schedule, especially when water savings are required quickly.
- Water savings may not be sustainable, especially when changes to customer water-using habits are involved.
- Some measures may not be publicly applicable, for example, although the use of gray water offers the potential for substantial water savings, it may not be popular with customers.

The RAST is generally established for each individual water efficiency measure included in the program; the sum of the various RASTs determines the overall WEP water savings target. Once the RAST is established for a measure, it will be applicable only to the specific set of circumstances (population, demographics, program schedule, public support, etc.) from which it was developed. The RAST associated with a measure will vary from municipality to municipality and, as such, there is no single "cookie cutter" approach to calculating these values—RAST values must be determined for each application on an individual basis.

Following is a list of some of the aspects that must be considered when establishing a measure's RAST.

- Cost/benefit ratio[18] of water efficiency measure to the customer
- Availability of incentives
- Public attitude toward water efficiency
- Household demographics
- Water rates and/or structures
- Expected building code requirements for new fixtures
- Water use by law enforcement protocol
- Expected advancements in plumbing fixture technology
- Expected changes in the costs of water-efficient fixtures

Once the RAST is established, the next phase in developing a WEP involves designing the implementation plan, described in the next section.

---

[18]Cost/benefit ratio: the cost of implementing a measure divided by the value of the resulting savings. A cost/benefit ratio of less than 1.0 indicates a cost-effective measure.

15.5 IMPLEMENTATION PLAN

The implementation plan considers and describes exactly how the WEP will be delivered and how the program's goals will be met. In a manner similar to the RAST, implementation plan will also vary from measure to measure and from municipality to municipality.

Although describing all of the potential elements that must be considered when designing an implementation plan is beyond the scope of this chapter, the following list identifies some of the more important elements.

- What is the implementation schedule, i.e., how quickly are the savings required?
- What capital works projects will be deferred or eliminated because of the WEP?
- How will the cost-effectiveness of the program be determined?
- If water-efficient fixtures are involved, will they be installed by professionals or self-installed?
- Will incentives be offered?
- How will you ensure that new fixtures are installed properly and achieve the maximum savings?
- What criteria will be used to approve plumbing fixtures and appliances?
- How will you ensure that water savings are maintained?
- How will removed plumbing fixtures be disposed?
- How will customer complaints be managed?
- Will additional staff be required?
- Are pilot programs required?
- Is monitoring required?
- Will public education be included?
- Will newspaper, radio, or TV promotion be included?
- What contingency plans are in place if savings are not achieved or exceeded?
- What effect will natural replacement of fixtures have on projected savings?
- What about the effect of "free riders" on the program costs?

As demonstrated by the list above, determining exactly how the WEP will be implemented is usually considerably more complex than determining why the WEP should be implemented. However, as stated earlier in this chapter, the most important aspect of any WEP is not the program itself, but the successful implementation of the program.

Many implementation issues are common to most plans. Four of the most important issues are described in the following paragraphs.

15.5.1 Natural Replacement

In time, all plumbing fixtures and appliances become old or their performance begins to deteriorate and they are replaced with newer units—though not necessarily with units that operate more efficiently. This replacement occurs even without incentives. By offering incentives or rebates, water efficiency programs often try to influence customers to select water-efficient fixtures or appliances. Knowing natural replacement rates allows a WEP designer to better estimate the water savings that can be accredited directly to the WEP implementation.

Example 5

If a residential toilet has an average life cycle of 25 years, the natural replacement rate for this fixture equals 4 percent per year (i.e., 1/25 = 0.04) and all toilets will theoretically be replaced with new units in 25 years.[19]

If water-efficient toilets are the only type of units available in the marketplace, then virtually all toilets will be water-efficient in 25 years even without offering incentives. If nonefficient toilets are still available, then there is a possibility that without incentives no water-efficient toilets will be installed within 25 years.

Free Riders

Customers who receive program incentives or rebates even though they would have participated in the program through natural replacement are considered "free riders," i.e., these participants increase the costs associated with implementing the program but do not increase the program effectiveness. For example, all participants in a program offering an incentive toward the purchase of a water-efficient toilet in an area where only water-efficient toilets are available are free riders.[20]

Incentives

Many water efficiency programs rely on the use of incentives to accelerate the adoption of a measure. Determining the optimum incentive amount, however, can require some research[21]—if the incentive is too low, the program will fail to meet its required participation targets; if it is too high, the program will cost more than it should. Incentive amounts are often based on targeted customer participation rates, overall program cost-effectiveness, and the urgency for water savings.

Be aware that changing the value of an incentive partway through a program may cause customer complaints, i.e., reducing incentives may offend later participants ("Why did they get more then than I will get now?"). On the other hand, increasing incentives may offend early participants ("Why did I get less than they are getting now, when I supported the program from the onset?").

Pilot Programs

Pilot programs are small-scale programs generally implemented immediately before a full-program rollout to verify design, implementation methodology, participation rates, etc. Because pilot programs usually include a significant level of monitoring, analysis, and evaluation, the unit costs are often considerably more than those of a full-program rollout.

---

[19]In practice, some toilets will be replaced before and some toilets will be replaced after the average life cycle is reached.

[20]In this situation, however, the incentive may be intended to accelerate the natural replacement rate of the nonefficient toilets.

[21]This research is often completed as part of a pilot program.

It is important that the design of the pilot program reflects that of the full-program rollout, e.g., rebate amounts, marketing, product types and qualities, etc.

15.6 MONITORING AND TRACKING

One of the most important aspects of any implementation plan is establishing the protocol that will be used to monitor and track the program results. Properly conducted program monitoring and tracking will determine if the efficiency measures are achieving their water savings targets (i.e., the RAST), if the program is on schedule, if program costs are on track, and if the water savings are being sustained. If proper monitoring is not performed, it will be impossible to assess the effectiveness of the WEP.

Because no two municipalities are exactly alike, monitoring programs must be designed to suit the specific conditions associated with each individual WEP. It is not possible in this chapter to outline all the parameters that should be considered, but some of the more common elements associated with program monitoring are outlined in the following section.

15.6.1 Water Audits

Some water efficiency programs attribute water savings to the conducting of water audits. Although a water audit can be an excellent tool to evaluate site conditions or even to determine the potential for reducing water demands, the audit itself does not save any water. It is quite possible to conduct an extensive and expensive water audit without saving any water. Perhaps the customer did not implement any of the measures identified through the audit, or perhaps there were simply no water saving opportunities identified. Water savings that do occur after a water audit is completed are the result of changes in water-using practices or equipment or both.

15.6.2 Water Meters

Water meters are important because:

- ❑ They help ensure equity (i.e., each customer pays for the water they receive).
- ❑ They can be used to help identify opportunities for water savings.
- ❑ They can be used to promote awareness.
- ❑ They provide a mechanism for measuring and tracking the effects of change.

However, water meters do not save water!

Based on the results of some poorly analyzed case studies, many water efficiency programs have mistakenly assigned water savings directly to the installation of water meters in homes that were previously billed on a flat-rate basis. All of the water savings realized in these homes are the result of other actions—such as changes in the customer's water-using habits, the installation of more efficient fixtures, or both. No water savings are attributable specifically to the meter.

396 CHAPTER 15 WATER EFFICIENCY PROGRAMS

Programs that include water savings because of meter installation run the risk of "double counting." For instance, a water efficiency program that has already considered the effects of installing efficient toilets, showerheads, and faucet aerators, as well as from improving customer habits (turning off faucets when not in use, using full loads for clothes and dishwashers, avoiding overwatering landscapes, etc.), would mistakenly overestimate the potential for water savings if they also include savings from installing a water meter.[22]

> The use of percentages as a performance indicator can be misleading.

15.6.3 The Use of Percentages versus Hard Values

Although percentages are commonly used to describe the distribution of a data set, the results can sometimes be misleading, as illustrated in the following example.

Example 6

The water demands of a single-family household are "premonitored" as part of a water efficiency program, and the data shown in Table 15.2 are collected.

The homeowner decides to take advantage of a municipal rebate and installs a new water-efficient clothes washer that uses only 60 percent of the water of his existing machine. No other changes are made to the home's water demands!

The home is again monitored and the data shown in Table 15.3 are collected.

TABLE 15.2 Premonitoring

Item	Demand (gpd)	Percentage
Toilet	60	30
Clothes washer	40	20
Shower	50	25
Faucet	50	25
Total	200	100

Source: Bill Gauley.

TABLE 15.3 Postmonitoring

Item	Demand (gpd)	Percentage
Toilet	60	32.6
Clothes washer	24	13.0
Shower	50	27.2
Faucet	50	27.2
Total	184	100

Source: Bill Gauley.

---

[22]Consider that no water savings would be expected from secretly metering a customer's water demands, nor would additional savings be expected by installing more than one meter on a customer's service.

The percentages illustrated in the tables seem to indicate the absurd conclusion that installing a water-efficient clothes washer will somehow increase the water demands associated with toilets, showers, and faucets. In reality, of course, the actual volumes of water associated with the other plumbing fixtures were not affected by the installation of the new clothes washer—only their percentage contribution to the reduced overall demand.

Example 7

For several years a small municipality was struggling with a high level of unaccounted-for water. Although they produced about 1.2 mgd (million gallons per day) of potable water, they only billed for about 1.0 mgd—a difference of about 17 percent.

Later, a brewery with an average demand of 0.8 mgd moved into town. Then the municipality produced 2.0 mgd and billed for 1.8 mgd. Their level of unaccounted-for water, at only 10 percent, seems to have been reduced.

This type of reasoning indicates the absurd conclusion that having a brewery relocate to your municipality will improve your distribution system's performance. In reality, of course, the actual volume of unaccounted-for water was unaffected by the relocation of the brewery.

15.6.4 The Hawthorne Effect

The Hawthorne effect is an initial improvement in a process caused by the obtrusive observation of that process. You should be aware of the Hawthorne effect when you are conducting a monitoring program. The Hawthorne effect occurs when program participants change their normal behavior due to the knowledge that their actions are being monitored. For this reason, monitoring programs often indicate actual conditions if they are implemented without the participants' knowledge.

Monitoring programs should be designed in such a way to reduce or eliminate the Hawthorne effect; i.e., bulk metering or other methods of blind testing should be used where possible.

Example 8

Information collected by data logging water meters that are located in outdoor meter pits (where participants may be oblivious to the monitoring) may provide a more accurate reflection of actual field conditions than data logging water meters located inside the home (where the participants may be aware of the monitoring program and, therefore, alter their behavior).

15.6.5 Diurnal Demand Curves

A diurnal curve is often used to illustrate demand rates versus time over a 24-hour period. The shape of the curve depends on the type of facility being monitored. Diurnal demand curves can be produced for individual homes, apartment buildings, subdivisions, industrial parks, or entire municipalities.

The data used to create the curves are obtained by data logging the building's water meter. Data are often collected for periods ranging from

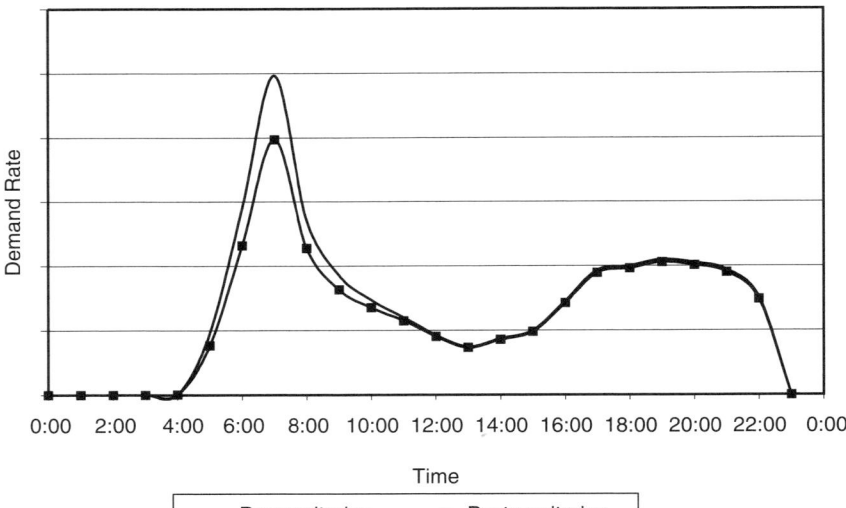

Figure 15.6 Diurnal curve of household water demands. (*Source: Bill Gauley.*)

24 hours to several days. The data are usually analyzed and then plotted to show average or "typical" results. It is important to remember that the amount of detail illustrated by the curve will depend on whether the demand data were data logged as instantaneous values or as average values and the frequency of collection.[23] Generally, the frequency of collection is based on the variability of the data, i.e., data that fluctuate significantly should be logged at a higher frequency.

When both "pre" and "post" data are collected, the water savings can be clearly illustrated by comparing the characteristics of the two diurnal curves. Figure 15.6 illustrates what might be expected when monitoring the water demands of a household before and after water-efficient toilets and showerheads are installed. The reduction in the morning's peak demands relates to the resulting water savings.

Diurnal curves can be used to illustrate information such as when people use water (bathing, showering, toilet use, etc.), the flow rate and duration of irrigation, the difference between weekday and weekend water demands, and the difference between summer and winter demands.

15.7 LOST REVENUES

Some municipalities may be concerned that implementing a water efficiency program will result in a reduction in water sales revenues. In fact, many water efficiency programs are implemented to allow a greater population to be serviced with the existing infrastructure; a larger population generally means a larger tax base, which will benefit the municipality.

---

[23]Using a data logging frequency of 1 h, for example, some data loggers will "turn on" every hour and log the parameter's value at that instant, while other data loggers will continuously monitor the parameter's value throughout the hour and then log the average of those values.

Programs that are implemented without a need (e.g., with no need to defer capital expansion projects) may, however, result in a reduction in revenues. The following points address this concern.

- ❏ Most water efficiency programs achieve demand reductions gradually over several years, providing ample time to implement small changes in water rates.
- ❏ Changes in summer weather patterns (i.e., cool wet summers versus hot dry summers) may have a more significant effect on annual water sales than the implementation of water efficiency programs.
- ❏ Reducing water production volumes will reduce costs associated with water treatment and pumping.

15.8 CONCLUSION

It is hoped that the material in this chapter will assist the growing number of persons developing, implementing, and monitoring water efficiency programs to better understand some of the important concepts associated with completing a successful program.

By developing a clear understanding of your WEP goals, the measures and methods that you will implement to achieve these goals, and the protocol that you will employ to monitor your results, it is hoped that water efficiency can become an even more important element of your future water demand planning.

Water is too valuable to waste; let's use it wisely.

CHAPTER

16

Using In-House Staff or a Contractor and Designing a Bid Document

Julian Thornton

> ***Case Study One***
> Water Leak Survey: A Performance-Based Approach
> *Vernon W. (Wes) Frye, P.E.*
> *Paul V. Johnson, P.E.*

16.1 INTRODUCTION

In this chapter we discuss how to prepare for interventions necessary in the field to resolve real and apparent loss problems located through the auditing and modeling phases discussed earlier in the book.

16.2 USING IN-HOUSE STAFF OR A CONTRACTOR

Most larger utilities have some kind of in-house expertise with knowledge of how to undertake a water audit or intervention against loss on either apparent or real losses. However, often these people have other duties, which make it hard for them to concentrate on the specialized tasks in an ongoing manner.

Many smaller utilities or industrial/commercial/institutional (ICI) systems do not have the in-house expertise and equipment necessary for a full audit and analysis of the water system.

If either the people cannot be dedicated to the job or they are not available, the decision is easy—a consultant or contractor should be employed to undertake the audit and subsequent intervention.

If staff are available, the following steps should be taken to organize a dedicated crew for loss control.

- ❑ Identify a team leader who will be full-time loss control supervisor.
- ❑ Identify the necessary test equipment for taking field measurements.
- ❑ Ensure that the equipment can be periodically tested for accuracy locally and is supported by a local supplier who can undertake repairs on a timely basis.
- ❑ Identify either a full-time or part-time team who can assist the team leader.
- ❑ Undertake detailed training on the methods and technologies chosen for the audit and intervention methods.
- ❑ Be prepared to give authority to the team.

> **I**f in-house staff is not available or cannot be fully committed, then a consultant or contractor should be used.

❏ Be prepared to give a budget to the team for annual testing and intervention.

Once the team has been identified and trained, they must identify where the faults in the current system lie and how best to resolve them. Usually the best way is the most cost-effective way!

Most audits and intervention tasks include the following:

❏ Master meter testing and repair
❏ Telemetry testing and repair
❏ Updating of system plans
❏ Sample testing and replacement of sales and revenue meters
❏ Leak detection and repair
❏ Reservoir and storage testing
❏ Selection of performance indicators
❏ Statistical analysis, modeling, and audit completion
❏ Pre- and postintervention monitoring

Other tasks which may arise as a result of the auditing and testing may include:

❏ Pressure management
❏ Level control
❏ Mains relining and rehabilitation
❏ Mains and service replacement

Some of the tasks listed above are quite time consuming and detailed, and have to be repeated frequently to ensure a sustainable and economic level of losses. Therefore in certain cases it may be preferable to use a specialized contractor or consultant to assist the in-house team or undertake the work in place of an in-house team.

The decision as to whether to utilize in-house staff or a contractor will really come down to time and money, as with most things! The other aspect to consider is the fact that a consultant or contractor will be specialized in the most-up-to-date methodologies and techniques of the field. Water company operators do not always have the opportunity to be exposed to latest technology if budgets do not allow them to attend conferences or seminars, and travel to see other utilities and discuss success and failure with others.

16.3 DESIGNING A BID DOCUMENT

16.3.1 Introduction

> The bid document must be clearly and carefully written to ensure that both the client and the contractor fully understand the project methodology, goals, and objectives.

If a decision has been made to call in a specialized contractor, careful planning must be undertaken prior to going to bid or negotiating directly with a contractor to ensure that both client and contractor understand exactly what the requirements and deliverables of the contract are. If everything is clearly spelled out in the beginning, it is much easier to select the best offer from several in the case of a bid, or to resolve any dispute which may later arise. If the bid document is not black and white there is room for speculation and uncertainty, which will inevitably waste time on both sides—the client and the contractor.

16.3.2 Important Factors to Consider

Obviously, each utility will have different requirements for a specialized contractor. Some utilities may require a full service and others may require specific tasks to be undertaken to complement the skills available from in-house personnel.

If the job is to be quite large and encompassing, it is a good idea to call a bidders meeting and present the overall situation clearly. Bidder packs should be distributed, with system condition and information reports.

System condition reports may contain some of the following information:

- ❑ System overview schematic if one is available.
- ❑ System plans (or a sample if there are many plans).
- ❑ Number of supply meters, type, and age.
- ❑ Number of sales meters, type, and age.
- ❑ If the above information is not available, clearly state that one of the objectives is to acquire this information.
- ❑ Topographical information.
- ❑ Storage information.
- ❑ Average system pressures.
- ❑ Schematic of supply zones if applicable.
- ❑ If zoning is not in place but is one of the desired deliverables, clearly state so.
- ❑ Length and type of mains and services.
- ❑ If the above information is not available, clearly state that this must be provided as one of the deliverables.
- ❑ Information on previous audits and water loss intervention.
- ❑ Estimated water losses.

A simple example follows.

XYZ utility comprises 800 mi of main supplying approximately 35,000 metered connections. The mains are primarily old cast iron mains laid around 1950, with some newer areas of PVC estimated to have been laid around 1980. Meters are between 10 and 30 years old. Meter maintenance has been sporadic. System plans are available in a 1:2000 scale but are not reliable in many cases. There are two main storage reservoirs, one old brick ground reservoir and one newer elevated concrete tank.

Water is supplied through the treatment plant in ZYZ Road. The treatment plant takes water from the XYZ River. The system is a direct pumped system with the two tanks balancing on the system.

There are two main pressure zones, A and B. Pressures in zone A range from 50 to 70 psi and in zone B from 40 to 100 psi. In addition to the water treatment plant, zone B receives water through a supply meter from ABC utility. Supply meters are Venturi type with 4–20-mA output to a telemetry system, which reports back to central control.

Topographical information is not available. Losses are unknown but are estimated at around 25 percent, of which it is felt that 70 percent is system leakage.

16.3.3 Project Goals

In all cases the bid document should clearly state the final objective of the contract. For example:

XYZ utility wishes to undertake a detailed water audit following AWWA M36 guidelines (or IWA audit guidelines). The main goal of the audit is to identify and rank, by cost to benefit, the best ways of reducing lost water or lost revenue in the system.

The audit will be complete, covering all aspects of potential loss, both real and apparent. Decisions as to the levels of loss and the potential benefits and costs of intervention will be based on real field measurements taken during the audit.

All data collection, testing, analysis, and recommendations for water loss control intervention will be the responsibility of the successful contractor.

Additionally, all system plans shall be updated, with separate 8.5 × 11 sheets for each pipe junction clearly showing valve positions triangulated from three fixed points. The new system plans shall be in ABC GIS format. GIS layers will include pipes, hydrants, valves, and detailed elevation contours.

The contract shall be conducted in three phases:

- Phase one: audit, system measurements, sample meter testing, and updating of plans
- Phase two: Leak detection and repair
- Phase three: Meter change-out and automated meter reading (AMR) system

The successful contractor shall go out to bid for leak detection services for phase two and meter replacement services for phase three of the project.

The contractor shall provide monthly progress reports and a detailed report at the end of phase one identifying all cost-to-benefit scenarios. The contractor shall provide a fixed sum for supervision of phases two and three.

16.3.4 Phase One Tasks

As a minimum, the contractor shall:

- Test all master and supply meters (see sample bid document below). Testing shall be to AWWA M6 recommendations.
- Perform demand analysis to identify classes of consumer and identify where most of the water supplied is being used, i.e., residential, commercial, industrial, institutional, agricultural, municipal, etc.
- Select a statistically accountable sample of sales meters and test them to allow analysis of the potential cost to benefit of meter change-out and meter resizing. The contractor shall comment upon and analyze the benefits of an AMR system.
- Update system distribution plans to an ABC GIS system as stated in the introduction.

- ❏ Perform hydraulic measurements to ascertain the level of real losses in the system and the benefits of leakage control. Hydraulic measurements will be defined as a minimum of 250 7-day pressure measurements at locations to be agreed at the start of the contract and 50 7-day flow measurements. Flows should be accurate to ±5 percent of real flow and pressures to 1 percent of real pressure. Equipment must be calibrated to a national standard volume or weight at the beginning and end of the contract and at two separate random occasions during the contract. Data will be manipulated in function to any drift recorded during the equipment testing.
- ❏ Identify the benefits of pressure management as a means of reducing and controlling real losses further (this task should include at least one pilot installation).
- ❏ Perform drop tests on the storage tanks to see if they are leaking.
- ❏ Identify potential losses from overflows.
- ❏ Provide a complete audit to the required guidelines, with accompanying ranking of loss recovery measures.
- ❏ Provide a complete cost-to-benefit analysis of all recommended measures, including an analysis of the potential benefits of AMR.
- ❏ Provide monthly progress reports.
- ❏ Provide a final report, including bids for leak detection and repair and meter change-out and AMR if applicable.
- ❏ Provide detailed training for utility staff on the measures taken during the audit.

The following is a sample bid document for a contractor to test supply meters, which are primary measuring devices such as orifice plates, Venturis, Dall and Pitot tubes, as in the case of our example above. Other tasks such as leak detection or meter replacement could be structured similarly.

16.3.5 Inspection and Testing of Primary Device

1. The contractor shall physically inspect and report on the visual condition of each primary device, the mechanical and hydraulic connections, and the suitability of the environmental housing, i.e., chamber, etc.
2. Using portable equipment provided by the utility, the contractor will measure the differential pressure (DP) in relation to the flow existing at the time of the test and will relate this back to the individual specifications for each device, which will be provided. The contractor will write a report pertaining to the accuracy of this DP.
3. If the DP does not match the flow/velocity, the contractor will suggest possible methods of recalibrating the primary device, i.e., cleaning, rodding, etc. If the primary device is past rehabilitation, the contractor will state this with detailed reasoning.

16.3.6 Inspection and Testing of DP Sensor and Converter

1. The contractor shall physically inspect and report on the visual condition, approximate age, and suitability of the electronic equipment,

stating serial number, make, and type of process, i.e., square root extractor, linear, etc.

2. After ascertaining the accuracy of the primary device, the contractor shall check both zero and span calibrations of the electronic equipment. A detailed report will be compiled identifying the conditions in which the calibration was found to be. The report will state the impact on potential metering error and what must be done to rectify a potential error.

3. If the device is found to be in error and can be calibrated, the contractor will perform the necessary calibration of both zero and span. The contractor will report on the calibration technique and values used. The contractor will then arrange with the utility to have the site reevaluated with an insertion meter for a final accuracy evaluation.

4. If the equipment is found to be in error and cannot be satisfactorily calibrated, the contractor will report on the reasons and make suggestions for the replacement of this equipment with other, more suitable equipment.

16.3.7 Transmission and Data Collection

1. The contractor will inspect all radio equipment, electrical connections, etc., to ensure that this equipment is transmitting and receiving proper data. The contractor will report on any faults or potential problems found.

2. In all cases, the contractor will collect data from the utility's files in the central control for the period of testing and in the case of recalibration, during this period. The contractor will analyze the archived data to ensure that it matches the findings in the field and that in the case of calibration, the data being collected after calibration is a true reflection of the real flow conditions in the field.

16.3.8 Replacement of Faulty Equipment

1. The contractor may provide a separate quotation for supply and services required to replace faulty equipment with new calibrated equipment.

2. If the contractor is called upon to install new equipment, it will be calibrated and tested in situ and a detailed report generated to identify settings, etc.

16.3.9 Contractor Experience and Requirements

1. The contractor must provide resumes of the instrument technicians who will undertake this service work. The technicians must have a minimum of 10 years relevant experience and be conversant with various makes and models of the above-mentioned equipment.

2. The contractor will be responsible for all test equipment required to undertake this work other than the portable DP meter and insertion meter, which will be supplied by the client. (If this is not the case, then the contractor should supply the insertion meter, etc.)

3. If the contractor is called upon to supply and install replacement equipment, the contractor must be prepared to provide all necessary equipment, fittings, connectors, etc., to ensure that the old equipment can be taken out of service and the new equipment fitted the same day, without delay or "downtime."
4. In addition to full reporting, the contractor will be required to provide "hands-on training" for client staff during all stages of the testing and or replacement and calibration.

16.4 SELECTING A CONTRACTOR

Once the bid specification has gone out and responses received, it is necessary to grade the responses, in a fashion that allows the best contractor to be selected for the job.

Many utilities put out a bid specification and then select the cheapest bidder. For a simple service this may be OK, but often for more detailed services it is not always the cheapest bidder that gives the best value for money.

One way of being more selective is to grade the responses on technical merit and price. To do this it is necessary to apply weighting to the skills and personnel of the bidding contractors and weighting to their financial bid. The bids are then compared on both technical and financial merit. Another popular way of financing a contract is on performance; this will be discussed later in this section.

> Ensure that you have a mechanism in place to allow you to select the best contractor for the job. The best is not always the cheapest.

16.4.1 Technical and Price Bids

Technical and price bids are usually made up of components similar to the following:

- Understanding of the problem
- Bidder's experience with similar projects
- Personnel experience with similar projects
- Additional innovation brought to the project
- Equipment to be used on the project
- Price

Each technical topic is assigned a weighted value; this value can then be divided by the price to give an overall weighted score.

Tables 16.1–16.3 show three scenarios. Bid One is the most technically competent in all respects, and in Table 16.1 Bid One is the winner. Table 16.2 shows how much cheaper the second bidder needs to be to win even though he is deemed to be less competent. Table 16.3 shows how much cheaper bidder three has to be to win even though he is much less competent than the other two bidders.

Technical and price bids are good ways of ensuring that the best bidder for the job gets the project; however, care should be taken to carefully analyze the data submitted and to be accountable for the decisions made. It is certain that somebody will complain and ask for justification!

TABLE 16.1 A Project with a Ceiling of $150,000 (Bid One Wins)

	Max. Weight Score	Bid One Allocated Score	Bid Two Allocated Score	Bid Three Allocated Score
Understanding of the problem	30	29	25	20
Bidder's experience with similar projects	25	25	24	20
Personnel experience with similar projects	25	24	23	19
Additional innovation brought to the project	5	3	3	0
Equipment to be used on the project	15	15	15	10
Total	100	96	90	69
Price	$150,000	$150,000	$145,000	$142,000
Maximum weighted score	66.67%	64.00%	62.07%	48.59%

Source: Julian Thornton.

TABLE 16.2 A Project with a Ceiling of $150,000 (Bid Two Wins)

	Max. Weight Score	Bid One Allocated Score	Bid Two Allocated Score	Bid Three Allocated Score
Understanding of the problem	30	29	25	20
Bidder's experience with similar projects	25	25	24	20
Personnel experience with similar projects	25	24	23	19
Additional innovation brought to the project	5	3	3	0
Equipment to be used on the project	15	15	15	10
Total	100	96	90	69
Price	$150,000	$150,000	$140,000	$142,000
Maximum weighted score	66.67%	64.00%	64.29%	48.59%

Source: Julian Thornton.

TABLE 16.3 A Project with a Ceiling of $150,000 (Bid Three Wins)

	Max. Weight Score	Bid One Allocated Score	Bid Two Allocated Score	Bid Three Allocated Score
Understanding of the problem	30	29	25	20
Bidder's experience with similar projects	25	25	24	20
Personnel experience with similar projects	25	24	23	19
Additional innovation brought to the project	5	3	3	0
Equipment to be used on the project	15	15	15	10
Total	100	96	90	69
Price	$150,000	$150,000	$145,000	$105,000
Maximum weighted score	66.67%	64.00%	62.07%	65.71%

Source: Julian Thornton.

It is a good idea when using a technical and price bidding structure to clearly state the weightings to be used in the bid documents. It is also a good idea to place a statement to the effect that "XYZ utility reserves the right of final judgment in the assignment of points for the technical criteria. By submission of the bid the bidder agrees to waive any rights to pursue financial claim for loss of earnings resulting from the loss of this bid."

16.4.2 Performance-Based Bids

Another way of ensuring quality for money is to go out to bid on a performance basis. In this scenario the bidder basically becomes a partner of the utility, sharing in the gain from reduced overheads or increased revenue streams. Obviously, if the bidder does not perform, there will be no payment for services rendered, or reduced payment if the risk is shared with the utility.

16.5 CASE STUDY ONE

Water Leak Survey: A Performance-Based Approach

Vernon W. (Wes) Frye, P.E., Assistant Director,
Metro Water Services, Nashville, Tennessee

Paul V. Johnson, P.E., Director of Water Services,
ADS Corporation, Marietta, Georgia

16.5.1 History and Background

Metro Water Services (MWS) provides water and sewer service to metropolitan Nashville and Davidson County, Tennessee, which encompasses an area of 533 mi^2 and has a population of over a half a million people. There are approximately 2400 mi of water main in the service area and almost 136,000 customers. Water is obtained from the Cumberland River and treated and pumped to the system by two treatment plants—Omohundro Drive, with a capacity of 90 mgd, and K. R. Harrington, with a capacity of 90 mgd. There are 45 reservoirs ranging in size from 50,000 gal to 52 million gal with a total capacity of 83 million gal. Water mains are sized up to 60 in, and the mains and service connections are comprised of galvanized, copper, lead, cast iron, ductile iron, reinforced concrete, PVC, and asbestos cement. The average daily pumpage through December 1999 was 82 mgd.

From 1972 until 1992, MWS had an annual contract with a firm to perform an ongoing leak survey program and to do additional studies as might be required. For the last 7 years the department has had a water usage team meet routinely to review the unaccounted-for water figures. The team was created when the numbers started to reach 30–35 percent. The team performed an in-house water audit based on AWWA guidelines and, rather than compare billed water versus pumped finished water, began to account for other losses such as breaks, street cleaning, reservoir draining, fires, etc. This resulted in lower unaccounted-for water numbers, but only to the level of 25–30 percent. Another initial recommendation was to review the source meters at the two treatment plants. These were Venturi meters, and recent testing had indicated a discrepancy in the Omohundro meters while the Harrington meters seemed to be within acceptable limits. The committee recommended that new Venturi meters be installed at the Omohundro facility, and these resulted in even lower unaccounted-for water figures.

While the numbers did improve, they were still of an unacceptable level. As a result, the committee recommended that the department undertake a leak survey, since it had been 7 years since a formal program had been in place.

16.5.2 Origin of the "Performance-Based" Approach

When MWS made a decision to undergo a leak survey, it was based on the high unaccounted-for water numbers that were being obtained on a monthly basis. At this same time, the purchasing agent for the metropolitan government was presenting a new approach for contracts. He referred to this approach as a "performance-based contract," in which the vendor, or bidder, submitted a proposal in which he would receive a reward or bonus obtained above a predetermined level of performance. The purchasing agent presented this method as a "win/win" contract in that both parties would benefit from increased performance. When asked how the additional performance and resulting bonus would be determined, the purchasing agent responded that the bidder would provide that information. This was a concept that was unclear at the time, but MWS felt it was worth further investigation. It quickly became apparent that there was nothing to lose with this concept and everything to gain, since we always had the option to reject all bids if they were unsatisfactory.

16.5.3 Development and Requests for Proposal (RFPs)

We began to develop the proposal by getting a number of key people together with representatives of the Purchasing Department. Prior to actual development, we found it necessary to "sell" the newcomers to the concept before meaningful progress could be made. Once this was done, we started to look at how the specifications should read. The purchasing agent kept telling us not to prepare details but to allow the submitters total latitude to tell us what they would do, how they would do it, and how they would be compensated. This was a rather tough concept to accept at first, but the more it was discussed, the more comfortable we felt about proceeding in that direction. We eventually arrived at the realization that we were going to permit the bidders to tell us how to do the work. It then became clear that this wasn't all that bad. Who would know better how one would perform than the one who would be doing the work? This was an enlightenment that really gave the process a large boost. With that in mind, it was decided that the advertisement would basically contain a brief history and very minimal instructions as to what work was to be completed. It also required that all submitters attend a pre-bid meeting for additional information.

Prior to the pre-bid meeting, Metro Purchasing received a number of questions. These were requested to be put in writing and answers were prepared. The questions and answers were then passed out at the pre-bid meeting for the benefit of all attendees. At that meeting there were a number of initial questions, such as what was the procedure, how much work needed to be done, how would they be compensated, and what were MWS's wishes? The response given to these was that MWS desired to reduce its unaccounted-for water rate and that we were expecting the bidders to tell us what needed to be done, how they were going to do it, how they would be compensated, what bonus plans they proposed, and expected time frames. This was a new concept to the bidders, and they continued to ask the same questions, but eventually it became obvious that we were expecting them to prepare the proposed contract. MWS would provide maps for review, data, and other information as necessary, but it was emphasized that all bidders would receive the same data so as not to give an advantage to one over the others.

It should be pointed out that the bidders included not only the traditional leak survey companies but also small, local independent firms and some that had not been in the leak survey business previously. Those Metro representatives attending the meeting all felt that the submittals would be most interesting and that there was no way to predict what might be received from the mix of bidders. The bidders were given 6 weeks to prepare their proposals and submit them to Purchasing. Prior to receipt of the proposals, an evaluation team was developed to review the submittals. The team consisted of employees in MWS who dealt directly with lost water issues and would be familiar with the terms and procedures included in the proposals. In addition, Purchasing had a representative on the team to ensure that the submittals were in compliance with Metro's purchasing procedures.

16.5.4 Review of Requests for Proposal (RFPs)

Once the RFPs were received in Purchasing, they were distributed to the members of the evaluation team to review. The team then met to discuss and possibly select a successful proposal. Each of the team members was given an evaluation sheet, divided into five parts, in which to rank the various areas of each bidder's proposal. The areas and a description are as follows.

Acceptability of Proposal

Is the proposal acceptable, potentially acceptable, or unacceptable? If potentially acceptable or unacceptable, indicate the apparent problems with the proposal.

Mistakes

Note any minor informalities or other mistakes made in the proposal. (Purchasing agent must determine whether a mistake is a minor informality, or whether it can be corrected.)

Scoring on Individual Evaluation Factors

1. Qualifications and experience (maximum of 34 points)
2. Merits of the business plan (maximum of 31 points)
3. Compensation (maximum of 30 points)
4. Small business participation (maximum of 5 points)

Total Score

Sum of scores on individual factors above.

Award

Determined by Metro Purchasing.

After each individual had rated the proposals, they met jointly to determine the successful bidder. At this meeting, each rater presented his scores for each bidder. These were then totaled and an average calculated to use to determine a "base ranking." At this time, individuals were asked to explain why they rated that area at a particular score. Once this was done, each rater was then asked if he concurred with the "average" value. This was an especially important step, since the individual's score could be lower or higher than the average score. This procedure was followed for each bidder's proposal and then the total numbers were listed. In our particular situation, there was a clear result of the findings of the raters.

16.5.5 Development of Contract and Implementation

Based on the findings, all bidders were notified as to whether they had been successful or not. A date was then set for the apparent successful bidder, ADS, to meet with the purchasing agent and representatives of MWS to finalize the contract. This was necessary, as the initial request for bids did not contain specific specifications and all parties needed to agree on the wording of the contract prior to its preparation and processing.

In their business plan, ADS had submitted a "fixed fee" based on miles of pipe surveyed and would receive a "bonus" after the guaranteed amount of leakage had been achieved. The guarantee was established at 1.25 mgd. The bonus was graduated as the amount of leakage increased. Other bidders had submitted proposals similar to this or had offered a flat fee based either on a monthly basis or amount of line surveyed.

Once the contract was signed, work began on the survey. Metro Water Services provided a single contact person to coordinate the activities of ADS. In addition, in order to provide timely results regarding the achievement of the guarantee, it was emphasized that the suspected leaks be repaired as quickly as possible. As the repaired leaks neared the guaranteed amount, the results of the survey started showing up on our water usage report, which is an indicator for our unaccounted-for water.

While not always done, it had been agreed to at the development stage that efforts would be made for both a Metro Water Services employee and an ADS employee to concur on the amount of leakage that had been discovered and repaired. It had been agreed that if a discrepancy occurred between the two, then a meeting would be held to arrive at an agreed figure. In addition, if there were any other problems that could not be worked out by the contract person and the ADS representative, then, once again, a meeting with additional representatives from both sides would be held to resolve the issue. It should be obvious at this point that this type of contract requires a willingness to negotiate, compromise, partner, or progress to "agree to agree." Even though the Purchasing Rules dictate standards that must be adhered to, the "agree-to-agree" process has worked extremely well when a situation requires it be applied.

ADS achieved the guaranteed amount approximately 10 months after the start of the contract. At that point, they continued to conduct their survey but, in addition to the fixed fee they had been receiving, they were also being paid a bonus.

16.5.6 Conclusion

At the time this particular type of contract was proposed, it was very hard to picture such a process ending up as a success story. As the contract progressed, it became clear to all involved that it was a very "workable" process in which all parties benefited. It is proceeding today with ADS receiving a bonus for their efforts and Metro Water Services achieving its goal of reducing unaccounted-for water and recovery of lost revenues.

There have been very few problems as the contract has progressed. It is felt this is due to the detailed work on the front end between Metro Water Services and ADS to ensure that all parties understood the business plan and agreed to the wording prior to its inclusion in the formal contract documents.

A very important part of this process was the "agree-to-agree" procedure. This has eliminated problems as they have occurred and hopefully will allow a smooth ending when this contract comes to completion. If you choose to enter into a contract of this type, you must be willing to negotiate, compromise, or partner. We continually refer to it as "agree to agree." If you choose to have everything in black and white and begin to dictate to the bidders how things should or will be done, then you have taken away a very important step from this process. You must remember that you asked the bidders to tell you how they would do the job; they do it every day, so who would know best how to perform?

The working relationship between Metro Water Services and ADS Environmental has definitely followed the agree-to-agree process and is a major reason for the success of this leak survey.

16.6 KEEP CONTRACTS SIMPLE

In most cases the best agreements and contracts are the simplest ones. In all cases the contractor and the client need to agree on a baseline for payment and a means of measuring increased efficiency over the baseline over time. An agreed value for reduced real (or apparent) losses and an agreed value for increased revenues need to be established.

The length of the agreement needs to be tailored around the amount of investment the contractor will make and the payback periods envisaged.

Obviously, these must be realistic or nobody would bid. It is in the utility's interest to allow the contractor to make money so that he will continue to provide good service.

A sample performance-based project structure follows.

1. Phase One (fixed rate, upfront payment, repaid from savings later): Perform source meter testing and a system water audit to an agreed format.
2. Phase Two (payment by performance)

    ❑ Set up temporary district meters and data log the district demands.
    ❑ Identify the minimum night flow period and establish reasonable nonrevenue water and real loss levels.
    ❑ Identify areas where leakage is evident and undertake "step testing" to quantify the leaks accurately.
    ❑ Pinpoint the leaks using sonic and correlation methods and report on the locations for a directed repair program.
    ❑ Meter testing and downsizing program.
    ❑ Detailed engineering report on the findings of the program with procedures for ongoing leakage control practices to be undertaken by utility staff.

In this simple case the baseline for recovered leakage could be the minimum night flow before and after leakage location and repair, assuming that consumption and pressure stays constant. A simple check would be for the contractor and utility supervisor to estimate each leak repaired for volume of losses. This could be compared back to the change in night flow. An agreement would have to be made if significant differences were encountered. Care should be taken to deal with backlog leakage, which may distort figures the first time.

Likewise, to calculate the gain in metered sales after testing and correct sizing, the agreement could be to use a weighted rolling average of the last 3 months of sales for the year of change-out against the previous year. This ensures that consumption differences from one month to the other do not confuse the issue and result in conflict. In the event that clients consume less after change-out because they start to instigate conservation measures, an agreement could be made whereby the utility pays a fixed charge to the contractor for a period of time to cover his costs. In this case the value of leakage recovered could be either:

❑ The purchase cost if water is being imported from a water supplier
❑ The variable production cost if water is being produced locally
❑ One or both of the above plus a value for deferral of capital expansion if leakage reduction helps to defer the construction of new water sources in the case that the utility is going to peak out.

The value of recovered revenue should be the sales cost of the water less any fixed charges.

16.7 SUMMARY

The ideas identified above are just an example of the myriad of possibilities for negotiation between utility and contractor. However, in summary, the best way to ensure a successful project for both parties is to state as clearly as possible, up front, the requirements and specifications of the contract. It also always helps to go into a project situation in a position of trust. As was the case in the Metro Water Services contract, an "agree to agree" basis is the healthiest one.

16.7.1 Checklist

- ❏ Bid documents must be clear and to the point.
- ❏ Bidders packages should be prepared with background data, however sketchy it may be.
- ❏ A bidders meeting is a good idea.
- ❏ Gray areas will cause confusion and can lead to dispute.
- ❏ Budgets must be realistic.
- ❏ Performance-based options can be negotiated.
- ❏ Realistic time frames must be negotiated to allow the contractor to take his money out.
- ❏ Good baseline data and performance indicators should be used to clearly identify the situation before and after intervention.
- ❏ Have an "agree to agree" clause, which should be adhered to by both parties.

CHAPTER

17

Understanding Basic Hydraulics

Julian Thornton

17.1 INTRODUCTION

We are about to embark on a major mathematical exercise to identify losses and their economic value, so we had better go back to school for a few pages and review some of the calculations, which we have either been bypassing or have forgotten! The following sections deal with some of the most frequently used calculations, tables, and transformations used in the water loss management and field-testing world.

17.2 PIPE ROUGHNESS COEFFICIENTS

All water pipes have a roughness factor, which plays a part in creating friction between the water running in the pipe and the pipe wall. Think about a glass pipe, which would be very slippery because of its high coefficient, as opposed to old rusted and encrusted cast iron pipe, which would have a very low coefficient. This roughness factor is very important when considering either computer modeling or pressure zoning and management. In cases where fire flows are critical, the roughness factor may indicate that a zone may not be shut in, due to poor hydraulic conductivity.

Additionally, the roughness factor can indicate pipes in poor condition and is often used by water system operators to identify which pipes should be earmarked for either replacement or some form of rehabilitation. Earlier in this book we dealt with the various types of rehabilitation techniques available and in use today, where and when to use them, and how they are best applied. We also discussed when to replace and when to repair or rehabilitate.

There are various methods of calculating roughness in a closed pipe, such as the Manning, Darcy-Weisbach, and Colebrook-White equations. However, the most common method for pressurized water pipes is through the use of the Hazen-Williams C factor. The Hazen-Williams formula can be found to the right.

In order to measure the C factor we must measure flows and pressures, levels and elevations, and pipe length and diameter. The following section identifies detailed methodology statement for this type of testing in the field.

A table showing average C factors for various pipes of various ages and diameters may be used as a first estimate when applying values for decision-making models, but such a table is not a substitute for actual values measured in the field.

- Continuity: $Q = V \times A$
- Hazen-Williams:
$$V = 1.318 \times C \times R^{0.63} \times S^{0.54}$$

418 CHAPTER 17   UNDERSTANDING BASIC HYDRAULICS

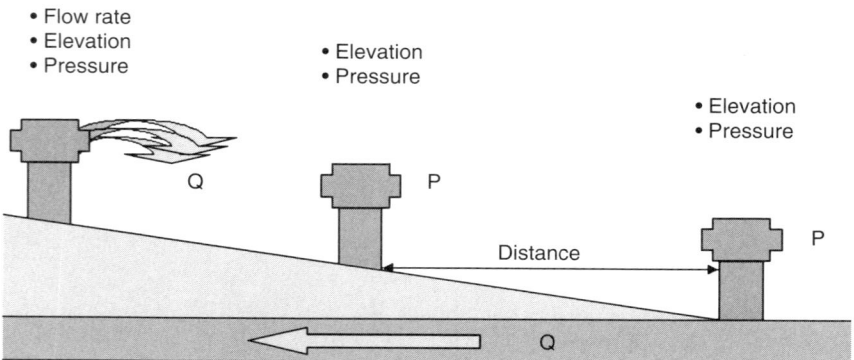

Figure 17.1 *C*-factor test. (*Source: Julian Thornton.*)

17.3 *C*-FACTOR TESTING IN THE FIELD

Accurate measurements of flow, pressure, level, diameter, and length must be undertaken. Figure 17.1 shows an example of a *C*-factor test laid out with two hydrants being used for pressure measurements and one for flow measurement. The spreadsheet in Table 17.1 shows a sample calculation using commercial software called Flowmaster, from Hastead Methods.

It is important to use calibrated equipment when undertaking this kind of testing, as the results are very sensitive, particularly to pressure. Further information on the equipment and calibration procedures used in this type of testing can be found in Chap. 6.

It is important that the operator has a basic feel for the results which will come out of the testing, so that unnecessary returns to the field can be avoided. After all, the idea of reducing losses and undertaking rehabilitation is to make a more efficient system! Fieldwork is a very important part of this process and can often be quite expensive when done properly either by contractor or by in-house staff. Proper planning and a good "ballpark" feel for what the results should be can help keep unnecessary work to a minimum.

While it is possible to have *C* factors higher than 130 and lower than 75, these cases are less likely, so by using this as a safety band the operator can query anything outside of these numbers while still on site and retest while the equipment is still set up. If the numbers are repeatable, then as long as they are not a long way off of the above recommendations, they may be realistic.

Age and pipe material play a large part in the *C* factor, with the worse conditions often being found on old, corroded, and untreated iron pipe. It is common to find that very old untreated iron pipes can be almost closed off by tuberculation, corrosion, and debris.

> ***C*** **factors should fall somewhere between 75 and 130.**

17.4 FIRE FIGHTING REGULATIONS

When considering changing pressures either by zoning or pressure management, as discussed in Chap. 12, it is important to remember that many countries—including the United States and Canada—have mandatory

TABLE 17.1 Sample C-Factor Calculation

Input	U.S. Units	Output	U.S. Units
Elevation @ 1	605.86 ft	Velocity	1.65 fps
Pressure @ 1	25.2 psi	Headloss	2.11 ft
Elevation @ 2	540.65 ft	Energy Grade @ 1	664.04 ft
Pressure @ 2	52.55 psi	Energy Grade @ 2	661.93 ft
Discharge	2325 gpm	Friction Slope	0.0004 ft/ft
Diameter	24 in		
Length	4710 ft		
Hazen-Williams Coef.	124.41		

Results calculated using Flowmaster.

flow and pressure levels for fire fighting demand. Although all countries do not have the same regulations, the author has had the opportunity to study several regulations and there is quite a lot of similarity among them. Most countries require a minimum flowing pressure of 20 psi or 15 m.

Before changing system pressures, it is also necessary to check the needs of local insurance underwriters in many cases. Often the insurance rating for a particular type of property is determined partially on local fire fighting capability, which is in many cases stated in terms of flow. In many cities the hydrant caps and bonnets are painted different colors to indicate the flow capacity of the hydrant.

In addition to understanding how to define hydrant needs for fire fighting, it is also important to undertake a demand analysis in any proposed area where zoning or pressure management may be undertaken, to identify the needs of internal building sprinkler systems. These systems can often be reset to accept lower inlet pressures, but consumers must be made aware of any potential change so that volumes may be properly calculated. Usually the benefit from pressure management is great enough that the cost of recalibrating these systems can be borne by the contract, which also ensures that the clients are happy.

17.5 FLOW TERMS

The term *flow* is used to describe the amount of water passing a point in pipework, perhaps a meter, in a certain time period. Flow is actually a moving volume of water (or any other substance). Flow can be recorded in many different units, usually depending on whether the country uses metric, imperial, or U.S. units of measurement. The most common units for flow measurement are gallons per minute (gpm, either imperial or U.S.), cubic feet per second (ft^3/s), and, in the case of large flows such as those found in bulk mains, transfer stations, and treatment plants, millions of gallons per day (mgd). Corresponding metric units are liters per second (L/s), cubic meters per hour (m^3/h), and megaliters per day (MLD).

During a water audit, it is often necessary to convert between different units of flow and also velocity. *Velocity* is the speed at which a liquid moves. Velocity itself does not tell you anything about how much water is flowing—only how quickly it is moving along the pipe.

CHAPTER 17 UNDERSTANDING BASIC HYDRAULICS

> $Q = V \times A$, where $Q =$ flow, $V =$ velocity, and $A =$ cross-sectional area.

Flow volume is calculated by multiplying the average velocity of the fluid by the cross-sectional area of the pipe in which it is flowing. The formula for calculating flow can be found to the left.

When calculating the cross-sectional area of the pipe we must make sure that we are using the same units for area and velocity, i.e., square feet and feet per second or square meters and meters per second.

Finally, to calculate flow, multiply the area by the velocity, or speed, of the liquid, which must be measured in the field.

Let's do some calculations of flow and velocity.

17.5.1 Example

A 6-in pipe has a velocity of 1 ft/s. How much flow does this represent in gallons per minute?

We use the equation $Q = V \times A$.

We fill in the missing information: $V = 1$ ft/s and $A = 0.785 \times 0.5 \times 0.5 = 0.196$ ft². So flow $Q = 1 \times 0.196$ ft³/s.

However, we want to express the flow in gallons per minute. We must multiply the volume in cubic feet per second by 7.48 to get a figure in gallons. Then we have to multiply by 60 to turn the time frame into minutes: $Q = 0.196 \times 7.48 \times 60 = 87.96$ gpm.

Let's now do the same calculation in metric units. We convert the above calculations as follows: $V = 0.3048$ m/s and $A = 0.785 \times 0.15 \times 0.15 = 0.0176625$ m². So flow $Q = 0.3048 \times 0.0176625 = 0.0053835$ m³/s.

We discussed earlier that metric units of flow are usually expressed as either liters per second or cubic meters per hour, so we should change our answer: $0.0053835 \times 1000 = 5.3835$ L/s, or 0.0053835×3600 (60 min of 60 s each in 1 h) $= 19.3806$ m³/h.

There are 3.78 L in a gallon, so, just to check our two calculations: $(5.3835/3.78) \times 60 = 85.45$ gpm. (The difference of around 1 percent is due to the rounding up or down of the decimal places.)

17.5.2 Types of Flow

Now that we have done some basic flow calculations, let us discuss the different types of flow which may be encountered. It is important to understand the flow conditions in a pipe, so that a suitable place for monitoring or testing can be selected.

Many problems in water system calculations are caused by incorrect siting of meters or test equipment. This is often the case because the individual who is responsible for the installation of the meter does not understand basic fluid dynamics. Other problems are caused by incorrect conversion of units, which is why it is important to understand the relationships between various types of units.

It is a good idea to write down the types of units currently in use in the water system you are going to audit and their respective conversions, before starting any field work or data analysis. (More information on data handling can be found in Chap. 5.) This way you can avoid costly mistakes later.

Although it is not necessary to understand fluid dynamics on a high level, a few basic ideas will come in handy. If we understand the basics, then when it comes time to troubleshoot we will be much better equipped to track down a problem.

If we do not understand the basics of fluid dynamics, we may end up fitting our test equipment at an unsuitable site. In this case our test data might be incorrect and we might input the wrong data to what might be a very important calculation.

The following sections cover the various types of flow. Although it is unusual to have to deal with all of these types of flow in a single water system, it is important to know that they exist.

Steady-State Flow

Flow is considered steady if, at a certain point in the pipe, the velocity of the water does not change with time but remains the same. Steady-state flows are not often found in field situations, but are sometimes used in simple modeling calculations.

Uniform Flow

Flow is considered to be uniform when the velocity does not change speed or direction from point to point. This condition can be found in transmission mains with long lengths of equal diameter and few restrictions such as butterfly valves or control valves. Most distribution-level mains experience nonuniform flow.

At the distribution level in the field, mains size is always changing, making velocity and pressure change constantly. In other situations water is passing through meters and control valves.

For our purposes, when measuring bulk flows on transmission systems, we should always record flow at a point where the flow is reasonably uniform. This is less likely to be possible when measuring distribution system flows.

Many manufacturers of portable equipment suggest a minimum of 10 times the pipe diameter upstream and 5 times the diameter downstream of any fitting or restriction. If in doubt, the 30/20 rule is a good bet; see Fig. 17.2. It is not good to install test equipment next to a permanent meter or valve, as the test equipment may then be less reliable than the equipment being tested.

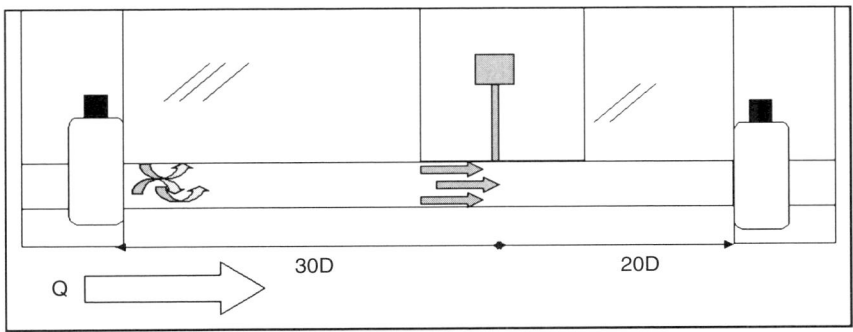

Figure 17.2 30/20 rule for installation of temporary measuring equipment. (*Source: Julian Thornton.*)

Laminar Flow

In a laminar flow condition the water moves along straight parallel paths, in layers or streamlines. The water velocity is not the same. Inside the pipeline, the layers of water closest to the pipe wall rub against the pipe walls and therefore travel more slowly. We discussed friction factors earlier in this chapter. We now know that by measuring flow and pressure we can actually calculate the internal condition of the pipe.

Laminar flow is not very common in water pipes and occurs only at very low velocities. Most utility situations have nonlaminar or turbulent flow. A term called a Reynolds number is used to calculate if flow is laminar or turbulent. Many pieces of flow monitoring equipment quote Reynolds numbers above or below which the equipment may be used. A Reynolds number is a function of velocity, diameter, and viscosity: $R = $ (velocity $\times$ diameter)/viscosity.

Most utility pipeline applications have Reynolds numbers in the hundreds of thousands. A Reynolds number of less than 2000 would indicate a Laminar flow condition.

Turbulent Flow

In a turbulent flow condition the water tumbles along in a more confused fashion, although this is the usual state for water flow in a water supply system. Figure 17.3 shows the difference between laminar and turbulent flow.

Although most water system flows are turbulent, the flow does still move in a forward direction. When monitoring flows we usually see that the velocity is greater toward the center of the pipe than it is at the sides. This is very important to understand, as we use different techniques to monitor flow in the field. Sometimes we use a single-point velocity meter, which requires finding the average velocity in the pipe and multiplying this by the cross-sectional area to find volumetric flow. At other times we use equipment which averages the velocity across the pipe diameter and calculate the average velocity for us. Either way, if we understand the basics we are better equipped to deal with anomalies.

It is important to note the law of conservation of mass, which states that material is neither created nor destroyed. So whatever flow enters a system must leave it (at either a consumer point or feed to another system, or leakage) or accumulate inside it (as in the case of storage tanks and reservoirs). Since water is incompressible, it cannot accumulate inside the system or individual pipes. This is why we have to have tanks or reservoirs inside a water system—to accumulate water. We usually refer to this as *storage*. The conservation of mass is the basis behind the water audit or balance. The water that enters a system is delivered to a customer, delivered to storage, or deliv-

Figure 17.3 Difference between laminar and turbulent flow. (*Source: Julian Thornton.*)

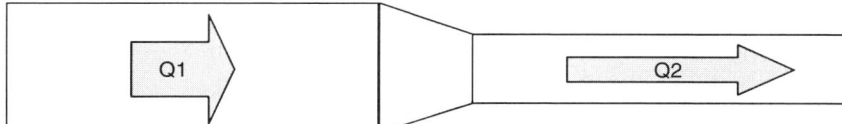

Figure 17.4 Conservation of mass. Q1 = Q2; however, the velocity will have increased. (*Source: Julian Thornton.*)

ered to a loss situation such as leakage or theft. It cannot simply disappear (even though in some audits it may seem that way for a while).

A gas system, however, is different. Storage is achieved by "line packing" or increasing the pressure of the gas in the line. Gas is compressible, so this is possible. This should be remembered if undertaking a tracer gas survey. Due to the fact that gas is compressible, a line break will cause a lot more damage when it is filled with gas than with water.

Figure 17.4 illustrates the law of conservation of mass.

Often it is necessary to measure velocity in a particular pipe. One of the main reasons is to decide if a site is suitable for a monitoring position. Sites with low velocities are not considered good places to monitor, as the equipment tends to stall out. Most flow monitoring equipment will state the stall-out velocity. This is often around 0.3 ft/s or 0.1 m/s. Flows calculated with velocities below these figures should be scrutinized for instability and potential error using the manufacturer's error curve and calibration certificate.

For example, we may know the flow and velocity leaving a treatment station or tank, and we want to undertake a flow balance to assess a system for leakage. Obviously, we want to reduce the inherent error in our monitoring equipment, so we try to locate the pipes with the highest velocity to monitor. Other pipes may be temporarily shut down during the period of testing. Larger pipes are often subjected to lower velocities, especially in newer systems where they may have been sized for future population growth.

17.6 PRESSURE TERMS

To understand pressure we need to think of containers of water resting on the ground. The weight of the water is 62.4 lb for every cubic foot; if we divide by 7.4 gal per cubic foot we find that we have 8.34 lb for every gallon. In metric units the corresponding weights and measures are 1 L = 1 kg, and 1 m^3, which is 1000 L, weights 1000 kg.

This weight resting on a surface exerts a force on that surface. That force is what we call pressure.

17.6.1 Example

One cubic foot of water resting on the ground exerts a force of 62.4 lb on that bottom square foot. If there are 2 ft^3 of water resting on a 1-ft^2 area, the pressure is 124.8 lb, and if there are 20 ft^2 resting on that bottom foot, then the pressure is 1248 lb. This is obviously important when we start thinking about storage and measuring pressures at tanks.

> ❑ Pressure is force per unit area.
> ❑ Pressure = weight × height

The same is true in metric units: 1 m³ of water resting on 1 m² of area exerts 1000 kg of pressure on that square meter, and 2 m³ of water resting on 1 m² of area exerts 2000 kg of pressure. If there are 20 m³ of water in a column resting on 1 m² of area, then the pressure is 20,000 kg.

During analysis of a water system, we often need to know the pressure at a particular point. Some of the reasons we may need to know are

- ❑ To calculate the amount of leakage occurring through a hole of a known size
- ❑ To calculate the amount of leakage through a varying hole or split
- ❑ To calculate the amount of water flowing from a fire hydrant during a test
- ❑ To figure out the condition of the main, when used in conjunction with a flow test
- ❑ To calibrate a computer model
- ❑ To see if a customer has sufficient pressure for supply

We just saw that there are 62.4 lb per square foot; however, water systems usually use pounds per square inch, or psi, as the denominator. We must therefore split our square foot into inches. There are 12 × 12 = 144 in in a square foot, so there are 62.4/144 = 0.433 lb per square inch. Therefore, for every foot of height (head), there is a pressure of 0.433 psi. Most people remember it the other way round, in terms of how many feet of water yield a pressure of 1 psi. So let us calculate this: 1/0.433 = 2.31 ft.

> **M**any systems that use metric units also use psi.

Metric calculations of height and weight (pressure) are generally easier, as the weight or pressure is stated as meters column of water. Therefore, if you have 20 m of water resting on 1 m² of area, you have a pressure of 20 m head. Sometimes this is stated in bar. One bar is 10 m head, so in this example 20 m head would be 2.0 bar. The metric system is quite easy to use, as the multipliers and dividers are all factors of 10 (e.g., 1 m of water = 1.42 psi; 1 bar of water pressure = 14.2 psi).

Static system pressure is due to the depth of water above the point of measurement; it has no relationship to the size of the water pipe or storage tank. Storage tanks of equal height but different shapes all provide the same pressure. See Fig. 17.5. Static system pressure within the distribution system is measured by taking the tank height plus the difference in elevation at the point of measuring in the system to the bottom of the tank. Static system pressure occurs only when there is very little flow in the system, which is usually at night. In some water systems the static pressure is never reached, because of high demands. As we discussed earlier in the chapter, the water flowing in a pipe is subjected to friction losses or headlosses. These headlosses change as flow conditions change within the system.

17.6.2 Gravity-Feed Systems

Often a water system will operate from static head in a tank and will gravity-feed into a system. Water is either pumped from a surface reservoir to a tank, so pumping costs do occur, or, as in the case of some mountain cities, water comes from a high-level natural reservoir or spring. See Fig. 17.6. This second type of system theoretically has the lowest operating costs and often the highest unaccounted-for water levels. The

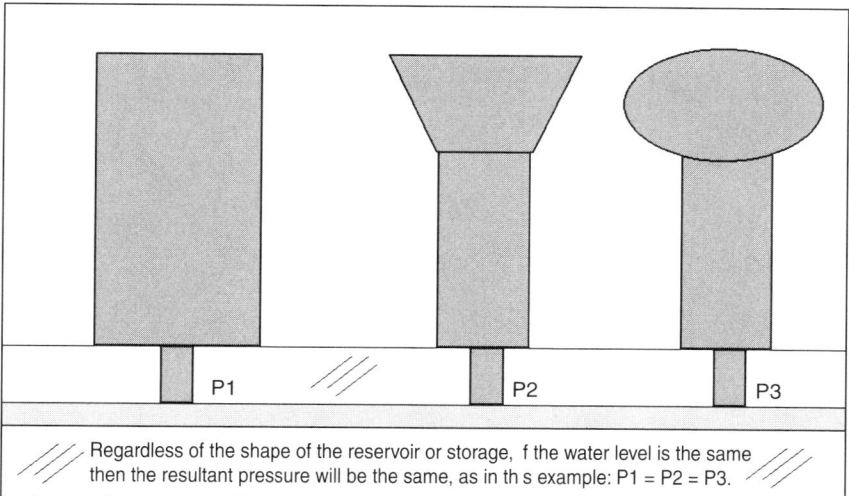

Figure 17.5 Storage tanks of equal height but different shapes all provide the same pressure. (*Source: Julian Thornton.*)

high unaccounted-for water levels are often the product of attitude, as the water is so cheap and pressures from a high-level reservoir run either unchecked or badly calibrated. These systems, however, often experience trouble in treating the large quantity of water passing through the station, particularly when turbidity is high, therefore making treatment costs high.

In many cases the only way to cost-justify a loss control project in a mountain community gravity fed without booster pumps is on deferral of capital costs for distribution system upgrades.

17.6.3 Pumped Systems

Pumps can also provide system pressure. The pumps actually provide lift or head. Pumped pressure is measured in the same way as static pressure.

Pumps are often used to lift water from a surface reservoir or well to a holding reservoir, standpipe, or water tower. Some systems, however, use direct pumped systems. See Fig. 17.7.

Most pumped water systems tend to have higher water prices than gravity-fed systems, particularly if the marginal cost of water is only calculated using the power and chemicals calculation. This of course could be untrue in cases where a system has extremely high treatment costs. An example is a system that uses reverse osmosis for treatment.

17.6.4 Pressure Measurements

Pressure measurements can be made in a number of ways. Some of the more common are

- ❏ Piezometers (level tube)
- ❏ Pressure gauges
- ❏ Pressure loggers

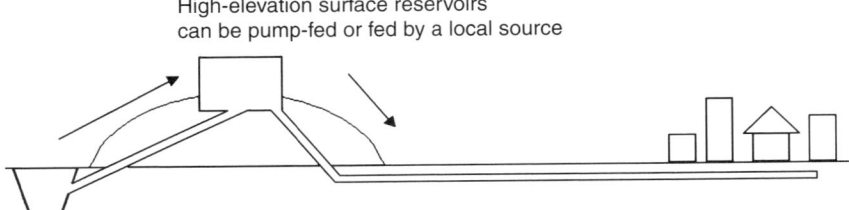

Figure 17.6 Gravity-fed system. (*Source: Julian Thornton.*)

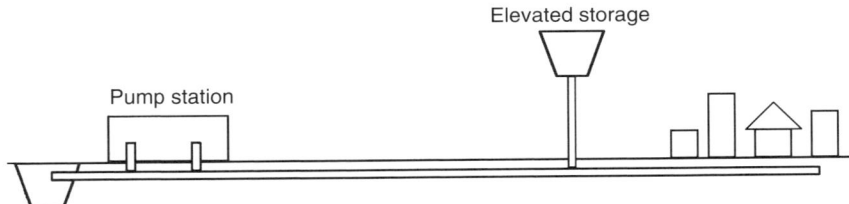

Figure 17.7 Pump-fed system. (*Source: Julian Thornton.*)

Most pressure measurements in water systems are made using *gauge pressure*, which is the difference between a given pressure and atmospheric pressure. *Absolute pressure* is the reading of pressure including atmospheric pressure. This type of reading may be used, for example, at a weather station. All of our measurements will be done using gauge pressure.

17.6.5 Effects of Pressure

Water hammer or *hydraulic shock* is the momentary increase in pressure which occurs in a dynamic (moving) water system, due to sudden change of direction or velocity of the water. Water hammer is often caused by incorrect operation or calibration of valves, pumps, or fire hydrants. A less severe form is often referred to as *surge*. This can be due to pressure fluctuations caused by natural changes in demand.

Many systems fit surge tanks, or surge anticipator valves (see Fig. 17.8), which provide pressure relief for a system. Reservoirs and storage tanks also help to vent unnecessary pressure. Reservoir storage is also often found to be an area of high leakage if not controlled properly. Most utilities are starting to fit pressure control systems, often consisting of fixed-outlet pressure-relief valves or altitude valves (see Fig. 17.9). In addition, pumped systems with high leakage levels often see benefits from installing surge anticipator valves. Some utilities are also seeing increased benefits from modulated pressure control. We dealt with this subject in more depth in Chap. 12.

> Water hammer and surge are often responsible for recurring leakage in a system.

17.7 SUMMARY

In this chapter we have covered some of the very basic concepts of flow and pressure and some of the effects of headloss. The following references were used during this research and are recommended reading for those who may wish to learn more about this topic.

Figure 17.8 Surge anticipator valve. (*Source: Watts ACV, Houston, Texas.*)

Figure 17.9 Altitude valve. (*Source: Watts ACV, Houston, Texas.*)

17.8 REFERENCES

General references regarding flow and pressure terms:

1. Hauser, B. A., *Practical Hydraulics Handbook,* Lewis Publishers, Chelsea, Mich., 1991.
2. Giles, R. V., Evett, J. B., and Liu, C., *Schaum's Outline of Theory and Problems of Fluid Mechanics and Hydraulics,* 3d ed., McGraw-Hill, New York, 1994.

CHAPTER

18

Articles and Case Studies

Julian Thornton

Article One
Water Loss Management in North America:
Just How Good Is It?
*Allan Lambert
Dale Huntington
Timothy G. Brown*

Case Study One
A U.S. Water Loss Case Study:
The Philadelphia Experience
George Kunkel

Case Study Two
The Loss Control in
São Paulo, Brazil
*Marcelo Salles
Holanda de Freitas
Francisco Paracampos*

Article Two
Accounting for Lost Water
Leslie M. Buie

Article Three
Plugging Billion-Gallon
Losses in Tennessee
Larry Counts

Article Four
A Holistic Approach to
Reducing Energy
Requirements for Centrifugal
Pumps and Pumping Systems
Horton Wasserman, P.E., D.E.E.

18.1 INTRODUCTION

Throughout this manual, case studies and articles have supported the text. The manual was designed to promote awareness of the state of water loss today and to stimulate interest in achieving better targets in the future. Various technologies and methodologies have been cited throughout the book.

The following articles and case studies discuss some of the efforts of practitioners and utilities to apply the methods stated and/or improve upon their previous practices.

Article One, by Allan Lambert, Dale Huntington, and Timothy G. Brown, discusses the new performance indicators recently published by the International Water Association (IWA) and their applicability to North American water loss management. (Further work in this area is being undertaken through recently approved AWWARF contract 2811, "Evaluating Water Loss and Planning Loss Reduction Strategies.")

Case Study One, by George Kunkel of the Philadelphia Water Department, details Philadelphia's efforts since the last century to combat water losses. Kunkel describes the evolution of a task team dedicated to loss management, culminating in a recent project to asses losses using the latest water balance and performance indicators from the IWA best practices manual.

Case Study Two, by Marcelo Salles Holanda de Freitas and Francisco Paracampos, details the efforts of one of the largest individual water undertakings in the world SABESP, São Paulo, Brazil, to identify and reduce water losses in a sustainable manner. Through necessity, SABESP has embarked upon large real and apparent loss reduction intervention projects in the field. SABESP is also currently reviewing its loss management practices and testing the IWA performance indicators as part of an international group of water companies that have volunteered to field-test the best practices.

Article Two is a detailed account by Leslie M. Buie of the efforts of DeKalb County, Georgia, to identify and reduce water losses in a sustainable manner.

Article Three is a detailed account by Larry Counts of the efforts of Johnson City, Tennessee, to become more efficient and accountable. This case study has been modified from its original 1997 form with data from continuing loss management efforts in 1998 and 1999.

Article Four, by Horton Wasserman of Malcolm Pirnie, covers some of the ways to reduce energy costs for pumping water in distribution systems. Energy conservation is often a direct or indirect benefit of a water loss reduction program and is becoming an issue in many countries.

These articles and case studies should be extremely valuable for readers who are considering improvements in their loss management practices, whether as a utility or municipality or as a consultant or contractor. Case studies are an extremely useful and effective way of justifying an investment in optimization, and articles from the literature are helpful, too. Many projects today include a literature search and current practice task. This chapter should provide useful information.

18.2 ARTICLE ONE

Water Loss Management in North America: Just How Good Is It?

Allan Lambert, International Water Data Comparisons Ltd., North Wales, United Kingdom
Dale Huntington, Huntington & Associates, Fallbrook, California
Timothy G. Brown, Heath Consultants Inc., Houston, Texas

18.2.1 Introduction

A 1996 overview of water production and delivery statistics in 469 U.S. water utilities showed metered water ratios averaging 84 percent, implying unaccounted-for water (UFW) of 16 percent of production, on average. The range of UFW was very large, from over 50 percent down to 1 percent, with a 13 percent modal (most frequently occurring) value. So how does this compare to other countries worldwide? An international data set of 27 systems in 20 countries, assembled by the International Water Association Task Force on Water Losses, which included two of the authors, showed a range of nonrevenue water (NRW) from 2 to 40 percent, with an average of 17.5 percent and a 13 percent modal value.

These overall statistics seem fairly similar. However, technical committees in many countries—including the American Water Works Association (AWWA) Leak Detection and Water Accountability Committee—have realized over recent years that expressing UFW or NRW as a percentage of production is only a very crude measure of true performance in managing water losses, because it is strongly influenced by differences in consumption.

The IWA Water Losses Task Force, working in conjunction with the IWA Task Force on Performance Indicators, has recently reviewed the traditional measures for comparing UFW/NRW/water losses. They recommended improved measures for comparing performance, applicable to a wide variety of different situations between, and within, countries. Key features of this internationally recommended approach are:

- ❑ A standard approach to water balance calculations used for performance comparisons
- ❑ Different NRW and water losses performance indicators for different purposes
- ❑ A component-based equation for predicting unavoidable annual real losses, taking key local system operating characteristics into account

These IWA Task Force recommendations were mentioned in a presentation to the AWWA Distribution Systems Symposium in Reno, Nevada, in 1999. As a result of the interest this generated, the Leak Detection and Water Accountability Committee encouraged a number of utilities to volunteer water audit and other data, to apply the approach to some North American supply systems. This presentation:

- Outlines the recommended IWA approach
- Compares the IWA equations for unavoidable real losses with approaches previously used to estimate unavoidable losses in North America
- Compares the recommended performance indicators for real losses for the volunteer North American supply systems with the international data set

18.2.2 An International Approach to Water Balance Calculations

Before attempting any performance comparisons, water balance data need to be reallocated into the IWA standard terminology and water balance components,[1] as shown in Fig. 18.1. These are based on international best practice, and include terms familiar to AWWA members such as "authorized consumption." However, the IWA has dropped the term "unaccounted-for water" in favor of "nonrevenue water," because there is no internationally accepted definition of UFW, and all components of the water balance can be accounted for using the above process.

All data are expressed as volumes per year. The calculation procedure is as follows:

- Obtain system input volume and correct for known errors.
- Obtain components of revenue water (right-hand column); calculate revenue water.
- Calculate nonrevenue water (system input − revenue water).

Own Sources	System Input (Allow for Known Errors)	Water Exported				Billed Water Exported
		Water Supplied	Authorized Consumption	Billed Authorized Consumption	Revenue Water	Billed Metered Consumption
						Billed Unmetered Consumption
Water Imported				Unbilled Authorized Consumption	Nonrevenue Water	Unbilled Metered Consumption
						Unbilled Unmetered Consumption
			Water Losses	Apparent Losses		Unauthorized Consumption
						Customer Metering Inaccuracies
				Real Losses		Leakage on Mains
						Leakage and Overflows at Storages
						Leakage on Service Connections up to Point of Customer Metering

Figure 18.1 International standard water balance. (*Source: "Water Loss Management in North America."*)

- Assess unbilled authorized consumption.
- Calculate authorized consumption (= billed + unbilled authorized consumption).
- Calculate water losses (= system input − authorized consumption).
- Assess components of apparent losses (right-hand column); calculate apparent losses.
- Calculate real losses (= water losses − apparent losses).
- Assess components of real losses from first principles (e.g., burst frequency/flow rate/duration calculations, night flow analysis, modeling) and cross-check with calculated volume of real losses.

Particular points to note, in comparison with the AWWA M36 audit procedure,[2] include the following:

- Authorized consumption is separated into billed and unbilled components, to allow both financial and operations performance indicators to be calculated.
- Authorized metered consumption does not include customer meter errors, which are part of apparent losses in the IWA water balance.
- Authorized consumption does not include volumes of losses from known (discovered) leaks, breaks, and storage tank leakage/overflows, or estimates of unavoidable losses—these are all part of "real losses" in the IWA methodology.
- An IWA standard water balance can easily be derived for components of an M36 water audit using the cross-referencing system in Table 18.1.

TABLE 18.1 Identifying IWA Water Balance Components from an M36 (2d ed., 1999) Audit Worksheet

IWA Water Balance Component	M36 Audit	M36 Description
System input volume	Line 4	Adjusted total water supply to the distribution system
Authorized metered consumption	Line 5	Uncorrected total metered water use
Authorized unmetered consumption	Line 12	Total authorized unmetered water
Apparent losses	Line 9	Corrected total metered water deliveries
	−Line 5	Uncorrected total metered water use
	+Line 14A	Accounting procedure errors
	+Line 14B	Unauthorized connections
	+Line 14H	Unauthorized use
Real losses	Line 14C	Malfunctioning dist. system controls
	+Line 14D	Reservoir seepage and leakage
	+Line 14E	Evaporation
	+Line 14F	Reservoir overflow
	+Line 14G	Discovered leaks
	+Line 16	Potential water system leakage

Source: "Water Loss Management in North America: Just How Good Is It?"

18.2.3 Recommended Performance Indicators for Nonrevenue Water and Real Losses

Table 18.2 shows the performance indicators (PIs) for nonrevenue water and real losses recommended by IWA (see Ref. 1) converted to North American units. The PIs are categorized by function and by level, defined as follows.

- *Level 1 (basic):* A first layer of indicators that provides a general management overview of the efficiency and effectiveness of the water undertaking.
- *Level 2 (intermediate):* Additional indicators, which provide better insight than the Level 1 indicators for users who need to go into more depth.
- *Level 3 (detailed):* Indicators that provide the greatest amount of specific detail, but are still relevant at the top management level.

Particular points to note with regard to Table 18.2 are as follows.

- *Fi36:* Percentage of nonrevenue water is the basic *financial* PI.
- *Fi37:* This detailed financial PI is very similar to the recommendation of the AWWA Leak Detection and Water Accountability Committee.[3]
- *WR1:* Real losses as a percent are unsuitable for assessing efficiency of management of distribution systems for control of real losses (because of the influence of consumption).

TABLE 18.2 IWA Recommended Performance Indicators for Nonrevenue Water and Water Losses

Function	Ref.	Level	Performance Indicator	Comments
Financial: Nonrevenue water by volume	Fi36	1 (Basic)	Volume of nonrevenue water as percentage of system input volume	Can be calculated from simple water balance
Financial: Nonrevenue water by cost	Fi37	3 (Detailed)	Value of nonrevenue water as percentage of annual cost of running system	Allows different unit costs for nonrevenue water components
Inefficiency of use of water resources	WR1	1 (Basic)	Real losses as a percentage of system input volume	Unsuitable for assessing efficiency of management of distribution systems
Operational: Real losses	p240	1 (Basic)	Gals/service line/day, when system pressurized	Best "traditional" performance indicator
Operational: Real losses	OP25	3 (Detailed)	Infrastructure leakage index	Ratio of current annual real losses to unavoidable annual real losses

Source: "Water Loss Management in North America: Just How Good Is It?"

- *Op24:* Gallons/service line/day is the most reliable of the traditional PIs for real losses, for all systems with service line densities of > 30/mi.

To improve on Op24, we need to take account of three key system-specific factors:

1. Average operating pressure
2. Location of customer meter relative to property line
3. Density of service connections

By expressing Op24 as gallons/service line/day/psi of pressure, the influence of pressure can be allowed for in intermediate (Level 2) comparisons.

An infrastructure leakage index (ILI) is a measure of how well the system is being managed for the control of real losses, at the current operating pressure. ILI is the ratio of current annual real losses to unavoidable annual real losses.

Unavoidable Annual Real Losses (UARL) are calculated from an equation[4] which takes into account average operating pressure, length of mains, number of service lines, and location of customer meters relative to the property line.

18.2.4 Previous North American Approaches to Estimating Unavoidable Losses

J. C. Smith's M.Sc. Thesis

J. C. Smith's M.Sc. thesis (see Ref. 4), written in 1987, provides a convenient summary of North American practice at the time. Smith found that the most commonly used method for estimating unavoidable leakage in North America was based on the Kuichling equation (published in 1887), which takes account of length of mains, number of pipe joints, hydrants and stop valves, and number of service connections. However, Smith's thesis concluded that this equation was "an inadequate and inappropriate method for estimating unavoidable leakage in modern distribution systems," because:

- It does not allow for the influence of pressure on leakage rates (Smith considers this to be the single most important factor influencing unavoidable leakage).
- The "drip rates" used by Kuichling may not be representative or valid.
- Kuichling's equation was developed for cast iron pipes, with jointing techniques (particularly on mains) much inferior to present practice.
- Kuichling's equation does not consider the effect of material type or age.
- Modern electronic leak detection equipment can locate hidden leaks which would have remained undiscovered 100 years ago.

Smith's definition of unavoidable leakage as "that part of system leakage which is lost but is not economical to locate and repair," includes:

- Small, nonsurfacing leaks which cannot be found by conventional leak detection methods (these are the only type which Kuichling considered "unavoidable")
- Leaks which do not have a high enough flow rate to make the cost of repair economical

Smith proposed an equation for calculating a threshold value of an individual leak flow rate (Q gal/min), given average repair cost and an assumed economic return period. However, he acknowledged that most of the repair costs are in excavation costs, that it is difficult to know the flow rate until the excavation has taken place, and once a leak has been excavated it is always sensible to repair it, whatever the size.

Smith then considered the two main alternatives to the Kuichling equation used in the United States. These are the "flat rate" method (an allowance per mile of mains), and the AWWA standard method for calculating allowable leakage on new water mains. He concluded that:

- The flat rate methods "neglect the effects of pressure, pipe materials, pipe size, system age, pipe location, frequency of service connections, and other system appurtenances."
- The AWWA standards provide reasonable methods of unavoidable leakage for new water mains, have the advantage of including for the effects of pressure, pipe material, and diameter, and pipe joint frequency, but do not allow for the effects of age or the frequency of connections or appurtenances to the system, so "they tend to provide overly conservative estimates for older pipe networks."

To try to overcome these deficiencies, Smith developed an equation for calculating an "unavoidable leakage index." by modifying the AWWA standard for new water mains. Factors were added for age, number of service connections, fire hydrants, valves and other system appurtenances, and pressure (assuming a square-root relationship between unavoidable leakage rate and pressure). Smith states that "Data is presently not available to test the assumptions used to develop the index equation…consequently some adjustment may be necessary to the equation before it can be uniformly applied to all water distribution systems. However, it does take account of the main factors which influence leakage…the only reasonable method available to a water utility manager to reduce unavoidable leakage is to reduce the system pressure and employ a comprehensive system maintenance and pipe replacement/rehabilitation program. The water customers often resist reduction of pressure."

It is not clear whether Smith was aware of published research in the United Kingdom[5] and Japan[6] which had shown that the relationship between leakage rate and pressure in sectors of distribution systems is, on average, approximately a linear (rather than a square-root) relationship.

Smith's conclusions were accepted and included almost word for word in the AWWA Research Foundation report *Water and Revenue Losses: Unaccounted-for Water*, in 1987.[7]

U.S. Practice, 1987–2000

Material provided to the authors by members of the AWWA Leak Detection and Water Accountability Committee, shows that "flat rate" assessments of unavoidable losses still predominate in U.S. practice. In 1997, Boston Water & Sewage Commission, using the same definition of unavoidable losses as Smith, considered that "the figures derived by Kuichling's formula and its derivatives, although dated, remain an accurate method to estimate the Commission's unavoidable leakage."

In a discussion of the AWWA Leak Detection and Water Accountability Committee of June 1997, on the subject of unavoidable and "noneconomic" leakage, which included reference to developing U.K. practice, the committee commented on the "wide array of terminology which seems to exist to describe a specific occurrence." In a later meeting, typical figures of 1000 to 3000 gal/mi/day were quoted. These appear to originate from a Report of AWWA Committee 4450D (Revenue Producing Water).[8] This 1957 report provides the definition of unavoidable leakage later used by Smith and states that "experience has shown that the value can range from 1000 to 3000 gpd per mile of pipe (main), depending on the age of the system, ground conditions, type of pipe and services, type of community, pressures and source of supply…a larger unaccounted-for figure is permissible where high pressures exist than under low-pressure conditions."

From the above, it appears that Smith's attempts to introduce a more rational approach to assessing unavoidable losses have not been taken up. This may be because the equation he proposed needs to be applied to many individual sections of different pipe diameter before the aggregate figure for unavoidable losses for a particular system can be calculated. Or it may be that as his equation had not been widely tested, practitioners are reluctant to depart from traditional "flat rate" methods. So North American practice in 2000 regarding assessment of unavoidable real losses can perhaps be summarized as follows.

- ❑ The AWWA definition of unavoidable losses includes not only undetectable small leaks, but also those which are considered uneconomic to repair.
- ❑ For over 40 years it has been recognized that several key factors are influential—notably pressure, density of service connections, age and type of pipe, etc.—but attempts by Smith and the AWWA Research Foundation to rationalize these in an equation do not appear to have been implemented to any significant extent.
- ❑ In North America, unavoidable real losses are usually expressed in gallons per mile of mains per day (although we now know from the work of the IWA Task Force on Water Losses that gallons per service line per day would be a better and more rational scaling factor).
- ❑ Flat rate allowances of between 1000 and 3000 gal/mi of mains/day are widely used, with no clear guidance as to how the key factors influence the achievable values
- ❑ No allowance is made for customer meter locations in southern states being located close to the property line, whereas in northern states they tend to be distant from the property line (with additional propensity for leakage on customers' private pipes upstream of the meter).

18.2.5 The IWA Water Losses Task Force Approach to Calculating Unavoidable Annual Real Losses (UARL)

The IWA approach is described in detail in the December 1999 issue of the IWA's *Aqua* magazine,[9] and can be seen as a natural development of previous North American attempts to take key local factors into account. The UARL component-based equation is based on auditable assumptions for break frequencies/flow rates/durations; it uses background and breaks estimates concepts[10] to calculate the unavoidable real losses for a system with well-maintained infrastructure, speedy good-quality repairs of all detectable leaks and breaks, and efficient active leakage control to locate unreported breaks.

The parameters used in the calculation, taken from Ref. 9 and converted to North American units, are shown in Table 18.3 (gal = U.S. gallon).

Table 18.4 shows these in a user-friendlier format for calculation purposes. The "UARL Total" values, in the units shown in Table 18.4, provide a rational yet flexible basis for predicting UARL values for a wide range of distribution systems, taking into account length of mains, number of service lines, location of customer meters relative to property line, and average operating pressure.

An important aspect of Table 18.4 is the value assigned to unavoidable "background" (undetectable) real losses, shown in column 2. These figures are based on international data, from analysis of night flows in sectors just after all detectable leaks and breaks have been located and repaired. This component of unavoidable real losses has, to the best knowledge of the authors, never previously been quantified in North American practice, yet it accounts for at least 50 percent of the unavoidable real losses components in Table 18.4.

There are many different ways to present the UARL equation. Figure 18.2 shows UARL in gallons per mile per day per psi of pressure (Y axis) plotted against density of service lines. The large variation with density of service lines shows why it is not recommended to use "per mile" for comparisons of real losses.

TABLE 18.3 Parameter Values Used for Calculation of Unavoidable Annual Real Losses (UARL) (all flow rates are at a reference pressure of 70 psi)

Infrastructure Component	Background (Undetectable) Losses	Reported Breaks	Unreported Breaks
Mains	8.5 gal/mi/h	0.20 breaks/mi/year at 50 gpm for 3 days duration	0.01 break/mi/year at 25 gpm for 50 days duration
Service lines, mains to property line	0.33 gal/service line/h	2.25/1000 service line/year at 7 gpm for 8 days duration	0.75/1000 service line/year at 7 gpm for 100 days duration
Service lines after property line (for 50-ft ave. length)	0.13 gal/service line/h	1.5/1000 service line/year at 7 gpm for 9 days duration	0.50/1000 service line/year at 7 gpm for 101 days duration

Source: "Water Loss Management in North America: Just How Good Is It?"

TABLE 18.4 Unavoidable Annual Real Losses in Component Format

Infrastructure Component	Background Losses	Reported Bursts	Unreported Bursts	UARL Total	Units
Mains	2.87	1.75	0.77	5.4	gal/mi mains/day/psi of pressure
Service lines, mains to property line	0.112	0.007	0.030	0.15	gal/service line/day/psi of pressure
Underground pipes between edge of street and customer meters	4.78	0.57	2.12	7.5	gal/mi u.g. pipe/day/psi of pressure

Source: "Water Loss Management In North America: Just How Good Is It?"

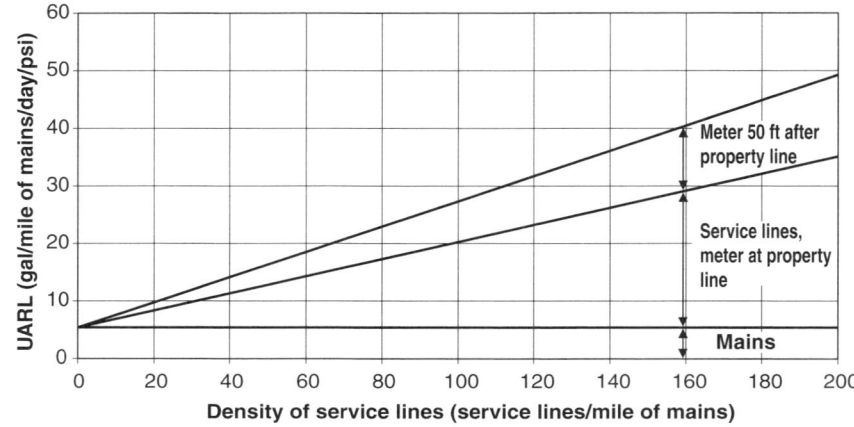

Figure 18.2 Unavoidable annual real losses (gal/mi of mains/day/psi) versus density of service lines. (*Source: "Water Loss Management in North America."*)

If the Table 18.4 UARL values are plotted as a graph of gallons per service line per day per psi of pressure versus density of service lines, the relatively flat curves show why "per service line" is preferred to "per mile" as the basic performance indicator, except at low densities of service lines. See Fig. 18.3.

18.2.6 Application to Data from North American Supply Systems

Three methods have been used to check the application of the IWA methodology to North American water supply systems.

1. *Comparison of IWA predicted range of unavoidable annual real losses with typical range quoted for U.S. systems.* Figure 18.2 can be used to calculate system-specific UARL in gallons per mile per day, by assuming upper and lower limits for density of service lines, customer meter location, and pressure.

❏ Lower limit: rural, meter 50 ft after property line; 30 service lines per mile; pressure 50 psi

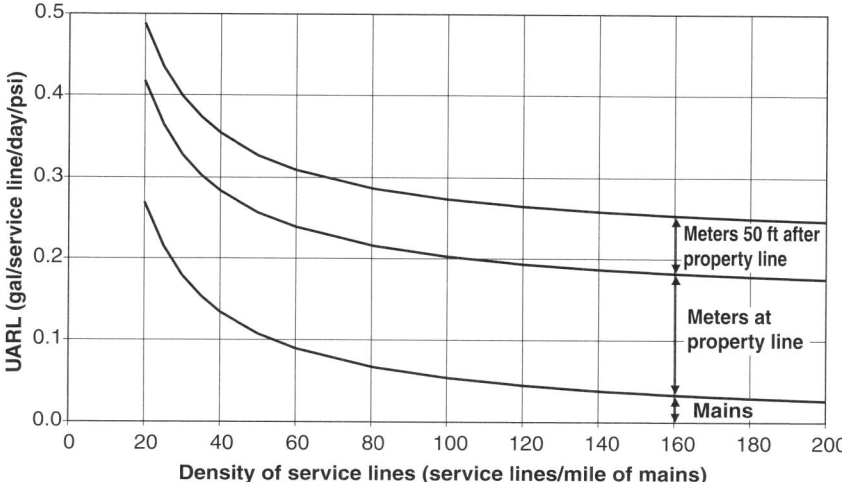

Figure 18.3 Unavoidable annual real losses (gal/service line/day/psi) versus density of service lines. (*Source: "Water Loss Management in North America."*)

- Unavoidable annual real losses: 12 gal/mi of mains/day/psi × 50 = 600 gal/mi/day
- Upper limit: urban, meter at property line, 160 service lines per mile, pressure 100 psi
- Unavoidable annual real losses: 40 gal/mi of mains/day/psi × 100 = 4000 gal/mile/day

Values predicted using the IWA equation span the typically quoted range of 1000 to 3000 gal/mi/day in the United States, but allow the estimates to be made on a system-specific basis.

2. Comparison of remaining background loss following leak detection surveys in small U.S. systems, with IWA unavoidable background loss predictions based on Table 18.4, column 2. Data so far are limited, but initial comparisons are encouraging, and more comparisons are sought.

3. Comparison of IWA recommended performance indicators for real losses for a number of North American systems, with the IWA international data set. Eleven North American water utilities volunteered data for this comparison. Analysis was complicated by each water balance being in a different format with different terminologies and lack of availability of average system pressures.

Seven of the systems provided sufficient data for the analysis as seen in Table 18.5. The values for nonrevenue water and the three new real losses performance indicators for the seven North American systems in Table 18.3 were entered alongside the values from the 27 systems from 20 countries in the IWA international data set. The North American systems are shown in black in the following bar charts.

The fact that almost all the seven North American systems in the data sample have nonrevenue water values greater than 16 percent (the average for the large 1996 sample referred to in the introduction to this article) means that these systems cannot be considered as a representative of North American performance in comparison to international performance: a wider range of system data would be needed for such a comparison.

TABLE 18.5 Summary of Statistics for Seven North American Systems

	A	B	C	D	E	F	G	Average
Traditional Statistics								
Metered water ratio (%)	81.2	84.7	67.6	76.0	78.0	78.6	66.5	76.1
Nonrevenue water (%)	18.8	15.3	32.4	24.0	22.0	21.4	33.5	23.9
Real losses (gal/mi/day)	3,652	3,558	4,975	13,378	13,960	13,105	20,686	10,473
Consumption (gal/service line/day)	337	551	358	466	772	756	444	526
Key System Factors								
Service line density/mi	86	67	40	104	85	80	145	87
Average meter location after property line (ft)	0	23	7	24	25	0	12	13
Average pressure, psi	75	75	85	70	72	65	55	71
Unavoidable Annual Real Losses (IWA Equation)								
gal/service line/day/psi	0.21	0.26	0.29	0.24	0.25	0.22	0.20	0.24
gal/service line/day	16.0	19.7	25.1	16.5	17.9	14.2	11.2	17.2
New PIs for Real Losses								
Basic: gal/service line/day	42	53	124	129	163	165	143	117
Intermediate: gal/service line/day/psi	0.57	0.70	1.46	1.84	2.27	2.53	2.59	1.71
Detailed: Infrastructure leakage index	2.66	2.69	4.96	7.80	9.13	11.63	12.69	7.37

Source: "Water Loss Management In North America: Just How Good Is It?"

When expressed as percent nonrevenue water (Fig. 18.4), the North American data bunches together in the "worst" 50th percentile of the extended data set; but when the new performance measures are used (see Figs. 18.5–18.7), a clearer discrimination of performance results, and it is seen that two of the North American systems are just within the "best" 50th percentile of the extended international data set.

In one of the North American systems, the IWA performance measures were calculated for three discrete subsystems, known to have very different leakage characteristics, and the individual ILI values ranged from 1.6 to over 10, reflecting the local situation reliably, but in greater detail than had previously been possible with the cruder, percent nonrevenue water approach. The methodology can be used for subsystems down to around 5000 service connections.

18.2.7 Concluding Comments

By expressing the outputs of the IWA task forces in units familiar to AWWA members, it is hoped that this article will stimulate further interest in the state of the art for water losses performance indicators. The methodologies may be viewed as a continuation of the work of North American researchers who have been seeking a more rational basis for

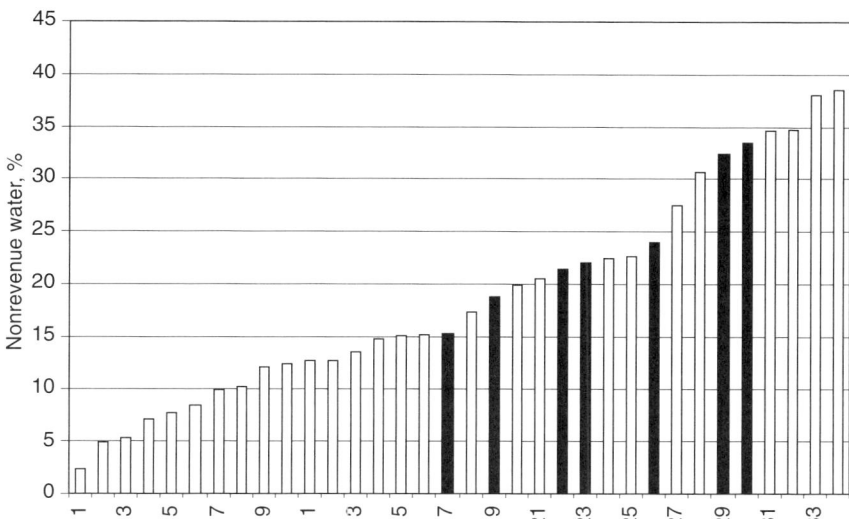

Figure 18.4 Nonrevenue water, percent of system input. (*Source: "Water Loss Management in North America."*)

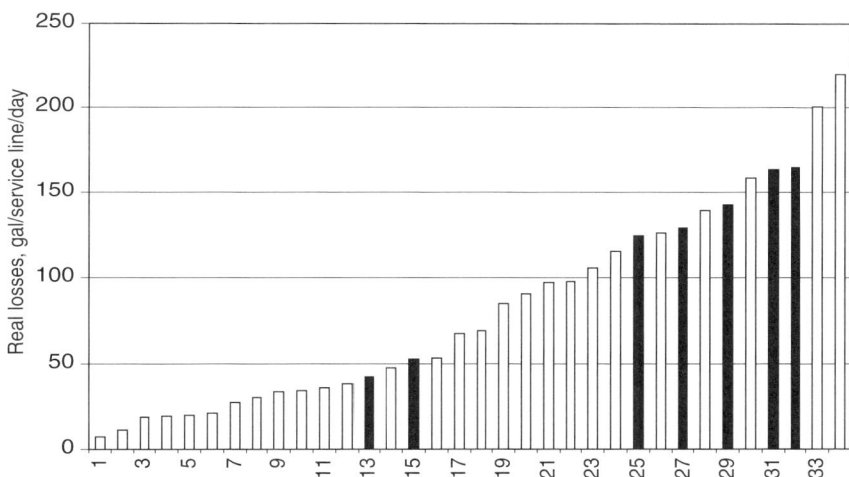

Figure 18.5 Real losses (gal/service line/day). (*Source: "Water Loss Management in North America."*)

comparisons, taking into account key system-specific factors, notable pressure, density of service lines, and location of customer meters relative to the property line.

There is no indication from the applications described in this article to suggest that the methodology is not applicable to North American systems, although further testing is of course recommended. Potential problems likely to be encountered in further applications include:

- ❑ The wide diversity of formats and terminology used for water balance calculations
- ❑ Training in a standard methodology for calculating average pressure

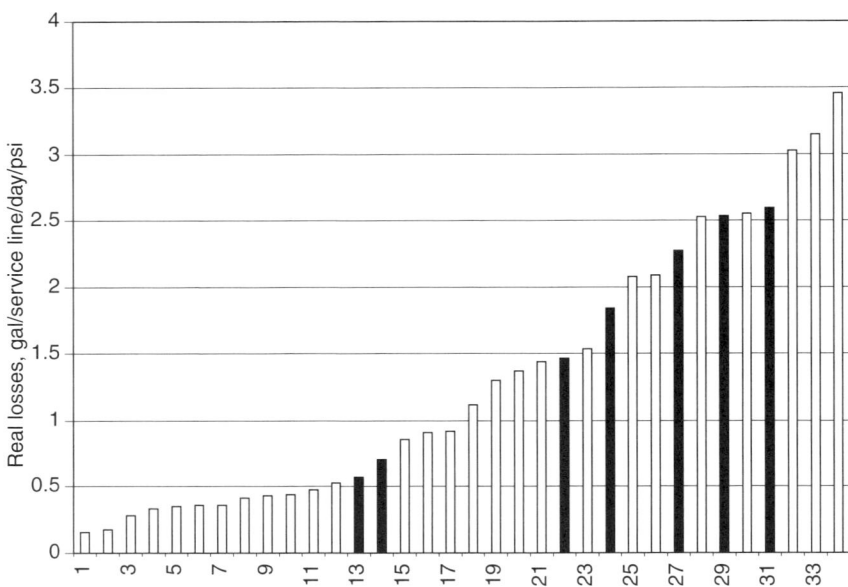

Figure 18.6 Real losses (gal/service line/day/psi). (*Source: "Water Loss Management in North America."*)

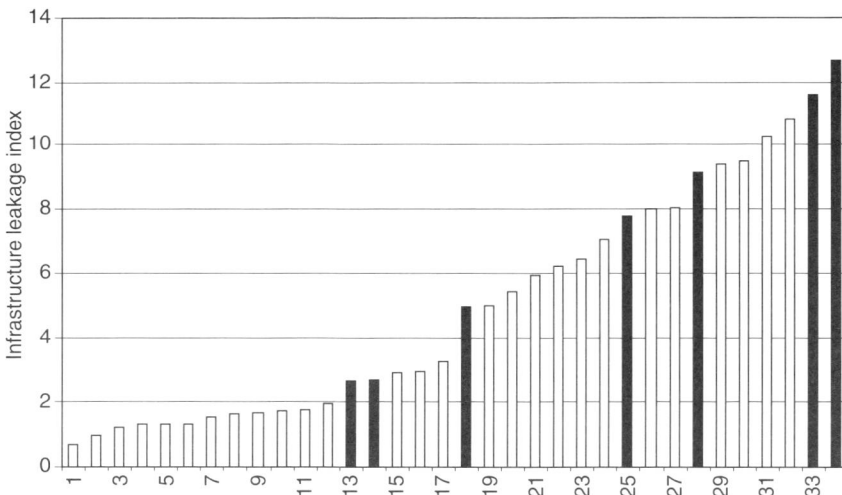

Figure 18.7 Infrastructure leakage index. (*Source: "Water Loss Management in North America."*)

❏ Both these problems, however, have been overcome in countries outside North America.

18.2.8 Acknowledgments

The authors acknowledge assistance from the North American water utilities, which provided data for the analysis, and support and encouragement from the AWWA Leak Detection and Water Accountability Committee.

18.3 CASE STUDY ONE

A U.S. Water Loss Case Study: The Philadelphia Experience

George Kunkel

18.3.1 Philadelphia's Water Supply: A History of Firsts

The City of Philadelphia has been a leader in water supply technology in the United States for over 200 years. In 1801 the fledgling city became the first in the young nation to construct two steam-driven pumping stations to bring Schuylkill River water to wooden tanks at the city's "Centre Square," where water was piped to 63 private homes, four breweries, and one sugar refinery. In 1815 a larger, improved system was commissioned, with the settling reservoir at "Fair Mount" to supply water to the growing city. By 1822 a dam, water-driven turbines, and Greek revival architecture were incorporated into the Fairmount Water Works. The site entertained many visitors, being widely recognized not only as an engineering marvel but also as a place of architectural splendor and beauty.

The distribution piping of these early systems consisted of bored wooden logs joined end to end by iron bands and caulking. The city's first water loss problem was realized when these pipes leaked badly and constantly. Philadelphia soon began to import British-made cast iron pipe to expand its water distribution system, and this material became the norm by 1832. The longevity of iron pipes—in use in Europe for hundreds of years—has been confirmed in Philadelphia, where several thousand feet of pipe segments installed in the 1820s still provide reliable service to this day.

While recognized for its historical significance as a center of government during the United States' birth as a nation, Philadelphia's expansion into a major city actually occurred during the industrial revolution of the nineteenth century as it evolved into a significant manufacturing center and bustling port. By 1900, however, the city's population of almost 1.3 million had also begun to degrade its two major river water sources: the Schuylkill and the Delaware. Again, Philadelphia demonstrated innovation by becoming one of the first large cities in the nation to construct water filtration plants, with five such facilities of various sizes commissioned between 1903 and 1911. At the time, Philadelphia's filtration system was the largest in the world. Philadelphia's readiness to apply emerging technology continued as it adopted cleaning and cement lining rehabilitation of water mains (1949), use of an analog computer, the McIlroy Fluid Network Analyzer (1956), and use of telemetry control of pumping stations—the forerunner to today's modern Supervisory Control and Data Acquisition (SCADA) systems (1958). More recently, the City of Philadelphia installed the largest automatic meter reading (AMR) system in the United States, with over 400,000 residential units outfitted in an initial phase between 1997 and 1999.

Philadelphia continues to meet today's complex challenges by providing a full range of water and wastewater services to a discerning public while maintaining a delicate balance with the natural environment.

Faced with increasing water quality and stormwater regulations, the Philadelphia Water Department (PWD) and Water Revenue Bureau (WRB) are further challenged by a contracting customer base, mounting infrastructure needs, and the fact that water loss in the city has traditionally been perceived as relatively high. The city has responded to these needs by developing a comprehensive capital planning and rehabilitation program focused on optimizing its assets and adopting best management practices for water delivery operations. At the start of the new millennium, Philadelphia continued its tradition of firsts by becoming the first water utility in the United States to explore the use of the progressive water loss management methods and technology developed internationally during the 1990s. This was carried out via the Leakage Management Assessment (LMA) Project, which was conducted by Bristol Water Services, Ltd. (BWS) and its subcontractor, International Water Data Comparisons, Ltd. (IWDC) of the United Kingdom.

18.3.2 Water Loss in Philadelphia

As with most communities in the early days of the United States, engineers focused on building and the development of the industrial potential that was evident in the young country. Water was critical to this end, and the coastal regions of the first states have always been blessed with abundant water resources. Engineers exploited these resources and were highly successful in creating a reliable water supply infrastructure. As growing communities or industry required more water, new wells or pumping stations were constructed. However, when John C. Trautwine was Philadelphia's Bureau Chief in 1898, water charges were assessed according to the type and number of plumbing fixtures in a business or home, rather than the actual quantity of water used. Greatly concerned about the enormous waste of water in the city, he felt that water meters would be the most successful way to encourage conservation. As a demonstration, he constructed and displayed the "Trautwine tank," which held 250 gal of water—the amount used by each Philadelphian every day at the time—in order to emphasize the significance of water use to the public. Accountability of water was again promoted with the installation of customer meters in the city throughout the next several decades. Despite these early displays of water conservation acumen, it is likely that the City of Philadelphia has never maintained a high level of water accountability in its operations. With water relatively available and inexpensive, Philadelphia's primary water supply goals were to provide a safe, sufficient supply of water for industrial, residential, and fire protection needs, and the city has continuously met these goals for over two centuries. Philadelphia completed the expansion of its distribution system and modernization of its water treatment plants by the mid-1960s, with an infrastructure capacity easily able to supply over 400 million gallons of high-quality water each day.

Philadelphia's population peaked at roughly 2.1 million people in the mid-1950s, with the average annual water delivery topping out at 377 MGD in 1957. Then began a slow, subtle shift in the demeanor and demographics of the city. Industry entered a gradual decline, as heavy manu-

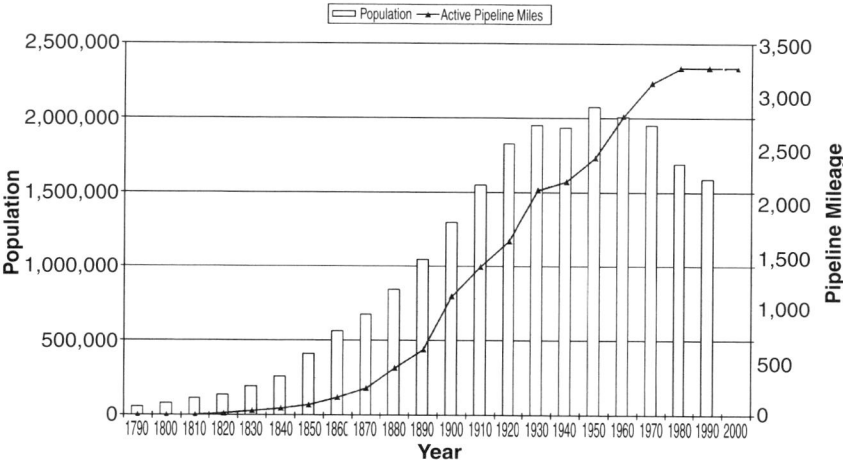

Figure 18.8 Population and water pipeline length. (*Source: City of Philadelphia.*)

facturing migrated away from major northeastern U.S. cities. The relocation of city dwellers to suburban areas furthered the decline in the city's population, which stood at 1.52 million as reported in the year 2000 Census. Yet, as shown in Fig. 18.8, its infrastructure size—three water treatment plants and 3200 mi of piping—remains roughly the same. With water export sales at 5 percent of its production, and only moderate additional sales potential believed to exist, the change in size of the customer base is projected to remain stable at best, or in continued slow decline for the next decade or longer. The city's water delivery, or water input to its distribution system, has also declined, reaching a record low in the city's modern history of just over 270 MGD for the fiscal year ending June 30, 2001 (FY2001). While water charges in Philadelphia have never been high by relative standards, the new demographics—with a larger portion of urban poor—have resulted in political pressure to keep water rates affordable.

Philadelphia assessed its water loss condition in limited and detailed fashion in 1975 and 1980 studies, respectively. A cursory review of delivery and billing data in 1975 asserted that the level of "unaccounted-for" water in the city appeared to be high and warranted further attention. Considerable attention was then given to the issue in 1980, when an Unaccounted-for Water Committee undertook a comprehensive, year-long study to identify sources of lost water in the city and propose actions to reduce losses and recoup revenue. A number of initiatives, including master meter calibration, expanded leak detection, and meter replacement, came about in the following years as a result of this endeavor. Unbilled water—defined as the difference between the amount of water delivery (system input) and customer billed usage (system output)—remained at levels well above 100 MGD, however, in the decade following this work.

Water loss took on a greater prominence for the city government in 1993, after a proposed 30 percent water rate (tariff) increase was roundly criticized and eventually reduced to single-digit increases totaling 7 percent over three years. The city's water loss standing was scrutinized and resulted in the formation of a permanent Water Accountability Committee to pursue

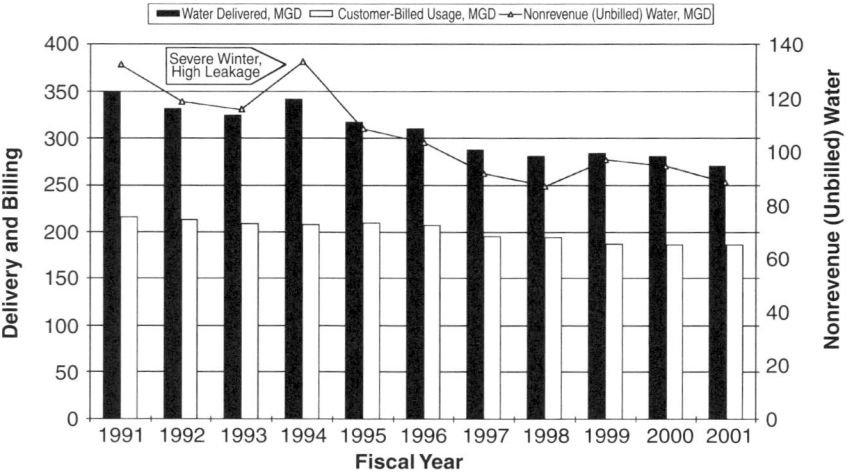

Figure 18.9 Water delivery and nonrevenue water. (*Source: City of Philadelphia.*)

water loss reductions. Further expansion of the main replacement and leak detection programs and a switch from quarterly to monthly billing were implemented shortly thereafter. These efforts made headway in bringing the city's excessive water losses under an initial degree of control. Prior to FY2000 the city measured water loss primarily by tracking unbilled water. Figure 18.9 reveals a notable decline in unbilled water after 1994. Unbilled water averaged almost 126 MGD from fiscal years 1990 to 1994, but realized a steady drop to an average of just less than 96 MGD for fiscal years 1995–2000. This success in cutting water loss is attributed to reductions in both real losses (leakage) and apparent losses (missed billings, meter error). It is believed that real losses have been reduced by a combination of stepped-up leak detection effort, improvements in leak repair job routing, and pipeline replacement. Apparent losses have been reduced by the use of new residential meters (installed with AMR), large meter right-sizing, missed billing recoveries, and metering/accounting of city-owned properties. While this improvement is significant, it is still understood by city managers that unbilled water—or nonrevenue water by International Water Association (IWA) terminology—over 90 MGD represents a large amount of water that is not being recovered.

18.3.3 In Search of Best Management Practices for Water Loss

During the 1990s Philadelphia's Water Accountability Committee sought to gain a better understanding of the nature of the seemingly high level of water loss occurring in Philadelphia's water supply system. The committee determined to identify current industry best practices and in 1995 began participation on the Leak Detection and Water Accountability Committee of the American Water Works Association (AWWA). The Philadelphia committee followed the recommendations of AWWA committee in assembling a water audit using a method that approximated that of AWWA publication M36, *Water Audits and Leak Detection*. After several years of gathering detailed information, the city produced its first distribution system water audit for the fiscal year ending June 30, 1996 (FY1996). In retrospect, a very basic water audit could have been issued one to two years earlier, and the author recommends this approach to any first-time auditors. The city continued with the modified M36 format through FY1999, but moved to use the new methodology compiled by the IWA as part of the LMA project for FY2000.

The years 1998–2001 were instrumental for the city's gaining awareness of the rapidly developing water loss technology and policy that was occurring internationally during the 1990s. As the author of this account assumed the role of chair of the AWWA Leak Detection and Water Accountability (LDWA) Committee in 1998, a movement was launched to raise awareness of water loss and explore the applicability of international water loss methods in North America. A number of international experts became members of the LDWA Committee, including Allan Lambert of IWDC, one of the world's foremost authorities on leakage management and also perhaps its most passionate disciple, having traveled to almost 50 countries in a several-year span to promote leakage management methods. It was also during this time that the IWA launched its effort to develop reliable performance indicators for water supply services. Mr. Lambert chaired the IWA Task Force on Water Loss, a five-country effort that included Timothy Brown of Heath Consultants as the AWWA representative from the United States. This group assembled the primary components of the IWA water balance methodology and performance indicators that were designed as a "best practice" approach to water auditing and comparison-making on an international scale. With these important developments occurring, Philadelphia again demonstrated its willingness to implement new technology by contracting with international experts on water loss management for an evaluation of its leakage standing.

18.3.4 Evaluating Real Losses: The Leakage Management Assessment Project

Early in the year 2000, Philadelphia initiated a contract funded at $50,000 with BWS to conduct the LMA project. In addition to the PWD and WRB, other project participants included IWDC, the subcontractor to BWS, and Pitometer Water Services (PWS), a division of Severn Trent Pipeline Services, Inc. PWS holds a regular contract with the city and performed field measurements to gather data for night flow analysis in four test areas at an approximate cost of $10,000, bringing the total project consultant cost to roughly $60,000.

The primary intention of the LMA was to gather information and critique the city's leakage management conditions with respect to the best practices being applied in water loss management throughout the worldwide water industry. Secondarily, the project provided opportunity for city personnel to become educated in the progressive leakage management methods in use internationally. The major steps of the project included:

1. Convert the Philadelphia Water Audit into IWA format (IWDC and PWD)

2. Obtain field measurements of water flow and pressure for night flow analysis in several test District Metered Areas (PWS and PWD)
3. Provide two presentations and a workshop on the methods of progressive leakage management technology to Philadelphia stakeholders (BWS and IWDC)
4. Conduct interviews with PWD personnel and gather data and information on PWD's leakage practices (BWS and IWDC)
5. Compile the project findings in a comprehensive report inclusive of recommendations for the city (BWS and IWDC)

The bulk of the work performed by the consultants in Philadelphia occurred from August to October 2000. The final report for the project was accepted in June 2001.

Philadelphia's Water Audit

BWS reviewed Philadelphia's water audit data for fiscal years 1997–1999, although all data from these audits had to be converted into approximate IWA format to allow comparisons with an international water utility data set that included information from more than 20 countries. During the course of the project, Philadelphia's Water Accountability Committee converted its routine water audit methodology to the IWA format and issued its audit for FY2000. For this period the WRB reported water billings of 185.8 MGD. The PWD reported that it treated and delivered to the distribution system an average of 280.5 MGD during the same period. The difference of these values represents nonrevenue water of 94.7 MGD, which is further broken down into 5.9 MGD of authorized unbilled usage (fire fighting, flushing), apparent (paper) losses of 18.6 MGD, and real losses (leakage) of 70.2 MGD. These losses exerted a financial impact to the city estimated at $16.7 million for the year. The summary of the FY2000 water audit in IWA format is shown in Fig. 18.10.

Conditions in Philadelphia

The City of Philadelphia operates one of the oldest water distribution systems in the United States. Approximately 60 percent of its pipeline is unlined cast iron installed between 1880 and 1930, with 6-in diameter being the most common water main size. The city's 3200 mi of distribution system piping extend across its 129-mi^2 area and provide water to approximately 460,000 customer accounts. Over the past three decades the city averaged 840 water main breaks, or bursts, per year, with half of the annual breaks occurring in the cold weather months of December through February. Additionally, in FY2000 the city documented 2368 leaks, with 1565, or 66 percent of the total, occurring on customer service connection piping. Accompanying Philadelphia's population decline of four decades has been a growing number of abandoned private properties. Left unmaintained, deteriorating service connection piping at many such properties have aggravated the trend of such leaks in the city. Distinctions in Philadelphia's terms "break" and "leak" do not follow with any standard terminology, so comparisons of leak and break data with that of other systems must be made carefully. In Philadelphia, customers are responsible to arrange for maintenance and repairs of leaks on their service connections, although the city has come to realize that this is not likely an efficient policy. Pressures in the water distribution system are typically between 40 and 70 psi, with an average citywide pressure taken as 55 psi. Small areas of the city are provided pressures in excess of 100 psi, with a high of 165 psi for a very small number of customer connections.

In addition to the large maintenance staff repairing several thousand main breaks and leaks each year, the PWD manages a significant capital program for infrastructure replacement. Approximately $25 million is expended each year to replace roughly 25 mi of pipeline, much of which is over 100 years old. This rate of replacement represents approximately 0.8 percent of the city's total pipeline mileage renewed each year. The primary criterion used to designate sections of pipe for replacement is the recent break or burst rate of the pipeline.

Leakage Management Assessment Approach and Findings

The LMA Project approach utilized by BWS included the four components of progressive leakage management shown in Fig. 18.11.

A comprehensive assessment of Philadelphia's *active leakage control* was conducted in a joint effort by all of the consultants. The assessment included analysis of data from four test District Metered Areas (DMAs) using the Bursts and Background Estimates (BABE) concept and an efficiency analysis of leak survey efforts.

The PWD has operated a focused leak detection and repair program since 1980. It maintains a Leak Detection Squad of roughly 20 employees, with crews performing leak surveys in search of "unreported leaks" on both day and night schedules. Leakage survey progress has typically been measured by the amount of system pipeline mileage covered per year. An annual survey goal of 1300 pipeline miles is used, which translates to roughly one-third of the total mileage each year, or a total system survey interval of three years. The Leak Detection Squad also consults to repair crews who have difficulty pinpointing "reported leaks" assigned to them for location and repair. The Leak Detection Squad utilizes leak correlators, noise loggers, and other leak noise sounding equipment to provide locations of leakage sources. All suspected leaks are then referred to various repair crews, or customers if leaks are determined to exist on private service connections. Costs to operate the 20-person Leak Detection Squad are approximately $900,000 per year, including personnel, vehicles, equipment, and training. The PWD employs in excess of 100 other employees engaged in making routine leak repairs in the water distribution system. The costs to employ this staff are roughly $4.5 million per year, although only a portion of their workload is leak repair, as they also perform general maintenance and replacement work on valves and hydrants, install new connections, and provide a variety of support functions.

The characteristics of the four test DMAs are shown in Table 18.6. These four areas were selected to provide a variety of conditions including average and high pressure, low and high rates of abandoned properties, and average and old infrastructure age. Closing pipeline valves to leave a single supply main in service to a discrete area of less than 1000 properties created the DMAs. DMA boundaries were also selected with the intent to keep the number of closed valves to a minimum. Twenty-four-hour flow

City of Philadelphia Annual Water Audit in International Water Association Format
Fiscal Year 2000: July 1, 1999 to June 30, 2000
(Water data shown in millions of gallons per day)

Corrected System Input Volume

Water Delivery	280.500
Master Meter Over-Registration	2.805
	277.695

Authorized Water Usage

Billed Metered	185.800
Billed Unmetered	0.000
Unbilled Metered	0.424
Unbilled Unmetered	2.570
	188.794

Water Losses 88.901

Apparent Losses

Customer Meter Under-Registration	2.008
Bypassed Flow to Fire System	0.100
Unauthorized Usage/Illegal Activities	6.150
Data/SCADA System Error	0.000
Customer Meter Malfunction	1.836
Meter Reading/Estimating Error	0.047
Accounts Lacking Proper Billing	2.500
City Properties	4.000
Billing Adjustments/Waivers	2.000
Apparent Losses	**18.641**

Real Losses

Operator Error (Tank Overflows)	0.000
Unavoidable Annual Real Loss	5.724
Recoverable Leakage	
Customer Service Line Leaks	38.633
Main Leaks	24.348
Main Breaks (Bursts)	0.067
Other	1.488
Real Losses	**70.260**

Fiscal Year 2000 Financial Data

$3,297 Apparent Losses per MG – Small Meter Accounts (5/8" and 3/4")
$2,904 Apparent Losses per MG – Large Meter Accounts (1" and Larger)
$2,828 Apparent Losses per MG for City Property Accounts
$3,120 Apparent Losses – Overall Average Customer Rate
$110.25 Real Losses – Marginal Cost per MG
$130,132 Real Loss Indemnity Costs – Added to Total

$151,181,693 Water Supply Operating Cost – Fiscal Year 1999 Data

Apparent Losses Cost: $13,750,897

Real Losses Cost: $2,965,228

Technical and Financial Performance Indicators for Water Supply System Losses

Technical Performance Indicator for Real Losses

Real Losses	70.240 MGD
Unavoidable Annual Real Losses	5.724 MGD
Infrastructure Leakage Index (ILI)	12.30
Recommended Long-Term Targets*	
Long-Term ILI	3.500
Target Long-Term Real Losses	20.000 MGD
Current Excess Real Losses	50.24 MGD
Value of Current Excess	
Real Losses	$2,027,259

* Recommended by Allan Lambert, International Water Data Comparisons, Ltd. An ILI of 3.5 represents the median value of an international data set drawn from over 20 countries. Specific Philadelphia economic influences may dictate a different long-term target ILI.

Financial Performance Indicator for Nonrevenue Water

Nonrevenue water is defined as real and apparent losses and unbilled authorized usage.

The nonrevenue cost ratio is the percentage of the annual cost of nonrevenue water over the annual running costs for the water system.

Nonrevenue water cost = $17,109 + $131,882 + $13,750,897 + $2,965,228 = $16,865,116

Nonrevenue cost ratio = ($16,865,116/$151,181,693) × 100 = 11.15%

Figure 18.10 Annual water audit in International Water Association format. (*Source: City of Philadelphia.*)

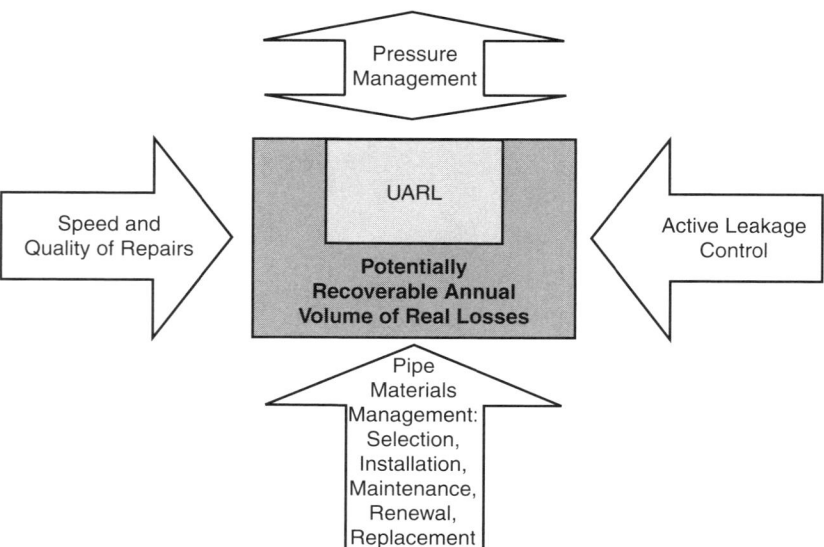

Figure 18.11 Four components of leakage management. (*Source: IWDC Ltd.*)

measurements were obtained by PPS via the use of insertion pitot rods (pitometers) on the supply main. A typical pitometer installation is shown in Fig. 18.12. The measurements were included in a BABE analysis and are shown in detail for DMA 4 in Fig. 18.13 and in summary for all DMAs in Table 18.7. The findings suggest, as believed to be true in many water systems, that Philadelphia's water leakage does not occur homogeneously throughout its system, but instead is concentrated in certain areas. The minimum night flow in two of the four DMAs was found to be negligible, at least to the degree discernible by the insertion pitometer. Using BABE analysis, leakage rates estimated at 15 equivalent service connection leaks (ESCL) and 54 ESCLs were derived for DMA 3 and DMA 4, respectively. These results indicate high amounts of active leakage.

Leakage flow rate and a leak's duration are fundamental to the BABE principles, and the PWD has refined its estimates of each as a result of the LMA. Water lost from a piping leak will mount based on the rate at which the water escapes through the leak and the amount of time that it is permitted to leak. For some leaks, such as those on certain types of plastic pipe, the leakage rate may increase with duration as the pipe failure expands. BABE concepts measure leakage duration in three component time periods: awareness, location, and repair. Awareness time (A) is the average time from the start of the leak until the utility becomes aware that the leak exists. Location time (L) is the average time that it takes the utility to pinpoint the source of the leak once it becomes aware that the leak exists. Repair time (R) is the time it takes to repair the leak once it has been located. Leakage losses can therefore be described by the following equation: leak volume = ALR time $\times$ flow rate.

Compared to data from several different countries, the leakage rates used by Philadelphia were found to be high for water main and service sizes less than 6 inches in diameter. The PWD has since moved to revise

TABLE 18.6 DMA Characteristics

	DMA 1	DMA 2	DMA 3	DMA 4
Inlet (gauging) point location	Tower, N. of Shurs	Roxborough, E. of Houghton	Crefeld, N. of Sunsct	Parkside, E. of Belmont
Inlet (gauging) point main diameter	6-in	6-in	8-in	10-in
Inlet (gauging) point elevation	103.3 ft	259.0 ft	346.0 ft	111.5 ft
DMA boundaries				
North	Shurs Lane	Dupont St.	Northwestern Ave.	Parkside Ave.
East	Manayunk Ave.	Henry Ave./Magdelena St.	Stenton Ave.	38th St.
South	Vassar Ave.	Walnut Lane	Norman Lane	Pennsgrove St.
West	Cresson St.	Ridge Ave.	Crefeld St.	Belmont Ave.
Water distribution plan number(s)	47, 48, 54	54, 55	79, 88	31
Pressure zone	Lower Roxborough Gravity	Upper Roxborough Gravity	Chestnut Hill	Belmont Gravity
Pipeline mileage	2.73	2.95	4.94	5.34
Weighted-average ground-level elevation	102.6 ft	236.8 ft	247.9 ft	100.0 ft
Critical point location	Hermit St., bending toward Manayunk Ave. in east limit of the DMA	Ridge Ave. and Conarroe St. in southwest limit of the DMA	Germantown Ave. and Norman Lane in east limit of the DMA	Girard Ave. and Belmont Ave. in southwest limit of the DMA
Elevation at critical point location	153.1 ft	274.6 ft	401.1 ft	122.0 ft
Appurtenance at average zone pressure (AZP) point	Hydrant on Sharp St., N. of Salaignac St.	Hydrant at Rector St. and Magdalena St.	Hydrant on Germantown Ave., S. of Hillcrest Ave.	Hydrant on Leidy St., W. of Girard Ave.
Average zone night pressure	97 psi	64 psi	96 psi	55 psi
Elevation at AZP point	100.4 ft	233.7 ft	256.0 ft	99.2 ft
Number of service connections	650	750	144	740
LMA project selection criteria	High pressure, residential	Average pressure, residential	High pressure, residential, low density of customer connections	Average pressure, high leakage, abandoned customer connections present

Figure 18.12 A typical pitometer installation. (*Source: Pitometer Water Services, a division of Severn Trent Pipeline Services, Inc.*)

the estimated leakage rates that are used in the tabulations of the annual water audit and other routine reports. These are shown in Table 18.8.

Leak survey records were also evaluated as part of the assessment of Philadelphia's active leakage control practices. BWS analyzed coverage frequency and records of survey (unreported) leaks located and repaired to determine whether the leakage survey frequency used by PWD is optimal. FY1999 data report that 17 main breaks and 125 service leaks were located by leak survey efforts and subsequently repaired. This is equivalent to

- 1 main break per 76 mi of main surveyed, or
- 1 service leak per 1528 connections, or
- 1 leak per 4 crew days

With a level of overall leakage believed to be relatively high, these unreported leak survey and repair findings appear to be relatively low. However, assessment of PWD's repair tracking system evidenced gaps in the routing of jobs to repair crews, creating the potential that a certain portion of suspected leaks identified during leak surveys are not being repaired in timely fashion, or at all. Also, it is likely that the time to repair many unreported leaks is lengthy and variable, due to the relatively bureaucratic process used to refer needed leakage repairs to customer-owners. Refining the repair tracking process to "close the gaps" and

Location:	Parkside – East of Belmont				Data Entry	Calculated	Default	FixedDefault	
Reference:	LMA4								
Plan No.:	31			Date Last Surveyed:	Sept. 1999				
	ASSUMPTIONS			Night Flow Measurement					
	Infrastructure Condition Factor:	3.00		Time:	0300 - 0400 hrs				
	N1 for Background Losses:	1.50		Date:	19-Sep-00				
	N1 for Breaks:	0.50		Method:	Insertion Meter in 10-inch Main				
	Estimated % of Toilet Tanks with Leaking Valves:	3.0%		Average Zone Pressure Point:	Leidy & Girard				
	Assumed Rate of Leakage of Toilet Tanks:	15	gph	Average Zone Night Pressure:		55	psi	Estimated	
	INFRASTRUCTURE			Minimum Inflow Rate over 1 hour:		361	gpm		
	Length of Mains, *mi*:	5.34		Minimum Inflow Rate over 1 hour:		21660	gph		
	No. of Service Connections:	740		Assessed Residential Night Use:		657	gph		
	No. of Plumbing Units:	770		Assessed Nonresidential Night Use:		83	gph		
	No. of Properties:	770		Leaking Toilet Tank Valves:		347	gph		
	No. of Abandoned Properties:	30		Background Loss, Mains:		95	gph		
	No. of Active Nonresidential Properties:	10		Background Loss, SC:		510	gph		
	No. of Active Residential Properties:	730		Background Loss, Plumbing:		113	gph		
				Background Loss & Assessed Night Use Total:		1804	gph		
	RESIDENTIAL NIGHT USE CALCULATION			Excess Night Flow:	19856		gph, Which May Include Exceptional Night Use.		
	Average No. of Persons per Res. Property:	3.00		Excess Night Flow:	331		gpm, Which May Include Exceptional Night Use.		
	% of People Using Toilet, 0200 to 0400 hrs:	6.0%	per hour	1 "Standard" Active Service Break =	6.9		gpm @ 70 psi		
	Average Toilet Tank Size, *gal*:	5.0		1 "Standard" Active Service Break =	6.1		gpm @ 55 psi		
	Assessed Residential Night Use, *gph*:	657		Excess Night Flow @ ICF of 3 =	54.1		"Standard" Active Service Breaks		
	Estimated Toilet Leakage, 3% at 15 gph:	347		or 1 "Standard" Active Service Break per	14		Service Connections		
	NONRESIDENTIAL NIGHT USE CALCULATION								
	Average. Nonresidential Night Use =	5	Times Average Residential Night Use (from British Data)						
	% Active, 0300 to 0400 hrs:	33%	Estimated from British Data						
	Assessed Nonres. Night Use, *gph*:	83							
	Assessed Res. & Nonres. Night Use, *gph*:	740							
	BACKGROUND LEAKAGE CALCULATION								
	Background Leakage Allowances for ICF* =	3.00							
	and Pressure Leakage Index N1 =	1.50		at AZNP of	70	psi	at AZNP of	55	psi
	On Mains* =	8.5	gal/mi/h		136	gph		95	gph
	On Service Connections to House Line* =	0.33	gal/SC/h		733	gph		510	gph
	Plumbing Systems** =	0.07	gal/PS/h		162	gph		113	gph
					1030	gph		718	gph
	* Taken from comparison of international data (Ref. 6).								
	**Value of 0.07 gph faucet drips taken from United Kingdom experience. (Source: BWS, IWDC.)								
	Toilet tanks with leaking bottom valves are pressure-independent, and allowed for in the residential night use calculation above.								
	The infrastructure condition factor (ICF) is the ratio of the actual background leakage to the unavoidable background leakage.								

Figure 18.13 BABE results of DMA 4. (*Source: City of Philadelphia.*)

reduce repair times would reduce leakage losses by limiting the duration that leaks are permitted to run. This would also provide a more reliable validation of the success of the current leak survey program.

Speed and quality of repairs were assessed by BWS and partially described above. The primary recommendations for Philadelphia are to review and potentially refine its policy, information handling, and job routing process to ensure that:

- Accountability exists: confirmed repairs address all unreported and reported leaks.
- Leak duration is minimized: currently times are tracked only for main break repairs. Consideration should be given to tracking the awareness–location–repair sequence for all leaks and breaks.
- An efficient and effective method is employed to address customer service connection leaks. The current method of referring leak repairs to customer/owners has known bureaucratic complexities and inefficiencies.

TABLE 18.7 DMA Results Summary Sheet

	DMA 1	DMA 2	DMA 3	DMA 4
Inlet (gauging) point location	Tower, N. of Shurs	Roxborough, E. of Houghton	Crefeld, N. of Sunset	Parkside, E. of Belmont
Inlet (gauging) point main diameter	6-in	6-in	8-in	10-in
Pressure zone	Lower Roxborough Gravity	Upper Roxborough Gravity	Chestnut Hill	Belmont Gravity
Pipeline mileage	2.73	2.95	4.94	5.34
Weighted-average ground-level elevation	102.6 ft	236.8 ft	247.9 ft	100.0 ft
Average zone night pressure	97 psi	64 psi	96 psi	55 psi
Elevation at AZP point	100.4 ft	233.7 ft	256.0 ft	99.2 ft
Number of service connections	650	750	144	740
LMA project selection criteria	High pressure, residential	Average pressure, residential	High pressure, residential, low density of customer connections	Average pressure, high leakage, abandoned customer connections present
Measured minimum inflow rate, gph	2,640	2,100	7,920	21,660
Assessed residential night use, gph	915	1,013	202	1,077
Background leakage losses, gph	1,387	853	480	718
Assessed use and background loss, gph	2,302	1,866	682	1,795
Excess night flow, gph	338	234	7,238	19,865
Excess night flow, gpm	5.6	3.9	120.6	331.1
Number of equivalent standard active service connection breaks at the average zone night pressure	0.7	0.6	14.9	54.1

TABLE 18.8 Estimated Leakage Rates: Before and After

| Type of Leak or Break | Diameter, in | Traditional PWD Leakage Flow Rates | | | | PWD Leakage Flow Rates @ 70 psi, Effective July 1, 2001 | | | | PWD Leakage Flow Rates @ 55 psi, Effective July 1, 2001 | | | |
| | | Unreported | | Reported | | Unreported | | Reported | | Unreported | | Reported | |
		gpm	MGD	gpm	MGD	gpm	MGD	gpm	MGD	gpm	MGD	gpm	MGD
Fire hydrant		3.5	0.005	3.5	0.005	3.5	0.005	3.5	0.005	3.1	0.004	3.1	0.004
Valve		6.9	0.010	6.9	0.010	6.9	0.010	6.9	0.010	6.1	0.009	6.1	0.009
Customer Service Lines													
Active	5/8	10.4	0.015	10.4	0.015	6.9	0.010	6.9	0.010	6.1	0.009	6.1	0.009
Active	3/4	17.3	0.025	17.3	0.025	6.9	0.010	6.9	0.010	6.1	0.009	6.1	0.009
Active	1	24.3	0.035	24.3	0.035	6.9	0.010	6.9	0.010	6.1	0.009	6.1	0.009
Active	2–4	34.7	0.050	34.7	0.050	13.9	0.020	13.9	0.020	12.3	0.018	12.3	0.018
Abandoned or vacant building	5/8	17.3	0.025	17.3	0.025	13.9	0.020	13.9	0.020	12.3	0.018	12.3	0.018
Abandoned or vacant building	1	31.2	0.045	31.2	0.045	13.9	0.020	13.9	0.020	12.3	0.018	12.3	0.018
Abandoned or vacant building	2–4	34.7	0.050	34.7	0.050	13.9	0.020	13.9	0.020	12.3	0.018	12.3	0.018
Water Mains													
Joint leak or repair band leak	6	10.4	0.015	10.4	0.015	10.4	0.015	10.4	0.015	9.2	0.013	9.2	0.013
Joint leak or repair band leak	8	17.3	0.025	17.3	0.025	17.3	0.025	17.3	0.025	15.3	0.022	15.3	0.022
Joint leak or repair band leak	10–48	27.8	0.040	27.8	0.040	27.8	0.040	27.8	0.040	24.6	0.035	24.6	0.035
Round (circumferential) crack	4	34.7	0.050			34.7	0.050	69.4	0.100	30.7	0.044	61.5	0.089
Round (circumferential) crack	6	55.5	0.080	55.5	0.080	55.5	0.080	111.1	0.160	49.2	0.071	98.4	0.142
Round (circumferential) crack	8	69.4	0.100	69.4	0.100	76.3	0.110	152.6	0.220	67.6	0.097	135.2	0.195
Round (circumferential) crack	10	83.3	0.120	83.3	0.120	93.8	0.135	187.6	0.270	83.1	0.120	166.2	0.239
Round (circumferential) crack	12	97.2	0.140	97.2	0.140	111.1	0.160	222.2	0.320	98.4	0.142	196.9	0.283
Longitudinal crack or split bell	6	69.4	0.100	69.4	0.100	69.4	0.100	138.9	0.200	61.5	0.089	123.1	0.177
Longitudinal crack or split bell	8	83.3	0.120	83.3	0.120	93.8	0.135	187.6	0.270	83.1	0.120	166.2	0.239
Longitudinal crack or split bell	10	97.2	0.140	97.2	0.140	111.1	0.160	222.2	0.320	98.4	0.142	196.9	0.283
Longitudinal crack or split bell	12	111.1	0.160	111.1	0.160	138.9	0.200	277.8	0.400	123.1	0.177	246.1	0.354

Quantities do not include the effects of different pipe materials.
Philadelphia's average water pressure is taken at 55 psi.
Unreported leaks: discovered during leak survey work. Reported leaks: all other leaks.

Reliably addressing leak repairs is critical to good leakage management. While methods such as leakage modeling, water auditing, and performance indicators have made considerable advances in managing leakage, the repair is usually the ultimate action that halts water lost through a leak. How well repair actions are managed is still a fundamental factor in how well water losses are managed in public water supply utilities.

BWS also assessed PWD's infrastructure management by reviewing its capital program for water main replacement and rehabilitation. BWS found that the program's current efforts have little impact on the reduction of real losses, although the program produces significant political and infrastructure benefits. At its current level of pipeline replacement, PWD is renewing roughly 0.8 percent of its total water main mileage each year. Opinions on the ideal replacement level vary, with some systems replacing up to 2.0 percent per year. PWD's primary driver for its replacement program has been to reduce the rate of main breaks to below 200 per 1000 mi of pipeline per year. However, PWD's rate of 259 main breaks per 1000 mi in FY2000 ranks it considerably less than the typical level of 320 breaks per 1000 miles encountered in systems in the United Kingdom. The PWD has lower average pressure (55 psi vs. 65 psi) and larger distribution mains (6-in versus 4-in) than most systems in the United Kingdom, factors which partially explain PWD's lower break rates.

Possible refinements to the strategies of the capital program main replacement initiative might produce a moderate additional reduction in leakage and help the primary objective of main break reduction. Currently, main replacement contractors only replace customer service connections from the main in the street center to the curb stop in the footway, roughly one-half of the length of the service connection. Consideration might be given to evaluate the costs and benefits of replacing the entire service line into customer properties. This might provide further leakage reduction at only a modest additional cost. Additionally, possible pressure management refinements might assist the reduction of main breaks by lowering average pressures and eliminating some of the pressure surges that occur in the distribution system—such as those that occur during pump operation changes at pumping stations. Another possible objective of the replacement program might be to reduce the total pipeline mileage in the distribution network. While Philadelphia's population and water demand dropped sharply in the latter decades of the twentieth century, the PWD's pipeline network remains largely the same 3200 mi that served earlier, larger generations of customers. Opportunities exist to eliminate redundant parallel mains in certain streets and sections of large-diameter transmission mains installed near reservoirs and pumping stations long decommissioned. Distribution system planning should move to ensure that only those assets that provide necessary service remain in the active system, while those that have become extraneous maintenance burdens and liabilities are removed from service. Finally, the PWD might extend water main rehabilitation funding by incorporating the use of "trenchless technology" methods into its infrastructure management toolkit.

Philadelphia operates a large, successful capital main replacement program that has achieved reductions in overall break rates and properly renewed portions of arguably the oldest water distribution system in the

United States. By incorporating relatively modest refinements in the program's strategies, multiple benefits of further leakage and main break reduction, and an optimally sized distribution network, could result.

Philadelphia likely has a moderate potential to improve leakage reductions and create other benefits by improved pressure management. Much of the city exists near sea level, with higher elevations reaching above 400 ft. The primary city center and other major concentrations of the customer population receive a moderate level of service of 35–45 psi. While the distribution network is divided into 11 typically large pressure zones, no permanent small management zones (DMAs) exist. While nighttime pressure reduction might be achievable in certain areas, a decision to implement permanent DMAs must first be implemented. Other refinements in pressure control, however, could be considered. PWD's Load Control Center monitors pressures at two or three locations in each of the 11 pressure zones via its SCADA system. Opportunities exist to analyze this data more closely in conjunction with pump operating protocols to determine if a reasonable amount of day or night "pressure shaving" might be accomplished with only relatively minor change in operating procedure. An example might be one of PWD's pressure zones where pumping rates are reduced during the nighttime hours. Providing reduced pumping for a slightly longer nighttime period could provide a benefit in reduced background leakage at virtually no additional cost, as long as the reduction does not compromise adequate levels of service. Exploring the extent of pressure surges in the distribution network might also lead to pressure management improvements that will further reduce the occurrence of main breaks. The PWD will soon be creating a hydraulic model of its distribution system, which should allow for analysis of pressure surges. Philadelphia may also have opportunity to reconfigure small portions of certain pressure zones that are served greater than 100 psi. Schemes to establish pressures in a 40–60-psi range might be investigated, but would likely need be planned in conjunction with main replacement work of the capital program. Like most water systems in the United States, Philadelphia's pressure levels were established conservatively high; therefore potential exists to lower pressures to reduce leakage and main breaks. With a relatively low, smooth topography, however, pressure management improvements are likely to be a supplementary improvement after the fundamental improvement opportunities of more efficient repairs and active leakage control.

By completing the FY2000 water audit in IWA format, Philadelphia is now able to apply the IWA performance indicators. The infrastructure leakage index (ILI) of 12.3 gives the city its first reliable comparative measure of its leakage standing relative to the best practices in the world. The value of the ILI reflects that the city's leakage is almost 13 times the technically low level achievable today. This level, while high, gives the city a reliable baseline from which to monitor improvements and make comparisons with other water utilities. With water relatively available and inexpensive, Philadelphia need not necessarily seek to attain technically low leakage levels, or an ILI close to 1.0. It should, however, seek to determine an appropriate economic leakage target that is based on the city's direct and indirect costs of water. While an economic assessment of leakage was not carried out during the LMA, IWDC did offer an ILI value of 3.5—the median value of the international data set of several dozen water utilities compiled by the IWA Task Force on Water Loss—as a potential initial target level for the PWD's planning.

The LMA project was highly successful in providing to the City of Philadelphia a clear understanding of its water loss standing relative to worldwide best practices. It also provided and explained the use of new tools and technology to manage leakage proactively to reduced levels. This information is assisting the planning efforts of the PWD and WRB to determine how best to apply the various methods now available to reduce water loss.

18.3.5 Addressing Apparent Losses

Philadelphia's FY2000 water audit indicates that the city's apparent losses of 18.6 MGD are just over one-quarter of its real losses of 70.2 MGD. Remarkably, however, apparent losses exert an impact of over $13.75 million annually on the city due to lost revenue potential, compared to real losses' $2.96 million impact as largely excess production costs. This stark difference occurs since apparent losses are valued at the retail cost charged to customers, which is almost always higher than the marginal cost of production used to valuate real losses. The marginal production cost and various retail costs used by the PWD and WRB in the city's water audit are given in Fig. 18.10. Since apparent losses represent service rendered without revenue recovered, these losses are usually highly cost-effective to recover—particularly since, for many water suppliers, they often occur in the fundamental operations of customer metering, meter reading, accounting, and billing.

Prior to 1997, the City of Philadelphia was greatly hampered in establishing a reliable evaluation of its apparent losses. Although the customer population was essentially fully metered, the city's customer billed usage included a suspected large degree of error, since manual meter reading efforts were often unsuccessful. Customer meters for most accounts in Philadelphia are located within customer buildings. Changing lifestyles have resulted in many residential properties being absent of any inhabitants during weekday business hours when manual meter reading was attempted. By the mid-1990s the WRB's meter reading success rate was in the mid-60s percent, and residential reads only the mid-30s percent. With quarterly meter reading cycles in use with a monthly billing interval, only 1 of every 7 water bills issued was based on an actual customer meter reading. In addition to compromising the accuracy of customer water usage data, estimated water bills resulted in frequent billing adjustments and high customer call/complaint volume.

Between 1997 and 1999, however, the City of Philadelphia successfully established the largest automatic meter reading (AMR) system in the United States with over 400,000 residential properties read remotely via radio transmission to vans patrolling regular meter reading routes. The project included the use of city plumbing contractors to install new customer meters manufactured by Badger Meter, Inc., and meter reading devices and services from Itron, Inc. With a primary intention of improving customer satisfaction with the billing process, it is hoped that AMR will ultimately also assist improvements in reducing apparent losses in the city. The installation phase and short-term subsequent period following installation has been one of considerable billing adjustment activity, since AMR was installed in many properties that had not been visited in years. Additionally, since 1999, the AMR project has continued in its second phase of industrial, commercial, and municipal property installations. As the installation transition passes, it is believed that the overall level of customer billed usage could rise slightly or hold stable (even as the city continues to lose population) as a result of AMR. More important, the number of customer billing complaints is expected to drop. In assisting water loss improvements, AMR is providing integrity to the assessment of customer billed usage as estimates are minimized and actual customer meter readings become the norm. Also, meter reading and billing are now on the same monthly schedule. Forthcoming improvements in the city's billing software will allow closer tracking of usage and billing trends. Directly assisting water loss reduction, the AMR system includes tamper-detection capabilities that thwart unauthorized usage by customers, a known problem in the city.

In employing AMR the PWD and WRB reorganized its metering and meter reading groups, since manual meter reading was no longer necessary. A "revenue protection" mission was added to the metering group, which now focuses on customer investigations as well as meter replacement and repair. With an entirely new customer meter population, considerable attention is now directed to investigate a large number of suspect accounts. Such accounts include "hard-to-install" holdouts from the initial AMR installation as well as the city's nonbilled accounts. The latter represent customer accounts that have had billing suspended for one of a number of administrative reasons. As nonbilled accounts grew without close monitoring over recent years, they came to represent a high potential for apparent losses. Often customers in nonbilled status would continue as such even after the end of the period of no water usage. The city's Revenue Protection group began to focus investigative work on suspect accounts late in FY1999. During its first full year of operation in FY2000, this group recovered revenue estimated at $2.07 million on unbilled water of 1.24 MGD. At its two-year anniversary its efforts had yielded over $4 million of recovered revenue. In the course of conducting its investigations, Revenue Protection has identified a number of gaps in the permitting, accounting, and overall information-handling procedures of the city, which have since been corrected. These gaps allowed many accounts to remain improperly in unbilled status when they had actually returned to water-using status.

Similar to the nonbilled accounts, the city accumulated a large number of suspect accounts of its own municipal buildings. Sometimes believed to be downplayed in importance since these accounts do not generate net revenue to the city, many municipal buildings have gone without water meters, meter maintenance, or meter reading. Many such properties have escaped listing in the city's billing system altogether, avoiding tracking of any water usage. The city's largest water treatment plant—and perhaps the city's largest water user at more than 2 MGD—fell into this category.

Several other plant and pumping facilities were found to lack meters, accounts, or both. Through the efforts of Revenue Protection and the continuing AMR project, investigations are ensuring that accounts exist for all customers, and include functional water meters and AMR devices. Automatic meter reading routes and billing procedures are being improved to ensure that all accounts are monitored effectively.

The PWD has also achieved success in one unusual source of lost water that has plagued older urban centers in the United States: fire hydrant abuse. With above-ground fire hydrants and large inner-city populations, hydrants have often been opened illegally as a means of heat relief during hot summer periods. In addition to high water loss, these dangerous events have worked to draw down distribution system pressures below safe levels to fight fires and protect against backflow. The PWD has achieved success in checking these phenomena by installing locally manufactured center compression locks (CCLs) on most of its fire hydrants. The device requires that a special adapter be used to open the hydrant by compressing an internal coil. The adapter must stay on the hydrant to keep it open. The adapter is removed to close the hydrant. Although some individuals have been able to defeat the CCL, they can only open one hydrant at a time and usually oblige the PWD by closing the hydrant (removing their makeshift adapter) when finished. This results in much less lost water than the pre-CCL era, when a single illegal wrench could be used to open numerous hydrants, which usually remained running at length before PWD personnel arrived to close them.

Reducing apparent losses is attractive since it offers high economic payback. In this way it "creates" previously uncaptured sources of funding and allows utilities to delay rate increases by equitably spreading costs among all users. The City of Philadelphia has made considerable headway in reducing apparent losses, but with over $10 million of such nonrevenue water still existing, much work remains.

18.3.6 Philadelphia's Water Loss Future

The City of Philadelphia has taken a leadership role with the American Water Works Association to raise awareness of water loss in the industry and the need to establish consistent reporting and loss reduction structures. Additionally, the city continued its tradition of water supply innovation by obtaining an assessment of its water loss standing by practitioners well versed in the progressive loss control methods developed internationally during the 1990s. Currently the results of that assessment are serving as the basis for planning that is occurring in the city to devise its long-term water loss strategy. The major recommendations of the LMA project include:

❑ Improve repair information systems and work routing to ensure accountability and reduce average leak run duration
❑ Evaluate options for more efficient repair of customer service connection leaks and replacement of complete connection piping lengths during water main replacement work

- Consider establishing district metered Areas (DMAs) in select or wide areas of the city to implement routine nighttime leakage measurement and leakage modeling
- Increase the coordination among groups performing leakage surveys, DMA measurements, leak repairs, and customer relations (service connection leaks) to better target high-leakage areas and focus all activities on creating measurable reductions in leakage
- Assess existing good water pressure data for opportunities to refine operating pressures and consider the use of advanced pressure management techniques for the limited areas of the city that experience pressure exceeding 100 psi
- Consider refinements of capital program objectives to reduce the overall total of active miles of pipeline and include trenchless technology to improve rehabilitation efficiency

The PWD is moving on several fronts to address the above recommendations. As of mid-year 2001, options to address customer service connections are being evaluated, including the use of customer insurance policies or outright subsidization of leakage repairs. Also, the PWD and BWS are launching an evaluation of the long-term feasibility and benefits of night flow monitoring.

The PWD has not modified its primary leakage management practices since the start of the LMA in August 2000. However, during June 2001, a second round of 24-hour measurements was obtained for the four test DMAs, with results shown in Table 18.9. These results indicate that the two areas (DMA 1 and DMA 2) with low leakage in 2000 still show low leakage. DMA 3 revealed a significant drop in leakage upon measurement in 2001, although it is not certain which leak repair action(s) is accountable for this reduction. DMA 4, the area of highest leakage in 2000, improved slightly with the 2001 measurement, but still evidences high leakage. The PWD plans to convene a focused, continued effort, with leak detection, flow measurement, leak repair, and customer relations groups meeting regularly to focus on DMA 4, with enhanced leak detection and repair efforts to drive down the high leakage. An assessment of current practices will be carried out to identify the nature of existing gaps in leakage management that have allowed this area to continue to experience high leakage over long periods of time.

In addition to major opportunities to make long-term reductions in leakage, the city will continue its efforts to reduce apparent losses and recoup lost revenue. The combined savings of recovered real and apparent losses in the past five years is believed to be much greater than the cost of the effort expended. This is confirming the notion that water loss recovery is a cost-effective undertaking. Philadelphia has a long history of taking progressive action to better its level of service to its customers. Its work on water loss management is a new and important chapter in this history and one that stands to influence a greater understanding of water loss in North America and the need to control it.

TABLE 18.9 Change in Night Flows

	DMA 1	DMA 2	DMA 3	DMA 4
Inlet (gauging) point location	Tower, N. of Shurs	Roxborough, E. of Houghton	Crefeld, N. of Sunset	Parkside, E. of Belmont
Inlet (gauging) point main diameter	6-in	6-in	8-in	10-in
Pressure zone	Lower Roxborough Gravity	Upper Roxborough Gravity	Chestnut Hill	Belmont Gravity
Pipeline mileage	2.73	2.95	4.94	5.34
Average zone night pressure	97 psi	64 psi	96 psi	55 psi
Number of service connections	650	750	144	740
Night Flow Measurements: Fall 2000				
Measured minimum inflow rate, gph	2,640	2,100	7,920.0	21,660
Excess night flow, gph	338	234	7,238	19,865
Excess night flow, gpm	5.6	3.9	120.6	331.1
Number of equivalent standard active service connection breaks at the average zone night pressure	0.7	0.6	14.9	54.1
Night Flow Measurements: Spring 2001				
Measured minimum inflow rate, gph	2,640	2,640	4,680	16,680
Excess night flow, gph	338	755	3,998	14,875
Excess night flow, gpm	5.6	12.6	66.6	247.9
Number of equivalent standard active service connection breaks at the average zone night pressure	0.7	1.9	8.2	40.5
Change in excess night flow	No change	8.7 gpm increase	54 gpm decrease	83.2 gpm decrease
Recommended action for pilot study	Monitor night flow at semi-annual or annual interval	Monitor night flow at semi-annual or annual interval	Monitor night flow quarterly; perform leak survey	Monitor night flow every 45 days; focus leak detection and repair

18.4 CASE STUDY TWO
The Loss Control Program in São Paulo, Brazil

Marcelo Salles Holanda de Freitas, SABESP
Francisco Paracampos, SABESP

18.4.1 Introduction

The metropolitan region of São Paulo, Brazil, has 17 million inhabitants settled in 800 km^2. The landscape is hilly, varying from 730 to 850 m above sea level. The São Paulo Water & Sewer Co., SABESP, supplies water and sanitation services through a distribution network of 26,000 km of mains, with 3 million connections for their customers and through bulk sales to six municipalities. The water system is totally metered, and consumers have individual building storage tanks.

Last year's [2000] average production was 62 m^3/s (1417 MGD); the figures for real losses were 15 percent, and for apparent losses were 18 percent.

The company has undergone an intense reorganization process during the last five years [1995–2000], aiming to achieve hard financial and operational targets and, so far, the results have fulfilled the expectations. In the last fiscal year (2000), a $250 million (U.S.) net profit was shown, which allowed a $400 million investment for 2001 to improve and maintain the systems.

18.4.2 The Loss Control Program

The key to our strategy was an extremely rational approach, which produced a significant recovery of lost water in a short time. The savings obtained in the short term funded the following stages of the program, longer-term structural programs. Additionally, SABESP adopted a more active position in daily operational routines and loss management activities.

Key Actions

1. Reduce the average pressure in the water distribution network—nearly 30 percent has pressures above 60 m head (88 psi).
2. Reduce response time to customer complaints about visible leakage.
3. To start on an intensive nonvisible leak detection program in the entire distribution network.
4. To implement rezoning in 15 sectors.
5. Renew and upgrade 1 million household meters in the system.
6. Replace 26,000 large consumer meters.
7. Reinforce antifraud actions.
8. Enhance the bulk metering system.

A brief discussion follows, explaining the implementation of each action.

Pressure Control

The water distribution network of São Paulo works on a gravity principle, and much of the system is made up of old piping (nearly 30 percent of pipes are older than 40 years). The city has experienced steady growth

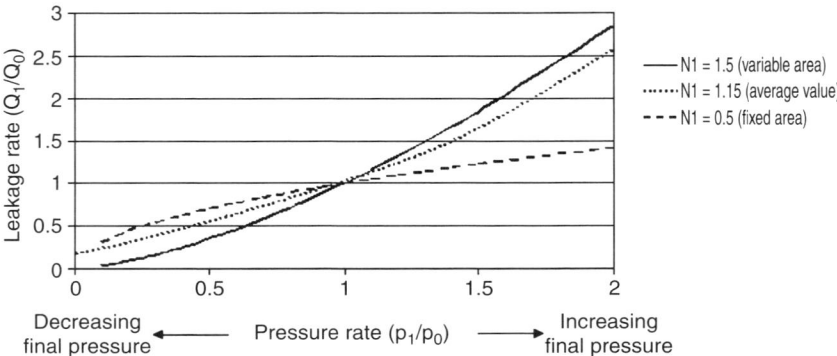

Figure 18.14 Relationship between leakage and pressure: $(Q_1/Q_0) = (p_1/p_0)^{N1}$, where Q_1 = final leakage, Q_0 = initial leakage, p_1 = final pressure, p_0 = initial pressure. (*Source: SABESP.*)

for decades, although recently a lower growth rate has been observed. However, 100,000 new service connections are still made annually.

The following picture summarizes the system. The distribution network usually has high velocity in old pipes, commonly undersized for the actual demand, and so significant headlosses occur during peak hours. Hence, an advanced pressure control strategy was compulsory to cope with such scenarios.

Another characteristic related to the São Paulo water supply system is the low overall distribution storage capacity (1,500,000 m³), although there is a positive impact as the customers have domestic roof tanks.

The pressure reduction program was essentially done with pressure-reducing valves (PRVs) and rezoning work in some sectors. Leakage varies with pressure as shown in Fig. 18.14. The key approach was to start the implementation of PRVs in large areas, regardless of the total head to be reduced in each area (Fig. 18.15 shows typical pressure ranges). SABESP has found that even a small amount of pressure reduction over a large area will provide excellent results. Undoubtedly, such a view drove SABESP to obtain the most significant savings with relatively low capital investment, due to the scale obtained with the overall program. It is worth remembering that a traditional concept for the application of PRVs is to seek higher pressure reduction in a smaller area. This was adopted in São Paulo when very critical points were addressed.

The evaluated savings for the 417 installed PRVs (Fig. 18.16 shows when the valves were installed) is 1.5 m³/s.

It is fair to report that some customers complained about having less pressure than previously. These complaints were handled case by case and, in some situations, such as very old four-story buildings without roof tanks, SABESP provided internal pumps.

The operation of the PRVs during the rotation supply event, which occurred in June/July 2000 as a consequence of a severe drought, proved hard to manage because of the low sensitivity of the hydraulic pilot to cover all of the flow range defined by the controller during a very restricted supply event; however, in most normal supply situations the installations work well.

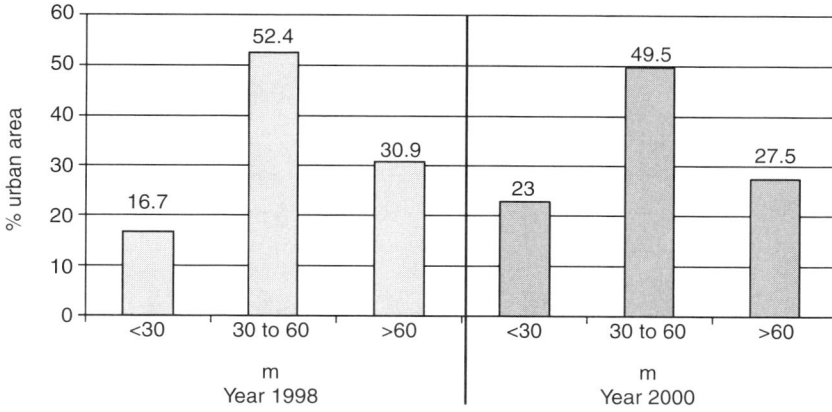

Figure 18.15 Pressure ranges for urban areas. (*Source: SABESP.*)

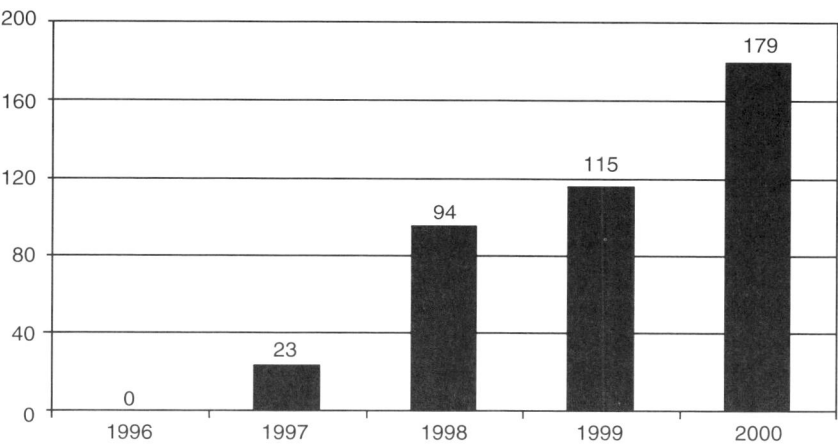

Figure 18.16 As of January 2001, 411 PRVs were installed. (*Source: SABESP.*)

Reaction Time

To shorten the time response to fix leakage, heavily dependent on contractors, a new organization was tailored, making just one contractor responsible for all stages of leak repair, including pavement reinstatement work.

The results of all initiatives brought down the 72 h average repair time, in 1994, to 24 h in 1999 and 13 h in 2000 (Fig. 18.17). Basically, it forced personnel to assure availability of teams 24 h a day, especially during peak demand hours. Training programs to create skilled gangs and a stricter technical specification of materials and practice is still under way (Fig. 18.18). Financially, this is a costly action, since it represents 50 percent of the annual budget for the real losses program.

Performance Figures

Monthly, 30,000 leaks are fixed; around 10 percent (roughly 3000) occur on main pipes and 90 percent on service connections, including stopcocks (36 percent).

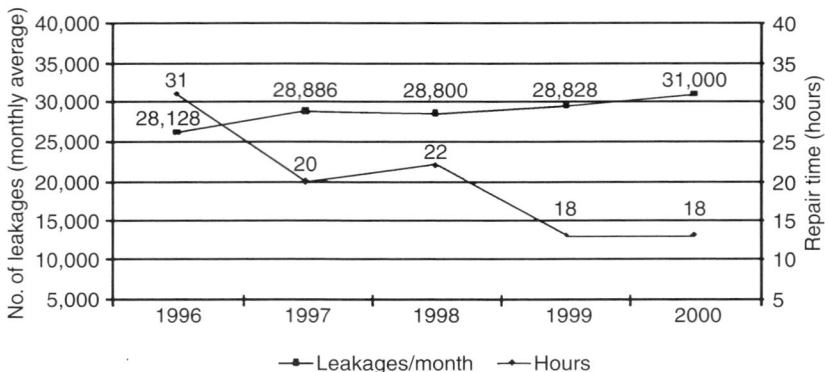

Figure 18.17 Leak repair response time. Distribution of leaks: 10 percent in mains, 90 percent in services (40 percent in meter mount, 40 percent in service pipe, 20 percent in inlet and adapters). (*Source: SABESP.*)

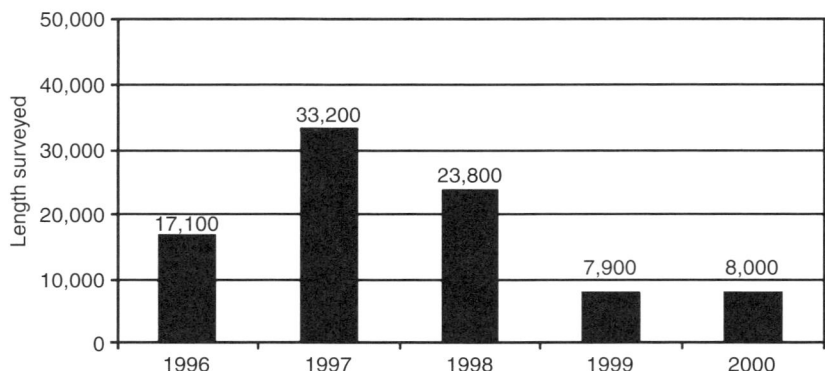

Figure 18.18 Main length surveyed for leakage and change in practice. New methods adopted: certification of leak detection professionals with ABENDE (Brazilian National Nondestructive Testing Association); field testing examination given by SABESP; enforcement of use of professionals in bidding process; lowest price as a bid criterion. (*Source: SABESP.*)

The original leak detection strategy was to cover the whole distribution network once. In the following years, the work should focus on the critical areas. These areas should be covered yearly and amount to 5000 km. Other points identified were to update the leak detection equipment and develop, in combination with contractors, a certification process, methods, practice, and personnel for assuring total efficiency.

The results have been positive. Contractors and SABESP teams, with an average 1.2 leaks/km, achieved good overall performance.

Financially, this action represents 10 percent of the annual budget for real loss control. Table 18.10 summarizes all of the actions scheduled for loss control.

Rezoning

As part of the pressure reduction program, it has been essential to rebuild main pipes in some sectors and zones. As said before, the intense

TABLE 18.10 Actions Scheduled for Loss Control

Action	Cost, US$ million	Volume recovered, millions m³ (2000–2004)
Real loss		
Reduce response time to customer complaints about visible leakage	5.5	63.5
Leakage repair	50	190.5
Start an intensive nonvisible leak detection program (entire distribution network)	11.5	131
Reduce the average pressure in the water distribution network	12.5	223
Implement rezoning in 15 sectors	12.5	15.7
Subtotal	92	623.7
Apparent losses		
To enhance the bulk metering system	1	
Renew and upgrade 1 million residential meters	16.5	101.9
Replace 26,000 large consumer meters	20	60.5
Reinforce antifraud actions	10	17.6
Correction of inclined residential meters	2	47
Subtotal	47.5	227
Total	139.5	850.7

Source: SABESP.

expansion of the city plus the presence of large sections of old pipes causes lack of pressure in some areas during peak demand hours and high pressures at night. Now new pressure zones are being implemented to lower the average pressure in the zones and minimize pressure at the critical nodes. The target is to rezone five new sectors per year in the whole metropolitan region. Some delay has occurred due to difficulties of rebuilding work in areas with dense traffic, as well as from budget constraints.

The expected volume recovered so far is 3.1 million m³/year. This kind of structural measure will be a continuing long-term strategy for SABESP.

Replace 1 Million Residential Meters

Inaccuracy of meters is thought to be responsible for a significant part of the apparent losses. The target is to renew 1 million residential meters, over three years (2000–2002), using a criterion of maximum possible volume to be recovered. So far, 3 m³/month is the average gain from 250,000 meters changed under this initiative (Fig. 18.19).

The strategy for maximizing this action is continually under development, and a new criterion for combining the type of meter and customer's average demand is being set.

In the present year [2002] US$8 million will be invested in new equipment, and 33 million m³ is the projected recovery.

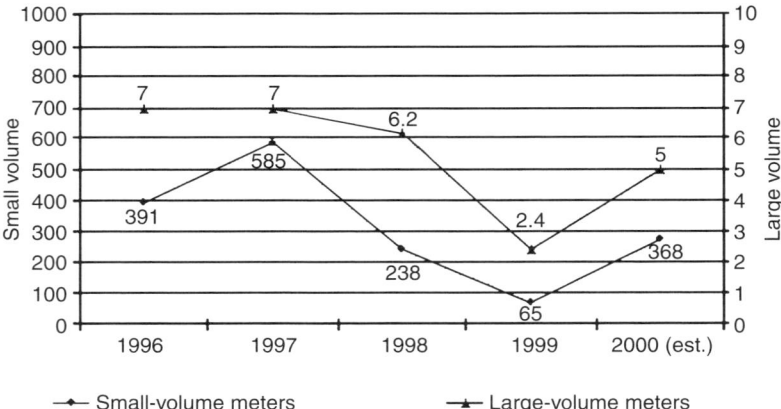

Figure 18.19 Revenue meter change-out. Goals: for small-volume meters, 700,000 change-out for 2001–2002; for large-volume meters, 26,000 change-out for 2001–2002. Key points: large-volume meters changed-out using performance-based contracting; 25 percent of meters changed-out to ISO class C; efficient change-out of small meters carried out in 2000. (*Source: SABESP.*)

Replace 26,000 Large Consumer Meters

The large customers are just 1 percent of the total connections and are responsible for 15 percent of total revenues. The analysis of all customer data, typical demand, characteristics of activity, and future trends has been evaluated as a preliminary stage for replacement of meters, under a performance contract between SABESP and a joint venture composed of suppliers and an engineering company.

The work is planned to be carried out over 24 months and had just begun in January 2001.

Reinforce Antifraud Actions

The water audit carried out three years ago [1997] indicated that nearly 13 percent of apparent losses are caused by fraud. The starting point for reducing fraud was to train internal teams for such action, and later on, contractors whose income was paid based on their performance.

Monthly, 4500 inspections are made, and 5 percent are confirmed as frauds, which has recovered 800 m^3 per event.

Enhance the Bulk Metering System

Many of the water treatment plant (WTP) meters are 20 years old, as are some zone meters, and not surprisingly, such equipment's accuracy is far from the expected 2 percent declared precision. The water audit pointed out that 3 percent of apparent losses are related to inaccuracy of large meters. However, initial checks showed larger figures.

In order to minimize this effect, 100 large meters, which represent 30 percent of the total, will be replaced during the next two years.

Reset Inclined Residential Meters

Some statistical checks regarding the position of residential meters pointed out that a significant number of meters (400,000) were not correctly set so a low efficiency range of metering occurs most of the time, and, depending on the volume through these meters, the projected apparent loss is considerable.

To minimize these effects, corrective actions are under way. The projected recovery for the period is 47 million m^3/year.

18.4.3 Conclusion

Table 18.10 summarizes the expected recovery for real and apparent losses and the investments necessary for the period 2000/2004. All the projections are based on several operational parameters or performance indicators. Assumptions had to be made and extrapolated as an average for the metropolitan region of São Paulo.

The key information to be taken from Table 18.10 is the positive relation (benefit/cost) in spite of the intensive capital necessary for the program.

The expected return on the investment ($139 million) from 2000 to 2004 is $272 million.

Some assumptions were made as part of the whole evaluation. One example is the variable used for rate of rise of background leakage when different water supply systems are considered within the metropolitan region. The supply systems cover different areas with different topography, age of pipes, and volumes.

The data used for the estimations were extrapolated from real measured field data; however, there is an element of error included in the estimates. SABESP is continually working to improve the level of field data recorded, the calculations made, and the overall loss management program.

18.5 ARTICLE TWO

*Accounting for Lost Water**

Leslie M. Buie

18.5.1 Summary

This article reports on efforts by DeKalb County (Georgia) to identify, measure, and reduce the amount of unaccounted-for water. This information may help those with similar concerns and interests to realize that this type of effort can benefit customers. DeKalb County used methodological and logical analysis of data that reflected the county's experience to resolve the issue of unaccounted-for water.

18.5.2 Background

DeKalb County could not account for almost 20 percent of the water pumped into its distribution system in 1989. Eight years later, in 1997, unaccounted-for water loss was reduced to about 11 percent, including

*Adapted from *Journal AWWA*, vol. 92, no. 7 (July 2000), by permission. Copyright © 2000, American Water Works Association.

7.6 percent that was lost but identified. When compared with 1997 cost and billing rates, the amount of water lost in 1989 represents about $8.9 million, or $24,400 per day. Compared with approximately $5.3 million worth of water lost in 1997, this number represents an annual reduction of $3.6 million, or a savings of $9800 per day. From 1986 through 1989, the annual average amount of lost water was 17.1 percent. DeKalb County was concerned about losses of this magnitude and wanted to meet the Georgia Environmental Protection Division's unaccounted-for water goal of 10 percent maximum. Thus, the county sought to identify and reduce the amount of unaccounted-for water.

DeKalb County's sole source of raw water is the Chattahoochee River. In 1997, the county withdrew more than 30.6 billion gal (115.8×10^6 m^3) of water from the river, an average of 83.9 MGD (0.32×10^6 m^3/day). More than 28.5 billion gal (107.9×10^6 m^3) were pumped into the distribution system. Of this amount, almost 25.4 billion gal (96.2×10^6 m^3) were billed to customers. The DeKalb County water system covers approximately 269 mi^2 (697 km^2) and serves more than half a million people. The system includes about 2450 mi (3942 km) of water main and 159,000 meters serving more than 260,000 customers. The existing water treatment plant has the capacity to treat 128 MGD (0.49×10^6 m^3/day). DeKalb County is redesigning its system to increase its capacity to 150 MGD (0.57×10^6 m^3/day). This additional capacity is scheduled to be available by 2003.

18.5.3 County First Tries to Analyze Unaccounted-for Water in 1991

DeKalb County first attempted to identify areas of unaccounted-for water in 1991 using guidance from AWWA and in-house resources. The country then compared its results with data from AWWA, water meter manufacturers, and neighboring municipal water facilities. A systematic meter testing program was initiated. The author reviewed and discussed billing records and practices with representatives from the department responsible for reading meters and collecting revenues. Data from other county departments and operations were reviewed, analyzed, and discussed with officials, and meters that measure water being pumped into the system were calibrated.

Analysis Occurs in Three Phases

The author approached the analysis in three phases. First, he identified possible causes for the unaccounted-for water. Second, he analyzed each potential cause and measured the water that could be attributed to it. Third, he determined opportunities for measuring or reducing the amount of lost water.

In the first phase, the author identified the following areas for further scrutiny: consumer meters, system supply meters, accounting and billing, construction consumption, fire department use, county department use, leaks, unauthorized use, and meter sizing.

Program Designed to Test Consumer Meters

The author designed and initiated a program to test consumer meters for accuracy. Although more than 94 percent of DeKalb County's meters are 0.75 in (18.7 mm), only about 50 percent of revenues are generated by this group of meters. In view of this fact, both small and large-size meters were included in the meter testing program. This testing program identified the largest single area of lost water. Annual losses resulting from the failure of meters to accurately measure water consumption from 1991 through 1997 ranged from 6.5 to 4.4 percent of water pumped into the system. If 1997 production, costs, and billing rates were used as a basis for illustration, meter inaccuracy alone would generate annual losses of 1.85 billion gal (7.0×10^6 m^3) at the 6.5 percent level and 1.26 billion gal (4.8×10^6 m^3) at the 4.4 percent level. The respective dollar amounts are $4.66 million and $3.15 million—an improvement of $1.51 million.

The actual volume (and the related cost effect) of lost water is the product of both the meter accuracy factors and the relative volumes of water measured by the respective meters. For example, the 0.75-in (18.7-mm) meters with a 97.6 percent accuracy factor generated 261.6 million gal (0.99×10^6 m^3) of lost water, whereas the 4-in (100-mm) meters with the accuracy factor of 93.2 percent (4.4 percentage points less) accounted for only 90.9 million gal (0.34×10^6 m^3)—just 35 percent of the loss from the 0.75-in (18.7-mm) meters. Sorting the test results into selected accuracy ranges and comparing the percentages of the tests conducted for each size meter within and over the ranges of accuracy obtained more insight into the variations of meter accuracy. Once the data are obtained and on file, these accuracy ranges can be varied to accommodate the specific needs of the entity.

Why Do Meters Fail to Provide Accurate Measurements?

Meters may fail to measure water flows accurately for many reasons. The reasons identified in the DeKalb County program include normal wear and tear in the aging process, damaged and broken registers, debris and solid materials that clog the meter mechanisms, broken seals that allow water damage to certain parts, bypass lines that fail or are left open, and damage by construction activities. Improperly sized meters that fail to measure flows accurately at certain rates also can contribute to lost water. DeKalb County performs some meter sizing analyses on a case-by-case basis, but does not have a formal right-sizing program. As a result, the county cannot conclusively determine the volume of water lost through improperly sized meters. However, the DeKalb County meter testing program partially compensates for this element by testing at flow rates below normal low flows for the larger meters. These low-flow accuracy rates are averaged in with those at the higher flow rates to arrive at the average meter accuracy. Thus, the element of improperly sized meters is partially compensated for in the meter testing procedures and the resulting average meter accuracy values. Any losses resulting from improper sizing not compensated for by including the low-flow test results would be included in the 3.42 percent of unaccounted-for water (see the discussion on leaks, unauthorized use, and meter sizing).

Small meters [up through 2 in (50 mm)] are tested in the meter shop. Meters that are 3 in (75 mm) and larger are field tested. All meters are tested at a minimum of three flow rates. The flow rates are selected to span the low to high flow ranges of the respective size meters. Additionally, some larger meters are tested at flows below the standard low flow. Although less than optimal water pressure has been a limiting factor in some of the high-flow ranges, the results are considered sufficiently reliable to determine a workable average accuracy for each size meter.

Meter-Testing Program Provides Data, Guidance

Results of the meter testing program provide valuable information and guidance in directing resources to repair or replace meters. In DeKalb County, analysis of meter accuracy by size was helpful in identifying those segments most in need. The downward trend in lost water experienced by the county is attributed in part to identifying losses caused by meter inaccuracy.

System Supply Meters Were Not Calibrated

The DeKalb County study found that the meters that measure water pumped into the distribution system (system supply meters) were producing questionable results. The system supply meters had not been calibrated, and they were found to be recording on the high side. No program or schedule was in place for calibrating the meters. Because these meters are now calibrated periodically, water pumped into the system is measured more accurately. DeKalb County now tries to have the meters tested by an independent contractor and the results systematically documented and recorded.

Additionally, the system supply meters were not operating for an extended period of time. During this period, readings from meters upstream in the treatment process were used to calculate the volume of water pumped into the system. These readings were adjusted by an estimated consumption-loss factor in the treatment process. After the supply meters were back in operation, the water pumped into the system for a 6-month period was measured by both the calculation method discussed earlier and the now functioning supply meters. The variance was found to be 2.8 percent of the true volume as measured by the meters. When this factor was applied to the volume calculated for the 5 months the supply meters were not operating, it was determined that water pumped into the system was overstated by 291.8 million gal (1.1×10^6 m^3), or more than 1.9 MGD (7000 m^3/day), overstating the amount of lost water by 2.8 percent for the 5-month period.

Each problem with the system supply meters had the effect of overstating the amount of lost water. Because small percentage variances at the production or supply point significantly affect the volume of lost water, accurate measurements at this point are critical.

Review of Accounting and Billing Identifies Some Lost Water

From a review of accounting records and discussions with county revenue collection officials, the author learned that there are accounts for which consumption is estimated. These are generally accounts for which meters cannot be read for one reason or another. In addition, a number of hardship abatements and some internal consumption accounts (i.e., county departments and operations that consume water but are not billed through the regular billing system) are not billed. By analyzing these accounts and this practice, a significant volume of previously unaccounted-for water was identified.

More than 99 percent of DeKalb County's water service accounts are billed on bimonthly cycles. These accounts consume about two-thirds of the total billed water. Because of this and variations in month-end backlogs, there is some distortion in the amount of lost water calculated on a month-to-month basis. Thus, a longer period (i.e., a 12- or 24-month moving average) provides a more accurate indicator of lost water.

Study Finds Several Areas of Unmeasured Consumption

This study identified a number of areas in which water consumption was not being metered or measured. Each area was initially researched and analyzed for indicators of water use. These areas are discussed as follows.

Fire Department

Water consumed to suppress fires or flush fire hydrants is not metered. For each fire suppression in which water was used, the fire department estimated the time the engine pumps were running and at what speed. Based on these data, water consumption was calculated for fire suppressions. To measure the amount of water used to flush fire hydrants, the estimated time hydrants are open during flushing was obtained. With that plus the size of the opening and the water pressure, the amount of water used for hydrant flushing was calculated.

Roads and Drainage

The county's roads and drainage department consumes water for washing streets, hydroseeding areas of new construction, and applying herbicide sprays. No records were available on the volume of water consumed in these operations. To obtain an accurate measurement, the author determined the number of tank trucks and the capacity of each used in these operations and then obtained the department's best estimate of the frequency and circumstances under which these tank trucks would be filled to calculate the volume of water consumed.

Water and Sewer

County records contained information on the number and frequency of line flushing, but the volume of water used was not measured. These records also did not show the volume of water consumed in cleaning access holes and streets around construction and repair sites. Thus, the author relied on best estimates for the average consumption for each of the procedures, which were provided by supervisors. These estimates were then applied to the frequency of the various operations to determine the volume of water consumed.

Private Development

Water used to test and flush new lines in private development projects is not metered or measured. The author identified the need for this information, and this consumption is now measured by first calculating the gallons per minute based on the size of the blow-off and the water pressure. This amount is then applied to a standard number of hours per project and multiplied by the number of projects to determine the unmeasured gallons used to test and flush new lines in private developments.

Leaks, Unauthorized Use, and Meter Sizing

Identifying and locating leaks and unauthorized uses of water and performing related repairs and corrective actions are ongoing efforts in DeKalb County. Meter right-sizing is addressed on a limited case-by-case basis, but in the absence of a systematically applied program, information on the full effect of improperly sized meters is not available. However, the meter testing program described earlier reflects some of this loss. Although DeKalb County has not completed a comprehensive measure of leakage, unauthorized use, or meter sizing, most of the remaining 3.4 percent of unaccounted-for water can be attributed to these elements.

18.5.4 County Benefits from Locating Lost Water

The AWWA Committee on Unaccounted-for Water has identified unmetered uses of water and recommends that no more than 1 percent of total annual pumpage be used to cover these uses. The categories AWWA identified are fires, hydrant flushing, street washing, sewer flushing, swimming pool filling, and construction and wrecking. According to the author's analyses, 1 percent adequately covers DeKalb County's consumption in these areas, and the 1 percent factor is now applied instead of repeatedly analyzing each category.

DeKalb County has identified and measured, with a reasonable degree of reliability, much of the water consumption which was previously unaccounted for. These analyses have contributed to courses of action designed to increase efficiency and reduce the costs of producing and distributing water to the citizens of DeKalb County. There is also an awareness of those areas in which water use could not be reduced, as well as the effects and related costs on water consumption. Because much of the previously lost water is now accounted for and the county has new and improved guidance for directing resources, the citizens benefit from cost savings.

18.6 ARTICLE THREE

*Plugging Billion-Gallon Losses in Tennessee**

Larry Counts, Superintendent of Water and Sewer, Johnson City (Tenn.) Water and Sewer Department

18.6.1 Introduction

The Johnson City (Tennessee) Water and Sewer Department made a commitment in 1992 to develop and manage the department with the goal of providing quality service and product to our customers. One area of improvement that we needed to make was to determine the amount of lost or unaccounted-for water in our distribution system.

*Adapted from *Opflow*, vol. 23, no. 12 (December 1997), by permission. Copyright © 1997, American Water Works Association.

When the question of unaccounted-for water came up over the past 25–30 years, no one knew how to properly address the issue. It appeared that numbers were pulled from thin air, and estimates ranged from 15 to 25 percent. In 1992 the department contracted with an engineering firm to perform a water supply and distribution study. The consultants determined that we had an unaccounted-for water percentage of 49.1 percent.

18.6.2 Team Established

After the report, an unaccounted-for water team was established to study this problem and develop a plan to reduce our water loss to below 25 percent. The initial team consisted of members from meter reading, engineering, water line maintenance, and our water plant. In 1996 our team expanded to include members from our fire department and upper-level management. Over the next four years we reduced our unaccounted-for water below the 25 percent goal established in 1992. In February 1997, our unaccounted-for water was 17.6 percent. The 1997 monthly average was 23.6 percent.

Johnson City incorporated in 1869, has a population of more than 50,000, covers 32 mi^2 (83 km^2), and is continuously growing.

The Johnson City Water and Sewer Department has 145 employees, three wastewater plants, and a water service complex that serves 27,000 accounts in portions of Washington, Carter, and Union counties. The system has 350 m (560 km) of sewer lines and 720 mi (1160 km) of water lines. There are 115 water and wastewater pumping stations. The average wastewater treatment is 14 MGD (53 ML/day) with a pumping capacity of 21.5 MGD (81.4 ML/day). The average water production is 15 MGD (57 ML/day), but the ability exists to produce 20 MGD (80 ML/day).

18.6.3 Charting a Strategy

The consultant's initial report was alarming, and the unaccounted-for water team brainstormed to develop a strategy of response. We considered our initial goal of reducing water loss to below 25 percent to be realistic. Through a series of meetings, the team developed a cause-and-effect chart broken down into four major categories: meters, leaks, unauthorized use, and unkept records. The following is a detailed breakdown of these categories.

1. Meters
   - Master meters
   - Covered meters
   - Meter leaks
   - Inoperable meters
   - Lost meter tap sheets
   - City and country buildings not metered
   - Meter accuracy
2. Leaks
   - Leaks not surfacing
   - Leaks surfacing

- Allowable joint leakage
- Reservoirs
- Hydrant leaks

3. Unauthorized use and water theft
   - Illegal taps from fire line
   - Illegal fill-ups—construction, agricultural
   - Vandalism
   - Meter manipulations—crossovers
   - Open or leaking bypasses
   - Illegal taps

4. Unkept records
   - Blow-offs and line flushing
   - Testing lines
   - Water use at pumping stations
   - Pressure and flows
   - Registers off meters
   - Sewer cleaning truck
   - Fire fighting
   - Hydrant testing—fire department
   - Equipment testing—fire department
   - Public works
   - Street and sidewalk cleaning
   - Records for interconnections with other systems
   - Inaccurate records
   - Swimming pool filling

All four of these major categories were broken down and scrutinized. All members of the team played an important role to help reduce our nonaccounted water. Following is a brief summary of how we improved each of the categories.

18.6.4 Meters

We contracted with a consulting firm to test our master meters at both our water plants. The results were not what we wanted to hear. Both master meters were registering too low—the Watauga plant by 15 percent and the Unicoi plant by 3.16 percent.

We located all our covered meters and repaired those that were inoperable. Meters were installed at any government-owned buildings that did not have meters. We replaced most of our residential meters that were more than 15 years old and initiated a program to change 2500 residential meters per year. Twice a year, we test half of our large meters and replace or repair them as needed.

18.6.5 Unauthorized Use and Theft

The water department has tried to educate the construction and agricultural communities concerning our water loss program. We have developed a program in which hydrant meters can be utilized through our meter system.

A meter lockout program is now in operation. Only authorized people have keys to unlock meters for service.

Bypasses around our large meters have been tagged to notify the public of the consequences of water theft. Tennessee law states: "It shall be unlawful for any person to injure or destroy any of the pipes, fixtures or other property of said company, or to turn on water, or to make any connection with the pipes or other fixtures after the same has been shut off, stopped, or disconnected by the company (TCA 65-2707). It shall also be unlawful for any person to take or use any water or other thing belonging to the company for any purpose without having previously contacted with the company thereof (TCA 65-2708)."

18.6.6 Unkept Records

Our record keeping pertaining to water loss needed a great deal of work. We had no idea how much water we were using for our flushing program, fire fighting, or hydrant testing. Nor did the department know the amount of water lost from breaks and leaks prior to repairs. Now, every time a line is flushed, the amount of water is estimated and put on the work order. We have supplied all our employees with charts that allow easy conversion to gallons per minute. The same procedure is followed with water line leaks and breaks.

When a leak or break is reported, a work order is issued to the appropriate crew leader. When the line is dug out and repairs are made, an estimate of water lost is calculated and turned in as accounted-for water. Our local fire department is reporting the amount of water used to fight fires or to flow hydrants. Public and private water line construction crews calculate the amount of water used to fill new lines, and we are working with rural volunteer fire departments to follow the same procedures.

18.6.7 Leaks

Leaks that surface have been covered by a routine procedure, and the amount of water was estimated and reported as described above. We knew that we had several water leaks and breaks that were not coming to the surface. A two-person crew with leak detection equipment surveyed some areas where we suspected problems. Finally, we determined that the job was too large for us to handle in-house. In September 1996, the city hired a consultant to help reduce our unaccounted-for water loss. The firm was charged with the following tasks:

1. Test the accuracy of source meters.
2. Conduct a meter consumption and block usage analysis of the system's 1- to 2-in (30- to 50-mm) sales meters.
3. Conduct site inspection of a representative sample of the system's 2-in (50-mm) sales meters.
4. Test the accuracy of 29 2-in (50-mm) sales meters.
5. Perform 24-hour flow profiles of 12 sales meters.
6. Analyze a 12-month history of meters that were tested.
7. Project the annual water and revenue loss due to inaccurate 2-in (50-mm) sales meters.

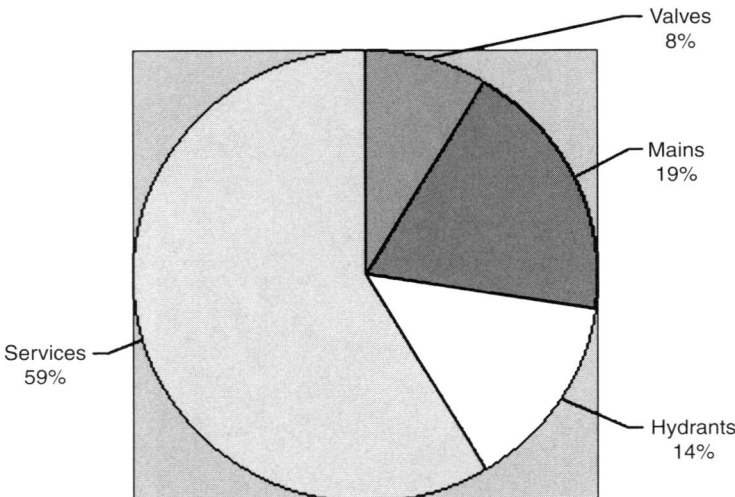

Figure 18.20 Percentage of leaks per leakage source. [*Source: Johnson City (Tennessee) Water and Sewer Department.*]

8. Identify deficiencies in the present meter testing program to improve accountability and revenue.
9. Conduct residential meter analysis.
10. Conduct zone flow measurements to determine minimum night flows in four zones.
11. Perform sonic leak detection to pinpoint leaks found during the zone flow measurements.
12. Perform a comprehensive sonic leak survey on areas of the system that were not included in the zone flow analysis.

The consulting firm completed its tasks in five months. The most significant information came from the sonic leak detection. There were 378 leaks pinpointed, totaling an estimated 1941 gpm (122 L/s). This equals more than 1 billion gal (4 billion L) per year. Based on current production costs of 20 cents per 1000 gal (4000 L), Johnson City was losing an estimated $204,000 per year to underground leaks. Figures 18.20 and 18.21 show the number of leaks per leakage source and the total estimated flow rates per leakage source in 1997.

Approximately 58 percent (220) of the 378 leaks have been repaired. A reduction in the water produced has been noted and can be seen in Fig. 18.22. This flow chart is for three years and can be compared to the rainfall chart for the same period.

The unaccounted-for water team set a new goal of lowering our unaccounted-for water to below 15 percent by the year 2001.

18.6.8 Year 2000 Update

As stated in the 1997 article "Plugging Billion-Gallon Losses in Tennessee," we set a goal for our unaccounted-for water, now referred to

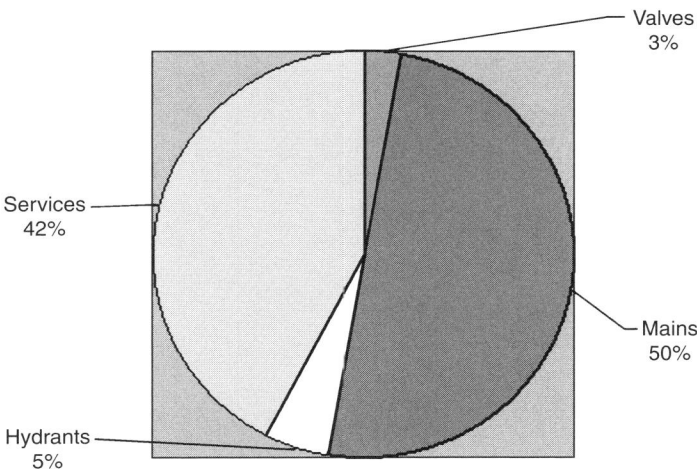

Figure 18.21 Percentage of total estimated gallons per minute per leakage source. [*Source: Johnson City (Tennessee) Water and Sewer Department.*]

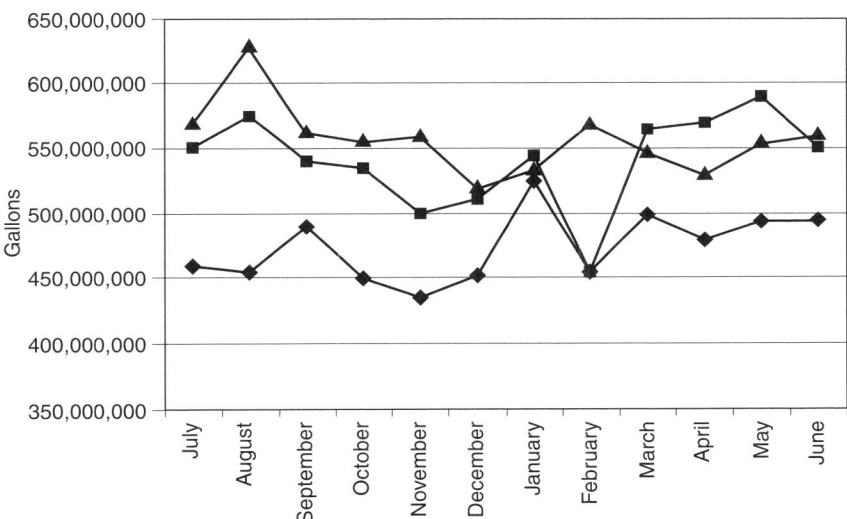

Figure 18.22 Reduction in water produced is seen from the three lines representing three years. [*Source: Johnson City (Tennessee) Water and Sewer Department.*]

as nonrevenue water, to be 15 percent by the year 2001. In 1992 we were at 49.1 percent nonrevenue; in 1997 we were at 23 percent; and in 1999 we averaged 16 percent. If the trend continues, we should meet our goal of 15 percent in 2001.

The Process

The process of maintaining gains in nonrevenue water is continuous. A plan must be developed to summarize progress on a schedule, preferably on a monthly basis. Failure to do this may lead to regression. It is important to keep vigilance on key areas.

The Survey

Since our initial leak survey in 1996, when 378 nonsurfacing leaks were discovered, amounting to more than 1 billion gal of water lost per year, we decided to survey 20 percent of our distribution system every year. This would allow us to cover our entire system in five years and not fall behind as had happened in the past.

The results of the 1998 survey were as follows:

- 57 leaks discovered
- 285 gpm
- 327,960 gpd
- 120M gpy

The results of the 1999 survey were as follows:

- 48 leaks discovered
- 294 gpm
- 423,360 gpd
- 154.5M gpy

See Figs. 18.23–18.26 for 1998 and then 1999 data for additional detailed information analysis on the leakage surveys data.

As time progresses and the cost of producing and delivering to the consumer potable, palatable water rises, the process of calculating and documenting the percentage of nonrevenue water will become more important. And hopefully you will carry with you at all times a sharp awareness of what this percentage is and what it represents in financial terms to your distribution system.

18.7 ARTICLE FOUR

A Holistic Approach to Reducing Energy Requirements for Centrifugal Pumps and Pumping Systems

Horton Wasserman, P.E., D.E.E., Malcolm Pirnie, Inc.,
White Plains, New York

18.7.1 Introduction

Energy requirements for pumping water are determined to a large extent by the pumping system configuration, components, and operation, as well as pump, motor, and drive efficiency. This article describes the relationships between the pumps, pump operation, and the distribution system and provides guidance on the use of these relationships to identify energy-saving opportunities.

18.7.2 General Equation for Pumping Energy

Energy requirements can be expressed simply by Energy = K [(flow × density × head)/(pump eff. × motor eff. × drive eff.)], where K is a conversion constant and eff. stands for efficiency.

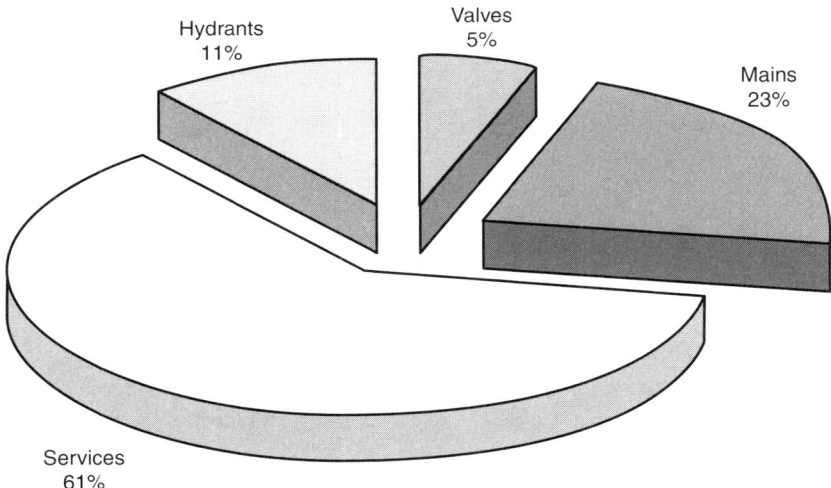

Figure 18.23 Percentage of leaks per leakage source in 1998. Leakage source and actual number of leaks: hydrants, 6; services, 35; valves, 3; mains, 13. [*Source: Johnson City (Tennessee) Water and Sewer Department.*]

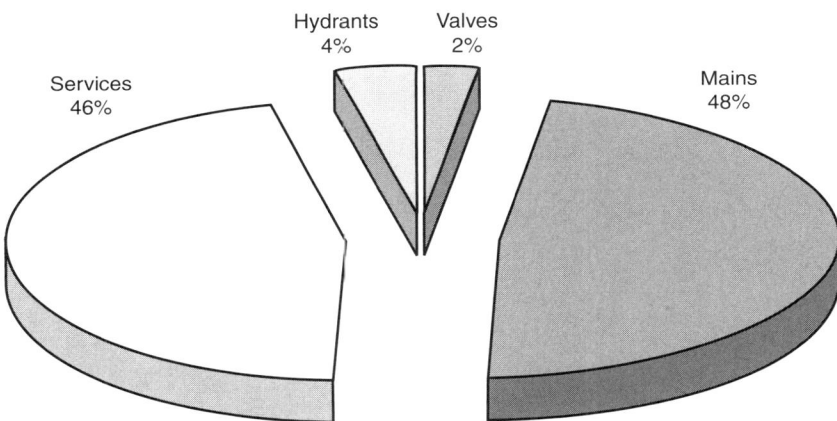

Figure 18.24 Percentage of total estimated gallons per minute per leakage source in 1998. Leakage source and actual estimated gallons per minute: hydrants, 8; services, 103.75; valves, 5; mains, 111. [*Source: Johnson City (Tennessee) Water and Sewer Department.*]

The keys to unlocking energy savings are either reducing the numerator or increasing the denominator or both. Following is a discussion of the opportunities for energy saving presented by each of the variables in the equation.

Flow

Methods for reducing flow are fairly obvious, but the energy savings can be dramatic. The methods include conservation, for which the many opportunities range from low-flow toilets and showers to more efficient filter backwash; correcting leakage; and reducing or eliminating bypass flows.

Bypass of flow from the pump discharge back to the suction sump using a bypass valve, at times in conjunction with a pump throttling valve,

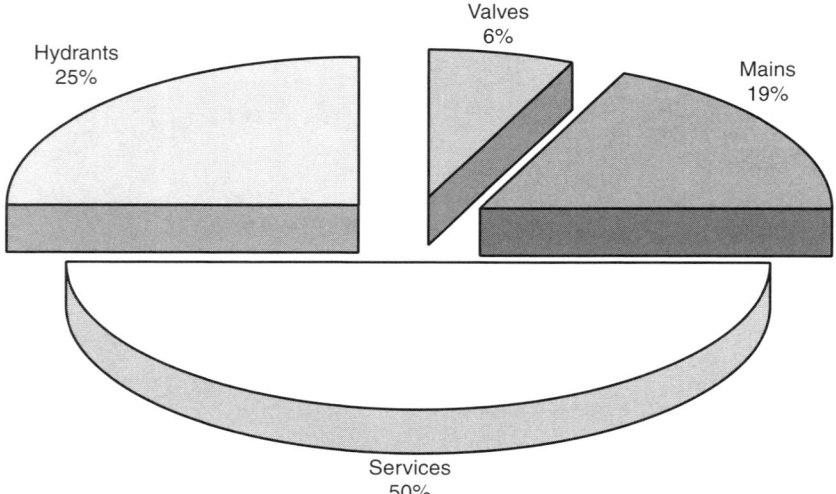

Figure 18.25 Percentage of leaks per leakage source in 1999. Leakage source and actual number of leaks: hydrants, 12; services, 24; valves, 3; mains, 9. [*Source: Johnson City (Tennessee) Water and Sewer Department.*]

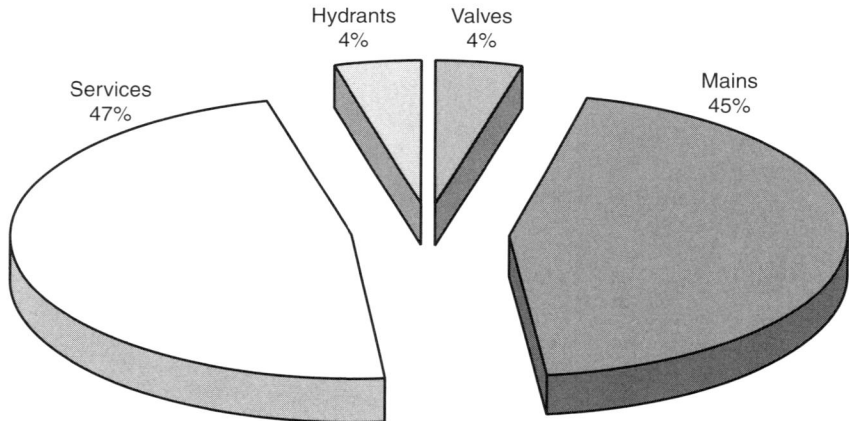

Figure 18.26 Percentage of total estimated gallons per minute per leakage source in 1999. Leakage source and actual estimated gallons per minute: hydrants, 11.15; services, 140.85; valves, 11; mains, 131. [*Source: Johnson City (Tennessee) Water and Sewer Department.*]

is a method to control flow rate over a wide range. This method was appropriate before the advent of high-energy costs and the availability of efficient variable-speed pump drives, which can reduce energy cost by pumping only the flow that is required for the application.

Density

Short of pumping on the moon, nothing can be done to reduce energy requirements by reducing density. While increasing temperature or aeration can reduce the density of water, less mass (pounds) of water would then be pumped.

Head

Techniques for reducing both static and dynamic head are numerous. They range from reducing friction loss to optimizing the design and operation of the pumping and distribution system.

Dynamic head can be reduced by

- Reducing velocity: Use larger piping or pump the same quantity of water over a longer period at a lower rate. Since friction loss is a function of the square of the velocity, pumping the same total quantity of water at a lower rate for a period that is twice as long can reduce energy requirements as much as 75 percent, assuming that there is no static head.
- Use of lower loss fittings: Use low angle reducers, Y fittings instead of T fittings, and lower loss check valves.
- Elimination of throttling to control pump delivery: Use variable-speed drives or multiple pumps.
- Reducing constrictions caused by air sediments: Install air valves and blow-offs.
- Increasing the C value of pipes: Use lined or plastic pipe, clean (pig) or clean and line existing piping.

Static head can be reduced by

- Installing siphons where appropriate
- Eliminating unnecessary air breaks
- Operating at higher sump elevations
- Operating at lower elevated storage elevations during minimum seasonal demand periods

18.7.3 System Types

The arrangement of the distribution system can have a significant impact on energy costs. Locating elevated storage beyond the center of load can result in a lower tank and, therefore, reduced static head requirements. In some cases, providing storage in the distribution system rather than at the treatment plant can reduce peak velocity requirements in the major transmission main and, therefore, friction losses.

Consideration should be given to the type of pumping system when looking for energy-saving opportunities. It is useful to characterize pumping systems as being open or closed, as each type requires a different approach to reducing head and, therefore, energy requirements.

An open pumping system can be defined as one in which the head is determined by the system head curve. The required pump head is the point at which the flow intersects the system curve and is independent of the shape of the pump curve. An example of an open system is pumping into a transmission main which supplies a reservoir or storage vessel. The strategy here is to reduce dynamic head by reducing peak velocities. Pumping the required volume of water for a longer time period and avoiding pumping at peak rates can accomplish this. This strategy may require more storage or more utilization of storage to equalize flow requirements, as well as multiple pumps or variable-speed drives. It is important to note that this strategy will not be as effective when applied to open systems having relatively high static heads, as there will be less dynamic head available for reduction.

A closed pumping system can be defined as one in which the head is determined by the pump and not the system curve. The pump head is the point at which the system demand intersects the pump curve and therefore is dependent on the shape of the pump curve. Examples of a closed system include distribution systems and plant water systems which do not have storage facilities. Significant energy reductions can be achieved for closed systems having a wide range of flow requirements by operating the pump(s) at close to their best efficiency points and avoiding operating at heads greater than that required to meet system requirements. This operation may require use of pumps with flat characteristic curves, which limit head variations in response to flow changes, multiple pumps, different-size pumps, or variable-speed drives.

18.7.4 Pump Efficiency

Selecting an efficient pump, minimizing hydraulic and mechanical losses, and sizing the pump so that it operates most efficiently most of the time can maximize pump efficiency.

Factors influencing the maximum available pump efficiency are the pump capacity, specific speed, and impeller trim.

- ❑ Capacity: Efficiencies greater than 90 percent are usually attainable for pumps with capacities exceeding 800 gpm. Maximum attainable efficiencies fall off rapidly to about 60 percent as capacity drops to 100 gpm. Therefore, in some circumstances, the use of fewer pumps may reduce energy requirements.
- ❑ Specific speed: The maximum attainable efficiency is achieved at a specific speed of about 3000 rpm. Efficiencies are reduced significantly at specific speeds which are substantially above or below 3000 rpm.
- ❑ Impeller trim: Usually the pump is most efficient when the largest-diameter impeller that is designed for the pump is installed. There may be a temptation to install a larger pump with a smaller-diameter (trimmed) impeller to suit current flow requirements, so that a larger impeller can be installed in the future when (and if) flow requirements increase. This may be appropriate but there is an energy cost, which must be considered.

18.7.5 Pump Losses

Pump losses can be divided into hydraulic and mechanical losses.

Hydraulic Losses

The efficiency of a diffuser (turbine)–type pump is reduced by hydraulic losses in the strainer (if used), suction inlet, column pipe, and discharge elbow. Manufacturer's curves for turbine pumps show bowl performance, which does not include these losses. Therefore, the pump bowl must produce more head at a higher energy cost to account for the hydraulic losses. Since the hydraulic losses are friction losses, they increase as the square of the velocity. Therefore, turbine pump efficiency is reduced as the flow increases. There may be opportunities to reduce these losses by increasing the size of the pump column and discharge head.

Mechanical Losses

Any device that produces heat consumes energy. Most pumps have bearings and packing or seals. Turbine-type pumps also have line shaft bearings and thrust bearings. Mechanical losses are generally small except for large-horsepower turbine pumps having long shafts. Line shaft bearing losses increase with the number of bearings (length of shaft). Thrust bearing losses increase with the hydraulic thrust and the weight of the rotating assembly. Except for making sure that the packing is not too tight, there is little that can be done to reduce mechanical losses. However, the mechanical losses result in an increase in the brake horsepower required to operate the pump, and therefore should be considered when evaluating alternatives.

18.7.6 Pump Sizing

The size of the pumps selected for each application will determine the actual operating efficiency.

General

A pump or a multiple-pump arrangement must be designed to meet the maximum system head and flow requirements. The maximum requirement for each pump can be defined as the maximum condition point (MCP). However, the pump or pumps usually must operate over a range of conditions. Often the actual operating head and flow requirements are significantly less than maximum. Since a centrifugal pump operates most efficiently at only one point on its head/capacity curve, which is defined as the design or best efficiency point (BEP), the challenge is to determine the condition at which the pumps will operate most often and select pumps which will be most efficient at that condition and still satisfy the requirements at the MCP. Energy requirements are determined by the pump efficiency at the operating point and not the BEP efficiency.

Systems with Variable-Speed Drives

The MCP is usually selected at the maximum pump speed. If the pump will usually operate at significantly less than maximum most of the time to satisfy reduced head and capacity requirements, the selected pump should have a BEP which is well to the left of the MCP. This will result in the pump being most efficient under the most frequent operating conditions. If the pump usually operates at speeds close to maximum and is rarely required to satisfy reduced head and capacity requirements, then the selected pump should have a BEP which is slightly to the left of the MCP. The analysis must consider energy requirements as well as the pump efficiency. Pump water horsepower requirements vary with the cube of the speed, and therefore the value of pump efficiency at higher speeds has a much greater impact on energy requirements than it does at lower speeds.

18.7.7 Motor Efficiency

Motor operating efficiency depends on two factors, load and design.

Load

Motor efficiency usually falls off when motors operate at less than 60 percent of full load. Motors should not be oversized for the application.

Design

Manufacturers can design motors to be more efficient than standard motors at an additional cost. The following should be considered.

- Energy-efficient motors: All two-, four-, and six-pole motors sold in the United States which are rated at 200 hp or less must meet energy-efficiency standards established by the Energy Policy Act of 1992 (EPACT).
- Premium-efficiency motors: Motors of the speed and sizes described above which have efficiencies 1–4 percent greater than that required by EPACT are available at additional cost.
- Large or lower-speed motors: Manufacturers of motors which have horsepower greater than 200 or more than six poles (slower than 1200 rpm) can usually provide motors (at additional cost) which have efficiencies which exceed their standard motor efficiency.
- Consideration should be given to selecting higher-efficiency motors when the capitalized value of the energy cost savings over the anticipated service life of the motor exceeds the additional cost for the higher-efficiency motor. The energy cost savings will depend on the difference in efficiency between the standard and the premium motor, the load on the motor (brake horsepower, not motor horsepower), anticipated motor operating time, and cost for energy (in kilowatts). The capitalized value of the increased efficiency of a motor which operates 10 percent of the time will be one-tenth that of a motor which operates continuously. Therefore, it may not be cost-effective to pay a premium for an efficient motor that seldom operates.

18.7.8 Drive (Variable-Speed) Efficiency

An analysis of the variable-speed drive (VSD) efficiency must consider the effect of varying the speed on the pump and motor operating efficiency, as well as energy losses in the drive itself. The effect on pump efficiency is dependent on the application.

VSDs are provided for pumps when it is necessary or desirable to vary flow in order to control a process (flow, pressure, level), avoid frequent start/stop operations, and reduce energy costs by pumping longer at a lower velocity. VSDs are an excellent alternative to throttling, but in some cases alternatives such as multiple constant-speed pumps, more storage, or pumps with a flatter curve may be a better solution. Following are application considerations.

Systems with a High Proportion of Static Head

VSDs applied to these systems will provide very little energy savings. However, flow control may be required for the process. Since the pump flow varies with the pump speed and the pump head varies with the square of the pump speed, a small speed change on systems with relatively high static head will produce a large flow change. The pump will then usually operate back on its curve at reduced pump efficiency in order to meet capacity requirements. Often use of multiple constant-speed pumps or several smaller variable-speed pumps instead of a large variable-speed pump will improve pump operating efficiency.

Closed Systems

Energy-saving alternatives to VSDs may include selecting pumps having relatively flat characteristic curves which show small changes in pump head as the capacity changes or multiple constant-speed pumps operating in sequence in response to flow requirements.

Many types of VSDs can be used to control pump speed. They include mechanical devices such as belt drives, fluid couplings, eddy current couplings, wound rotor motors, direct-current motors, and variable-frequency drives (VFDs) applied to AC motors. Since the latter is currently the drive most often applied to pumps in our industry, the following is limited to VDFs.

- ❏ VDF efficiency: Manufacturers provide efficiency data. However, the data may not include the energy needed to operate coolant pumps, air conditioning equipment, or fans, as may be required for large drives under certain conditions. This energy requirement reduces the net efficiency of the drive.
- ❏ Motor efficiency: VDFs produce harmonic currents, which reduce motor efficiency. These currents can be controlled to some extent by installing output reactors or filters, which require energy, although minimal, for operation.
- ❏ Overall efficiency: When comparing VDFs to alternatives for reducing energy, consideration should be given to the overall efficiency of the complete drive, including the VDF and the motor, as well as the effect of the speed change on the operating efficiency of the pump. If a variable-speed system is installed for the sole purpose of reducing head by 10 percent, there are no energy savings.

18.7.9 Conclusion

A holistic approach, which looks at all the components of the pumping system as well as the operation, can be used to identify appropriate alternatives for reducing energy requirements.

18.8 REFERENCES

1. Alegre, H., Hirner, W., Baptista, J., and Parena, R., *Performance Indicators for Water Supply Services,* Manual of Best Practice, International Water Association, 2000.
2. American Water Works Association, Manual M36, *Water Audits and Leak Detection,* 2d ed., AWWA, 1999.

3. AWWA Leak Detection and Water Accountability Committee, "Committee Report: Water Accountability," *Journal AWWA,* July 1996.
4. Smith, J. C., Estimating Unavoidable Leakage in Water Distribution Systems, M.Sc. thesis, Department of Civil Engineering, Brigham Young University, Provo, Utah, April 1987.
5. National Water Council Standing Technical Committee, Report No. 26, Technical Working Group on Waste of Water, Leakage Control Policy and Practice, 1980.
6. Ogura, *Japan WaterWorks Association Journal,* May 1981.
7. AWWA Research Foundation, *Water and Revenue Losses: Unaccounted-for Water,* 1987.
8. AWWA Council Report 4450D, 1957.
9. Lambert, A., Brown, T. G., Takizawa, M., and Weimer, D., "A Review of Performance Indicators for Real Losses from Water Supply Systems," *AQUA,* December 1999.
10. Lambert, A. O., Myers, S., and Trow, S., *Managing Water Leakage: Economic and Technical Issues,* Financial Times Energy Publications, 1998.
11. U.S. Army Corps of Engineers, Philadelphia Water Supply Infrastructure Study, May 1985.
12. Morgan Stanley & Co., Statement, City of Philadelphia, Water and Wastewater Revenue Bonds, Series 1995.
13. American Water Works Association, Manual M36, *Water Audits and Leak Detection,* 2d ed., 1999.
14. International Water Association, *Performance Indicators for Water Supply Services,* Manual of Best Practice, IWA, 2000.
15. Bristol Water Services Limited, Final Report, City of Philadelphia Leakage Management Assessment Project, 2001
16. Lambert, A., Huntington, D., and Brown, T. G. "Water Loss Management in North America: Just How Good Is It?" paper presented to Workshop on Progressive Developments in Leakage and Water Loss Management, AWWA, Distribution Systems Symposium, New Orleans, September 2000.

APPENDIX

Water Accountability

> **Article One**
> Water Accounting for Management and Conservation
> *Janice A. Beecher and John E. Flowers*
>
> **Article Two**
> Committee Report: Water Accountability
> *AWWA Leak Detection and Water Accountability Committee*
>
> **Article Three**
> Conducting a Water Audit
> *Chapter 2 of AWWA Manual M36*

A.1 INTRODUCTION

The purpose of this appendix is to illustrate various discussions on water accountability in North America today. Article One was published in 1999 by the American Water Works Association (AWWA) and discusses some of the confusion in the marketplace today with regard to methodology and terminology. The article also reinforces the need for good accounting and supply-side conservation. It is one of the more recent articles to be published which demonstrates growing awareness of the need to improve both performance and the indicators we use to manage losses. Article Two is a Leak Detection and Water Accountability (LDWA) Committee report, and Article Three is a reprint of Chap. 2 of AWWA Manual M36, which provides basic audit guidelines.

There is currently increased interest in North America in adopting the recent IWA guidelines and performance indicators as a standardized approach to better water loss management. The American Water Works Association Research Foundation has recently let a contract which will involve testing of the IWA methods at three North American utilities. Readers are encouraged to follow this progress and, if applicable, adopt this new approach.

A.2 ARTICLE ONE

*Water Accounting for Management and Conservation**

Janice A. Beecher[†]
John E. Flowers[†]

A.2.1 Introduction

Today, the commodity water systems deliver has greater value than ever before. Extraction, treatment, storage, and pumping all add value to the water resource. Given mounting infrastructure costs and growing constraints on water resources, water managers must strive to account for all the water that travels from source to end users.

Conservation is a key rationale for improving methods for tracking water. Conservation on the supply side can be particularly effective because the supply side is under utility managers' direct control, and water savings

---

*Adapted from *Opflow*, vol. 25, no. 5 (May 1999), by permission. Copyright © 1999, American Water Works Association.
[†]Janice A. Beecher, Ph.D, is the principal of Beecher Policy Research, Inc. John E. Flowers is the program director for the U.S. Environmental Protection Agency's Water Alliances for Voluntary Efficiency program.

translate directly to cost savings (without adversely affecting revenues). In *Water Conservation Plan Guidelines,* published by the U.S. Environmental Protection Agency (USEPA) in August 1998, water accounting and loss control, along with universal source and end-user metering, are discussed as basic and essential conservation measures for water systems of all sizes.*

The guidelines' approach to water accounting is similar to existing practices common to many systems. The emphasis on accounting, however, shifts attention from any single standard or percentage to the process of tracking water through the water delivery system and identifying potential areas for improvement.

A.2.2 Confusing Terms and Standards

Numerous imprecise terms are used to represent the difference between the water that is withdrawn from the source and water that is eventually distributed to end users. For example, the terms "water losses" and "unaccounted-for water" are sometimes used interchangeably. But not all unaccounted-for water is lost; some might be given away or used for authorized purposes. The term "uncompensated usage" has been used to include water used by public authorities, water used for flushing and other maintenance purposes, leakage, and uncollected accounts from customers.

A 1987 study by Lynn P. Wallace for the American Water Works Association Research Foundation made this distinction between "account" and "nonaccount" water:

- Account water is all water for which an account exists; the water is metered, and the account is billed.
- Nonaccount water is the sum of all water produced or purchased by a water utility that is not covered by the term "account water."

This proposed nomenclature, which is adopted in the guidelines' accounting system, has not really caught on in the water industry. For the most part, the industry uses the similar term "unaccounted-for water" to mean leaks and other kinds of avoidable losses. However, the measurement of unaccounted-for water can be confusing because the numerator and the denominator used to calculate the percentage are not obvious. Is the percentage amount supposed to represent all water not metered and sold, or only water lost through leaks? How the percentage is calculated makes a meaningful difference.

On top of the confusion about terms is confusion about standards. Any single standard (expressed in terms of volume or a percentage) for unaccounted-for water will not be valid, realistic, or appropriate for many water systems. Many system characteristics—such a size, age, and service population density, physical terrain, soil characteristics, and pipe materials—will affect leakage rates. Systems also have different production-cost profiles against which the cost effectiveness of leak detection and control programs can be evaluated.

In 1996, the AWWA's Leak Detection and Accountability Committee recommended 10 percent as a benchmark for unaccounted-for water,

---

*http://www.epa.gov/OWM/genwave.htm#guideline

replacing a 15 percent standard that apparently was based more on folklore than analysis. Even the 10 percent benchmark has not achieved consensus in the water industry.

The AWWA committee suggested that "regardless of the water system's size, water loss should be expressed in terms of actual volume, not as a percentage." This volumetric measure, the committee points out, is essential for estimating the monetary value of losses. The volume of lost water can be multiplied by the unit cost of water production (on the retail rate) to estimate the cost of the lost water. From an economics perspective, the true value of losses is the *marginal* or *incremental* unit cost of production (that is, the cost of producing the next increment of supply). Incremental or marginal costs more accurately reflect water's value, which will increase as supply alternatives become scarcer. Reducing leakage and loss can help systems avoid high supply-side operating and capital costs.

A clearer system of water accounting could be used to improve data collection, and evaluate and eventually establish a standard (or set of standards for the different water subaccounts) based on a sound empirical understanding and agreement about the basic purpose and concepts of water accounting.

> Water loss should be expressed in terms of volume, not as a percentage.

A.2.3 A Water Accounting System

All water systems, even smaller systems, should implement a basic system of water accounting. AWWA Manual M36, *Water Audits and Leak Detection,* provides guidance for this process. Water system managers should try to track water throughout the system—from water sources to end users—and identify areas that may need attention, particularly large volumes of nonaccount water.

A system of water accounting should be based on experience and observation. Thus, metering and audits play an important role in implementing a system of water accounting. A system of water *accounting* is essential to water *valuation,* that is, placing a monetary value on losses.

Nonaccount water includes water that is *metered but not billed,* as well as *all unmetered water.* Unmetered water may be authorized for certain utility purposes, such as operation and maintenance, and for certain public uses such as fire hydrant maintenance. Unmetered water also includes unauthorized uses, including losses from accounting errors, malfunctioning distribution system controls, thefts, inaccurate meters, or leaks. In some cases, nonaccount water may represent losses such as meter inaccuracy and theft is associated with specific customer accounts. Some unauthorized uses may be identifiable. When they are not, these unauthorized uses constitute *unaccounted-for water.*

Implementing a system of water accounting is a necessary first step in developing strategies for loss control, as well as for metering nonaccount water. A system of water accounting is provided in Fig. A.1. This system for tracking water begins with total water produced and ends with unaccounted-for water. In this system, "unaccounted-for water" has a more literal meaning. Figure A.2 provides a worksheet that water system managers can use to account for water.

APPENDIX A WATER ACCOUNTABILITY

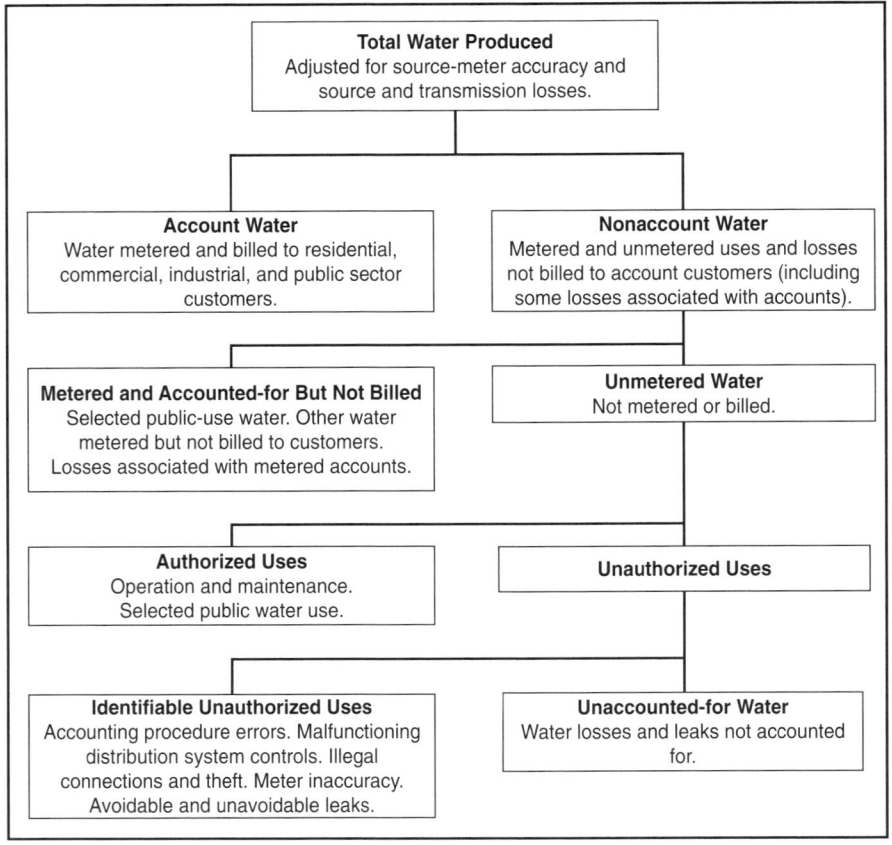

Figure A.1 Water accounting system. (*Source: "Water Accounting for Management and Conservation."*)

A.2.4 Additional Strategies

The USEPA's *Water Conservation Plan Guidelines* identify a number of management strategies that can be used in conjunction with a system of water accounting. These conservation measures can be highly effective in reducing nonaccount, unaccounted-for, and lost water. Key strategies include the following.

- *Repair known leaks:* Water lost produces no revenues for the utility. The cost of water leakage can be measured in terms of the operating costs associated with water supply, treatment, and delivery. Repairing larger leaks can be costly, but it also can produce substantial savings in water and expenditures over the long run.
- *Universal metering:* Source water, service connections, and all water provided free of charge for public use should be metered, to provide the most accurate usage data. If source water is unmetered, usage can be estimated by multiplying the pumping rate by the time of operation based on electric meter readings.
- *Analysis of nonaccount water:* Nonaccount water use should be analyzed to identify potential revenue-producing opportunities, as well as recoverable losses and leaks. Some utilities might consider charging for water previously given away for public use

Line	Item	Volume (gal)	% of Amount in Line 1
1	**Total Source Withdrawals and Purchases**		**100%**
2	*Adjustments to source water supply [a]*		
2A	Adjustment for source meter error (+ or −)		
2B	Adjustment for change in reservoir or tank storage (+ or −)		
2C	Adjustment for transmission line losses (−) [a]		
2D	Adjustments for other source contributions or losses (+ or −) [a]		
3	Total adjustments to source water (add lines 2A through 2D)		
4	**Adjusted Source Water (subtract line 3 from line 1)**		%
5	*Metered water sales*		
5A	Metered residential sales		
5B	Metered commercial sales		
5C	Metered industrial sales		
5D	Metered public sales		
5E	*Other metered sales*		
6	Total metered sales (add lines 6 through 8)		
7	Adjustment for meter reading lag time (+ or −)		
8	Adjustment for meter errors (+ or −) [a]		
9	Adjusted total meter sales (add lines 6 through 8)		
10	**Nonaccount Water (subtract line 9 from line 4)**		%
11	Metered and accounted-for but not billed		
11A	Public-use water metered but not billed		
11B	Other water metered but not billed		
12	*Authorized unmetered water: operation and maintenance*		
12A	Main flushing		
12B	Process water at treatment plant		
12C	Water quality and other testing		
13	*Authorized unmetered water: public use*		
13A	Storm drain flushing		
13B	Sewer cleaning		
13C	Street cleaning		
13D	Landscaping in large public areas		
13E	Firefighting, training, and related maintenance		
14	*Other authorized metered use*		
14A	Swimming pools		
14B	Construction sites		
14C	Other unmetered uses		
15	Total authorized unmetered water (add lines 11A through 14C)		
16	**Total Unauthorized Losses (subtract line 15 from line 10)**		%
17	*Identifiable water losses and leaks*		
17A	Accounting procedure errors [a]		
17B	Malfunctioning distribution system controls		
17C	Illegal connections and theft		
17D	Meter inaccuracy		
17E	Unavoidable water leaks		
17F	Avoidable water leaks		
18	Total identifiable water losses and leaks (add lines 17A through 17F)		
19	**Unaccounted-for Water (subtract line 18 from line 16)**		%

[a] Methodology subject to industry and regulatory standards.

Figure A.2 Worksheet. (*Source: "Water Accounting for Management and Conservation."*)

and increasing efforts to reduce illegal connections and other forms of theft.
- *System audit:* A system audit can identify and measure authorized metered and unmetered uses and provide a more accurate analysis of nonaccount water. The AWWA's Manual M36 is an excellent resource for information on conducting water audits.
- *Leak detection and repair strategy:* Systems also should institute a comprehensive leak detection and repair strategy. This strategy may include regular on-site testing using computer-assisted leak detection equipment, a sonic leak detection survey, or another acceptable method for detecting leaks along water distribution mains, valves, services, and meters. Divers can be used to inspect and clean storage tank interiors.
- *Automated sensors/telemetry:* Remote sensor and telemetry technologies can be used for ongoing monitoring and analysis of source, transmission, and distribution facilities. These sensors and monitoring software can alert operators to leaks, fluctuations in pressure, problems with equipment integrity, and other concerns.
- *Loss prevention program:* Periodic pipe inspections, cleaning, lining, and other maintenance efforts can improve distribution system performance and prevent leaks and ruptures from occurring. Utilities might also consider methods for minimizing water used in routine water system maintenance procedures in accordance with other applicable standards.

A.2.5 Conclusion

As water costs and prices rise, the benefits of water accounting and loss control are becoming more obvious. Shifting the focus from a single performance standard to a more refined system of accounting could be very beneficial to the water industry. The USEPA's *Water Conservation Plan Guidelines* introduces terms that clearly identify where all the water produced goes, and by using this system of water accounting, managers can track water throughout their systems and identify opportunities for improvement. As experience with water accounting grows, this system can be fine-tuned and adopted to provide additional guidance for reducing avoidable leaks and losses.

The report that follows was published on behalf of the AWWA National Leak Detection and Water Accountability Committee in 1996. The committee has since started to adopt recent IWA performance indicators as discussed in Chaps. 4 and 18 and is currently awaiting the results of the upcoming AWWA Research Foundation (AWWARF) study to determine which performance indicators are the correct ones for North American conditions.

While the committee report discusses losses as percentages, which is not the form the author prefers, the report does state that utilities should state their losses as a volume and put a cost on the loss. This is extremely important for any utility and is still very much a current concept. The report later justifies why percentages are not the correct indicator for water system losses.

A.3 ARTICLE TWO

Committee Report: Water Accountability*

AWWA Leak Detection and Water Accountability Committee[†]

A.3.1 Introduction

Often, decision makers in the water supply field are satisfied when they can account for 85 percent of the water they produce. Recognizing the problem of lost or non-revenue-producing water and desiring to find solutions for member utilities, the AWWA's Distribution and Plan Operations Division asked the Leak Detection and Water Accountability Committee to write this report, which recommends that because of increasing demand and higher operational costs, the goal for lost or non-revenue-producing water should be less than 10 percent. The report also proposes that certain guidelines should be followed when the goal of 10 percent is not met.

Over the past several years, it has not been unusual to hear statements from water utilities throughout the country such as "AWWA says that 15 percent unaccounted-for water is acceptable" or "Our water loss is pretty close to the AWWA guidelines of 15 percent." In fact, AWWA has never adopted a policy or issued guidelines to the effect that 15 percent unaccounted-for water is acceptable. AWWA's Distribution and Plant Operations Division asked the National Committee on Leak Detection and Water Accountability to determine how this impression arose, to research the issue of unaccounted-for water, and to issue guidelines and recommendations that specifically address unaccounted-for water and effective water loss management for water utilities.

A.3.2 1957 Report Identified as Source of Figure

Apparently, the source of the frequently heard statement that AWWA accepts a 15 percent rate of unaccounted-for water is a committee report presented at the 1957 AWWA annual conference in Atlantic City, New Jersey, and subsequently published in *Journal AWWA*.[1] The committee report states that "unaccounted-for water may vary from 10–15 percent in a well operated system where the consumption is between 100 and 125 gpcd (379 and 473 L/d)." Since that article was published 39 years ago, two areas of water loss management—operating costs and technological resources—have undergone dramatic changes.

A.3.3 Operating Cost Increases

Virtually all costs of producing and distributing potable water have increased dramatically over the past 30 or 40 years—treatment plant

---

*Adapted from *Journal AWWA*, vol. 88, no. 7 (July 1996), by permission. Copyright © 1996, American Water Works Association.

[†]David A. Liston (chair), Timothy G. Brown (vice-chair), F. S. Brainard, Jr., Donald E. Britt, John P. Corless, Jr., Reed G. Craft, William A. Finger, John G. Hock, Chris J. Kleinert, William E. Luta, Keith J. Nelson, Glen C. Phipps, Keith Wadsworth, Dean A. Wheadon, L. Harvey Wicklund, and Glennon N. Zelch.

expansions and improvements, development of additional water supplies (pumping costs), labor at all staff levels, regulatory compliance, restoration expenses, and so on. As the total cost of operation rises, the cost of unaccounted-for water also rises at a corresponding rate.

A.3.4 Technology to Reduce Water Loss

Because of increasing costs of production, distribution, and unaccounted-for water, many technological advances aimed at reducing water loss have been developed. These include leak detection and pinpointing instruments, more accurate metering devices, instrumentation to test meter accuracy, rate-of-flow recording for meter sizing and typing, and data collection. In addition, a wide range of techniques and methodologies provide practical application of these advanced technologies to identify losses within a water system and to implement cost-effective corrective action. Because of these significant advances, the AWWA's Leak Detection and Water Accountability Committee recommends the goal for unaccounted-for water should be less than 10 percent.

A.3.5 Method for Determining "True" Unaccounted-for Water

The basic steps for quantifying the amount of water loss within a water system are as follows (*regardless of the water system's size, water loss should be expressed in terms of actual volume, not as a percentage*).

1. Accurately determine the amount of water being produced or purchased and delivered to the distribution system for a 13-month period of operation. The production quantities are used to establish the base number against which all other calculations in the water accountability process will be made. It is therefore imperative that the production quantities be accurate. This requires annual accuracy testing of source meters.
2. Determine the total amount of water sales for the same period of operation as measured by all meters in the system. This includes estimated accounts.
3. Subtract the total amount of water sold from the total amount of water purchased or produced.
4. Identify and quantify all other categories of water use in the system. It is recommended that all water use in the various categories be metered, so the water can be accurately accounted for instead of ending up in the unaccounted-for water category where it does not belong. If actual metering is not possible, every effort should be made to accurately estimate each type of water use to determine realistic usage quantities for each category.

    The various categories of water use in a water system include bulk water sales (including construction), known leakage, tank (storage facility) drainage, storage tank overflows, line flushing, fire protection, bleeding or blow-off done during the winter or for taste and odor episodes, and municipal uses (sewer cleaning,

street cleaning, golf course, parks and recreational facilities, hydrant flow tests, unknown miscellaneous uses, and all other nonrevenue uses).
5. Subtract the total quantity of water use for the same period of operation for all of the identified categories in step 4 from the quantity of water remaining after step 3.
6. The quantity of water that remains is the water system's amount of unaccounted-for water. True unaccounted-for water consists of the following: unidentified leakages, meter inaccuracies, theft, underestimated accounts, improperly typed and sized meters, meter reading errors, and accounting errors.

A.3.6 Express Water Loss in Terms of Volume

Regardless of the water system's size, water loss should be expressed in terms of actual volume, not as a percentage. This is necessary for the utility to be able to determine the true annual cost of unaccounted-for water. Consider the following example.

A water utility produces 2 MGD (7.6 ML/day) and has a true unaccounted-for water rate of 20 percent. The utility adds a large-volume user that uses 0.5 MGD (1.9 ML/day), which increases the production to 2.5 MGD (9.5 ML/day). What happens to the 20 percent unaccounted-for water? It becomes 16 percent. Has the utility actually reduced its water loss and the associated costs of the loss?

Don't be misled by percentages. Measure performance with respect to unaccounted-for water strictly by comparing the volume of water lost with the volume that was lost in prior years. The "percentage unaccounted" so often used, although it is a convenient yardstick of comparison, can be misleading.

> **D**on't be misled by percentages—they can be deceiving!

A.3.7 Convert Water Loss to Dollar Loss

The amount of water loss is more meaningful than the percentage of unaccounted-for water. When the total volume of unsold water is known, the utility can place a value on that water and determine the cost effectiveness of implementing corrective action.

The simplest way to estimate the potential financial loss is to make two assumptions:

❏ All water loss results from underground pipe leakage.
❏ All water loss results from underregistering water meters.

Usually the least amount of financial loss would be related to underground leakage, because that amount of the loss depends on the direct production costs associated with producing that amount of water. Three components make up direct production costs: cost of raw water, energy costs (electricity), and treatment costs (chemicals). Therefore, the total volume of underground lost water is multiplied by the unit production rate (excluding labor) to determine the approximate financial loss to the utility.

Of course, the cost of underground leakage would be of greater value if leakage repairs eliminated the need for plant expansion.

Usually the most expensive water loss in the distribution system is caused by both underregistration of water meters and theft of water. This water loss has the highest potential value because it is "sellable" at the retail water rate. The total water loss volume related to underregistration and theft should be multiplied by the retail rate to determine the approximate lost revenue.

Experience dictates that total water loss in a system does not result from one cause but from several. Generally, a utility can split the difference between financial loss from leakage and from metering. The utility can then estimate how much money is being lost because of unaccounted-for water. The actual split will vary from one utility to another and will be determined by the age of meters, water quality, system pressure, age of pipe, and pipe material. For instance, if a utility has excellent water quality (e.g., minimal buildup of sand or minerals) and an aggressive meter maintenance program, it will tend to weigh the cost factors toward production costs rather than retail rate. An example of determining the dollar value of unaccounted-for water is

- ❑ Total daily production: 1 MGD (3.8 ML/day)
- ❑ Total known usage: 0.8 MGD (3 ML/day)
- ❑ Difference: 0.2 MGD (0.8 ML/day)
- ❑ Production costs: $0.30/1000 gal ($0.08/1000 L)
- ❑ Average retail rate: $2.50/1000 gal ($0.70/1000 L)

To determine the minimum lost revenue, multiply 0.2 MGD (0.8 ML/day) of unmetered water by the production cost. If all unmetered water were lost through leakage, the direct cost to the utility would be $21,900.

To determine the maximum amount of financial loss to the water system, multiply the 0.2 MGD (0.8 ML/day) by the retail rate; the result is $182,500 per year. If all unmetered losses occurred in the area of underregistering water meters, the financial loss attributable to that condition would be nearly nine times that of the loss attributable to leakage.

If the utility knows what is causing distribution system water losses, it may want to weight the cost factors toward either leakage or metering. For instance, it may be determined that metering is a greater problem than leakage by a factor of 2:1. The approximate cost of lost water in the system would then be $130,000 per year. When wastewater revenue loss is added to this example, the effect on the system is amplified. For many systems, this could be a significant loss.

A.3.8 Weigh the Costs

After the utility has determined the annual cost (or cost range) of unaccounted-for water, management can make a more informed decision concerning the cost effectiveness of corrective action. For example, if a utility is losing $100,000 per year because of unaccounted-for water and it has an aggressive meter accuracy testing and repair program, it can be reasonably sure most of the loss is attributable to leakage. If leak detection and pinpointing survey of the distribution system will cost about $10,000, it is likely that such a survey will be cost effective. Likewise, if a utility is losing $100,000 per year in unaccounted-for water and it has recently conducted a comprehensive leakage detection and pinpointing survey, it can reasonably conclude that most of the loss is attributable to meter inaccuracies or underregistering. If a testing and repair program will cost about $20,000, it would be cost effective.

Regardless of the size of the water utility, determining the cost of loss should be conducted on a case-by-case basis. Each water system has unique characteristics and variables that must be considered when the cost of water is calculated for any given system—e.g., the quantity and the quality of the raw water, the number and size of commercial and industrial meters, the extent of pumping required (energy costs), and treatment costs.

Today's water system managers are faced with a variety of challenges to be met and problems to be solved. Drought, contamination, lack of available funding sources, increased regulations for water quality and monitoring, and aging distribution systems are among some of the issues that confront water utilities.

As the cost of producing and distributing potable water continues to escalate, it will be important for water systems managers to implement effective water loss management programs. Excessive amounts of water loss or unaccounted-for water will not be tolerated by regulatory agencies or the general public as water rates continue to increase.

It is fortunate that the necessary technologies, expertise, and methodologies are available to identify and substantially reduce lost water and to reduce unaccounted-for water to a more acceptable and realistic level. As the twenty-first century gets under way, the goal for unaccounted-for water should be less than 10 percent.

A.3.9 Reference

1. "Revenue-Producing versus Unaccounted-for Water," *Journal AWWA* 49:12:1587 (Dec. 1957).

A.3.10 Bibliography

Brown, T. G., "Basic Leak Detection Is Necessary for Any System," *Opflow* 11:10:1 (Oct. 1985).

Brown, T. G., "The Tangible and Intangible Benefits of Leakage Control," *Proc. 1986 AWWA Distribution System Symposium,* Minneapolis, Minn.

Hock, J. G., "A Comprehensive Approach to the Control of Unaccounted-for Water," *Proc. 1989 AWWA Distribution Systems Symposium,* Dallas, Tex.

"Leak Detection Programs Save Water, Money: Twelve Helpful Hints for Getting Started in Leak Detection," *Opflow* 17:12:1 (Dec. 1991).

A.4 ARTICLE THREE

Conducting a Water Audit

Chapter 2 of AWWA Manual M36, Water Audits and Leak Detection, 2d ed., 1999; reprinted by permission

---

AWWA MANUAL M36

Chapter 2

Conducting a Water Audit

This chapter outlines the basic steps involved in conducting a water audit (see Figure 2-1). Instructions for accomplishing each step are included. In some cases, more than one method is described for accomplishing a particular step, thus allowing for the various configurations of water systems. Each water supplier may choose the technique best suited to the system under study.

BEFORE STARTING

A number of decisions should be made before beginning a water audit. While discussion of all factors is beyond the scope of this book, three factors that influence the reliability of the study are discussed in the following paragraphs.

Water Audit Worksheet

A sample water audit worksheet is included in this chapter (Figure 2-2). This chapter describes each step of the audit, including what information is needed, how to get that information, and how to enter it on the worksheet. Examples accompany the instructions to show how the data are gathered and how they are entered on the worksheet. Figure 2-2 uses data from examples in this chapter; appendix A provides a blank water audit worksheet.

Instructions for entering information on the worksheet appear in blue.

Set a Study Period

A water audit is a study over time. Choose a time period that allows analysis and evaluation of total system water use. One month or even six months is too short a time to give an overall picture of water flow through the system (see Figure 2-3). A 12-month study period is recommended. Most utility records are kept by the calendar or fiscal year; either system makes 12 months of data available. However, a calendar

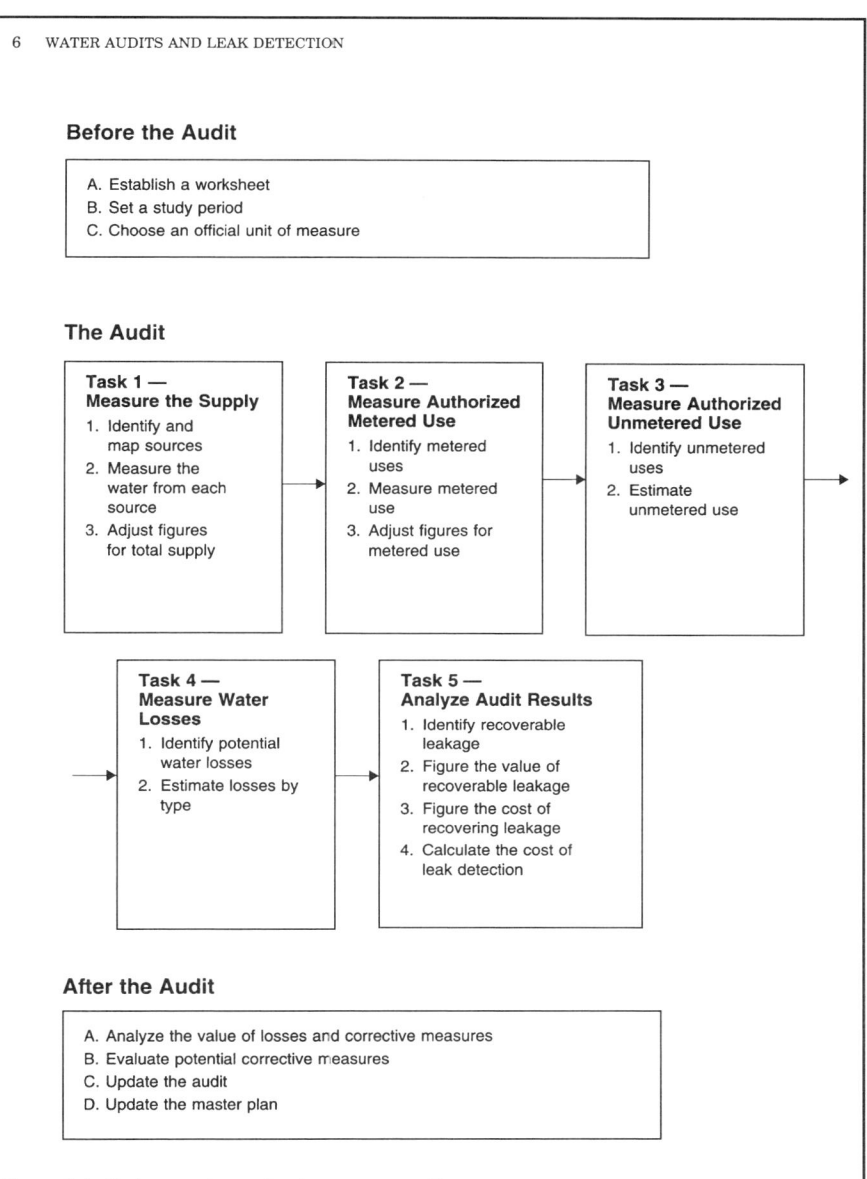

Figure 2-1 Basic steps in conducting a water audit

APPENDIX A WATER ACCOUNTABILITY

CONDUCTING A WATER AUDIT 7

WATER AUDIT WORKSHEET

For: County Water Co.
Audit Study Period: Jan. 1 – Dec. 31 1998

Line	Item	Subtotal	Total Cumulative	Units*
Task 1—Measure the Supply				
1	Uncorrected total water supply to the distribution system (total of master meters)		3,672.36	mil gal
2A–C	Adjustments to total water supply			
2A	Source meter error (+ or –)	+82.90		mil gal
2B	Change in reservoir and tank storage (+ or –)	+0.83		mil gal
2C	Other contributions or losses (+ or –)	0		mil gal
3	Total adjustments to total water supply (add lines 2A, 2B, and 2C)		+ 83.73	mil gal
4	Adjusted total water supply to the distribution system (add lines 1 and 3)		3,756.09	mil gal
Task 2—Measure Authorized Metered Use				
5	Uncorrected total metered water use	3,258.0		mil gal
6	Adjustments due to meter reading lag time (+ or –)	+0.20		mil gal
7	Metered deliveries (add lines 5 and 6)		3,258.20	mil gal
8A–C	Total sales meter error and system-service meter errors (+ or –)			
8A	Residential meter error	134.33		mil gal
8B	Large meter error	29.97		mil gal
8C	Total (add lines 8A and 8B)		164.30	mil gal
9	Corrected total metered water deliveries (add lines 7 and 8C)		3,422.50	mil gal
10	Corrected total unmetered water (subtract line 9 from line 4)		333.59	mil gal
Task 3—Measure Authorized Unmetered Use				
11A	Firefighting and firefighting training	9.70		mil gal
11B	Main flushing	1.60		mil gal

*Units of measure must be consistent throughout the worksheet. The particular unit used (that is, gallons, millions of gallons, acre feet, cubic feet, cubic metres, or other unit) is left to the user.

Figure 2-2 Completed water audit worksheet for County Water Company (continued)

8   WATER AUDITS AND LEAK DETECTION

Line	Item	Water Volume Subtotal	Total Cumulative	Units*
	Measure authorized unmetered use (continued)			
11C	Storm-drain flushing	0.50		mil gal
11D	Sewer cleaning	0.65		mil gal
11E	Street cleaning	1.75		mil gal
11F	Schools	0		mil gal
11G	Landscaping in large public areas:			
	Parks	40.00		mil gal
	Golf courses	120.00		mil gal
	Cemeteries	3.00		mil gal
	Playgrounds	5.40		mil gal
	Highway median strips	0.65		mil gal
	Other landscaping	0.50		mil gal
11H	Decorative water facilities	0 (metered)		mil gal
11I	Swimming pools	0 (metered)		mil gal
11J	Construction sites	0 (metered)		mil gal
11K	Water quality and other testing (pressure-testing pipe, water quality, etc.)	0 (metered)		mil gal
11L	Process water at treatment plants	0.07		mil gal
11M	Other unmetered uses	0		mil gal
12	Total authorized unmetered water (add lines 11A through 11M)		183.82	mil gal
13	Total water losses (subtract line 12 from line 10)		149.77	mil gal
Task 4—Measure Water Losses				
14A	Accounting procedure errors	11.63		mil gal
14B	Unauthorized connections	0.33		mil gal
14C	Malfunctioning distribution system controls	0		mil gal
14D	Reservoir seepage and leakage	+ 0.07		mil gal
14E	Evaporation	9.78		mil gal

*Units of measure must be consistent throughout the worksheet. The particular unit used (that is, gallons, millions of gallons, acre feet, cubic feet, cubic metres, or other unit) is left to the user.

Figure 2-2 Completed water audit worksheet for County Water Company (continued)

506 APPENDIX A WATER ACCOUNTABILITY

		Water Volume		
Line	Item	Subtotal	Total Cumulative	Units*
	Measure identified water losses (continued)			
14F	Reservoir overflow	0		mil gal
14G	Discovered leaks	21.88		mil gal
14H	Unauthorized use	0		mil gal
15	Total identified water losses (add lines 14A through 14H)		43.69	mil gal

Task 5—Analyze Audit Results

Line	Item	Subtotal	Total Cumulative	Units*
16	Potential water system leakage (subtract line 15 from line 13)		106.08	mil gal
17	Recoverable leakage (multiply line 16 by 0.50)		53.04	mil gal

Line	Item	Dollars per Unit of Volume
18A–B	Cost savings	
18A	Cost of water supply	$ 690/mil gal
18B	Variable operation and maintenance costs	$ 153/mil gal
19	Total costs per unit of recoverable leakage (add lines 18A and 18B)	$ 843/mil gal

Line	Item	Dollars per Year
20	One-year benefit from recoverable leakage (multiply line 17 by line 19)	$ 44,712.72
21	Total benefits from recovered leakage (multiply line 20 by 2)	$ 89,425.44
22	Total costs of leak detection project	$ 24,215.00
23	Benefit-to-cost ratio (divide line 21 by line 22)	3.69

Prepared by:

Name John Smith

Title Distribution Manager   Date 3/27/99

*Units of measure must be consistent throughout the worksheet. The particular unit used (that is, gallons, millions of gallons, acre feet, cubic feet, cubic metres, or other unit) is left to the user.

Figure 2-2 Completed water audit worksheet for County Water Company (continued)

10 WATER AUDITS AND LEAK DETECTION

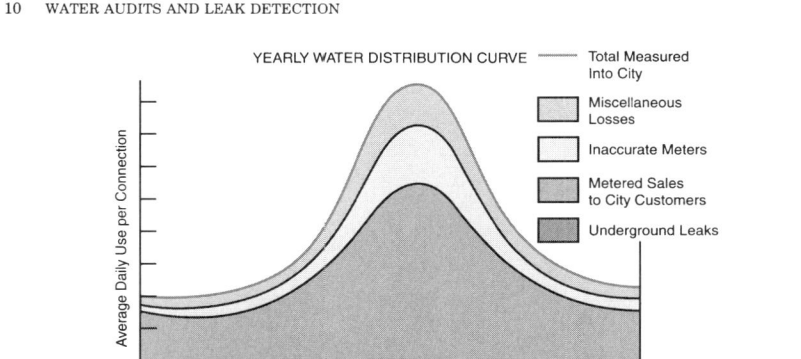

Figure 2-3 One calendar year is recommended for the audit study period because it includes seasonal variations

year is recommended to reduce the effects of lag time in meter reading, and it is long enough to include seasonal variations.

Choose a Unit of Measure

The same unit of measure should be used throughout the water audit. Most utilities in North America record total flow measurements in acre-feet, cubic feet, or gallons. Choose a unit of measure that suits the utility and use the same unit of measure throughout the audit.

In this book, the unit of measure is gallons. By way of comparison, 1 acre-ft equals 43,560 ft^3 or 325,851 gal.*

After establishing the water audit's official unit of measure, note the unit of measure used on each measuring device (that is, meter, weir, Parshall flume, and so on). Also note the conversion factor to be used when reading the device.

TASK 1—MEASURE THE SUPPLY

This task tells how much water enters the distribution system and where it comes from.

Step 1-1 Identify and Map Sources

1-1A Identify sources. Identify all water sources that supply the distribution system, including interconnections with other systems and intermittent sources or emergency supplies. Make a list of these sources.

---

*Refer to appendix F for US customary units to metric conversions.

CONDUCTING A WATER AUDIT 11

Table 2-1 Source measuring devices for County Water Company

	Water Source		
	Source 1	Source 2	Source 3
Type of measuring device	Venturi	Propeller	Venturi
Identification number (may be serial number)	0000278-A	8759	OC-16
Frequency of reading	Daily	Weekly	Daily
Type of recording register	Dial	Dial	Builder type M
Units registers indicate	100,000 gal	Gallons	Cubic feet
Multiplier (if any)	1.0	1.0	100.0
Date of installation	1950	1968	1955
Size of conduit	24 in.	8 in.	11.5 in.
Frequency of testing	Annual	2 years	4 months
Date of latest calibration	4/1/91	8/21/91	1/15/92

1-1B Map the system. Find an existing map of the distribution system. The map should show the principal mains of the entire delivery system. A scale of 1 in. to 400 ft makes the map legible and easy to work with.

If no map of the distribution system exists, use aerial photos or a city or county map and draw the distribution mains on a transparent overlay.

If no suitable map can be found, a hand-drawn one can be used.

1-1C Plot the sources. First, choose a symbol to represent each type of water source, including aqueduct turnouts; wells; surface diversions, such as lakes, streams, or reservoirs; interconnections; and emergency sources. Then, draw the symbol on the map according to where the source is located.

Step 1-2 Measure the Water From Each Source

1-2A Identify measuring devices. Visit each source and note what type of measuring device is used (for example, meter, Parshall flume, weir, or stream gauge). (See appendix B for information on types of measuring devices and meters.) Note basic information about the measuring device, including the type, identification number, frequency of reading, type of recording register, unit of measure (and conversion factor, if necessary), multiplier, date of installation, size of conduit, frequency of testing, and date of last calibration. Using that information, construct a table similar to Table 2-1.

1-2B Record total water from each source. Record how much water was produced by each source for each month and for the entire audit period (see Table 2-2).

Most meters have some type of register, or totaling device. Registers may be round-reading or direct-reading. Round-reading registers have a series of small dials with pointers, registering cubic feet, or gallons, in tens, hundreds, thousands, and ten thousands. Direct-reading registers have one large sweep hand for testing and a direct-reading dial that shows total units of volume.

Calculate the total water produced from all water sources during the study period. Enter the amount on line 1 of the worksheet.

12   WATER AUDITS AND LEAK DETECTION

Table 2-2 Total water supply for County Water Company (uncorrected)*

Month	Source 1 Turnout 41 mil gal	Source 2 Well Field mil gal	Source 3 City Intertie mil gal	Total for Sources 1, 2, and 3 mil gal
January	0	130.34	104.27	234.61
February	0	195.51	65.17	260.68
March	0	260.68	0	260.68
April	130.34	130.34	0	260.68
May	265.57	97.76	0	363.33
June	299.78	0	81.46	381.24
July	303.04	0	84.72	387.76
August	325.85	0	89.61	415.46
September	293.27	32.59	32.59	358.45
October	130.34	32.59	97.76	260.69
November	130.34	0	130.34	260.69
December	130.34	0	97.76	228.10
Yearly Total	2,008.87	879.81	783.68	3,672.36†

*Study period is one calendar year.
†This is the total supply from all three sources for the year. Note that it is an *uncorrected* figure; if necessary, it will be adjusted.

Step 1-3 Adjust Figures for Total Supply

Figures for the total water supply, based on readings from source meters and measuring devices, are raw data. The raw data must be adjusted for a number of factors, including (1) meter inaccuracies, (2) changes in reservoir and storage levels, (3) nonmetered sources, and (4) losses that occur before water reaches the distribution system. These adjustments are made in the steps described below; they are recorded on lines 2A, 2B, and 2C on the worksheet.

1-3A Verify meter accuracy. Although most production sources are measured by meters, some are measured by other devices, such as Parshall flumes or weirs. Supply figures (like those used in Table 2-2) are based on readings of these measuring devices. Any error in any measuring device must be discovered and corrected; incorrect supply data invalidates the entire water audit.

To be sure meters are accurate, compare the results of meter tests to applicable AWWA standards. If a meter measures incorrectly and the error exceeds the standard for its category, repair and recalibrate the meter to function within standard limits.

If a meter has not been tested within the last 12 months, test the meter.

Possible causes of meter error. If source meters are inaccurate, inspect each one in the field. Normal wear is not the only cause of inaccurate meter readings. Check to be sure the meter is the right type and size for the application and that it is installed correctly. (See appendix B for types and uses of meters as well as sizing parameters and installation guidelines.) Check the size against manufacturers' recommended ranges. Be sure the meter is level; most meters are not designed for sloped or vertical operation. Inspect the meter to see if hard-water encrustation is interfering with its measurement.

Also check to verify that the proper registers were selected and installed correctly. Finally, be sure the register is read correctly. Have an employee other than the regular meter reader make a special reading of master meters, or have an employee accompany the meter reader to verify sample readings. Check to be sure the meter is read correctly, recorded correctly, and the correct conversion factor is used.

Check venturi meters. Check venturi meters for blockages in the throats of the meters. Test the primary device with a pitot rod. Testing the meter with a pitot rod shows whether or not the installation is adequate for nonturbulent flows. The meter's primary device should be tested at different flow ranges. If pressure deflection for appropriate flows is adjusted without checking the venturi itself, the meter may still record flows erroneously.

Testing meters. There are four ways meters may be tested. (Meter testing, surveillance, and calibration are discussed in appendix C.) Meter testing methods are listed here in order of effectiveness, with the most effective first.

1. Test the meters in place. Some pipes may need to be replaced to make this possible.

2. Compare meter readings with readings of a calibrated meter installed in series with the original meter.

3. Record meter readings for a given flow over a specified time period. Remove the meter and replace it with a calibrated meter. Record readings from the calibrated meter for the same flow over the same period; compare the readings.

4. Test the meter at a meter testing facility.

Meters can be tested with portable equipment. Pump efficiency flow testing can be used to check meters; it is sometimes provided free of charge by electric utilities. Some utilities use an averaging rod meter or anubar to test meters, but results may be off by as much as 10 percent. A standard single-point pitot rod must be used for accurate results.

Meter testing may be done by an outside agency. Consultants, meter manufacturers, and special testing laboratories offer testing services.

1-3B Adjust supply totals. Adjust the monthly and annual supply data from Table 2-2 for meter error. To do this, divide the uncorrected metered volume (UMV) by the measured accuracy of the meter (a percentage expressed as a decimal) and subtract the UMV as follows:

$$\frac{\text{uncorrected meter volume}}{\text{percent accuracy}} - \text{uncorrected meter volume}$$

$$= \text{corrected meter volume} \qquad (\text{Eq 2-1})$$

Table 2-3 shows how to adjust the supply totals from Table 2-2 to yield the adjusted measurements.

Enter the total adjustments due to meter error on line 2A of the worksheet.

1-3C Adjust reservoir and tank storage. If source meters are located upstream of reservoirs and storage tanks, then stored water must be accounted for in the water audit. Generally, water flowing out of storage is replaced; as the "replacement" water flows from the source into storage, it is measured as supply into the system. If the reservoirs have more water at the end of the study period than at the beginning, then the increased storage is measured by the source meters but not delivered to consumers. Such increases in storage should be subtracted from the metered supply. Conversely, if there is a net reduction in storage, then the decreased amount of stored water should be added to the metered supply. Table 2-4 shows how to figure the change in storage volume.

Enter the changes in reservoir and tank storage on line 2B of the worksheet.

14 WATER AUDITS AND LEAK DETECTION

Table 2-3 Total water supply for County Water Company (adjusted for meter error)

Source	Yearly Total: Uncorrected Metered Volume (UMV)* *mil gal*	Meter Accuracy (MA) *percent*	Meter Error Calculation $\dfrac{UMV}{MA\dagger} - UMV$	Meter Error *mil gal*	Corrected Metered Volume‡ *mil gal*
1	2,008.87	95	(2,008.87/0.95) − 2,008.87	+105.73	2,114.60
2	879.81	100	(879.81/0.00) − 879.81	+00.0	879.81
3	783.68	103	(783.68/1.03) − 784	−22.83	760.85
Total adjustments due to meter error..........................				+82.90	

*Based on Table 2-2.
†A percentage, written as a decimal (95 percent = 0.95).
‡The corrected meter volume for sources 1, 2, and 3 is 3,755.26 mil gal; note that this is 82.90 mil gal greater than the total supply given for these sources in Table 2-2. This is a way to double-check your arithmetic. The new total is not recorded on the worksheet—the "total adjustment due to meter error" is. This is only one of three adjustments that must be made to the raw data given in Table 2-2.

Table 2-4 Changes in reservoir storage for County Water Company

Reservoir	Start Volume *gal*	End Volume *gal*	Change in Volume *gal*
Apple Hill	32,350	36,270	+3,920
Cedar Ridge	278,100	240,600	−37,500
Monument Road	978,400	318,400	−660,000
Davis	187,300	55,300	−132,000
Total change in reservoir storage..			−825,580 gal

Remember: *Decreases* in storage are *added* to the supply; storage *increases* are *subtracted* from the supply.

1-3D Other adjustments. Some water supplies may be subject to other types of contributions or losses. For example, there may be an additional source that enters the water system between the source meter and the finished water system. This could result from infiltration into an open channel. Likewise, losses may be introduced through an unlined or open channel. These additions or losses should be accounted for as "other contributions or losses" on the worksheet.

Enter other contributions or losses on line 2C of the worksheet.
Remember: Always use the same unit of measure.
Add lines 2A, 2B, and 2C; enter the sum on line 3 of the worksheet.
Add line 3 to line 1; enter the sum on line 4 of the worksheet.

TASK 2—MEASURE AUTHORIZED METERED USE

Authorized water is any water used for all uses approved by the utility. Most authorized use is metered, but some is not. Metered water, usually sold to consumers, includes industrial, commercial, residential, agricultural, governmental, and other uses. Unmetered water is used for irregular or mobile public purposes, such as street cleaning or firefighting. All unmetered uses should have meters installed if possible.

CONDUCTING A WATER AUDIT   15

This task tells how much water goes to metered deliveries. The next task discusses unmetered deliveries.

Remember: To be accurate, the water audit must be consistent. Be sure to use the same 12-month study period and the same unit of measure for studying consumption as was used to study supply.

Step 2-1 Identify Metered Uses

2-1A Identify metered accounts. Identify all users who should have meters. Accounts can be identified by meter serial number, connection number, assessor's parcel number, street address, or account number. Assign each account to a meter-reading route.

Be sure to include all accounts for which data on metered use are available, even if the account is not billed. Remember, too, to take into account water provided to other agencies.

2-1B Describe meters. For all active accounts, list meters according to identification number and size of meter. Sort by type of use, including industrial, commercial, residential, agricultural, wholesale transfers, and other. This can help to identify accounts that represent larger volumes of sales and, therefore, greater potential earnings (see Table 2-5).

Check carefully to be sure all information is correct. Consider the possibility of accounting procedure errors, improper computer programming, incorrect meter reading, and unauthorized use. When data on metered use are unavailable or incorrect, water losses are possible. Task 4 gives methods for estimating potential water losses due to incorrect data on metered use.

Step 2-2 Measure Metered Use

2-2A Figure total (uncorrected) water use for each size of meter. Add the water consumption for all accounts and connections for each size of meter by month (or other billing period) and for the entire study period (see Table 2-6).

Calculate the total water consumed through all meters during the audit period. Enter the amount on line 5 of the worksheet.

Remember: Use the same unit of measure that was used for measuring supply in task 1.

2-2B Adjust for lag time in meter readings. Corrections must be made to metered use data when the source-meter reading dates and the customer-meter reading dates do not coincide with the beginning and ending dates of the audit study period.

Adjusting for one meter route. For example, a utility is studying one calendar year, January 1 through December 31. Source meters are read on the first day of each month and customers' meters are read on the 10th day of each month. The goal is to calculate the amount of water supplied and consumed for the calendar year.

Source meters. No correction is made for source meters, because their reading usually occurs on the days that the study period begins and ends. If the last reading (December 31) was a day late (January 1), then the water supplied for January 1 should be subtracted from the total water use read.

Customer meters. Because customer meter readings do not coincide neatly with the study period, a correction must be made. The best way to account for changes in the number of customers and in use patterns is to prorate water use for the first and last billing periods within the study period.

The first billing period has only 10 days that actually occur in the study period. Yet the billing information represents 31 days of use. If sales for that December 11

16 WATER AUDITS AND LEAK DETECTION

Table 2-5 Water consumption by meter size

Meter Size *in.*	Number of Meters	Percent of Total Meters	Percent of Metered Consumption
5/8	11,480	94.1	70.1
3/4	10	0.08	0.1
1	338	4.4	2.8
1 1/2	124	1.0	2.8
2	216	1.8	11.7
3	15	0.12	6.4
4	7	0.05	2.0
6	6	0.05	2.5
Total	12,196	100.00	100.0

Table 2-6 Total metered water use (uncorrected)

	Type of Use				
Month	Residential *mil gal*	Industrial *mil gal*	Commercial *mil gal*	Metered Agriculture *mil gal*	Totals for All Meters *mil gal*
January	146.6	35.8	8.1	0	190.5
February	162.9	35.8	8.1	0	206.8
March	162.9	35.8	8.1	0	206.8
April	179.2	39.1	8.1	24.4	250.8
May	211.8	42.4	8.1	57.0	319.3
June	228.1	48.9	8.1	74.9	360.0
July	260.3	48.9	8.1	57.0	374.3
August	266.5	48.9	8.1	74.9	398.4
September	228.1	45.6	8.1	65.2	347.0
October	162.9	35.8	8.1	0	206.8
November	162.9	35.8	8.1	0	206.8
December	146.6	35.8	8.1	0	190.5
Yearly Total	2,318.8	488.6	97.2	353.4	3,258.0

through January 10 period are 33.204 mil gal, the amount applicable to the study period is

$$33.204 \text{ mil gal} \times \frac{10 \text{ days}}{31 \text{ days}} = 10.711 \text{ mil gal} \quad \text{(Eq 2-2)}$$

Thus, only 10.711 mil gal of the use read on January 10 applies to the study period.

At the end of the study period, there are 21 days not included in the billing data collected on December 10. Use for the last 21 days in December is obtained from the following month's billing. If sales for that month are 36.66 mil gal, the amount applicable to the study period is

$$36.66 \text{ mil gal} \times \frac{21 \text{ days}}{31 \text{ days}} = 24.83 \text{ mil gal} \quad \text{(Eq 2-3)}$$

CONDUCTING A WATER AUDIT 17

Thus, 24.83 mil gal is added to the use read on December 10.

Adjusting for many meter routes. The preceding discussion describes the basic method for correcting lag time in meter reading when all customers' meters are read on the same day. That seldom happens, however. Usually, meters are assigned to different routes and read on different days. Therefore, a meter lag correction should be used for each meter reading route, particularly if each customer's meter is read on the same date each month.

A meter lag correction can involve a number of steps. In our example, County Water Company has three meter routes, each with its own reading date. The study period is one calendar year, and the consumption is prorated for each meter route or book. Meters are read bimonthly: route A on the first of the month, route B on the 10th of the month, and route C on the 20th of the month (see Figure 2-4).

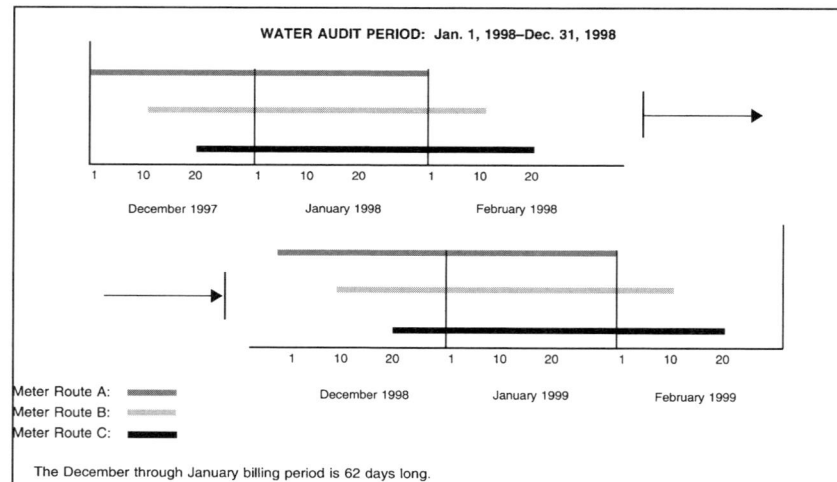

The December through January billing period is 62 days long.

Route	Date Read	Sales	Adjustment
A	2/1/98	4.0 mil gal	31/62 = 2.0 mil gal
B	2/10/98	3.3 mil gal	21/62 = 1.1 mil gal
C	2/20/98	3.6 mil gal	11/62 = 0.6 mil gal

Total adjustment to eliminate 1997 sales from study period . . . . –3.7 mil gal

This amount appeared on the February billing, but the water was used the previous December.

Route	Date Read	Sales	Adjustment
A	2/1/99	4.2 mil gal	31/62 = 2.1 mil gal
B	2/10/99	3.3 mil gal	21/62 = 1.1 mil gal
C	12/20/99	3.9 mil gal	11/62 = 0.7 mil gal

Total adjustment to add 12/98 sales to study period . . . . +3.9 mil gal

This amount did not appear on the final bill for the year; it is prorated from the bill on which it appears.

Net adjustment . . . . +0.20 mil gal

Figure 2-4 Detailed meter lag correction

18 WATER AUDITS AND LEAK DETECTION

The uncorrected total metered use (from step 2-2A, Table 2-6) is based on bills issued during the study period. But, because of the bimonthly billing schedule, these bills would not include all water used during the year. Some water shown as used in the first billing period (issued in February) actually occurred in the preceding December. The last set of bills, issued in November and December, would not include water used in December. Two corrections need to be made. First, water used in the month preceding the study period must be subtracted from consumption figures. Second, water used in the final month of the study period must be added.

Figure 2-4 shows how to adjust sales figures for meter lag time. Many utilities combine accounting and billing procedures into a computerized format to make this procedure easier and quicker.

Prorate water sales figures to adjust for lag time in meter reading. Enter the net adjustment on line 6 of the worksheet.

Add all metered deliveries. Enter the sum on line 7 of the worksheet.

Step 2-3 Adjust Figures for Metered Use

Because there are so many customer meters, it is not practical to inspect and test every one each year. Instead, annual inspections and testing should include all meters greater than 2 in. in diameter, along with a random sample of smaller meters.

2-3A Check for proper installation. Review the utility's practices on meter selection, sizing, and installation to see whether or not present practices permit accurate operation. If they do not, revise the practices as necessary so that meters will operate correctly. (See appendices B and C for more information.)

Commercial and industrial meters produce a much larger share of revenue per account than do residential meters. Commercial and industrial accounts should be inspected for proper selection, sizing, and installation. In addition, inspect and test all large meters before they are used. Not all new meters are accurate.

2-3B Test residential meters. Test a random sample of residential meters— 50 to 100 is a good number. Residential meters may be tested on a test bench or sent to the factory or a consultant for testing. (For more information, see appendix C of this book as well as Chapter 6 of *Water Meters—Selection, Installation, Testing, and Maintenance*, AWWA Manual M6.)

Meter replacement programs. Many utilities are involved in meter replacement programs. For those utilities, to calculate meter error for the entire system, the random sample of meters must include some of the meters being replaced. Test a representative sample of the new residential meters before putting them to use.

2-3C Calculate total sales meter error. Total sales meter error includes meter errors from all meter sizes, including residential, commercial, and industrial.

Calculate residential meter error. Residential meters are tested for low, medium, and high flows. The results, expressed as a percentage of accuracy, are used to calculate the total meter error at average flow rates. Tables 2-7 through 2-9 demonstrate how to use existing meter test data to calculate total residential meter error. The data in the table are based on Table 2-6.

Calculate large meter error. Tables 2-10 through 2-12 show how to use existing meter test data to calculate total large meter error. The mean registration data in Table 2-10 are used to calculate the meter error for large meters.

One of the benefits of a water audit is the potential increase in revenue resulting from testing and repairing large meters (performed as part of the audit). One can estimate the amount of revenue to be gained by repairing the meters; detailed instructions are given in appendix C.

Calculate total sales meter error. Total sales meter error includes meter errors from all meter sizes. In short, total sales meter error = residential meter error + large meter error. Using the data given in Tables 2-6 through 2-11, the total sales meter error for County Water Company is

$$134.33 \text{ mil gal} + 29.97 \text{ mil gal} = 164.30 \text{ mil gal} \quad \text{(Eq 2-4)}$$

Add lines 8A and 8B on the worksheet. Enter the sum on line 8C.
Add lines 7 and 8C on the worksheet. Enter the sum on line 9.
Subtract line 9 from line 4.

TASK 3—MEASURE AUTHORIZED UNMETERED USE

The volume of unmetered water must be carefully estimated to produce an accurate audit. This task includes descriptions of ways to measure water used for unmetered purposes. In selecting the best procedure for a given situation, consider the difficulty of gathering information; the degree of precision necessary; the availability of measuring equipment and skilled personnel; and the need for hiring consultants, buying more equipment, or training employees. When it is evident that a particular water use is very low, a rough estimate could replace a complete, detailed calculation.

It is recommended that all uses be metered, even if the customer is not billed for the use.

Procedures for Estimating Usage

Most unmetered water use can be estimated using either the batch or discharge procedure. Which procedure is best for a given use depends on how the water is applied. In some cases, the best way to estimate use is to adapt consumption figures from a similar, metered facility.

Batch procedure. When water is transported in a tank truck or container of some sort, use the batch procedure. Multiply the volume of the tank or other container by the number of times it is filled from the distribution system. This yields the volume of water delivered from the distribution system. Careful record keeping is necessary for accurate estimates.

Discharge procedure. When water is applied directly from a pipe, as in a sprinkler system, use the discharge procedure. Multiply the rate of water discharge by the total time it flows. This yields the volume of water delivered from the distribution system. The discharge rate may vary and the application period will vary in length and frequency. Again, careful record keeping is necessary for accurate estimates.

Comparison procedure. For some facilities and areas, such as schools, swimming pools, construction sites, or golf courses, consumption figures may be adapted from similar facilities, provided they are alike in size, hours and type of use, landscaping, and most other details. Any differences must be accounted for. For example, at a construction site, work habits are important. If the crew at a metered site turns off water between uses while the crew at an unmetered site lets the water run continuously, the borrowed consumption figures will have to be adjusted considerably.

Remember: No matter what procedure is used, be sure to use the same unit of measure used for other parts in the audit.

20 WATER AUDITS AND LEAK DETECTION

Table 2-7 Weighting factors for flow rates related to volume percentages for ⅝-in. × ¾-in. water meters*

Percent of Time		Range gpm	Average gpm	Percent Volume†
15	Low	0.50–1.0	0.75	2.0
70	Medium	1–10	5.00	63.8
15	High	10–15	12.50	34.2

*Based on information from Tao, Penchin, "Statistical Sampling Technique for Controlling the Accuracy of Small Meters," *Journal AWWA*, 6:296 (1982).
†Percent volume refers to the proportion of water consumed at the specified flow rate, as compared to the total volume consumed at all rates. In this example, only 2.0 percent of the total water consumed occurs at the low-flow range of approximately 0.5–1 gpm.

Instead of using the percentage of volumes shown here, you may compute your own percentage volume data. Using special dual-meter yokes and recording meters, you can determine the actual flow rates for your water meters.

Table 2-8 Meter testing data from a random sample of 50 meters for County Water Company

Test Flow Rates		Mean Registration percent
Low flow	(0.25 gpm)	88.8
Medium flow	(2.0 gpm)	95.0
High flow	(15.0 gpm)	94.0

Table 2-9 Calculation of residential water meter error

Percent Volume* (%V)	Total Sales Volume† (Vt) mil gal	Volume at Flow Rate (Vf) (%V × Vt) mil gal	Meter Registration (R)‡ percent	Meter Error (ME) ME = Vf/(0.01R) − Vf mil gal	Meter Error (ME) mil gal
2.0	2,318.8	46.38	88.8	[(46.38/0.888) − 46.38]	5.85
63.8	2,318.8	1,479.39	95.0	[(1,479.39/0.95) − 1,479.39]	77.86
34.2	2,318.8	793.03	94.0	[(793.03/0.94) − 793.03]	50.62
Total residential meter error (line 8A)					134.33

*From Table 2-7.
†Based on residential water sales data in Table 2-6.
‡From Table 2-8.

Table 2-10 Volume percentages for large meters for County Water Company*

Flow Rates	Percent of Volume Delivered
Low	10
Medium	65
High	25

*For this example, assume flow recordings were made for 24 h in July and February to indicate the percent of volume delivered by large meters at low-, medium-, and high-flow rates.

Table 2-11 Meter test data for large meters for County Water Company

Meter ID Number	Size in.	Meter Type	Date of Installation	Manufacturer	Test Date	Mean Registration at Various Flow Rates: (Designated as Percent of Registration)		
						Low	Medium	High
XYZ001	3	Turbine	6/83	Sensus	10/91	89	85	100
X00ZAA	3	Turbine	6/83	Sensus	10/91	70	88	98
NB123	4	Displace	7/80	Sparling	10/91	95	99	102
NB456	6	Compound	9/77	Sparling	10/91	98	92.5	102
AA002	6	Propeller	5/66	Hersey	10/91	98	98	103
Sum of mean registrations						450	462.5	505
Mean registration for five meters tested						90	92.5	101

Table 2-12 Calculation of large water meter error

Percent Volume (%V)*	Total Sales Volume† (Vt) mil gal	Volume at Flow Rate (Vf) mil gal	Meter Registration (R)‡ percent	Meter Error (ME) ME = Vf/(0.01)R − Vf mil gal	Meter Error (ME) mil gal
10	488.6	48.86	90.0	[(48.86/0.90) − 48.86]	5.4
65	488.6	317.59	92.5	[(317.59/0.925) − 317.59]	25.75
25	488.6	122.15	101.0	[(122.15/1.01) − 122.15]	−1.21
Total meter error for large meters (line 8B)					29.97

*From Table 2-10.
†From Table 2-6.
‡From Table 2-11.

Step 3-1 Identify Unmetered Uses

Many water utilities provide some authorized, but unmetered, use of water. This is typically labeled "authorized, unaccounted-for water." Most often, unmetered consumers are scattered throughout the service area at public buildings, open-space public areas, and special facilities designed to protect the public. Unmetered uses may include firefighting and training; flushing mains, storm drains, and sewers; street cleaning; schools; landscaping/irrigation in large public areas; wintertime bleeders; decorative water facilities; swimming pools; and construction sites. Water used for water quality and other testing, as well as process water at treatment plants, is also included in this category.

When conducting the audit, check to be sure these uses are unmetered. If they are metered, they should be handled as metered use, discussed in task 2.

Step 3-2 Estimate Authorized Unmetered Use

To estimate total unmetered use, break down the total into various uses.

3-2A Firefighting and training. This is defined as water drawn from hydrants, fire-sprinkler systems, and other water sources dependent on the piped water distribution system. It may be used for fire suppression, testing fire equipment, flushing sprinkler systems, or hazardous-materials reduction performed by public-safety crews. It may also be used to train firefighters, airport personnel, and other public-safety employees and volunteers. This category excludes water drawn from ponds, rivers, or other water supplies not connected to the piped water distribution system.

To estimate this use, check fire department records on training, flushing, and fire suppression. Many fire departments use more water for training than for fighting fires. Where flowmeters on standby fire systems show water use, the maintenance superintendent of the building may have fire or test records.

Some fire departments require a "run report" whenever a unit responds to a call. A survey of run reports for all fire calls made during the audit study period in the water service area should yield an excellent estimate of the amount of water used by the fire department. Remember to eliminate calls to locations where the water used came from water supplies not connected to the distribution system.

Estimates of other firefighting uses, such as sprinkler systems (including their testing), require calculation of the flow of the system and the duration of operation. For this calculation, the discharge procedure is used. To acquire the raw data needed for the calculation, survey and inspect meters at schools, stores, apartments, industrial sites, lumberyards, warehouses, and other similar locations. The more complete the survey, the more accurately the final estimate will reflect water used in testing, and in leaky or incorrectly connected sprinkler systems.

In our example, there are four fire companies in the service area. None of them make run reports. However, their logs show a total of 10 structural fires and a 5-day wildfire (for which water was airlifted from an open reservoir), plus 8 days (48 work hours) of training in which water was used. Estimates of water use are 6.5 mil gal for firefighting and 3.2 mil gal for training. Water used for fighting the wildfire is not included because it was not drawn from the distribution system.

Add fire-department use and other use to determine total use for firefighting and training. Enter the sum on line 11A of the worksheet.

3-2B Flushing mains. This water is vented from the distribution system, frequently to storm drains, to clean the system of contaminants and debris.

Many utilities with standard flushing procedures maintain logs that include the location of the main or blowoff and the length of time it flowed. Some utilities meter the amount of water released. (If this use is metered, the total should be included in the uncorrected total metered water use in task 2.)

Estimating water used for flushing mains requires a series of discharge estimates. For each location flushed, multiply the flow rate by the duration of the discharge. For instance, 50 gpm released for 30 min yields 1,500 gal. If the discharge rate is not constant, calculate the volume by figuring the area under a curve (see Figure 2-5).

In our example, mains were flushed through relief valves and hydrants on 18 occasions (54 h total), accounting for about 1.6 mil gal.

Enter the total volume used for flushing mains on line 11B of the worksheet.

3-2C Flushing storm sewers. This is water from the distribution system discharged through fire hydrants to flush and clean storm drains, culverts, or catch basins.

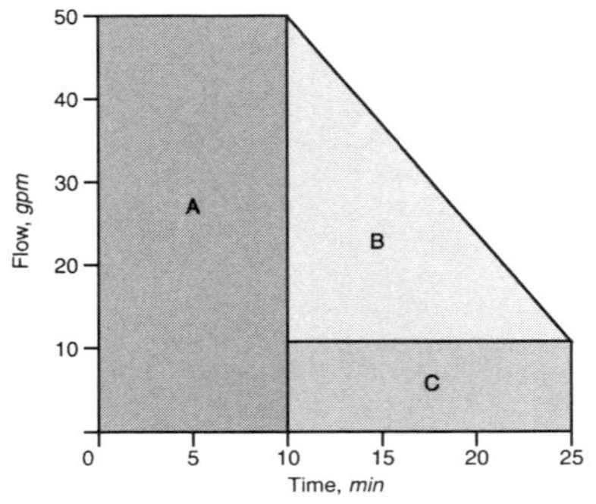

CONDUCTING A WATER AUDIT 23

The discharge flow was constant for 10 min at 50 gpm, then uniformly reduced to 10 gpm over the next 15 min, and then was shut off.

Volume A = 50 × 10 = 500 gal
Volume B = 0.5 × (50 − 10) × (25 − 10) = 300 gal
Volume C = 10 × 15 = 150 gal
Total volume = 950 gal

Figure 2-5 Calculation of water volume from variable-rate discharge

To estimate water used for flushing storm drains, contact the department responsible for the work. Ask for logs of cleaning activity, including the number of trucks used, their capacities, how frequently they were filled, and portable meter readings. Use the discharge method if water is supplied directly from the piped distribution system. If water is transported by truck, use the batch method.

In our example, the county department of wastewater treatment estimated about 0.50 mil gal of water was used to clear congested storm drains.

Enter the total volume used for flushing storm drains on line 11C of the worksheet.

3-2D Flushing sanitary sewers. This is water from the distribution system discharged through fire hydrants to flush and clean sanitary sewers. This category also includes water used in sewage treatment plants for treatment processes and maintenance. This water should be metered and kept track of accordingly.

Use the procedure described in step 3-2C, for flushing water.

In our example, the county department of wastewater treatment estimated about 60 days' work using jet vactor and releases from hydrants for an estimated 0.65 mil gal of water.

Enter the total volume used for flushing sewers on line 11D of the worksheet.

3-2E Street cleaning. This water is used to clean roadways. It may be released directly from fire hydrants or sprayed from trucks, sweepers, or other equipment. It includes water used for cleaning park walkways, boat ramps, bus stops, parking areas, and bike paths.

Estimates are usually made with the batch method. Check with local street department to find out the number of trucks or other equipment used, their capacities, how many days they were used during the study period, and how many times a day they were filled. Also identify the types of equipment used to wash

24 WATER AUDITS AND LEAK DETECTION

Table 2-13 Estimate of water volumes used by tank trucks

Vehicle	Capacity gal	Number of Refills per Day	Number of Days Used per Year	Volume per Vehicle per Year gal
A	200	× 5	× 200	= 200,000
B	500	× 10	× 150	= 750,000
C	2,000	× 2	× 200	= 800,000
Total annual use				1,750,000

recreational vehicle ramps and paths and the frequency of their use. Table 2-13 shows how to calculate total street-cleaning estimates.

Volumes used in direct cleaning from hydrants must be estimated using the discharge procedure.

Enter the total volume used for street cleaning on line 11E of the worksheet.

3-2F Schools. Water is used in schools for domestic sanitation, heating, and air conditioning. This designation may also apply to schoolyards and playgrounds that are supplied by school water services.

Estimate schools' use by comparing them with metered schools having similar landscaped areas and use characteristics, including the number of students and faculty, hours of use, and recreational facilities.

In our example, all schools are metered, so there is no unmetered use; a zero is entered on the worksheet.

Enter the total volume used for unmetered schools on line 11F of the worksheet.

3-2G Landscaping in large public areas. This water is used to irrigate parks, golf courses, cemeteries, playgrounds, highway median strips, and similar areas.

As with schools and other public areas, the easiest method of estimating this total water use may be to compare use with metered landscaped areas having similar use, watering schedules, size, and plant growth.

If records have been kept, it may be possible to calculate the actual amount of water applied to the landscaped areas. This would include runoff from misapplied watering. Ask the people who maintain the areas for information on the frequency and duration of watering. Remember that watering schedules vary at different times of the year.

For median strips and other landscaped areas watered by tank trucks, use the batch procedure for estimating volume.

For unmetered sprinkler systems, the discharge method can be used. Essential factors are (1) the discharge rate at each supply pipe to an irrigated area and (2) the total amount of time water is applied at each area. Obviously, time-controlled irrigation systems make the calculation easier. When figuring the amount of time water is applied, remember to use the total time the service is discharging, rather than the period for one lateral. Figure 2-6 demonstrates how to estimate the volume used for landscape irrigation. Figure 2-7 gives the landscape irrigation figures for the example, County Water Company.

Enter the total volume used for each category of unmetered landscaping on the lines under line 11G of the worksheet.

3-2H Decorative facilities. This water is used for cleaning and maintaining water quality in pools, fountains, and other decorative facilities.

EXAMPLE ESTIMATE OF PRIVATE LANDSCAPE WATERING

A single 2-in. service provides irrigation water to 4½-acre Sunnyslope Park at the rate of 160 gpm. Each of three laterals provides equal amounts of water and is controlled by a common timer.

Lateral A operates from 1:00 a.m. to 3:00 a.m. Lateral B operates from 3:00 a.m. to 5:00 a.m. Lateral C operates from 5:00 a.m. to 7:00 a.m. The system irrigates according to the following schedule:

May and September	Every third day
June	Every second day
July and August	Daily

How much water is applied from May through September?

Here's how to work out the answer:

The service supplies 160 gpm or 9,600 gallons per hour (160 × 60). It operates 6 hours each day the park is watered. During those 6 hours, 9,600 gallons per hour × 6 hours = 57,600 gallons of water applied.

The number of watering days must now be calculated:

Month	Days in Month	Frequency of Watering	Number of Days Watered
May	31	Every third day	11
June	30	Every second day	15
July	31	All days	31
August	31	All days	31
September	30	Every third day	11
Total			99 days

The total amount of water applied during the five-month period is

$$57{,}600 \text{ gpd} \times 99 \text{ days} = 5{,}702{,}400 \text{ gal}$$
$$= 762{,}353 \text{ ft}^3$$
$$= 5.7 \text{ mil gal}^*$$

*The final answer must be given in the audit's official unit of measure.

Figure 2-6 Estimating landscape irrigation

The major causes of water loss from open-air, standing bodies of water are evaporation, water drained from a pool during maintenance, water used for cleaning, and leaks.

Evaporative loss. Appendix D explains how to calculate evaporative loss.

Pool drainage. To estimate water loss from pool drainage, use the following equation:

$$V \times F = V_w \qquad \text{(Eq 2-5)}$$

Where:

V = volume of pool at the time it is drained (the volume of the pool when full minus the amount that is likely to have evaporated)
F = frequency of pool draining
V_w = volume of water loss due to drainage

Cleaning. To estimate the water lost in cleaning, ask maintenance workers about pool volumes and the frequency of cleaning and flushing.

26 WATER AUDITS AND LEAK DETECTION

EXAMPLE OF ESTIMATED PUBLIC LANDSCAPE IRRIGATION

Landscaping in large public areas is not metered. Three estimating methods were used.

1. Parks, Playgrounds, and Cemeteries.
For parks, playgrounds, and cemeteries comparisons were made with Mayfair City parks, playgrounds, and cemeteries, which are metered. Landscape was irrigated 2 mil gal/acre for parks and playgrounds and 1.5 mil gal/acre for cemeteries.

Type of Area	Total Area in Service District acres	Annual Water Rate mil gal/acre	Total Yearly Water Use mil gal
Park	20.0	2.0	40.0
Playground	2.7	2.0	5.4
Cemetery	2.0	1.5	3.0

2. Golf Courses.
Water use at the 62-acre municipal golf course was estimated by measuring the rate of flow and determining the length of watering periods.

Typical water application: 10 h/day
250 days/year
800 gpm

(10 h/day) (60 min/h) (250 days/year) (800 gpm) = 120 mil gal/year

3. Median Strips.
Highway median strips are watered by a large tanker truck carrying 500 gal, refilled 10 times a day, used 130 days a year.

(500) × (10) × (130) = 0.65 mil gal/year

4. Miscellaneous.
A 25-acre memorial park is watered, but specific quantities are not recorded. Water use is estimated at 0.5 mil gal and included as "Other" on the worksheet.

Figure 2-7 County Water Company landscape irrigation figures

For an unmetered source, ask how much time maintenance work requires after the pool is drained. Ask if the hose or refill pipe is left running during that time. Determine flow rates for the appropriate outlet, refill pipe, or hose, and calculate the volume used. If the source is a hose bibb from a metered facility, no further calculation is needed.

Leaks. To estimate leakage, subtract the average amount that should be lost to evaporation from the normal water volume. The difference is leakage.

Add water lost to evaporation, drainage, cleaning, and leaks. Add losses by type of facilities (for example, parks or buildings) within the service area.

In our example, County Water Company has no unmetered decorative facilities.

Enter the total volume used on line 11H of the worksheet.

3-2I Swimming pools. Here, water is used to maintain volume and water quality, including cleaning filters, decks, and walkways, and operation of sanitary and drinking water facilities associated with swimming pools. If concessionnaires tap into unmetered water intended for the pool, that use should be included, too.

Many pools are metered. In that case, their use is already counted as a metered use, under task 2. If they are not metered, estimate the volume of use from information provided by operations and maintenance staff. Useful estimates can be made by comparing water use with metered pools of similar size and use.

In our example, County Water Company has no unmetered swimming pools. Enter the volume used for all *unmetered* pools on line 11I of the worksheet.

3-2J Construction sites. Water is delivered, principally through hydrants, to trucks for controlling road dust, site preparation, landscaping, temporary domestic use, and materials processing (for example, mixing concrete).

To estimate total use, use consumption data from metered construction sites for similar projects. Data may be obtained from regulatory water agencies. Compare the practice of shutting off supply at unmetered sites with the practices at metered sites and compensate for the difference.

It is recommended that all contractors be required to use a portable meter and report the readings.

In our example, County Water Company has no unmetered construction sites. Enter the total volume used on line 11J of the worksheet.

3-2K Water quality and other testing. This water is used to test distribution system output to meet public health standards and to test meters and new mains.

Estimate water used by contacting operations staff to determine testing frequency as well as duration and volumes of water used. Amounts probably vary with each user.

In our example, County Water Company uses less than 500 gal/year for testing. The amount is so small it is disregarded. Enter the total volume used on line 11K of the worksheet.

3-2L Process water at treatment plants. This is water lost—not recycled—after washing filters or draining sedimentation basins at source water treatment plants. If the meter is located after the treatment plant rather than before the plant, disregard this factor.

Estimate this water use by contacting plant operations staff and checking records.

In our example, operators for County Water Company do not keep specific records. It is estimated that a total of 0.07 mil gal is released to waste after filter backwash each year.

Enter the total volume on line 11L of the worksheet.

3-2M Other. An unmetered use may not fit any of the categories described here. In that case, determine the best means for estimating the total volume used.

In our example, there are no miscellaneous unmetered uses. Enter the total volume on line 11M of the worksheet.

Add all unauthorized water uses (lines 11A through 11M). Enter the sum on line 12 of the worksheet.

TASK 4—MEASURE WATER LOSSES

This task shows how much water is lost from the distribution system, and how much of that is lost to leaks.

As we have seen, most water that passes through a distribution system is accounted for, that is, records show how much is used for what purpose. Accounted-for water could be defined as water that is either metered or used for an authorized, unmetered use. On the other hand, some water use is neither metered nor authorized. This water is considered lost from the system. The water does not produce revenue and is not available for beneficial uses. To determine how much water is lost by the system, subtract the volume of authorized, unmetered water from the corrected total unmetered water.

Subtract line 12 from line 10. Enter the difference on line 13 of the worksheet.

Step 4-1 Identify Potential Water Losses

Most water losses can be attributed to the following causes:

- accounting procedure errors
- unauthorized connections
- malfunctioning distribution-system controls
- reservoir seepage and leakage
- evaporation
- reservoir overflow
- unauthorized water use
- discovered leaks
- other leaks

The total volume of water lost to unknown leaks can be determined by first accounting for all other losses and subtracting them from the total water loss.

Step 4-2 Estimate Losses

This step identifies types of water losses and methods of estimating their volume.

4-2A Accounting procedure error. Water may seem to be lost from the system because of overlapping billing cycles (see discussion of meter lag time), misread meters, improper calculations, or computer-programming errors. These types of losses are paper mistakes that can be identified by careful, step-by-step review of record keeping and the computer process. For example, if the water meter registers in cubic feet and the water bill is in gallons, an incorrect conversion factor would introduce an error.

The entire billing and accounting procedure, from meter reading to printing billing statements, should be reviewed. In some agencies, the accounting and billing functions are scattered among several departments or organizational units. Such a fragmented arrangement can lead to miscommunication and error.

The task of reviewing billing and accounting can be simplified by checking a representative sample of accounts. A sampling of accounts from each meter-route book should be checked for accuracy. To do this, perform the following steps:

1. Determine the number of accounts to be checked from each book or route.
2. Choose a random sample of accounts to be checked.
3. Check meter readings. Have an employee other than the regular meter reader read meters for the identified accounts, or send another employee to accompany the meter reader to verify the readings. Both readings should be taken on the same day, if possible. The purpose is to determine whether or not the register is read and recorded properly and if the conversion factor is used properly. Compare the water-use volume for both meter readings. Calculate the billing manually.
4. Compare the total water use with total billed amounts. They should agree. If they don't, find out the reason for the discrepancy.
5. If meter-book readings show a substantial difference from billed amounts, review the normal billing process step by step, line by line.

In the example, four meter readers were accompanied on their rounds for a half-day each. This procedure sampled about 800 connections. In addition, the billing department's computer procedures were reviewed. The utility discovered that 0.5 percent of the 10,200 residential meters were inoperable or misread. The water loss was estimated at 0.228 mil gal per year for each inaccurate meter reading. The following equation shows the total water loss for the 1-year audit study period.

$$0.005 \times 10,200 \times 0.228 \text{ mil gal} = 11.628 \text{ mil gal} \qquad \text{(Eq 2-6)}$$

Enter the total loss due to accounting errors on line 14A of the worksheet.

4-2B Unauthorized connections. Identify active connections when unauthorized water use is not included in the accounting and billing process. Unauthorized use of water is usually accidental. For example, taps may be made to unmetered fire lines in a building (the installer may have been unaware that the lines are reserved for fire control); connections classified as inactive may be in use; or meters may not be read or the readings may not be entered into the accounting system. Occasionally, unauthorized water use may be deliberate. A customer may tap into the main or someone else's service to avoid paying for water. Good billing software should disclose improper practices, both deliberate and accidental.

The first step in measuring water lost to unauthorized connections is to determine which accounts are active and which are inactive. Connections listed as inactive should be confirmed. For some inactive accounts, the meter is removed or locked in the "off" position. If meters have not been deactivated, the meter reader should check the inactive connections periodically to see if they have been used. Telephone calls or visits to the premises may help meter readers keep abreast of changes in the use of these connections.

To identify unauthorized use, compare all active and inactive accounts with all locations that could receive water. One way to do this is to write the meter-identification numbers on the distribution system map and check that each parcel has a meter. A second method is to use an aerial photograph to identify places of water use and, where possible, correlate them with account numbers on a map. A third method is to check each meter-identification number against the assessor's parcel number. Any parcel without an account number should be investigated. (Keep in mind that some water uses may not have an account number or may not be metered.) A fourth method involves the measurement of flow into the suspected facility. A portable metering device is installed on the feed main upstream of the facility's meter. The flow is measured and recorded for a 24-h period. Results are compared to readings of the facility's meter. A discrepancy between the two amounts reveals that either the facility's meter is registering inaccurately or that water is being diverted through a bypass.

In our example, one small shop was using water illegally at an annual rate of 0.326 mil gal/year.

Enter the total volume for illegal connections on line 14B of the worksheet.

4-2C Malfunctioning distribution system controls. Water loss may result from improper application, malfunctioning, or improperly set system controls.

The basic steps for determining the volume of loss remain the same: determine (1) the rate of loss; (2) the length of time during which the loss occurred; and (3) the frequency of loss.

Valves. Valves are the controlling devices in a water distribution system. They are used for both isolation and control functions. Isolation valves are usually manually operated, while control valves function automatically. Many valves have indicators that show the position of the valve; valves without indicators can be retrofitted. Indicators make it easy to inspect valves to be sure they are positioned properly.

30   WATER AUDITS AND LEAK DETECTION

All control valves fail sooner or later. Each installation that uses automatic system controls should be inspected to determine whether or not the valve is working. Also check to see that the valve has been set properly and whether or not its size and design are suitable for its intended purpose.

Different kinds of valves must be checked for different problems.

Altitude control valves. Altitude-control valves can cause a tank or reservoir to spill if the valve is broken or set improperly. The valve is normally set to prevent the tank from overflowing.

Pressure-relief valves. If the pressure-relief valve is set too low for the system's range of pressures, then each time the pressure reaches the high range the valve will cause water to spill. Sometimes an unnecessary spill occurs because a pressure-regulating valve has been readjusted but the pressure-relief valve was not reset.

Pressure-reducing, pressure-sustaining, and pressure-maintaining valves. If any one of these valves is improperly set, it can cause an altitude-control valve, a pressure-relief valve, or a surge-control valve to spill water.

Surge-relief valves. If these valves are set too low, they can spill water as a blowoff to the atmosphere or into a tank or drain or back to the suction well of a pump.

Pump-discharge valves. When the pump-discharge valve fails, it acts like a check valve that is partially open. This may allow water to discharge from the distribution system down the well.

In this example, County Water Company uses an updated system schematic to methodically check all controls. All operated correctly.

Enter the total loss from malfunctioning distribution system controls on line 14C of the worksheet.

4-2D Reservoir seepage and leakage. This is the loss from linings, bottoms, or walls of storage tanks or ponds. Water loss is estimated by closing the inflow and outflow of the reservoir and noting the change in storage level over several days. From this information, the rate of seepage can be calculated. Because water may leak only at certain elevations, the seepage test should be performed several times at successively lower surface-water elevations. Clearwells (finished water storage) are also subject to seepage and should be inspected regularly.

In our example, each of County Water Company's closed reservoirs was individually valved off from the system for 24 h. One tank showed a change in elevation. The loss was calculated to be 192 gal/day, or 0.07 mil gal/year.

Enter the total loss from reservoir seepage and leakage on line 14D of the worksheet.

4-2E Evaporation. Most reservoirs that store treated water are covered and lined, which reduces evaporation significantly. Some clearwells and reservoirs are open to the atmosphere and subject to evaporation. Losses can be calculated by measuring the surface area and applying the proper evaporation data for the area (see appendix D).

In our example, Crystal Lake is an open reservoir with 5 acres of surface area. Estimated evaporation in excess of rainfall is 6 ft/year. Total annual loss is

$$6 \text{ ft} \times 5 \text{ acres} = 30 \text{ acre-ft, or } 9.776 \text{ mil gal/year} \qquad (Eq\ 2\text{-}7)$$

Enter the total volume lost to evaporation on line 14E of the worksheet.

4-2F Reservoir overflow. This occurs most often because the altitude control valve is faulty or missing. To calculate total loss, both the periods of overflow and the overflow discharge rate must be determined. (Overflow does not include water discharged to the distribution system.) If discharge is not directly measured (for

CONDUCTING A WATER AUDIT 31

Table 2-14 Rate of leak losses

Diameter of Hole in.	Total Loss gal/day	Number of Days Leak Existed	Loss gal
1	243,000	30	7,290,000
1	243,000	30	7,290,000
0.25	15,200	180	2,736,000
0.25	15,200	300	4,560,000
Total loss from all leaks			21,876,000

example, by a stream gauge below the discharge point), reservoir overflow is calculated by subtracting reservoir outflow to the distribution system from the inflow to the reservoir. For open reservoirs, evaporation losses should be deducted from the reservoir inflow before attempting to calculate overflow.

In our example, no reservoirs overflowed.

Enter the total volume lost to reservoir overflow on line 14F of the worksheet.

4-2G Discovered leaks. Losses from leaks that are found and repaired can be measured to determine the rate of loss and the total volume lost during the life of the leak. Methods of estimating leak rates, with tables, are described in chapter 4.

In our example, four leaks were repaired. Two were large leaks believed to be short-lived (30 days), due to their disruptive nature. The other two were assumed to have been active the entire year until their discovery. Leak rates were not estimated in the field. The rates were estimated from the size of the hole in the pipe, using Greeley's formula as shown in Table 2-14. All mains involved operated at 50 psi.

Enter the total loss from discovered leaks on line 14G of the worksheet.

4-2H Unauthorized water use. Most unauthorized water use occurs when individuals vandalize fire hydrants or open the hydrants to fill water trucks. Comparing construction permits with temporary-use billings shows where large amounts of water are being taken without metering and billing.

In this example, there was no unauthorized water use.

Enter the volume lost to unauthorized water use on line 14H of the worksheet.

Add all the identified water losses (lines 14A through 14H). Enter the sum on line 15 of the worksheet.

TASK 5—ANALYZE AUDIT RESULTS

Audit results may indicate loss problems resulting from faulty metering, unauthorized taps, leaking reservoirs, or leaking mains and services. To determine which corrective efforts are cost-effective, first estimate the value of the recoverable leakage and the cost of recovering it. If the value of the recoverable water exceeds the cost of recovering it, then carry out a metering and unauthorized tap program and a leak detection and repair program.

This section shows how to figure the benefits and costs of a leak detection survey.

Step 5-1 Identify Recoverable Leakage

Not all leakage is recoverable. To determine what portion may be recovered, first find out how much leakage there probably is, then figure what percentage of that can be recovered.

32   WATER AUDITS AND LEAK DETECTION

5-1A Potential leakage. Generally, more water is lost than can be accounted for. This missing water is called potential leakage. It is easily calculated by subtracting identified losses from total losses.

Subtract line 15 from line 13. Enter the difference on line 16 of the worksheet.

5-1B Recoverable leakage. Not all leaks can be detected and repaired. Experience indicates that about 25 to 75 percent of all potential losses can be recovered.* To calculate the recoverable leakage, multiply the potential leakage by 0.50.

Multiply line 16 by 0.50. Enter the total on line 17 of the worksheet.

Step 5-2 Figure the Value of Recoverable Leakage

Saving water saves money. What cost savings would be achieved if the leakage was prevented? There are two types of cost savings: (1) the cost of purchasing the water and (2) variable operations and maintenance costs associated with storing, treating, and delivering the water. Both of these costs vary with the amount of water going into the distribution system. Both exclude fixed costs. The cost savings equal the value of recoverable leakage.

5-2A Purchase cost. This is what it costs the utility to buy water from another water supplier. Recovering leakage reduces the amount of water purchased. Usually, the most effective cost reduction results from reducing the amount of water purchased or produced from the most expensive source of supply.

In our example, the purchase price of water from the CC Aqueduct is $690/mil gal.

Enter the cost-per-unit of water from the utility's most expensive supply on line 18A of the worksheet.

5-2B Operations and maintenance. Operations and maintenance costs are paid to treat and pressurize water in the system. If the utility pumps water from its own wells, then recovering leakage will reduce the amount of energy needed for pumping.

NOTE: Only operation and maintenance costs that vary with the amount of water delivered are to be included here. Fixed costs that do not vary with the amount of water delivered should not be included.

In our example, $137.7/mil gal is spent to purchase power, while $15.3/mil gal is spent on water treatment. Variable operations and maintenance costs total $153/mil gal.

Enter the unit costs of variable operations and maintenance on line 18B of the worksheet.

5-2C Total cost per unit. The total cost per unit is the sum of costs for purchase and costs for operations and maintenance.

Add lines 18A and 18B. Enter the sum on line 19.

Step 5-3 Figure the Cost to Recover Leakage

This step is accomplished in two parts.

5-3A One-year benefit. To determine the benefit of leak repair over one year, multiply the recoverable leakage by the unit cost of recoverable leakage.

Multiply line 17 by line 19. Enter the product on line 20 of the worksheet.

---

*According to the California Department of Water Resources' *Statewide Leak Detection Grant Program Final Report* (1988). For 47 agencies, the average recoverable leakage was 48 percent of the unidentified water losses indicated by audits.

5-3B Two-year benefit. The average lifetime of a leak before it requires repair is estimated to be two years, depending largely on pipe and soil conditions and the extent of a leak detection program. The total benefit, then, accrues over those two years. To compute total benefits from recoverable leakage, multiply the 1-year benefit by 2.

Multiply line 20 by 2. Enter the product on line 21 of the worksheet.

In some situations, it may be appropriate to include additional benefits from increased leak detection and repair. For example, a more active detection program may allow investments in new resources or delayed construction of a treatment plant, resulting in substantial financial savings from deferred capital spending.

Step 5-4 Calculate the Cost of Leak Detection

The cost of conducting a leak detection survey can be estimated by preparing a leak detection and repair plan. Chapter 4 provides suggestions for how to prepare the plan. Prepare the plan now to get an idea of the cost for leak detection.

Enter the cost for leak detection on line 22 of the worksheet.

NOTE: The cost of leak repair is not included. Since leaks are continually discovered and repaired in the normal course of utility operations, leaks found in the leak detection program would be repaired eventually. If leaks are repaired as part of the leak detection program, the utility avoids the expense of repairing them as they are discovered accidentally. Savings on future repair costs are often overlooked when estimates of savings from leak detection are made, but they can be nearly as great as the cost of repairing leaks as part of the program. The real cost of repairing leaks in a program is generally very small.

For example, when the average life of the leak is 24 months and the real interest rate is 3 percent, benefits from avoided future repair costs amount to 94 percent of the cost of repairing leaks at the time of the leak detection program. In other words, the real cost is only 6 percent of the cost of repairing the leaks found in the program. To simplify the calculation, the cost of repairs has been assumed to be 0.

To determine the benefit–cost ratio, divide total benefits from recoverable leakage by the total costs of the leak detection program.

Divide line 21 by line 22. Enter the total on line 23 of the worksheet.

If the benefit–cost ratio is greater than 1.0, then the benefits of leak detection are greater than the costs, and the program should be implemented.

AFTER THE AUDIT

Outlined here are steps management can take to correct the unproductive elements in the water supply system.

Analyze the Value of Losses and Corrective Measures

Evaluations of corrective measures should be based on cost, feasibility, and savings. The water audit tells the utility where the greatest losses occur. This information allows the utility to set priorities. In setting priorities, managers need to incorporate local constraints.

The choice of corrective actions is influenced by a number of factors, including the following:

- where losses occur
- how much water is lost in each problem area
- what action is needed to reduce water loss

- the cost of reducing water loss
- savings that will result from reducing water loss (this is based on the benefit–cost ratio)
- timetable for implementing the measures

Evaluate Potential Corrective Measures

Corrective measures include performing a leak detection survey and leak repair program, replacing mains that have a history of serious leaks, exercising valves annually, and implementing corrosion-control procedures.

Based on our example, County Water Company, the following recommendations might be made: conduct a leak detection program; test customer meters when received from the manufacturer; check accounts for stopped meters; recommend that public parks, playgrounds, cemeteries, and golf courses be metered; have the leak repair crew measure leak discharge when leaks are uncovered; and perform water audits annually.

Update the Audit

Annual updates provide information to help managers adjust priorities and monitor progress made on system maintenance. Equally important, the update can identify new areas of system losses; this helps managers establish new annual maintenance goals. Updating a water audit is usually less expensive than the original audit.

In some cases, it may be worthwhile to install permanent flowmeters in distribution zones. Continuous monitoring allows detection teams to be sent to areas of highest leakage and provides clear results of detection and repair.

Update the Master Plan

The utility's master plan can be used to set priorities and schedule corrective actions to maintain the distribution system. Managing a distribution system requires current information on the system's delivery capacity, maintenance, and water quality. An updated master plan, supplemented by a water audit and leak detection program, supplies that current information by providing the following:

- identification of problem areas and areas of potential water savings
- analysis of water and cost savings achieved by corrective action
- feasibility analysis of corrective actions based on cost and organizational constraints
- analysis of improved water system efficiencies resulting from past and proposed corrective actions
- analysis of greater system efficiencies versus expansion of resources, treatment plant, or distribution system
- projected water needs
- an implementation schedule for corrective actions
- updated maps showing the system's physical relationships and characteristics

APPENDIX

B

Meter Installation and Testing

Chapter 4, Meter Installation, and Chapter 5, Testing of Meters—Test Procedures and Equipment, of AWWA Manual M6, Water Meters—Selection, Installation, Testing, and Maintenance, *4th ed., 1999; reprinted by permission*

AWWA MANUAL M6

Chapter 4

Meter Installation

INTRODUCTION

Water meters for customer service are installed, or set, in two ways: indoor and outdoor settings. In an indoor setting, the meter is installed inside the customer's premises, usually in the basement. In an outdoor setting, the meter is installed underground in a pit or meter box, which is usually located at the curb end of the service line. Historically, indoor settings have been used primarily in northern states where severe winter weather may cause frost damage, and outdoor settings have been used in warm, temperate climates.

INDOOR VERSUS OUTDOOR SETTINGS

The advantages of outdoor settings include meter location at point of delivery to customer; elimination of a separate curb box; reduction in meter damage from water heater failure; and no need to enter the customer's home for reading, inspection, and replacement. Disadvantages of outdoor settings include high costs when frost protection is required; reading difficulty; high maintenance due to flooding or snow; damage due to vandalism; liability exposure from the public for tripping accidents; and possible pit modifications due to grade changes.

The advantages of indoor settings include potentially lower installation costs; reduced damage and maintenance through elimination of exposure to outdoor conditions; and, for some utilities, the opportunity for customer contact via the meter reader. Disadvantages of indoor settings include missed readings because of entry problems; hot-water damage; basement flooding due to frost-protective-bottom fracture or miscellaneous leaks; and consumer complaints about meter readers.

Remote Registers

The advent of inexpensive and reliable remote-meter-reading devices, which can be read manually or automatically, has added installation capabilities that modify and, in some cases, eliminate the previously accepted advantages and disadvantages of indoor versus outdoor settings. For example, it is now possible to read meters set indoors from outside the house and eliminate entry or call-back problems. It is also possible to read

40   WATER METERS—SELECTION, INSTALLATION, TESTING, AND MAINTENANCE

outside meters installed in pits without removing meter pit lids by using aboveground, mounted remote or automatic reading devices mounted in pit lids.

With the flexibility of remote registers, the utility manager can set meters to achieve minimum installation costs as well as minimum meter-reading costs. Because of the economic advantages of bringing indoor meter readings to the outside of the house, most remote-reading devices have initially been installed in northern areas. Remote-reading devices are now offered by all domestic meter manufacturers and are described in chapter 9. These devices offer an alternative in areas where outdoor pit settings have traditionally been used. The decision can now be made on the basis of overall economics, considering initial investment, installation cost, reading cost, and maintenance expense. Elimination of pit hazards, public injury liability, and pit maintenance should also be considered.

METER INSTALLATION

In addition to special considerations for indoor and outdoor meter settings, large meter settings and turbine meter installations also have special requirements. This section includes general considerations applicable to all meter installations and their related special needs.

General Considerations

Although standard specifications exist for meters, valves, pipe, and tubing, there are no standards for meter settings; however, there are certain principles that should be observed. Specific problems and questions related to installation can best be addressed by consulting manufacturers of meters or setting and testing equipment. Over the years, a wealth of experience has been accumulated to provide meter settings that ensure optimum meter performance and service accessibility, along with ease of installation and low cost. Installation hardware is available in a wide variety of sizes, types, and materials to meet virtually any installation preferences or requirements.

Basic requirements of an acceptable meter installation are as follows:

1. Position meter in horizontal plane for optimum meter performance.

2. Locate meter so that it is readily accessible for reading, servicing, and/or testing.

3. Provide leak tight, permanent setting to ensure that the meter can be removed from service without negatively affecting customer's plumbing.

4. Provide for permanent electrical grounding that does not use the meter to prevent accidental shock to meter service personnel.

5. Protect meter from freezing and other conditions that could damage the installation.

6. Provide high-quality inlet shutoff valve to allow meter maintenance. Location of meter may also dictate a meter valve on outlet side to prevent water draining back when meter is removed.

7. Provide a minimum loss of pressure.

8. Consider public safety and design installation to prevent accidents.

To avoid future operating problems, all open connections should be capped whenever a meter is removed from its setting for any length of time. A meter idler can be used in place of caps to provide the same protection. Also, meters should be protected from heat and direct sunlight during storage and transit prior to installation or after removal.

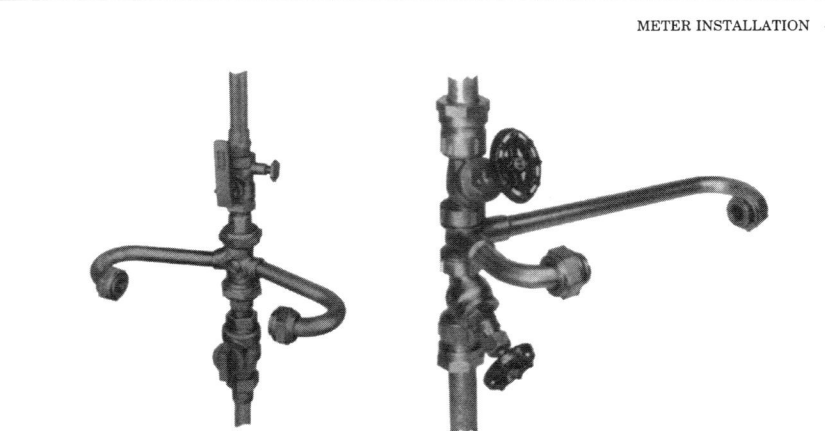

Figure 4-1 Two types of meter setters for vertical indoor piping

On all indoor settings, it is highly important that electrical continuity be maintained through the water line. Most utilities require electrical bonding around meters to prevent accidental electrocution of service personnel changing meters. If the meter setting itself does not provide a continuous electrical circuit when the meter is removed, a permanently bonded electrical grounding strap should be provided. Electrical grounding is a requirement specified by the National Electrical Code, and all service and installation personnel should be advised of this safety requirement. Most commercially available, prepared meter settings provide a continuous metallic circuit, even when the meter is removed from the line. AWWA opposes the grounding of electrical systems to pipe systems conveying drinking water to a customer's premises. Two types of commercially available meter settings are shown in Figure 4-1.

In the United States and several European countries, ample protection from backflow and backsiphonage is required in single family dwellings, as well as at commercial and industrial sites. The water supplier should be familiar with AWWA C510, *Standard for Double Check Valve Backflow-Prevention Assembly* and AWWA C511, *Standard for Reduced-Pressure Principle Backflow-Prevention Assembly* and the requirements of local, state, and federal authorities, whichever takes precedence. This ensures that proper consideration will be given to the meter installation design, thereby providing for the required backflow preventers.

Indoor Settings

The installation of a small meter in a basement or utility room is a relatively simple job. However, improper and unsatisfactory settings do occur because of the absence of applicable standards. When indoor meters are to be read directly, a drawing should be made showing basic installation requirements. This drawing should include minimum and maximum elevations above the floor, the type of recommended connections, required valving, and minimum access space required for reading and service. The drawing should specify that the meter is to be located in the supply line as near as practicable to the point where it enters the basement or building. If a holding device is not used, the drawing should require an electrical ground connection across the meter.

42   WATER METERS—SELECTION, INSTALLATION, TESTING, AND MAINTENANCE

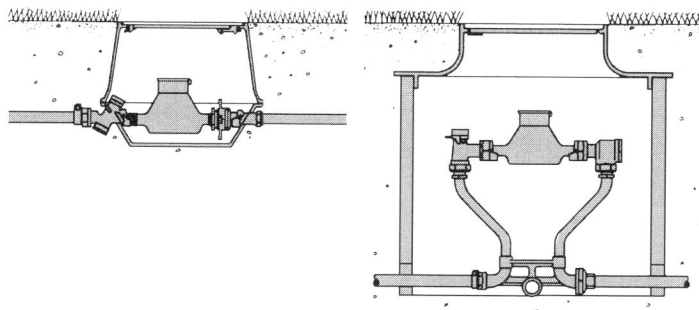

Figure 4-2   Outdoor meter settings with integral yoke (left) and meter yoke (right)

This standard drawing should be furnished to applicants for service, and compliance with the drawing should be one of the conditions for the utility's agreement to serve the premises. This type of drawing provides assurance of uniformly proper installation for the mutual benefit of the utility and the customer.

Outdoor Settings

Many factors influence the design, materials, installation details, and overall performance of an outdoor water-meter installation. Considerations include soil conditions, groundwater level, maximum frost penetration, and accessibility for ease of reading and service. Regardless of pit depth, the meter itself should be located at a depth from the surface that makes reading and accessibility convenient. It is also important to provide 2 in. (50 mm) to 4 in. (100 mm) of clearance between the service piping and the bottom of the meter pit to avoid any damage or strain that may occur if the meter box settles after installation. Consideration should also be given to the location of the curb stop or service control valve. This valve may be made an integral part of the meter setting or may be located elsewhere and housed in a separate curb box. Figure 4-2 illustrates good outdoor meter-setting practice.

The wide variation in ground frost penetration throughout the country makes it impossible to detail a universally practical outdoor setting. In areas where frost penetration is more than a few inches, serious consideration for frost protection is required. Knowledge of local conditions is necessary to select a pit of sufficient size and depth to provide frost protection. Differences in rate of frost penetration, depending on soil conditions, add considerable complications to this problem. Experience with outdoor installations under a given set of conditions is the best guide for avoiding freezing problems. Suggestions for meter pit design, including recommended size and depth, can be obtained from meter-box manufacturers.

Once sufficient experience has been established for meter-setting standards, it is recommended that a drawing be prepared as a guide for further meter settings. This drawing should specify (1) that the meter pit be located as close as possible to the utility line, which is the point of customer delivery; (2) that the lid of the pit or box be placed flush with the ground surface; (3) that no portion of the riser piping or meter be less than 1 in. (25 mm) to 2 in. (50 mm) from any portion of the meter box (more if required for frost protection); (4) the distance belowground surface where the meter spuds or

METER INSTALLATION 43

Table 4-1 Large meter installation guides for compound meters and class I and II turbine meters

Types of Fittings	Distance Upstream in Pipe Diameters	Distance Downstream in Pipe Diameters
Tees and crosses	10	5
Elbows and reducers	10	5
Tees and crosses with strainer or straighteners upstream	5	5
Elbows and reducers with strainer or straighteners upstream	5	5
Angle strainer	5	3
Basket strainer	5	3
Gate valve	1 to 3	1 to 3
Butterfly valve	5	5
Plug valve	5	5
Check valve	*	5
Pressure regulator	*	5
Test tee and plug	*	3
Saddle	*	3

* These fittings are not recommended for use upstream of a water meter. Under extreme high pressures, when it is necessary to protect the meter, pressure regulators may be considered for upstream use only after consulting with the meter manufacturer.

couplings are to be located; (5) the dimensions of the meter box to be used for each size meter; and (6) the location of the curb stop or service-control valve.

Large Meter Settings

Large meter settings, although made less frequently than small meter settings, are relatively expensive and require considerable preliminary planning. Large meters are heavy and removal for service or testing can be costly and time-consuming. Provisions for fire service must also be given serious consideration.

Some utilities have adopted the practice of installing meters in manifolds of two equal branches with meters one size smaller than the main line. These installations consist of compound, turbine, multi-jet, or positive displacement meters. Caution should be exercised to ensure both branches contain meters of the same size, type, model, and manufacturer so that the water flow is balanced and performance life of the meters is equal. A manifold installation provides assurance of continuous service, because a metered, alternative water flow is available during maintenance or emergency situations. Figures 4-3 and 4-4 show examples of various manifold installations using good installation practice and fire-service provisions.

If a manifold contains two or more branches or if there are meters of different types or sizes, the installation requires considerable care in its design. Flow regulating valves (not shutoff valves) are required for pressure loss adjustments to ensure proper water distribution through the various branches. The meter manufacturers or other engineering experts on hydraulics must be consulted because of the technical issues involved.

Check valves may be necessary in some manifolds to prevent recirculation or to improve the low flow registration accuracy of the manifold system. Use of check valves requires careful consideration and should be designed by a professional.

To determine minimum distance to install subject fittings from class I and/or II turbine meters, multiply the nominal pipe diameter of the installation by the appropriate number found in Table 4-1.

Custom-built meter setting devices in 1½-in. (40-mm) and 2-in. (50-mm) sizes with built-in bypass systems and valves with locking arrangements are now available.

44 WATER METERS—SELECTION, INSTALLATION, TESTING, AND MAINTENANCE

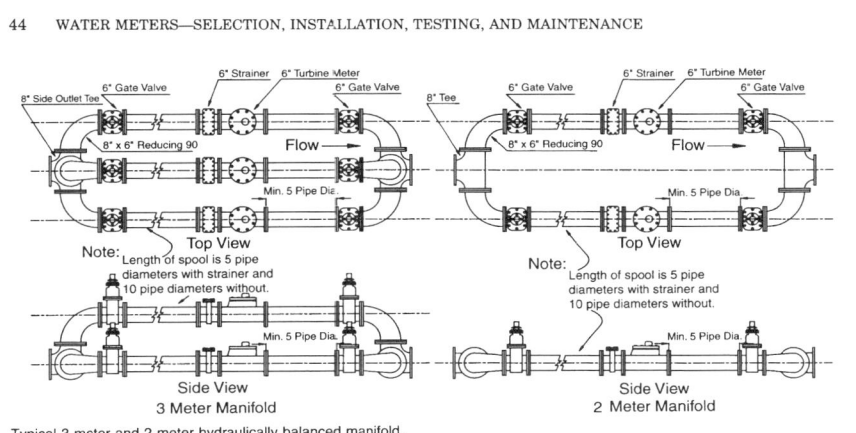

Typical 3 meter and 2 meter hydraulically balanced manifold.

Figure 4-3   Manifold of large meters

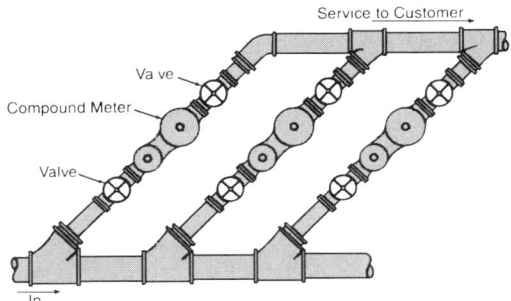

Figure 4-4   Diagram of manifold of three large meters

These devices offer a uniformity of meter settings and facilitate meter maintenance as well as save space in a typical installation. Figure 4-5 shows a typical available device.

Many utilities prefer single-unit installations for large meters and find them very satisfactory. When a single large meter is installed, a bypass circuit should be provided so that meter maintenance can be accomplished without service interruptions. When large meters are installed in a vault, provision should be made for at least 20 in. (500 mm) of clearance to the vertical vault walls and at least 24 in. (600 mm) of head space from the highest point on the meter to the vault cover. Also, when a large meter installation is planned, it is essential that practical testing requirements be carefully considered in the meter and vault layout. Test valves should be installed to permit volumetric field tests, and provisions should be made for discharging test water.

The size, type, and meter brand may have a variety of size test ports for field testing. If the test port provided on the water meter is not adequate to produce the desired test flow, downstream test tees should be incorporated into the installation.

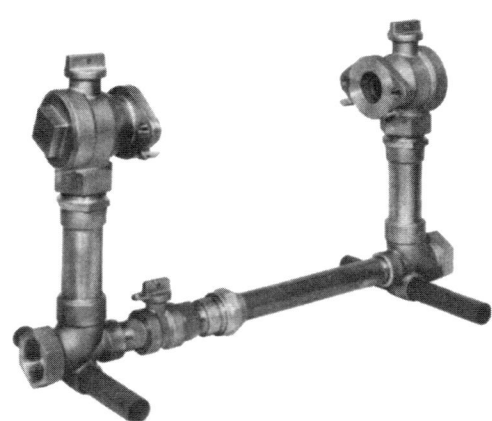

Typical flanged water meter setter complete with angle ball valves, bypass with lockable shutoff valve, and support brackets to hold meter in place during meter gasket replacement.

Figure 4-5 Flanged water meter setter

Satisfactory large meter settings can be designed in a variety of ways, depending on specific requirements, code specifications, and individual preferences. However, it should be noted that these settings represent sizeable investments that will provide long and satisfactory service if adequate planning is done in advance. Valuable advice and installation recommendations can be obtained by contacting the meter manufacturers; this advice should be sought in the preliminary planning stages of a new or unusual installation.

Class I or Class II Turbine-Meter Installations

For optimum life and best accuracy, inferential meters work efficiently when there is a swirl-free, uniform-flow-velocity profile in the pipe immediately upstream of the meter. Because turbine meters are inferential meters, certain precautions or good practices are recommended, including the following.

Class I and class II turbine meters should be positioned horizontally. Piping should be arranged to ensure the meter remains full of water at all times. Elbows, reducers, tees, and crosses installed without a strainer upstream from the meter should be no closer than 10 pipe diameters of straight pipe of the same nominal diameter on the meter upstream and five diameters downstream. As a result of their increased flow capacity, smaller turbine meters are often used to replace larger-sized meters of other types. When pipe reducers are required in such cases, it is important that only gradual or tapered concentric reducers be used. Care should be exercised that the piping flange gaskets are centered and not protruding into the main flow stream.

Avoid installing check valves or pressure-regulating devices upstream of the meter. When check valves or pressure-regulating devices are required in the piping system, they should be installed downstream of the meter at a minimum of five pipe diameters. When backflow prevention devices are required, they should be installed downstream at a minimum of five pipe diameters. Full-opening ball or gate valves are preferred for the meter set's isolation valves. Butterfly or plug valves may also be used

46 WATER METERS—SELECTION, INSTALLATION, TESTING, AND MAINTENANCE

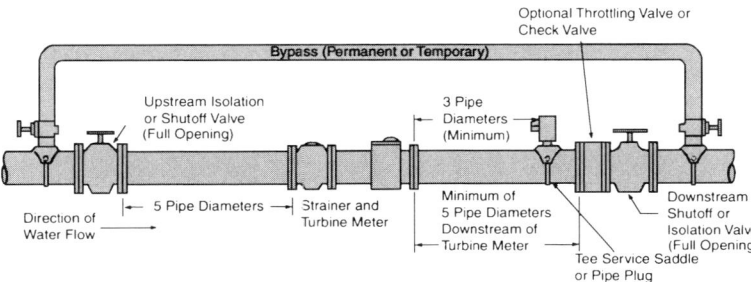

Figure 4-6 Optimum turbine meter installation

as isolation valves, if they are located at a minimum of five pipe diameters upstream and downstream of the meter. If the piping system requires a throttling valve, it should always be downstream of the meter at a minimum of five pipe diameters. It is recommended that a bypass arrangement (either permanent or temporary) be provided to permit uninterrupted customer service during periodic testing or routine maintenance in single-set meter installations. To facilitate periodic testing, provision should be made for a test tee or plug at a minimum of three pipe diameters downstream of the meter.

A rigid, flat-plate or Z-plate strainer is recommended to protect the turbine metering element from debris carried in the flow stream. The effective open area of the strainer element should be at least twice the open pipe area of the meter inlet. When angle or regular basket strainers are used, they should be installed at a minimum of five pipe diameters upstream. When the piping system is also used for fire service, only a Factory Mutual Laboratories (FM)-approved or Underwriters Laboratories (UL)-listed meter should be used. Some meters require approved fire-service strainers as part of the metering package. Fire-service-rated strainers must be installed upstream of the meter. When the piping installation, by necessity, creates a flow swirl in the upstream piping, a flow straightener should be used. At least two types are available: one type incorporates a concentric tube bundle, and the other uses a system of vanes. Either type of straightener can be installed integrally in the meter or immediately upstream. If a flow straightener is not used, the run of straight pipe immediately upstream of the meter should be increased at a minimum of 10 pipe diameters.

Caution should be exercised to avoid entrained air in the meter piping. This is most critical during meter startup when large slugs of entrained air could cause damage to the meter's internal measuring mechanism. Slowly filling the meter piping with a small bleed valve is good practice with the upstream isolation valve open and the downstream isolation valve closed. If possible, the small air-bleed valve should be located at a high point in the surrounding meter piping. The test opening, if valved, is used for this function.

The installation guidelines described are considered good practices for any meter installation and are repeated again in this section because of their importance in turbine meter installations. Figure 4-6 illustrates many of the suggested installation criteria for class I and class II turbine meters.

AWWA MANUAL M6

Chapter 5

Testing of Meters—Test Procedures and Equipment

INTRODUCTION

A water meter, like any other mechanical device, is subject to wear and deterioration and, over a period of time, loses its peak efficiency. How long water meters retain their overall accuracy depends on many factors, such as the quality of the water being measured, rates of flow and total quantity, and chemical buildup and abrasive materials carried by the water. The only way to determine whether a specific meter is operating efficiently is to test it. Establishing a meter maintenance program is very difficult, as it involves repetitive testing. From the individual customer's viewpoint, meters should be tested to protect the customer against meter inaccuracy that could result in overcharges from over registration. This matter is also of concern to utility management. Experience shows, however, that the greater concern of a water utility should be the inequities and revenue loss that result from under-registration of meters.

The economic advantage of meter maintenance programs has been recorded in many articles, but most of these programs have represented concentrated efforts to rehabilitate meters after a long period of nonmaintenance. These programs are of little value in answering the question of how often meters should be tested. Unfortunately, there can be no single answer, as the economic result depends on factors such as rates charged for water; the effects of waters of different qualities on meters; and the cost of removing, testing, repairing, and installing meters. A proper economic balance should be attained. If meters are not adequately maintained, the utility loses revenue. Conversely, if the cost of a meter maintenance program is more than the loss of revenue incurred if the meters were not tested, the overall result is economic waste, and the utility's customers incur the unnecessary expense.

48 WATER METERS—SELECTION, INSTALLATION, TESTING, AND MAINTENANCE

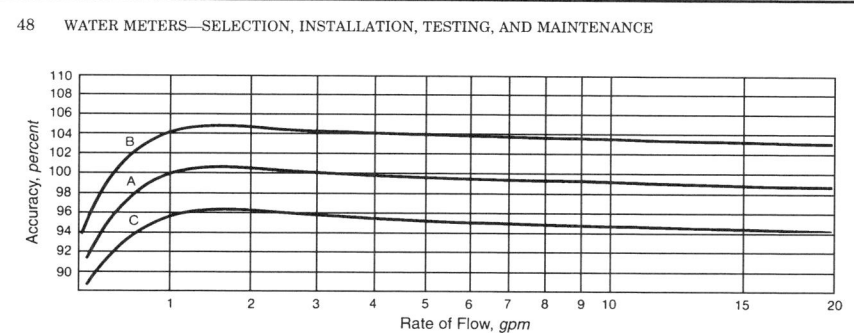

Highest registration occurs at approximately 7 to 10 percent of maximum capacity. Altering of change-gear ratio shifts curve vertically but does not affect shape of curve.

Figure 5-1 Effect of modifying change-gear combinations in typical ⅝-in. (15-mm) meter

ACCURACY LIMITS

Accuracy limits are established to ensure that water meters record as accurately as possible. Meters have an inherent variation of 2 to 3 percent in registration over the entire range of flows, except very low flows just above those that the meter will not register. As an example, a ⅝-in. (15-mm) water meter in good condition will register within the following limits: 95 percent or higher at ¼-gpm (0.06-m^3/h) flow, a rise to a maximum of 101.5 percent at 2 gpm (0.45 m^3/h) (usually 10 percent of rated meter capacity), and then a falling off on a flat curve to not less than 98.5 percent at 20 gpm (4.5 m^3/h). This is the rated meter capacity for a ⅝-in. (15-mm) meter (refer to Figure 5-1 for an illustration of a typical performance curve).

It may not be economically feasible to repair older meters to meet the accuracy requirements for new meters. The water utility should carefully weigh all costs to make a specific determination. For this reason, separate accuracy limits are shown in Table 5-3 for new and repaired meters on the minimum-flow test. The limits set for repaired meters represent those that require good meter-shop procedures. Meter repair work is not acceptable if repaired meters do not register at least 90 percent on this test. A higher percentage is recommended for desirable shop-quality standards.

In many cases, meter manufacturers have provided revenue-maintenance programs in the form of replacement measuring-chamber assemblies, wherein the accuracy of the repaired meter can be restored to the same level as that of a new meter.

Determining Accuracy Limits for Meter Types

For displacement, multi-jet, propeller, and turbine meters, a typical method of establishing test-flow-percent accuracy is the algebraic sum of 15% of the low flow results, 70% of the intermediate flow results, and 15% of the maximum flow results. For compound and fire-service meters, the normal test-flow-percent accuracy should be one third of the algebraic sum of the accuracy results at the maximum test-flow rate of the main line meter and the maximum and intermediate test-flow rates of the bypass meter.

Accuracy Limits for Removal From Service

Meters with determined accuracy limits beyond those shown in Table 5-1 should not remain in service unless repaired. Determining the optimum number of years a meter should remain in service between tests is achieved by testing 5 percent of those meters

Table 5-1 Required accuracy limits for compliance with guidelines

Meter Type (all sizes)	Accuracy Limits as Found by Testing *percent*	
	Normal Test Flow Rates	Minimum Test Flow Rates
Displacement	96–102	80–102
Multi-jet	96–102	80–104
Propeller and turbine	96–103	Not applicable
Compound and fire service	95–104	Not applicable

next scheduled or past due for periodic testing under an existing testing schedule. If the results of these tests fall within the accuracies shown in Table 5-1, it is assumed that the remaining meters will produce the same average test results. This procedure should be followed each year until the test results indicate that a longer time interval between tests would still produce results compliant with the accuracy limits in Table 5-1. At this time, the remaining meters should be removed and tested as part of the periodic test program. At this point the optimum number of years for periodic testing can be determined. As long as all factors remain unchanged, it can be assumed that the periodic testing period will remain constant.

Statistical sample testing in a meter distribution system is an alternative method for determining the optimum number of years a meter should remain in service, especially residential meters ⅝ in. (15 mm) through 1 in. (25 mm) in size. Sample testing is a cost-effective management approach to determine the variables affecting meter performance and to monitor the overall accuracy of the metering system. Using established statistical methods, a random selection of meters determined by the year the meter was installed will provide data on the entire metering system. Weighting the statistical sample information with the system demand information determines the service-life decision.

The sample-test database, sorted by the purchase year and by the consumption rate, can be used to identify or window groups of meters with similar accuracy problems. This information will be used as a criteria to select changeout based on performance rather than age or type. This ensures that the money used to exchange meters is spent on the poorest performing meters within the system.

Calibration

In a mechanical-drive meter and in many magnetic-drive meters, the gear train includes two changeable gears allowing the ratio between the motion of the measuring element and the register to be adjusted for maximum accuracy of registration. When gears are changed, the effect on the performance curve is to raise or lower it, as shown in Figure 5-1. Changed gears are no substitute for good meter-repair standards. Gearing changes should be based on tests made at the high point on the performance curve to ensure that the meter registers near optimum accuracy.

The following description demonstrates how the displacement of water in the measuring chamber is transmitted to the meter register and converted to standard units of measure. Although the maximum speed of ⅝-in. (15-mm) meters is limited to 435 piston revolutions or nutations/ft^3 (15,400 rev/m^3) by 1995 AWWA standards, most meters have measuring speeds considerably below this maximum. Furthermore, meters are manufactured with three to five gear-reduction stages. In the example that follows, a composite figure of element's speed of 310 nutations/ft^3 (10,950 rev/m^3) is used in combination with a four-gear reduction train.

50   WATER METERS—SELECTION, INSTALLATION, TESTING, AND MAINTENANCE

The volume of the measuring-chamber assembly of this composite meter is, therefore, approximately $1/310$ ft^3, and 310 revolutions of the disc spindle are required to cause the smallest hand of the meter register to make one complete revolution and record 1 ft^3.

The intermediate gear train consists of four sets of reduction pinions and gears, each driving pinion has seven teeth and each driven gear has 28 teeth. The meter design (or trial) gears for registration in cubic feet consists of a driver gear with 24 teeth and a driven gear with 29 teeth.

In any gear train, the change in speed is the product of the number of teeth of the driver gears divided by the number of teeth of the driven gears. Therefore, in this meter, the reduction of the 310 revolutions of the disc spindle to one revolution of the 1-ft^3 index of the register is done as follows:

$$310 \times \frac{7}{28} \times \frac{7}{28} \times \frac{7}{28} \times \frac{7}{28} \times \frac{24}{29} = \frac{7,440}{7,424}$$

$$= 100.215\% \text{ (Curve A in Figure 5-1)}$$

This same meter geared for registration in gallons requires only a different set of change gears. For example, if a 24-tooth driver gear and 39-tooth driven gear are used, and the number of revolutions of the disc spindle are increased from 310 to 413$1/3$, an increase of one third, as 10 gal equals 1.3368 ft^3, then

$$413\frac{1}{3} \times \frac{7}{28} \times \frac{7}{28} \times \frac{7}{28} \times \frac{7}{28} \times \frac{24}{39} = \frac{9,920}{9,984}$$

$$= 99.359\%$$

Thus, the smallest index of the meter register records 10 gal for each complete revolution.

The following computations illustrate the effect of changing gear ratios above and below the example. For simplicity in computation, the driver gear only is changed. The shifting effect on the performance curves is shown in Figure 5-1.

$$310 \times \frac{7}{28} \times \frac{7}{28} \times \frac{7}{28} \times \frac{7}{28} \times \frac{25}{29}$$

$$= 104.391\% \text{ (Curve B in Figure 5-1)}$$

$$310 \times \frac{7}{28} \times \frac{7}{28} \times \frac{7}{28} \times \frac{7}{28} \times \frac{23}{29}$$

$$= 96.039\% \text{ (Curve C in Figure 5-1)}$$

These computations illustrate a percentage spread much wider than would be used in good meter-repair practice. If both gears are changed, a much finer adjustment is obtainable.

These examples show that one basic design is used for meters recording in various units of measure by mere substitution of the change gears and register dial plate. This feature is of value from a production cost standpoint but is not the only reason for changing gears. These gears are also used to compensate for differences in meter accuracy resulting from manufacturing tolerances, wear, or other service conditions.

Multi-Jet Meters—Regulating Device

Contrary to the positive-displacement meters, where the ratio between the register and the measuring chamber is changed mechanically, in the multi-jet meters changes are made hydraulically while the gear train is left untouched.

TESTING OF METERS—TEST PROCEDURES AND EQUIPMENT 51

Calibration is done by a variable-port regulating device incorporated along an internal bypass channel in the meter. Usually the device will be in the form of a screw, accessible from the outside without the need to disassemble the meter or the register and protected by a sealed plug.

Turbine Meters—Regulating Device

Most turbine meter designs have an adjusting vane that can be used for a simple hydraulic calibration. Adjustment is done with a screwdriver. Changing the position of the adjusting vane will affect the entire accuracy curve, but it might not shift the whole curve by the same amount. Most manufacturers mark the initial setting of the adjusting vane. The adjusting vane slot is covered by a cap and protected by a seal wire.

TESTING NEW METERS

All new meters should be tested for accuracy of registration at flow rates and test-flow quantities in accordance with Table 5-3 before they are placed in service. This procedure ensures the water utility that the new meter is accurate and that a complete history is available when the meter is eventually brought back to the maintenance shop for inspection or maintenance, or if a customer challenges the meter's accuracy.

During the procurement process, specifications that require the manufacturer to provide certified meter test results are advantageous. Certified test results may easily be transferred to a database thus establishing the complete history of a meter. The database can be referenced for inspection or maintenance work, or to address customer concerns. A statistical sample testing of new meter shipments to verify accuracy and to maintain confidence in manufacturer test results is an efficient cost alternative to testing every new meter.

Program Coordination

To start a program of periodic testing, it is necessary to set an arbitrary time in which to complete the work. Also, it is desirable to select a period of years that coincides with the best estimate of the frequency with which meters should be tested. In this way, the work load is leveled out, and approximately the same number of meters will be due for testing each time. If, for example, a utility with 10,000 meters in service sets up a program for testing meters on a 10-year cycle, the utility has to remove approximately 900 meters each year. This amount is less than 10 percent of the number in service, as there are always meters that will not remain in service for the full period but will be removed for other reasons.

In order to provide for even work flow, both in the changing of meters and the shop work, the number of orders required should be prepared on a daily basis. Assuming 250 working days in a year, in order to complete the periodic testing of 900 meters per year, the testing of approximately four meters each day would be necessary in addition to other required work. If, therefore, four orders are written each day, the progress of the program may be reviewed at any time by a count of the number of incomplete orders for changing meters for periodic test and a check to see if the shop work is completed without a backlog of meters.

Although the testing of 10,000 meters may seem a staggering job, it is surprising that once the work is started on a systematic basis, the additional work is absorbed and soon becomes routine. As actual test results of meters removed from service are accumulated, experience is obtained as to how long it takes, on the average, for meters to lose sensitivity on the low flows. The length of time meters are permitted to remain in service can be adjusted on the basis of known results.

52 WATER METERS—SELECTION, INSTALLATION, TESTING, AND MAINTENANCE

It is generally considered advisable to provide for more frequent testing of large meters, because an error in their registration affects revenue to a much greater extent. Furthermore, current and compound meters may under or over register to a much greater degree than positive-displacement meters.

If enough 3-in. (80-mm) and larger meters are installed, the repair and testing of these larger meters may be delegated to one particular person or crew. They will develop special skills that are necessary for the effective maintenance of larger meters. A survey of the largest utilities in the United States determined that the testing period for the larger meters is conducted on a yearly basis. However, the surprisingly large variation in test periods indicates a need for close study of this important policy. In any meter-testing program, accurate and readily available records are essential. A formal, ongoing meter record program should be established as an initial step in the program. Electronic data processing has proven to be a highly effective tool in maintaining an effective meter record program.

When any testing program is considered, some general observations are pertinent. An initial service period should be assumed, whether it is 30, 25, 20 years, or less. Older meters and those carrying the heaviest volume should be given priority, because volume measured is definitely related to meter wear. Magnetic-drive meters should have a longer maintenance period than mechanical drive meters.

Probably the best advice that can be given regarding a meter testing program is to be alert to and study all phases of the metering field; there is no substitute for experience in determining the best procedure for any one utility. Although a metered system is the best known for equitably spreading the cost of water service, serious inequities and injustices can occur unless all meters are maintained at a high, uniform level of efficiency and unless every reasonable effort is made to prevent inequities from occurring.

Test Procedures

There is no phase of water-utility operation that has been handled in so many different ways as that of testing water meters. The closest approach to standard test procedures has been the accuracy requirements contained in the various AWWA meter standards, and these have been widely used as a basis for establishing individual testing methods. With the exception of specific rates set forth for testing meters on a minimum flow, the specifications make no provision for the number of different rates of flow on which meters should be tested; what these various rates of flow should be; or the quantities of water to be used in running such tests. It is understandable that the meter specifications do not contain these provisions, as the accuracy requirements are primarily a warranty to the buyer that meters purchased in accordance with the specifications will register within certain accuracy limits. The confusion and wide variance in testing procedures result from the fact that the testing of water meters in ordinary shop practice is primarily concerned with meters that are not new but that have been removed from service and repaired. Each individual has had to begin with the information available and develop testing procedures. Under such circumstances, it is not difficult to understand the reason for the widely divergent procedures that have developed over the years, many of which do not produce reasonable answers on overall operational ability of the meters tested.

Although many state regulatory commissions have adopted regulations concerning frequency of meter tests, it should be noted that any arbitrary time interval applied to several localities, each with its own unique local conditions, is not economically feasible for all. Table 5-2, compiled in 1995 and showing data as of 1994, lists these regulations. It must be recognized, however, that the very existence of such regulations has often resulted in better maintenance of meters. It is inexplicable why meter maintenance is, in too many instances, considered of secondary importance. Only when meters

TESTING OF METERS—TEST PROCEDURES AND EQUIPMENT 53

Table 5-2 State public service commission regulations for periodic testing of water meters as of Nov. 30, 1994*

			Meter Size in. (mm)								Meter Size in. (mm)			
			5/8 (15)	3/4 (20)	1 (25)	1 1/2 (40)	2 (50)	3 (80)	4 (100)	6 or larger (150)	5/8 (15)	3/4 (20)	1 (25)	1 1/2 (40)	2 (50)
State	Effective Year	Rule Number	Interval Between Tests Years								Registration Between Tests 1,000 ft³ (28 m³)				
Alabama	(2)	W-17	10	8	6	4	4	3	2	1					
Alaska	1986	6.02	10	8	6	4	4	4	4	4	200	300	400		
Arizona (1)															
Arkansas															
California (3)	1967	103	20	20	15	10	10	10	10	10					
Colorado†	1949	21	5	5	5	4	4	3	2	1					
Connecticut	1966	16-11-88	8	8	8	4	4	3	2	1					
Delaware	1980	2076	15	15	10	10	3	3	3	1					
Florida															
Georgia (1)															
Hawaii															
Idaho (1)															
Illinois	1975	I.A.C.83 Part 600	6	6	4	4	4	4	4	4	100	300	300		
Indiana	1988	170 LAC 6-1	10	10	8	6	4	4	4	4	100	150	300		
Iowa (4)	1986	21.6(10)													
Kansas (1)															
Kentucky	1992	KAR 5:066	10	10	10	4	4	2	1	1					
Louisiana (1)															
Maine (3)	1987	Chap. 620	8	8	8	6	6	4	2	1	100	150	300		
Maryland	1968	6.9.2	10	8	6	4	4	2	2	1					
Massachusetts (1)															
Michigan (5)	1963	R460.13601	10	8	6						100	150	300		
Minnesota (1)															
Mississippi (1)															
Missouri	1968	42	10	6	4	4	4	4	4	4	200	300	400		
Montana	(2)	9	10	10	10	10	10	10	10	10					
Nebraska (1)															
Nevada (1)															
New Hampshire (2)		605.04(6)	10	10	4	4	4	2	1	1					
New Jersey	(2)	14:9-3.2	10	8	6	4	4	4	4	4	750	1,000 (6)	2,000 (6)		
New Mexico (1)															
New York	1981	16 NYCRR Part 500.1	15	15	5	4	4	3	2	1	1,500 (6)	2,000 (6)	2,000 (6)	3,000 (6)	7,000 (6)
North Carolina (7)	1952	R7-32				4	4	3	2	1					
North Dakota (1)															
Ohio (4)	1991	4901:1-15-21													
Oklahoma (4)	1971	165:65-7-11	10	10	10	6	6	4	2	2					
Oregon (1)															
Pennsylvania	1963	7	10	8	6	4	4	4	4	4	100	150	300		
Rhode Island	1966	Schedule	10	10	10	10	10	2	1	1					
South Carolina (1)															
South Dakota (1)															
Tennessee (8)	(2)														
Texas †															
Utah †	1937	R746-300-3													
Vermont															
Virginia (1)															
Washington	1971	480-110-161	10	8	6	4	4	2	2	2					
West Virginia	1948	38	7	7	5	4	4	2	1	1					
Wisconsin (4)	1994	PSC 185	10	10	10	4	4	2	2	1					
Wyoming	1979	Sec. 608	10	8	6	4	4	4	4	4	100	150	300		

*Data are condensed for simplification. Detailed or specific current information should be obtained from the state regulatory agency. Testing intervals included in this tabulation are statutory and are not recommended periods.
†No response was received from current survey. Data from old survey are included in table.
(1) No regulations for periodic testing having been adopted.
(2) Effective date not furnished.
(3) A utility may submit a different periodic test plan for meters to the commission for approval.
(4) As often as necessary to comply with commission accuracy standards.
(5) Meters larger than 1 in. (25 mm) should be tested following recommendations in AWWA standards.
(6) Registration is in units of 1,000 gal.
(7) Applies only to compound, velocity, and fire-line meters.
(8) Each utility sets its own schedule.

are formally recognized as the only means by which revenue is equitably obtained to operate the water system will the necessary time and study be given to the question of how often it is necessary to test meters for the most efficient and economic results.

Table 5-3 includes recommended data for testing cold-water meters by the use of volumetric tanks or weight scales. Accuracy standards for new meters are contained in the latest revisions of the following AWWA standards: C700, C701, C702, C703, C704, C708, and C710.

54 WATER METERS—SELECTION, INSTALLATION, TESTING, AND MAINTENANCE

Table 5-3 Test requirements for new, rebuilt, and repaired cold-water meters*

	Maximum Rate (All Meters)			Intermediate Rate (All Meters)				Minimum Rate (New and Rebuilt)				Minimum (Repaired)
Size in.	Flow† Rate gpm	Test Quantity†† gal	ft³	Accuracy Limits percent	Flow** Rate gpm	Test Quantity†† gal	ft³	Accuracy Limits percent	Flow Rate gpm	Test Quantity†† gal	ft³	Accuracy Limits percent	Accuracy Limits percent (min.)
				Displacement Meters (AWWA C700 and C710)									
5/8	15	100	10	98.5–101.5	2	10	1	98.5–101.5	1/4	10	1	95–101	90
5/8 × 3/4	15	100	10	98.5–101.5	2	10	1	98.5–101.5	1/4	10	1	95–101	90
3/4	25	100	10	98.5–101.5	3	10	1	98.5–101.5	1/2	10	1	95–101	90
1	40	100	10	98.5–101.5	4	10	1	98.5–101.5	3/4	10	1	95–101	90
1½	50	100	10	98.5–101.5	8	100	10	98.5–101.5	1½	100	10	95–101	90
2	100	100	10	98.5–101.5	15	100	10	98.5–101.5	2	100	10	95–101	90
3	150	500	50	98.5–101.5	20	100	10	98.5–101.5	4	100	10	95–101	90
4	200	500	50	98.5–101.5	40	100	10	98.5–101.5	7	100	10	95–101	90
6	500	1,000	100	98.5–101.5	60	100	10	98.5–101.5	12	100	10	95–101	90
				Multi-jet Meters (AWWA C708)									
5/8	15	100	10	98.5–101.5	1	10	1	98.5–101.5	1/4	10	1	97–103	90
5/8 × 3/4	15	100	10	98.5–101.5	1	10	1	98.5–101.5	1/4	10	1	97–103	90
3/4	25	100	10	98.5–101.5	2	10	1	98.5–101.5	1/2	10	1	97–103	90
1	35	100	10	98.5–101.5	3	10	1	98.5–101.5	3/4	10	1	97–103	90
1½	70	100	10	98.5–101.5	5	100	10	98.5–101.5	1½	100	10	97–103	90
2	100	100	10	98.5–101.5	8	100	10	98.5–101.5	2	100	10	97–103	90
				Class I Turbine Meters (AWWA C701)									
3/4	30	100	10	98–102					1	10	1	98–102	—
1	50	100	10	98–102					1.5	10	1	98–102	—
1½	100	200	20	98–102					12	100	10	98–102	—
2	160	500	50	98–102					16	100	10	98–102	—
3	350	1,000	100	98–102					24	100	10	98–102	—
4	600	1,500	200	98–102					40	400	50	98–102	—
6	1,250	4,000	500	98–102					80	1,000	100	98–102	—
8	1,800	7,000	900	98–102					140	1,000	100	98–102	—
10	2,900	10,000	1,300	98–102					225	1,000	100	98–102	—
12	4,300	15,000	2,000	98–102					400	1,000	100	98–102	—
				Class II Turbine Meters (AWWA C701)									
1½	120	500	50	98.5–101.5					4	100	10	98.5–101.5	—
2	160	500	50	98.5–101.5					4	100	10	98.5–101.5	—
3	350	1,000	100	98.5–101.5					8	100	10	98.5–101.5	—
4	630	1,500	200	98.5–101.5					15	100	10	98.5–101.5	—
6	1,400	4,000	500	98.5–101.5					30	1,000	100	98.5–101.5	—
8	2,400	7,000	900	98.5–101.5					50	1,000	100	98.5–101.5	—
10	3,800	10,000	1,300	98.5–101.5					75	1,000	100	98.5–101.5	—
12	5,000	15,000	2,000	98.5–101.5					120	1,000	100	98.5–101.5	—
14	7,500	20,000	2,500	98.5–101.5					150	1,000	100	98.5–101.5	—
16	10,000	30,000	4,000	98.5–101.5					200	1,000	100	98.5–101.5	—
18	12,500	40,000	5,000	98.5–101.5					250	1,000	100	98.5–101.5	—
20	15,000	40,000	5,000	98.5–101.5					300	1,000	100	98.5–101.5	—
				Propeller Meters (AWWA C704)									
2	100	300	40	98–102					35	200	25	98–102	90
3	250	800	100	98–102					40	200	25	98–102	90
4	500	1,500	200	98–102					50	250	30	98–102	90
6	1,200	2,500	300	98–102					90	500	60	98–102	90
8	1,500	3,000	400	98–102					100	500	60	98–102	90
10	2,000	4,000	500	98–102					125	500	60	98–102	90
12	2,800	6,000	800	98–102					150	750	100	98–102	90
14	3,750	8,000	1,000	98–102					250	1,000	130	98–102	90
16	4,750	10,000	1,300	98–102					350	1,500	200	98–102	90
18	5,625	12,000	1,600	98–102					450	2,000	250	98–102	90
20	6,875	15,000	2,000	98–102					550	2,500	300	98–102	90
24	10,000	20,000	2,500	98–102					800	4,000	500	98–102	90
30	15,000	30,000	4,000	98–102					1,200	6,000	800	98–102	90
36	20,000	40,000	5,000	98–102					1,500	7,500	1,000	98–102	90
42	28,000	40,000	5,000	98–102					2,000	10,000	1,300	98–102	90
48	35,000	50,000	6,000	98–102					2,500	12,500	1,500	98–102	90
54	45,000	60,000	8,000	98–102					3,200	16,000	2,000	98–102	90
60	60,000	70,000	9,000	98–102					4,000	20,000	2,500	98–102	90
66	75,000	80,000	11,000	98–102					4,750	25,000	3,000	98–102	90
72	90,000	90,000	12,000	98–102					5,500	28,000	3,500	98–102	90

TESTING OF METERS—TEST PROCEDURES AND EQUIPMENT

Table 5-3 Test requirements for new, rebuilt, and repaired cold-water meters*—continued

	Maximum Rate (All Meters)				Change Over Point (All Meters)				Minimum Rate (New and Rebuilt)				Minimum (Repaired)
Size in.	Flow† Rate gpm	Test Quantity†† gal	ft³	Accuracy Limits percent	Flow** Rate gpm	Test Quantity†† gal	ft³	Accuracy Limits percent	Flow Rate gpm	Test Quantity†† gal	ft³	Accuracy Limits percent	Accuracy Limits percent (min.)
					Compound Meters (AWWA C702)§								
2	160	400	50	97–103				90–103				95–101	90
3	320	1,000	100	97–103				90–103				95–101	90
4	500	1,500	200	97–103				90–103				95–101	90
6	1,000	3,000	400	97–103				90–103				95–101	90
8	1,600	4,000	500	97–103				90–103				95–101	90
10	2,300	4,000	500	97–103				90–103				95–101	90
					Fire-Service Type (AWWA C703) (Test at intermediate rate not necessary.)§								
				Accuracy Limits percent									
				Type 1 / Type 2									
3	350	700	100	97–103 / 98.5–101.5									90
4	700	1,500	200	97–103 / 98.5–101.5				Not Less Than 85%				Not Less Than 95%	90
6	1,600	3,000	400	97–103 / 98.5–101.5									90
8	2,800	5,000	700	97–103 / 98.5–101.5									90
10	4,400	9,000	1,200	97–103 / 98.5–101.5									90

*A rebuilt meter is one that has had the measuring element replaced with a factory-made new unit. A repaired meter is one that has had the old measuring element cleaned and refurbished in a utility repair shop.

†These are suggested test flows and test quantities. Testing for high rates of flow can be achieved by testing the meter at 25% of the meters rating if the manufacturer's original test certificate indicates a linear curve between 25% and 100% of the rated flow range.

††Quantity should be one or more full revolutions of the test hand but not less than 3 min running.

§The bypass meter should be tested in accordance with the appropriate test requirements for the type of meter used.

**As this rate varies according to manufacturer, it should be determined for each type of meter tested.

Metric Conversions: in. × 25.4 = mm, gal × 0.003785 = m³, gpm × 0.2268 = m³/h, ft³ × 0.02831 = m³.

The primary reason for meter tests is to ensure that the cost of water service is equitably distributed among all customers. Unless all meters register within defined limits of accuracy, equitable cost distribution does not occur. For this reason, regulatory commissions have established definitive meter-accuracy requirements and the frequency of tests for water meters. A review of such commission requirements indicates that each agency varies slightly in test frequencies, so it is important to be familiar with local regulations.

In addition, loss of revenue to the water utility will occur if the meters are not maintained efficiently. Unfortunately, meters may under-register for long periods without complete stoppage. It is necessary to test meters periodically to minimize this loss of revenue. The accuracy of displacement meters, inferential meters, multi-jet or current meters, is also subject to change while they are in service, and they may either under or over register. The period of time for which water meters retain overall accuracy is variable and depends on the characteristics, quality, and volume of the water being measured. The rates charged for water service also have a distinct bearing on how frequently meters should be tested. It is difficult to determine the economic balance between the cost of more frequent testing and potential loss in revenue caused by meter under-registration. Proper meter tests are necessary, however, in any such evaluation. Unless meter-testing procedures reflect overall operational ability and the same procedures are followed consistently, changes in meter accuracy after periods of service cannot be determined.

Finally, it would be advantageous if everyone spoke the same language. There are many seminars held annually on meter department operation and test procedures, but there can be no common ground for comparison unless the results are based on one method or standard.

Elements of a Meter Test

Accuracy denotes the comparison between the indicated value of water delivered of the meter being tested as compared to the true value determined by the test system. In the water industry, the meter's accuracy is most commonly called the *percent registration* (% registration). A water meter that has 101% accuracy at a given flow rate indicates that 1% more water has been delivered to the proving system than the testing system indicates (this error is called 1% fast). Conversely, a meter that has a 99% registration will indicate 1% less water than that delivered to the prover (i.e., 1% slow).

Flow range is the maximum and minimum flow rates established by the manufacturer of the meter being tested wherein the meter will provide the stated accuracy with acceptable life. The range is sometimes expressed as a ratio where the maximum flow rate is compared to the minimum flow rate. A meter capable of being accurate from 20 gpm (4.5 m^3/h) to 200 gpm (45 m^3/h) would have a 10:1 flow range as an example.

A meter's characteristic accuracy curve (also called a performance curve) is a continuous curve drawn between all accuracy points over the meter's flow range that is representative of all replicates of that particular meter when new. Most water meter manufacturers' literature contains a performance curve for their particular meter model. Generally speaking, all meters of the same general type (disc, piston, multi-jet, compound, or turbine) will exhibit the same characteristic accuracy curve.

Repeatability is the measure of deviation of a series of accuracy tests from the test's mean value where all of these accuracy tests were conducted under exactly identical conditions. This term can be applied to either the test system or the meter under test. Most new water meters tested on a precision test stand in the main portion of their characteristic accuracy curve have repeatabilities of ± 0.1%.

Precision usually refers to the combined sources of all errors in a test system which may include data presentation and operator interpretation errors as well as the combined proving system repeatability errors of each component. In addition to these sources of error, a test system might have a bias where the test system indicated proof quantity is displaced from the true value at one or all proof quantities.

The following are the three basic elements of a meter test.

1. The number of different rates of flow over the operating range of a meter required to determine overall meter efficiency.

2. The quantities of water necessary at the various test rates to provide reasonable resolution of meter registration accuracy.

3. Accuracy limits that meters must meet on the different rates to be acceptable for use.

Test Rates

The three rates of flow necessary to properly test displacement, compound, and propeller water meters are maximum, intermediate, and minimum. At least one additional test, preferably more, is necessary within the changeover range of flows of compound and fire-service meters to determine overall operational efficiency and accuracy of registration.

Tests for full-flow accuracy do not need to be made at the "safe maximum capacity" rate shown in the appropriate AWWA standard. Registration curves of water meters show that meters in good operating condition follow a general pattern of registration. The specific profile of the accuracy curve can be different for each type of meter. Usually, there will be an intermediate point of maximum registration above the low-flow-metering zone. Depending on the size and type of meter, this point may vary between 3 to 10 percent of the rated meter capacity. At rates above that of maximum registration, the accuracy curve is fairly flat so that there is little difference in accuracy over a wide range of flows. Selection of the maximum rate of flow at which meters are tested is, therefore, not of major importance. Maximum-rate test flows of 25 percent or more, if desired by the owner, of rated capacity are practical, because meters are seldom operated at rated capacity. This lower test rate is advantageous with multiple testing of small meters and is possible with pressures and testing equipment usually available.

The intermediate rate of flow should be at or near the high point of registration to ensure against over registration of the meter on any rate of flow and, therefore, should be approximately 10 percent of rated capacity. The minimum-rate-flow test discloses operational ability and proficiency of meter repair more than either the maximum- or intermediate-flow tests. All three, however, are necessary to evaluate overall meter accuracy. Test rates of flows should be measured in actual units, such as gallons per minute. Rates based on size of orifice are not reliable because of possible enlargement of the orifice from wear or changes in pressure in the supply line.

Test Quantities

The quantity of water required to provide acceptable accuracy depends on the accuracy and resolution of the test equipment, as well as the desired accuracy of the test results. Commercial calibrated tanks with visually read height scales can provide overall equipment error of less than 0.3% if the scale reading error is small compared to the final height of the collected water. Measuring tanks based on weight can provide even higher accuracy (e.g., 0.25%) if the accuracy of the weighing system is appropriate for the amount of collected water to be weighed. In both systems the use of digital equipment can reduce instrument reading error to zero and improve overall system accuracy, although the inherent accuracy of the load cell or other weighing device would still produce some testing error.

When choosing the quantity of water to be collected for any test, the maximum possible volume or weight measurement error should be compared to the final volume or weight of the collected water. For example, assume a maximum testing error (0.25%) is desired for a volumetric tank that has a 100-gal capacity at a water column height of 50 in. (2 gal per in.). Assume, also, that the maximum scale error is 0.05 in. (including reading error). To achieve a maximum error of 0.25%, the minimum volume of water that must be collected is 40 gal, assuming a constant diameter tank.

$$\frac{0.05 \text{ in. (Scale and Reading Error)}}{0.0025 \text{ (Minimum Desired Accuracy)}} = 20 \text{ in. Column Height}$$

$$20 \text{ in.} \times \frac{2 \text{ gal}}{\text{in.}} = 40 \text{ gal}$$

Since most volumetric tanks have scales that read directly in gallons or cubic feet, the appropriate height scale error should be stated in gallons or cubic feet. Since the volumetric reading is simply a height reading converted to volume from overall column height, the required collection quantity is exactly the same as above.

Max volume error:

$$\frac{0.05 \text{ in.} \times 2 \text{ gal}}{\text{in.}} = 0.1 \text{ gal}$$

$$\frac{0.1 \text{ gal}}{0.0025} = 40 \text{ gal (Desired accuracy)}$$

58 WATER METERS—SELECTION, INSTALLATION, TESTING, AND MAINTENANCE

Table 5-4 Percentage registration tables for test quantities other than 10, 100, or 1,000 gal or ft^3

Meter Reading ft^3 or gal	Percent of Actual Volume	Meter Reading ft^3 or gal	Percent of Actual Volume	Meter Reading ft^3 or gal	Percent of Actual Volume	Meter Reading ft^3 or gal	Percent of Actual Volume	Meter Reading ft^3 or gal	Percent of Actual Volume
19.0	95.0	29.0	96.6	39.0	97.5	49.0	98.0	59.0	98.3
1	95.5	1	97.0	1	97.7	1	98.2	1	98.5
2	96.0	2	97.3	2	98.0	2	98.4	2	98.6
3	96.5	3	97.6	3	98.2	3	98.6	3	98.8
4	97.0	4	98.0	4	98.5	4	98.8	4	99.0
5	97.5	5	98.3	5	98.7	5	99.0	5	99.1
6	98.0	6	98.6	6	99.0	6	99.2	6	99.3
7	98.5	7	99.0	7	99.2	7	99.4	7	99.5
8	99.0	8	99.3	8	99.5	8	99.6	8	99.6
19.9	99.5	9	99.6	9	99.7	9	99.8	9	99.8
				Actual Volume*					
20.0	100.0	30.0	100.0	40.0	100.0	50.0	100.0	60.0	100.0
1	100.5	1	100.3	1	100.2	1	100.2	1	100.1
2	101.0	2	100.6	2	100.5	2	100.4	2	100.3
3	101.5	3	101.0	3	100.7	3	100.6	3	100.5
4	102.0	4	101.3	4	101.0	4	100.8	4	100.6
5	102.5	5	101.6	5	101.2	5	101.0	5	100.8
6	103.0	6	102.0	6	101.5	6	101.2	6	101.0
7	103.5	7	102.3	7	101.7	7	101.4	7	101.1
8	104.0	8	102.6	8	102.0	8	101.6	8	101.3
20.9	104.5	9	103.0	9	102.2	9	101.8	8	101.5
21.0	105.0	31.0	103.3	41.0	102.5	51.0	102.0	61.0	101.6

*For 100 division dial, move decimal one place to right.

Metric Conversion: gal × 0.004546 = m^3.

For a weight-based collection system, a similar result would be:

Required Accuracy = 0.25%
Maximum Weight
Measurement Error = 0.5 lb

$$\frac{0.5 \text{ lb (Measurement Error)}}{0.0025 \text{ (Minimum Desired Accuracy)}} = 200 \text{ lb}$$

$$\frac{200 \text{ lb Water}}{8.34 \text{ lb/gal}} = 24 \text{ gal}$$

Where large quantities of water are used for testing large meters, it is more expedient to employ gravimetric means involving the use of electronics with load cells. These procedures make possible testing "on the fly," thereby eliminating the effects of valve opening and closing, particularly at high flows.

Table 5-4 is intended for ascertaining the percentage of true quantity indicated by the meter, when the meter is run for a measured tank volume of 20, 30, 40, 50, or 60 units. Examples: If the test tank is filled to a 30-gal mark and the meter reads 29.4 gal, the meter is registering 98 percent of the true quantity passed through it. If the meter has a 100-gal test hand, the tank is filled to 400 gal, and the meter reads 407 gal, it is registering 101.7 percent, i.e., ($^{407}/_{400}$) × 100 = 101.7 percent.

It must be stressed that the test volumes associated with the various meter sizes, types, and flow rates indicated in Table 5.3 be followed to minimize measurement system error. Compressed volumetric testing at lesser volumes can result in significant errors from:

- non-calibrated regions of volumetric tank
- sight glass gauge readability
- meter registration test circle resolution

Testing Interval

Ongoing meter-testing programs have long been advocated by AWWA and are required by most public utility commissions. It is in the best interest of both the utility and the customer that testing of meters be part of an ongoing maintenance program.

The chemical and physical characteristics of water are the most important factors affecting the performance of a meter. Because the characteristics of water vary throughout the country, an arbitrary number of years is not the criterion to use for determining the length of time between tests. From an economic standpoint, a meter should remain in service until it ceases to register within accepted accuracy limits. Because meter testing programs are costly, prudent management dictates that meters should be left in service as long as practical. Because of these variable factors, it is recommended that a utility's own test results be used to determine the length of time its meters should remain in service between tests.

The following guidelines were established so that a utility could establish its own periodic meter-test intervals based on its historical data. The guidelines are designed to be flexible so that different test intervals may be established for different types of meters or for different manufacturers' meters. Time intervals may even be established for new meters that differ from those for repaired meters. Different time intervals may also be established for meters in areas supplied with water from different sources if those sources result in water-quality variances.

Meters will be considered to be in compliance with these guidelines if both of the following conditions are met:

1. Ninety-five percent of the meters scheduled for tests on a periodic basis are actually tested. (It must be recognized that 100 percent of the meters scheduled for test cannot always be tested through no fault of operational procedures.)

2. At least ninety-five percent of the meters actually tested register results within the accuracy limits shown in Table 5-1 for both normal and minimum test-flow rates. These accuracy limits are determined prior to any adjustment or repair of the meter after it has been in service for a period of time. Only meters tested in conjunction with the periodic testing program should be used in computing average periodic accuracy results.

Statistical Sample Testing

Sample testing is an alternative method to evaluate the performance and service life of $5/8$-in. (15-mm) through 1-in. (25-mm) meters. In addition to providing insight into variables affecting meter performance, this method also allows the utility to monitor the overall accuracy of its meters. Sampling is a useful management tool in addressing metering activities, as well as, identifying problems with specific meters.

Most utilities' metering programs involve a changeout policy once a meter has reached a certain age or predetermined life expectancy. These programs assume that all newly purchased meters perform better than the older meters in the system. If this assumption is wrong then the utility may be exchanging meters that are more accurate than those remaining, therefore producing a loss of revenue. Statistical sample testing identifies the poorest performing meters and allows the utility to exchange meters based on performance.

Statistical data is used and accepted throughout the business world as an excellent tool for making informed management decisions. Information developed from sample testing will provide the utility with data in which trends analysis can be made and performance levels for specific meters can be identified.

The first steps to implement a sample testing program is to understand and identify the established statistical methods. This may be accomplished through educational training or by referencing basic statistical methods in textbooks.

After developing a basic understanding of statistical sampling, the following steps should be implemented:

1. Determine desired confidence level.
2. Determine appropriate sample lot size for the confidence level.
3. Determine criteria for testing (size, age, volume, type).
4. Randomly select and retrieve.
5. Test and document.
6. Analyze test data.
7. Report with recommendations.

The following is an example that identifies the benefits of sample testing and performance-based meter changeouts versus annual changeout based on age. In this example, a 95 percent confidence level is achieved. The appropriate lot size to achieve a 95 percent confidence level was determined to be a minimum of 70 meters for a particular set year, regardless of total population size. The criteria used in this example to sample test meters was the year in which the meter was put in service, "set year." A minimum of 70 meters from each sample lot (set year) is randomly pulled from field service and is returned to the meter test facility for shop testing.

Testing standards identified in this manual should be followed and adhered to. Extreme care should be used when retrieving and testing sample metering to best duplicate field accuracy/conditions. The example data includes an overall weighted average determined by assigning weights such as 15 percent–70 percent–15 percent to the low, intermediate, and high test results, respectively.*

Based on established statistical methods, reports such as Table 5-5 can be developed from the sample data. If these data were used to exchange meters based on performance, the meters set in 1982 would be scheduled for exchange. If the exchange criteria were based on age, the utility would exchange meters more accurate than others in the system, therefore not being cost-effective.

Although confidence levels are lost, the sample data can be queried by purchase year (Table 5-6) and by consumption (Table 5-7). Reviewing the data in different ways will identify the group or type of meters that are least accurate. In Table 5-6, purchase year 1982 is again identified as the least accurate group of meters in the sample data.

Statistical testing is a management tool. As utility managers strive to make good management decisions, data to assist the decision making is essential. Statistical testing provides the data to make cost-effective management decisions.

---

* Statistical sampling technique for controlling the accuracy of small water meters. Penchih Tao. *Journal AWWA*. June 1982.

Table 5-5 1989 sample test meters test results—set date

Set Date	Flow Rates			Overall Weighted Average
	Low	Intermediate	High	
1988	94.72	100.63	90.30	99.54
1987	97.66	100.90	99.63	100.22
1986	86.27	100.98	99.58	98.56
1985	88.86	100.68	98.97	98.65
1984	72.11	100.08	98.61	95.61
1983	79.34	100.02	97.95	96.61
1982	70.09	98.83	98.28	94.43
1981	71.76	100.34	98.82	95.83
1980	71.70	101.03	99.59	96.43
1979	86.20	100.96	99.43	98.52
1978	78.84	100.93	99.68	97.43
1977	88.85	99.90	99.68	98.21
1976	89.07	98.88	99.11	97.44
1975	82.24	100.50	99.46	97.61
1974	90.35	101.24	99.66	99.37
1973	89.50	101.38	99.94	98.38

Set Year	Number of Meters Tested	Slow Meters (accuracy 98.5 percent)				Accurate Meters (accuracy 98.5–101.5 percent)				Fast Meters (accuracy 101.5 percent)				Entire Population	
		Number of Meters	Percentage of Total	Average	Standard Deviation	Number of Meters	Percentage of Total	Average	Standard Deviation	Number of Meters	Percentage of Total	Average	Standard Deviation	Average	Standard Deviation
88	79	3	3.80	83.38	1.78	76	96.20	100.18	0.51	0	0.00	0.00	0.00	99.54	3.27
87	79	1	1.27	84.07	0.00	78	98.73	100.43	0.51	0	0.00	0.00	0.00	100.22	1.90
86	79	12	15.19	88.18	5.56	66	83.54	100.38	0.51	1	1.27	103.15	0.00	98.56	4.93
85	77	13	16.88	91.12	6.13	64	83.12	100.18	0.63	0	0.00	0.00	0.00	98.65	4.27
84	77	31	40.26	89.17	6.70	46	59.74	100.04	0.68	0	0.00	0.00	0.00	95.67	6.84
83	78	31	39.74	91.67	8.87	47	60.26	99.86	0.58	0	0.00	0.00	0.00	96.61	6.90
82	78	37	47.44	88.40	13.73	40	51.28	88.84	0.67	1	1.28	101.56	0.00	94.43	11.07
81	80	34	42.50	90.53	5.99	46	57.50	99.74	0.66	0	0.00	0.00	0.00	95.83	6.02
80	77	24	31.17	87.81	4.46	53	68.83	100.32	0.85	0	0.00	0.00	0.00	96.42	6.34
79	79	12	15.19	88.61	4.86	67	84.61	100.29	0.67	0	0.00	0.00	0.00	98.52	4.64
78	79	21	26.58	88.76	5.43	54	68.35	100.49	0.64	4	5.06	101.55	0.00	97.43	5.95
77	80	15	18.75	87.63	20.37	62	77.50	100.60	0.61	3	3.75	101.66	0.10	98.21	10.20
76	75	14	18.67	85.99	21.87	59	78.67	100.01	0.72	2	2.67	101.97	0.24	97.44	10.95
75	77	21	27.27	91.32	7.99	55	71.43	99.92	0.72	1	1.30	102.60	0.00	97.61	5.72
74	79	12	15.19	93.54	4.82	63	79.75	100.32	0.71	4	5.06	101.97	0.27	99.37	3.19
73	25	2	8.00	89.62	4.00	22	88.00	100.14	0.65	1	4.00	102.25	0.00	99.38	3.18

Total Number of Meters Tested—1,198
Total Number of Slow Meters—283   Percent of Total—24
Total Number of Accurate Meters—898   Percent of Total—75
Total Number of Fast Meters—17   Percent of Total—1

Sample ⅝ in. (15 mm) and ⅝ in. × ¾ in. (15 mm × 20 mm) meters by set year

Table 5-6 1989 sample test meters test results—purchase date

Purchase Date	Low	Flow Rates Intermediate	High	Overall Weighted Average
1988	95.93	100.59	99.36	99.71
1987	94.77	100.73	99.48	99.65
1986	94.21	100.98	99.70	99.77
1982	63.28	98.71	98.30	93.34
1981	77.91	100.52	98.59	96.84
1980	87.84	100.62	98.84	97.09
1979	78.82	101.03	99.49	97.61
1978	84.71	100.98	99.62	98.34
1977	88.41	99.98	99.71	98.21
1974	88.38	99.54	99.11	97.80
1973	91.41	101.13	99.72	99.46

Purchase Year	Number of Meters Tested	Slow Meters (accuracy 98.5 percent)				Accurate Meters (accuracy 98.5–101.5 percent)				Fast Meters (accuracy 101.5 percent)				Entire Population	
		Number of Meters	Percent- age of Total	Average	Standard Deviation	Number of Meters	Percent- age of Total	Average	Standard Deviation	Number of Meters	Percent- age of Total	Average	Standard Deviation	Average	Standard Deviation
88	51	1	1.96	83.22	0.00	50	98.04	100.04	0.49	0	0.00	0.00	0.00	99.71	2.38
87	75	3	4.00	82.88	1.22	72	96.00	100.35	0.49	0	0.00	0.00	0.00	99.65	3.47
86	85	4	4.71	85.66	0.27	81	95.29	100.47	0.52	0	0.00	0.00	0.00	99.77	3.18
82	112	68	60.71	89.03	11.14	44	39.29	99.99	0.68	0	0.00	0.00	0.00	93.34	10.21
81	122	42	34.43	91.23	6.16	80	65.57	99.78	0.63	0	0.00	0.00	0.00	96.84	5.40
80	214	61	28.50	89.53	7.37	152	71.03	100.08	0.70	1	0.47	103.15	0.00	97.09	6.22
79	76	17	22.37	88.03	4.27	59	77.63	100.38	0.72	0	0.00	0.00	0.00	97.61	5.56
78	115	22	19.13	89.22	5.56	89	77.39	100.34	0.65	4	3.48	101.55	0.00	98.34	5.11
77	86	16	18.60	87.50	19.73	66	76.74	100.59	0.60	4	4.65	101.63	0.10	98.21	9.95
74	120	21	17.50	86.77	18.76	93	77.50	100.01	0.75	6	5.00	102.14	0.28	57.80	9.38
73	23	5	21.74	96.30	2.43	17	73.91	100.27	0.77	1	4.35	101.55	0.00	99.46	2.14

Total Number of Meters Tested—1,079
Total Number of Slow Meters—260   Percent of Total—24
Total Number of Accurate Meters—803   Percent of Total—74
Total Number of Fast Meters—16   Percent of Total—1

Sample ⅝ in. (15 mm) and ⅝ in. × ¾ in. (15 mm × 20 mm) meters by purchase year

Table 5-7 1989 sample test meters test results—consumption

Cons. Rng	Per 1,000 Gallons Consumption	Number of Meters Tested	Flow Rates Low	Flow Rates Intermediate	Flow Rates High	Overall Weighted Average
01	0–100	120	89.42	100.36	98.87	98.50
02	100–200	144	86.03	100.81	99.31	98.37
03	200–300	148	86.79	100.70	99.22	97.49
04	300–400	120	78.81	99.72	99.12	98.49
05	400–500	98	79.61	100.47	99.01	97.13
06	500–600	76	81.18	99.43	99.03	96.63
07	600–700	75	72.75	100.73	99.25	96.31
08	700–800	90	87.02	100.87	99.28	98.55
09	800–900	52	84.56	101.02	99.25	98.28
10	900–1,000	61	81.88	100.08	99.40	97.67
11	1,000–1,100	42	83.87	98.73	99.40	96.60
12	1,100–1,200	9	80.61	100.88	99.38	97.61
13	1,200–1,300	37	89.58	101.35	99.48	99.31
14	1,300–1,400	30	89.58	101.19	99.22	100.04
15	1,400–1,500	24	96.36	101.00	99.22	97.67
16	1,500–1,600	21	87.41	99.53	99.26	97.67
		58				

Cons. Rng	Slow Meters (accuracy 98.5 percent)				Accurate Meters (accuracy 98.5–101.5 percent)				Fast Meters (accuracy 101.5 percent)				Entire Population	
	Number of Meters	Percent-age of Total	Average	Standard Deviation	Number of Meters	Percent-age of Total	Average	Standard Deviation	Number of Meters	Percent-age of Total	Average	Standard Deviation	Average	Standard Deviation
01	14	11.67	88.44	10.59	105	87.50	100.19	0.54	1	0.83	103.15	0.00	98.50	5.99
02	23	15.97	87.98	5.18	121	84.03	100.34	0.59	0	0.00	0.00	0.00	98.37	5.01
03	32	21.62	87.83	5.33	115	77.70	100.15	0.68	1	0.68	101.55	0.00	97.49	5.68
04	37	30.83	87.97	13.72	81	67.50	126.26	0.75	2	1.67	101.64	0.09	96.49	9.53
05	34	34.69	91.26	6.04	63	64.29	100.22	0.69	1	1.02	101.55	0.00	97.13	5.59
06	25	32.89	89.52	16.40	51	67.11	100.11	0.66	0	0.00	0.00	0.00	96.63	10.66
07	25	33.33	88.21	5.19	48	64.00	100.30	0.76	2	2.67	101.90	0.35	96.31	6.50
08	21	23.33	93.15	5.68	67	74.44	100.15	0.74	2	2.22	101.88	0.33	98.55	4.11
09	11	21.15	91.30	5.99	41	78.85	100.16	0.74	0	0.00	0.00	0.00	98.28	4.60
10	18	29.66	91.66	5.59	43	70.49	100.18	0.74	0	0.00	0.00	0.00	97.67	4.98
11	9	21.43	83.18	24.83	32	76.19	100.20	0.70	1	2.38	102.17	0.00	96.60	13.40
12	11	29.73	91.68	5.83	25	67.57	100.06	0.67	1	2.70	101.55	0.00	97.61	5.03
13	5	16.67	93.79	5.22	23	76.67	100.25	0.78	2	6.67	102.20	0.40	99.31	3.36
14	4	16.67	93.76	4.94	19	76.17	100.20	0.72	1	4.17	102.25	0.00	99.21	3.25
15	2	9.52	96.15	0.00	18	85.71	100.38	0.61	1	4.76	101.55	0.00	100.04	1.40
16	13	22.41	89.76	11.58	43	74.14	99.87	0.73	2	3.45	101.76	0.14	97.67	6.97

Total Number of Meters Tested—1,196   Percent of Total—24
Total Number of Slow Meters—284   Percent of Total—75
Total Number of Accurate Meters—895   Percent of Total—1
Total Number of Fast Meters—17

64 WATER METERS—SELECTION, INSTALLATION, TESTING, AND MAINTENANCE

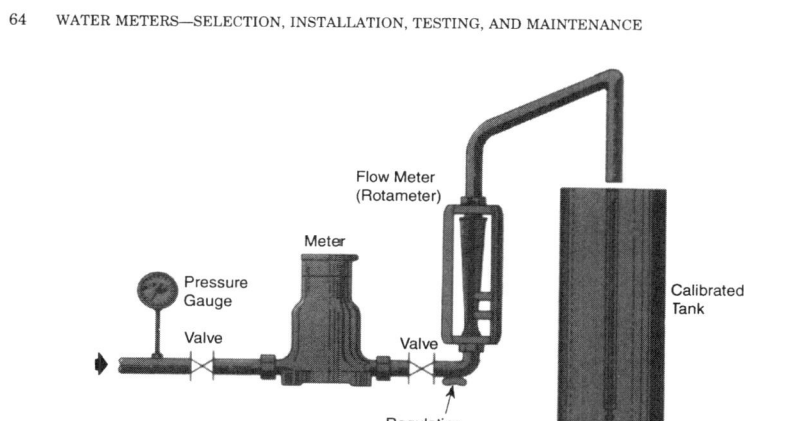

Figure 5-2   Basic requirements for a volumetric meter-testing assembly

Test Equipment

Equipment required to test a water meter may be very simple (Figure 5-2). A rotameter is a useful flow-measuring device for precise control in meter testing. It consists of a tapered, calibrated glass tube in which a stainless-steel rotor is free to move up and down in the center of the tube, guided by a stainless steel guide. The instrument is positioned above a flow-regulating valve. Water passing up the tapered tube raises the rotor to a stable position. The rate of flow is read across the top of the rotor on a scale of figures etched in the body of the glass tube. Rotameters are commercially available for meter-repair-shop testing apparatus. Figure 5-3 illustrates basic features of a rotameter.

If a large number of meters is to be tested, complete test equipment is commercially available. Also, the various meter-testing components may be purchased individually from commercial manufacturers, and an organization can fabricate test equipment for any particular application.

Meters may be tested singly or in groups, and the equipment selected should be based on the work load. When reasonable judgment is exercised in the selection of test equipment, the cost is quickly repaid and more accurate results are usually obtained. Requisites of a good meter test facility are:

- a test bench on which meters may be quickly and securely held
- an inlet valve
- a quick-closing valve on bench discharge and a flow-regulating valve
- a rate-of-flow indicator
- one or more tanks, preferably calibrated for volumetric testing, or a tank set on platform scales for testing by weight (calibrated tanks, of course, must be installed in a plumb position to avoid inaccuracies of calibration)
- adequate lighting of test area

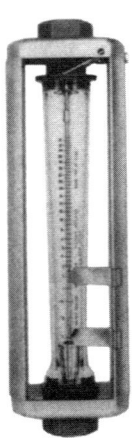

Figure 5-3 Rotameter rate-of-flow indicator

- cabinets for systematic storage of change gears, hand tools, and records
- an ample supply of water so that pressure will fluctuate as little as possible
- convenient and accessible location, and carefully designed shop layout to increase efficiency

Basic equipment for testing ⅝-in. (15-mm), ¾-in (20-mm), and 1-in. (25-mm) meters consists of the following:

- a 10-ft^3 or 100-gal (0.5-m^3) capacity tank, preferably calibrated for volumetric testing
- a 1-ft^3 or 10-gal (0.05-m^3) capacity tank, similarly equipped
- a test bench provided with necessary fittings, including a control inlet valve, a quick-acting rate control valve on the discharge, and a rate-of-flow indicator. This bench may be of single-meter capacity or designed for simultaneous testing of a number of meters (Figure 5-4).

Similar equipment should be provided for testing meters larger than 1 in. An additional tank, 100 ft^3 or 1,000 gal (5 m^3) or larger, should be provided for testing 3-in. (80-mm) and larger meters. A rate-of-flow indicator of larger capacity is also essential. For high flow rates for testing large meters, a Venturi meter with a calibrated manometer or an electronic flow meter to indicate rates of flow is recommended. Test benches (Figure 5-5) for 1½-in. (40-mm) and 2-in. (50-mm) meters are available and are recommended if sufficient testing of meters in these sizes is required. Piping should be provided for testing 3-in. (80-mm) and larger meters, if such sizes are shop-tested (Figure 5-6). This testing is accomplished by making a piping arrangement with space for the longest meter and an assortment of filler pieces for inserting meters of shorter length (Figures 5-7). This arrangement is frequently used for 1½-in. (40-mm) and 2-in. (50-mm) meters, instead of a special test bench, particularly when few meters of these

66 WATER METERS—SELECTION, INSTALLATION, TESTING, AND MAINTENANCE

The test bench can test ⅝-in. (15-mm) or ⅝-in. (15-mm) × ¾-in. (20-mm) meters, or 1-in. (25-mm) meters. The unit features a rate-of-flow indicator and an electronically actuated shutoff valve, with two calibrated tanks. This style bench features hydraulically or mechanically operated clamping mechanisms.

Figure 5-4 Test benches for small meters

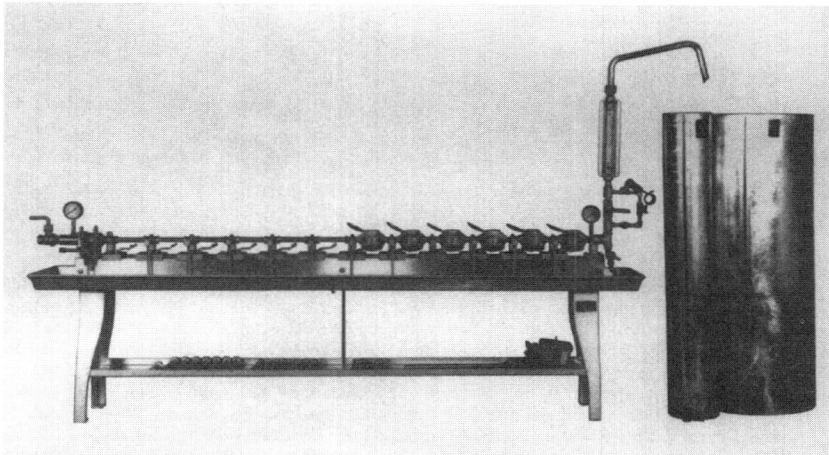

This multiple-unit test bench provides for effective testing of larger meters. Note free access to bench for handling heavy meters.

Figure 5-5 Test bench for 1½-in. (40-mm) and 2-in. (50-mm) meters

562 APPENDIX B METER INSTALLATION AND TESTING

This test bench is designed to test 3-in. (80-mm) through 12-in. (300-mm) water meters of different types. Multiple meters can be tested depending on their size and length. A hydraulic system is used to clamp the meters in place, and the bench is operated using a programmable gravimetric system. All water is recirculated.

Figure 5-6 Testing of large meters

This test facility includes two 50-hp (37,300-W) pumps rated at 1,250 gpm (284 m^3/h) each. It includes two 5,000-gal or 500-ft^3 (19-m^3) tanks in addition to several smaller tanks all based on gravimetric systems.

Figure 5-7 Large meter testing facility

68   WATER METERS—SELECTION, INSTALLATION, TESTING, AND MAINTENANCE

sizes are in use. For 3-in. (80-mm) and larger testing equipment, a surge tank or air chamber at the tester inlet may be advantageous to eliminate water hammer that could result from quickly stopping a larger test stream.

Procedure

Although the following description covers the actual steps in testing a single $5/8$-in. (15-mm) meter, the only differences for larger meters are the rates of flow and test quantities used. Test equipment is the same as that previously described.

1. Clamp the meter securely to the test bench. (Do not tighten more than necessary to make a watertight connection, as there is a possibility of distorting the meter housing or extruding the washer into the waterway.)

2. Open the register-box cover.

3. Open the discharge valve first, then open the inlet valve gradually and run the water to waste until the entrapped air is cleared. This process also ensures a full discharge line to the tank.

4. Shut off the discharge valve.

5. Check 100-gal or 10-ft^3 (0.5-m^3) tank discharge to ensure that the tank is empty and then close the tank drain valve. (Tank discharge should be to an open drain so that any possible leakage of the tank drain valve can also be observed.)

6. Revolve the meter register to set the test hand at the zero mark. In so doing, revolve the register backward beyond the zero mark and then reverse its direction to bring the test dial hand forward to the mark. This procedure takes up any possible backlash in the gearing. An option to setting the register to zero, is to simply record the register starting position, providing that the backlash is removed.

7. Sealed registers on some magnetic-drive meters require a special test ring that is set on top of the register box and can be rotated so that the test hand starts at zero on the test dial.

8. Open the test bench discharge valve as rapidly as prudently possible to the desired rate for the maximum-flow test (15 gpm [3.4 m^3/h]). Continue the flow at this rate until the meter test hand has made 10 complete revolutions, then stop the test at the starting mark. Read the meter accuracy from the scale of the calibrated tank (Figure 5-8) or, if the tank is not calibrated, by weighing the water in the tank. Trickling or bumping water through the meter to accomplish an exact volume of measurement is not recommended.

9. Record the results of this test on the record form.

10. If the meter being tested is a repaired meter and its accuracy does not fall within prescribed limits, refer to the change-gear chart and replace change gears with a set that should cause the meter to register within required limits, or use calibration adjustments or new parts to bring the unit within limits. Repeat the test as outlined. (Do not depend on the theoretical change in meter accuracy from changing gears. Be sure that the meter tests correctly by rerunning the test after any change in gears is made.)

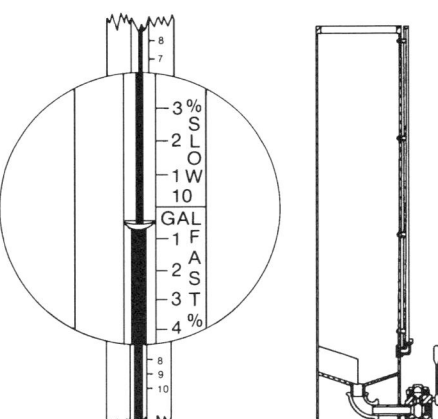

The calibration scale shows the amount by which the tested meter is over or under registering, in percent. The reading illustrated is ½ percent fast (meter registers 100.5 percent of true quantity passed through it).

Figure 5-8 Detail of calibration scale of testing tank and sectional view of tank with outlet valve

11. If a repaired meter is being tested and is found to register within the limits on the maximum-flow test rate, repeat steps 3 through 8 for the intermediate (2 gpm [0.5 m^3/h]) and minimum (¼ gpm [0.06 m^3/h]) tests, discharging the test water into the 10-gal or 1-ft^3 [0.05-m^3] tank. The test quantity for these rates is only 10 gal or 1 ft^3 (0.05 m^3), and the meter register test hand completes only one full revolution.

12. If, however, the test is to determine the condition of a meter removed from service, it is recommended to rearrange the order of test flows. The preferable order in this event is the minimum-, intermediate-, and, finally, the maximum-flow test. If a meter that has been in service for some length of time is tested first on the high rate of flow, there is a possibility of freeing the disc and thereby obtaining a false impression of the meter's condition on lower rates. Conversely, a meter should be tested as soon as possible after removal from service to prevent the drying of deposits in the measuring chamber, as this condition tends to give an adverse impression of the meter's condition while it was in service.

13. After the three separate tests have been run, the meter has been fully tested and may be removed from the test bench after the inlet valve of the test bench has been closed and pressure has been released by a partial opening of the discharge valve.

14. Once the test is concluded, the meter should be drained, the register sealed, and the dust caps placed on the meter spuds before the meters are put in storage.

The calibration of a tank is based on its rated capacity, and if lesser quantities are used in testing, as is sometimes done, the calibrated percentage errors are not valid.

70 WATER METERS—SELECTION, INSTALLATION, TESTING, AND MAINTENANCE

For example: a 10-gal tank may be used as a 1-gal tank by lowering the water level to the 9-gal mark and running water to the 10-gal mark. This procedure is not recommended but is sometimes carried out to save testing time. Under such conditions, each 1 percent error on the calibration strip becomes 10 percent.

It should be noted that a rate-of-flow indicator is affected by water temperature below the 1-gpm (0.2-m³/h) rate. Most rate-of-flow indicators are calibrated for 70°F (21°C), and viscosity changes due to temperature affect the reading.

Rate-of-flow indicators are usually equipped with a pointer to indicate true rates at the minimum-flow points. They are raised for low-temperature water and lowered for high-temperature water. A calibrated cylinder and a stopwatch are effective calibrating instruments for the ¼- and ½-gal points; or the following formula may be used to verify the actual rate of flow during a standard accuracy test:

$$RT = V$$

Where:
  R = rate of flow, gpm
  T = time of flow, min
  V = volume of test tank, gal

Example: It takes 4 min to fill a 1-gal calibrated cylinder:

$$R = \frac{V}{T} = \frac{1}{4} = \frac{1}{4} \; gpm$$

This check on the accuracy of low-flow calibration is required when a marked change in water temperature of the test water occurs.

Multi-Jet Meters

The same procedure is also used for multi-jet meters, the only difference being that the registers are not rotatable, and special test dials are not applicable. After the meter has been thoroughly flushed with water until the entrapped air is cleared, the flow is stopped, and a reading of the meter is taken, then the test bench discharge valve is opened and the filling of the calibration tank commences. Flow is stopped when the meter shows the appropriate quantity, as required by the standard. The accuracy of the meter at that specific flow is then read from the scale of the calibrated tank.

Multiple-Meter Testing

Multiple testing of meters is identical to the testing of one meter, except for one important factor. In multiple-meter testing, it is not possible to use the meter test dials for determining the test quantity, as each meter has a slightly different accuracy. The test quantity is discharged into the tank, and the flow is stopped when the scale on the tank indicates that the exact quantity has been delivered. Each meter's accuracy is determined by reading its register. The arithmetic of the test must be clearly understood when meters are tested in this manner. In the tests described for testing one ⅝-in. (15-mm) meter, only two quantities were used, 100 gal or 10 ft³ (0.5 m³) for the maximum-flow test and 10 gal or 1 ft³ (0.05 m³) for the intermediate- and minimum-flow tests. One full revolution of the test hand of a ⅝-in. (15-mm) meter registers either 10 gal or 1 ft³, and the test-hand circle is divided into 10 equal parts. Therefore, on the maximum-flow test, when the test hand makes 10 full revolutions, each subdivision of the test dial circle is a 1 percent variation. On the intermediate- and minimum-flow tests of 10 gal or 1 ft^3, the test hand makes only one revolution, and each subdivision represents a 10 percent variation.

When ⅝-in. (15-mm) meters are tested in groups and the exact test quantity is discharged into the tank, the percentage of error between divisions of the test dial depends on the quantity, i.e., 1 percent for 100 gal or 10 ft^3 and 10 percent for 10 gal or 1 ft^3. Furthermore, it must be determined whether the meter is fast or slow. If, in group testing ⅝-in. (15-mm) meters on a quantity of 100 gal, the test hand of one meter has made 10 full revolutions plus one additional mark (1 gal), the meter has registered 101 gal when only 100 gal were delivered; and the meter is registering 101 percent of true quantity, or 1 percent fast. Similarly, if the test hand of another meter has made 10 revolutions minus two marks (2 gal), it registers only 98 gal when 100 gal are being delivered and, therefore, is registering 98 percent of true quantity or 2 percent slow.

When 10 gal or 1 ft^3 are used in multiple testing, each division of the test dial circle represents 10 percent error, and if a meter register shows on the test that the test hand has only moved $9/10$ of the full circle, it is registering only 90 percent. Unless meter registers with center-sweep test hands or special test registers are used, it is necessary to interpolate between the marks for more accurate readings.

Although concern is often expressed over the time involved in testing meters on low rates of flow and large test quantities, multiple testing of meters reduces this factor on a unit basis. Furthermore, a relatively low-cost, automatic, electric cutoff valve can be installed in the test bench unit, which eliminates the need for someone to stand by to ensure against overrun of the test. This attachment, shown in Figure 5-9, consists of an electric valve with a strainer, a relay with transformer, one low-voltage electrode for each tank, and the necessary wiring and switches. When the water in the tank rises to the test volume level (at which point the lower end of the low-voltage electrode is positioned), contact is made, and a current path is formed through the relay. This causes

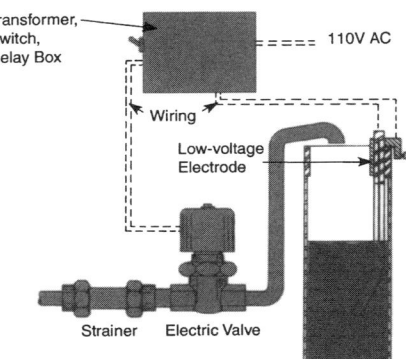

This device automatically stops the flow of water when the testing tank contains the preset quantity of water. This feature eliminates the need for someone to stand watch to ensure against overrun of the test.

Figure 5-9   Automatic cutoff valve

This facility can test 144 meters ¾-in. × 1-in. (20-mm × 25-mm) in size or 120 meters 1-in. (25-mm) in size simultaneously. Two rows at a time are tested on fast flows and all are tested in series on lower flows. The 10-ft³ (0.5-m³) tank in the background measures flow volumetrically.

Figure 5-10 Large multiple-test table

the relay contacts to open, which de-energizes the solenoid circuit of the electric valve and causes it to close and stop the flow. Low-voltage electrodes can also be used to ring a warning bell or other monitoring device when the tank is almost full. (See Figure 5-10 for a large multiple-testing operation.)

When multiple-meter tests are made on multi-jet meters, two suitable conditions should be provided: (1) a constant, nonpulsating flow; and (2) intermediate coupling pieces between the meters having the same bore diameter as that of the meter and having a length between three and five times that of the diameter.

FIELD TESTING

Larger meters are frequently ignored as long as they continue to record consumption. These meters may be few in number, but they represent a significant amount of revenue for a water system. Typically, the largest 10 percent of the meters measure 40 to 60 percent of a system's consumption.

If these larger meters are so important to a water system's financial health, why are they not maintained to provide peak performance? The answer is that larger meters are harder to work on and repair; spare parts are expensive; assemblies are sometimes complex; and it takes more knowledge and better trained repair personnel to maintain them than for smaller meters.

Maintenance is delayed, because meter installations are frequently crowded, or piping compromises are necessary. There is no bypass piping to continue supplying water to the customer during a meter accuracy test, or it is difficult to dispose of the

TESTING OF METERS—TEST PROCEDURES AND EQUIPMENT 73

test water. In addition, working space around the meter may be restricted and unsafe, causing the system's management to be concerned about the liability, and the span of control of the testing personnel.

Nonetheless, because these larger meters are so important, their operating condition must be monitored on a systematic and timely basis. One alternative approach is for qualified test personnel to test these meters on-site.

On-Site Testing Rationale

On-site testing of the larger meters is one way to ensure their proper operation; another is to test the meters at the water system's facilities. There are certain advantages to testing the meters at the water system's facilities, and in some cases it would be prudent. In the majority of situations, on-site testing is more economical in time and resources. From a technical standpoint, the piping in a meter installation can have a definite influence on a meter's accuracy, and this irregularity can be detected by the on-site tests.

The diagram in Figure 4-6 of this manual specifies correct meter installations. Properly operating isolation valves and a straight pipe, a minimum of 5 pipe diameters in front of the meter and a minimum of 3 pipe diameters downstream of the test tee, are critical in attaining engineered accuracy and performance.

One acceptable method of maintaining proper performance of certain types of larger meters is to replace the operating components and assemblies while leaving the meter body in place. If the measuring and registration functions are within one integral assembly, accuracy tests are not required after installation. If separate assemblies are involved, a final on-site accuracy test is appropriate. Without accuracy testing of separate assemblies, the potential loss of significant revenue increases.

Because of the revenue these large meters produce, a formal on-site meter testing program should be part of every water system's maintenance operation. Increased revenue and water accountability gains will offset the initial investment and continuing costs of the testing programs. Actual case histories at Boston, Mass., East Orange, N.J., Columbia, S.C., and other utilities, have shown such returns to be both immediate and dramatic.[*]

What's Involved With On-Site Testing

Setting up and maintaining an on-site meter testing program involves a number of factors. The meter pit must have adequate and safe space in which personnel can maneuver. Occupational Safety and Health Administration (OSHA) requirements must be adhered to, and everyone in the testing program should be clearly instructed on its guidelines. The piping arrangement around the meter must include some method to positively isolate the meter, while still maintaining an adequate flow to the end user. Temporary or permanent by-pass piping needs to be installed.

Some larger meters have built-in test plugs, others do not. For installations that require test outlets, the outlets can be fabricated in a number of ways. Service saddles and reducing tees are the most frequently used approaches. These need to be installed according to the recommendations of the meter manufacturer and located so that the connecting hose to the on-site tester is correctly located downstream of the meter.

A prerequisite to testing is that the isolating, downstream shut-off valve provides positive shut-off. If the valve is not completely operational and there is leakage when the valve is in the off position, the test will not be accurate.

---

[*] Data is available from Schlumberger Industries.

74 WATER METERS—SELECTION, INSTALLATION, TESTING, AND MAINTENANCE

To assist the testing operation, a short length of pipe should be permanently attached to the test outlet, along with a shut-off valve that can be locked into position. This will save considerable time and effort in preparing the meter site for testing.

Finally, and most importantly, the personnel assigned to perform the tests must be properly trained and have the appropriate test equipment. There are certain techniques and procedures that must be properly followed when using the test equipment. The consequences of discharging large volumes of water at high flow rates must be understood, appreciated, and accounted for. Improper use of the equipment can be harmful to the operators, the meter, and the surrounding area.

The techniques for performing the tests, selecting the appropriate test flow rates, determining the accuracies, and reaching conclusions, must be known and carefully followed to obtain valid test results. Meter testers should be considered specialists.

Types of Test Equipment

Residential meters, meters up to 1 in. (25 mm) in size, can be tested on-site in several ways to determine accuracy. The test equipment (Figure 5-12) and methods for determining the accuracy of these meters are not applicable to testing the larger meters. The larger meters require specialized test equipment that can handle a wide range of flow rates and provide accurate, valid data. These devices can either be purchased as a manufactured assembly or fabricated by the water utility.

The equipment for the larger meters (Figure 5-11) is available as portable test packages, installed on trailers, or mounted in a van (Figures 5-13a and 5-13b) or pickup truck.

Figure 5-11 Field testing flanged meters

TESTING OF METERS—TEST PROCEDURES AND EQUIPMENT 75

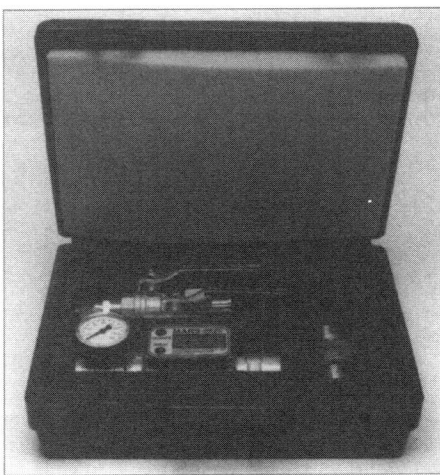

Figure 5-12 Field-test apparatus for small meters

Regardless of the style, these testers all contain certain basic elements that are required to properly test turbine, compound, and propeller meters.

Because of the wide flow ranges involved, a tester includes at least two, and sometimes three, meters of varying capacities. A shutoff valve is located downstream of each meter to control the flow rate for the various tests. A pressure gauge is required to check both the line pressure and the residual pressure at the tester. Sometimes resettable registers and/or flow raters are included to reduce the time required to complete a test.

Flexible hoses are required to connect the test equipment to the test connection of the meter being tested. Because of the pressures and hydraulic forces, the hoses must be in good condition and positioned as straight as possible between the two meters. For the larger testers, it is important that the tester itself be anchored to a vehicle, or a hold-down device, because the tester will want to move when the flow is shut off.

The master meters used on the testers must be capped, protected, and handled with care when not in use. They should also be recalibrated periodically to ensure accurate measurement.

Test Procedure

Before testing, it is necessary to know what the typical accuracy curve of the meter is for each specific brand, model, and size of meter being tested. This information can be obtained from the meter manufacturer's literature. A chart can be prepared that lists the flow rates at which each meter should be tested to properly assess its operational condition. Table 5-3 will suffice in the absence of factory data.

76 WATER METERS—SELECTION, INSTALLATION, TESTING, AND MAINTENANCE

The mobile meter test van is capable of testing meter sizes 3 in. (80 mm) to 12 in. (300 mm) without removing the meters from service. The unit is self-contained, computer driven, and fully automatic.

Figure 5-13a Mobile meter test van

This is a view of the automatic control panel inside the test van.

Figure 5-13b Mobile test van automatic control panel

For positive displacement meters, chapter 2 of this manual provides the three flow rates (low, mid, and high) which apply to all meter brands. For turbine and propeller meters, either this manual or the meter literature should be consulted. For compound meters, it is important to know where "cross-over" is located so that it can be specifically tested. For this information, the manufacturer's accuracy curves are preferable. (Different brands of compounds have various cross-over flow rates. A discussion of these cross-over flow rates is beyond the scope of this manual.)

The various suppliers of large meter test equipment provide detailed procedures for conducting the accuracy tests. In general, the tester is hooked up and the line flushed. The flow rate should be set to flush the testing apparatus, and the leak test should be conducted by observing the meter flow indicator.

Considerations When Testing

When testing a meter on-site, compare the accuracy of the meter in question to another, calibrated meter. The calibrated meter has its own performance characteristics and is not 100 percent accurate across its entire flow range, unless it is electronically corrected to level the accuracy curve. Other than at the very low flows, where the acceptable accuracy is 95 percent (minimum), there should not be a concern over differences of 1 or 2 percent.

Once the testing begins, the testing order is from the low flows to the higher flows. Experience has shown that when most meters begin to wear or lose accuracy, it occurs at the lower flows rather than the higher. If a meter is performing accurately up through the lower 25 percent of its capacity, it will typically test accurately through the rest of the range. This is especially true on the very large meters.

Another item to consider is what to do with the water after it has gone through the tester. A sudden high flow rate, such as one thousand gallons per minute, could reduce the supply pressure available and disturb any debris in the service line to the customer. It could also reduce the water supply in a nearby portion of the system, or it could cause considerable damage if the water is not discharged properly. Sheet ice on the street is a major life-threatening hazard. Great care should be taken to prevent water from freezing on the road surface or walking surfaces near the test site.

When setting up the tests, no test should be less than one minute long, and the meter's sweep hand should make at least one complete revolution. The residual pressure on the tester should never be less than 20 psi (140 kPa) when running a high-flow test. Also, for safety, the tester should not be operated on lines with static pressure exceeding 80 psi (550 kPa) unless provisions are made to secure the tester.

Detailed records are important to monitor trends in performance, along with the accuracies obtained at the various flow rates. Record the meter's registration before and after the testing, so the customer is not charged for the water used during the test.

The local utility must determine how inaccurate a meter can become before repairs or replacement can be justified. A cost-versus-benefit consideration is required, based on a number of factors important to the local water system. The considerations should also include the sizing and selection of the replacement meter. Another factor should be the updating of the meter installation for future on-site testing.

APPENDIX

C

Types of Flowmeters

Chapter 4 of AWWA Manual M33, Flowmeters in Water Supply, *1989; reprinted by permission*

AWWA MANUAL M33

Chapter 4

Types of Flowmeters

The most common flowmeters used in today's water treatment facilities are discussed in this chapter. These include the Venturi and modified Venturis, orifice plate, magnetic, turbine and propeller, sonic, vortex, and averaging Pitot flowmeters.

VENTURI FLOWMETERS

Accuracy*: ±0.75 percent of rate
Repeatability†: ±0.25 percent
Rangeability‡: 4:1 and up to 10:1
Size range: 1–120 in. (25 mm–3 m)
Head loss: Low
Relative cost: Medium

The most commonly used flow-measurement devices are the differential-pressure flowmeters, including the Venturi meter, or Venturi. Their popularity is largely due to a combination of flexibility, simplicity, ease of installation, and reliability.

In the Venturi, a defined constriction (throat) within the meter body causes an increase in flow velocity at the constriction, resulting in a corresponding decrease in pressure at that point (Figure 4-1). The difference in pressure between the upstream pressure connection (piezometer) and the pressure connection at the constriction (throat) is proportional to the square of the flow. Flow is then calculated from the square root of the measured pressure differential.

---

*Accuracy: The degree to which an indicated value matches the actual value of a measured variable. In process instrumentation accuracy is the degree of conformity of an indicated value to a recognized accepted standard value, or ideal value.

†Repeatability: The closeness of agreement among a number of consecutive measurements of the output for the same value of the input under the same operating conditions, approaching from the same direction, for full-range traverses.

‡Rangeability: The ratio of the maximum flow rate to the minimum flow rate of a meter.

10 FLOWMETERS

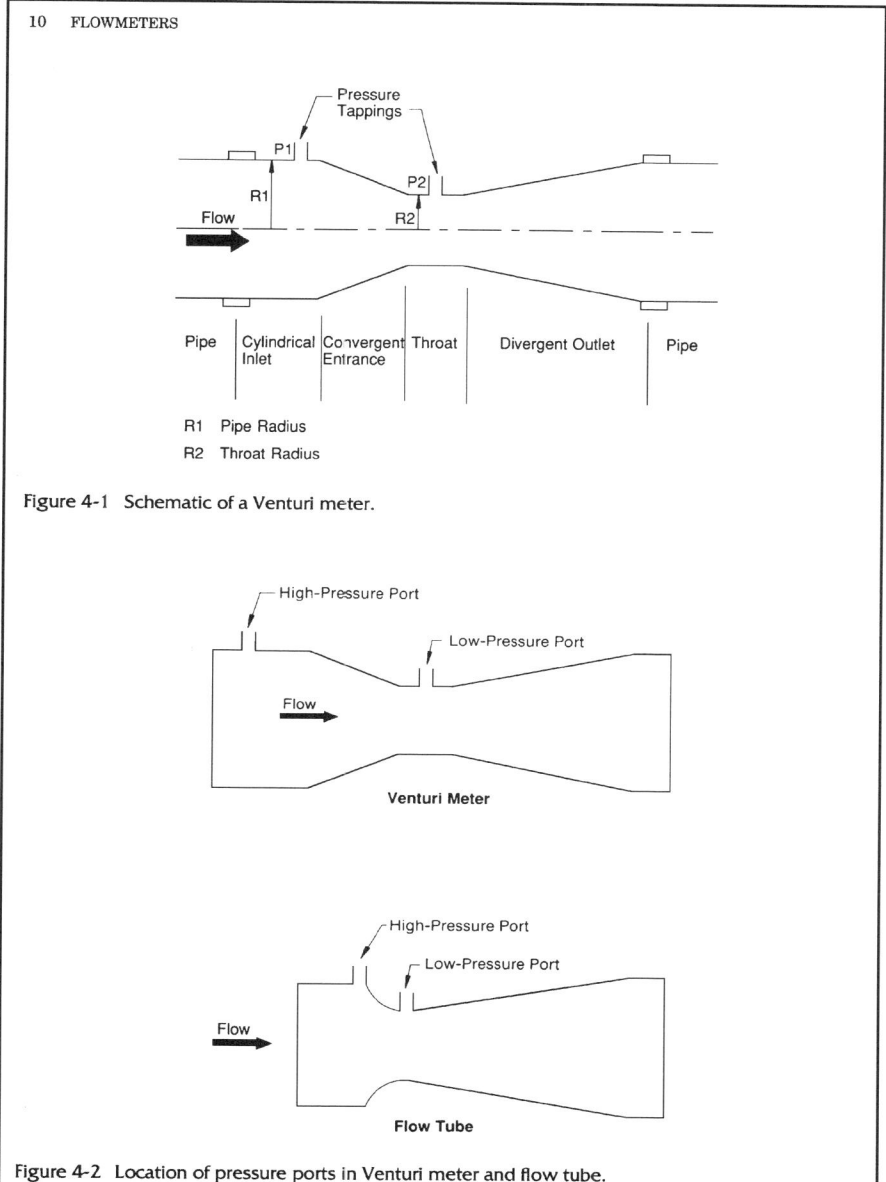

Figure 4-1 Schematic of a Venturi meter.

Figure 4-2 Location of pressure ports in Venturi meter and flow tube.

To obtain an electrical signal for flow measurement, the pressure at the ports must be transferred to a differential-pressure transducer. The transducer is typically a flexible diaphragm with a small chamber; one of the two pressure lines from the meter tube is connected to each side. The pressures each exert a force on the diaphragm and cause deflection of the sides. To sense the deflection of the diaphragm, a strain gauge, or a variable capacitance or a variable reluctance device is built into the transducer. This device generates an electrical signal proportional to the pressure difference across the diaphragm. Direct-reading differential pressure may be detected and scaled using a U-tube manometer or double-bellows gauge.

It is important to note that the major distinction between the Venturi and modified forms of this device, such as flow tubes and flow nozzles, is that the pressure connections are pure static measurements taken at points where the direction of fluid flow is not changing (see Figure 4-2). The shape of the approach to the throat conditions the flow pattern in the throat. With this hydraulic profile conditioning built into the body inlet structure, only the shortest lengths of straight pipe on the inlet side are necessary, when compared to other primary flowmeters. For the same pressure differential and the same size pipe and throat constriction, the Venturi yields 60 percent less head loss than the orifice plate meter.

The accuracy of a Venturi is ±0.75 percent of actual flow. This is exclusive of any additional tolerances introduced by instrumentation used with the meter to measure the differential pressure. Contrary to much of the published data, the Venturi has no range limitations up to the speed of sound. However, the instrumentation does have definite limits. Some instruments can provide a 10:1 range, several 8:1, and many only 4:1.

The Venturi meter has a low permanent head loss and does not obstruct the flow of suspended matter. The coefficient of discharge of the classic Venturi is 0.983, which remains constant regardless of pipe or throat size. The Venturi has the most comprehensive documentation regarding its coefficient, with hundreds of thousands of laboratory test points taken in every conceivable piping configuration.

The Venturi requires the greatest laying length of all differential-pressure meters and is, therefore, the heaviest flow tube as well. Venturis are commonly constructed from cast iron and naval bronze.

Installation

Before installing a Venturi, there are factors to consider in the selection process.

- Select a meter with the highest differential pressure, while considering how much head loss can be tolerated. A high differential provides the greatest energy for driving the instrumentation, which improves the range and accuracy.
- Select a meter with the smallest throat size in any given line size, again considering head loss. In the literature and manufacturer's tables, throat size is defined in terms of the beta ratio—the ratio of throat diameter to pipe diameter.
- Review the upstream piping configuration. A Venturi is not affected by the downstream configuration except for a slight increase in head loss.
- Consider future expansion of the facility, which may increase required flow.

Installation of a Venturi is critical to the accuracy of the differential-pressure measurement. Errors in computing flow from the differential pressure may become unacceptable if the piping arrangements create distorted flow. Swirls or vortices that affect meter accuracy can be produced by a projecting gasket, misalignment, or a burr on a pressure tap.

12 FLOWMETERS

Preferably, the Venturi should be installed with its axis horizontal, and the fluid should enter the tube with a fully developed velocity profile that is free from swirls and vortices. In a horizontal installation, the pressure port tappings must not be at the bottom, subject to clogging, nor at the top, subject to air bubble trapping. The preferred location is on the side in the horizontal plane of the centerline.

Maintenance

As a flowmeter with no moving parts, the Venturi would normally require less attention than, for example, a turbine meter. However, there can be extensive piping, fittings, and valves in the pressure-differential assembly. Lines may clog and corrosion may begin to appear. Periodic disassembly, inspection, and cleaning should be practiced. The pressure sensors, in particular, should be removed and inspected. While it is especially important to inspect the throat of the Venturi for debris or deposits, the high fluid velocity usually scours the throat.

Manufacturers provide step-by-step instructions for checking out the meter components. The procedures include disassembly, inspection and testing, parts replacement, and reassembly. The emphasis is on the differential-pressure unit and electrical housing. Instructions for meter zero and span adjustment are also included. A troubleshooting guide may be provided with symptoms, potential sources, and recommended corrective action. To aid the user, illustrated drawings, schematic diagrams, and parts lists are provided in the manufacturer's manual. Figure 4-3 provides an example of a troubleshooting guide for a differential-pressure transducer.

Advantages and Disadvantages

The following are advantages of the Venturi meter:
- life expectancy of a Venturi with manometer is documented to be greater than 50 years (this may not be true of modern instrumentation in general, although some earlier instruments were good for 30 years or more); materials of construction are well documented for long life;
- simplicity of construction, including no moving parts;
- no sudden change in contour and no sharp corners;
- relatively large pressure recovery in the recovery cone, resulting in low head loss and leading to substantial power savings where large flows are metered; and
- heavily documented in the literature as an acceptable type of flowmeter.

The following are disadvantages of the Venturi tube:
- larger units are costly to purchase and install;
- largest and heaviest of the differential-pressure meters; and
- differential pressure is not linear with flow rate, requires square root extraction, and reduces rangeability.

MODIFIED VENTURIS

Several flowmeters are derived from the Venturi and operate on the same principle. They are classified as differential-pressure producers, and include flow tubes, or "low loss" (proprietary) flow tubes. The object of any modification to the Venturi is to achieve shorter laying length, less cost, and/or higher head recovery. The length of the throat in the flow tubes is much shorter than in Venturi meters. This may result in a less stable flow pattern through the throat, which will affect accuracy.

578 APPENDIX C TYPES OF FLOWMETERS

Troubleshooting—Differential-Pressure Transducer

SYMPTOM: High Output

POTENTIAL SOURCE AND CORRECTIVE ACTION

1. **Primary Element**
   Check for restrictions at primary element.
2. **Impulse Piping**
   Check for leaks or blockage.
   Check that blocking valves are fully open.
   Check for entrapped gas in liquid lines and for liquid in dry lines.
   Check that density of fluid in impulse lines is unchanged.
   Check for sediment in transmitter process flanges.
3. **Transmitter Electronics Connections**
   Make sure bayonet connectors are clean and check the sensor connections.
   Check that bayonet pin #8 is properly grounded to the case.
4. **Transmitter Electronics Failure**
   Determine faulty circuit board by trying spare boards.
   Replace faulty board.
5. **Sensing Element**
   See Sensing Element Checkout Section.
6. **Power Supply**
   Check output of power supply.

SYMPTOM: Erratic Output

POTENTIAL SOURCE AND CORRECTIVE ACTION

1. **Loop Wiring**
   Check for adequate voltage to the transmitter.
   Check for intermittent shorts, open circuits, and multiple grounds.
   NOTE: Do not use over 100 V to check the loop.
2. **Process Fluid Pulsation**
   Adjust electronic damping pot (4–20 mA DC only).
3. **Impulse Piping**
   Check for entrapped gas in liquid lines and for liquid in dry lines.
4. **Transmitter Electronics Connections**
   Check for intermittent shorts or open circuits.
   Make sure that bayonet connectors are clean and check the sensor connections.
   Check that bayonet pin #8 is properly grounded to the case.
5. **Transmitter Electronics Failure**
   Determine faulty board by trying spare boards.
   Replace faulty circuit board.

SYMPTOM: Low Output or No Output

POTENTIAL SOURCE AND CORRECTIVE ACTION

1. **Primary Element**
   Check installation and condition of element.
   Note any changes in process fluid properties that may affect output.
2. **Loop Wiring**
   Check for adequate voltage to transmitter.
   Check for shorts and multiple grounds.
   Check polarity of connections.
   Check loop impedance.
   NOTE: Do not use over 100 V to check the loop.
3. **Impulse Piping**
   Check that pressure connection is correct.
   Check for leaks or blockage.
   Check for entrapped gas in liquid lines.
   Check for sediment in transmitter process flange.
   Check that blocking valves are fully open and that bypass valves are tightly closed.
   Check that density of fluid in impulse piping is unchanged.

Adapted from Rosemount Inc., Eden Prairie, Minn.

Figure 4-3 Excerpt from differential-pressure transducer troubleshooting procedures.

14 FLOWMETERS

Although similar to the Venturi, the flow tube has a relatively short transition section from the inlet to the throat (see Figure 4-2), and a recovery cone of gradually increasing diameter. The flow tube causes significantly less pressure drop than the orifice plate flowmeter (discussed later), but it is costlier. With its shortened inlet section, the flow tube is usually less expensive than the Venturi for the same line size. The body structure does not provide hydraulic profile conditioning, and the flow tubes are more sensitive to upstream flow disturbances.

A very low head loss, lower than that for a comparable Venturi with the same beta ratio, has been achieved by the "low-loss" flow tube. The head loss is a percentage of the differential, based on throat size.

Because the materials of construction for the flow tube are the same as those of the Venturi, long life may be anticipated. Some flow tubes will lose their discharge coefficient at relatively low and high Reynolds numbers, but only the former is important in the water market. The accuracy is ±1 percent of the flow rate, and the range, like the Venturi, is limited only by the associated instruments.

Another variation on the Venturi meter is the insert flow tube. The convenience of the unit is based on its short laying length, single mounting flange with no inlet cone upstream of the flange, and low head loss. Both the high- and the low-pressure metering taps are built into the flange.

ORIFICE PLATE FLOWMETERS

Accuracy: ±1/4 to ±2 percent of full scale
Repeatability: ±0.25 percent
Rangeability: 4:1
Size range: All
Head loss: Medium
Relative cost: Low

The orifice plate flowmeter consists of a thin plate with a hole in it (Figure 4-4). The standard plate is a circular disc, usually of stainless steel, 1/8–1/2 in. (3–13 mm) thick, and containing an orifice with a sharp leading edge. The hole is usually concentric with the pipe into which the plate is inserted and perpendicular to the axis

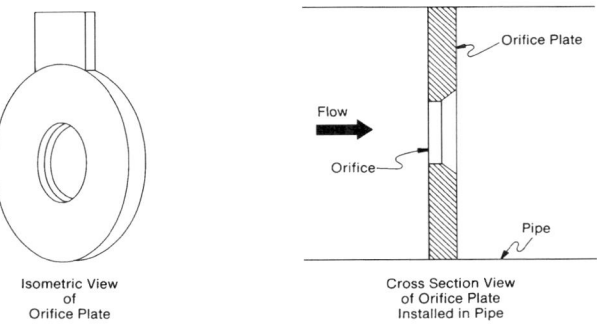

Figure 4-4 Schematic of an orifice plate flowmeter.

of flow. The plate is installed in a pipeline between two flanges, and the differential pressure across the plate is measured. Because of its simplicity, low cost, ease of installation, and reasonable accuracy, the orifice plate is among the most common primary elements for metering flow.

Most orifice plate meters for clear water are made with a circular orifice concentric with the pipe. In special applications, notably for flowing solids, the orifices may also be eccentric or segmental. Passage of entrained solids is permitted if the hole is tangent to the inside surface of the bottom of a horizontally laid pipe.

As in a typical differential-pressure flowmeter, the pressure upstream (approximately one pipe diameter from the plate) is compared with the pressure downstream at the point of narrowest streamflow. The narrowest section of flow is called the *vena contracta*, and is taken to be one half pipe diameter downstream from the plate. These points define where to locate the pressure taps.

Numerous eddies form downstream from the plate between the pipe wall and the *vena contracta*, causing kinetic energy to be dissipated as heat. This accounts for the medium to high head loss associated with the orifice plate meter.

The orifice plate size can be fabricated to accommodate any pipe size. Inherent accuracy, independent of pipe diameter errors and the differential-pressure sensor, may be $\pm 1/4$ percent to ± 2 percent of full scale.

Orifice plates are also available as part of integrated pipe assemblies, which include the plate, pipe length, pressure taps, and, in some cases, the differential-pressure sensor and signal transmitter.

Installation

The orifice plate is usually mounted between a pair of flanges. Manufacturers may extend the plate to include a tab above the edge of the pipe flange. The tab, suitable for a nameplate, may contain pertinent data on the specific installation and may identify the upstream side. To prevent errors in flow measurement, the gaskets should not protrude across the plate face beyond the inside pipe wall. Typically, the orifice plate also requires a straight run of smooth flow before and after the plate. Pressure taps must be installed perpendicular to the pipe wall. For horizontal pipe runs, the pressure taps should be in the horizontal plane of the pipe centerline. Burrs and intrusions at the tap must be removed.

Maintenance

As a differential-pressure flowmeter with no moving parts, the orifice plate meter requires maintenance similar to that for the Venturi. The orifice should be visually checked periodically to be sure its dimensions and sharp leading edge are unaffected by the flow. Degeneration of the sharp edge can result in significant errors in the measurement. Solids may collect behind the plate at the bottom and gases may be trapped at the top, which can be cleared by removal of the plate. Some manufacturers provide special mounting devices to allow the orifice plate to be inserted and removed from the pipe without interrupting the flow, which simplifies maintenance.

The pressure taps should also be examined for possible obstruction. The pressure-differential assembly should be periodically checked.

Advantages and Disadvantages

The following are advantages of orifice plate flowmeters:
- lowest cost differential-pressure meter;
- economically manufactured to very close tolerances;

16 FLOWMETERS

- easy to install and replace; and
- no moving parts.

The following are disadvantages of orifice plate flowmeters:

- high permanent pressure loss through orifice plate can cause significant power costs;
- volume flow is nonlinear;
- rangeability is lower than with linear-output flowmeters; and
- long upstream and downstream straight pipe runs are required.

MAGNETIC FLOWMETERS

Accuracy: ±0.5 percent of rate
Repeatability: ±0.25 percent
Rangeability: 10:1
Size range: 0.1–120 in. (2.5 mm–3 m)
Head loss: None
Relative cost: High

In a magnetic flowmeter (Figure 4-5), a magnetic field is generated around an insulated section of pipe. Water passing through the magnetic field induces a small electric current that is proportional to the velocity of the water flow. The electric current is measured and converted to a numerical indication of water flow. The amount of the voltage produced is approximately 0.05 V/ft/s for the velocity of the fluid, and the output is essentially linear. The liquid moving through the meter must have sufficient ion content to provide a conductivity of 20 μS/cm. Water treatment plant flows meet this conductivity requirement.

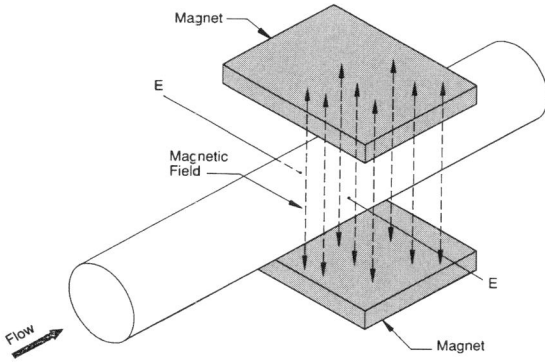

Figure 4-5 Schematic of a magnetic flowmeter.

The meter's accuracy is ±0.5 percent of the actual flow rate; it can be used over a velocity range of 0.2–35 ft/s (0.06–11 m/s). The range of a magnetic flowmeter is 10:1. For clean water, the allowable velocity range is from 1 ft/s (0.3 m/s) minimum to 10 ft/s (3 m/s) maximum.

The head loss across a magnetic flowmeter is the loss that is caused by the length of pipe forming the body of the meter. Head loss would be increased if it were necessary to reduce pipe diameter and incorporate a smaller size meter in order to raise the flow velocity in the meter to an acceptable level.

A magnetic meter has no constriction in its cross-sectional area, so the flow profile is not affected. Hence, whatever profile error may exist in the piping is carried through the metering area. However, area averaging makes the magnetic flowmeter less sensitive than other types of flowmeters to profile changes. The magnetic meter does not have a discharge coefficient definable by its hydraulic shape. Consequently, all magnetic meters should be calibrated, even those of the same line size.

Sizes of magnetic flowmeters range from 0.1 to 120 in. (2.5 mm to 3 m) in diameter. Following are some of the obvious advantages for this type of meter. The magnetic meter has a short laying length, and is much lighter in weight than a Venturi meter. The meter is constructed of stainless steel; therefore, it has a long life. In addition, the solid-state electronics that generate the flow signal should also have a long life.

Installation

A magnetic flowmeter should be installed with the electrodes located at the ends of a horizontal diameter rather than a vertical diameter. This will ensure continued electrode immersion even when air bubbles are present in the water flow.

The following installation steps should be taken in order for the meter to function properly:

- select a meter that has a reasonable velocity at maximum flow (on the order of 5 ft/s [1.5 m/s] and a velocity at the minimum flow not less than 1 ft/s [3 m/s]);
- provide suitable upstream piping length, as recommended by the meter manufacturer;
- ensure proper grounding isolation between the meter body and the pipeline, to avoid transient voltage interference;
- provide 120 V AC power input at about 100 W (models are available for 220 V); and
- use electrical bonding strips to bypass the currents around the meter, if the meter is installed in a pipe that is part of an electrogalvanic-corrosion prevention system.

Maintenance

Manufacturers' troubleshooting and maintenance procedures for magnetic flowmeters are frequently documented on logic flowcharts. These charts include functional statement blocks and "Yes–No" arrows that indicate the next step in the checking sequence. By using the chart, the meter repair person is directed to the problem sources, diagnostics, and recommendations for remedial action. Flowcharts may be provided for meter installation, and primary, converter, and power supply checks. Figure 4-6 (page 18) is an example of a troubleshooting flowchart used to check meter installation.

18 FLOWMETERS

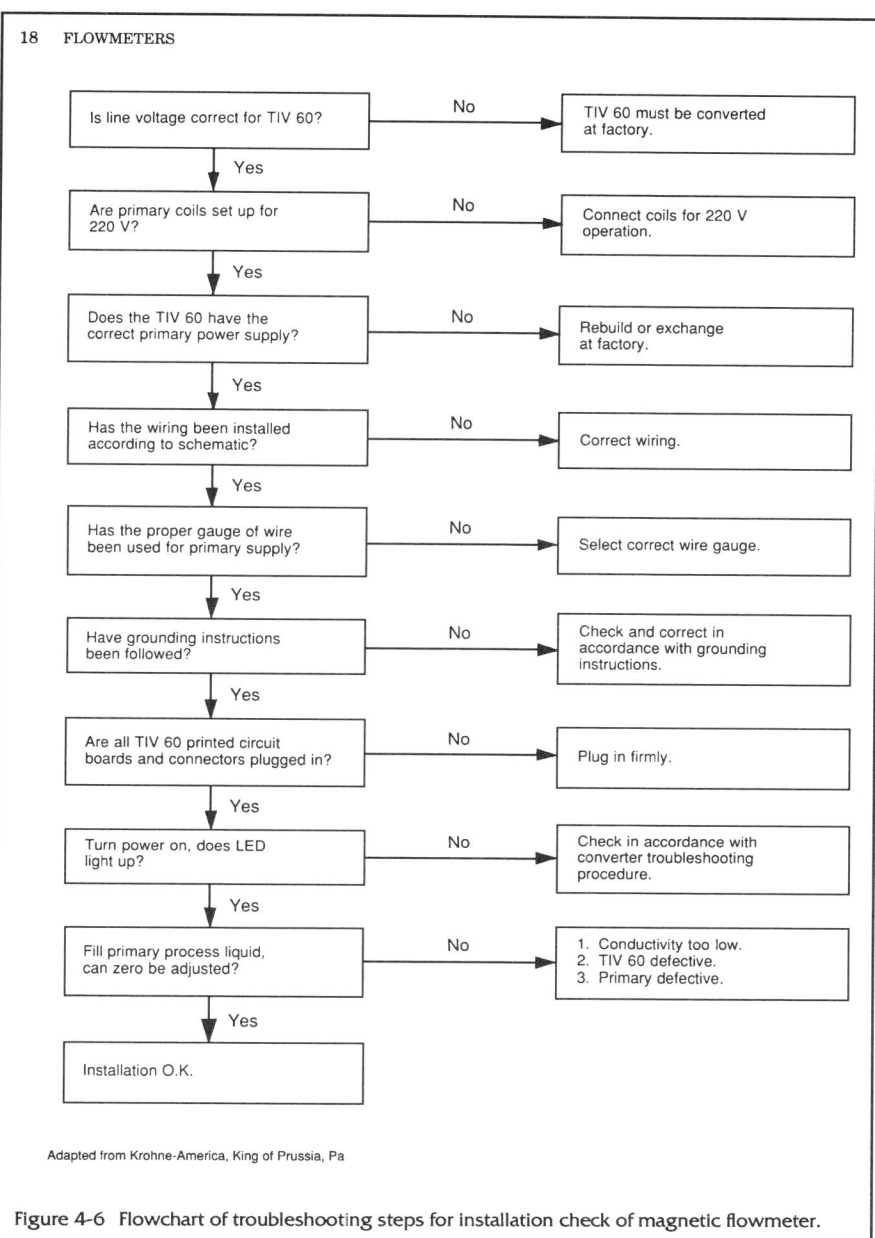

Adapted from Krohne-America, King of Prussia, Pa

Figure 4-6 Flowchart of troubleshooting steps for installation check of magnetic flowmeter.

Some manufacturers provide special equipment for testing, troubleshooting, and recalibrating meters. Instructions and detailed diagrams are provided for disassembly and replacement of components. If a meter is considered to be not repairable in the field and the instructions indicate a warning against tampering with the sealed portion of the unit, no routine maintenance is recommended. Service or installation problems should be referred to the manufacturer's service or field office.

Electrode cleaning. Deposits and incrustations, including calcium carbonate, on the meter's electrodes can impair accuracy, unless the deposits have the same conductivity as the fluid. Such deposits can cause variable or high resistance between the electrodes, thereby introducing errors. Methods for cleaning the electrodes vary among the manufacturers. The following are examples of what may apply to most magnetic flowmeter electrodes:

- electrodes may be removed for inspection, cleaning, and replacement, without removing the meter from the line;
- conical electrodes are available that extend into the flow stream where the scouring effect of the liquid velocity is more likely to inhibit coating;
- continuous ultrasonic vibrations will prevent deposition of particulate and crystalline materials on the electrodes. This is not effective for grease deposits; and
- bypass piping to maintain flow may be included in the installation to allow for periodic inspection and cleaning of the meter's inner wall.

Recent advances in electronic signal conditioning have significantly reduced the effects of electrode fouling. Manufacturers now state that among the newer meters cleaning is not necessary, except under unusual circumstances.

Advantages and Disadvantages

The following are advantages of magnetic flowmetering devices:

- no obstruction to flow (unless meter spool size is reduced from pipe size and thereby causes head loss);
- minimum effective head loss, essentially that of the straight pipe replaced by the meter; magnetic flowmeters are highly suitable for applications where low head loss is essential;
- available over a wide range of sizes from 0.1 to 120 in. (2.5 mm to 3 m) in diameter;
- bi-directional, therefore, suitable for measuring reverse flows;
- linear output;
- variations in fluid density, viscosity, pressure, and temperature have little effect on performance; and
- unless very severe, upstream nonsymmetrical flow patterns and flow disturbances do not seriously affect the flow measurement.

The following are disadvantages of magnetic flowmetering devices:

- metered liquid must have an electrical conductivity of 20 µS/cm or greater (this is not a problem with finished drinking water);
- for smaller pipe sizes, the meters become relatively bulky and expensive;
- high accuracy is expensive, and each meter must be individually calibrated in a water test circuit; and

20 FLOWMETERS

- the meter is sensitive to the geometry and electric properties of the flow tube and magnetic core, and to variations in the coil supply current.

TURBINE AND PROPELLER FLOWMETERS

Accuracy: ±0.5 to 2 percent of rate
Repeatability: ±0.02 percent
Rangeability: 10:1
Size range: $3/16$–24 in. (5–600 mm)
Head loss: High
Relative cost: High

In turbine and propeller flowmeters, flowing water strikes rotor blades, which rotate at a rate proportional to the flow velocity. The rotor (turbine wheel) of a turbine meter generally fills the cross section of the pipe and is mounted to spin freely between two central bearings supported in the pipe wall (Figure 4-7). The rotor (propeller) of a propeller meter (see Figure 4-7) is mounted on bearings at the downstream end of the pipe. The tapered propeller nose faces upstream into the flow and is mounted to spin freely in line with the pipe axis.

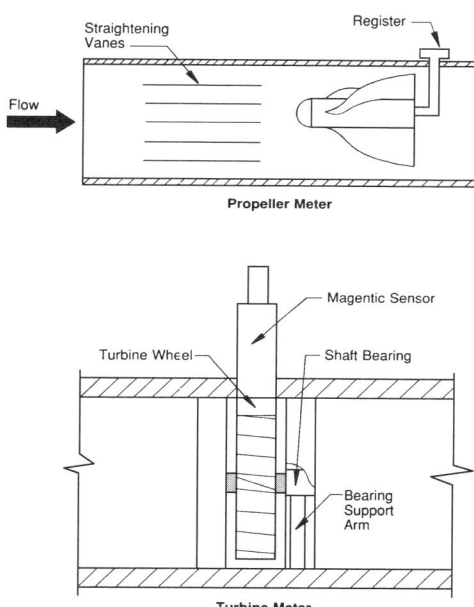

Figure 4-7 Schematic of turbine and propeller flowmeters.

Within given limits of flow rate and fluid viscosity, the rotor speed and volumetric flow rate maintain a linear relationship. A magnetic proximity sensor in the meter transduces the rotor velocity to an equivalent frequency signal. Hence, the rotation of the rotor blades causes a known number of cycles per unit volume. The meter coefficient is the calibration K factor, which is a known number of pulses for a given volume measured. The K factor is typically constant over a 10:1 flow range within a linearity tolerance of ±0.25. For identical meters, the K factor may vary as a result of manufacturing tolerances. Hence, each meter should be calibrated for its own specific K factor value.

According to meter manufacturers, turbine meter accuracy varies from ±0.5 percent to ±2 percent of flow. Among the manufacturers of propeller meters, there is agreement on accuracy of ±2 percent of actual flow. At low flow rates, accuracy is reduced to ±5 percent. The rangeability in larger meter sizes varies with good repeatability at 10:1, and can go up to 25:1.

The orientation and configuration of the meter blade profile are important to the application of the turbine meter. Straight-bladed meters may be less affected by variations in velocity, while helical-bladed meters are generally less affected by variations in viscosity.

Installation

Turbine and propeller flowmeters may be installed with a strainer to prevent solids from interfering with the rotor mechanism. Because these meters are affected by upstream configurations that cause swirls or velocity fluctuations, manufacturers frequently provide or recommend built-in straightening vanes upstream in the pipe. These are installed to minimize the effect of profile irregularities and to smooth flow entering the meter. Straight pipe lengths of at least five pipe diameters upstream and two pipe diameters downstream are recommended.

When selecting a meter for a specific application, care should be taken to ensure that the maximum flow rating will not be exceeded, except for short periods. Running over the maximum speed for extended periods will increase bearing wear and shift the meter's coefficient (K).

Location near a point of chemical injection should be avoided, and the electronic mechanisms that generate the pulses should be protected from electromagnetic influence.

Maintenance

A turbine or propeller flowmeter will demonstrate long life providing periodic mechanical maintenance is performed. Factory maintenance programs are available for meter testing, maintenance, and recalibration. Periodic inspection, calibration, and service should be performed at least once a year.

Propeller meters are generally manufactured for ease of disassembly and extraction of the metering unit from the pipe body. No special tools are required for maintenance, according to many manufacturers. When the metering unit is removed, a cover plate is commonly installed to continue line service. During service and disassembly, the complete metering assembly should be examined for wear and corrosion. Parts should be cleaned, and worn or damaged parts replaced. Troubleshooting procedures are provided by manufacturers, covering problems, causes, and recommended corrective actions. An excerpt from the troubleshooting procedures for a turbine meter is shown in Figure 4-8 (page 22).

22 FLOWMETERS

Troubleshooting—Turbine Meter

INTRODUCTION

Turbine meter system malfunctions are usually restricted to two areas: electrical/electronic, or mechanical.

When a malfunction or an apparent malfunction occurs, the electrical and electronic systems should first be thoroughly checked in accordance with the manufacturer's recommended procedures, prior to checkout of the turbine meter. Only when the source of the malfunction cannot be found in the electrical or electronic systems should the turbine meter be inspected.

SYMPTOM: Fluid Delivery Greater Than Indicated on Totalizer

NOTE: Verify that the proper "K" factor value is entered in the electronic readout device.

1. **Possible Cause**—Foreign material collected on rotor or bearings. When foreign material collects on or in the bearings or if material wraps around the rotor (such as strands of PTFE tape), the angular velocity of the rotor will be reduced. This allows more fluid to pass through the meter than the pulse train indicates.

   Corrective Action—Remove the meter from the line and visually inspect internally. If foreign material is present, remove the material. If no foreign material is found, disassemble the meter in accordance with the instructions of Paragraph 4-3. Clean parts according to Paragraph 4-1. Reassemble and reinstall the meter in the line.

   If no foreign material is found, and cleaning does not eliminate the problem, check bearing wear as performed in the next paragraph.

2. **Possible Cause**—Excessive bearing wear. Excessive bearing wear will lower the angular velocity of the rotor and rotation may stop completely. Effect is the same as with foreign material.

   Corrective Action—Replace bearings or bushing assembly and journal. Recalibrate as required in accordance with Paragraph 4-2.

SYMPTOM: Fluid Delivery Less Than Indicated on Totalizer

NOTE: Verify that the proper "K" factor value is entered in the electronic readout device.

1. **Possible Cause**—Ground loop in electrical circuit.

   Corrective Action—See manufacturer's installation manual.

2. **Possible Cause**—Gasification of liquid in meter. ITT Barton recommends that a back pressure be applied to the system. This back pressure should be twice the magnitude of the net pressure loss through the meter plus twice the magnitude of the vapor pressure of the liquid being metered. If this back pressure is not maintained, gasification within the meter can occur, causing an over-spin of the rotor. This will indicate a greater than actual delivery and cause damage to the bearings.

   Corrective Action—Provide a back pressure at the meter by using accepted design means and practices.

3. **Possible Cause**—Entrained gas or bubbles in metered liquid. If the fluid has a significant amount of entrained gas or gas bubbles that are released in the reservoir, over-registration can occur. Turbine meters measure the actual volume of the liquid. If a unit volume is expanded by entrained gas or bubbles, this expansion will be registered. Gas bubbles may cavitate the rotor and produce effects similar to those caused by gasification of liquid in the meter (see previous paragraph).

   Corrective Action—Eliminate entrained gases or bubbles by use of an air eliminator or other acceptable method.

Adapted from *Turbine Meter Troubleshooting Procedures* (ITT Barton, City of Industry, Calif.)

Figure 4-8 Excerpt from turbine meter troubleshooting procedures.

Advantages and Disadvantages

The following are advantages of turbine and propeller flowmeters:

- the very large sizes have excellent short-term repeatability, and can be extremely accurate when calibrated periodically;
- output flow signal is directly proportional to a pulse train with high linearity over a broad range of flow rates, at least 10:1 for large meters;
- head loss is moderate, decreasing with larger sizes;
- meter size is the same as the diameter of pipe in which it is installed; and
- flow is not blocked if the meter seizes up.

The following are disadvantages of turbine and propeller flowmeters:

- systematic mechanical maintenance and lubrication are required;
- expensive, particularly for use in large-diameter pipe;
- wear or fouling of meter surfaces will gradually deteriorate the calibration, making periodic recalibration necessary to maintain high accuracy. Corrosive liquids, liquids of poor lubricating quality, and liquids with a high proportion of suspended solids will cause bearing problems;
- calibration factor is sensitive to changes in viscosity of the flowing liquid;
- sensitive to flow disturbances and swirls; and
- bearing friction is detrimental to performance of smaller meters.

SONIC FLOWMETERS

Accuracy: ±1 to ±2.5 percent of rate
Repeatability: ±0.25 percent
Rangeability: 20:1
Size range: 0.125–120 in. (3.1 mm–3 m)
Head loss: None
Relative cost: High

In a sonic (or ultrasonic) flowmeter, a pair of transceivers (transmitter–receiver) are positioned diagonally across the meter body, as shown in Figure 4-9 (page 24). The transceivers transmit and receive an ultrasonic pulse in the direction of flow, followed by a return pulse against the direction of flow. In a flowing liquid, the speed of the pulse directed downstream is increased by the speed of the stream; when directed upstream, the speed of the pulse is slowed by the streamflow. The time difference between the two is a function of fluid velocity and, by computation, the rate of flow. The transit time difference and the allowance for pipe diameter are converted by a microprocessor circuit in the meter to a standard output signal for volume flow. This sonic meter, used to measure the flow of clean water, is known as a "time-of-flight," "transit-time," or "through-transmission" meter. The sonic meter accuracy is ±1 percent of rate over a 10:1 range. Accuracy must be checked in the manufacturer's literature, since some manufacturers cite accuracy in terms of percent full scale.

Because sonic flowmeters use microprocessor circuits to generate and record flow, special digital features can be incorporated into the unit. Light-emitting diodes (LEDs) can indicate circuit operation or alarms, and provide readouts. The user can select output units, ranges, totalizing, and the type of signal transmission.

24 FLOWMETERS

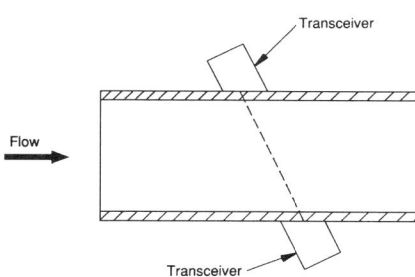

Figure 4-9 Schematic of an ultrasonic time-of-flight flowmeter.

The head loss of the meter is no more than that of an equivalent length of pipe. In cases of low flow rates, a less-than-line-size meter may be used, which will introduce some additional head loss.

Since there is no constriction, as in a Venturi or flow tube, there may be no flow conditioning. Therefore, any profile error caused by the piping configuration is carried through the meter and picked up by its velocity-sensing system. However, flow straighteners may alleviate the problem of turbulence. Some sonic meters use a single-path sensor system that requires more careful layout of upstream piping. To minimize profile sensitivity, some meters have multiple-path sensing to average out the fluid profile error, and thus increase metering accuracy.

Since the meter does not have a discharge coefficient that is defined by its hydraulic shape, all units, even of the same line size, must be calibrated.

The meter body is constructed of materials that have a long life. Because sonic flowmeters have been on the market for a relatively short time, the amount of performance data associated with field tests is not extensive. Consequently, the life cycles of the transceivers and related electronics over long periods of time have yet to be proven by experience.

One type of ultrasonic flowmeter is the Doppler, which requires only one transceiver to send pulses into the flow. This meter depends on particles of solids or entrained bubbles in the fluid to reflect energy pulses back to the transceiver. Consequently, this type of meter is not suitable for potable water.

Installation

A sonic meter requires approximately 10 pipe diameters of straight run upstream and 5 pipe diameters downstream for proper performance. This may differ with varying piping configurations and manufacturers. The meter should not be located near a point where there is a sudden pressure drop that might release minute quantities of gas from the liquid. The meter cannot function accurately when gas bubbles are present. Piping should be isolated from noise and vibration that might interfere with the sound propagation of the meter. The meter requires a power source and may use from 8 to 50 W of power.

The meter is available in two forms: a spool piece with integral transceivers or a transceiver assembly for clamp-on mounting outside an existing pipe.

Maintenance

Maintenance of sonic flowmeters is minimal. Sonic flowmeters should have a long operational life because of obstructionless flow, solid-state circuitry, and no moving parts.

Built-in electronic self-diagnostics generate circuit tests and display functions, and plug-in modular construction permits rapid replacement of defective parts.

Maintenance procedures are executed through the microprocessor, checking and adjusting the 4 and 20 mA output levels, and the level of the transmitted and received signal. A low signal level may indicate incorrect transducer installation, obscured sonic path (bubbles, solids), sedimentation, or cabling fault.

Advantages and Disadvantages

The following are advantages of sonic flowmeters:

- no obstruction to flow, thus, no head loss;
- not restricted to use with conductive liquids (as are magnetic flowmeters);
- clamp-mounted meters do not jeopardize pipe wall structure;
- clamp-mounted meters do not interrupt process flow during maintenance or replacement;
- no mechanical moving parts;
- linear output over a wide range (100:1);
- adaptable to a wide range of pipe diameters;
- low installation and operating costs; and
- bi-directional flow is allowable.

The following are disadvantages of sonic flowmeters:

- sensitive to change in fluid composition;
- high solids content or entrained bubbles distort and block propagation of sound waves;
- measures mean velocity across a diameter, which is not the same as the weighted mean velocity;
- sensitive to flow-velocity profile; accuracy can be impaired by changes in pipe wall roughness and by changes from laminar to turbulent flow;
- accuracy impaired by upstream and downstream flow disturbances, such as elbows and valves, which affect the velocity profile;
- positioning of the opposing transceivers is critical to ensure signal interception; and
- in clamp-mounted use, the presence of sound absorptive or scattering scale or coating on the inner walls of the pipe may prevent the meter from working (This is not true when transceivers are mounted through the wall on a spool piece.).

VORTEX FLOWMETERS

Accuracy: ±0.75 percent of rate
Repeatability: ±0.15 percent

26 FLOWMETERS

Rangeability: 40:1
Size range: 1–10 in. (25–250 mm)
Head loss: Medium
Relative cost: High

In a vortex flowmeter, a nonstreamlined or bluff body, the vortex shedding element obstructs and splits the flow through the pipe forcing two streams around the barrier, and creating vortices downstream in the flow. These vortices are caused by the swirling of the fluid into the low-pressure area behind the body (see Figure 4-10). The vortices alternately rotate in opposite directions, with the spacing between them proportional to the fluid flow velocity. This also creates an oscillating pressure variation from side to side of the immersed vortex device.

Numerous methods are available for measuring the vortex-train frequency or the frequency of the pressure oscillations. In all cases, external electronics convert the frequency signal into a standard analog value that is proportional to the flow velocity or a pulse train suitable for input to a totalizer.

Accuracy of vortex flowmeters is ±0.75 percent of rate, with a repeatability of better than ±0.15 percent. Flow rate is approximately 15 ft/s (5 m/s), with a range of 40:1. The flow range is a function of pipe size, being dependent on the Reynolds number and on cavitation in the pipe.

Head loss is somewhat higher than that of an unobstructed pipe, because of the presence of the vortex element. The added loss is equivalent to about 4 psi (28 kPa) at maximum flow.

The frequency/flow characteristic in the pipe is a function only of the shape of the body. Consequently, a generic coefficient can be used for all meters having the same body profile, regardless of pipe size.

Pipe sizes used with vortex flowmeters range from 1 to 10 in. (25 to 250 mm), and larger sizes can be used. The maximum size of pipe that can be used depends on the pulse frequency per unit volume. This limit exists because, in larger pipes, the full-scale frequency may be too low to enable the signal-conditioning electronics to make acceptably accurate measurements.

The vortex meter has no moving parts, and has only the spool piece and body exposed to the fluid. It is typically constructed of stainless steel, resulting in relatively low maintenance requirements.

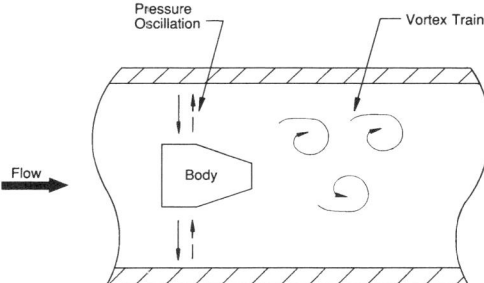

Figure 4-10 Schematic of a vortex flowmeter.

Installation

The vortex meter requires a fully developed flow profile, which means that upstream disturbances must be minimized. An upstream straight pipe length of at least 10 diameters is desirable, and flow straighteners may be needed where severe disturbances are present. A length of five diameters of straight pipe downstream should also be used, to minimize the effects of disturbances on the vortex train. Severe vibration of the pipe caused by "noisy" pumps and valves and the continued presence of bubble streams can affect meter accuracy by introducing false signals into the sensing elements. These problems must be considered before installation of the meter.

Maintenance

The meter is not expected to fail if properly installed and operated under normal conditions. If failure occurs, flowcharts of troubleshooting instructions and related diagrams are provided by manufacturers to isolate and remedy the problem. An example of a flowchart for troubleshooting is shown in Figure 4-11 (page 28).

Parts are usually removable, according to detailed procedures found in the instruction manuals, and replaceable in the field. These include the vortex shedding element, the sensor assembly, and output signal electronics. The electronics can be replaced without interrupting the pipe flow.

Because it has no moving parts, there is virtually no maintenance associated with the installed fixed assembly of meter spool and vortex shedder. This assumes that installation is meticulously executed according to manufacturer instructions regarding orientation, alignment, piping connections, ambient conditions, power connections, wiring, and flow range applications.

Detailed instructions for maintenance usually refer to electronic adjustments, such as zero, span, noise balance, and minimum measurable velocity. However, all such adjustments, while explained in the meter manual, are set in the factory and should not be altered after proper installation.

Advantages and Disadvantages

The following are advantages of vortex flowmeters:
- low head loss;
- no moving parts;
- long-lived construction;
- high turndown ratio (the ratio of the maximum design flow rate to the minimum design flow rate); and
- simplicity in design and installation.

The following are disadvantages of vortex flowmeters:
- sensitive to flow-profile distortion;
- affected by pipe vibration and bubble streams;
- not bi-directional; and
- limited range of pipe size.

AVERAGING PITOT FLOWMETERS

Accuracy: $\pm 1/2$ to ± 5 percent of full scale
Repeatability: 0.5 percent

APPENDIX C TYPES OF FLOWMETERS **593**

28 FLOWMETERS

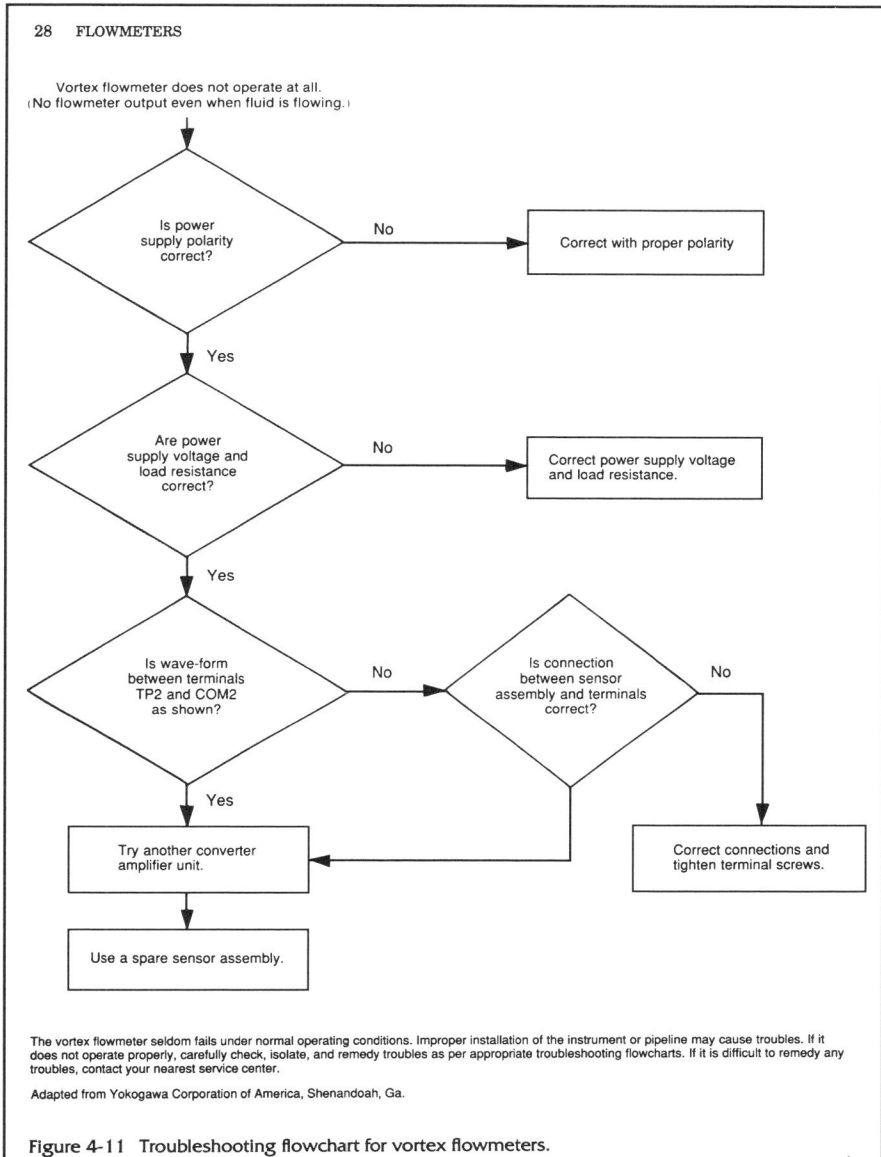

The vortex flowmeter seldom fails under normal operating conditions. Improper installation of the instrument or pipeline may cause troubles. If it does not operate properly, carefully check, isolate, and remedy troubles as per appropriate troubleshooting flowcharts. If it is difficult to remedy any troubles, contact your nearest service center.

Adapted from Yokogawa Corporation of America, Shenandoah, Ga.

Figure 4-11 Troubleshooting flowchart for vortex flowmeters.

Rangeability: 4:1
Size range: ½–96 in. (13 mm–2.4 m)
Head loss: Low
Relative cost: Low

The averaging Pitot flowmeter consists of an insertion tube, or probe, that is placed through the pipe cross section (Figure 4-12). Multiple ports face upstream into the flow to provide sampled pressures at selected points along the pipe diameter. The multiple pressures are sensed as an interpolated average by the internal tube to provide an averaged pressure over the pipe cross section. A port or ports facing downstream register static pressure. The device produces a differential pressure between the averaged velocity head ports and the static head port(s). As with the Venturi flowmeter, the fluid velocity is proportional to the square root of the differential between the resultant upstream pressure and the static pressure.

Accuracy for a Pitot meter is $\pm\frac{1}{2}$ percent to ± 5 percent of full scale. The meter's range is greatly limited by the sensitivity of the differential-pressure sensor and the results obtainable for low differential. The meter, by nature, is velocity-profile sensitive. Placing it in different planes at the same point in a pipe may result in different readings. Laboratory calibration is not valid for field conditions unless the pipe configuration is duplicated.

According to manufacturers, the shape of the insertion tube generally will cause most foreign material to flow around the probe rather than accumulate on it. Ordinarily, material that does affect the probe does not significantly affect meter performance unless, in extreme cases, the ports are completely obstructed or buildup changes the outside shape.

Installation

The insertion tube must be installed into the pipe according to the manufacturer's instructions. Deviation from perpendicular to the axis of the pipe in any direction will

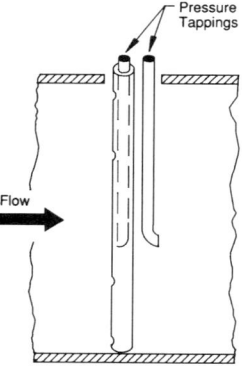
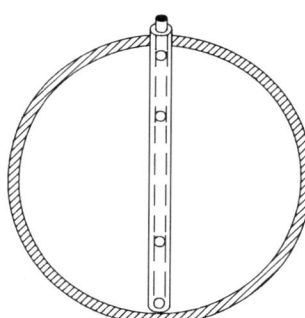

Figure 4-12 Schematic of an averaging Pitot flowmeter insertion tube.

30 FLOWMETERS

affect the sensed pressures. Misalignment of the tube from the diameter beyond 3° in the pipe cross-sectional plane or 5° upstream or downstream out of the crosssectional plane will cause significant error. Rotation of the tube beyond 3° from strict upstream/downstream will also cause errors in the sensed pressure.

It is also important to install the correct size insertion tube in the correct size pipe. The location of the ports on the insertion tube is based on the pipe's inside diameter and wall thickness. Installed in the proper size line with the proper fittings, the tube's sensing ports will be at their correct locations, and the meter will respond with the designed accuracy. If the tube is inserted in an incorrect line size, it can be expected to provide a repeatable signal, but must be recalibrated to yield accurate flow measurements.

The location of tube ports is based on a fully developed turbulent flow, with a velocity profile that is consistent across the pipe in all directions. The averaging of pressures will not be correct in an inconsistent flow profile, and errors in flow measurement will result. Upstream devices, such as valves, elbows, and pipe diameter changes, influence the flow profile. Therefore, as with other meters, sufficient lengths of pipe must be provided upstream of the insertion tube to allow the turbulent flow profile to develop. Flow straighteners may be used to reduce the necessary length of straight pipe upstream. Tables of recommended upstream and downstream straight pipe lengths are provided by manufacturers. Upstream lengths vary from 7 to 24 pipe diameters without straighteners, and 3 to 9 pipe diameters with straighteners. Downstream straight pipe lengths vary from three to four pipe diameters.

Maintenance

The probe should be removed and cleaned periodically. Probe removal does not require shutting down the system. Sensing ports and internal passages can be cleaned using external pressure without being removed. Manufacturer's recommendations should, of course, be followed in cleaning the probe. The sensor design will handle most flow conditions without clogging. Nevertheless, if the fluid is contaminated, periodic purging of the internal passages may be necessary. Designs can be provided to facilitate purging.

Advantages and Disadvantages

The following are advantages of averaging Pitot flowmeters:
- removable without shutting down system;
- nonconstricting design produces low head losses;
- meter cost is low;
- easily installed at any time by making a wet tap in the pipeline; and
- materials of construction and the nature of the parts provide for long life.

The following are disadvantages of averaging Pitot flowmeters:
- any leaks in the instrument lines or connections will significantly affect the meter accuracy, because the flow measurement depends on the comparison of two pressures generated by the flow past the device, and the pressure change may be small; and
- potential for calibration error exists due to misalignment of the insertion tube.

APPENDIX

D

Demand Profiling for Optimal Meter Sizing*

D.1 INTRODUCTION

Precise, customer-specific demand profiles are used to generate valuable types of water use data. A demand profile consists of rate-of-flow data describing water use versus time. Such data are typically gathered directly from a utility customer's existing meter installation using specialized flow recorders that attach to meters and log water usage per unit of time. See Figs. D.1 and D.2.

Figure D.1 Data being captured from an existing meter. (*Source: F. S. Brainard and Company.*)

*This appendix was provided by permission of Brad Brainard of F. S. Brainard and Company and will eventually form part of the new AWWA M22.

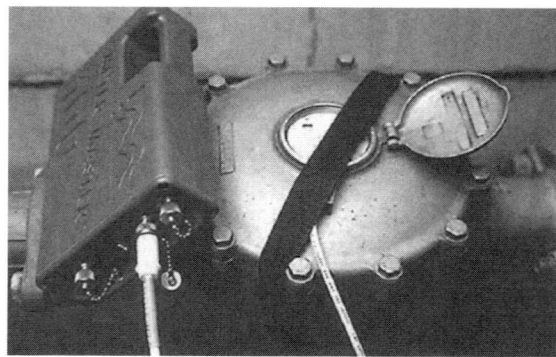

Figure D.2 Data being captured from an existing meter. (*Source: F. S. Brainard and Company.*)

Demand profiles generated from existing meters provide data essential for making a variety of critical decisions. Data logged from water meters is more accurate than other measures because a water meter represents the most precise way to measure actual water use. Flow recorders accomplish their mission without interrupting the accurate registration of the water meter and, typically, without altering the existing meter configuration. In a small number of cases, adapters are required but are easily installed.

Applications for customer demand profiles may be grouped into three general categories: (1) meter sizing and maintenance, (2) water use audits, and (3) cost of service studies. While only the first application is discussed in detail here, it is worth remembering that the same data gathered for meter sizing purposes has other important applications and can benefit a variety of utility divisions, including distribution, metering, conservation, customer service, engineering, and finance. In the case of water use audits, demand profiles assist with conservation programs, leak detection, customer service, and hydraulic modeling. In the case of cost-of-service studies, demand profiles are used to obtain data regarding the variability of use by residential, commercial, industrial, and wholesale customer class groups. Because the same data can be used in support of all these applications, it is important when collecting the data to consider all of the potential applications for which the data may be of value presently or in the future. For example, if a cost-of-service study or hydraulic model requires only hourly demand data, you may still choose to store the data in 10–60-s increments so that the same data can be used for meter sizing and maintenance programs.

In general, a demand profile should accurately provide peak flow data and the percentage and volume of water used in critical flow ranges. Critical flow ranges include, as a minimum, flow below the specified accuracy range of a meter, flow at the crossover range in a compound meter setting, and high flow. The objective is to size the meter properly for maximum accountability and revenue recovery without adversely effecting pressure levels or fire flow requirements. It is also important to consider meter maintenance costs. It may be that a 6-in turbine meter could better serve a customer with constant flows of 600 gpm than a 4-in turbine meter because, while both would accurately measure the flows, the 6-in turbine would experience less degradation from wear and tear. The most obvious direct benefit of proper meter sizing is the accurate measurement of water use; the more closely a meter is matched to a customer's usage pattern, the more water will be accounted for and billed. What is often not quite so obvious is the potential size of revenue gains associated with proper meter sizing.

Tim Edgar, in *The Large Water Meter Handbook,* illustrates this potential revenue gain by the case of a 100-unit apartment building with a 4-in turbine meter. The actual monthly consumption was 500,000 gal, but much of that volume was at low flow rates. Because the turbine meter was not accurate at flow rates less than 12 gpm, 15 percent of the volume went unrecorded and unbilled in both water and, as is very often the case, sewer charges. The result was a revenue loss of $1700.00 per year (at $3/1000 gal for combined water and sewer). As Edgar points out, if a utility has 100 such incorrectly sized meters, those 100 meters would cost a utility over $1 million in lost revenue over six years.[1]

As an example, the Boston Water and Sewer Commission began a downsizing program in 1990. John Sullivan, Boston's Director of Engineering, reported in presentations to the American Water Works Association that, between August 1990 and April 1992, the city had accounted for an additional 113,784 ft^3 of water per day (0.8 MGD). With just the meters downsized in the first year of the program, Boston anticipated the total increase in revenue over five years from combined water and sewer billings to be $6.8 million (1991 dollars). These savings would only be realized in systems with many oversized turbine meters.

While the most direct benefit of proper meter sizing is increased revenue and accountability, meters offer a distribution system much more valuable than just revenue enhancement. Any decision made by a utility related to water usage can only be as good as the consumption data collected from meters. In general, demand profiles provide valuable data to improve distribution system design, performance, and management. In addition to finding ways to increase accounted-for water levels and revenue, demand profiles help to identify service size requirements, clarify meter maintenance requirements, define water use characteristics for conservation programs, enhance customer satisfaction and awareness, improve hydraulic models, and establish equitable and justifiable rate structures. Additionally, with increased water scarcity and cost, conservation has become an important industry issue. For many utilities, conservation has become the most cost-effective means to improve water resource availability. All of these distribution system design, performance, and management objectives are dependent on the capability of a system's meters to account for usage as accurately as possible, which can only occur as a consequence of sizing meters properly for each and every application.

D.2. RECORDER DESIGN

D.2.1 Theory of Operation

Demand profiles are generated with electronic flow recorders. The portable flow recorders discussed here are also referred to as demand

Figure D.3 Sensor attached to meter with Velcro strap. (*Source: F. S. Brainard and Company.*)

profilers, demand recorders, and data loggers. The devices pick up data from either the meter's internal drive magnets or the meter's pointer movement and store the data for later downloading into a desktop or handheld computer for analysis. These recorders can be moved from one meter site to the next with minimum effort and operate with standard meters, thereby eliminating the need for special registers. Typically, the magnetic or optical sensor is either strapped to the outside of a meter using Velcro or heavy-duty tape or is integral to an adapter located between the meter body and the existing register. See Fig. D.3.

Because of potential adverse operating conditions (meter pits, temperature extremes, rough handling, public access), recorders should be submersible, durable, and securable. In order to provide extended data storage capability in remote locations, recorders should also offer substantial battery life. This section describes current technology for demand profiling. As new technologies evolve in this field, they should be evaluated in order to promote this area of knowledge and capability.

D.2.2 Recording Methods

Flow recorders using magnetic pickups sense the magnetic field generated by the magnetic coupling of a water meter's internal drive magnets and convert the magnetic flux change into a digital pulse that is logged into memory and later downloaded into a PC for analysis. Optical pickup devices sense the meter pointer passing beneath the sensor and also store the signal as digital pulses to be later downloaded. Each pulse is associated with a known volume of water. The principal advantage of a magnetic pickup is the higher resolution of data made possible by the rotation speed of a meter's magnets. In almost all cases, the drive magnets inside a meter rotate much faster than the sweep hand (pointer) on the register's dial face. In small meters, the number of magnet rotations per unit of time can be as high as approximately 30 per second at 20 gpm. At this rate, the magnets are rotating 900 times as fast as the sweep hand. In the case of turbine meters, the rotation speed of the magnets can vary greatly, from approximately 800 times the speed of the sweep hand to the same speed as the sweep hand. Available adapters can substantially increase the resolution of the data on many of the slower-magnet-speed meters by isolating an additional magnet with a higher rotation speed. Optical and mechanical adapters are available to enable compatibility with the older gear-driven meters, which preceded magnetic-drive meters.

D.2.3 Installing Magnetic Sensors

Because most meters have the magnetic coupling directly under the register, it is typically easy to pick up a reliable signal by placing the sensor on the side of the register. Almost without exception, the magnetic coupling is directly under the register in the case of all 2-in and smaller positive displacement and multijet meters. If the magnetic coupling is not directly under the register, it is typically in the center of the turbine rotor in the middle of the flow. In this case, the magnetic sensor must be placed on the side of the meter body in order to be as close to the drive magnets as possible. As discussed above, adapters are required for some meters, such as gear-driven meters. See Fig. D.4.

If the magnetic coupling is under the register but the register has shielding on the sides, the sensor may have to be located directly on top of the register in order to circumvent the shield. Because the recorder's magnetic sensor is essentially picking up the electromagnetic noise generated by a water meter, the sensor can be susceptible to picking up noise generated by other sources of electromagnetic noise such as motors, generators, and alarm systems. The recorder's sensing circuitry should be designed to consistently pick up the magnetic signal generated by a water meter's drive magnets, while minimizing the potential for picking up electromagnetic noise from other sources.

Figure D.4 Sensor picks up pulses from register magnets. (*Source: F. S. Brainard and Company.*)

D.2.4 The Recorder's Data Storage Capacity

It is essential that a recorder have adequate data storage capacity in order to enable the recorder to store a substantial amount of data. As discussed in greater detail in Sec. D.3.3, flow data must be logged into memory in small time increments if accurate maximum and minimum flow rate data is to be ensured. The potential factor of difference in the observed maximum flow rate between a 10-s and a 60-s data storage interval monitoring the exact same flow is 6:1. The potential factor of difference in the observed maximum flow rate between a 10-s and a 300-s (5-min) data storage interval monitoring the exact same flow is 30:1. In other words, if a solitary flow usage of 200 gal occurred for just 10 s at a rate of 1200 gpm, whereas the 10-s data storage interval could detect this high flow rate of 1200 gpm, the 300-s data storage interval would observe a maximum flow rate of just 40 gpm because the 200 gal would be averaged over 5 min rather than averaged over 10 s.

Obviously, this difference could have serious ramifications for a meter size selection. Frequently, users choose to store data for one week when assessing the size of a commercial/industrial user's meter, in order to ensure that a representative sample of flow data is gathered. If a user is to store 10-s data for one week, the recorder must be able to continuously store a minimum of 60,480 intervals of data. For other applications, such as cost-of-service studies and hydraulic modeling, a smaller data storage capacity is required than for meter sizing; however, if the data are to be used most efficiently, the storage capacity should provide for high-resolution data so that the data may be used effectively for the various applications.

D.3 RECORDING DATA

D.3.1 Length of Record

As discussed above, many recorder users choose to store data from commercial/industrial sites for one week because certain high-rate water uses (e.g., a cleaning operation at a factory) may occur on only a particular day each week. It is important to discuss water usage with a customer prior to storing data, if possible, to ensure that the duration of the recording period is sufficient to get a representative sample of flow data. In the case of multitenant residential or hotels/motels, 24 hours of data may be sufficient as long as the data are collected during hot weather in the case of residential and high occupancy in the case of hotels/motels. Essentially, it is best to make some effort to understand a user's water use characteristics in order to select the optimum length of the data storage period. Experience with different types of users over time will also provide an indication of the optimum record length for different classes of users. The record length is critical and should be determined on a case-by-case basis.

D.3.2 Customer's Water Use Habits

Data should be recorded during a period in which the user experiences typical peak, average, and minimum flow rates and for duration sufficient to capture those rates. For example, it would not be appropriate to record data at a school or factory during a vacation period. Similarly, as mentioned above, you would want to record data for at least a week at an industrial site if there were evidence that the customer performed different operations on different days of the week. Seasonal cycles are as important to consider as weekly ones. Weather at different times of the year may substantially alter demand patterns. If a user uses a lot more water on a hot summer day, it is important to record data on such a day in order to capture peak flow data.

The personnel performing an analysis should anticipate potential changes in demand patterns. At a residential development, it would be important to consider the number of additional units currently under construction. It is also important to resurvey a user if the type of use changes. Commercial lease space can have a high rate of turnover. A warehousing or distribution company with substantially lower water usage could replace a bottling company. If the meter is not resized, the new user will be the beneficiary of a lot of free water.

D.3.3 The Recorder's Data Storage Interval

The data storage interval is the period of time over which a flow recorder counts pulses before that interval's pulse count is logged into memory. The interval determines the resolution of the raw data file from which all subsequent graphs and reports are generated: the shorter the interval, the greater the detail possible in subsequent graphs and reports. For example, a data storage interval of 10 s allows accurate data analysis for periods of 10 s or longer. The user selects the data storage interval before the recorder goes into the field. As long as the graph/report generating software allows for adjustment of the time interval over which maximum and minimum flow rates are calculated (see Sec. D.4.2), the data storage interval should be kept short, e.g., 10 s.

Keeping the data storage interval short is particularly important in order to provide sufficient data resolution to determine maximum flow rates accurately. In order to ensure the accurate identification of a maximum flow rate, the data storage interval cannot exceed 50 percent of the duration of a maximum flow event. For example, if an industrial customer has a particular operation which occurs just once each 30 min, lasts 30 s, and uses 500 gal of water (i.e., a demand of 1000 gpm), identification of the 1000-gpm flow rate can only be assured if data are logged into memory at least once each 15 s. If the data storage interval is between 15 and 30 s, there is an increasing likelihood that the maximum flow rate will be understated due to the possibility that no data storage interval begins and ends within the 30-s event. If the data storage interval is more than 30 s, the likelihood becomes a certainty. In this particular example, a data storage interval of 15 s or less would show the 1000-gpm flow rate. On the other hand, if the data storage interval is 15 min (900 s), the maximum flow rate would appear as only 33 gpm, because all that is known is that a total of 500 gal was used during a 15-min period, and 500 gal divided by 15 min is 33 gpm. If the data storage interval were 5 min, a maximum flow rate of 100 gpm would be indicated. A lower maximum flow rate would be indicated if the 500-gal usage was divided between two 5-min data storage intervals. As can be seen, a serious meter sizing error can easily be made if the recorded data are not stored at a level of resolution sufficient to capture the actual maximum flow rate.

As another example, let's say that a small manufacturing company has an operation which periodically uses 250 gal of water for 10 s (which equates to a rate of 1500 gpm) in addition to its other uses. This scenario is simulated graphically in Figs. D.5 to D.7. In each case, the same data was used to create each graph; the only difference is the data storage interval, which, in this case, is also the interval used for maximum and minimum flow rate calculations. In the case shown in Fig. D.5, a data storage interval of 10 s is used and a true maximum flow rate of 1520 gpm is identified. In the case shown in Fig. D.6, the data storage interval is 60 s and the calculated maximum flow rate is reduced to 280 gpm. In the case shown in Fig. D.7, the data storage interval is 300 s (5 min) and the true maximum flow rate disappears into the rest of the data.

Although the above examples exaggerate normal circumstances, they are intended to illustrate the potential for meter sizing errors if one ignores the importance of data resolution.

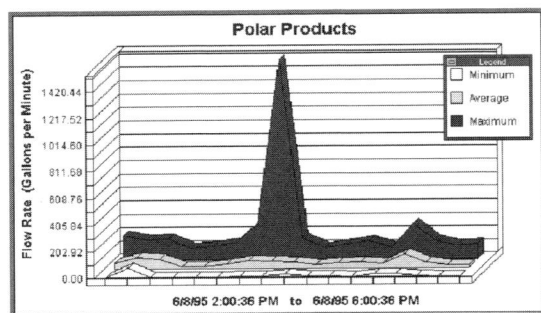

Figure D.5 Data storage interval of 10 s (Meter-Master Model 100 program). (*Source: F. S. Brainard and Company.*)

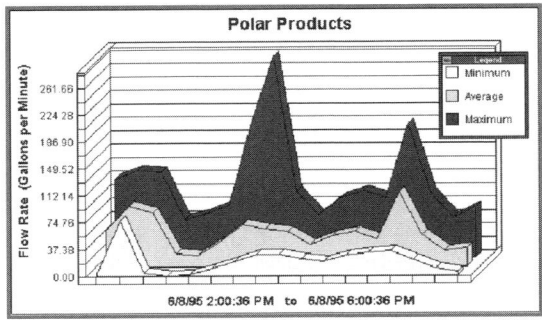

Figure D.6 Data storage interval of 60 s (Meter-Master Model 100 program). (*Source: F. S. Brainard and Company.*)

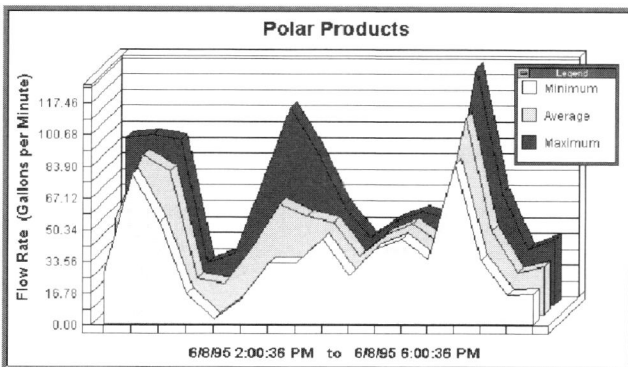

Figure D.7 Data storage interval of 300 s (5 min) (Meter-Master Model 100 program). With this interval, the true maximum flow rate disappears into the rest of the data. (*Source: F. S. Brainard and Company.*)

It should be noted that there are disadvantages to making the data storage interval too small. This interval defines the size of the downloaded data file and the length of time you can record before running out of memory. The same test recorded with a 5-s interval will take up to six times more memory than one stored with a 30-s interval. Furthermore, larger files take longer to download and to generate graphs and reports. Generally, a 10-s interval provides adequate detail and recording time for most applications. If you are making a long recording and a 10-s interval would use up all of the logger's memory before the recording is completed, lengthen the data storage interval. Another problem with too short an interval is discussed in Sec. D.3.4 and in Sec. D.4.2. Briefly, if too short an interval is used on a meter with slow-moving drive magnets (or sweep hand, in the case of optical sensors), skewing (exaggeration) of maximum and minimum flow rates can occur because there is too little data for accurate calculations. A recorder's operating instructions should identify such meters so that care is taken when selecting intervals for data presentation. Software design can improve the integrity of downloaded data by intelligently interpreting pulse data in order to minimize the potential for exaggerated maximum and minimum flow rates.

D.3.4 The Meter's Pulse Resolution

The meter's pulse resolution is defined as the number of pulses generated that equate to a unit of liquid measure. For magnetic pickups, the resolution is the number of meter magnet poles (as the magnets rotate) which equate to a unit of liquid measure. It is desirable that the internal magnets revolve as fast as possible without degrading the reliability of the meter; accordingly, the higher the number of magnet poles per unit of measure, the better. Faster magnets generate more pulses, which translates into greater data accuracy. For optical pickups, the same considerations apply to the speed of sweep-hand rotation. Therefore, it is important to have some knowledge concerning the speed at which a meter generates pulses. A flow recorder's operating instructions should provide guidance in this area.

The pulse resolution (or factor) is especially important when determining maximum and minimum flow rates. The issues are very similar to those discussed in Sec. D.3.3. Concerning maximum flow rates, if a magnet (or sweep hand) is rotating slowly, it is possible that a large, short-term usage could take place without any evidence of its occurrence. For example, if a 6-in turbine meter (meter "a") generates just one magnetic pulse for each 500 gal while another 6-in turbine (meter "b") generates one pulse for each 2 gal, the 250-gal usage at 1500 gpm described in the preceding section might not even be identified at all by a recorder attached to meter "a," while meter "b," with fast-moving magnets, would have provided 125 pulses to the recorder. Furthermore, if the recorder attached to the meter with the slow-moving magnets did detect one pulse within a 10-s interval, it might be erroneously assumed that 500 gal were used during that 10-s interval, which would equate to a flow rate of 3000 gpm because, if one pulse is logged in 10 s, this is the equivalent of 6 pulses per minute, and 6 pulses per minute multiplied by 500 gal per pulse equals 3000 gal per minute. Accordingly, a meter with fast-moving magnets can provide continuously accurate data throughout the flow ranges, whereas a meter with slow-moving magnets cannot. Likewise, using optical sensors, the faster the rotation of the sweep hand, the more accurate the resultant data will be. However, an optical sensor would have to detect numerous pulses per revolution of a sweep hand in order to approach the substantial level of accuracy achievable by a magnetic sensor.

Minimum flow rates identify leakage rates and affect the selection of turbine versus compound meters in larger applications. In order to ensure the accurate identification of minimum flow rates, as with maximum flow rates, a user must know which meters have slow-moving drive magnets. For example, if a meter's magnets are providing just one pulse for each 20 gal, and the current flow rate is a steady rate of just 5 gpm, only 1 pulse will be generated each 4 min. If one observes the data in time increments smaller than once each 4 min, the flow rate will appear to vary between zero and some amount greater than the actual flow rate of 5 gpm. As an illustration, if a 1-min time interval is used for observing the data, the flow rate will appear to equal zero for 3 of each 4 min and 20 gpm for 1 of each 4 min because each pulse, equaling 20 gal, will appear just once each 4 min when a steady flow rate of 5 gpm is occurring. If a 4-min time interval is used to observe the data, it will appear as if a steady flow rate of 5 gpm is occurring.

Figures D.8 and D.9 represent the scenario just described. Both graphs were generated from the exact same data, but the time increments used to view the data are 1 and 4 min, respectively. Each pulse from the meter equals 20 gal, and they were spaced 4 min apart (except during the initial interval shown). Software design can help by evaluating the data to determine the likelihood that raw pulse data should be averaged over longer periods of time because the pulse distribution indicates the presence of a constant flow rate.

Unless you are actually at the meter site watching the meter at the time of the event, it is not possible to know with certainty whether a periodic use of 20 gal is occurring or a steady flow rate of 5 gpm is occurring. If each pulse from the meter equals a smaller amount of water, such as 1 gal, the true picture will be much clearer.

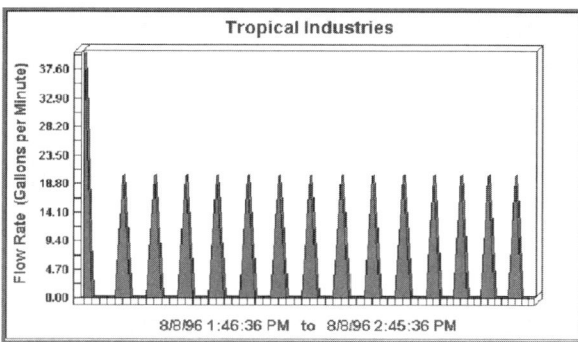

Figure D.8 Pulse resolution is extremely important (Meter-Master Model 100 program). (*Source: F. S. Brainard and Company.*)

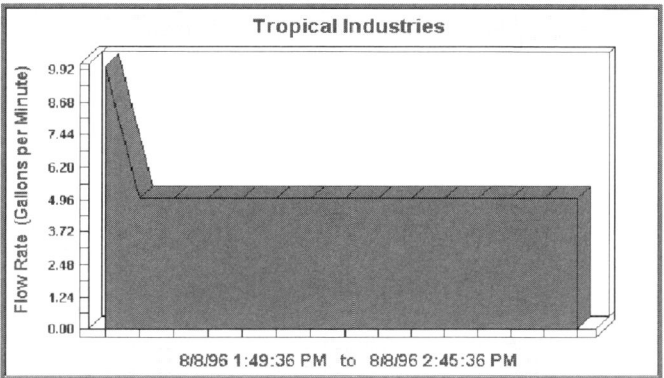

Figure D.9 Pulse resolution is extremely important (Meter-Master Model 100 program). (*Source: F. S. Brainard and Company.*)

The key to getting accurate flow data is generating a sufficient number of pulses per time interval. In the case of magnetic pulses, all 2-in and smaller positive-displacement and multijet meters provide good pulse resolution, such that the data can reasonably be observed in time increments as small as 10 s. Because some turbine meters have magnets that rotate relatively slowly, the minimum time increment necessary for observing minimum (low) flow rate data, such as leakage rates, may be as long as 5, 10, or more min, unless the software can interpret the data intelligently. Adapters which increase the magnetic pulse resolution are useful in determining accurate flow rate data because the flow data may be accurately viewed in smaller time increments, which minimizes the need to interpret the data using potentially inaccurate assumptions. The same considerations apply when using optical sensors. Meters with rapidly rotating sweep hands will provide more accurate flow data than meters with slow-moving sweep hands. Slow-moving sweep hands will not allow an optical sensor to achieve the resolution needed for accurate maximum and minimum flow rate calculations, unless the optical sensor generates numerous pulses per revolution of the sweephand, e.g., 50.

D.3.5 Meter Accuracy

When one uses a flow recorder, it is assumed that the meter to which it is attached is accurate. A flow recorder cannot determine meter accuracy, but it can determine the accurate meter configuration for a meter site. Because a flow recorder is only as accurate as the meter to which it is attached, routine meter testing is important when using recorders to determine the appropriate meter size. Because most meter inaccuracy involves underregistration of usage, a flow record on an underregistering meter can result in selection of an undersized meter.

Ideally, a meter should be tested for accuracy, and repaired/recalibrated if testing indicates that it is not accurate, prior to recording data for meter sizing purposes. As discussed in Sec. D.5.3, a demand profile performed in conjunction with a flow test may indicate that all of the flow is occurring in an accurate range of the meter, even though the meter is not accurate throughout the flow ranges. If this is the case, the meter does not need to be repaired/recalibrated because no accountability or revenue is currently being lost.

Flow recorders should be considered a valuable companion tool as part of a meter test program. As referred to in the previous paragraph, a flow recorder can identify the percentage of flow in low, medium, and high flow ranges. With this information, testing can be focused on the ranges in which most of the usage is occurring, and unnecessary and costly repairs can sometimes be avoided. If a flow record indicates that all of the flow at an oil refinery or brewery is occurring in a high flow range, it is not relevant whether or not the meter is accurate at low and medium flows.

D.4 CREATING REPORTS/GRAPHS

D.4.1 Verifying Data Accuracy

One of the principal advantages of recording flow data directly from water meters rather than using alternative technologies, such as ultrasonic devices, is that the resultant flow data is based on and may be verified against the meter's registration. Graphs and reports generated from the data may be used with confidence because the accuracy is based on the premise that a water meter is the most accurate and reliable means to measure potable water use. However, if the accuracy of the data generated with a flow recorder is not verified by comparing the total volume observed by the flow recorder to the total volume registered by the water meter itself during the data storage period, this key advantage is lost.

Verification of data accuracy is critical and is accomplished (1) by requiring the user to enter the beginning and ending meter readings when downloading data and (2) by having an accurate meter magnetic pulse factor database so that the total volume registered by the meter may be compared to the total volume registered by the flow recorder. This procedure also requires the operator to take special care, when making a record of the meter readings, that the numbers are accurate and include digits down to the decimal. In order to read a meter down to the decimal, a digit for all rotating dials and painted on "zeros" must be read.

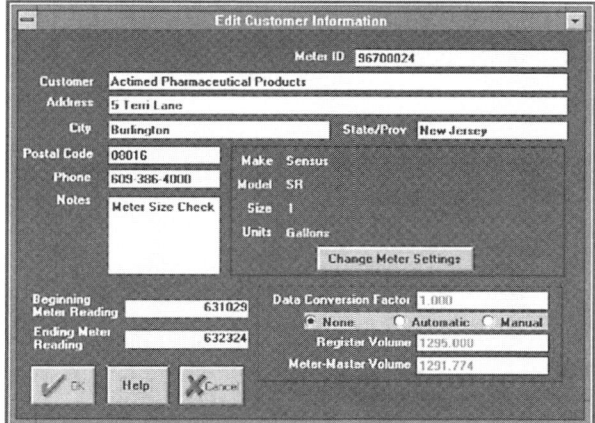

Figure D.10 Meter-Master Model 100 program. (*Source: F. S. Brainard and Company.*)

The sample software screen shown in Fig. D.10 requires the user to compare the meter's register volume to the flow recorder's observed volume. The numbers should either be extremely close or differ by an explicable margin. In this case, the electronically recorded total of 1291.774 gal compares favorably with the water meter's registered volume of 1295 gal during the same period. The software calculates register volume by subtracting the meter's beginning register reading from the ending register reading. This example is based on a magnetic pickup; it calculates the recorder volume by multiplying the total magnetic pulse count for the entire recording period times the magnetic pulse factor for that meter in the software's database. An explicable difference between the two total volumes would include differences due to change gears used in some meters for calibration purposes. Because the change gears are used to speed up and slow down the register to match the activity below in the meter's chamber, the recorder's volume could differ from the register's volume by as much as 15 percent even though both the meter and the recorder may have functioned 100 percent accurately. The software screen shown includes an automatic "data conversion factor" option so that the recorder's volume can automatically be calibrated to match the meter's volume 100 percent in such cases.

D.4.2 Data Resolution and the "Max–Min" Interval

Data resolution refers to the time intervals over which volume and maximum, average, and minimum flow rates are calculated. The sample software screen shown in Fig. D.11 displays volume, maximum, minimum, and average flow data in a grid format. In this case, the volume interval is the time interval represented by each line of data. A volume interval of 300 s will provide volume data as well as maximum, minimum, and average flow rates for each 5 min of the survey. When creating a report or graph, the longer (larger) the volume interval, the shorter the report and the fewer the points plotted on a graph.

To compute a flow rate, the software calculates the number of pulses per unit of time selected. For example, if a 10-s Max–Min interval is selected,

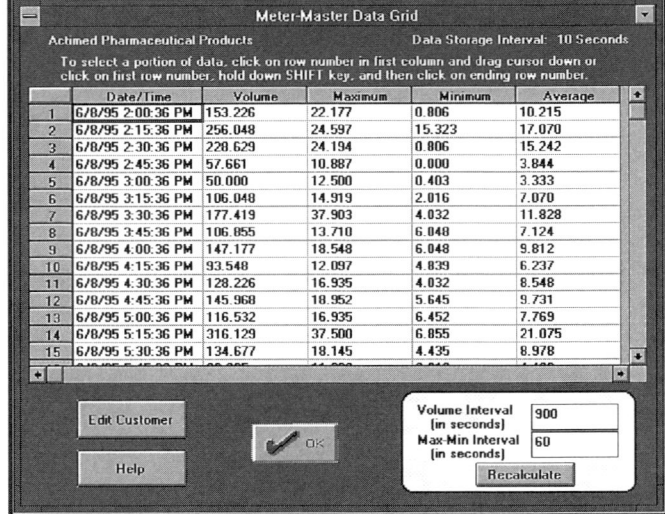

Figure D.11 Meter-Master Model 100 program. (*Source: F. S. Brainard and Company.*)

each volume interval is divided into 10-s increments, and the increments with the largest and the smallest pulse counts are converted to per-minute maximum and minimum flow rates, respectively. This represents one widely used method for calculating flow rate data. The considerations for selecting the Max–Min interval are similar to those related to the data storage interval; however, besides ensuring that the interval is sufficiently short for accurate flow rate calculations, one must ensure that the interval selected is not too short. In general, the maximum flow rate gets both larger and more accurate as the Max–Min interval gets smaller, until a point, after which the maximum flow rate exceeds reality. The point after which the maximum flow rate exceeds reality is a function of the meter's pulse resolution. The faster the magnet or sweep hand, the smaller the Max–Min interval can be without skewing the data. Similarly, the minimum flow rate gets both smaller and more accurate as the Max–Min interval gets smaller, until a point, after which the minimum flow rate becomes smaller than reality. Again, the rotation speed of the meter's magnet or sweep hand is the determining factor. It is important that a recorder's operating instructions provide guidance concerning this issue. If you familiarize yourself with those meters that generate few pulses per time interval and the volumetric equivalents of each pulse for such meters, the selection of appropriate volume and Max–Min intervals for viewing the data becomes more apparent. As mentioned previously, the software can also be designed to minimize the potential for exaggeration as the Max–Min interval is shortened.

Selection of the time intervals for viewing the data depends, in part, on the application. As discussed above it is important to consider the type of usage profile typically generated by each class of user. Usage at multifamily residential locations, for example, typically does not differ substantially in small time increments. Demand typically ramps steadily up and down in the morning and evening, which allows for longer time intervals when viewing the data. On the other hand, an industrial user may have

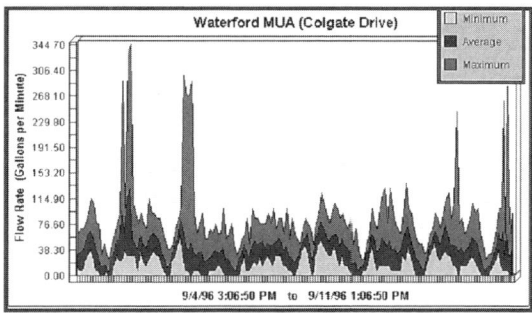

Figure D.12 Max./avg./min. graphs (Meter-Master Model 100 program). (*Source: F. S. Brainard and Company.*)

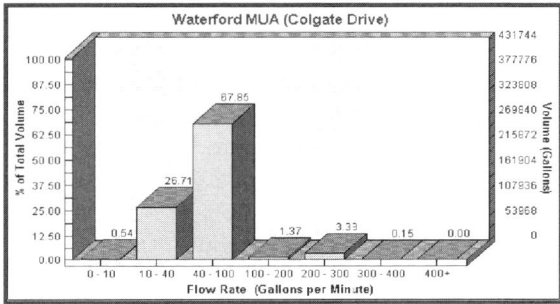

Figure D.13 Rate-versus-volume graph (Meter-Master Model 100 program). (*Source: F. S. Brainard and Company.*)

high-volume wash cycles with short duration, requiring shorter time intervals for accurate maximum flow rate calculations.

D.4.3 Graph/Report Presentation Options

Software can present data in an endless variety of formats and styles. Generally, the software should provide options to view volume data, max/avg./min flow rate data, and rate-versus-volume data. The sample graphs shown in Figs. D.12 and D.13 display max/avg./min flow rate data and rate-versus-volume data. The max/avg./min graph is useful for identifying instantaneous maximum and minimum flow rates and the duration of events. The rate-versus-volume graph is useful for meter sizing and maintenance programs because it shows the percentage and volume of water being used in various flow ranges.

D.5 USING DEMAND PROFILES TO SIZE AND MAINTAIN METERS

D.5.1 Summary of Meter Sizing Benefits

The use of demand profiles for meter sizing applies to all users. Although relatively standard meter size and water use patterns characterize single-family residential customers, outdoor residential water use can differ

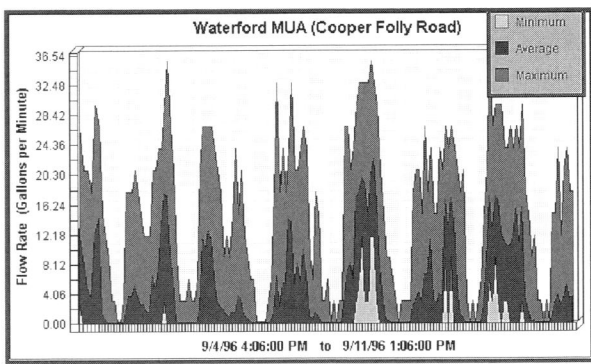

Figure D.14 The specified accuracy range for an 8-in turbine meter is approximately 40–3500 gpm (Meter-Master Model 100 program); the flow rate never exceeded 40 gpm at this site. (*Source: F. S. Brainard and Company.*)

substantially, requiring meter sizes larger than the norm. For users other than single-family residential, each customer generates a unique demand profile, and the meter should be sized accordingly. Although generic demand data can be developed for various customer class groups based on demographic and business type information, the cost of gathering customer-specific demand data is minimal when compared to the revenue and community relations benefits associated with maximizing meter accuracy and water use accountability.

Multifamily residential meters (e.g., apartment buildings) are the most consistently oversized meters because of both traditional fixtures count methods of sizing and the advent of efficient, low-volume fixtures. The graph displayed in Fig. D.14 is from an 8-in wholesale connection serving a small residential community. Although the specified accuracy range for an 8-in turbine meter is approximately 40–3500 gpm, the flow rate never exceeded 40 gpm at this site. Accordingly, the customer received a lot of free water. Replacement of the meter with one that is properly sized and configured will substantially increase both accounted-for water levels and revenue. The revenue gain is enhanced by higher sewer charges, which are typically a function of water charges. Although in some cases a smaller customer surcharge based on the meter size means less revenue to a utility in the short run, the overall water service cost to a community is reduced as a consequence of the lower capital costs associated with smaller meters.

Proper meter sizing has positive spillover effects with other programs. For example, a cost-of-service study in support of a rate structure design can only be fair and equitable if all of the sample sites have properly sized meters. Leak detection efforts are undermined if a meter is oversized, because low flows are needlessly undetectable and the meter's pulse resolution is less than it would be with a smaller meter. Similarly, hydraulic models, conservation efforts, and other programs all benefit from accurate use registration, which is dependent on proper meter sizing.

D.5.2 Compound versus Turbine Decisions

Many utilities experience shifting philosophies concerning the application of compound versus turbine meters. Compound meters are more expensive and have higher maintenance costs, but they register accurately through a broader range of flows. By comparison, turbine meters are less expensive to purchase and maintain, but offer a smaller accuracy range. For each meter application there is an optimum solution, and a demand profile will enable you to make the correct decision in each instance. If a compound meter is installed when a turbine is more appropriate, excessive maintenance costs and problems can be expected, and the utility will spend money unnecessarily. Conversely, if a turbine is installed when a compound is more appropriate, registration will be lost, and, once again, the utility will unnecessarily lose money and accountability.

A rate-versus-volume graph like that in Fig. D.13 enables a user to determine the amount of flow occurring in the crossover range of a compound meter setting that week. In this crossover range, there is a substantial drop in the level of accurate use registration by the meter setting because the turbine side of the compound setting is just starting to move, and, consequently, all flow through the turbine is below its accuracy range. If there is a meaningful amount of flow in the crossover range, an alternative compound meter size or a single meter setting should be considered.

D.5.3 Meter Maintenance Considerations

Another related use of demand profiles is meter maintenance programs, especially large meter maintenance programs. Some utilities consider demand profiles when making meter test, repair, and/or replace decisions because the demand data enables the utility to perform an accurate cost/benefit analysis of these three maintenance options on a case-by-case basis. For example, if a 10-in turbine meter tests 100 percent accurate in a high flow range, 90 percent accurate in a medium flow range, and 80 percent accurate in a low flow range, the conventional wisdom would average the three accuracies, which would equal 90 percent, and recommend repair. However, if a demand profile indicates that the flow rate never drops below 1000 gpm, the in-service meter accuracy for the subject application would equal 100 percent because all flow is occurring in a high flow range. With the advantage of a demand profile, costly and unnecessary service interruption and repair costs can be avoided and appropriate maintenance programs can be devised. Proper check valve operation in a compound meter setting can also be evaluated by ensuring that the turbine side does not move unless the small side exceeds a specified flow rate.

Water meters, like any piece of machinery, have optimum performance ranges, and projected test requirements can be related to a user's demand profile. If a 4-in meter is constantly being driven at a flow rate close to its high-end performance rating, more frequent repair requirements can be anticipated.

D.6 REFERENCE

1. Edgar, T., *Large Water Meter Handbook,* Flow Measurement Publishing, Dillsboro, N.C., 1995, pp. 41–42.

APPENDIX

E

Pipe Properties

Appendix E2 of Piping Handbook, *7th ed., edited by M. L. Nayyar, McGraw-Hill, New York, 2000; reprinted by permission*

616 APPENDIX E PIPE PROPERTIES

TABLE E2.1 Principal Properties of Commercial Pipe (in U.S. Customary Units)

Nominal pipe size, outside diam. (in)	Schedule number* a	Schedule number* b	Schedule number* c	Wall thickness (in)	Inside diam. (in)	Inside area (in²)	Metal area (in²)	Ft² outside surface per ft	Ft² inside surface per ft	Weight per ft (lb)†	Weight of water per ft (lb)	Moment of inertia (in⁴)	Elastic section modulus (in³)	Radius gyration (in)	Plastic section modulus (in³)
½ 0.840	…	…	5S	0.065	0.710	0.396	0.158	0.220	0.186	0.538	0.171	0.012	0.029	0.275	0.039
	…	…	10S	0.083	0.674	0.357	0.197	0.220	0.177	0.671	0.155	0.014	0.034	0.269	0.048
	40	Std	40S	0.109	0.622	0.304	0.250	0.220	0.163	0.851	0.132	0.017	0.041	0.261	0.059
	80	XS	80S	0.147	0.546	0.234	0.320	0.220	0.143	1.088	0.101	0.020	0.048	0.251	0.072
	160	…	…	0.187	0.466	0.171	0.383	0.220	0.122	1.304	0.074	0.022	0.053	0.240	0.082
	…	XXS	…	0.294	0.252	0.050	0.504	0.220	0.066	1.714	0.022	0.024	0.058	0.220	0.096
¾ 1.050	…	…	5S	0.065	0.920	0.665	0.201	0.275	0.241	0.684	0.288	0.025	0.047	0.349	0.063
	…	…	10S	0.083	0.884	0.614	0.252	0.275	0.231	0.857	0.266	0.030	0.057	0.343	0.078
	40	Std	40S	0.113	0.824	0.533	0.333	0.275	0.216	1.131	0.230	0.037	0.071	0.334	0.100
	80	XS	80S	0.154	0.742	0.432	0.435	0.275	0.194	1.474	0.188	0.045	0.085	0.321	0.125
	160	…	…	0.218	0.614	0.296	0.570	0.275	0.161	1.937	0.128	0.053	0.100	0.304	0.154
	…	XXS	…	0.308	0.434	0.148	0.718	0.275	0.114	2.441	0.064	0.058	0.110	0.284	0.179
1 1.315	…	…	5S	0.065	1.185	1.103	0.2553	0.344	0.310	0.868	0.478	0.0500	0.0760	0.443	0.102
	…	…	10S	0.109	1.097	0.945	0.413	0.344	0.2872	1.404	0.409	0.0757	0.1151	0.428	0.159
	40	Std	40S	0.133	1.049	0.864	0.494	0.344	0.2746	1.679	0.374	0.0874	0.1329	0.421	0.187
	80	XS	80S	0.179	0.957	0.719	0.639	0.344	0.2520	2.172	0.311	0.1056	0.1606	0.407	0.233
	160	…	…	0.250	0.815	0.522	0.836	0.344	0.2134	2.844	0.2261	0.1252	0.1903	0.387	0.289
	…	XXS	…	0.358	0.599	0.2818	1.076	0.344	0.1570	3.659	0.1221	0.1405	0.2137	0.361	0.343
1½ 1.900	…	…	5S	0.065	1.770	2.461	0.375	0.497	0.463	1.274	1.067	0.1580	0.1663	0.649	0.219
	…	…	10S	0.109	1.682	2.222	0.613	0.497	0.440	2.085	0.962	0.2469	0.2599	0.634	0.350
	40	Std	40S	0.145	1.610	2.036	0.799	0.497	0.421	2.718	0.882	0.310	0.326	0.623	0.448
	80	XS	80S	0.200	1.500	1.767	1.068	0.497	0.393	3.631	0.765	0.391	0.412	0.605	0.581
	160	…	…	0.281	1.338	1.406	1.429	0.497	0.350	4.859	0.608	0.483	0.508	0.581	0.744
	…	XXS	…	0.400	1.100	0.950	1.885	0.497	0.288	6.408	0.412	0.568	0.598	0.549	0.921
2 2.375	…	…	5S	0.065	2.245	3.96	0.472	0.622	0.588	1.604	1.716	0.315	0.2652	0.817	0.347
	…	…	10S	0.109	2.157	3.65	0.776	0.622	0.565	2.638	1.582	0.499	0.420	0.802	0.560
	40	Std	40S	0.154	2.067	3.36	1.075	0.622	0.541	3.653	1.455	0.666	0.561	0.787	0.761
	80	XS	80S	0.218	1.939	2.953	1.477	0.622	0.508	5.022	1.280	0.868	0.731	0.766	1.018
	160	…	…	0.343	1.689	2.240	2.190	0.622	0.422	7.444	0.971	1.163	0.979	0.729	1.430
	…	XXS	…	0.436	1.503	1.774	2.656	0.622	0.393	9.029	0.769	1.312	1.104	0.703	1.667

E.14

3 3.500	... ... 40 80 160 ...	... ... Std XS ... XXS	5S 10S 40S 80S	0.083 0.120 0.216 0.300 0.437 0.600	3.334 3.260 3.068 2.900 2.626 2.300	8.73 8.35 7.39 6.61 5.42 4.15	0.891 1.274 2.228 3.02 4.21 5.47	0.916 0.916 0.916 0.916 0.916 0.916	0.873 0.853 0.803 0.759 0.687 0.602	3.03 4.33 7.58 10.25 14.32 18.58	3.78 3.61 3.20 2.864 2.348 1.801	1.301 1.822 3.02 3.90 5.03 5.99	0.744 1.041 1.724 2.226 2.876 3.43	1.208 1.196 1.164 1.136 1.094 1.047	0.969 1.372 2.333 3.081 4.128 5.118
4 4.500	... ... 40 80 120 160 ...	... ... Std XS ... ... XXS	5S 10S 40S 80S	0.083 0.120 0.237 0.337 0.437 0.531 0.674	4.334 4.260 4.026 3.826 3.626 3.438 3.152	14.75 14.25 12.73 11.50 10.33 9.28 7.80	1.152 1.651 3.17 4.41 5.58 6.62 8.10	1.178 1.178 1.178 1.178 1.178 1.178 1.178	1.135 1.115 1.054 1.002 0.949 0.900 0.825	3.92 5.61 10.79 14.98 18.96 22.51 27.54	6.40 6.17 5.51 4.98 4.48 4.02 3.38	2.811 3.96 7.23 9.61 11.65 13.27 15.29	1.249 1.762 3.21 4.27 5.18 5.90 6.79	1.562 1.549 1.510 1.477 1.445 1.416 1.374	1.620 2.303 4.312 5.853 7.242 8.415 9.968
6 6.625	... ... 40 80 120 160 ...	... ... Std XS ... ... XXS	5S 10S 40S 80S	0.109 0.134 0.280 0.432 0.562 0.718 0.864	6.407 6.357 6.065 5.761 5.501 5.189 4.897	32.2 31.7 28.89 26.07 23.77 21.15 18.83	2.231 2.733 5.58 8.40 10.70 13.33 15.64	1.734 1.734 1.734 1.734 1.734 1.734 1.734	1.677 1.664 1.588 1.508 1.440 1.358 1.282	5.37 9.29 18.97 28.57 36.39 45.30 53.16	13.98 13.74 12.51 11.29 10.30 9.16 8.17	11.85 14.40 28.14 40.5 49.6 59.0 66.3	3.58 4.35 8.50 12.23 14.98 17.81 20.03	2.304 2.295 2.245 2.195 2.153 2.104 2.060	4.628 5.647 11.280 16.600 20.718 25.176 28.890
8 8.625	... ... 20 30 40 60 80 100 120 140 ... 160	... ... Std ... XS ... ... XXS ...	5S 10S ... ... 40S ... 80S	0.109 0.148 0.250 0.277 0.322 0.406 0.500 0.593 0.718 0.812 0.875 0.906	8.407 8.329 8.125 8.071 7.981 7.813 7.625 7.439 7.189 7.001 6.875 6.813	55.5 54.5 51.8 51.2 50.0 47.9 45.7 43.5 40.6 38.5 37.1 36.5	2.916 3.94 6.58 7.26 8.40 10.48 12.76 14.96 17.84 19.93 21.30 21.97	2.258 2.258 2.258 2.258 2.258 2.258 2.258 2.258 2.258 2.258 2.258 2.258	2.201 2.180 2.127 2.113 2.089 2.045 1.996 1.948 1.882 1.833 1.800 1.784	9.91 13.40 22.36 24.70 28.55 35.64 43.39 50.87 60.63 67.76 72.42 74.69	24.07 23.59 22.48 22.18 21.69 20.79 19.80 18.84 17.60 16.69 16.09 15.80	26.45 35.4 57.7 63.4 72.5 88.8 105.7 121.4 140.6 153.8 162.0 165.9	6.13 8.21 13.39 14.69 16.81 20.58 24.52 28.14 32.6 35.7 37.6 38.5	3.01 3.00 2.962 2.953 2.938 2.909 2.878 2.847 2.807 2.777 2.757 2.748	7.905 10.636 17.540 19.311 22.210 27.448 33.050 38.326 45.013 49.745 52.778 54.230

E.15

APPENDIX E PIPE PROPERTIES

TABLE E2.1 Principal Properties of Commercial Pipe (in U.S. Customary Units) *(Continued)*

Nominal pipe size, outside diam. (in)	Schedule number* a	b	c	Wall thickness (in)	Inside diam. (in)	Inside area (in²)	Metal area (in²)	Ft² outside surface per ft	Ft² inside surface per ft	Weight per ft (lb)†	Weight of water per ft (lb)	Moment of inertia (in⁴)	Elastic section modulus (in³)	Radius gyration (in)	Plastic section modulus (in³)
	...	...	5S	0.134	10.482	86.3	4.52	2.815	2.744	15.15	37.4	63.7	11.85	3.75	15.103
	...	...	10S	0.165	10.420	85.3	5.49	2.815	2.728	18.70	36.9	76.9	14.30	3.74	18.489
	20	...	...	0.250	10.250	82.5	8.26	2.815	2.683	28.04	35.8	113.7	21.16	3.71	27.568
	...	...	...	0.279	10.192	81.6	9.18	2.815	2.668	31.20	35.3	125.9	23.42	3.70	30.597
	30	...	...	0.307	10.136	80.7	10.07	2.815	2.654	34.24	35.0	137.5	25.57	3.69	33.490
10	40	Std	40S	0.365	10.020	78.9	11.91	2.815	2.623	40.48	34.1	160.8	29.90	3.67	39.381
10.750	60	XS	80S	0.500	9.750	74.7	16.10	2.815	2.553	54.74	32.3	212.0	39.4	3.63	52.573
	80	...	...	0.593	9.564	71.8	18.92	2.815	2.504	64.33	31.1	244.9	45.6	3.60	61.246
	100	...	...	0.718	9.314	68.1	22.63	2.815	2.438	76.93	29.5	286.2	53.2	3.56	72.384
	120	...	...	0.843	9.064	64.5	26.24	2.815	2.373	89.20	28.0	324	60.3	3.52	82.939
	140	...	...	1.000	8.750	60.1	30.6	2.815	2.291	104.13	26.1	368	68.4	3.47	95.396
	160	...	...	1.125	8.500	56.7	34.0	2.815	2.225	115.65	24.6	399	74.3	3.43	104.695
	...	...	5S	0.156	12.438	121.4	6.17	3.34	3.26	20.99	52.7	122.2	19.20	4.45	24.744
	...	...	10S	0.180	12.390	120.6	7.11	3.34	3.24	24.20	52.2	140.5	22.03	4.44	28.443
	20	...	...	0.250	12.250	117.9	9.84	3.34	3.21	33.38	51.1	191.9	30.1	4.42	39.068
	30	...	...	0.330	12.090	114.8	12.88	3.34	3.17	43.77	49.7	248.5	39.0	4.39	50.917
	...	Std	40S	0.375	12.000	113.1	14.58	3.34	3.14	49.56	49.0	279.3	43.8	4.38	57.445
12	40	...	...	0.406	11.938	111.9	15.74	3.34	3.13	53.53	48.5	300	47.1	4.37	61.886
12.750	...	XS	80S	0.500	11.750	108.4	19.24	3.34	3.08	65.42	47.0	362	56.7	4.33	75.073
	60	...	...	0.562	11.626	106.2	21.52	3.34	3.04	73.16	46.0	401	62.8	4.31	83.543
	80	...	...	0.687	11.376	101.6	26.04	3.34	2.978	88.51	44.0	475	74.5	4.27	100.078
	100	...	...	0.843	11.064	96.1	31.5	3.34	2.897	107.20	41.6	562	88.1	4.22	119.717
	120	...	...	1.000	10.750	90.8	36.9	3.34	2.814	125.49	39.3	642	100.7	4.17	138.396
	140	...	...	1.125	10.500	86.6	41.1	3.34	2.749	139.68	37.5	701	109.9	4.13	152.508
	160	...	...	1.312	10.126	80.5	47.1	3.34	2.651	160.27	34.9	781	122.6	4.07	172.399

APPENDIX E PIPE PROPERTIES

14 14.000	...	...	5S	0.156	13.688	147.20	6.78	3.67	3.58	23.0	63.7	162.6	23.2	4.90	29.900
	...	...	10S	0.188	13.624	145.80	8.16	3.67	3.57	27.7	63.1	194.6	27.8	4.88	35.867
	10	...	...	0.250	13.500	143.1	10.80	3.67	3.53	36.71	62.1	255.4	36.5	4.86	47.271
	20	...	...	0.312	13.376	140.5	13.42	3.67	3.50	45.68	60.9	314	44.9	4.84	58.467
	30	Std	...	0.375	13.250	137.9	16.05	3.67	3.47	54.57	59.7	373	53.3	4.82	69.633
	40	...	...	0.437	13.126	135.3	18.62	3.67	3.44	63.37	58.7	429	61.2	4.80	80.416
	...	XS	...	0.500	13.000	132.7	21.21	3.67	3.40	72.09	57.5	484	69.1	4.78	91.167
	60	...	...	0.562	12.876	130.2	23.73	3.67	3.37	80.66	56.5	537	76.7	4.76	101.545
	...	...	...	0.593	12.814	129.0	24.98	3.67	3.35	84.91	55.9	562	80.3	4.74	106.660
	...	...	...	0.625	12.750	127.7	26.26	3.67	3.34	89.28	55.3	589	84.1	4.73	111.889
	80	...	...	0.687	12.626	125.2	28.73	3.67	3.31	97.68	54.3	638	91.2	4.71	121.869
	...	...	...	0.750	12.500	122.7	31.2	3.67	3.27	106.13	53.2	687	98.2	4.69	131.813
	...	...	...	0.875	12.250	117.9	36.1	3.67	3.21	122.66	51.1	781	111.5	4.65	150.956
	100	...	...	0.937	12.126	115.5	38.5	3.67	3.17	130.73	50.0	825	117.8	4.63	160.166
	120	...	...	1.093	11.814	109.6	44.3	3.67	3.09	150.67	47.5	930	132.8	4.58	182.519
	140	...	...	1.250	11.500	103.9	50.1	3.67	3.01	170.22	45.0	1127	146.8	4.53	203.854
	160	...	...	1.406	11.188	98.3	55.6	3.67	2.929	189.12	42.6	1017	159.6	4.48	223.931
16 16.000	...	...	5S	0.165	15.670	192.90	8.21	4.19	4.10	28.00	83.5	257	32.2	5.60	41.375
	...	...	10S	0.188	15.624	191.7	9.34	4.19	4.09	32.00	83.0	292	36.5	5.59	47.006
	10	...	...	0.250	15.500	188.7	12.37	4.19	4.06	42.05	81.8	384	48.0	5.57	62.021
	20	...	...	0.312	15.376	185.7	15.38	4.19	4.03	52.36	80.5	473	59.2	5.55	76.798
	30	Std	...	0.375	15.250	182.6	18.41	4.19	3.99	62.58	79.1	562	70.3	5.53	91.570
	...	...	...	0.437	15.126	179.7	21.37	4.19	3.96	72.64	77.9	648	80.9	5.50	105.872
	40	XS	...	0.500	15.000	176.7	24.35	4.19	3.93	82.77	76.5	732	91.5	5.48	120.167
	...	...	...	0.562	14.876	173.8	27.26	4.19	3.89	92.66	75.4	813	106.6	5.46	134.002
	60	...	...	0.625	14.750	170.9	30.2	4.19	3.86	102.63	74.1	894	112.2	5.44	147.826
	...	...	...	0.656	14.688	169.4	31.6	4.19	3.85	107.50	73.4	933	116.6	5.43	154.542
	...	...	...	0.687	14.626	168.0	33.0	4.19	3.83	112.36	72.7	971	121.4	5.42	161.201
	80	...	...	0.750	14.500	165.1	35.9	4.19	3.80	122.15	71.5	1047	130.9	5.40	174.563
	...	...	...	0.843	14.314	160.9	40.1	4.19	3.75	136.46	69.7	1157	144.6	5.37	193.866
	...	...	...	0.875	14.250	159.5	41.6	4.19	3.73	141.35	69.1	1193	154.1	5.36	200.393
	100	...	...	1.031	13.938	152.5	48.5	4.19	3.65	164.83	66.1	1365	170.6	5.30	231.383
	120	...	...	1.218	13.564	144.5	56.6	4.19	3.55	192.29	62.6	1556	194.5	5.24	266.745
	140	...	...	1.437	13.126	135.3	65.7	4.19	3.44	223.64	58.6	1760	220.0	5.17	305.750
	160	...	...	1.593	12.814	129.0	72.1	4.19	3.35	245.11	55.9	1894	236.7	5.12	331.993

E.17

620 APPENDIX E PIPE PROPERTIES

TABLE E2.1 Principal Properties of Commercial Pipe (in U.S. Customary Units) *(Continued)*

Nominal pipe size, outside diam. (in)	Schedule number* a	b	c	Wall thickness (in)	Inside diam. (in)	Inside area (in²)	Metal area (in²)	Ft² outside surface per ft	Ft² inside surface per ft	Weight per ft (lb)†	Weight of water per ft (lb)	Moment of inertia (in⁴)	Elastic section modulus (in³)	Radius gyration (in)	Plastic section modulus (in³)
18 18.000	...	...	5S	0.165	17.670	245.20	9.24	4.71	4.63	31.00	106.2	368	40.8	6.31	52.486
	...	...	10S	0.188	17.624	243.90	10.52	4.71	4.61	36.00	105.7	417	46.4	6.30	59.649
	10	...	...	0.250	17.500	240.5	13.94	4.71	4.58	47.39	104.3	549	61.0	6.28	78.771
	20	...	...	0.312	17.376	237.1	17.34	4.71	4.55	59.03	102.8	678	75.5	6.25	97.624
	...	Std	...	0.375	17.250	233.7	20.76	4.71	4.52	70.59	101.2	807	89.6	6.23	116.508
	30	...	...	0.437	17.126	230.4	24.11	4.71	4.48	82.06	99.9	931	103.4	6.21	134.824
	...	XS	...	0.500	17.000	227.0	27.49	4.71	4.45	93.45	98.4	1053	117.0	6.19	153.167
	40	...	...	0.562	16.876	223.7	30.8	4.71	4.42	104.75	97.0	1172	130.2	6.17	170.954
	...	...	...	0.625	16.750	220.5	34.1	4.71	4.39	115.98	95.5	1289	143.3	6.15	188.763
	...	...	...	0.687	16.626	217.1	37.4	4.71	4.35	127.03	94.1	1403	156.3	6.13	206.029
	60	...	...	0.750	16.500	213.8	40.6	4.71	4.32	138.17	92.7	1515	168.3	6.10	223.313
	...	...	...	0.875	16.250	207.4	47.1	4.71	4.25	160.04	89.9	1731	192.8	6.06	256.831
	80	...	...	0.937	16.126	204.2	50.2	4.71	4.22	170.75	88.5	1834	203.8	6.04	273.078
	100	...	...	1.156	15.688	193.3	61.2	4.71	4.11	207.96	83.7	2180	242.2	5.97	328.496
	120	...	...	1.375	15.250	182.6	71.8	4.71	3.99	244.14	79.2	2499	277.6	5.90	380.904
	140	...	...	1.562	14.876	173.8	80.7	4.71	3.89	274.23	75.3	2750	306	5.84	423.335
	160	...	...	1.781	14.438	163.7	90.7	4.71	3.78	308.51	71.0	3020	336	5.77	470.386
20 20.000	...	...	5S	0.188	19.634	302.40	11.70	5.24	5.14	40	131.0	574	57.4	7.00	71.869
	...	...	10S	0.218	19.564	300.6	13.55	5.24	5.12	46	130.2	663	66.3	6.99	85.313
	10	...	...	0.250	19.500	298.6	15.51	5.24	5.11	52.73	129.5	757	75.7	6.98	97.521
	...	...	...	0.312	19.376	294.9	19.30	5.24	5.07	65.40	128.1	935	93.5	6.96	120.947
	20	Std	...	0.375	19.250	291.0	23.12	5.24	5.04	78.60	126.0	1114	111.4	6.94	144.445
	...	...	...	0.437	19.126	287.3	26.86	5.24	5.01	91.31	124.6	1286	128.6	6.92	167.273
	30	XS	...	0.500	19.000	283.5	30.6	5.24	4.97	104.13	122.8	1457	145.7	6.90	190.167
	...	...	...	0.562	18.876	279.8	34.3	5.24	4.94	116.67	121.3	1624	162.4	6.88	212.403
	40	...	...	0.593	18.814	278.0	36.2	5.24	4.93	122.91	120.4	1704	170.4	6.86	223.412
	...	...	...	0.625	18.750	276.1	38.0	5.24	4.91	129.33	119.7	1787	178.7	6.85	234.701
	...	...	...	0.687	18.626	272.5	41.7	5.24	4.88	141.71	118.1	1946	194.6	6.83	256.354
	...	...	...	0.750	18.500	268.8	45.4	5.24	4.84	154.20	116.5	2105	210.5	6.81	278.063
	60	...	...	0.812	18.376	265.2	48.9	5.24	4.81	166.40	115.0	2257	225.7	6.79	299.140

E.18

APPENDIX E PIPE PROPERTIES

		80			0.875	18.250	261.6	52.6	5.24	4.78	178.73	113.4	2409	240.9	6.77	320.268
		100			1.031	17.938	252.7	61.4	5.24	4.70	208.87	109.4	2772	277.2	6.72	371.343
		120			1.281	17.438	238.8	75.3	5.24	4.57	256.10	103.4	3320	332	6.63	449.564
		140			1.500	17.000	227.0	87.2	5.24	4.45	296.37	98.3	3760	376	6.56	514.500
		160			1.750	16.500	213.8	100.3	5.24	4.32	341.10	92.6	4220	422	6.48	584.646
					1.968	16.064	202.7	111.5	5.24	4.21	379.01	87.9	4590	459	6.41	642.442
22 22.000				5S	0.188	21.624	367.3	12.88	5.76	5.66	44	159.1	766	69.7	7.71	89.446
				10S	0.218	21.564	365.2	14.92	5.76	5.65	51	158.2	885	80.4	7.70	103.435
		10			0.250	21.500	363.1	17.16	5.76	5.63	58	157.4	1010	91.8	7.69	118.271
		20	Std		0.375	21.250	354.7	25.48	5.76	5.56	87	153.7	1490	135.4	7.65	175.383
		30	XS		0.500	21.000	346.4	33.77	5.76	5.50	115	150.2	1953	177.5	7.61	231.167
					0.625	20.750	338.2	41.97	5.76	5.43	143	146.6	2400	218.2	7.56	285.638
					0.750	20.500	330.1	50.07	5.76	5.37	170	143.1	2829	257.2	7.52	338.813
		60			0.875	20.250	322.1	58.07	5.76	5.30	197	139.6	3245	295.0	7.47	390.706
		80			1.125	19.750	306.4	73.78	5.76	5.17	251	132.8	4029	366.3	7.39	490.711
		100			1.375	19.250	291.0	89.09	5.76	5.04	303	126.2	4758	432.6	7.31	585.779
		120			1.625	18.750	276.1	104.02	5.76	4.91	354	119.6	5432	493.8	7.23	676.034
		140			1.875	18.250	261.6	118.55	5.76	4.78	403	113.3	6054	550.3	7.15	761.602
		160			2.125	17.750	247.4	132.68	5.76	4.65	451	107.2	6626	602.4	7.07	842.607
24 24.000				5S	0.218	23.564	436.1	16.29	6.28	6.17	55	188.9	1152	96.0	8.41	123.301
				10S	0.250	23.500	434	18.65	6.28	6.15	63.41	188.0	1316	109.6	8.40	141.021
		10			0.312	23.376	430	23.20	6.28	6.12	78.93	186.1	1629	135.8	8.38	175.080
			Std		0.375	23.250	425	27.83	6.28	6.09	94.62	183.8	1943	161.9	8.35	209.320
		20			0.437	23.126	420	32.4	6.28	6.05	109.97	182.1	2246	187.4	8.33	242.657
			XS		0.500	23.000	415	36.9	6.28	6.02	125.49	180.1	2550	212.5	8.31	276.167
		30			0.562	22.876	411	41.4	6.28	5.99	140.80	178.1	2840	237.0	8.29	308.788
					0.625	22.750	406	45.9	6.28	5.96	156.03	176.2	3140	261.4	8.27	341.576
		40			0.687	22.626	402	50.3	6.28	5.92	171.17	174.3	3420	285.2	8.25	373.490
					0.750	22.500	398	54.8	6.28	5.89	186.24	172.4	3710	309	8.22	405.563
		60			0.968	22.064	382	70.0	6.28	5.78	238.11	165.8	4650	388	8.15	513.800
		80			1.218	21.564	365	87.2	6.28	5.65	296.36	158.3	5670	473	8.07	632.768
		100			1.531	20.938	344	108.1	6.28	5.48	367.40	149.3	6850	571	7.96	774.131
		120			1.812	20.376	326	126.3	6.28	5.33	429.39	141.4	7830	652	7.87	894.044
		140			2.062	19.876	310	142.1	6.28	5.20	483.13	134.5	8630	719	7.79	995.313
		160			2.343	19.314	293	159.4	6.28	5.06	541.94	127.0	9460	788	7.70	1103.215

E.19

TABLE E2.1 Principal Properties of Commercial Pipe (in U.S. Customary Units) *(Continued)*

Nominal pipe size, outside diam. (in)	Schedule number* a	Schedule number* b	Schedule number* c	Wall thickness (in)	Inside diam. (in)	Inside area (in²)	Metal area (in²)	Ft² outside surface per ft	Ft² inside surface per ft	Weight per ft (lb)†	Weight of water per ft (lb)	Moment of inertia (in⁴)	Elastic section modulus (in³)	Radius gyration (in)	Plastic section modulus (in³)
26 26.000	...	...	...	0.250	25.500	510.7	19.85	6.81	6.68	67	221.4	1646	126.6	9.10	165.771
	10	...	...	0.312	25.376	505.8	25.18	6.81	6.64	86	219.2	2076	159.7	9.08	205.891
	...	Std	...	0.375	25.250	500.7	30.19	6.81	6.61	103	217.1	2478	190.6	9.06	246.258
	20	XS	...	0.500	25.000	490.9	40.06	6.81	6.54	136	212.8	3259	250.7	9.02	325.167
	...	...	...	0.625	24.750	481.1	49.82	6.81	6.48	169	208.6	4013	308.7	8.98	402.513
	...	...	...	0.750	24.500	471.4	59.49	6.81	6.41	202	204.4	4744	364.9	8.93	478.313
	...	...	...	0.875	24.250	461.9	69.07	6.81	6.35	235	200.2	5458	419.9	8.89	552.581
	...	...	...	1.00	24.000	452.4	78.54	6.81	6.28	267	196.1	6149	473.0	8.85	625.333
	...	...	...	1.125	23.750	443.0	87.91	6.81	6.22	299	192.1	6813	524.1	8.80	696.586
28 28.000	...	...	...	0.250	27.500	594.0	21.80	7.33	7.20	74	257.3	2098	149.8	9.81	192.521
	10	...	...	0.312	27.376	588.6	27.14	7.33	7.17	92	255.0	2601	185.8	9.78	239.197
	...	Std	...	0.375	27.250	583.2	32.54	7.33	7.13	111	252.6	3105	221.8	9.77	286.195
	20	XS	...	0.500	27.000	572.6	43.20	7.33	7.07	147	248.0	4085	291.8	9.72	378.167
	30	...	...	0.625	26.750	562.0	53.75	7.33	7.00	183	243.4	5038	359.8	9.68	468.451
	...	...	...	0.750	26.500	551.6	64.21	7.33	6.94	218	238.9	5964	426.0	9.64	557.063
	...	...	...	0.875	26.250	541.2	74.56	7.33	6.87	253	234.4	6865	490.3	9.60	644.018
	...	...	...	1.000	26.000	530.9	84.82	7.33	6.81	288	230.0	7740	552.8	9.55	729.333
	...	...	...	1.125	25.750	520.8	94.98	7.33	6.74	323	225.6	8590	613.6	9.51	813.023
30 30.000	...	...	5S	0.250	29.500	683.4	23.37	7.85	7.72	79	296.3	2585	172.3	10.52	221.271
	10	...	10S	0.312	29.376	677.8	29.19	7.85	7.69	99	293.7	3201	213.4	10.50	275.000
	...	Std	...	0.375	29.250	672.0	34.90	7.85	7.66	119	291.2	3823	254.8	10.48	329.133
	20	XS	...	0.500	29.000	660.5	46.34	7.85	7.59	158	286.2	5033	335.5	10.43	435.167
	30	...	...	0.625	28.750	649.2	57.68	7.85	7.53	196	281.3	6213	414.2	10.39	539.388
	40	...	...	0.750	28.500	637.9	68.92	7.85	7.46	234	276.6	7371	491.4	10.34	641.813
	...	...	...	0.875	28.250	620.7	90.06	7.85	7.39	272	271.8	8494	566.2	10.30	742.456
	...	...	...	1.000	28.000	615.7	91.11	7.85	7.33	310	267.0	9591	639.4	10.26	841.333
	...	...	...	1.125	27.750	604.7	102.05	7.85	7.26	347	262.2	10653	710.2	10.22	938.461

APPENDIX E PIPE PROPERTIES

32 32.000	⋯ 10 20 30 40 ⋯ ⋯ ⋯ ⋯	⋯ ⋯ Std XS ⋯ ⋯ ⋯ ⋯ ⋯	⋯ ⋯ ⋯ ⋯ ⋯ ⋯ ⋯ ⋯ ⋯	0.250 0.312 0.375 0.500 0.625 0.688 0.750 0.875 1.000 1.125	31.500 31.376 31.250 31.000 30.750 30.624 30.500 30.250 30.000 29.750	779.2 773.2 766.9 754.7 742.5 736.6 730.5 718.3 706.8 694.7	24.93 31.02 37.25 49.48 61.59 67.68 73.63 85.52 97.38 109.0	8.38 8.38 8.38 8.38 8.38 8.38 8.38 8.38 8.38 8.38	8.25 8.21 8.18 8.11 8.05 8.02 7.98 7.92 7.85 7.79	85 106 127 168 209 230 250 291 331 371	337.8 335.2 332.5 327.2 321.9 319.0 316.7 311.6 306.4 301.3	3141 3891 4656 6140 7578 8298 8990 10372 11680 13023	196.3 243.2 291.0 383.8 473.6 518.6 561.9 648.2 730.0 814.0	11.22 11.20 11.18 11.14 11.09 11.07 11.05 11.01 10.95 10.92	252.021 313.299 375.070 496.167 615.326 674.652 732.563 847.893 961.333 1072.898
34 34.000	⋯ 10 ⋯ 20 30 40 ⋯ ⋯ ⋯ ⋯	⋯ ⋯ Std XS ⋯ ⋯ ⋯ ⋯ ⋯ ⋯	⋯ ⋯ ⋯ ⋯ ⋯ ⋯ ⋯ ⋯ ⋯ ⋯	0.250 0.312 0.375 0.500 0.625 0.688 0.750 0.875 1.000 1.125	33.500 33.376 33.250 33.000 32.750 32.624 32.500 32.250 32.000 31.750	881.2 874.9 867.8 855.3 841.9 835.9 829.3 816.4 804.2 791.3	26.50 32.99 39.61 52.62 65.53 72.00 78.34 91.01 103.67 116.13	8.90 8.90 8.90 8.90 8.90 8.90 8.90 8.90 8.90 8.90	8.77 8.74 8.70 8.64 8.57 8.54 8.51 8.44 8.38 8.31	90 112 135 179 223 245 266 310 353 395	382.0 379.3 376.2 370.8 365.0 362.1 359.5 354.1 348.6 343.2	3773 4680 5597 7385 9124 9992 10829 12501 14114 15719	221.9 275.3 329.2 434.4 536.7 587.8 637.0 735.4 830.2 924.7	11.93 11.91 11.89 11.85 11.80 11.78 11.76 11.72 11.67 11.63	284.771 354.093 424.008 561.167 696.263 763.575 829.313 960.331 1089.333 1216.336
36 36.000	⋯ 10 ⋯ 20 30 40 ⋯ ⋯ ⋯ ⋯	⋯ ⋯ Std XS ⋯ ⋯ ⋯ ⋯ ⋯ ⋯	⋯ ⋯ ⋯ ⋯ ⋯ ⋯ ⋯ ⋯ ⋯ ⋯	0.250 0.312 0.375 0.500 0.625 0.750 0.875 1.000 1.125	35.500 35.376 35.250 35.000 34.750 34.500 34.250 34.000 33.750	989.7 982.9 975.8 962.1 948.3 934.7 920.6 907.9 894.2	28.11 34.95 42.01 55.76 69.50 83.01 96.50 109.96 123.19	9.42 9.42 9.42 9.42 9.42 9.42 9.42 9.42 9.42	9.29 9.26 9.23 9.16 9.10 9.03 8.97 8.90 8.89	96 119 143 190 236 282 328 374 419	429.1 426.1 423.1 417.1 411.1 405.3 399.4 393.6 387.9	4491 5565 6664 8785 10872 12898 14903 16851 18763	249.5 309.1 370.2 488.1 604.0 716.5 827.9 936.2 1042.4	12.64 12.62 12.59 12.55 12.51 12.46 12.42 12.38 12.34	319.521 397.384 475.945 630.167 782.201 932.063 1079.768 1225.333 1368.773
42 42.000	⋯ ⋯ 20 30 40 ⋯ ⋯ ⋯ ⋯	⋯ Std XS ⋯ ⋯ ⋯ ⋯ ⋯ ⋯	⋯ ⋯ ⋯ ⋯ ⋯ ⋯ ⋯ ⋯ ⋯	0.250 0.375 0.500 0.625 0.750 1.000 1.250 1.500	41.500 41.250 41.000 40.750 40.500 40.000 39.500 39.000	1352.6 1336.3 1320.2 1304.1 1288.2 1256.6 1225.3 1194.5	32.82 49.08 65.18 81.28 97.23 128.81 160.03 190.85	10.99 10.99 10.99 10.99 10.99 10.99 10.99 10.99	10.86 10.80 10.73 10.67 10.60 10.47 10.34 10.21	112 167 222 276 330 438 544 649	586.4 579.3 572.3 565.4 558.4 544.9 531.2 517.9	7126 10627 14037 17373 20689 27080 33233 39181	339.3 506.1 668.4 827.3 985.2 1289.5 1582.5 1865.7	14.73 14.71 14.67 14.62 14.59 14.50 14.41 14.33	435.771 649.758 861.167 1070.013 1276.313 1681.333 2076.554 2461.500

E.21

TABLE E2.1 Principal Properties of Commercial Pipe (in U.S. Customary Units) *(Continued)*

Notes: The following formulas were used in the computation of the values shown in the table:

Weight† of pipe per foot (pounds)	$= 10.6802 t(D - t)$
Weight of water per foot (pounds)	$= 0.3405 d^2$
Square feet outside surface per foot	$= 0.2618 D$
Square feet inside surface per foot	$= 0.2618 d$
Inside area (square inches)	$= 0.785 d^2$
Area of metal (square inches)	$= 0.785(D^2 - d^2)$
Moment of inertia (inches⁴)	$= 0.0491(D^4 - d^4)$
	$= A_M R_g^2$
Elastic section modulus (inches³)	$= \dfrac{0.0982(D^4 - d^4)}{D}$
Plastic section modulus	$= \dfrac{(D^3 - d^3)}{6}$
Radius of gyration (inches)	$= 0.25 \sqrt{D^2 + d^2}$

A_M = area of metal (square inches)
d = inside diameter (inches)
D = outside diameter (inches)
R_g = radius of gyration (inches)
t = pipe wall thickness (inches)

* Schedule numbers: Standard weight pipe and Schedule 40 are the same in all sizes through 10 in; from 12 in through 24 in, standard weight pipe has a wall thickness of ⅜ in. Extra-strong weight pipe and Schedule 80 are the same in all sizes through 8 in; from 8 in through 24 in, extra-strong weight pipe has a wall thickness of ½ in. Double extra-strong weight pipe has no corresponding schedule number.
  a: ANSI/ASME B36.10 Steel Pipe Schedule Numbers.
  b: ANSI/ASME N36.10 Steel Pipe Nominal Wall Thickness Designations.
  c: ANSI/ASME B36.19 Stainless Steel Pipe Schedule Numbers.

† The ferritic stainless steels may be about 5 percent less and the austenitic stainless steels about 2 percent greater than the values shown in this table which are based on weights for carbon steel.

APPENDIX E PIPE PROPERTIES

TABLE E2.1M Principal Properties of Commercial Pipe (Metric Data)

d_n Nom dia	Schedule		D Outside dia	t Wall thick	d Inside dia	A_i Inside area	A_m Metal area	S_o Outside surf	S_i Inside surf	w_p Pipe wt.	w_w Water wt.	I Mom of inert	z_e Elast sec mod	R_g Rad of gyr	z_p Plast sec mod
mm			mm	mm	mm	cm²	cm²	m²/m	m²/m	kg/m	kg/m	cm⁴	cm³	cm	cm³
3	10S		10.3	1.2446	7.811	0.479	0.354	0.0324	0.0245	0.277	0.048	0.037	0.072	0.323	0.103
	Std	40	10.3	1.7272	6.846	0.368	0.465	0.0324	0.0215	0.364	0.037	0.044	0.086	0.309	0.129
	XS	80	10.3	2.413	5.474	0.235	0.598	0.0324	0.0172	0.468	0.024	0.051	0.099	0.292	0.155
6	10S		13.7	1.651	10.398	0.849	0.625	0.0430	0.0327	0.489	0.085	0.116	0.169	0.430	0.241
	Std	40	13.7	2.235	9.23	0.669	0.805	0.0430	0.0290	0.630	0.067	0.137	0.200	0.413	0.298
	XS	80	13.7	3.023	7.654	0.460	1.014	0.0430	0.0240	0.794	0.046	0.156	0.228	0.392	0.354
10	10S		17.145	1.651	13.843	1.505	0.804	0.0539	0.0435	0.629	0.151	0.244	0.285	0.551	0.398
	Std	40	17.145	2.311	12.523	1.232	1.077	0.0539	0.0393	0.843	0.123	0.303	0.354	0.531	0.513
	XS	80	17.145	3.2	10.745	0.907	1.402	0.0539	0.0338	1.098	0.091	0.359	0.418	0.506	0.633
15	5S		21.336	1.651	18.034	2.554	1.021	0.0670	0.0567	0.799	0.255	0.498	0.467	0.698	0.641
	10S		21.336	2.108	17.12	2.302	1.273	0.0670	0.0538	0.997	0.230	0.596	0.558	0.684	0.782
	Std	40	21.336	2.769	15.798	1.960	1.615	0.0670	0.0496	1.265	0.196	0.711	0.667	0.664	0.962
	XS	80	21.336	3.734	13.868	1.510	2.065	0.0670	0.0436	1.617	0.151	0.836	0.783	0.636	1.174
		160	21.336	4.75	11.836	1.100	2.475	0.0670	0.0372	1.938	0.110	0.921	0.863	0.610	1.342
	XXS		21.336	7.468	6.4	0.322	3.254	0.0670	0.0201	2.547	0.032	1.009	0.946	0.557	1.575
20	5S		26.67	1.651	23.368	4.289	1.298	0.0838	0.0734	1.016	0.429	1.020	0.765	0.886	1.035
	10S		26.67	2.108	22.454	3.960	1.627	0.0838	0.0705	1.273	0.396	1.236	0.927	0.872	1.275
	Std	40	26.67	2.87	20.93	3.441	2.146	0.0838	0.0658	1.680	0.344	1.541	1.156	0.848	1.634
	XS	80	26.67	3.912	18.846	2.790	2.797	0.0838	0.0592	2.190	0.279	1.864	1.398	0.816	2.046
		160	26.67	5.537	15.596	1.910	3.676	0.0838	0.0490	2.878	0.191	2.193	1.645	0.772	2.529
	XXS		26.67	7.823	11.024	0.954	4.632	0.0838	0.0346	3.626	0.095	2.411	1.808	0.721	2.938
25	5S		33.401	1.651	30.099	7.115	1.647	0.1049	0.0946	1.289	0.712	2.081	1.246	1.124	1.666
	10S		33.401	2.769	27.863	6.097	2.665	0.1049	0.0875	2.086	0.610	3.151	1.887	1.087	2.605
	Std	40	33.401	3.378	26.645	5.576	3.186	0.1049	0.0837	2.494	0.558	3.635	2.177	1.068	3.058
	XS	80	33.401	4.547	24.307	4.640	4.122	0.1049	0.0764	3.227	0.464	4.396	2.632	1.033	3.817
		160	33.401	6.35	20.701	3.366	5.396	0.1049	0.0650	4.225	0.337	5.208	3.119	0.982	4.732
	XXS		33.401	9.093	15.215	1.818	6.944	0.1049	0.0478	5.436	0.182	5.846	3.501	0.918	5.624
32	5S		42.164	1.651	38.862	11.862	2.101	0.1325	0.1221	1.645	1.186	4.318	2.048	1.434	2.711
	10S		42.164	2.769	36.626	10.536	3.427	0.1325	0.1151	2.683	1.054	6.681	3.169	1.396	4.305
	Std	40	42.164	3.556	35.052	9.650	4.313	0.1325	0.1101	3.377	0.965	8.104	3.844	1.371	5.316
	XS	80	42.164	4.851	32.462	8.276	5.686	0.1325	0.1020	4.452	0.828	10.063	4.773	1.330	6.792
		160	42.164	6.35	29.464	6.818	7.145	0.1325	0.0926	5.594	0.682	11.815	5.604	1.286	8.230
	XXS		42.164	9.703	22.758	4.068	9.895	0.1325	0.0715	7.747	0.407	14.198	6.734	1.198	10.529
40	5S		48.26	1.651	44.958	15.875	2.418	0.1516	0.1412	1.893	1.587	6.573	2.724	1.649	3.588
	10S		48.26	2.769	42.722	14.335	3.957	0.1516	0.1342	3.098	1.433	10.275	4.258	1.611	5.737
	Std	40	48.26	3.683	40.894	13.134	5.158	0.1516	0.1285	4.038	1.313	12.899	5.345	1.581	7.335
	XS	80	48.26	5.08	38.1	11.401	6.891	0.1516	0.1197	5.395	1.140	16.283	6.748	1.537	9.516

626 APPENDIX E   PIPE PROPERTIES

TABLE E2.1M Principal Properties of Commercial Pipe (Metric Data) (Continued)

d_n Nom dia	Schedule	D Outside dia	t Wall thick	d Inside dia	A_i Inside area	A_m Metal area	S_o Outside surf	S_i Inside surf	w_p Pipe wt.	w_w Water wt.	I Mom of inert	z_e Elast sec mod	R_g Rad of gyr	z_p Plast sec mod
mm		mm	mm	mm	cm²	cm²	m²/m	m²/m	kg/m	kg/m	cm⁴	cm³	cm	cm³
	160	48.26	7.137	33.986	9.072	9.220	0.1516	0.1068	7.219	0.907	20.078	8.321	1.476	12.191
	XXS	48.26	10.16	27.94	6.131	12.161	0.1516	0.0878	9.521	0.613	23.635	9.795	1.394	15.098
	—	48.26	13.335	21.59	3.661	14.631	0.1516	0.0678	11.455	0.366	25.560	10.593	1.322	17.056
	—	48.26	15.875	16.51	2.141	16.151	0.1516	0.0519	12.645	0.214	26.262	10.884	1.275	17.983
50	5S	60.325	1.651	57.023	25.538	3.043	0.1895	0.1791	2.383	2.554	13.106	4.345	2.075	5.685
	10S	60.325	2.769	54.787	23.575	5.007	0.1895	0.1721	3.920	2.357	20.780	6.890	2.037	9.180
	Std 40	60.325	3.912	52.501	21.648	6.933	0.1895	0.1649	5.428	2.165	27.713	9.188	1.999	12.470
	XS 80	60.325	5.537	49.251	19.051	9.530	0.1895	0.1547	7.461	1.905	36.124	11.977	1.947	16.677
	160	60.325	8.712	42.901	14.455	14.126	0.1895	0.1348	11.059	1.446	48.379	16.039	1.851	23.429
	XXS	60.325	11.074	38.177	11.447	17.134	0.1895	0.1199	13.415	1.145	54.579	18.095	1.785	27.315
	—	60.325	14.275	31.775	7.930	20.652	0.1895	0.0998	16.168	0.793	60.002	19.893	1.705	31.242
	—	60.325	17.45	25.425	5.077	23.504	0.1895	0.0799	18.402	0.508	62.955	20.872	1.637	33.850
65	5S	73.025	2.108	68.809	37.186	4.696	0.2294	0.2162	3.677	3.719	29.550	8.093	2.508	10.605
	10S	73.025	3.048	66.929	35.182	6.701	0.2294	0.2103	5.246	3.518	41.092	11.254	2.476	14.935
	Std 40	73.025	5.156	62.713	30.889	10.993	0.2294	0.1970	8.607	3.089	63.662	17.436	2.406	23.796
	XS 80	73.025	7.01	59.005	27.344	14.538	0.2294	0.1854	11.382	2.734	80.089	21.935	2.347	30.665
	160	73.025	9.525	53.975	22.881	19.002	0.2294	0.1696	14.876	2.288	97.928	26.820	2.270	38.696
	XXS	73.025	14.021	44.983	15.892	25.990	0.2294	0.1413	20.348	1.589	119.49	32.726	2.144	49.734
	—	73.025	17.145	38.735	11.784	30.098	0.2294	0.1217	23.564	1.178	128.54	35.204	2.067	55.218
	—	73.025	20.32	32.385	8.237	33.645	0.2294	0.1017	26.341	0.824	134.19	36.752	1.997	59.243
80	5S	88.9	2.108	84.684	56.324	5.748	0.2793	0.2660	4.500	5.632	54.15	12.183	3.069	15.883
	10S	88.9	3.048	82.804	53.851	8.221	0.2793	0.2601	6.436	5.385	75.84	17.061	3.037	22.475
	Std 40	88.9	5.486	77.928	47.696	14.376	0.2793	0.2448	11.255	4.770	125.58	28.251	2.956	38.227
	XS 80	88.9	7.62	73.66	42.614	19.458	0.2793	0.2314	15.233	4.261	162.09	36.466	2.886	50.490
	160	88.9	11.1	66.7	34.942	27.130	0.2793	0.2095	21.240	3.494	209.44	47.119	2.779	67.644
	XXS	88.9	15.24	58.42	26.805	35.267	0.2793	0.1835	27.610	2.680	249.43	56.114	2.659	83.871
	—	88.9	18.415	52.07	21.294	40.777	0.2793	0.1636	31.925	2.129	270.52	60.859	2.576	93.572
	—	88.9	21.59	45.72	16.417	45.654	0.2793	0.1436	35.743	1.642	285.15	64.151	2.499	101.173
90	5S	101.6	2.108	97.384	74.485	6.589	0.3192	0.3059	5.158	7.448	81.56	16.055	3.518	20.870
	10S	101.6	3.048	95.504	71.636	9.437	0.3192	0.3000	7.388	7.164	114.68	22.575	3.486	29.614
	Std 40	101.6	5.74	90.12	63.787	17.286	0.3192	0.2831	13.533	6.379	199.27	39.226	3.395	52.810
	XS 80	101.6	8.077	85.446	57.342	23.731	0.3192	0.2684	18.579	5.734	261.39	51.455	3.319	70.823
	XXS	101.6	16.154	69.292	37.710	43.363	0.3192	0.2177	33.949	3.771	409.89	80.686	3.074	119.348
100	5S	114.3	2.108	110.084	95.179	7.430	0.3591	0.3458	5.817	9.518	116.94	20.462	3.967	26.537
	10S	114.3	3.048	108.204	91.955	10.653	0.3591	0.3399	8.340	9.196	164.94	28.861	3.935	37.735
	—	114.3	4.775	104.75	86.179	16.430	0.3591	0.3291	12.863	8.618	246.83	43.189	3.876	57.317

E.24

APPENDIX E PIPE PROPERTIES

E.25

	Std	40	114.3	6.02	102.26	82.130	20.478	0.3591	0.3213	16.033	8.213	301.05	52.677	3.834	70.656
	XS	80	114.3	8.56	97.18	74.173	28.436	0.3591	0.3053	22.262	7.417	400.02	69.995	3.751	95.920
		120	114.3	11.1	92.1	66.621	35.988	0.3591	0.2893	28.175	6.662	484.63	84.800	3.670	118.676
			114.3	12.7	88.9	62.072	40.537	0.3591	0.2793	31.736	6.207	531.22	92.952	3.620	131.782
		160	114.3	13.487	87.326	59.893	42.715	0.3591	0.2743	33.442	5.989	552.36	96.652	3.596	137.892
	XXS		114.3	17.12	80.06	50.341	52.267	0.3591	0.2515	40.920	5.034	636.16	111.314	3.489	163.356
			114.3	20.32	73.66	42.614	59.994	0.3591	0.2314	46.970	4.261	693.31	121.315	3.399	182.271
			114.3	23.495	67.31	35.584	67.025	0.3591	0.2115	52.474	3.558	737.06	128.970	3.316	198.056
125	5S		141.3	2.769	135.762	144.760	12.051	0.4439	0.4265	9.435	14.476	289.20	40.934	4.899	53.148
	10S		141.3	3.404	134.492	142.064	14.747	0.4439	0.4225	11.545	14.206	350.72	49.642	4.877	64.743
	Std	40	141.3	6.553	128.194	129.070	27.740	0.4439	0.4027	21.718	12.907	631.07	89.324	4.770	119.077
	XS	80	141.3	9.525	122.25	117.379	39.432	0.4439	0.3841	30.871	11.738	860.37	121.779	4.671	165.690
		120	141.3	12.7	115.9	105.501	51.309	0.4439	0.3641	40.170	10.550	1071.0	151.595	4.569	210.719
		160	141.3	15.875	109.55	94.257	62.553	0.4439	0.3442	48.973	9.426	1249.8	176.894	4.470	251.075
	XXS		141.3	19.05	103.2	83.647	73.164	0.4439	0.3242	57.280	8.365	1400.0	198.155	4.374	287.014
			141.3	22.225	96.85	73.670	83.141	0.4439	0.3043	65.091	7.367	1524.9	215.834	4.283	318.791
			141.3	25.4	90.5	64.326	92.484	0.4439	0.2843	72.406	6.433	1627.5	230.357	4.195	346.663
150	5S		168.275	2.769	162.737	208.000	14.398	0.5287	0.5113	11.272	20.800	493.1	58.607	5.852	75.858
	10S		168.275	3.404	161.467	204.766	17.631	0.5287	0.5073	13.804	20.477	599.3	71.232	5.830	92.544
			168.275	5.563	157.149	193.961	28.437	0.5287	0.4937	22.263	19.396	942.2	111.980	5.756	147.342
	Std	40	168.275	7.112	154.051	186.389	36.009	0.5287	0.4840	28.191	18.639	1171.4	139.219	5.704	184.847
	XS	80	168.275	10.973	146.329	168.171	54.226	0.5287	0.4597	42.454	16.817	1685.4	200.310	5.575	271.961
		120	168.275	14.275	139.725	153.334	69.063	0.5287	0.4390	54.070	15.333	2065.0	245.426	5.468	339.522
		160	168.275	18.237	131.801	136.436	85.962	0.5287	0.4141	67.300	13.644	2454.6	291.738	5.344	412.570
	XXS		168.275	21.946	124.383	121.510	100.887	0.5287	0.3908	78.985	12.151	2761.0	328.151	5.231	473.444
			168.275	25.4	117.475	108.388	114.009	0.5287	0.3691	89.258	10.839	3001.0	356.683	5.131	523.970
			168.275	28.575	111.125	96.987	125.410	0.5287	0.3491	98.184	9.699	3187.4	378.829	5.041	565.461
200	5S		219.075	2.769	213.537	358.127	18.817	0.6882	0.6708	14.732	35.813	1100.7	100.483	7.648	129.566
	10S		219.075	3.759	211.557	351.516	25.427	0.6882	0.6646	19.907	35.152	1474.0	134.563	7.614	174.292
			219.075	5.563	207.949	339.629	37.315	0.6882	0.6533	29.214	33.963	2127.8	194.251	7.551	253.665
		20	219.075	6.35	206.375	334.507	42.437	0.6882	0.6483	33.224	33.451	2402.6	219.336	7.524	287.441
		30	219.075	7.036	205.003	330.074	46.870	0.6882	0.6440	36.694	33.007	2637.0	240.738	7.501	316.465
	Std	40	219.075	8.179	202.717	322.754	54.190	0.6882	0.6369	42.425	32.275	3017.3	275.456	7.462	363.968
		60	219.075	10.312	198.451	309.313	67.631	0.6882	0.6235	52.949	30.931	3693.3	337.175	7.390	449.792
	XS	80	219.075	12.7	193.675	294.604	82.340	0.6882	0.6084	64.464	29.460	4400.2	401.708	7.310	541.595
		100	219.075	15.062	188.951	280.407	96.536	0.6882	0.5936	75.578	28.041	5049.8	461.010	7.233	628.052
		120	219.075	18.237	182.601	261.877	115.067	0.6882	0.5737	90.086	26.188	5849.4	534.012	7.130	737.642
		140	219.075	20.625	177.825	248.357	128.587	0.6882	0.5587	100.671	24.836	6398.4	584.126	7.054	815.203
		160	219.075	23.012	173.051	235.201	141.743	0.6882	0.5437	110.970	23.520	6904.6	630.343	6.979	888.677
			219.075	25.4	168.275	222.398	154.546	0.6882	0.5287	120.994	22.240	7370.8	672.906	6.906	958.236
			219.075	28.575	161.925	205.930	171.014	0.6882	0.5087	133.887	20.593	7932.2	724.149	6.811	1044.792

APPENDIX E: PIPE PROPERTIES

TABLE E2.1M Principal Properties of Commercial Pipe (Metric Data) (Continued)

d_n Nom dia	Schedule	D Outside dia	t Wall thick	d Inside dia	A_i Inside area	A_m Metal area	S_o Outside surf	S_i Inside surf	w_p Pipe wt.	w_w Water wt.	I Mom of inert	z_e Elast sec mod	R_g Rad of gyr	z_p Plast sec mod
mm		mm	mm	mm	cm²	cm²	m²/m	m²/m	kg/m	kg/m	cm⁴	cm³	cm	cm³
250	5S	273.05	3.404	266.242	556.729	28.836	0.8578	0.8364	22.576	55.673	2621.2	191.993	9.534	247.519
	10S	273.05	4.191	264.668	550.166	35.399	0.8578	0.8315	27.714	55.017	3199.3	234.337	9.507	302.978
	—	273.05	5.563	261.924	538.817	46.748	0.8578	0.8229	36.599	53.882	4182.7	306.372	9.459	398.094
	20	273.05	6.35	260.35	532.361	53.204	0.8578	0.8179	41.654	53.236	4733.1	346.684	9.432	451.763
	30	273.05	7.798	257.454	520.583	64.982	0.8578	0.8088	50.874	52.058	5719.9	418.966	9.382	548.826
	Std 40	273.05	9.271	254.508	508.738	76.828	0.8578	0.7996	60.148	50.874	6690.2	490.036	9.332	645.349
	XS 60	273.05	12.7	247.65	481.690	103.875	0.8578	0.7780	81.324	48.169	8822.0	646.179	9.216	861.533
	80	273.05	15.062	242.926	463.488	122.077	0.8578	0.7632	95.574	46.349	10191.0	746.454	9.137	1003.653
	100	273.05	18.237	236.576	439.574	145.991	0.8578	0.7432	114.296	43.957	11909.5	872.331	9.032	1186.168
	120	273.05	21.412	230.226	416.294	169.272	0.8578	0.7233	132.523	41.629	13495.1	988.470	8.929	1359.143
	—	273.05	22.225	228.6	410.434	175.131	0.8578	0.7182	137.110	41.043	13880.6	1016.705	8.903	1401.933
	140	273.05	25.4	222.25	387.949	197.616	0.8578	0.6982	154.714	38.795	15309.1	1121.341	8.802	1563.289
	160	273.05	28.575	215.9	366.097	219.468	0.8578	0.6783	171.822	36.610	16620.3	1217.382	8.702	1715.683
	—	273.05	31.75	209.55	344.879	240.687	0.8578	0.6583	188.434	34.488	17820.8	1305.313	8.605	1859.372
	—	273.05	38.1	196.85	304.342	281.223	0.8578	0.6184	220.170	30.434	19915.0	1458.708	8.415	2121.655
300	5S	323.85	3.962	315.926	783.902	39.817	1.0174	0.9925	31.172	78.390	5093.7	314.570	11.311	405.454
	10S	323.85	4.572	314.706	777.859	45.859	1.0174	0.9887	35.903	77.786	5844.7	360.948	11.289	466.104
	20	323.85	6.35	311.15	760.380	63.339	1.0174	0.9775	49.588	76.038	7984.3	493.084	11.228	640.218
	30	323.85	8.382	307.086	740.646	83.072	1.0174	0.9647	65.037	74.065	10341.3	638.650	11.157	834.390
	Std	323.85	9.525	304.8	729.660	94.058	1.0174	0.9576	73.638	72.966	11626.7	718.031	11.118	941.379
	40	323.85	10.312	303.226	722.144	101.574	1.0174	0.9526	79.523	72.214	12495.1	771.659	11.091	1014.118
	XS	323.85	12.7	298.45	699.575	124.144	1.0174	0.9376	97.192	69.957	15048.5	929.349	11.010	1230.249
	60	323.85	14.275	295.3	684.885	138.833	1.0174	0.9277	108.692	68.489	16666.8	1029.292	10.957	1369.066
	80	323.85	17.45	288.95	655.747	167.971	1.0174	0.9078	131.505	65.575	19775.4	1221.267	10.850	1640.027
	—	323.85	19.05	285.75	641.303	182.415	1.0174	0.8977	142.813	64.130	21266.2	1313.335	10.797	1772.143
	100	323.85	21.412	281.026	620.274	203.444	1.0174	0.8829	159.276	62.027	23377.3	1443.711	10.720	1961.840
	—	323.85	22.225	279.4	613.117	210.601	1.0174	0.8778	164.879	61.312	24079.7	1487.092	10.693	2025.678
	120	323.85	25.4	273.05	585.565	238.153	1.0174	0.8578	186.450	58.557	26707.9	1649.398	10.590	2267.947
	140	323.85	28.575	266.7	558.646	265.072	1.0174	0.8379	207.525	55.865	29158.9	1800.766	10.488	2499.205
	—	323.85	31.75	260.35	532.361	291.357	1.0174	0.8179	228.104	53.236	31441.0	1941.699	10.388	2719.710
	160	323.85	33.325	257.2	519.557	304.162	1.0174	0.8080	238.128	51.956	32512.8	2007.891	10.339	2825.182
350	5S	355.6	3.962	347.676	949.381	43.768	1.1172	1.0923	34.266	94.938	6765.7	380.524	12.433	489.929
	10S	355.6	4.775	346.05	940.521	52.628	1.1172	1.0872	41.202	94.052	8098.1	455.460	12.405	587.746
	—	355.6	5.334	344.932	934.454	58.695	1.1172	1.0836	45.952	93.445	9003.4	506.375	12.385	654.472
	—	355.6	5.563	344.474	931.974	61.175	1.1172	1.0822	47.894	93.197	9371.7	527.091	12.377	681.683
	10	355.6	6.35	342.9	923.477	69.672	1.1172	1.0773	54.547	92.348	10626.3	597.656	12.350	774.646

APPENDIX E PIPE PROPERTIES

	—	355.6	7.137	341.326	915.018	78.131	1.1172	1.0723	61.169	91.502	11863.8	667.256	12.323	866.759
	20	355.6	7.925	339.75	906.588	86.561	1.1172	1.0674	67.769	90.659	13085.8	735.987	12.295	958.142
	—	355.6	8.738	338.124	897.931	95.218	1.1172	1.0623	74.546	89.793	14328.9	805.902	12.267	1051.541
	Std 30	355.6	9.525	336.55	889.590	103.559	1.1172	1.0573	81.076	88.959	15515.3	872.628	12.240	1141.100
	40	355.6	11.1	333.4	873.016	120.133	1.1172	1.0474	94.052	87.302	17840.1	1003.381	12.186	1317.833
	—	355.6	11.913	331.774	864.521	128.628	1.1172	1.0423	100.703	86.452	19014.7	1069.441	12.158	1407.764
	XS	355.6	12.7	330.2	856.338	136.811	1.1172	1.0374	107.110	85.634	20135.3	1132.469	12.132	1493.984
	60	355.6	15.062	325.476	832.011	161.138	1.1172	1.0225	126.155	83.201	23403.7	1316.294	12.052	1747.856
	—	355.6	15.875	323.85	823.718	169.431	1.1172	1.0174	132.647	82.372	24496.3	1377.743	12.024	1833.553
	80	355.6	19.05	317.5	791.732	201.417	1.1172	0.9975	157.689	79.173	28608.1	1609.006	11.918	2160.063
	100	355.6	23.8	308	745.062	248.087	1.1172	0.9676	194.228	74.506	34315.6	1930.010	11.761	2624.718
	120	355.6	27.762	300.076	707.218	285.931	1.1172	0.9427	223.855	70.722	38689.1	2175.988	11.632	2990.990
	140	355.6	31.75	292.1	670.122	323.027	1.1172	0.9177	252.898	67.012	42754.9	2404.665	11.505	3340.638
	160	355.6	35.712	284.176	634.258	358.891	1.1172	0.8928	280.976	63.426	46477.6	2614.039	11.380	3669.605
400 5S		406.4	4.191	398.018	1244.218	52.957	1.2767	1.2504	41.460	124.422	10709.7	527.051	14.221	678.025
10S		406.4	4.775	396.85	1236.926	60.248	1.2767	1.2467	47.168	123.693	12149.4	597.902	14.201	770.272
	10	406.4	6.35	393.7	1217.368	79.807	1.2767	1.2368	62.481	121.737	15969.2	785.885	14.146	1016.360
	20	406.4	7.925	390.55	1197.965	99.209	1.2767	1.2270	77.671	119.797	19698.4	969.409	14.091	1258.541
Std 30		406.4	9.525	387.35	1178.414	118.760	1.2767	1.2169	92.977	117.841	23395.5	1151.355	14.036	1500.599
XS 40		406.4	12.7	381	1140.094	157.080	1.2767	1.1969	122.978	114.009	30465.5	1499.286	13.927	1969.218
	60	406.4	16.662	373.076	1093.165	204.010	1.2767	1.1721	159.719	109.316	38805.6	1909.727	13.792	2532.479
	80	406.4	21.4412	363.5176	1037.867	259.307	1.2767	1.1420	203.012	103.787	48183.0	2371.209	13.631	3180.791
	100	406.4	26.187	354.026	984.376	312.798	1.2767	1.1122	244.889	98.438	56790.9	2794.826	13.474	3791.705
	120	406.4	30.937	344.526	932.255	364.919	1.2767	1.0824	285.695	93.226	64740.3	3186.038	13.320	4371.222
	140	406.4	36.5	333.4	873.016	424.158	1.2767	1.0474	332.074	87.302	73250.5	3604.847	13.141	5010.459
	160	406.4	40.462	325.476	832.011	465.164	1.2767	1.0225	364.177	83.201	78814.1	3878.645	13.017	5440.481
450 5S		457.2	4.191	448.818	1582.091	59.645	1.4363	1.4100	46.696	158.209	15301.4	669.354	16.017	860.107
10S		457.2	4.775	447.65	1573.867	67.869	1.4363	1.4063	53.135	157.387	17366.7	759.699	15.997	977.443
	10	457.2	6.35	444.5	1551.795	89.941	1.4363	1.3964	70.415	155.180	22856.6	999.853	15.942	1290.849
	20	457.2	7.925	441.35	1529.879	111.857	1.4363	1.3865	87.573	152.988	28231.1	1234.956	15.887	1599.844
Std 30		457.2	9.525	438.15	1507.775	133.961	1.4363	1.3765	104.878	150.777	33574.3	1468.691	15.831	1909.259
	30	457.2	11.1	435	1486.173	155.563	1.4363	1.3666	121.790	148.617	38720.9	1693.826	15.777	2209.458
XS		457.2	12.7	431.8	1464.388	177.348	1.4363	1.3565	138.846	146.439	43835.9	1917.581	15.722	2510.002
	40	457.2	14.275	428.65	1443.100	198.636	1.4363	1.3466	155.512	144.310	48761.2	2133.034	15.668	2801.532
	60	457.2	19.05	419.1	1379.514	262.222	1.4363	1.3166	205.294	137.951	63043.5	2757.807	15.506	3659.509
	80	457.2	23.8	409.6	1317.683	324.054	1.4363	1.2868	253.702	131.768	76314.6	3338.348	15.346	4475.070
	100	457.2	29.362	398.476	1247.083	394.653	1.4363	1.2519	308.974	124.708	90723.7	3968.667	15.162	5383.123
	120	457.2	34.925	387.35	1178.414	463.322	1.4363	1.2169	362.735	117.841	103977.6	4548.450	14.981	6242.017
	140	457.2	39.675	377.85	1121.320	520.416	1.4363	1.1871	407.434	112.132	114426.0	5005.513	14.828	6937.385
	160	457.2	45.237	366.726	1056.268	585.468	1.4363	1.1521	458.363	105.627	125698.5	5498.624	14.653	7708.342

E.27

TABLE E2.1M Principal Properties of Commercial Pipe (Metric Data) (Continued)

500	5S		508	4.775	498.45	1951.345	75.489	1.5959	59.101	195.135	23897.7	940.853	17.792	1209.260
	10S		508	5.537	496.926	1939.431	87.404	1.5959	68.428	193.943	27586.5	1086.081	17.766	1398.006
		10	508	6.35	495.3	1926.760	100.075	1.5959	78.349	192.676	31484.9	1239.563	17.737	1598.112
	Std	20	508	9.525	488.95	1877.672	149.162	1.5959	116.779	187.767	46345.8	1824.637	17.627	2367.082
	XS	30	508	12.7	482.6	1829.218	197.616	1.5959	154.714	182.922	60638.8	2387.354	17.517	3116.336
		40	508	15.062	477.876	1793.582	233.252	1.5959	182.613	179.358	70912.3	2791.821	17.436	3661.096
		60	508	20.625	466.75	1711.038	315.797	1.5959	247.238	171.104	93932.8	3698.142	17.247	4902.169
			508	22.225	463.55	1687.657	339.178	1.5959	265.543	168.766	100256.3	3947.100	17.193	5248.361
		80	508	26.187	455.626	1630.452	396.383	1.5959	310.329	163.045	115360.9	4541.769	17.060	6085.257
		100	508	32.537	442.926	1540.825	486.010	1.5959	380.497	154.082	137979.1	5432.247	16.849	7367.108
		120	508	38.1	431.8	1464.388	562.447	1.5959	440.340	146.439	156258.5	6151.908	16.668	8431.313
		140	508	44.45	419.1	1379.514	647.320	1.5959	506.787	137.951	175466.0	6908.112	16.464	9580.820
		160	508	49.987	408.026	1307.575	719.260	1.5959	563.109	130.757	190848.8	7513.732	16.289	10527.913
550	5S		558.8	4.775	549.25	2369.360	83.110	1.7555	65.067	236.936	31889.7	1141.363	19.588	1465.722
	10S		558.8	5.537	547.726	2356.230	96.240	1.7555	75.347	235.623	36827.3	1318.084	19.562	1694.966
		10	558.8	6.35	546.1	2342.261	110.209	1.7555	86.283	234.226	42050.0	1505.013	19.533	1938.150
	Std	20	558.8	9.525	539.75	2288.106	164.364	1.7555	128.680	228.811	62004.3	2219.193	19.423	2874.067
	XS	30	558.8	12.7	533.4	2234.585	217.885	1.7555	170.582	223.459	81266.5	2908.608	19.313	3788.219
			558.8	15.875	527.05	2181.697	270.772	1.7555	211.988	218.170	99852.9	3573.835	19.204	4680.862
			558.8	19.05	520.7	2129.443	323.027	1.7555	252.898	212.944	117779.9	4215.447	19.095	5552.253
		60	558.8	22.225	514.35	2077.822	374.648	1.7555	293.312	207.782	135062.3	4834.011	18.987	6402.648
		80	558.8	28.575	501.65	1976.480	475.989	1.7555	372.652	197.648	167757.8	6004.218	18.773	8041.472
		100	558.8	34.925	488.95	1877.672	574.798	1.7555	450.009	187.767	198062.3	7088.844	18.563	9599.384
		120	558.8	41.275	476.25	1781.398	671.072	1.7555	525.383	178.140	226095.2	8092.168	18.355	11078.432
		140	558.8	47.625	463.55	1687.657	764.813	1.7555	598.773	168.766	251972.9	9018.356	18.151	12480.663
		160	558.8	53.975	450.85	1596.449	856.021	1.7555	670.179	159.645	275808.8	9871.469	17.950	13808.127
600		10	609.6	6.35	596.9	2798.299	120.343	1.9151	94.217	279.830	54748.3	1796.203	21.329	2310.964
	Std	20	609.6	9.525	590.55	2739.077	179.565	1.9151	140.581	273.908	80843.9	2652.360	21.219	3430.214
	XS		609.6	12.7	584.2	2680.489	238.153	1.9151	186.450	268.049	106111.3	3481.341	21.108	4525.651
		30	609.6	14.275	581.05	2651.660	266.981	1.9151	209.020	265.166	118343.5	3882.662	21.054	5060.300
			609.6	15.875	577.85	2622.534	296.108	1.9151	231.823	262.253	130568.0	4283.725	20.999	5597.532
		40	609.6	17.45	574.7	2594.020	324.622	1.9151	254.147	259.402	142404.6	4672.068	20.945	6120.590
			609.6	19.05	571.5	2565.213	353.429	1.9151	276.700	256.521	154231.5	5060.090	20.890	6646.112
	5S		609.6	5.537	598.526	2813.565	105.077	1.9151	82.265	281.356	479830.7	1572.531	21.358	2020.505
			609.6	22.225	565.15	2508.525	410.117	1.9151	321.081	250.852	177119.3	5811.002	20.782	7671.646
		60	609.6	24.587	560.426	2466.763	451.879	1.9151	353.776	246.676	193653.4	6353.456	20.702	8419.783
		80	609.6	30.937	547.726	2356.230	562.412	1.9151	440.313	235.623	236075.8	7745.267	20.488	10369.359
		100	609.6	38.887	531.826	2221.417	697.225	1.9151	545.858	222.142	285184.5	9356.446	20.225	12685.870
		120	609.6	46.025	517.55	2103.757	814.885	1.9151	637.974	210.376	325680.9	10685.069	19.992	14651.104

650	140	609.6	52.375	504.85	2001.777	916.865	1.9151	1.5860	717.814	200.178	358998.2	11778.156	19.788	16310.639
	160	609.6	59.512	490.576	1890.181	1028.460	1.9151	1.5412	805.182	189.018	393560.2	12912.079	19.562	18078.760
	—	660.4	6.35	647.7	3294.873	130.477	2.0348	2.0348	102.151	329.487	69775.7	2113.133	23.125	2716.552
	10	660.4	7.925	644.55	3262.903	162.448	2.0747	2.0249	127.181	326.290	86459.2	2618.390	23.070	3374.093
	Std	660.4	9.525	641.35	3230.584	194.766	2.0747	2.0149	152.482	323.058	103159.0	3124.138	23.014	4035.523
	20	660.4	12.7	635	3166.929	258.421	2.0747	1.9949	202.318	316.693	135565.4	4105.554	22.904	5328.634
	XS	660.4	15.875	628.65	3103.907	321.443	2.0747	1.9750	251.658	310.391	167014.0	5057.965	22.794	6596.139
	—	660.4	19.05	622.3	3041.519	383.832	2.0747	1.9550	300.502	304.152	197523.9	5981.949	22.685	7838.294
	—	660.4	22.225	615.95	2979.764	445.587	2.0747	1.9351	348.850	297.976	227114.1	6878.076	22.577	9055.357
	—	660.4	25.4	609.6	2918.642	506.709	2.0747	1.9151	396.702	291.864	258803.0	7746.912	22.469	10247.582
	—	660.4	28.575	603.25	2858.154	567.197	2.0747	1.8952	444.059	285.815	283609.3	8589.017	22.361	11415.227
700	—	711.2	6.35	698.5	3831.984	140.612	2.2343	2.1944	110.085	383.198	87328.4	2455.805	24.921	3154.914
	10	711.2	7.925	695.35	3797.500	175.096	2.2343	2.1845	137.082	379.750	108264.6	3044.561	24.866	3919.915
	Std	711.2	9.525	692.15	3762.629	209.967	2.2343	2.1745	164.384	376.263	129243.7	3634.526	24.810	4689.995
	XS 20	711.2	12.7	685.8	3693.906	278.690	2.2343	2.1545	218.186	369.391	170021.2	4781.247	24.700	6197.165
	30	711.2	15.875	679.45	3625.817	346.779	2.2343	2.1346	271.493	362.582	209681.5	5896.554	24.590	7676.682
	—	711.2	19.05	673.1	3558.362	414.234	2.2343	2.1146	324.304	355.836	248245.3	6981.026	24.480	9128.801
	—	711.2	22.225	666.75	3491.539	481.057	2.2343	2.0947	376.619	349.154	285732.9	8035.235	24.372	10553.779
	—	711.2	25.4	660.4	3425.351	547.245	2.2343	2.0747	428.439	342.535	322164.7	9059.750	24.263	11951.871
	—	711.2	28.575	654.05	3359.795	612.801	2.2343	2.0548	479.762	335.980	357560.6	10055.134	24.156	13323.334
750	5S	762	6.35	749.3	4409.632	150.746	2.3939	2.3540	118.019	440.963	107602.6	2824.216	26.717	3626.052
	10	762	7.925	746.15	4372.635	187.743	2.3939	2.3441	146.984	437.263	133458.8	3502.857	26.662	4506.642
	Std	762	9.525	742.95	4335.209	225.169	2.3939	2.3341	176.285	433.521	159392.3	4183.524	26.606	5393.628
	XS 20	762	12.7	736.6	4261.420	298.958	2.3939	2.3141	234.054	426.142	209870.9	5508.421	26.496	7131.247
	30	762	15.875	730.25	4188.264	372.114	2.3939	2.2942	291.328	418.826	259060.7	6799.494	26.385	8839.163
	40	762	19.05	723.9	4115.741	444.637	2.3939	2.2742	348.106	411.574	306983.9	8057.321	26.276	10517.633
	—	762	22.225	717.55	4043.852	516.526	2.3939	2.2543	404.389	404.385	353662.5	9282.480	26.167	12166.913
	—	762	25.4	711.2	3972.596	587.782	2.3939	2.2343	460.175	397.260	399118.1	10475.541	26.058	13787.259
	—	762	28.575	704.85	3901.973	658.405	2.3939	2.2144	515.465	390.197	443372.5	11637.069	25.950	15378.927
800	—	812.8	6.35	800.1	5027.817	160.880	2.5535	2.5136	125.953	502.782	130794.5	3218.368	28.513	4129.964
	10	812.8	7.925	796.95	4988.306	200.391	2.5535	2.5037	156.886	498.831	162286.8	3993.277	28.458	5134.272
	Std	812.8	9.525	793.75	4948.327	240.370	2.5535	2.4936	188.186	494.833	193898.9	4771.134	28.402	6146.424
	XS 20	812.8	12.7	787.4	4869.470	319.226	2.5535	2.4737	249.923	486.947	255506.8	6287.076	28.291	8130.878
	30	812.8	15.875	781.05	4791.247	397.450	2.5535	2.4537	311.163	479.125	315642.1	7766.783	28.181	10083.580
	40	812.8	17.475	777.85	4752.068	436.629	2.5535	2.4437	341.837	475.207	345395.9	8498.915	28.126	11055.669
	—	812.8	19.05	774.7	4713.657	475.039	2.5535	2.4338	371.909	471.366	374328.4	9210.836	28.071	12004.789
	—	812.8	22.225	768.35	4636.701	551.996	2.5535	2.4138	432.158	463.670	431589.2	10619.813	27.962	13894.759
	—	812.8	25.4	762	4560.378	628.319	2.5535	2.3939	491.911	456.038	487447.8	11994.286	27.853	15753.746
	—	812.8	28.575	755.65	4484.688	704.008	2.5535	2.3740	551.168	448.469	541927.2	13334.824	27.745	17582.007
850	—	863.6	6.35	850.9	5686.538	171.014	2.7131	2.6732	133.887	568.654	157100.1	3638.261	30.309	4666.651
	10	863.6	7.925	847.75	5644.513	213.039	2.7131	2.6633	166.788	564.451	194993.2	4515.823	30.254	5802.806

E.29

TABLE E2.1M Principal Properties of Commercial Pipe (Metric Data) (*Continued*)

d_n Nom dia			D Outside dia	t Wall thick	d Inside dia	A_i Inside area	A_m Metal area	S_o Outside surf	S_i Inside surf	w_p Pipe wt.	w_w Water wt.	I Mom of inert	Z_e Elast sec mod	R_g Rad of gyr	Z_p Plast sec mod
mm	Schedule		mm	mm	mm	cm²	cm²	m²/m	m²/m	kg/m	kg/m	cm⁴	cm³	cm	cm³
	Std		863.6	9.525	844.55	5601.981	255.571	2.7131	2.6532	200.087	560.198	233057.7	5397.354	30.198	6948.382
	XS		863.6	12.7	838.2	5518.057	339.495	2.7131	2.6333	265.791	551.806	307321.2	7117.211	30.087	9196.058
		20	863.6	15.875	831.85	5434.767	422.785	2.7131	2.6133	330.999	543.477	379915.9	8798.423	29.977	11409.935
		30	863.6	17.475	828.65	5393.034	464.518	2.7131	2.6033	363.671	539.303	415874.7	9631.187	29.921	12512.862
		40	863.6	19.05	825.5	5352.110	505.442	2.7131	2.5934	395.711	535.211	450867.0	10441.571	29.867	13590.269
	—		863.6	22.225	819.15	5270.087	587.465	2.7131	2.5734	459.927	527.009	520199.6	12047.234	29.757	15737.316
	—		863.6	25.4	812.8	5188.697	668.855	2.7131	2.5535	523.647	518.870	587938.3	13615.987	29.648	17851.332
	—		863.6	28.575	806.45	5107.940	749.612	2.7131	2.5335	586.872	510.794	654108.0	15148.401	29.540	19932.573
900	—		914.4	6.35	901.7	6385.796	181.148	2.8727	2.8328	141.821	638.580	186715.7	4083.894	32.105	5236.113
		10	914.4	7.925	898.55	6341.258	225.687	2.8727	2.8229	176.690	634.126	231823.0	5070.493	32.050	6512.244
	Std		914.4	9.525	895.35	6296.172	270.772	2.8727	2.8128	211.988	629.617	277163.1	6062.184	31.994	7799.502
		20	914.4	12.7	889	6207.181	359.763	2.8727	2.7929	281.659	620.718	365706.4	7998.827	31.883	10326.788
	XS		914.4	15.875	882.65	6118.824	448.120	2.8727	2.7729	350.834	611.882	452372.6	9894.413	31.773	12818.226
		30	914.4	19.05	876.3	6031.100	535.844	2.8727	2.7530	419.513	603.110	537188.3	11749.525	31.663	15274.073
		40	914.4	22.225	869.95	5944.009	622.935	2.8727	2.7330	487.696	594.401	620180.1	13564.744	31.553	17694.585
	—		914.4	25.4	863.6	5857.552	709.392	2.8727	2.7131	555.383	585.755	701374.3	15340.646	31.444	20080.017
	—		914.4	28.575	857.25	5771.728	795.216	2.8727	2.6931	622.575	577.173	780797.1	17077.802	31.335	22430.627
1050	—		1066.8	6.35	1054.1	8726.790	211.551	3.3515	3.3116	165.623	872.679	297383.2	5575.237	37.493	7141.147
	Std		1066.8	9.525	1047.75	8621.965	316.376	3.3515	3.2916	247.691	862.196	442100.1	8288.341	37.382	10647.836
		20	1066.8	12.7	1041.4	8517.773	420.568	3.3515	3.2717	329.263	851.777	584209.6	10952.561	37.271	14112.276
	XS		1066.8	15.875	1035.05	8414.214	524.127	3.3515	3.2517	410.339	841.421	723743.2	13568.488	37.160	17534.723
		30	1066.8	19.05	1028.7	8311.289	627.052	3.3515	3.2318	490.919	831.129	860732.1	16136.710	37.050	20915.433
		40	1066.8	25.4	1016	8107.339	831.002	3.3515	3.1919	650.592	810.734	1127201	21132.369	36.830	27552.668
	—		1066.8	31.75	1003.3	7905.922	1032.419	3.3515	3.1520	808.281	790.592	1383862	25944.162	36.612	34026.029
	—		1066.8	38.1	990.6	7707.039	1231.302	3.3515	3.1121	963.987	770.704	1630959	30576.656	36.395	40337.565

Index

AADD (average annual day demand), 385–387
Above-ground valve installation, 302–303
Absolute pressure, 426
Account water, 492
Accountability, 2, 26–27
  audits for (*See* Audits)
  AWWA committee report on (*see* American Water Works Association (AWWA)
  for management and conservation:
    accounting systems in, 493
    conclusion, 496
    introduction, 491–492
    strategies in, 494–496
    terms and standards in, 492–493
Accounting systems:
  for conservation, 493
  in DeKalb County, 474–475
  water losses from, 20–23, 525
Accuracy:
  of acoustic detection techniques, 250–252
  in demand profile recording, 608–609
  of meter reader reports, 376
  in meter testing, 543–546, 551, 572
  of meters:
    in audits, 509, 515–518
    magnetic, 582
    pitot, 594
    turbine and propeller, 586
    Venturi, 576
    vortex, 591
Acoustic detection techniques, 177, 179–180
  accuracy of, 250–252
  for boundary valves, 207–208
  case study, 190–192
    field-scale experiments in, 196–198
    initial experiments in, 194–196
    leak detection in, 199–201
    leak identification in, 192
    pipe coating in, 201–202
    pipe loops in, 193
    pipeline evaluations in, 198
    research in, 194
    summary, 202
    test chambers in, 193
  equipment for, 103–105
  in Halifax, 237–238
  in Johnson City, 480
Acoustic emission (AE) technology, 192

Active leakage management, 171–172
  case studies:
    acoustic technology for, 190–203
    ground-penetrating radar, 227–233
    Heath Consultants, 183–185
    Severn Trent process, 211–216
    technology application, 235–239
    tracer gas process, 216–226
  correlator surveys in, 185–189
  flow analysis in, 203–208
  infrared testing in, 233
  leak detection in, 177–183
  mapping in, 172–177
  noise logger surveys in, 210–211
  in Philadelphia system, 449
  reservoir leakage in, 233–235
  step testing in, 208–210
ACVs [*see* Automatic control valves (ACVs)]
Adjusting vanes for turbine meters, 546
AE (acoustic emission) technology, 192
Age in C-factor testing, 418
Air compressors, 219, 222–223
Air conditioning needs, 368
Air release valves, 305
Air scouring, 329
Al Jalazon Refugee Camp Water Network, 307
Altitude valves, 426–427
  losses from, 527
  for pressure control, 110, 112
  for reservoir and tank control, 287, 289
Ambient noise and listening sticks, 103
American Water Works Association (AWWA):
  accountability report by:
    1937 report, 497
    costs in, 500–501
    dollar losses in, 499–500
    introduction, 497
    operating cost increases in, 497–498
    technology in, 498
    unaccounted-for water in, 498–499
    volume losses in, 499
  audit instructions by, 28, 36, 118
  for statistics, 49
  lost water value estimates, 26
  main break estimates, 15
  upgrading cost estimates, 8
  UARL estimate standards by, 435
AMR (automatic meter reading) systems:
  benefits of, 168, 375

AMR (automatic meter reading) systems (*Cont.*):
  for illegal usage detection, 23
  in Philadelphia system, 443, 446, 460–462
Analog values, 96–97
Anchorage in valve installation, 302
Ancient water systems, 4
Anode beds, 319–322
Antifraud actions (*see* Unauthorized water usage)
Apartments, meter sizing in, 364–365
Apparent losses, 2, 5, 161–164
  billing systems in, 16, 168, 375–378
  causes of, 15–17
  in England vs. United States, 31
  meters in, 342–345
    benefits of, 369–375
    field testing, 353–357
    large, 345–353
    master meter calibration, 341–342
    production meter testing, 166
    reading of, 168
    replacement of, 168
    sales meter testing, 166–167
    sizing, 124, 167, 357–369
    specifications, 167
  in nonrevenue water, 39
  in Philadelphia system, 460–462
  revenue recovery in, 169
  unauthorized usage in, 17, 168
Aqualog device, 213
Attitudes, 27
Audits, 19
  authorized use in:
    metered, 511–516
    unmetered, 118, 516–524
  for Boston, 52, 56–57
  confidence factors in, 153
  for conservation, 496
  corrective measures after, 530–531
  demand profiles for, 598
  economics in, 154–158
  in efficiency programs, 395
  instructions for, 28, 36, 118
  international standard, 38–40
  for losses, 35–36
  meter accuracy in, 509, 515–518
  for meter reader reports, 376
  for Philadelphia, 52, 54–55, 448
  preparation for, 502–507
  results analyses in, 528–530

633

Audits (*Cont.*):
  sensitivity analyses in, 154
  source measurements in, 508–509
  spreadsheet models for, 118–124
  supply measurements in, 507–508
  total supply figures in, 509–511
  water loss measurements in, 524–528
  worksheets for, 502, 504–506
Australia, water programs in, 7
Authorized consumption, 40
  in audits:
    metered, 511–516
    unmetered, 118, 516–524
Automated billing systems, 375
Automated sensors and telemetry for conservation, 496
Automatic control valves (ACVs), 269–271, 284, 294
  installation points for, 285–286
  multiple valve sectors in, 286
  for pressure reduction, 284–285
  for reservoir and tank control, 286–291
Automatic meter reading (AMR) systems:
  benefits of, 168, 375
  in Philadelphia system, 443, 446, 460–462
  for unauthorized usage detection, 23
Automation in optimization programs, 10
Average annual day demand (AADD), 385–387
Average velocity point measurements, 66
Averaging mode in pulse counting, 94–95
Averaging pitot meters, 592–595
AWWA (*see* American Water Works Association (AWWA)

BABE (breaks and background estimates) concepts:
  in Philadelphia system, 449, 452
  for real losses, 136–140
Backfill in tracer gas process, 218
Background data for audits, 118–120
Background losses:
  in BABE model, 136–137, 139–140
  in leakage calculations, 272–273
  in UARL estimates, 437–438
Balancing flows, worksheets for, 61
Balancing levels, worksheets for, 61–62
Balancing pressures, worksheets for, 61
Ball valves:
  and pressure, 262
  for reservoir and tank control, 287–288
Base demands, 383–384
Base sewerage flows, 384–385
Batch procedure for usage estimates, 516
Bearings in pumps, 487
Benefits and costs (*see* Costs and benefits calculations)
BEP (best efficiency point) in pumps, 487

Berea-Alexander Park supply district case study:
  background, 279–280
  district and pipe condition in, 280–281
  final design in, 282
  preinstallation investigation and plan in, 281–282
  results of, 282–283
  selection criteria in, 280
Best efficiency point (BEP) in pumps, 487
Bid documents, 402
  data collection requirements in, 406
  equipment replacement in, 406
  experience requirements in, 406–407
  factors in, 403
  inspection and testing in, 405–406
  project goals in, 404
  tasks in, 404–405
Bids in contractor selection, 407–409
Billed authorized consumption, 40
Billing systems, 375
  apparent losses from, 16, 168, 375–378
  in DeKalb County, 474–475
  errors in, 20–21, 525
  meter reader report audits in, 376
  in optimization programs, 10
  in Philadelphia system, 460
Blue Pages, 161
Boreholes in Cheadle Water Works project, 336, 338
Boring bars, 182
Boston, audit data for, 52, 56–57
Boston Housing Authority (BHA), 363–364
Boston Water and Sewer Commission (BWSC), 358, 599
Boundary valve controls, 207–208, 269
Brazil:
  São Paulo (*see* São Paulo case study)
  water programs in, 7
Breaks and background estimates (BABE) concepts:
  in Philadelphia system, 449, 452
  for real losses, 136–140
Bulk metering systems, 470–471
Bulk supply meters, 108–109
Bunker Hill housing development, 364
Bursts, 14
  and pipe replacement, 328
  water mains, 29
Bursts and Background Estimates (BABE) concept:
  in Philadelphia system, 449, 452
  for real losses, 136–140
Bypassed meters, identifying, 378
Bypasses:
  in Johnson City, 479
  for large meters, 345–346
  for meter testing, 344
  with pumps, 483–484
  for turbine meters, 541
  for valve installation, 300, 302

C-factor testing, 417–418
Cables, maintaining, 110
Calibration, 98–99
  of level sensors, 63
  of meters, 563–564
    in DeKalb County, 474
    master meters, 341–342
    in meter testing, 544–545
    ultrasonic, 76–77
  of pressure sensors, 63, 101–102
Calibration K factor, 586
Canada, water programs in, 7
Capacity of pumps, 486
Capital costs and programs:
  limitations of, 381
  in meter sizing, 368
  in Philadelphia system, 449, 458
Carriage concerns for portable equipment, 110
Cast iron ring fractures, 257–258
Castle Donington DMA, 214
Cathodic protection systems, 318–321
Cavitation concerns in valve sizing, 294–296
Cement pipe linings, 329–330
Center compression locks (CCLs), 462
Chamber valve installation, 302–303
Characteristic accuracy curves, 551
Charts:
  for flow, 83–84
  for pressure measurement, 101–102
Cheadle Water Works project, 330–331
  background, 331
  benefits in, 338–339
  conclusions, 340
  organization of, 332–333
  progress in, 338
  proposals in, 333–338
Cleaning pipes, 328
Clearwell losses, 527
Clicks in valve changes, 208
Clocks for flow charts, 84
Close-fit pipe lining, 328
Closed pumping systems, 486
Closed valves, 208
Coatings in acoustics, 201–202
Coefficient of discharge in Venturi meters, 576
Coefficients in BABE modeling concepts, 138–140
Coherence analysis, 199
Commercial meters, sizing, 356, 365–366
Comparative flows in testing, 98
Comparative pairs of meters, 107
Comparison procedure for usage estimates, 516
Compound meters, 19, 85, 89, 91
  crossover flow rates for, 353
  in demand profiles, 612–613
  flow data for, 93
  installing, 347–348
  in meter sizing case study, 360
  testing, 355

Compressors:
  in Gasophons, 217
  in tracer gas process, 219, 222–223
Computer modeling, 18
Condition reports, 403
Condominiums, meter sizing in, 364–365
Conduit testing (see Tracer gas process)
Confidence factors in audits, 153
Connection leaks, 32
Conservation, 2, 26
  accountability for (see Accountability)
  demand profiles for, 598
  Heath Consultants case study, 183–185
  in leak repair, 243
  in meter sizing, 368–369
  pressure management for, 262–263
  for pump energy savings, 483
Conservation of mass, 422–423
Construction:
  in Cheadle Water Works project, 335
  costs:
    in costs-to-benefits calculations, 156–157
    in economics of leak repair, 252
  unmetered use for, 524
Consultants in Johnson City, 479–480
Consumption analysis model, 124–131
Contamination risk reductions, 252
Contractors:
  bid documents for (see Bid documents)
  contracts for, 412–414
  vs. in-house staff, 401–402
  in performance-based approach case study, 409–413
  selecting, 407–409
Control methods, 6–7
Control valves:
  losses from, 526–527
  in meter testing, 349
  in pressure management [see Automatic control valves (ACVs)]
Controllers in pressure management, 114–115, 297–300
Cooling uses in base demands, 383
Corporation stop leaks, 257, 258
Corrective measures, value of, 531
Correlator surveys, 185–189
Corrosion, 315–316
  controlling, 165–166
  electrolytic (see Stray current environment case study)
  flow measurement affected by, 76
Cost per unit of recoverable leakage, 118, 121
Cost-to-intervene calculations, 156
Cost-to-usage relationships, 17
Costs:
  in audits, 529
  increases in, 497–498
  of leak detection and repair, 246, 248–249, 252, 530
  in water loss, 499–501

Costs and benefits calculations:
  in audits, 123
  in conservation, 263
  demand profiles for, 598
  in economics of lost water, 155–156
    in 100 percent supplied systems, 156–157
    graphs for, 158–159
    in rotational supply systems, 157–158
  in meter replacement, 352–353, 572
  in pressure management, 275–279, 283–284
Cover factor in helium-spread patterns, 224–225
Crack leaks in testing, 192
Cracking for pipe replacement, 328
Critical flow ranges, demand profiles for, 598
Critical nodes in pressure management, 274
Cross-correlation, 199–200
Cross-sectional pipe area, 420
Crossover flow rates for compound meters, 353
Cubic meters per hour, 419
Cultural attitudes, 27
Current anode beds, 319–322
Current in stray current environments (see Stray current environment case study)
Curtis Hall Municipal Building, 367
Customer issues:
  in Cheadle Water Works project, 338–339
  meter benefits in, 369–375
  pressure management for, 265
Customer meter sizing model, 124
Customer service piping leaks, 14

Data calibration forms, 63, 102
Data collection and management:
  in bid documents, 406
  in Cheadle Water Works project, 333
  in leak repair case study, 244
  in meter sizing, 357, 369
  worksheets for, 59–62
Data loggers, 81–83, 100–101
  for demand profiles, 600
  for insertion meters, 72–74
  for leak noise, 107
Data transmission methods, 85
Databases in meter sizing, 361–362
Dead-end testing, 206–208
Dead-weight tests, 98–99, 101
Decision page for meter sizing, 124
Decorative facilities, unmetered use for, 521
Deferred expenditures:
  in costs-to-benefits calculations, 157
  in leak repair economics, 252
DeKalb County:
  analysis in, 472

DeKalb County (Cont.):
  accounting and billing in, 474–475
  meters in, 472–474
  unmeasured consumption in, 475–476
  background, 471–472
  benefits in, 476
  summary, 471
Demand analysis and control:
  in efficiency programs, 382–389, 397–398
  in flow analysis, 206
  in reservoir and tank control, 288–290
Demand-based pressure controllers, 115, 298–299
Demand profiles:
  benefits of, 611–612
  compound vs. turbine decisions in, 612–613
  maintenance considerations in, 613
  recorders for:
    data storage capacity of, 602
    magnetic sensor installation, 601
    recording methods, 600–601
    theory of operation, 599–600
  recording data in:
    customer habits in, 602–603
    data storage interval in, 603–605
    length of record in, 602
    meter accuracy in, 608
    pulse resolution in, 605–607
  reports and graphs in:
    accuracy in, 608–609
    presentation options for, 611
    resolution in, 609–611
Demographics, 27
Density in pump general energy equation, 484
Desalination, 6
Design in pump motor efficiency, 488
Desktop study in flow analysis, 204–205
Diaphragm valves, 110, 113, 294
Diaphragms in Venturi meters, 576
Dielectric polyethylene encasements, 325–326
Differential pressure meters and transducers, 574
  averaging functions in, 66
  in hydrant meters, 78–81
  in Venturi meters, 576
Diffusers, hydrant, 79
DIPRA (Ductile Iron Pipe Research Association), 317
Direct-feed systems, 16
Discharge:
  of test water, 349, 352
  in usage estimates, 516
  in Venturi meters, 576
Discounts, 17
Discovered leaks in evaluating water losses, 40–44
Diseases, water-borne, 6, 190

636 INDEX

Distribution systems:
 historical, 3
 losses from, 526
 pressure management for, 263–265
District metered areas (DMAs), 18
 establishment of, 211
 leakage measurements in, 19
 night flow in, 31–32
 in Philadelphia system, 449, 452–453, 456, 459, 463
Diurnal demand curves, 397–398
Doppler meters, 74–75, 589
Drainage, water usage for, 475
Drive efficiency in pumps, 488–489
Drought situation scenario, 147–149
Dry holes in leak repair case study, 250
Ductile Iron Pipe Research Association (DIPRA), 317
Ductile iron pipes in stray current environments (see Stray current environment case study)
Dynamic head in pump general energy equation, 485

Economics of lost water, 5
 costs and benefits calculations in, 155–156
  in 100 percent supplied systems, 156–157
  graphs for, 158–159
  prices and usage in, 154–155
  in rotational supply systems, 157–158
 leak repair case study, 241–242
  analysis in, 244–245
  background, 242–243
  conclusions, 252–254
  data collection in, 244
  discussion, 252
  leak types in, 247–250
  objectives of, 243–244
  results of, 245–247
  sonic accuracy in, 250–252
 revenue effects:
  in apparent loss control, 169
  in costs-to-benefits calculations, 157
  in efficiency programs, 398–399
  in optimization programs, 10
  from pressure management, 266–267
  from rotational supplies, 156
  from unaccounted-for water, 357
 in Severn Trent case study, 212
Economics page for meter sizing, 124, 127
Edgar, Tim, 599
Efficiency and efficiency programs, 4, 381
 benefits of, 382
 implementation plans in, 393–395
 monitoring and tracking in, 395
  audits in, 395
  diurnal demand curves in, 397–398
  Hawthorne effect in, 397
  meters, 395–396
  percentages vs. hard values in, 396–397

Efficiency and efficiency programs (Cont.):
 of motors, 488
 pressure management for, 263–265
 of pumps and pump drives, 486, 488–489
 revenue effects in, 398–399
 system demand components in, 382–383
  average annual day demand, 385–387
  base demands, 383–384
  base sewerage flows, 384–385
  maximum summer/peak day demands, 387–389
  summary, 389
 targets in, 389–390
  realistically achievable savings, 392
  theoretical maximum savings, 390–392
Elbows in leak detection, 201–202
Electrical continuity:
 through meters, 536
 in stray current environment, 317–318
Electricity costs in water loss, 499
Electrode cleaning, 584
Electrolytic corrosion (see Stray current environment case study)
Electronic geophones, 104–105
Electronic meters, 87, 560
Electronic self-diagnostics, 590
Emergency situations, valves for, 297–298
Energy costs in water loss, 499
Energy-efficient pump motors, 488
Energy equation for pumps, 482–485
Engineering costs, 156–157
England, 7, 29–32
Entrained air, 541
Environmental perspective, 28
Environmental Protection Agency (EPA), 28
Environmental regulation, 7
Epoxy pipe linings, 329
Equipment and techniques, 65
 for acoustic leak detection, 103–105
 in bid documents, 406
 calibration, testing, and dead-weight tests, 98–99
 for meter testing, 107–109, 559–563, 569–570
 output readings from:
  analog values in, 96–97
  pulse recordings, 88–96
  summary, 97–98
 permanent, 84–87, 89
 portable (see Portable equipment)
 for pressure:
  controlling, 110–115
  measuring (see Pressure and pressure management)
 for tracer gas process, 105, 216–217
Errors:
 accounting and billing, 20–23, 525
 measurement, 19–20
 meter, 515–518

Estimated average nodes in pressure management, 274
Estimated read syndrome, 168
Estimates:
 in audits, 153
 by AWWA, 8, 15, 26
 in leakage repair economics, 250–251
 for real losses, 136–140
Evaluating water losses, 35–36
 audits for, 38–40
 inadequacy of, 36–37
 IWA recommendations for, 46–49
 performance indicators for, 44–46
 published statistics for, 49–51
 standards for, 37–38
 summary, 51–52
 unavoidable losses in, 40–44
Evaporation losses, 527
Event recording, 95
Excavation in tracer gas process, 226
Expanding power law in pressure management, 270
Experience requirements in bid documents, 406–407
Explosive decompression danger in tracer gas process, 219–220

Fairmount Water Works, 443
FAVAD concepts, 118, 128–136
Ferrule leaks, 258
Fidelia Way housing development, 364
Field meter testing, 353–357, 567–568
Fill type in tracer gas process, 218
Filters:
 in geophone surveys, 181
 in testing, 99
Filtration in Philadelphia system, 443
Financial effects (see Economics of lost water)
Financial performance indicators, 47
Finished water temperature (FWT) as leak predictor, 254–255
 analysis in, 256
 conclusions, 257
 operational observations in, 255–256
Fire fighting:
 in DeKalb County, 475
 parallel valves for, 296–297
 in pressure management, 266
 regulations for, 418–419
 unmetered use for, 519
Fire hydrant abuse, 462
Fire line meters:
 installing, 349
 testing, 355
Fittings, maintaining, 110
Fixed and variable area discharge (FAVAD) concepts, 118, 128–136
Fixed outlet control, 269–271, 426–427
Fixed paths in leakage calculations, 271–272
Flat-rate charges, 17

Flat rate method for UARL estimates, 435–436
Flexible hoses in meter testing, 570
Float control valves, 110, 112, 113
Flow charts, portable, 83–84
Flow range in meter testing, 551
Flow Search device, 362–363
Flow straighteners for sonic meters, 589
Flow tubes in Venturi meters, 577–578
Flow valves:
  for pressure control, 110, 113
  for reservoir and tank control, 288–290
Flowmaster software, 418
Flows and flow rates, 419–420
  in active leakage management, 203–208
  in BABE model, 139
  balancing, 61
  demand profiles for (see Demand profiles)
  example, 420
  in leakage, 15
  loggers for, 81–83, 100–101
    for demand profiles, 600–601
    for insertion meters, 72–74
    for leak noise, 107
  measuring:
    portable ultrasonic meters for, 76–77
    in pressure management, 274
  and meters:
    compound, 93, 353
    propeller, 90
    sizing, 361–363
    testing, 347, 350, 551–552, 565
    turbo, 88
  in pump general energy equation, 483–484
  testing equipment for, 363
  types of, 420–423
  units for, 59–60
  in valve installation, 304
Flushing lines and sewers, unmetered use for, 519–520
Flushometers, 368
Formats for data collection worksheets, 62
Fort Drum, 198, 200–201
Foundry condominium complex, 365
Fountain sound in leak frequency, 179
4–20-mA output, 85, 96, 100
Free riders, 394
Frequency:
  filters for, 179
  meters for, 85–86
  for pressure loggers, 100
  recording, 88, 90–94
Frequency of leakage, pressure management for, 273
Friction sound in leak frequency, 179
Frost in helium-spread patterns, 225–226
FWT (finished water temperature) as leak predictor, 254–255
  analysis in, 256
  conclusions, 257
  operational observations in, 255–256

G-Man device, 369–375
Gallons per minute (gpm), 419
Galvanic cathodic protection systems, 318–319
Galvanic corrosion, 317
Gas process (see Tracer gas process)
Gasophon, 216–217
Gauge pressure, 426
Gears in meters, 544–545
Genesis Meter Module (GMM) radio transponder, 337
Geographic Information System (GIS), 107, 173–175
Geography in water loss, 27
Geophones, 104–105, 180–183
Germany, water programs in, 7
GIS (Geographic Information System), 107, 173–175
Glaciers, 3
Global Positioning System (GPS), 173–174
GMM (Genesis Meter Module) radio transponder, 337
Government action in water loss, 28–29
Government-owned properties, accounting for, 22
GPR [see Ground-probing radar (GPR)]
GPS (Global Positioning System), 173–174
Graphs for meter sizing, 124, 126
  accuracy in, 608–609
  presentation options for, 611
  resolution in, 609–611
Gravity-feed systems, 424–425
Greater Johannesburg Metropolitan Council (GJMC), 279–280
Ground-probing radar (GPR), 176
  case study:
    future of, 232–233
    growth in use of, 228–230
    history of, 227–228
    method in, 230–231
    present state in, 230
    pros and cons of, 231–232
Ground surveys, 180–183
Grounding cells, 317
Grounding of meters, 536
Groundwater level in helium-spread patterns, 225
Guaranteed storage, pressure management for, 264

Half-cell electrodes, 323
Halifax Regional Municipality, 235–239
Hawthorne effect, 397
Hazen-Williams C factor, 417–418
Head losses:
  in magnetic meters, 582
  in pump general energy equation, 485
  in São Paulo system, 466
  in valve installation, 300–301
  in valve sizing, 294–296
Health risks, 6, 156, 190

Heat in pumps, 487
Heath Consultants case study, 183–185
Helium (see Tracer gas process)
HGLs (hydraulic grade lines), 61
High water levels in Cheadle Water Works project, 338
Holiday homes, tracking, 377
Hoses for meter testing, 345
Hot tapping, 68–70
Human failings, 11–12
Hydrant diffusers, 79
Hydrant leaks and surveys:
  for leak detection, 177–180
  in leak repair case study, 247, 249–250
Hydrant meters, 78–81
Hydrant needs for fire fighting, 419
Hydraulic ACVs, 284
Hydraulic capacity in reservoir and tank control, 291
Hydraulic connections:
  leaks in, 257, 260
  in valve installation, 302
Hydraulic dead-weight testers, 101
Hydraulic grade lines (HGLs), 61
Hydraulic shock, 426
Hydraulics:
  C-factor testing, 417–418
  demand profiles for, 598
  fire fighting regulations, 418–419
  flow, 419–423
  pipe roughness coefficients in, 417
  pressure, 423–426
  in pressure management, 264–265, 274–275
  in pump losses, 486–487
Hydrologic cycle, 3
Hydrostatic testing, 218

ICI meters, 17, 108
ILI (infrastructure leakage index):
  as performance indicator, 47–50, 52
  in Philadelphia system, 459–460
  in technology application case study, 239
Illegal actions (see Unauthorized water usage)
Impact of lost water, 4–5
Impact sound in leak frequency, 179–180
Impeller trim of pumps, 486
Impressed current anode beds, 319–322
In-house staff vs. contractors, 401–402
In-place calibration, 77
Incentives in efficiency programs, 393–394
Incremental unit cost of production, 493
Indoor meter installation:
  vs. outdoor, 534–535
  settings, 536–537
Industrial, commercial, and institutional (ICI) population, 17, 108
Industrial customers:
  meter sizing for, 356
  reservoir and tank control for, 290

638 INDEX

Infrared testing, 233
Infrastructure:
　in BABE modeling concepts, 138
　in Philadelphia system, 458–460
Infrastructure leakage index (ILI):
　as performance indicator, 47–50, 52
　in Philadelphia system, 459–460
　in technology application case study, 239
Insert flow tubes in Venturi meters, 579
Insertion meters, 66–68
　data logger fitting for, 72–74
　hot tapping with, 68–70
　installation process, 70–72
　profiling with, 72
Inspection in bid documents, 405–406
Installation:
　of meters (see Meter installation)
　of pressure management equipment, 274–275
　of valves (see Valves)
Institutional properties in meter sizing, 366
Insulating pipes, 317
Insurance ratings, 419
Internal wells, 378
International standard audits, 38–40
International Water Association (IWA), 31, 35–37, 41
　unavoidable real loss approach by, 437–438
　water loss evaluation recommendations for, 46–49
Intervals:
　in demand profiles:
　　in recording, 603–605
　　in reports, 609–611
　in meter testing, 554
Invensys Metering Systems, 93–94
Investigations in meter sizing case study, 362
Iron pipes in stray current environments (see Stray current environment case study)
Irrigation:
　in base demands, 383
　unmetered use for, 523
Isolation valves:
　losses from, 526–527
　in pressure management, 294
IWA (International Water Association), 31, 35–37, 41
　unavoidable real loss approach by, 437–438
　water loss evaluation recommendations for, 46–49

James Michael Curley Recreational Center, 366, 368
Japan, water programs in, 7
Jerusalem Water Undertaking (JWU), 307–308

Johannesburg:
　Berea-Alexander Park supply district, 279–283
　leakage detection in, 227–230
Johnson City:
　introduction, 476–477
　leaks in, 479–480
　meters in, 478
　record keeping in, 479
　strategy for, 477–478
　team for, 477
　unauthorized use and theft in, 478–479
　update, 480–482
Joints for stray currents, 317–318, 325

K factor, 586
Kuichling equation, 9, 434–436

Labor costs, 246
Lag time in meter readings, 512, 514
Lambert, Allan, 136
Laminar flow, 422
Landscaping, unmetered use for, 521–523
Large flows, parallel valves for, 296–297
Large meters:
　bypass lines for, 345–346
　installation of, 538–540
　test outlets for, 346
　testing, 346–353, 562
Large Water Meter Handbook, 599
Leak Detection and Water Accountability Committee, 49
Leak Detection Squad, 449
Leak noise correlation, 106, 185–189
Leak noise loggers, 107
Leak plugs in testing, 192
Leakage:
　active leakage management (see Active leakage management)
　causes of, 14
　detection and location of:
　　acoustic (see Acoustic detection techniques)
　　acoustic surveys for, 177, 199–201
　　for conservation, 496
　　demand profiles for, 598
　　geophone surveys for, 180–183
　　hydrant surveys for, 177–180
　　leak noise correlators for, 106, 185–189
　　in leak repair economics case, 252
　　in meter sizing, 368
　　in real loss control, 164
　　in Severn Trent case study, 213
　　in tracer gas process, 226
　　visual surveys for, 177–178
　in evaluating water losses, 40–44
　leak survey case study:
　　conclusion, 413
　　contract development in, 412–413
　　history and background, 409–410
　　requests for proposals in, 410–412
　　pipes (see Cheadle Water Works project)

Leakage (*Cont.*):
　pressure management for, 261–262, 270–273
　rate estimates for, 250–251, 606–607
　repairing, 241
　　economics of (see Economics of lost water)
　　leak types in, 257–260
　　in real loss control, 164–165
　　safety in, 257
　　water temperature as predictor, 254–257
　　types of, 247–250, 257–260
Leakage Management Assessment (LMA) Project, 444, 447–460
Least-cost curves, 212
Legislation for water quality, 28
Lesotho Highlands Water Scheme, 229–230
Level drop tests, 233
Levels:
　balancing, 61–62
　calibrating, 63
　sensors for, 100–101
　units for, 60
Lexan test probes, 221
Lift-and-shift strategy, 214
Line flushing, water usage for, 475
Line shaft bearing losses, 487
Line valves, 269
Linings for pipes, 328–330
Listening sticks, 103–104, 179
Liters per second (L/s), 419
LMA (Leakage Management Assessment) Project, 444, 449, 452–460
Loads in pump motor efficiency, 488
Local support in pressure management, 294
Location of leaks (see Leakage)
Lockout program, 479
Loggers, 81–83, 100–101
　for demand profiles, 600
　for insertion meters, 72–74
　for leak noise, 107, 210–211, 213
Logging (see Loggers; Recording data)
Logistics in Cheadle Water Works project, 337
Louisville Water Co., 254–257
Low-flow meters, 167
Low flow rates in meter sizing, 361
Low loss flow tubes in Venturi meters, 577
Lower-speed motors, 488

M36 audits, 28, 36, 118
Magnetic meters, 85, 92, 581–582
　advantages and disadvantages of, 584–585
　averaging functions in, 66
　insertion meters, 66, 69
　installing, 582
　maintaining, 582–584

INDEX

Magnetic pickup devices, 600–601, 605–607
Mains:
   in Cheadle Water Works project, 334, 338
   finished water temperature in, 254
   flushing, 519
   in leak repair case study, 247, 249–250
   in Philadelphia system, 448, 458
   replacement of, 166–167, 327
   in tracer gas process, 218–220, 222–223
   for valve installation, 300–301
Maintenance:
   costs of:
     in costs-to-benefits calculations, 157
     demand profiles for, 598, 613
     in pressure management, 294
   of meters:
     magnetic, 582–584
     pitot, 595
     turbine and propeller, 586–587
     Venturi, 577
     vortex, 592
   of pressure measurement equipment, 109–110
   of valves, 284, 306–307
Malaysia, water programs in, 7
Manual M36, 28, 36, 118
Manual read systems, 85, 168
Mapping:
   in active leakage management, 172–177
   in audits, 507–508
Marginal unit cost of production, 493
Master meters, 18
   calibration of, 341–342
   in meter testing, 570
   testing, 108–109, 166, 342
   types of, 341
Master plans, updating, 531
Materials:
   in leak detection and repair program, 246
   in tracer gas process, 105, 216–217
Max-Min interval, 609–611
Maximum condition point (MCP) in pumps, 487
Maximum summer/peak day demands, 387–389
McIlroy Fluid Network Analyzer, 443
Measurements:
   in audits, 507–509, 524–528
   errors in, 19–20
   pressure (see Pressure and pressure management)
Mechanical dead-weight testers, 101
Mechanical geophones, 104
Mechanical listening sticks, 103–104
Mechanical losses in pumps, 487
Mechanical meters, 87, 89
Megaliters per day (MLD), 419
Metal-based pipes, 166

Meter calibration, 563–564
   in DeKalb County, 474
   master meters, 341–342
   in meter testing, 544–545
   ultrasonic, 76–77
Meter efficiency test worksheets, 351
Meter installation:
   checking, 515
   general considerations, 535–536
   indoor settings, 536–537
   indoor vs. outdoor, 534–535
   insertion meters, 70–72
   introduction, 534
   settings for, 537–540
   turbine meters, 540–541
   ultrasonic meters, 76
Meter Master device, 82
   in customer relations case study, 369–375
   in meter sizing case study, 362–364
Meter pits, 568
Meter readings:
   AMR systems:
     benefits of, 168, 375
     for illegal usage detection, 23
     in Philadelphia system, 443, 446, 460–462
   in apparent loss control, 168
   auditing, 376
   lag time in, 512, 514
   measurement errors in, 20
   meter type in, 377
Meter sizing, 356–357
   in apparent loss control, 167
   case study, 357
     background, 357–358
     conclusions, 367–369
     past practices, 360–361
     project approach, 361–363
     results, 363–367
     unaccounted-for water in, 358–360
   in DeKalb County system, 473, 476
   demand profiles for (see Demand profiles)
   model for, 124
Meter testing, 166–167, 343–345, 510
   accuracy in, 543–546, 551, 572
   for apparent losses, 166–167, 353–357
   considerations in, 572
   in DeKalb County system, 472–474
   elements in, 551
   equipment for, 107–109, 559–563, 569–570
   field, 353–357, 567–568
   flow rates in, 551–552
   intervals in, 554
   introduction, 542–543
   large meters, 346–353
   master meters, 108–109, 342
   multi-jet meters, 545–546, 565
   multiple meters, 565–567
   on-site, 343–344, 568–572
   in pressure management, 107–109

Meter testing (*Cont.*):
   procedures for, 547–551, 563–565, 570–572
   program coordination for, 546–547
   quantities in, 552–554
   statistical sampling in, 554–558
   turbine meters, 346, 355, 546
Metered use in audits, 511–516
Meters:
   accuracy of (see Accuracy)
   for apparent losses (see Apparent losses)
   averaging pitot, 592–595
   calibration of (see Meter calibration)
   characteristics of, 86, 87, 89
   in Cheadle Water Works project, 336–337
   comparative pairs of, 107
   compound (see Compound meters)
   for conservation, 494
   in customer relations case study, 369–375
   data transmission methods in, 85
   in DeKalb County system, 472–474
   in efficiency programs, 395–396
   hydrant, 78–81
   importance of, 18–19
   insertion (see Insertion meters)
   installation of (see Meter installation)
   in Johnson City, 478
   magnetic (see Magnetic meters)
   measurement errors in, 19–20
   orifice plate, 579–581
   output correlation in, 85–86
   pulse recording in, 89, 91
   reading (see Meter readings)
   in São Paulo case study, 469–471
   in Severn Trent, 331
   sizing (see Meter sizing)
   sonic, 588–590
   testing (see Meter testing)
   tilted, 357
   turbine and propeller (see Turbine and propeller meters)
   Venturi (see Venturi meters)
   vortex, 590–593
Metro Water Services (MWS), 409–413
Millions of gallons per day (mgd), 419
Minimum flow rates in demand profile recording, 606
Mitigation of stray current, 324–326
MLD (megaliters per day), 419
Modeling water losses, 117–118
   audit spreadsheet models, 118–124
   commercial models, 140–141
     scenario one, 141–146
     scenario three, 147–150
     scenario two, 146–147
   consumption analysis model, 124–131
   customer meter sizing model, 124
   pressure analysis and FAVAD concepts in, 128–136
   real losses, 136–140

640 INDEX

Modulation speed in valve installation, 305–306
Monitoring points in pressure management, 273–274
Motor efficiency in pumps, 488
Multi-jet meters, 545–546, 565
Multifamily residential meters, 612
Multiple leaks, tuned location method for, 200–201
Multiple meters, testing, 565–567
Multiple valve sectors with ACVs, 286
Municipal properties in meter sizing, 366–367
Museum of Fine Arts, 366
MWS (Metro Water Services), 409–413

National Awards Program for Energy Innovation, 184
National Leakage Initiative, 2, 7, 30
National Risk Management Laboratory, 191
Natural replacement in efficiency programs, 393–395
Network analysis models, 117
New-technology leak detection equipment, 106–107
New Zealand, water programs in, 7
Night flows, 18–19
  in DMAs, 31–32
  in flow analysis, 204–205
  in Philadelphia system, 463–464
Noise:
  leak noise correlation, 106, 185–189
  and listening sticks, 103
  with magnetic sensors, 601
Noise logging, 107
  in active leakage management, 210–211
  in Severn Trent case study, 213
Nonaccount water, 492, 494
Nonbilled accounts, 461
Nondestructive evaluation methods, 191
Nonrevenue water (NRW), 37–38, 50
  in AWWA committee report, 498–499
  components of, 39
  performance indicators for, 46–50, 433–434
  revenue losses from, 357
  statistics for, 49
  in technology application case study, 236
  in water balance, 431
Nonvisible leakage, detecting, 164
North America, water loss management in (*see* Water loss management)
Northeastern University, 366

Oak Ridge Reservation, 198, 200–201
Office of Water Services (OFWAT), 30, 46
On-site operations:
  meter testing, 343–344, 568–572
  valve installation investigation, 300

100 percent supplied systems, costs and benefits calculations in, 156–157
One-year benefits in audits, 529
Open pumping systems, 485
Open valves, 208
Operating cost increases, 497–498
Operational pipeline evaluations, 198
Operations and maintenance costs in audits, 529
Optical devices:
  in demand profile recording, 600–601, 605–607
  switch sensors, 87
Optimization programs, 8–11
Orifice plate meters, 579–581
Outdoor meter installation:
  vs. indoor, 534–535
  settings for, 537–538
Output correlation in meters, 85–86
Output readings, 89, 91
  averaging mode in, 94–95
  in demand profile recording, 605–607
  event recording in, 95
  frequency recording, 88, 90–94
  state recording, 96
Overflows:
  in evaluating water losses, 40–44
  in pressure management, 273
  in real loss control, 165
  testing for, 233–235
Overhead:
  in leak detection and repair program, 246
  in optimization programs, 10
Overlap in meter readings, 376

Paper mistakes, 525
Parallel installations, 293, 296–298
Paths in leakage calculations, 271–272
Patroller device, 213
Pavement in leak detection and repair program, 246
Payments to contractors, 414
Peak day demands, 387–389
Penalties, 17
Per-capita use in pressure management, 266–268
Percent registration in meter testing, 551
Percentages, limitations of, 44–46, 396–397, 499
Performance-based bids, 409–413
Performance curves in meter testing, 551
Performance in optimization programs, 8–9
Performance indicators:
  for evaluating water losses, 44–46
  IWA recommendations for, 439
  for nonrevenue water and real losses, 46–50, 433–434
  in Philadelphia system, 459
Performance Indicators for Water Supply Services, 31, 35
Permalog device, 211, 213–215

Permanent equipment, 84–87, 89
Personnel requirements in tracer gas process, 219
Perspectives on water loss, 25
  in England and Wales, 29–32
  future, 32–33
  in United States, 25–29
Philadelphia system, 443–444
  audit data for, 52, 54–55, 448
  best management practices in, 446–447
  future prospects, 462–464
  real loss evaluation in, 447–448
    approach and findings in, 449–460
    audit for, 448
    conditions in, 448–449
    remediation in, 460–462
    water loss in, 444–446
Philadelphia Water Department (PWD), 444
Pigging for pipes, 329
Pilot programs, 394–395
Pilot rods, 66–67
Pinhole leaks, 259
Pipes and pipelines, 315
  in acoustic technology case study, 193, 198, 201–202
  in Berea-Alexander Park system, 280–281
  bursts in, 14
    and pipe replacement, 328
    water mains, 29
  in C-factor testing, 418
  coatings for, 201–202
  corrosion in, 76, 165–166, 315–316
  customer service leaks, 14
  in leakage case study (*see* Cheadle Water Works project)
  for meter testing, 560, 568
  for pitot meters, 595
  properties of, 615–632
  rehabilitation methods for, 328–330
  replacement of, 166–167, 327–328, 330
  roughness coefficients for, 417
  in São Paulo system, 466
  in stray current environments (*see* Stray current environment case study)
  in tracer gas process, 218
  for vortex meters, 591
Piston residential meters, 85, 91
Piston valves, 110, 114, 294
Pitometer installations, 452, 454
Pitot meters, 592–595
Plans:
  in active leakage management, 172–173, 175, 177
  in efficiency programs, 393–395
  in tracer gas process, 218
  updating, 531
Plastic-based pipes, 166
Plunger bars in gasophons, 217
Pneumatic rock drills, 217
Polyethylene encasements, 325–326

INDEX **641**

Population growth, 1, 27
Portable equipment, 65–66
　flow charts, 83–84
　flow loggers, 81–83
　hydrant meters, 78–81
　insertion meters, 66–74
　pressure measurement, 99–103
　step testers, 85
　storage and carriage concerns, 110
　ultrasonic meters, 74–78
Potential leakage in audits, 529
Pounds per square inch (psi) for pressure, 424
Power factor in leakage calculations, 272
Precision in meter testing, 551
Premium-efficiency pump motors, 488
Prepaid systems, 169
Pressure and pressure management, 15, 261, 423
　balancing, 61
　in Berea-Alexander Park supply district, 279–283
　calibrating, 63
　for conservation, 262–263
　controllers for, 114–115
　in correlator surveys, 188
　cost-to-benefit calculations for, 275–279, 283–284
　effects of, 426
　for efficient distribution, 263–265
　example, 423–424
　fire flow concerns in, 266
　from fire hydrant abuse, 462
　flow measurements in, 274
　gravity-feed systems, 424–425
　installation locations for, 274–275
　for leakage reductions, 261–262, 270–273
　measurements and measurement equipment for, 99–102, 274, 425–426
　　acoustic leak detection, 103–105
　　dead-weight testers, 101
　　maintaining, 109–110
　　meter testing, 107–109
　　new-technology, 106–107
　　portable charts, 101–102
　　portable loggers, 100–101
　　portable sensors, 103
　monitoring points in, 273–274
　nonpayment issues in, 263
　overflow control in, 273
　in Philadelphia system, 459
　pumps in, 269, 425
　in Ramallah system, 307–312
　in real loss control, 165
　reservoir filling problems from, 268
　revenue losses from, 266–267
　in São Paulo system, 465–467
　sectorization in, 268–269
　units for, 60
　valves for (*see* Valves)

Pressure and pressure management (*Cont.*):
　in Venturi meters, 574–575
　in water loss models, 128–136
Pressure-dependent demand, 143, 146
Pressure gauges for meter testing, 344
Prices:
　in contractor selection, 407–409
　in lost water economics, 154–155, 243
　world, 5
Priorities:
　in BABE modeling concepts, 136–140
　in Cheadle Water Works project, 333
　of losses, 171
Private development, unmeasured consumption in, 475
Process water at treatment plants, unmetered use for, 524
Product costs in costs-to-benefits calculations, 156–157
Production meters (*see* Master meters)
Profiles:
　demand (*see* Demand profiles)
　velocity, 66–70, 72
Project goals in bid documents, 404
Propeller meters (*see* Turbine and propeller meters)
Property damage savings in leak repair, 252
PRVs (pressure-reducing valves), 110, 143, 145, 270
　losses from, 527
　in São Paulo system, 466
Psi (pounds per square inch) for pressure, 424
Public health concerns, 6, 156, 190
Public housing, 363–364
Public image and relations:
　in leak repair economics case, 252
　in rotational supplies, 156
Public uses and losses, 4
Pulse output meters, 85–86
Pulse recording, 89, 91
　averaging mode in, 94–95
　in demand profile recording, 605–607
　event recording in, 95
　frequency recording, 88, 90–94
　state recording in, 96
Pumped systems, 425
Pumps:
　efficiency of, 486, 488–489
　general energy equation for, 482–485
　introduction, 482
　losses from, 161, 486–487, 527
　in meter sizing, 367–368
　in pressure management, 269
　sizing, 487
　system types, 485–486
Purchase cost in audits, 529
PWD (Philadelphia Water Department), 444

Qpmax and qpmin in valve sizing, 294–296
Quality problems in rotational supplies, 156
Quality testing, unmetered use for, 524
QualServe program, 254
Quantities in meter testing, 552–554
Quantity and values of water losses, 17
　accounting errors in, 20–23
　measurement errors in, 19–20
　metering in, 18–19

Radio-read systems, 337–338
Ramallah case study:
　background, 307–308
　financing solutions in, 311–312
　investigations in, 308–310
　problem definition, 308
　proposed solutions in, 310–311
　results in, 312
RASTs (realistically achievable savings targets), 392
Rate of flow (*see* Flows and flow rates)
Raw data from meter readings, 377
Real losses, 2, 5
　causes of, 13–15
　controlling, 161–164
　　active leakage management (*see* Active leakage management)
　　corrosion control in, 165–166
　　leak detection in, 164
　　leak response in, 164–165
　　mains replacement in, 166–167
　　overflow reductions in, 165
　　pressure management in, 165
　　service replacement in, 167
　　zoning in, 165
　in England vs. United States, 30–31
　modeling, 136–140
　in nonrevenue water, 39
　performance indicators for, 46–50, 433–434
　in Philadelphia, 447–448
　　approach and findings in, 449–460
　　audit for, 448
　　conditions in, 448–449
　　unavoidable annual real losses:
　　　IWA estimates, 437–438
　　　in performance, 48
　　　previous estimates, 434–436
　　　statistics for, 49–51
Realistically achievable savings targets (RASTs), 392
Rebates in efficiency programs, 393
Recording data, 97
　customer habits in, 602–603
　data storage interval in, 603–605
　in Johnson City, 479
　length of record, 602
　loggers for (*see* Loggers)
　meter accuracy in, 608
　pulse resolution in, 605–607

Recording data (*Cont.*):
  recorders for:
    data storage capacity of, 602
    magnetic sensor installation, 601
    pulse recording (*see* Pulse recording)
    recording methods, 600–601
    theory of operation, 599–600
  usage by time, 369–375
Recoverable leakage in audits, 529
Reducing tees, 344
Reed switch sensors, 87
Regulations:
  environmental, 7
  fire fighting, 418–419
Regulators in Gasophons, 217
Rehabilitation methods for pipes, 328–330
Remote meters, 534–535
Remote-node pressure controllers, 115
Repairs, 241
  economics of (*see* Economics of lost water)
  leak types in, 257–260
  in real loss control, 164–165
  safety in, 257
  in tracer gas process, 226
Repeatability in meter testing, 551
Replacement:
  in bid documents, 406
  in efficiency programs, 393–395
  in leak detection and repair program, 246
  of meters, 168, 352–353, 572
  of pipes, 166–167, 327–328, 330
  of revenue meters, 356–357
Reported leaks, 29
  in BABE model, 136–137
  in UARL estimates, 437–438
  vs. unreported leaks, 14
Reports:
  in demand profiles:
    accuracy in, 608–609
    presentation options for, 611
    resolution in, 609–611
    system condition, 403
Requests for proposals (RFPs), 410–412
Requirements, water, 2–3
Reservoir and tank control, 286–287
  altitude valve controls for, 287, 289
  ball valve controls for, 287–288
  in Cheadle Water Works project, 336, 338
  demand control in, 288–290
  for industrial customers, 290
  leakage in:
    in audits, 118, 122, 510
    losses from, 527
    testing for, 233–235
  in pressure management, 268, 274, 424–425
  in real loss control, 165
  seasonal variations in, 291
  for sectors with weak hydraulic capacity, 291

Residential End Use Study (REUS), 390
Residential meters, 85, 91–92
  multifamily, 612
  sizing, 356–357
  testing, 108
Residual pressure in meter testing, 352
Resolution:
  in demand profile reports, 609–611
  of sensors, 100–101
Response time:
  in real loss control, 164–165
  in São Paulo system, 467–468
  (*See also* Economics of lost water)
Restoration of terminated service, 23
Restroom usage, model for, 124–131
REUS (Residential End Use Study), 390
Revenue and revenue losses (*see* Economics of lost water; Nonrevenue water (NRW))
Revenue meters, replacing, 356–357
Revenue Protection group, 461
Reynolds number, 422
Rezoning in São Paulo system, 468–469
RFPs (requests for proposals), 410–412
RMS (root-mean-square) signal transition, 194–195
Roads, water usage for, 475
Rolling diaphragm valves, 294
Rome, water system in, 4
Root-mean-square (RMS) signal transition, 194–195
Rotameters, 559–560
Rotating chains and rods for pipes, 329
Rotating meters, 20
Rotating store mode, 97
Rotational supply systems:
  costs and benefits calculations in, 157–158
  negative impacts from, 156
  scenario for, 147–149
Roughness coefficients, 417
Rubber-gasketed joints, 317, 325

Sacrificial anodes, 318
Safe Drinking Water Act, 190
Safety:
  in leak repair, 257
  in meter testing, 345, 568, 572
Sales meter testing, 166–167
Salt water, 3
Sampling:
  in meter testing, 554–558
  in SCADA systems, 97
São Paulo case study:
  introduction, 465
  loss control program in, 465–467
  antifraud actions in, 470
  meters in, 469–471
  performance data in, 467–468
  reaction time in, 467
  rezoning in, 468–469

SCADA (Supervisory Control and Data Acquisition):
  for pressure management, 300
  sampling in, 97
  in technology application case study, 237–238
Schools:
  meter sizing in, 365
  unmetered use for, 521
Scraper trowels for pipes, 329
Seasonal variations, reservoir and tank control for, 291
Sectorization in pressure management, 268–269
Seeding of clouds, 6
Sensitivity analyses, 154
Sensors and sensor locations:
  automated, 496
  calibration of, 63, 101–102
  for flow measurement, 76, 82–83
  in leak detection, 200
  for pressure, 100–103
  in sonic meters, 589
Series installations, 293, 297–298
Service saddles, 344
Services:
  in Cheadle Water Works project, 334, 338
  leaks in, 247, 249–250
  replacement of, 167, 327
Severn Trent system:
  achievements in, 215
  background, 211–212
  conclusion, 215
  leakage detection in, 213
  Permalog device in, 213–215
  pipe leakage in (*see* Cheadle Water Works project)
  strategies in, 212–213
Sewers, flushing, 519–520
Sextus, Julius Frontinius, 44
Shielding:
  magnetic sensors, 601
  pipes from stray currents, 317
Shutoff valves, 294
Signal difference for leak detection, 199–200
Single-point velocity meters, 66, 72, 74
Sizing:
  meters (*see* Meter sizing)
  pumps, 487
  valves, 292, 294–296
Sleeve valves, 294
Slip lining, 328
Small leaks in testing, 192
Soil factors:
  in geophone surveys, 181
  in leak detection, 201
  in tracer gas process, 221–222, 224–225
Sonic detection (*see* Acoustic detection techniques)
Sonic meters, 588–590

Source measurements in audits, 508–509
Source meters, testing, 108–109
South Africa, water programs in, 7
Space limitations in meter sizing, 368
Specific speed of pumps, 486
Speed of leak repair:
    in real loss control, 164–165
    in São Paulo system, 467–468
    (See also Economics of lost water)
Spray linings for pipes, 329–330
Spreadsheet models for audits, 118–124
Standards:
    in accountability, 492–493
    in audits, 38–40
Start-up procedures in valve installation, 303–305
State recording, 96
Static head in pump general energy equation, 485
Static system pressure, 424
Statistical models, 275–279
Statistical sampling, 554–558
Statistics for evaluating water losses, 49–51
Steady-state flow, 421
Step testing:
    in active leakage management, 208–210
    portable testers, 85
    in Severn Trent case study, 213
Stewardship, 11–12
Stop tap bashing, 213
Storage:
    in demand profile recorders, 602
    capacity of, 602
    intervals for, 603–605
    in flow, 422–423
    in portable equipment, 110
Storage tanks (see Reservoir and tank control)
Store till full mode, 97
Storm sewers, flushing, 519–520
Straightening vanes, 586
Strainers:
    in testing, 99
    for turbine meters, 541, 586
Stray current environment case study, 316–317
    cathodic protection systems in, 318–321
    conclusions, 326–327
    electrical discontinuity in, 317–318
    impressed current anode beds in, 319–322
    mitigation in, 324–326
    pipeline route investigation in, 322–324
Street cleaning, 520–521
Study periods in audits, 502, 507
Subsurface interface radar, 227
Sullivan, John, 599
Summer demands, 383, 387–389

Supervisory control and data acquisition (SCADA) systems:
    for pressure management, 300
    sampling in, 97
    in technology application case study, 237–238
Supplier losses:
    apparent, 15–17
    problems in, 13
    quantity and values of water in, 17–23
    types of, 13–15
Supply measurements in audits, 507–508
Supply nodes in pressure management, 273
Surfaces in geophone surveys, 182
Surge anticipator valves, 426–427
Surge-arrestor valves, 264–265
Surge-relief valves, 527
Surges, 264–265, 426–427
Sustainability of water savings, 383–384
Sustaining valves:
    for demand control, 288–290
    setting for, 305
Swimming pools, unmetered use for, 523
Swing-gate principle in step testers, 84
SWT metering, 336–338
System condition reports, 403
System data in BABE modeling concepts, 138
System demand components, 382–383
    average annual day demand, 385–387
    base demands, 383–384
    base sewerage flows, 384–385
    maximum summer/peak day demands, 387–389
    summary, 389
System maps, 507–508
System surveys, 237

Tanks:
    in audits, 510
    for meter testing, 560
    in residential situations, 262
    (See also Reservoir and tank control)
Tapping sleeves, 70–71
Task Force on Water Loss, 35
Teamwork in Cheadle Water Works project, 334–335
Technical bids, 407–409
Technology application case study, 235–236
    details of, 236–239
    unaccounted-for water in, 236
Tees in leak detection, 201–202
Temperature as leak predictor, 254–255
    analysis in, 256
    conclusions, 257
    operational observations in, 255–256
Tennessee, Johnson City (see Johnson City)
Tennessee Energy and Water Conservation Program, 183–185

Terminology:
    in accountability, 492–493
    variances in, 37–38
Test benches, 560–561
Test chambers, 193
Test hole placement in tracer gas process, 220
Test outlets for large meters, 346
Test plugs, 344, 346, 568
Testing:
    in bid documents, 405–406
    DP sensors, 405–406
    equipment, 98–99
    meters (see Meter testing)
    for reservoir leakage, 233–235
Theft (see Unauthorized water usage)
Theoretical maximum savings, 390–392
30/20 rule, 421
Throttled line valves, 269
Through-transmission meters, 588
Tilted meters, 357
Time-based pressure controllers, 114–115, 298–299
Time intervals:
    in demand profiles:
        in recording, 603–605
        in reports, 609–611
    in meter testing, 554
Time-of-flight meters, 588
Time periods, units for, 60
Time permitted to run in leakage, 15
Time response:
    in real loss control, 164–165
    in São Paulo system, 467–468
    (See also Economics of lost water)
Timed frequency of usage, 369–375
Total cost per unit, 529
Total sales meter errors, 515–518
Total supply figures, 509–511
Tracer gas process:
    equipment and materials in, 105, 216–217
    field operations in, 219–226
    helium and helium injection in, 217
    in Gasophons, 217
    in mains, 222–223
    pinpointing, 224–226
    preparation for, 219–220
    spread patterns in, 224
    prerequisites in, 217–219
    storage in, 423
    summary, 216
Training:
    for meter testing, 344
    unmetered use for, 519
Transceivers, 588–589
Transducers, 576
Transferability, 185
Transient waves, 264–265, 426–427
Transit time meters, 75, 77, 588
Transmission methods, 85
Trautwine, John C., 444
Trautwine tanks, 444

644 INDEX

Treatment costs, 499
Trenchless pipe replacement, 166, 327–328
Tube ports for pitot meters, 595
Tuned linear leak location, 196–201
Turbine and propeller meters, 85, 89, 585–586
   advantages and disadvantages of, 588
   in demand profiles, 612–613
   flow data for, 88, 90
   hydrant meters, 78
   insertion meters, 66, 68
   installing, 347, 540–541, 586
   maintaining, 586–587
   specifications for, 86
   testing, 346, 355, 546
   time pulse frequency in, 87
Turbulent flow, 422–423
Two-phase flow in leak detection, 201
Two-year benefits in audits, 530

UARL (unavoidable annual real losses), 41–42
   calculating, 42–44
   IWA estimates, 437–438
   in performance, 48
   previous estimates, 434–436
   statistics for, 49–51
Ultrasonic meters, 74–76, 588–590
   applications for, 78–79
   for flow measurement, 76–77
   installing, 76
   selecting, 77–78
   for transit time meter readings, 77
Unaccounted-for water [see Nonrevenue water (NRW)]
Unaccounted-for Water Task Force, 358–360
Unauthorized water usage, 22–23
   in apparent loss control, 17, 168
   in DeKalb County, 476
   in Johnson City, 478–479
   losses from, 528
   in meter sizing case study, 369
   in São Paulo, 470
Unavoidable annual real losses (UARL), 41–42
   calculating, 42–44
   IWA estimates, 437–438
   in performance, 48
   previous estimates, 434–436
   statistics for, 49–51
Unavoidable leakage index, 435
Unbilled water usage, 40
   in meter sizing case study, 358–359
   in nonrevenue water, 39
   in Philadelphia system, 445–446
Uncompensated usage, 492
Underground utilities:
   in helium-spread patterns, 225
   in tracer gas process, 219
Undetectable losses in UARL estimates, 437–438

Uniform flow, 421
United Kingdom, 7, 29–32
United States, water loss perspectives in, 25–29
   vs. England, 29–32
   future, 32–33
Units of measure:
   in audits, 507
   in data collection, 59–61
Universal metering for conservation, 494
Unmetered use, 118, 516–517
   in DeKalb County, 475–476
   estimating, 519–524
   identifying, 378, 518
Unregistered connections (see Unauthorized water usage)
Unreported leaks and breaks, 14, 29
   in BABE model, 136–137
   in UARL estimates, 437–438
Updating master plans, 531
Urban Watershed Research Facility (UWRF), 190, 193
Utilities in tracer gas process, 219

Vacant properties, 377
Vacuum conditions, 148
Valuation, accounting for, 493
Valves:
   for demand control:
    installation of:
     air release valves for, 305
     anchorage in, 302
     automatic control valves, 285–286
     chamber vs. above-ground, 302–303
     commissioning, 303
     headloss concerns in, 300–301
     hydraulic connections in, 302
     location in, 300
     modulation speed in, 305–306
     stability in, 306
     start-up procedures, 303–305
   open and closed, 208
   in pressure management:
    ACVs for, 269–271, 284–285, 294
    controllers for, 297–300
    for hydraulic impact, 264–265, 426–427
    losses from, 526–527
    maintaining, 306–307
    multiple valve sectors in, 286
    parallel, 296–297
    for reservoir and tank control, 286–291
    in São Paulo system, 466
    selecting and sizing, 291–300
    series, 297
    settings for, 304–305
    throttled line valves, 269
    types of, 110–114
   in water loss models, 143, 145
   in step testing, 209–210
   surge anticipator, 426–427
Variable-frequency drives (VFDs), 489

Variable paths in leakage calculations, 271–272
Variable-speed drives (VSDs), 487–489
Variations, parallel valves for, 296–297
Velocity:
   calculation of, 186–187
   in flow, 419–422
   profiles for, 66–70, 72
Velocity meters, 85, 92
Vena contracta in orifice plate meters, 580
Venturi meters, 574–576
   advantages and disadvantages of, 577
   checking, 510
   installing, 576–577
   maintaining, 577
   modified, 577–579
   testing, 560
Verification of demand profile data, 608–609
Vertical leaks, 201
VFDs (Variable-frequency drives), 489
Visual leak surveys, 177–178
Volume:
   in flow, 420
   units for, 60
   for water loss, 492–493
Volumetric tests, 98
Vortex meters, 590–593
VSDs (variable-speed drives), 487–489

Wales, 7
Water accountability [see Accountability; American Water Works Association (AWWA); Audits]
Water Audits and Leak Detection, 28, 36
Water balance, 19
   international approach to calculations, 431–432
   standard for, 38–39
Water-borne disease, 6, 190
Water conservation (see Accountability; Conservation)
Water cycle, 3
Water efficiency programs (WEPs) (see Efficiency and efficiency programs)
Water hammer, 77, 426
Water loss case studies and articles:
   DeKalb County (see DeKalb County)
   Johnson City (see Johnson City)
   Philadelphia (see Philadelphia system)
   pumps, 482–489
   São Paulo (see São Paulo case study)
Water loss management, 2
   conclusion, 440–442
   introduction, 430–431
   IWA data in, 438–440
   needs and requirements for, 8–10
   performance indicators for, 433–434
   unavoidable loss estimates in:
    IWA approach, 437–438
    previous approach, 434–436
   water balance calculations in, 431–432

INDEX

Water loss measurements in audits, 524–528
Water Revenue Bureau (WRB), 444
Water temperature as leak predictor, 254–255
   analysis in, 256
   conclusions, 257
   operational observations in, 255–256
Water utility organization, 27
Weak hydraulic capacity, reservoir and tank control for, 291
Weather conditions in tracer gas process, 219
Weighting factors:
   in contractor selection, 407–408
   for meter errors, 517
Weldolets, 70–71
WEPs (Water efficiency programs) (*see* Efficiency and efficiency programs)
Westchester Joint Water Works, 241–254
White, Hylton, 227
Wooden logs in Philadelphia system, 443
Worcestershire operational area, 214
WRB (Water Revenue Bureau), 444

Zero-flow calibration, 76–77
Zero shut tests, 206
Zero-volume accounts, 377
Zones:
   in flow analysis, 203–206
   in real loss control, 165
   in technology application case study, 237–238

ABOUT THE AUTHOR

Julian Thornton is involved in numerous professional water-related activities. He specializes in water loss management and has practiced in many locations around the world, including the United States, Canada, the United Kingdom, Africa, the Middle East, and Asia. He has supported American Water Works Association committees for the last 10 years and is currently Chair of the New Technology Subcommittee for the AWWA National Leak Detection and Water Accountability Committee. Mr. Thornton may be contacted at watloss@attglobal.net.